Fields and Waves in Electromagnetic Communications

Fields and Waves in Electromagnetic Communications

Nemai Chandra Karmakar
Monash University, Clayton Campus, Melbourne, Victoria, Australia

WILEY

Published by John Wiley & Sons, Inc., Hoboken, New Jersey.
Published simultaneously in Canada.

Library of Congress Cataloging-in-Publication Data Applied for:

Hardback ISBN: 9781119472193

Cover Design: Wiley
Cover Image: © ImageFlow/Shutterstock
Author photo: Courtesy of Nemai Chandra Karmakar

Set in 9.5/12.5pt STIXTwoText by Straive, Pondicherry, India

The book is dedicated to the great teachers of electromagnetics in the 20th and 21st centuries.

Special gratitude to:

i) Late Professor Dr. Abdul Matin, Bangladesh University of Engineering and Technology, for developing my interests in advanced electromagnetics with his finest pedagogy in late 1980s
ii) Professor Dr. Arun Bhattacharyya, The University of Saskatchewan, for giving me the mentorship on fundamental research in the computational electromagnetics in early 1990s
iii) Professor Dr. John Ness, Managing Director, Mitec Limited and Adjunct Professor, The University of Queensland, for giving me the exposures to hands-on applications of complex electromagnetic theories in the early to mid-1990s
iv) Late Professor Dr. Marek Bialkowski, The University of Queensland, for giving me the opportunities for independent research on applied electromagnetics from mid to late 1990s

Contents

Preface

Electromagnetics (EM) deals with the abstract natures of the EM fields and waves and their interactions with matters. It is the most abstract discipline in the entire electrical and electronics engineering field. The theories of EM also influence other fields such as particle and quantum physics, optics and thermodynamics just to name a few. The textbook has presented a holistic approach to the advanced EM (AEM) field and wave theory and modern applications. The uniqueness of the textbook is that each topic is supported by the most recent technological developments, analogies, similes, anecdotes and emerging applications of EM in a novel pedagogy. The chapter outlines of the book are provided in the table. The textbook can be studied for a semester or two in the modern university. The hours that covers the topic in a 12- to 14-week semester are given in the table as well. The hours are indicative.

Chapter 1 (4 hrs)	Uniform Plane Wave I	Deals with the fundamentals of plane wave fields, near and far-field concepts, polarizations of electromagnetic fields
Chapter 2 (6 hrs)	Wave Propagation in Homogeneous, Non-Dispersive Lossy Media	Deals with advanced topics on wave propagation in lossy media such as good conductors and dielectrics
Chapter 3 (6 hrs)	Wave Propagation in Dispersive Media	Deals with phase and group velocities and pulse broadening in dispersive media, and polarization of electromagnetic waves
Chapter 4 (8 hrs)	Wave Propagation in Dispersive Media – Advanced Topic[1]	Deals with the atomic model and the classical electron theory of matters, wave propagation in conductors, dielectric and charged particles such as plasma
Chapter 5 (8 hrs)	Reflection and Transmission of Normal Incidence	Deals with generic theory of transmission and reflection at discontinuities and detailed theory of normal incidence
Chapter 6 (8 hrs)	Oblique Incidence Theory	Deals with the fundamental concepts of oblique incidence with coordinate systems, conducting and dielectric boundaries, ray trace model, impedance concept, and Maxwell's field analysis
Chapter 7 (4 hrs)	Wave Propagation in Lossy Media	Deals with generic theory of wave propagation in lossy media and emerging applications in biomedical engineering, telecommunications and EMI/EMC
Total 44 hrs		

[1]Readers will comprehend why circular polarized signals propagating through plasma are used in satellite communications.

Experimental learning is the prime to gaining a solid foundation in AEM. Each important abstract theory is supported with many examples, problems, review questions, analogies, and anecdotes from the built environment and nature so that students can correlate the complex AEM processes with such objects, artifacts and natural phenomena. At the same time, such abstract theories are provided with detailed derivations so that a beginning student can understand each step of the developed theory. Therefore, the textbook has both depth and breadth of the covered AEM topics.

In conclusion, the textbook has presented comprehensive AEM materials so that when a student leaves university to pursue his/her career in the field, s/he competently would deal with the design and decision-making problems related to the technical issues of wireless and guided electromagnetics. This type of AEM knowledge and transferable skills are highly sought after in the modern industry. Therefore, this book provides graduating engineering students and experienced engineers with the enhanced learning experience and outcomes of the most recent and state-of-the-art wireless communications technologies which utilize the AEM discipline.

Finally, prudent feedback from a PhD student on the textbook: *Regarding the chapter, the abstract ideologies on electromagnetics were excellently demonstrated with the simple and real-life examples which I found the most interesting part of reading. Besides, the in-detail analysis and discussion on the fundamentals were really helpful for the readers. The final integration of the real-world applications attached to the topics is also very useful and quite relevant to the topic of discussion.*

I am pretty much confident that this book will be really helpful in the field of electromagnetics which will directly benefit our groups' present and future students.

I wish you all the best with this book and feel very lucky to assist you in this megaproject. – Shahreen Hasan, PhD student, Monash Microwave, Antenna, RFID and Sensor (MMARS) Laboratory, Monash University, 1 July 2020.

Therefore, the main motto of the textbook is to give the practicing engineers and would-be practicing engineers advanced knowledge on AEM and skills to solve problems associated with the state-of-the-art applications of AEM. Utmost care has been given and research has been performed to find the emerging and cutting edge applications of the AEM in the range of digital design, signal integrity, biomedical, mining and geoscience, telecommunications, power engineering, electronics, signal processing, antenna technology, RFID, wireless sensors and microwave active and passive design, atmospheric engineering, remote sensing, satellite communications, precision agriculture, food safety and security, climate change, driverless cars, vehicle to vehicle communications, radar engineering, optical communications, defense, reconnaissance, tactical engineering, and optical engineering. These vast applications of AEM and their depictions of fundamental theories enrich the textbook. Not only the university students and educators but all the decision-making leaders in many government departments and nonprofit organizations (NGOs) would be benefited from the book.

Nemai Chandra Karmakar
Monash University,
Clayton Campus, Melbourne, Victoria,
Australia
March 2023

Acknowledgments

Monash University has a long tradition of teaching and research in advanced electromagnetics (AEM). I express my sincere gratitude to my previous and current colleagues who taught the topic at various levels and enlightened me in the subject matters. They were great teachers and researchers in the field.

For the scholarship in teaching and leadership in education, I got the inspiration to write the manuscript from Monash University. I express my sincere appreciation to Monash University, and particularly, the Department of Electrical and Computer Systems Engineering (ECSE) for creating an opportunity for this scholarly pursuit of writing a reference book on advanced electromagnetics. This pursuit also provides an Australian contribution to AEM pedagogy in the modern university. I also acknowledge Australian Research Council (ARC) Discovery and Linkage Project Grants. Many of the data, resources and research themes produced in the textbook were drawn from ARC funded research projects. Therefore, the textbook has become a research monograph of the emerging technologies in chipess RFID, wireless communications, microwave biomedical engineering, remote sensing smart antennas and satellite communications just to name a few.

I express my sincere thanks to Ms. Megan Deacon, Copyright Office, Monash University, for her continuous guidance for all copyright and ethical issues of scholarly publications. From the beginning to the end of the preparation of the manuscript, she was so kind to guide me in the copyright issues of the manuscript preparation process. Alongside, I also express my sincere appreciation to my friend Mr. Prodip Roy, Library Officer, (Digital Collections), RMIT University Library, for his valuable guidance in copyright issues for the manuscript.

My sincere thanks to Mr. Brett Kurzman, Editor, Professional Learning, Wiley, for his encouragement to write a textbook on Advanced Electromagnetics, formalizing the review process, and accepting the proposal. I must appreciate his patience and guidance for the manuscript. It was so nice to meet him in many international antenna and microwave symposia and exchange pedagogical ideas and thoughts on the textbook. He was so kind to donate relevant resources to help me prepare the manuscript. Thanks to Ms. Sarah Lemore, the former managing editor K&L Content Management, Wiley. She provided valuable guidance in all aspects of the manuscript preparation. I must appreciate her nice cooperation and patience for the repeated delays in the preparation of the manuscript due to my busy schedules in teaching and research at Monash University. Also I thank the Wiley Content Refinement Specialist, Mr. Kavin Shanmughasundaram, Chennai, India for his patience and hard work to compose and edit the proof of the manuscript.

My PhD students are the lifeblood of my teaching and research, and inspiration for the textbook. I am tremendously indebted to their contributions in all aspects of the manuscript. Without countless bits of help and sacrifice, patience, and encouragement, the project could not have been completed. Therefore, it was big teamwork like any of my big research projects with my PhD students at my research laboratory. I have no language to express my hearty gratitude to them for their invaluable and timely contributions and their sacrifice. Trong Ho helped me from the beginning to the end of the manuscript. He was my inspiration and walked along with me without any hesitation all the time. He not only helped me to edit, draw many of the images, and format many of the chapters with great enthusiasm. I am indebted to his unique contributions to the manuscript. Guanghui Ma took huge responsibilities for the final logistics to prepare the manuscript, appropriate acknowledgments, and citations of all figures, tables and contents, and uploading the materials in Wiley provided OneDrive. He spent countless hours for the preparation of the manuscript and led the team of PhD students in the book project. Jiewei Feng, Farah Bilawal, Shahreen Hassan, Huu Nguyen, Javad Aliasgari Mosleh Abadi, Parya Fathi, Jiewei Feng, Jose Arganaras, Mahabub Alam, Kim Trinh, Likitha Lasantha and Shahed Khan contributed tremendously toward the preparation and editing of the manuscript. Without their help, the book project would not come to fruition. Shahreen indebted me to write a judicious reflection after reviewing and editing Chapter 3. Trong Ho. Guanghui Ma and Farah Bilawal checked many problems of the chapters. I really indebted to Farah Bilawal for preparation of index terms of the manuscript in the final stage of the proof setting.

I am highly indebted to Late Professor, M.A. Islam for his compiled textbook titles Electromagnetic Theory, BUET (formerly EPUET), Dhaka, Bangladesh, 1969. It was a textbook in the School of EEE in BUET in 1970s. Many of the derivations and theme of advanced EM has been borrowed with due acknowledgements.

Finally, I would like to acknowledge the support provided by my loving wife Shipra, daughters Antara and Ananya, and dogs Navy and Lilac. They always gave me a pleasant surrounding while I was busy writing the manuscript. I acknowledge my elder brother Mr. Hirendra Nath Karmakar who was an inspiration and blessing for me to complete a big project like this.

In the end of December 2022, I did final proof editing and corrections staying in my in-laws house in Merul (Badda), Dhaka, Bangladesh. During this time, my beloved mother-in-law Mrs. Shefali Rani Das took great care of my wellbeing. Without her affection and blessing in the final stage of the manuscript proofing, it would not be possible to publish the book on time. I gratefully acknowledge her support.

Nemai Chandra Karmarkar
Monash University,
Clayton Campus, Melbourne, Victoria
Australia
January 2023

About the Companion Website

This book is accompanied by a companion website

www.wiley.com/go/fieldsandwavesinelectromagneticcommunications

The website includes:

- This website material contains solution manual.

1

Uniform Plane Wave

1.1 Introduction to Uniform Plane Wave

With the emergence of new wireless technologies, electromagnetic wave propagation in nonconventional media has forced to revisit the classical electromagnetic theories in dynamic conditions. In this chapter, a few very important classical electromagnetic theories are examined in the light of wave propagations in time varying cases. The emerging applications are RF/microwave/millimeter-wave printed electronics on nonconventional materials, and study of propagation of electromagnetic waves in very complex composite media. The chapter first defines the uniform plane wave in the light of far field radiation from a dipole antenna followed by the uniform plane wave concept. Derivation of electromagnetic wave equations with the aid of Maxwell's equations and one-dimensional wave equation are presented next. The wave is characterized with its phase velocity meaning the velocity of wave propagation and the wave impedance, which is the function of the constitutive parameters of the medium. The above derivations are extended to the time harmonic field wave equation and the refractive index of medium and dispersion of medium. The time harmonic wave solution and polarization of the uniform plane wave are presented next. Finally, the energy balance equation which is called Poynting theorem is presented in detail for a few special cases followed by the conclusion of the chapter.

Maxwell unified all classical electromagnetic principles and laws into four potent equations. These equations are so universal that they can be used to characterize electromagnetic field wave equations for any frequency, for the static as well as the dynamic field conditions. In this regard, an electromagnetic field can be defined as: (i) every function that satisfies Maxwell's equations, which is finite, continuous,[1] and single valued, and (ii) propagates through homogeneous, linear, and isotropic medium. In this chapter, we shall develop the uniform plane wave concept of the electromagnetic fields using Maxwell's equations for any frequency in a medium as stated above. In this chapter, we examine this aspect in depth.

In solving the field wave equations, we need to deal with the complex phasor vector field quantities which are the functions of both space and time in double derivatives. In these situations, these vector functions are not easily solvable because of the following reasons: (i) no field solution is possible to get directly from scalar functions. Here, the functions are dependent on time and spatial coordinates, and (ii) the complexity of analyses of the vector functions increases as we move from one coordinate system to another coordinate system. As for example, the complexity of vector analysis of the electromagnetic fields increases as we move from the rectangular

1 Recall the boundary conditions that we have derived in the previous chapters for the dynamic fields.

Fields and Waves in Electromagnetic Communications, First Edition. Nemai Chandra Karmakar.

Companion website: www.wiley.com/go/fieldsandwavesinelectromagneticcommunications

coordinate system to the cylindrical coordinate system to the spherical coordinate system. There are also other seven coordinate systems in the vector analysis with which the electromagnetic fields are resolvable, but we only confine ourselves in the three main coordinate systems. Also note that electromagnetic field propagation in bounded and unbounded media behave differently. In unbounded medium, the antenna radiates spherically, whereas in bounded structure such as rectangular metallic waveguide we solve electromagnetic field problems with the rectangular wave equations. In optical fibers, we analyze the electromagnetic fields using cylindrical coordinate systems. With the simplest one, the rectangular coordinate system, we shall develop the electromagnetic field wave theory in this coordinate system in the chapter. Then gradually we define field solutions for more complex coordinate systems in cylindrical and spherical coordinate systems. (iii) It is also not possible to solve 3D scalar wave equations in arbitrary coordinate systems by the use of separation of variable method.[2] That is why, we keep our wave field analysis into simple configurations such as rectangular, circular, and spherical coordinate systems.[3]

Figure 1.2 illustrates how wireless communications networks established by the time varying electromagnetic field waves harness each and every device in the system for efficient wireless communications. As we have seen in the big picture of wireless communications shown in Figure 1.1, the wireless channel is established with the wave propagation. In this chapter, we shall study more

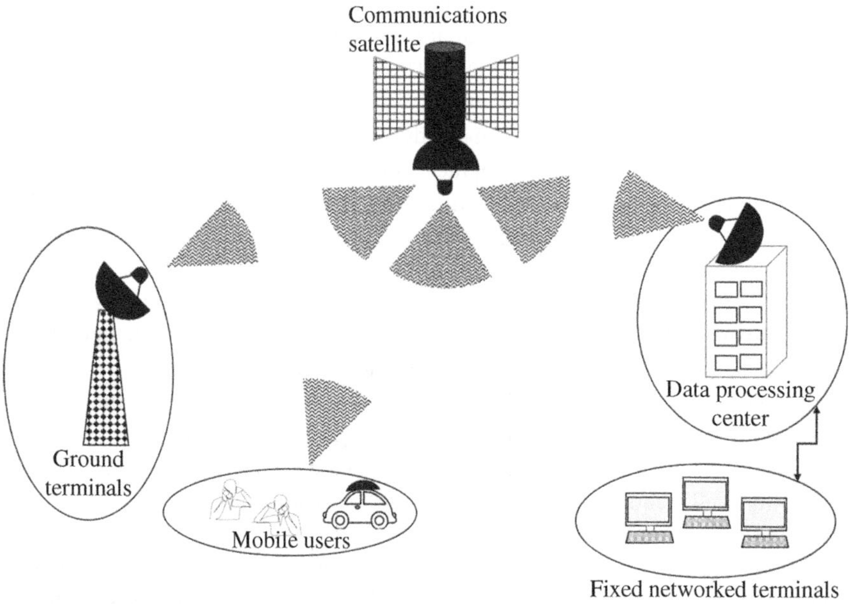

Figure 1.1 Wireless communications networks showing how the electromagnetic waves harness each and every device in the system for efficient wireless communications.

---

2 It is a method to solve a vector field which is a function of multiple independent variables. In this method we derive the original function as the product of two functions of individual independent variables and then we solve the function one by one. Finally, we multiply the individual solutions to get the final solution.

3 Islam, M.A. (1969). *Electromagnetic Theory*. Dacca, East Pakistan: EPUET.

details of the wireless channel augmented with the theory of uniform plane wave propagation. A brief description of the topics covered in the chapter are as follows:

i) Definition of Uniform Plane Waves
In the definition of uniform plane waves in this chapter, the very beginning of the book, we already have given a very good understanding of the uniform plane wave. We have learned that in a uniform plane wave, the electric field $\boldsymbol{E}$, the magnetic field $\boldsymbol{H}$ and, the Poynting vector $\boldsymbol{P} = \boldsymbol{E} \times \boldsymbol{H}$, which is also called the power flow density vector in (W/m^2), are orthogonal to each other, and the wave propagates orthogonally with respect to the directions of the two fields.
ii) Derivation of Field Wave Equations
After the proper definition of the uniform plane wave, we shall derive the electromagnetic plane wave equations. As we know, in a uniform plane wave the phase front is like a flat surface on which the phase of the signal is the same. From the derivation of the field wave equations, we shall now learn how to derive electromagnetic wave equations using simple differential equations. Uniform plane wave equation is a double derivative equation of both spatial and time coordinates. We shall see that the vector wave equation has two solutions, forward and reverse traveling wave fields.
iii) Wave Impedance
Every plane wave has its intrinsic wave impedance which is defined as the ratio of electric field and the magnetic field intensities at any point in space. The wave impedance is a function of the constitutive parameters such as medium permittivity ε (F/m), permeability μ (H/m), finite conductivity σ (S/m), and angular frequency ω (rad/m). We shall define the intrinsic wave impedance of the lossless and lossy media in the chapter. You may recall the wave impedance of the plane wave in free space is 377 Ω.
iv) Polarizations of Uniform Plane Waves
The polarizations of uniform plane waves is a significant parameter that regulates the efficacy of electromagnetic communications. The polarization of the field wave is the spatial orientation of the electric field vector with respect to time. We shall learn more in details of various polarizations of the field waves and their significance in the wireless communications.
v) Poynting Theorem
And finally, Poynting Theorem, that tells us the power flow density vector $\boldsymbol{P} = \boldsymbol{E} \times \boldsymbol{H}$ (w/m^2). Poynting Theorem is an energy balance equation that tells us that in a homogeneous, linear, and isotropic medium, the power input is equal to the electric and magnetic energy stored plus the power dissipated as the Ohmic loss as it propagates real power as the plane wave. The time average power is defined as

$$\boldsymbol{P}_{ave} = \frac{1}{2} Re\,(\boldsymbol{E} \times \boldsymbol{H}^*) \tag{1.1}$$

Finally, in this chapter, we shall solve many interesting problems as we go along with the theories on the uniform plane wave.

This chapter provides the very important concept of medium properties derived from the wave equation. There, we shall define the attenuation and phase constants, and the loss tangent for lossy medium. In defining those parameters, we shall go in deeper in defining the propagation of wave in good conductors and good dielectrics. These concepts are critically important in designing any practical RF and microwave circuits.

The aim of this chapter is to develop the concept of electromagnetic waves, and how they propagate in free space and in different complex media. It helps readers to understand the meaning of the polarization of plane waves and how introduction of a metallic or dielectric slab in their propagation paths affects their behaviors.[4] The reader will also obtain the concept of 3D and 2D vector plots of electromagnetic field distributions in various conditions via some illustrations generated with Computer Simulation Technologies (CST) Microwave Studio™. A few practical examples and worked out problems of the plane wave propagation is also provided to enhance understanding in a real-world scenario.

In this section, we investigate the different aspects of plane waves: how they look like, how they behave when encountered with different obstacles such as buildings, walls inside a room, on the roads, and in the free space when wireless propagations happen. Recall Maxwell's two curl equations, Faraday's and Ampere's laws that tell us that for a plane wave to be generated, a source of time varying electromagnetic fields must exist. According to Ampere's circuital law:

$$\nabla \times \boldsymbol{H} = \boldsymbol{J}_c + \frac{\partial \boldsymbol{D}}{\partial t} \tag{1.2}$$

In other words, if the electric field, $\boldsymbol{E}(t)$, varies at a point in space resulting in a time-varying electric flux density, $\boldsymbol{D}(t)$, it produces a time-varying magnetic field in the region around that point instantaneously. On the other hand, Faraday's law states that:

$$\nabla \times \boldsymbol{E} = -\frac{\partial \boldsymbol{B}}{\partial t} \tag{1.3}$$

The law tells us about the induction of electric field $\boldsymbol{E}(t)$ due to the time varying magnetic field $\boldsymbol{B}(t)$. As for example if a time varying magnetic field cuts a conductor it generates an induced current that produces the induced electromotive force V_{emf} or electric potential or curling electric field. A good example is an antenna that cuts a time varying magnetic field and as a result induces V_{emf}. We amplify the received signal and process it for meaningful information in our radio receiver. This means we are always encountering Faraday's law in telecommunications engineering. Receiving a call via a mobile phone is a good example of Faraday's law in action.

1.2 Fundamental Concept of Wave Propagation

We consider at some point in space there is a radiating point source. Good example is an infinitesimal dipole antenna, also called a Hertzian dipole. This antenna isotopically radiates EM fields radially outward in all directions. We usually study dipole antennas as the first lesson of the antenna analysis. Assume the antenna carries a time varying uniform current density $\boldsymbol{J}(t)$ in (A/m). It is a legitimate guess because we assume a very tiny antenna on which the time varying current exists.

Review Questions 1.1: Introduction to Uniform Plane Wave
Q1. Explain how does the EM field waves propagate from one point to another point in space? What is the mechanism? What fundamental laws they follow?

4 This extra exercise helps develop concepts of wave incidence, transmission and reflection at the obstacles of the wave. We shall develop the concept of reflection and refraction of the wave at the discontinuities of media and derive the reflection and transmission coefficients when we shall cover the normal and oblique incidence cases. Also, the concept paves the way to understand the multipath fading which is a common problem of wireless mobile communications when the signal reflects from and penetrates buildings, walls, trees, and roads.

Let us observe the dipole antenna in Figures 1.2a and 1.3a. It is a source of EM wave propagation. By Ampere's circuital law, due to the current density $\boldsymbol{J}(t)$, which is present on the dipole, a time varying magnetic field, $\boldsymbol{H}(t)$, is induced in the surrounding region. By Faraday's law of the electromagnetic induction, the time varying magnetic field induces a time varying electric field $\boldsymbol{E}(t)$.

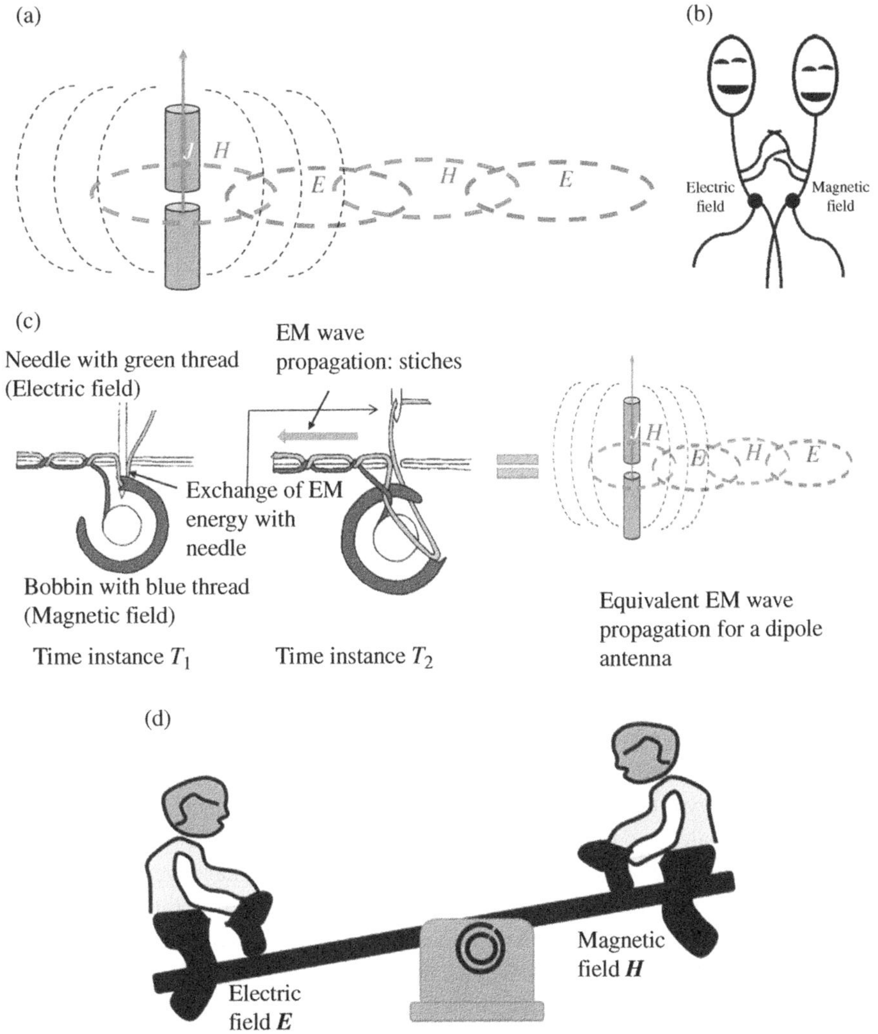

Figure 1.2 Depiction of electromagnetic wave propagation (a) wave propagation due to the time varying current $\boldsymbol{J}(t)$ that creates the time varying magnetic field $\boldsymbol{H}(t)$ by Amperes law and then the time varying magnetic field creates the time varying electric field $\boldsymbol{E}(t)$ by Faraday's law. Then the oscillating fields alternate between the time varying electric and magnetic fields as they propagate away from the source $\boldsymbol{J}(t)$, the dipole, (b) a depiction of exchange between the electric and magnetic field with an analogy of tap dancers who exchange their positions of their legs in a rhythmic manner which resembles the oscillating frequency and exchange of energy between the electric and magnetic fields. *Source:* Ulaby, F.T. (2005). *EM for Engineers*. Upper Saddle River, NJ, USA: Pearson Prentice Hall, (c) a sewing machine's sewing mechanism has analogy of the electromagnetic field wave propagation, and (d) exchange of electric and magnetic energy in instantaneous electromagnetic field wave is equivalent to seesaw.

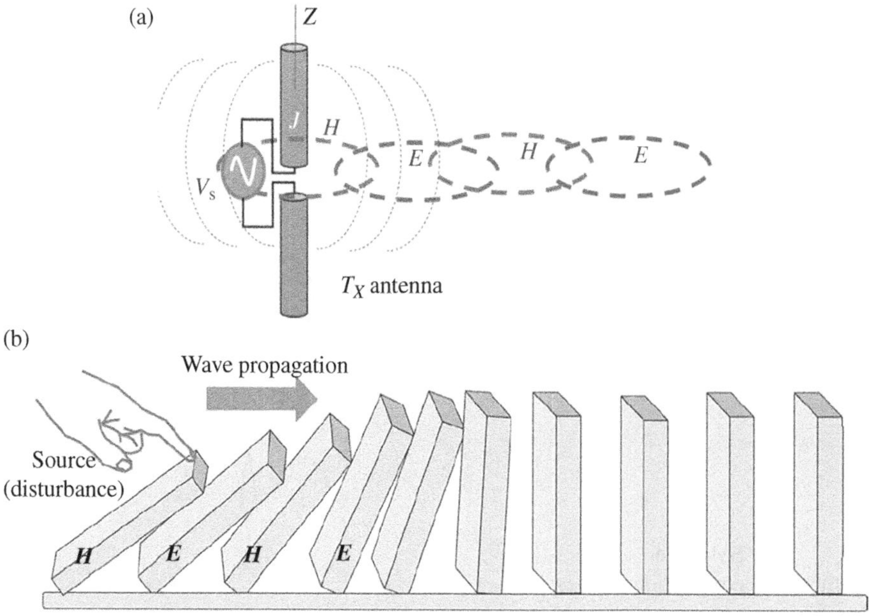

Figure 1.3 (a) Electromagnetic wave propagation via a dipole antenna connected with a generator, and (b) analogy of electromagnetic wave propagation as the domino effect of a series of vertical bricks made of alternate electric and magnetic fields. Here the figure push is the generator with alternating source voltage $V_s(t)$ that creates the induced current $\boldsymbol{J}(t)$ on the dipole.

Energy is passing back and forth between electric and magnetic fields as they travel away from the source point at the speed of light. Note that away from the dipole there is no original current source $\boldsymbol{J}(t)$ present, but still the electromagnetic wave fields carry real power in the direction of propagation due to these two laws in action.

It is an interesting analogy of the series of bricks vertically placed at a uniform distance like a domino. A slight push to the first brick propagates the mechanical energy to the next brick, and then after a moment, the energy is transferred to the third brick and the energy is propagating on and on. Now if we put the analogy of the domino effect in the electromagnetic system, the first push is equivalent to the source of the oscillation; here the current source that is generated by an oscillator that oscillates at an angular frequency $\omega = 2\pi f(\text{rad/s})$ where, f is the operating frequency in Hz.[5] Note that the source (hand) is sitting in its original position and does not need any further push but the signal is propagating indefinitely without any source present along with the propagation. This is the foundation for wireless propagation in which no source needs to be present in free space.

These source free oscillating fields away from the source in free space constitute the propagating waves of electromagnetic energy. As shown in Figure 1.3a, the electromagnetic energy exchanges hands between the electric and magnetic energies, and travels away from the source point creating electromagnetic wave propagation. And this alternating or periodic oscillation constitutes *the time*

5 See also the video of domino brick and the wave propagation due to push of a series of brick. It also tells that the wave energy propagates with finite velocity. The velocity is called the phase velocity of the wave propagation.

harmonic fields. We shall discuss the time harmonic fields after derivation of the generic wave equations. The sinusoidally varying electromagnetic fields are called the time harmonic fields. A signal waveform can be transformed into the sinusoidally varying field using Fourier series expansion. This assumption simplifies the electromagnetic field analysis.

Review Questions 1.2: Mechanism of Uniform Plane Wave

Q1. Discuss different analogies of the different mechanisms of creations of the uniform plane wave.

Q2. What is the analogy of seesaw in electromagnetic wave propagation?

1.3 Plane Wave Concept

As we discussed in this chapter, the wave radiates spherically, but at a remote distance away from the source the spherical wave resembles a uniform plane wave,[6] UPW in short. The distance is called the far field distance in the antenna nomenclature. Usually the far field distance is defined as $R \geq (2D^2/\lambda)$, where D is the largest dimension of the radiating source, here is the active part of the antenna[7] and λ is the wavelength in free space. As you know the wavelength λ is inversely proportional to the operating frequency f, therefore, the far field is also a function of the operating frequency. As the frequency goes up the wavelength gets shorter and the far field distance is also getting larger. However, due to the resonant nature of the antenna, the effective dimension D of the antenna is also proportional to the wavelength λ, hence the antenna is getting shorter almost linearly with the frequency. And the far field distance varies with the squared terms of D. Therefore, in general, the far field distance is almost inversely proportional to the operating frequency.

The far field distance R represents the distance at which the phase error between a perfect plane wave and the antenna's wavefront is equal to 22.5° or less. This phenomenon of the far field is shown in Figure 1.4. That means we can assume at this distance; the field's phase front resides on a plane as shown by the dotted line in Figure 1.4a. As we move further away from the source, the wave front becomes planar and then we can assume the wave is a uniform plane wave. We also defined that for a uniform plane wave, the electric and magnetic fields are oriented orthogonally to each other as well as the direction of wave propagation.

Figure 1.4b illustrates a uniform plane wave generated with a horn antenna. The illustration is generated from a 3D animation of CST Microwave Studio. The detailed field orientation and the direction of the wave propagation are shown in Figure 1.5. Both $\boldsymbol{E}$ and $\boldsymbol{H}$ fields are sinusoidal/time harmonic in nature. The electric (along the x-axis) and magnetic fields (along the y-axis) are orthogonal to each other and transverse to the direction of propagation (propagating along the z-axis) as

---

6 It follows the surface area of a sphere $S = 4\pi R^2$, where R is the radius of the sphere. As R increases, the spherical surface looks like a flat surface. Similarly, since the Earth surface is too large compared to a normal human, it looks like a flat surface to us.

7 The active part of the antenna is called effect length for a dipole antenna and effective area for a patch, a horn and a reflector antennas. Using the diagonal distance D of a rectangular horn aperture is considered in the calculation of the far field.

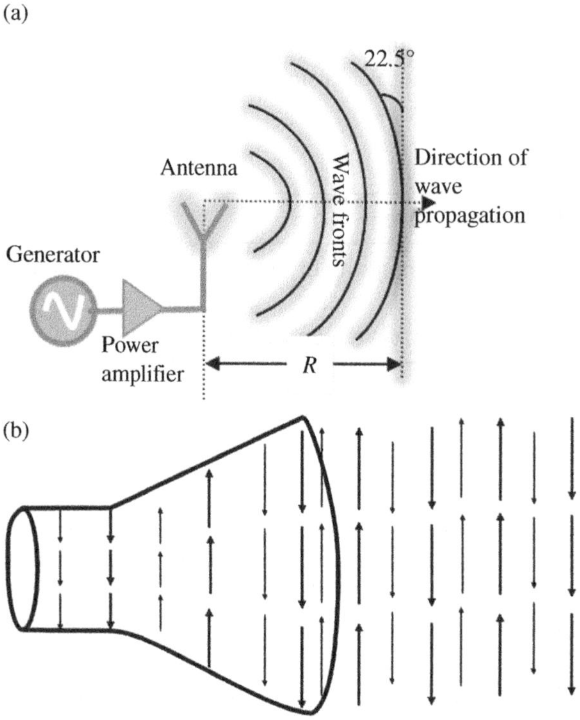

Figure 1.4 (a) Definitions of near field (spherical) wave and the far field (flat) waves at a distance from a radiating antenna, and (b) CST Microwave Studio rendering of plane wave creation from a horn antenna. Observe the wave front becomes planar as it moves away from the horn antenna.

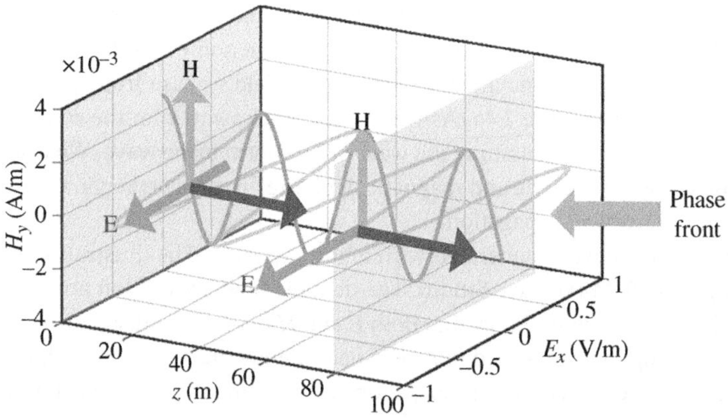

Figure 1.5 Definitions of near field (spherical) wave and the far field (flat) waves at a distance from a radiating antenna.

depicted in Figure 1.5. The power is propagating along the z-axis and is defined with the power density vector $\boldsymbol{P} = \boldsymbol{E} \times \boldsymbol{H}$. The phase on which the electric and magnetic fields reside is called the phase front or the wave front. The direction of the wave propagation follows the right-hand rule: $\boldsymbol{P}_z = \boldsymbol{E}_x \times \boldsymbol{H}_y$. Therefore, the uniform plane wave is also called the transverse electromagnetic (TEM) wave. The magnitude and phase of the field vectors in a UPW are equal at every point on the wavefront. For simplicity we analyze the field wave problem in rectangular coordinate. We usually work on the spherical coordinate for the antenna analysis.

You can watch some YouTube video in the following link.[8] You may find many videos and lecture notes on this topic on YouTube to enhance your understanding of UPW. You may also use free trial version of CST Microwave Studio™ animations software downloadable from website upon request from the vendor to obtain the visualization of the abstract nature of EM field in 3D animation via www.cst.com. In the next section, we shall derive the field wave equation using Maxwell's equations.

Review Questions 1.3: Introduction to Uniform Plane Wave

Q1. What is the far field of an electromagnetic wave?
Q2. Why do we need to consider the effective dimensions of the sources in calculating the far field distance?
Q3. Explain the salient features of a uniform plane wave.
Q4. What is the definition of the phase front? What is wavefront?
Q5. What is the far field distance of a radiating source? How do you determine the far field distance for a physical source?
Q6. The antenna operates at 1 GHz frequency and has the largest dimension of 210 mm. Calculate the far field distance.
Q7. Why a uniform plane is also called a TEM wave?

Example 1.1: Introduction to Uniform Plane Wave
The electric field $\boldsymbol{E}(z, t) = 10\cos(10^8\pi t - 4\pi z)\boldsymbol{a}_x$ (V/m) and the magnetic field is $\boldsymbol{H}(z, t) = 26.53\cos(10^8\pi t - 4\pi z)\boldsymbol{a}_y$ (mA/m). Calculate the power density or Poynting vector $\boldsymbol{P}$ in (W/m^2).

Answer: $\boldsymbol{P} = \boldsymbol{E} \times \boldsymbol{H} = 10\cos(10^8\pi t - 4\pi z)\boldsymbol{a}_x \times 26.53 \times 10^{-3}\cos(10^8\pi t - 4\pi z)\boldsymbol{a}_y = 377\cos^2(10^8\pi t - 4\pi z)\boldsymbol{a}_z$ (W/m^2).

Example 1.2: Uniform Plane Wave
Calculate the wave impedance of the wave in Example 1.1

Answer: The wave impedance is the ratio of the x-directed electric field and y-directed magnetic fields.

$$\eta = \frac{E_{xo}}{H_{yo}} = \frac{10\left(\frac{\text{V}}{\text{m}}\right)}{26.53\left(\frac{\text{mA}}{\text{m}}\right)} = 377\ (\Omega).$$

8 http://www.walter-fendt.de/html5/phen/electromagneticwave_en.htm (accessed 05 December 2017)
http://www.edumedia-sciences.com/en/media/222 (accessed 05 December 2017)
https://www.youtube.com/watch?v=c_IUic_oatY (accessed 05 December 2017)

Example 1.3: Derivation of Electromagnetic Wave Equation
Now we are going to derive the wave equation from Maxwell's equations. Recall Maxwell's equations in point form.

$$\nabla \cdot \boldsymbol{E} = \frac{\rho_v}{\varepsilon} \tag{1.4}$$

$$\nabla \cdot \boldsymbol{H} = 0 \tag{1.5}$$

$$\nabla \times \boldsymbol{E} = -\mu \frac{\partial \boldsymbol{H}}{\partial t} \tag{1.6}$$

$$\nabla \times \boldsymbol{H} = \boldsymbol{J_c} + \varepsilon \frac{\partial \boldsymbol{E}}{\partial t} \tag{1.7}$$

In far away from the source, we legitimately consider the source free region where the field wave propagates with real energy. Therefore, in free space we assume $\rho_v = 0$; therefore, Gauss law for electric field case reduces to $\nabla \cdot \boldsymbol{E} = 0$.

Now, consider a homogeneous, linear and isotropic medium, which has time invariant constitutive parameters permittivity (ε), and permeability (μ) and the electromagnetic wave propagates in this medium. Then Gauss law for electric and magnetic cases are:

$$\nabla \cdot \boldsymbol{E} = \frac{\rho_v}{\varepsilon} = 0 \text{ and } \nabla \cdot \boldsymbol{H} = 0.$$

The Gauss law for magnetic case is zero due to the fact that there is no magnetic source in reality, irrespective of the distance. Also, at far away from the source, the conduction current $\boldsymbol{J_c} = \sigma \boldsymbol{E}$ is zero. Using the above two source free assumptions in the free space, we can write the two curl equations of Maxwell:

$$\begin{aligned} \nabla \times \boldsymbol{E} &= -\mu \frac{\partial \boldsymbol{H}}{\partial t} \\ \nabla \times \boldsymbol{H} &= \varepsilon \frac{\partial \boldsymbol{E}}{\partial t} \end{aligned} \tag{1.8}$$

The symbols have their usual physical meanings. We shall be using these two source-free curl equations in deriving EM wave equations.

Review Questions 1.4: Introduction to Uniform Plane Wave

Q1. Explain why do we consider J and ρ_v both zero in the derivation of the electromagnetic wave equation?
Q2. Show that in a source free region $\nabla \times \boldsymbol{E} = -\mu(\partial \boldsymbol{H}/\partial t)$ and $\nabla \times \boldsymbol{H} = \varepsilon(\partial \boldsymbol{E}/\partial t)$.

Next, take curl on both sides of Faraday's law, we obtain,

$$\nabla \times \nabla \times \boldsymbol{E} = \nabla \times \left(-\mu \frac{\partial \boldsymbol{H}}{\partial t} \right) \tag{1.9}$$

Observe on the right side of the above equation, the curl is a position derivative in x, y, and z that is acting on a time derivative $\partial/\partial t$. A wave is a two independent dimensional entity with positions (x, y, z) and time (t). Now using the identity

$$\nabla \times \nabla \times \boldsymbol{E} = \nabla \cdot \boldsymbol{E} - \nabla^2 \boldsymbol{E} \tag{1.10}$$

on the left-hand side and taking μ out of $\nabla\times$ in the right-hand side as it is not varying with space,[9] yields

$$\nabla \cdot \boldsymbol{E} - \nabla^2 \boldsymbol{E} = -\mu \frac{\partial(\nabla \times \boldsymbol{H})}{\partial t} \tag{1.11}$$

Now if you substitute Ampere's law in point form $\nabla \times \boldsymbol{H} = \sigma \boldsymbol{E} + \varepsilon(\partial \boldsymbol{E}/\partial t)$ in the right-hand side and do the time derivative on it then we get the complete expression for the field wave equation as follow:

$$\nabla \cdot \boldsymbol{E} - \nabla^2 \boldsymbol{E} = -\mu\sigma \frac{\partial \boldsymbol{E}}{\partial t} - \mu\varepsilon \frac{\partial^2 \boldsymbol{E}}{\partial t^2} \tag{1.12}$$

Since in the source free region, no volume charge density ρ_v is present[10] hence $\nabla \cdot \boldsymbol{E} = 0$. Taking the source free assumption in free space, we obtain the generic wave equation for the electric field:

$$\nabla^2 \boldsymbol{E} - \mu\sigma \frac{\partial \boldsymbol{E}}{\partial t} - \mu\varepsilon \frac{\partial^2 \boldsymbol{E}}{\partial t^2} = 0 \tag{1.13}$$

where the Laplacian operator ∇^2 is a double derivation operator in spatial coordinates and is defined in the rectangular coordinate as:

$$\nabla^2 = \nabla \cdot \nabla = \frac{\partial^2}{\partial x^2} + \frac{\partial^2}{\partial y^2} + \frac{\partial^2}{\partial z^2} \tag{1.14}$$

Likewise, we can derive the generic wave equation for the magnetic field:

$$\nabla^2 \boldsymbol{H} - \mu\sigma \frac{\partial \boldsymbol{H}}{\partial t} - \mu\varepsilon \frac{\partial^2 \boldsymbol{H}}{\partial t^2} = 0 \tag{1.15}$$

In the above two wave equations of $\boldsymbol{E}$ and $\boldsymbol{H}$, we observe that $\boldsymbol{E}$ and $\boldsymbol{H}$ are the functions of both space (x, y, z) and time t. This wave equation is called the *Helmholtz wave equation*.[11] Similar equation is derived for $\boldsymbol{H}$. This derivation of $\boldsymbol{H}$ is left as an exercise.

Review Questions 1.5: Introduction to Uniform Plane Wave

Q1. Derive the field wave Equation (1.15) for the time varying magnetic field from Maxwell's equations in free space.

Laplacian operator ∇^2 can be decomposed into independent three spatial dimensions x, y, and z. Therefore, each equation can be broken up into their individual spatial components: E_x, E_y, E_z, H_x, H_y, and H_z. This yields the corresponding six one dimensional field wave equations in x, y, and z directions. Each of the six equations is a second-order differential equation. Remember we shall no more use partial differential equation as we are working with one dimensional independent spatial

9 We assume the medium is linear, isotropic, and homogeneous.

10 As discussed above in free space away from the source there is not active source like volume charge density ρ_v. as well as current density J present.

11 Helmholtz was the teacher of Heinrich Hertz and both worked in the same laboratory in Germany.

Table 1.1 One-dimensional electromagnetic field wave equations.

1-D electric field wave equations		1-D magnetic field wave equations	
$\frac{d^2\mathbf{E}}{dz^2} = \mu\sigma\frac{\partial \mathbf{E}}{\partial t} + \mu\varepsilon\frac{\partial^2 \mathbf{E}}{\partial t^2}$	(1.16a)	$\frac{d^2\boldsymbol{H}}{dx^2} = \mu\sigma\frac{\partial \boldsymbol{H}}{\partial t} + \mu\varepsilon\frac{\partial^2 \boldsymbol{H}}{\partial t^2}$	(1.17a)
$\frac{d^2\mathbf{E}}{dy^2} = \mu\sigma\frac{\partial \mathbf{E}}{\partial t} + \mu\varepsilon\frac{\partial^2 \mathbf{E}}{\partial t^2}$	(1.16b)	$\frac{d^2\boldsymbol{H}}{dx^2} = \mu\sigma\frac{\partial \boldsymbol{H}}{\partial t} + \mu\varepsilon\frac{\partial^2 \boldsymbol{H}}{\partial t^2}$	(1.17b)
$\frac{d^2\mathbf{E}}{dy^2} = \mu\sigma\frac{\partial \mathbf{E}}{\partial t} + \mu\varepsilon\frac{\partial^2 \mathbf{E}}{\partial t^2}$	(1.16c)	$\frac{d^2\boldsymbol{H}}{dx^2} = \mu\sigma\frac{\partial \boldsymbol{H}}{\partial t} + \mu\varepsilon\frac{\partial^2 \boldsymbol{H}}{\partial t^2}$	(1.17c)

variable only. These equations can be solved in terms of one-dimensional position and time. The solution is an equation defining forward and reverse travelling field waves in these three respective coordinates. One example for the physical meaning of the field wave propagation of $\boldsymbol{E_x}$ and $\boldsymbol{H_y}$ fields along z-direction is shown in Figure 1.5. Table 1.1 summarizes the one-dimensional field wave equations for both electric and magnetic fields.

The above equations for the $\boldsymbol{E}$ and $\boldsymbol{H}$ fields are for the case of variations in one spatial coordinate only. As we have seen from the depiction of the uniform plane waves of the $\boldsymbol{E}$ and $\boldsymbol{H}$ fields components in Figure 1.5, that the uniform plane wave is transverse to the direction of the wave propagation, z. According to the illustration of the wave propagation along z-direction, the z-directed electromagnetic field components must be zero. Therefore, for the uniform plane wave we shall solve for the two transverse field components along x- and y-directions assuming that the direction of wave propagation is along z-axis. Let us derive the field solutions for $\boldsymbol{E_x}$ field component first. The scalar field wave equation for $\boldsymbol{E_x}$ is:

$$\frac{\partial^2 \boldsymbol{E_x}}{\partial z^2} = \mu\sigma\frac{\partial \boldsymbol{E_x}}{\partial t} + \mu\varepsilon\frac{\partial^2 \boldsymbol{E_x}}{\partial t^2} \tag{1.18}$$

We observe that $\boldsymbol{E_x}$ is a function of z and t only. To solve for it, we resort the method of separation of variables. In this method we can decompose a function into the product of the two functions of individual independent variables separately as follow[12]:

$$E_x(z, t) = f(z)g(t) \tag{1.19}$$

where $f(z)$ is a function of z only and $g(t)$ is a function of t only.

Substitution for the Equation (1.19) in the 1D field wave Equation (1.18) above, we obtain

$$\frac{\partial^2 f(z)g(t)}{\partial z^2} = \mu\sigma\frac{\partial f(z)g(t)}{\partial t} + \mu\varepsilon\frac{\partial^2 f(z)g(t)}{\partial t^2} \tag{1.20a}$$

$$g(t)\frac{\partial^2 f(z)}{\partial z^2} = \mu\sigma f(z)\frac{\partial g(t)}{\partial t} + \mu\varepsilon f(z)\frac{\partial^2 g(t)}{\partial t^2} \tag{1.20b}$$

Now dividing both sides by $f(z)g(t)$, we obtain

$$\frac{1}{f(z)}\frac{d^2 f(z)}{dz^2} = \frac{\mu\sigma}{g(t)}\frac{dg(t)}{dt} + \frac{\mu\varepsilon}{g(t)}\frac{d^2 g(t)}{dt^2} = -k^2 \text{ a constant} \tag{1.20c}$$

12 Similar derivation is also available in Islam, M.A. (1969). *Electromagnetic Theory*. Dacca, East Pakistan: EPUET.

k is a constant due to the fact that both sides are the singular functions of the respective independent variables z on the left-hand side and t on the right-hand side, respectively. The individual equations of the singular function are:

$$\frac{d^2f(z)}{dz^2} + k^2f(z) = 0 \tag{1.20d}$$

$$\frac{d^2g(t)}{dt^2} + \frac{\sigma}{\varepsilon}\frac{dg(t)}{dt} + \frac{k^2}{\mu\sigma}g(t) = 0 \tag{1.20e}$$

The general solution for $f(z)$ is:

$$f(z) = Ae^{-jkz} + Be^{jkz} \tag{1.21a}$$

where A and B are coefficients and usually represent the amplitudes of the forward and reverse functions of $f(z)$. For equation containing $g(t)$, we assume similar solution

$$g(t) = Ce^{-jpt} \tag{1.21b}$$

and substitute Equation (1.21b) in the above Equation (1.20e) then we obtain,

$$Ce^{-pt}\left(p^2 - \frac{\sigma}{\varepsilon}p + \frac{k^2}{\mu\sigma}\right) = 0 \tag{1.22}$$

Since Ce^{-pt} cannot be zero, therfore

$$p^2 - \frac{\sigma}{\varepsilon}p + \frac{k^2}{\mu\sigma} = 0 \tag{1.23}$$

In the above equation we observe that p and k are not independent of each other. Therefore, if one of them, for example p is specified then k can also be calculated. Now we combine solutions for $f(z)$ and $g(t)$ in the above two equations, we obtain the complete solution for E_x,

$$E_x(z, t) = f(z)g(t) = \left(Ae^{-jkz} + Be^{jkz}\right)\left(Ce^{-jpt}\right) \tag{1.24}$$

or,

$$E_x(z, t) = E_{1x}e^{(-jkz-pt)} + E_{2x}e^{(jkz-pt)} \tag{1.25}$$

where $E_{1x} = AC$ and $E_{2x} = BC$ are used.

In a similar mathematical manipulation, we can obtain the solution for the other transverse field components E_y,

$$E_y(z, t) = E_{1y}e^{(-jkz-pt)} + E_{2y}e^{(jkz-pt)} \tag{1.26}$$

Now combining the individual complete we obtain the complete solution for the transverse electric field:

$$E(z, t) = E_1e^{(-jkz-pt)} + E_2e^{(jkz-pt)} \tag{1.27}$$

Likewise, we can derive the solution for the magnetic field,

$$H(z, t) = H_1e^{(-jkz-pt)} + H_2e^{(jkz-pt)} \tag{1.28}$$

We can easily develop the relationship between E_1, E_2, H_1 and H_2 and show that the field quantities are orthogoal to each and the direction of wave propagation.

Review Questions 1.6: Introduction to Uniform Plane Wave
Q1. Explain why we need to decouple 3D electromagnetic field wave equations into discrete 1D spatial field wave equation. **Q2.** Sketch the procedure to calculate the 1D spatial field wave equation.

Exercise 1.1: Derivation of Uniform Plane Wave
Q1. Derive Equation (1.28) from Equation (1.17a).

1.4 One Dimensional Wave Equation Concept

In the beginning of the chapter we mentioned that the vector field analysis is a complex and challenging process, and it is not possible to get vector field solution using the conventional mathematical process. The complexity of the analysis increases with the coordinate systems. In this regard, the field components are decomposed into their individual one dimensional field components and the assumptions of variations are made to simplify the process. This process we have just witnessed in the above field wave analysis. Therefore, the simplest approach to this field analysis is to find the field solution in one dimensional wave equations. In this section, we shall develop the concept of one dimensional field wave equation and the relevant field solutions. This theory is also supported by the practical system. In guided structures such as transmission lines, waveguides and optical fibers; and radiation of waves via antennas, we always look for solutions for one dimensional fields, and develop the appropriate excitation mechanism to support the theory.[13]

Let us have a look at the relation of the field components and the direction of propagation as we have derived in the above section. Assume that the wave propagates along z-axis only and no variations along x and y axes. Let us assume the medium is lossless, and the region is charge free as we have assumed before. To solve the problem we take Maxwell's equations again as follows:

From Maxwell's equation,

$$\begin{bmatrix} \boldsymbol{a}_x & \boldsymbol{a}_y & \boldsymbol{a}_z \\ 0 & 0 & \dfrac{\partial}{\partial z} \\ E_x & E_y & 0 \end{bmatrix} = -\mu \frac{\partial}{\partial t}\left(\boldsymbol{a}_x H_x + \boldsymbol{a}_y H_y + \boldsymbol{a}_z H_z\right) \tag{1.29}$$

Simplification leads to the following differential equations:

$$\frac{\partial E_y}{\partial z} = \mu \frac{\partial H_x}{\partial t} \tag{1.30}$$

$$\frac{\partial E_x}{\partial z} = -\mu \frac{\partial H_y}{\partial t} \tag{1.31}$$

13 We have witnessed such excitation mechanism during the development of the field wave propagation via a dipole antenna. As we move along applications of the wave equation in specific cases such as waveguides, optical fibers and antennas, we shall encounter various specific excitation mechanisms.

$$0 = \mu \frac{\partial H_z}{\partial t} \tag{1.32}$$

From (1.2) with the conduction current $J_c = 0$ and $\boldsymbol{D} = \varepsilon \boldsymbol{E}$ in 3D coordinates we obtain,

$$\begin{bmatrix} \boldsymbol{a}_x & \boldsymbol{a}_y & \boldsymbol{a}_z \\ 0 & 0 & \dfrac{\partial}{\partial z} \\ H_x & H_y & 0 \end{bmatrix} = \varepsilon \frac{\partial}{\partial t} \left(\boldsymbol{a}_x E_x + \boldsymbol{a}_y E_y + \boldsymbol{a}_z E_z \right) \tag{1.33}$$

Simplification leads to the following differential equations:

$$\frac{\partial H_y}{\partial z} = -\varepsilon \frac{\partial E_x}{\partial t} \tag{1.34}$$

$$\frac{\partial H_x}{\partial z} = \varepsilon \frac{\partial E_y}{\partial t} \tag{1.35}$$

$$0 = \varepsilon \frac{\partial E_z}{\partial t} \tag{1.36}$$

From the above derivations we observe that E_z and H_z must either be zero or time independent to satisfy the last bottom two equations. Since we are investigating the time varying field quantities, they must be zero. This confirms that if we assume the field quantities vary only in one direction then both $\boldsymbol{E}$ and $\boldsymbol{H}$ must be transverse to the direction of the field variation. In the present case the variation is assumed to be in z-direction only, and hence both electric and magnetic fields must have only x and y components. This is the fudamental definition of *the transverse electromagnetic field waves*.

Now we shall work on the field wave equations of the transverse wave that propagates only in the z direction. We start with

$$\frac{\partial E_y}{\partial z} = \mu \frac{\partial H_x}{\partial t} \tag{1.37}$$

and differentiate with respect to z to obtain the field wave equation as follow:

$$\frac{\partial^2 E_y}{\partial z^2} = \mu \frac{\partial H_x}{\partial z \partial t} \tag{1.38}$$

Similarly for

$$\frac{\partial H_x}{\partial z} = \varepsilon \frac{\partial E_y}{\partial t} \tag{1.39}$$

with partial derivate with respect to t we obtain,

$$\frac{\partial^2 H_x}{\partial t \partial z} = \varepsilon \frac{\partial^2 E_y}{\partial t^2} \tag{1.40}$$

Now combining Equations (1.38) and (1.40), we obtain

$$\frac{\partial^2 E_y}{\partial z^2} = \mu\varepsilon \frac{\partial^2 E_y}{\partial t^2} \tag{1.41}$$

Equation (1.41) is the field wave equation of E_y. In the above derivation we assume that the time varying magnetic field is a continuous function of both t and z. Therefore, the differentiation with respect to t and z can be done in any order that yields the same results.

Likewise we can derive the field wave equations for of E_x, H_x, and H_y. as follows:

$$\frac{\partial^2 E_x}{\partial z^2} = \mu\varepsilon \frac{\partial^2 E_x}{\partial t^2} \tag{1.42}$$

$$\frac{\partial^2 H_x}{\partial^2 z} = \mu\varepsilon \frac{\partial^2 H_x}{\partial t^2} \tag{1.43}$$

$$\frac{\partial^2 H_y}{\partial^2 z} = \mu\varepsilon \frac{\partial^2 H_y}{\partial t^2} \tag{1.44}$$

Exercise 1.2: Derivations of One-Dimensional Wave Equation

Q1. Derive the one-dimensional field wave equations from the relevant Maxwell's Equations (1.42)–(1.44).

We can generalize all the above one dimensional scalar field wave equations in the following form:

$$\frac{\partial^2 f}{\partial z^2} = \frac{1}{v^2}\frac{\partial^2 f}{\partial t^2} \tag{1.45}$$

where $v = \frac{1}{\sqrt{\mu\varepsilon}}$ is the velocity of wave propagation.

The wave of this type is called *the wave equation of vibrating string*. Figure 1.6 illustrates the waves for the stretched vibrating string: (a) stretched string, (b) fundamental mode ($n = 1$), (c) second harmonic ($n = 2$), and (d) third harmonic ($n = 3$).

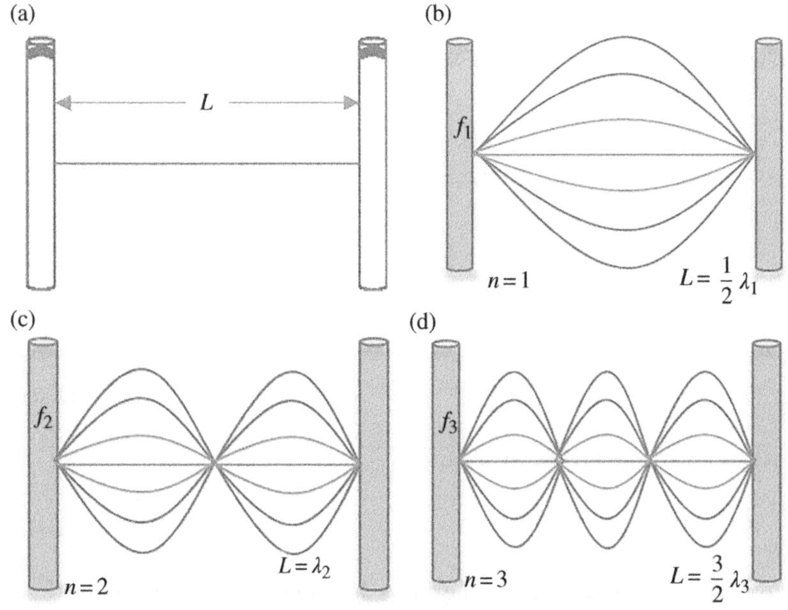

Figure 1.6 Illustration of string waves: (a) stretched string at equilibrium, (b) disturbance of fundamental mode frequency with wavelength λ, (c) 2nd harmonic, and (d) 3rd harmonic.

We solve for the one dimensional wave equation using separation of variable method and understand the salient feature of the one dimensional wave such as its two solutions of forward and reverse travelling natures, the velocity of wave propagation and definition of the uniform plane wave.

As per the method of separation of variables, we assume $f = Z(z)T(t)$ where, $Z(z)$ and $T(t)$ are the functions of z only and t only, respectively. Substitution for $f = Z(z)T(t)$ in the above Equation (1.45) and diving both sides by v^2 we obtain,

$$\frac{d^2Z(z)}{dz^2} + \frac{c^2}{v^2}Z(z) = 0 \tag{1.46}$$

$$\frac{d^2T(t)}{dt^2} + c^2T(t) = 0 \tag{1.47}$$

Here c^2 is the separation variable. The solutions for the above two double derivative equations are:

$$Z(z) = A_1e^{-jcz/v} + A_2e^{jcz/v} \tag{1.48}$$

$$T(t) = Ce^{jct} \tag{1.49}$$

and the complete solution for $f(z, t)$ is obtained by the product of the two above Equations (1.48) and (1.49):

$$f(z,t) = \left(A_1e^{-jcz/v} + A_2e^{jcz/v}\right)Ce^{jct} = Ae^{jc(t+(z/v))} + Be^{jc(t-(z/v))} \tag{1.50}$$

Review Questions 1.7: Derivation of General Wave Equation
Q1. Derive the wave equation of vibrating strings. **Q2.** What is the velocity of the wave?

We find that the solution has two parts: (i) one with the function of $(t-(z/v))$ and can be deifned as $F_1(t-(z/v))$, and (ii) second one with the function of $(t+(z/v))$ and that can be defined as $F_2(t+(z/v))$. These functions must satisfy the generic wave Equation (1.45).

The function $F_1(t-(z/v))$ represents the forward travelling wave along +ve z-direction and $F_2(t+(z/v))$ represents the reverse travelling wave in the −ve z-direction, both are with the same velocity (m/s). Figure 1.7 illustrates the forward and reverse travelling wave function $F_t(t-(z/v))$ and $F_r(t+(z/v))$ in the sinusoidal form.

1.5 Wave Motion and Wave Front

We have defined the wave velocity $v = \left(1/\sqrt{\mu\varepsilon}\right)$ in a linear, isotropic, and homogeneous medium with the constitutive parameters ε and μ. Now we understand the physical significance of the wave motion and the wave front. These two concepts are vital for the uniform plane wave theory.

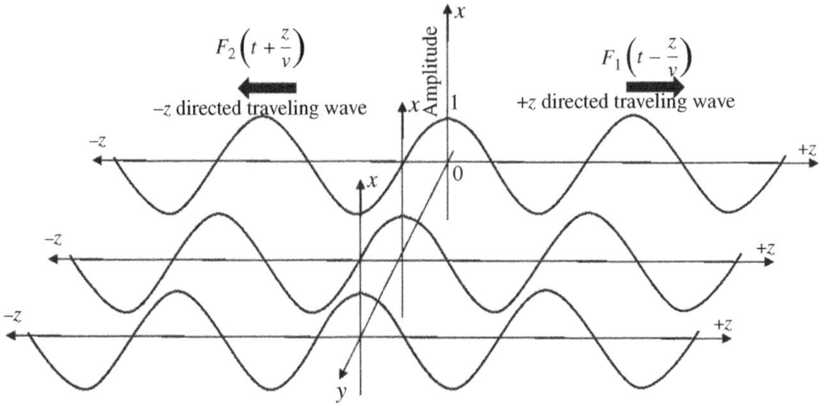

Figure 1.7 Concept of the field solution of the plane wave propagation with forward and reverse travelling waves on *xz*-plane with the propagation direction along *z*-axis.

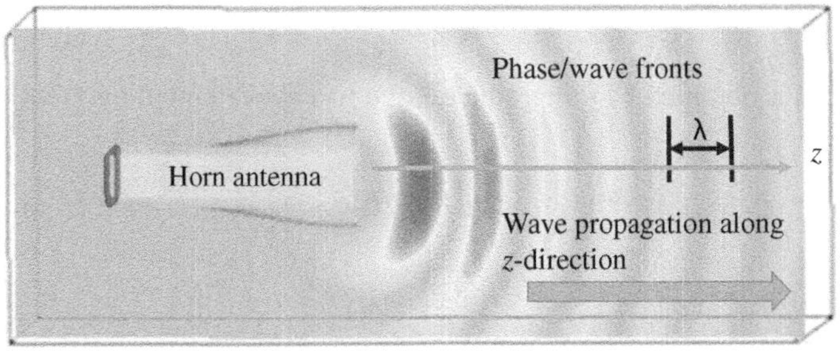

Figure 1.8 Concept of the wave front of the EM wave as it propagates away from a horn antenna (CST Microwave Studio rendition).

The wave motion is defined as the uniform propagation speed at which the constant *phase front*, also called *wave front*, of the wave energy travels. Each surface of the constant phase is called *the wave front.* Figure 1.8 illustrates such wave front propagates from a horn antenna as generated using CST Microwave Studio. As you may observe in the CST rendition of the wave propagation of an electromagnetic wave, for each wave there is a *characteristic distance* between two successive wave fronts of the same family of the wave front of which all members are simultaneously in the same phase. This characteristic distance is called *the wavelength.* The wavelength is related with the frequency of operation as $\lambda = v/f$ where f is the frequency of wave propagation.

Example 1.4: Derivation of Electromagnetic Wave Equation
If the horn antenna is propagating an electromagnetic wave at 9 GHz in air with $\varepsilon = 8.854 \times 10^{-12}$ (F/m) and $\mu = 4.475 \times 10^{-7}$ (H/m). Calculate the wavelength of the wave.

Answer:

$$\lambda = \frac{v}{f} = \frac{c}{f} = \frac{3 \times 10^8\ (\text{m/s})}{9 \times 10^9\ (\text{Hz})} = 3.3\ (\text{cm})$$

The above wave front is also called *the wave train*. This is a continuous sequence of wave fronts of which those intervals of the constant wavelength λ, and also called constant time period $T = 1/f$, and they are always in the same phase. As you can also see that the wave fronts near the antenna aperture is spherical and as the wave propagates away from the antenna aperture it become planar. In all cases the wave fronts are in the same phase. Depending on the shapes of the wave front they have different names. As for example, the wave fronts near the antenna consist of surfaces of concentric spherical surfaces. In this case, they are called *spherical waves*. This wave is also called *the near field* wave in antenna nomenclature.[14] In a case, if we can make a wave front in cylindrical shape then the wave is called *cylindrical wave*. Therefore, depending on the shapes of the wave fronts we define the electromagnetic waves.

As the wave propagates away from the antenna, the field which is planar in shape is called *the far field* wave in the antenna nomenclature. The far field wave is also called *plane wave* meaning the phase front of the wave consisting of a plane surface as shown in Figure 1.8. If the amplitude of the field vectors remains constant over each plane which is perpendicular to the direction of propagation, then we call it as a *uniform plane wave* or in short UPW.

Review Questions 1.8: Introduction to Uniform Plane Wave

Q1. Define wavelength of a time varying electromagnetic propagating signal
Q2. What is wave motion?
Q3. Show that a uniform plane wave has two solutions, one is forward travelling wave and one is reverse travelling wave.
Q4. Define uniform plane wave.

1.6 Phase Velocity of UPW

In the preceding section, we have examined that the function $F_1(t-(z/v))$ is a solution of an electromagnetic wave equation that travels at a velocity v. Now we shall develop the physical concept of the wave velocity in this section. We all have examined that the wave equation has two solutions $F_1(t-(z/v))$ and $F_2(t+(z/v))$, the forward and reverse travelling wave functions, respectively. The wave function has two independent variables, the distance z along z-axis and time t. As the wave travels, the time increases with the distance irrespective of the direction of the wave propagation. To understand the physical concept appropriately, let us take an example of a car traveling with a certain uniform speed v along z-axis as shown in Figure 1.9. Assume that the location of the car is $z = z_1$ at an instant of time $t = t_1$. As the car travels with time, the location of the car is $z = z_2$ at the instant

14 The near field waves are of two types: (i) the spherical waves which are very close to the antenna aperture are called non-radiating near field wave, and (ii) the spherical waves which are in between the non-radiating near field wave and the far field waves are called radiating near field waves. In the first case of the waves, the non-radiating reactive energy is stored in the waves and does not contribute in the real power radiation. The radiating near field starts detaching the radiative field's energies in real power. However, the phase front of the wave remains curvilinear.

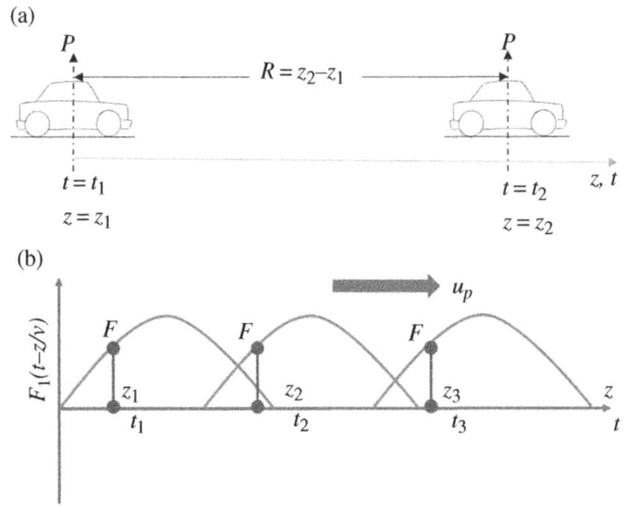

Figure 1.9 Concept of the phase velocity (a) analogy of car travelling a distance R at a velocity $v = ((z_2 - z_1)/(t_2 - t_1)) = (R/(t_2 - t_1))$, and (b) the constant phase point F of a wave function $F_1(t-z/v)$ travelling along z-axis with a phase velocity u_p.

of time $t = t_2$. By this time, the car travels a distance R with a uniform velocity v. Therefore, the magnitude of velocity of the car along z-axis is:

$$v = \frac{z_2 - z_1}{t_2 - t_1} = \frac{R}{t_2 - t_1} \text{ (m/s)} \tag{1.51}$$

If we want to measure the velocity of the car precisely, we need to measure the distance and time precisely as well. To measure the distance in precision details, we need to fix a point on the car as shown in Figure 1.9a. At the time instant t_1, the location point P on the car is always fixed as for the time instants t_2, and t_3 etc. The fixed-point P on the car is equivalent to the fixed or constant phase or the initial state of a uniform plane wave.[15]

Now look at the plots of the wave function $F_1(t-(z/v))$ at different locations along the z-axis as the time progresses from t_1 to t_2 to t_3. At these time instances, the wave travels from z_1 to z_2 to z_3 along the z-axis, respectively. In the three progressive plots of the wave functions $F_1(t-(z/v))$ we set a fixed reference amplitude position F on the plots and set the three distances at the three distance–time instances (z_1, t_1), (z_2, t_2), and (z_3, t_3), respectively, as shown in Figure 1.9b. This means as time increases, the same plot moves to a distance along the positive z-axis. The two independent variables are z and t that are denoted along the horizontal axis. In all cases, we assume the velocity v to be constant at all time. The above examination reveals that as time t increases the distance z must increase in order that $(t-(z/v))$ remains constant at all time.

15 The phase of the signal is such a concept that needs to be measured precisely. We may ignore small deviation of amplitude but not the phase. We hinted this issue in the development of the concept of the retarded potential. In the antenna radiation theory, we shall consider the minute phase variation very seriously, and include minute phase variation in the calculation. The direction of the radiated wave propagation and beamforming of the antenna are highly dependent on the phase excitation of individual antenna elements.

This means the solution of the wave function $F_1(t-(z/v))$ is constant as long as $(t-(z/v))$ is constant. And in this case, we designate $F_1(t-(z/v)) = F$ for all of the three-time instances and travelled distances.

Now to determine the velocity of the wave propagation, we focus our attention to the reference point F on the plot of $F_1(t-(z/v))$. We have observed that the fixed amplitude point F is moving with the plot of $F_1(t-(z/v))$ along the z-direction. Its height is equivalent to the phase state of the wave function and must remain constant all the time. Hence the argument $(t-(z/v))$ must remain constant all the time.

Now we are in a position to define the velocity of the wave in the physical sense as we can measure precisely the distance and time based on the above explanation. We set a reference point F on the plots of the wave function that is propagating along the z-axis at different time instances with a uniform velocity v. The velocity of the wave function is obtained simply taking the derivative of distance z with respect to time t. This means the velocity of the constant phase point F is:

$$v = \frac{dz}{dt} \tag{1.52}$$

This means the velocity with which a point of the constant phase of the wave is propagating is simply $v = (dz/dt)$. Thus, we conclude that a wave function $F_1(t-(z/v))$ that moves at a fixed velocity v along the positive z-direction. Likewise we can also show that $F_2(t+(z/v))$ is moving along the negative z-direction at a velocity $v = (dz/dt)$ to keep the argument $(t+(z/v))$ fixed. Here, as t increases z must increase negatively with a proportionality constant $1/v$ to keep the phase of the wave function fixed.

We have already defined $= \left(1/\sqrt{\mu\varepsilon}\right)$ (m/s). The velocity is known as *the phase velocity* of the uniform plane wave, and in many instances, it is denoted as u_p, as shown in Figure 1.9b.

Example 1.5: Derivation of Phase Velocity
Show that the phase velocity has unit (m/s).

Answer:
Permittivity has unit F/m = $(coulomb)^2/\{\text{N}.(\text{m})\}^2$, permeability has unit H/m = V.s/(A.m), *energy* = N.m = V.A.s, *coulomb* = A.s

Answer:

$$\mu\varepsilon = \frac{\text{V.s}}{\text{A.m}} \times \frac{(coulomb)^2}{\text{N.}(\text{m})^2} = \frac{\text{V.s.}(\text{A.s})^2}{\text{A.}(\text{V.s.A}).(\text{m})^2} = \frac{(\text{s})^2}{(\text{m})^2}$$

$$\therefore v = \frac{1}{\sqrt{\mu\varepsilon}}\left(\frac{\text{m}}{\text{s}}\right)$$

Thus, v has the dimension of velocity (m/s), and is independent of frequency, but dependent on the constitutive parameters μ and ε. It represents the velocity of the phase or the original state of the signal. It is important to point out that it is not the velocity of energy of the wave. We shall discuss the energy velocity, which is called the *group velocity*, during developing the concept of uniform plane wave for time harmonic cases.

Review Questions 1.9: Phase Velocity to Uniform Plane Wave

Q1. Define phase velocity. What is the physical meaning of the phase velocity of the electromagnetic wave?
Q2. Who described that the light wave was an electromagnetic wave?
Q3. Explain the similarities of the electromagnetic and light waves in terms of the phase velocity
Q4. What is group velocity?

Exercise 1.3: Phase Velocity to Uniform Plane Wave
Q1. Show that in free space the phase velocity of the electromagnetic wave is $v = 3 \times 10^8$ (m/s).

Hints: $\varepsilon \cong (1/36\pi) \times 10^{-9}$ (F/m) and $\mu = 4\pi \times 10^{-7}$ (H/m).

Q2. A coaxial cable has a Teflon filling which has $\varepsilon_r = 2.3$ and $\mu_r = 1$. Calculate the electromagnetic signal's phase velocity in the coaxial cable. Compare the velocity with respect to the free space phase velocity.

Answer:
1.978×10^8 (m/s); 66% of free space velocity.

In free space and in vacuum, the electromagnetic wave travels with the velocity equal to the speed of light 3×10^8 (m/s).[16] We therefore, both light and electromagnetic plane waves travel with the same velocity in vacuum. This tells us that light wave and electromagnetic waves are similar in nature. They are similar in many ways such as: (i) both waves are transverse to the direction of wave propagation, (ii) in free space both types of waves have exactly similar behavior, and (iii) the properties of reflection, refraction, and transmission of both wave types have similarities not only in free space or vacuum but also in any medium. These similarities laid the foundations for the electromagnetic wave theory of light wave by Maxwell. This textbook covers the theory of both electromagnetic and light waves for the advanced level applications of microwave passive design such as metallic waveguides, antennas, optical waveguides, and optical fibers, respectively.

So far, we have examined the wave motion of the plane wave. From the examination we have understood a few salient features of the plane wave: (i) a plane wave can travel in both −ve and +ve z-directions, (ii) the wave travels with a uniform velocity v, (iii) the reference phase point on the wave is always constant, and (iv) the velocity of the wave is independent of frequency of operation, but depends on the constitutive parameters of the medium of propagation.

Now let us go back to the wave equation of a particular wave function say the z-directed travelling electric field component E_x:

$$\frac{\partial^2 E_x}{\partial^2 z} = \mu\varepsilon \frac{\partial^2 E_x}{\partial t^2} \tag{1.53}$$

16 Maxwell has discovered the phenomena of the similarity between the electromagnetic and light waves. Therefore, covering the light wave theory in terms of the electromagnetic phenomena and electromagnetic field solutions is a salient feature of the book compared to other traditional textbooks on electromagnetism.

Using the definition of the phase velocity $v = \left(1/\sqrt{\mu\varepsilon}\right)$, we can rewire the equation as

$$\frac{\partial^2 E_x}{\partial z^2} = \frac{1}{v^2}\frac{\partial^2 E_x}{\partial t^2} \tag{1.54}$$

Based on our previous examination of the solution for E_x we can write the general solution of the one dimensional wave equations as

$$E_x = f_1\left(t - \frac{z}{v}\right) + f_2\left(t + \frac{z}{v}\right) \tag{1.55}$$

where, f_1 and f_2 are two arbitrary functions of $(t - (z/v))$ and $(t + (z/v))$, that travel along +ve and −ve z-directions, respectively. In electromagnetic nomenclature we denote the forward travelling wave as

$$E_x^+(z, t) = f_1\left(t - \frac{z}{v}\right) \tag{1.56a}$$

and the reverse travelling wave as

$$E_x^-(z, t) = f_2\left(t + \frac{z}{v}\right) \tag{1.56b}$$

Therefore, we can write the complete solution as:

$$E_x(z, t) = E_x^+(z, t) + E_x^-(z, t) \tag{1.56c}$$

Likewise we can develop the magnetic wave field solutions from the wave equation

$$\frac{\partial^2 H_y}{\partial^2 z} = \frac{1}{v^2}\frac{\partial^2 H_y}{\partial t^2} \tag{1.57}$$

for H_y field component as:

$$H_y(z, t) = H_y^+(z, t) + H_y^-(z, t) \tag{1.58}$$

In the next section we shall use the above equation to develop the relationship between the corresponding magnetic field component to develop the concept of wave impedance. Again we shall resort to Maxwell's equations in the development.

Exercise 1.4: One-Dimensional Wave Equation

Q1. Show that the general solution of the one-dimensional wave equation for the magnetic field is:

$$H_y(z, t) = H_y^+(z, t) + H_y^-(z, t)$$

1.7 Wave Impedance

The concept of wave impedance is somewhat different than the impedance that we are used to in the theory of alternating current (AC) circuit. In the AC circuit, the impedance is fixed for a particular frequency, and is defined as the ratio of the voltage and current waveforms. The impedance of an AC circuit is a complex quantity and is comprised of the real and reactive circuit elements. In free space electromagnetic wave propagation, we define the intrinsic wave impedance that relates the time varying electric and magnetic fields at any point in space. The impedance is a function of the constitutive parameters of the medium such as σ, ε, and μ. In lossless, homogeneous, and linear medium, the intrinsic impedance is a real quantity and does not change with frequency.

In this section, we shall develop the physical understanding of *the wave impedance*, also called *the intrinsic wave impedance*. As coined by Maxwell in his legendary theory, the electromagnetic wave carries real power as it propagates along the direction of propagation. Take the example of a radar as shown in Figure 1.10. The wave from the radar's transmitter travels with the speed of light and hits a target at a certain distance in free space. As the electromagnetic wave hits the target, it experiences a different impedance than that for the free space. Therefore, the reflection of the electromagnetic wave is due to the fact that the intrinsic impedance of the object is different from that of the free space. The intrinsic impedance is defined by the ratio of the electric and magnetic field intensities at a point in space. Therefore, we understand that the reflection of the electromagnetic wave is due to the change in the intrinsic impedance of the medium in which it propagates. The reflected signal from the target starts propagating in the reverse direction and the radar's antenna receives the returned echo of the target.[17] The sensitive radar receiver processes the weak received signal and provides useful information of the target. Usually, a radar provides the target's distance from the radar, the speed and direction of travel of the target, and the bearing of the target.

Now let us derive the intrinsic wave impedance using Maxwell's equation that we derived in Section 1.6:

$$\frac{\partial E_x}{\partial z} = -\mu \frac{\partial H_y}{\partial t} \tag{1.31}$$

Substitution for

$$E_x(z,t) = E_x^+(z,t) + E_x^-(z,t) = f_1\left(t - \left(\frac{z}{v}\right)\right) + f_2\left(t + \left(\frac{z}{v}\right)\right) \tag{1.59}$$

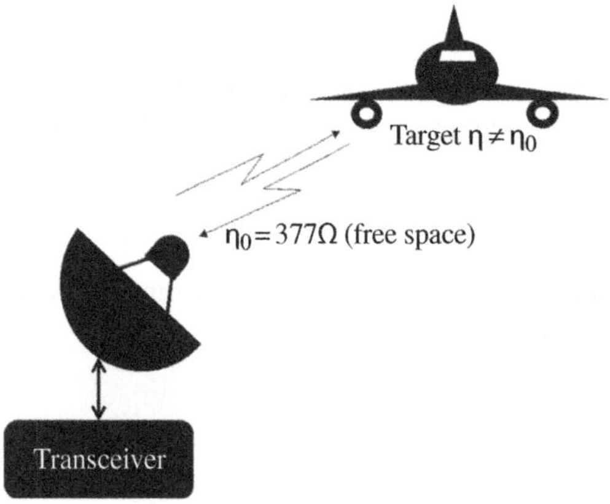

Figure 1.10 A radar hits a target in free space and the returned echo from the target is received by the radar receiver. The reflection is due to the fact that the intrinsic impedance of the object is different than that for the free space.

17 In previous section we solved wave equation and found the general solution has two components: the forward travelling wave function $F_1(t-(z/v))$ and the reverse travelling wave function $F_2(t+(z/v))$. The reflected signal is the reverse travelling wave in the radar example.

in the left-hand side of the above equation yields,

$$\frac{\partial E_x(z,t)}{\partial z} = -\frac{1}{v}f_1'\left(t-\left(\frac{z}{v}\right)\right) + \frac{1}{v}f_2'\left(t+\left(\frac{z}{v}\right)\right) = -\mu\frac{\partial H_y}{\partial t} \tag{1.60}$$

Now integrating Equation (1.60) with respect to,[18] we obtain:

$$H_y(z,t) = H_y^+(z,t) + H_y^-(z,t) = \frac{1}{\mu v}E_x^+(z,t) - \frac{1}{\mu v}E_x^-(z,t) \tag{1.61}$$

As can be seen, H_y has two parts: (i) a forward travelling wave component $H_y^+(z,t)$, and (ii) a reverse travelling wave component: $H_y^-(z,t)$. Linking the respective forward and reverse travelling waves of both electric and magnetic field components, we can define the intrinsic wave impedance η as follow:

$$\eta = \frac{E_x^+(z,t)}{H_y^+(z,t)} = -\frac{E_x^-(z,t)}{H_y^-(z,t)} = \frac{1}{\mu v} = \sqrt{\frac{\mu}{\varepsilon}}\ (\Omega) \tag{1.62}$$

The intrinsic wave impedance is always positive and has dimension of ohms. It is very useful in wave analysis. In free space, the intrinsic wave impedance is:

$$\eta = \sqrt{\frac{\mu_0}{\varepsilon_0}} = \sqrt{\frac{4\pi\times 10^{-7}}{\frac{1}{36\pi}\times 10^{-12}}} \cong 377\ (\Omega) \tag{1.63}$$

Review Questions 1.10: Wave Impedance

Q1. What is intrinsic wave impedance?
Q2. What is the value of intrinsic wave impedance in vacuum?
Q3. Show that the unit of intrinsic wave impedance is Ohm.[19]

Exercise 1.5: One-dimensional Wave Equation
Q1. Derive the wave impedance using the following Maxwell's equation: $(\partial E_y/\partial z) = \mu(\partial H_x/\partial t)$

1.8 Time Harmonic Field Wave Equations

In the preceding chapter we have already introduced the time harmonic fields and its significance. In engineering applications, the sinusoidally varying or the time harmonic fields are of special significance. In laboratory and in the field, we measure such time varying field waveforms using oscilloscopes. Even if the field variation is not purely sinusoidal, any field function can be decomposed

18 Here we neglect any constant of integration which can give rise to at best a static field.
19 Use the proper dimensions of permittivity and permeability.

with Fourier series expansion as the combinations of purely sinusoidally varying field functions. In addition to these, the time harmonic field expression in the phasor vector formed with the aid of Euler's theorem simplifies the field analyses of complex problem so greatly that it becomes an integral tool for practical field solutions in the electromagnetic theory.

Now we go back to the Helmholtz generic wave equations for the electric and magnetic fields that we derived in the beginning of the chapter:

$$\nabla^2\boldsymbol{E} - \mu\sigma\frac{\partial \boldsymbol{E}}{\partial t} - \mu\varepsilon\frac{\partial^2 \boldsymbol{E}}{\partial t^2} = 0 \tag{1.13}$$

$$\nabla^2\boldsymbol{H} - \mu\sigma\frac{\partial \boldsymbol{H}}{\partial t} - \mu\varepsilon\frac{\partial^2 \boldsymbol{H}}{\partial t^2} = 0 \tag{1.15}$$

In time harmonic field analysis, Helmholtz simplified the time harmonic case by simply substituting the time derivative $\partial/\partial t = j\omega$ in the above generic field wave equations. How we are getting this? Do you have a clue?

Euler's theorem defines:

$$e^{j\omega t} = cos\,(\omega t) + j\,sin(\omega t) \tag{1.64}$$

Therefore, if we take time derivative of $e^{j\omega t}$, we obtain

$$\frac{de^{j\omega t}}{dt} = j\omega e^{j\omega t};\ \frac{d^2e^{j\omega t}}{dt^2} = (j\omega)(j\omega)e^{j\omega t} = -\omega^2 e^{j\omega t} \tag{1.65}$$

This gives us the substitution of generic time varying field quantities to time harmonic case $(\partial/\partial t) = j\omega$. Thus, Maxwell's equations in source free region can be transformed into their time harmonic expressions as follows:

Time-varying field expression		Time-harmonic field expression	
$\nabla\cdot\mathbf{E} = 0$	(1.66a)	$\nabla\cdot\boldsymbol{E}_s = 0$	(1.67a)
$\nabla\cdot\boldsymbol{H} = 0$	(1.66b)	$\nabla\cdot\boldsymbol{H}_s = 0$	(1.67b)
$\nabla\times\boldsymbol{E} = -\mu\frac{\partial \boldsymbol{H}}{\partial t}$	(1.66c)	$\nabla\times\boldsymbol{E}_s = -j\omega\mu\boldsymbol{H}_s$	(1.67c)
$\nabla\times\boldsymbol{H} = \varepsilon\frac{\partial \boldsymbol{E}}{\partial t}$	(1.66d)	$\nabla\times\boldsymbol{H}_s = j\omega\varepsilon\boldsymbol{E}_s$	(1.67d)

We define the time harmonic field expressions with a subscript s where $s = j\omega$ is implied as shown above. Now let us go back to the two generic field wave equations and substitute $\partial/\partial t = j\omega$ then we obtain,

$$\nabla^2\boldsymbol{E} - j\omega\mu\sigma\boldsymbol{E} + \omega^2\mu\varepsilon\boldsymbol{E} = \nabla^2\boldsymbol{E} - j\omega\mu(\sigma + j\omega\varepsilon)\boldsymbol{E} = 0 \tag{1.68}$$

$$\nabla^2\boldsymbol{H} - j\omega\mu\sigma\boldsymbol{H} + \omega^2\mu\varepsilon\boldsymbol{H} = \nabla^2\boldsymbol{H} - j\omega\mu(\sigma + j\omega\varepsilon)\boldsymbol{H} = 0 \tag{1.69}$$

With the substitution, the previous generic wave equations with time derivatives become such a simplified form without any derivative or fraction, they just become a straight equation in one line. Therefore, substitution for $\partial/\partial t$ with $j\omega$ simplifies the wave equations significantly. In short, we can write the above equations as

$$\nabla^2\boldsymbol{E} - \gamma^2\boldsymbol{E} = 0 \tag{1.70}$$

$$\nabla^2\boldsymbol{H} - \gamma^2\boldsymbol{H} = 0 \tag{1.71}$$

The general solutions of the electric and magnetic field wave equations can be easily derived using the exponential form as follows:

$$\boldsymbol{E}_{xs} = E_0^+ e^{-\gamma z} + E_0^- e^{\gamma z} \tag{1.72}$$

and,

$$\boldsymbol{H}_{ys} = H_0^+ e^{-\gamma z} + H_0^- e^{\gamma z} = \frac{E_0^+}{\eta} e^{-\gamma z} - \frac{E_0^-}{\eta} e^{\gamma z} \tag{1.73}$$

where

$$\gamma = \sqrt{j\omega\mu(\sigma + j\omega\varepsilon)} = \alpha + j\beta \tag{1.74}$$

is called the propagation constant with its unit (1/m). And the two components of γ are: α is called *the attenuation constant* (Np/m) and β is called *the phase constant* (rad/m).[20] Both are real positive numbers. For a lossy medium with finite conductivity σ, the propagation constant γ is a complex number. If the medium is lossless, then α is zero and $\gamma = j\beta$ is fully imaginary that represents the oscillatory nature and represents uninterrupted propagation of the wave. Now we derive the expressions for α and β in terms of the constitutive parameters.

Taking square of the propagation constant in (1.75), we obtain:

$$\gamma^2 = \alpha^2 - \beta^2 + j2\alpha\beta = j\omega\mu(\sigma + j\omega\varepsilon) = -\omega^2\mu\varepsilon + j\omega\mu\varepsilon \tag{1.75}$$

Separating the real and imaginary parts of the above equation, respectively, we obtain

$$\alpha^2 - \beta^2 = -\omega^2\mu\varepsilon \tag{1.76a}$$

and

$$2\alpha\beta = \omega\mu\sigma \tag{1.76b}$$

Again, taking the square of the absolute value of γ in (1.75), we obtain,

$$|\gamma|^2 = \alpha^2 + \beta^2 = \omega^2\mu\varepsilon\sqrt{1 + (\sigma/\omega\varepsilon)^2} \tag{1.77}$$

Adding Equations (1.76a) and (1.77) we obtain:

$$2\alpha^2 = \omega^2\mu\varepsilon\left\{\sqrt{1 + \left(\frac{\sigma}{\omega\varepsilon}\right)^2} - 1\right\} \tag{1.78}$$

$$\therefore \alpha = \omega\sqrt{\frac{\mu\varepsilon}{2}\left[\sqrt{1 + \left(\frac{\sigma}{\omega\varepsilon}\right)^2} - 1\right]} \tag{1.79}$$

Subtracting Equation (1.76a) from Equation (1.77) and with simple manipulation as above we obtain:

$$\beta = \omega\sqrt{\frac{\mu\varepsilon}{2}\left[\sqrt{1 + \left(\frac{\sigma}{\omega\varepsilon}\right)^2} + 1\right]} \tag{1.80}$$

---

20 The phase constant β (rad/m) is also called the wave number. The physical meaning of the wave number of a wave is that how many radians per meter the wave rotates in the phase plane where one wavelength λ represents 2π angular rotation. Therefore, β increases with frequency as well as with the constitute parameter ε and μ.

Example 1.6: Wave Propagation in Lossy Medium
Given, the operating frequency $f = 1$ GHz, the conductivity of the medium $\sigma = 0.1$ (S/m), relative permittivity $\varepsilon_r = 2.2$, and the relative permeability $\mu_r = 1$, calculate the attenuation constant α in (Np/m) and the propagation constant β in (rad/m).

Answer:
We shall work out the problem step by step. First angular frequency $\omega = 2\pi f$. Then $\mu\varepsilon/2$, $(\sigma/\omega\varepsilon)$, $\sqrt{1+(\sigma/\omega\varepsilon)^2}$ and finally, we substitute all the calculated numerical values of the given parameters into expression for $\alpha = \omega\sqrt{(\mu\varepsilon/2)\left[\sqrt{1+(\sigma/\omega\varepsilon)^2}-1\right]}$ and

$$\beta = \omega\sqrt{\frac{\mu\varepsilon}{2}\left[\sqrt{1+\left(\frac{\sigma}{\omega\varepsilon}\right)^2}+1\right]}.$$

This yields $\alpha = 22.6$ (Np/m), the unit of attenuation constant. (Np/m) is in natural log scale because we are dealing with exponential form of the wave equation e^{α}. *Np* means at what distance the amplitude diminishes by $1/e$ of its original value. 1 Np = 8.686 dB.
The angular frequency is derived as $\omega = 2\pi f = 2\pi \times 10^9$ (rad/s)
Now we calculate the parameters in the equations of α and β one by one

$$\frac{\mu\varepsilon}{2} = \frac{4\pi\times 10^{-7}\times\left(\dfrac{10^{-9}}{36\pi}\right)\times 2.2}{2} = 1.22\times 10^{-17}$$

$$\frac{\sigma}{\omega\varepsilon} = \frac{0.1}{2\pi\times 10^{-9}\times\left(\dfrac{10^{-9}}{36\pi}\right)} = 1.8;\ \sqrt{1+\left(\frac{\sigma}{\omega\varepsilon}\right)^2} = 2.06$$

$$\therefore \alpha = 2\pi\times 10^9\sqrt{1.22\times 10^{-17}\left[\sqrt{2.06}-1\right]} = 22.6\,(\text{Np/m})$$

1 Np = 8.686 dB

$$\therefore \alpha = 22.6\times 8.686 = 196.3\,(\text{dB/m})$$

The phase constant is: $\beta = 2\pi\times 10^9\sqrt{1.22\times 10^{-17}\left[\sqrt{2.06}+1\right]} = 38.4\,(\text{rad/m})$
As stated before, the phase constant is also called *the wave number.*

α and β in Lossless Medium
For loss less medium $\sigma = 0$, this simplifies the calculation significantly as follows:

$$\alpha = \omega\sqrt{\frac{\mu\varepsilon}{2}\left[\sqrt{1+\left(\frac{0}{\omega\varepsilon}\right)^2}-1\right]} = 0 \tag{1.81}$$

$$\beta = \omega\sqrt{\frac{\mu\varepsilon}{2}\left[\sqrt{1+\left(\frac{0}{\omega\varepsilon}\right)^2}+1\right]} = \omega\sqrt{\mu\varepsilon} = \frac{2\pi f}{u_p} = \frac{2\pi}{\lambda} \tag{1.82}$$

$$\therefore \beta = \frac{\omega}{u_p} = \frac{2\pi}{\lambda} \tag{1.83}$$

The phase velocity is:

$$u_p = \frac{\omega}{\beta} \tag{1.84}$$

Substitutions for the value of angular frequency $\omega = 2\pi \times 10^9$ (rad/s) and the free space wave velocity

$$u_0 = \frac{1}{\sqrt{\mu_0 \varepsilon_0}} = 3 \times 10^8 \text{ (m/s)} \text{ yield:}$$

$$\beta = \frac{2\pi \times 10^9}{\frac{3 \times 10^8}{\sqrt{2.2}}} = 31.06 \text{ (rad/m)}$$

Review Questions 1.11: Wave Propagation in Lossy Medium

Q1. What is α? What is β? Can you explain them in terms of exponential form of these quantities?

Q2. We do not like a lossy medium as a wave losses its power due to the finite conductivity of the media. Explain from the above examinations of propagation and attenuation constants, and wave impedance, how we can know if a medium is lossy and how much power we are losing.

Q3. Explain the significance of lossy and low loss media in RF/microwave/mm-wave active and passive circuit design.

Q4. Explain if you prefer to use a low cost but highly lossy FR4 laminate for microwave design.

Q5. As a microwave design engineer, explain how do you modify the laminate for efficient design with low cost FR4 laminate.

1.8.1 Summary of Propagation Constant

1) If we put α, a real number as the exponent with its negative value along with the distance z, $e^{-\alpha z}$; this represents exponentially decaying factor as the wave propagates along the positive z-direction. Thus, α is called *the attenuation constant.*
2) If we put $j\beta$, an imaginary number as an exponent with its negative value along with the distance z, $e^{-j\beta z}$, this represents uninterrupted oscillating factor as the wave propagates infinitely along the z-direction. This is an oscillating quantity with phase velocity u_p, time period T, and wavelength λ. β is called *the phase constant* or *the wave number.*
3) The phase constant is also called *the wave number*, because it tells how many radians a wave cycles in one meter. This is a very important concept for practical design. Therefore, development of this concept of propagation constant, attenuation and phase constant is very significant in practical design exercises.
4) The propagation constant β is very important concept in wave propagation as it conveys many important pieces of information about the characteristics of the wave in the specific media and guided structures. We shall cover these elements in the subsequent chapters.

5) The concepts of the attenuation constant α and the phase constant β are integral parts of any wave type.
6) For simplicity, we, in most cases, neglect the attenuation constant α and only work on the phase constant β. For loss less media, the conductivity σ is zero, then the propagation constant γ reduces to $j\beta$, where $\beta = \omega\sqrt{\mu\varepsilon} = 2\pi/\lambda\ (\text{rad/m})$. This means the wave does not attenuate as it travels. We use β in every calculation of plane wave, transmission lines, waveguides, optical fibers and antennas.
7) Finally, the phase velocity, u_p, which is also called the velocity of wave propagation, or you can say the velocity of the wave front is $u_p = \omega/\beta$. This is another important expression that we shall be using in the above occasions.

1.9 Refractive Index of Medium and Dispersion

The refractive index of an optical medium is a very important design parameter in optical fiber design. It is considered to be one of the most important characteristic parameters of wave propagation in optical spectra. Maxwell hypothesized the concept of the light waves as electromagnetic waves from the definition of the refractive index n. The refractive index n of a medium is defined as the ratio of the phase velocity of light in vacuum to the phase velocity of light in that medium. Thus, the refractive index is defined mathematically as:

$$n = \frac{u_0}{u_p} = \sqrt{\frac{\mu\varepsilon}{\mu_0\varepsilon_0}} = \sqrt{\frac{\mu_0\mu_r\varepsilon_0\varepsilon_r}{\mu_0\varepsilon_0}} = \sqrt{\mu_r\varepsilon_r} \tag{1.85}$$

Most optical media are dielectric media, and the relative permeability $\mu_r \cong 1$, and the refractive index is defined as:

$$n = \sqrt{\varepsilon_r} \tag{1.86}$$

For the above example with the relative permittivity $\varepsilon_r = 2.2$, and the relative permeability $\mu_r = 1$, the refractive index is $n = \sqrt{\varepsilon_r} = 1.48$. Note for most optical materials $\mu = \mu_0$, therefore, (1.86) is a valid expression. On the basis of the expression of n, the refractive index, Maxwell hypothesised that the light wave and the electromagnetic wave have the same wave phenomena.

As stated above, the refractive index of an optical medium is the key parameter in designing optical fibers in which the refractive index is gradually changed for avoidance of modal dispersion. See the two illustrations of two differently refractive index graded optical fibers in Figure 1.11a and observe how by changing the refractive index profile inside an optical fiber the signal propagation changes its profile. Also observe how the incoming pure impulse is distorted at the end of the optical fibers. The profiling of the refractive index inside the optical fiber helps reduce distortion of the light wave inside the optical fiber.

Likewise, our eyes are conceived as the optical devices that communicate optically with the outer world. Figure 1.11b shows the refractive indices of various parts of a human eye. With various eye diseases such as myopia, hyperopia, and astigmatism, the refractive indices of various layers of eyes change and cause focusing problem of optical signals. The rays of lights with a sound eye with 20/20 vision and unfocused ray with eye diseases are shown in Figure 1.11b. Using appropriate lenses in reading glasses we can correct the eyesight errors.

As you may observe in Figure 1.11a that with the change of the refractive index profile, the pulse distortion at a certain distance in the optical fiber can be controlled. This distortion of pulses due to the refractive index change is called *dispersion* and the medium in which the refractive index

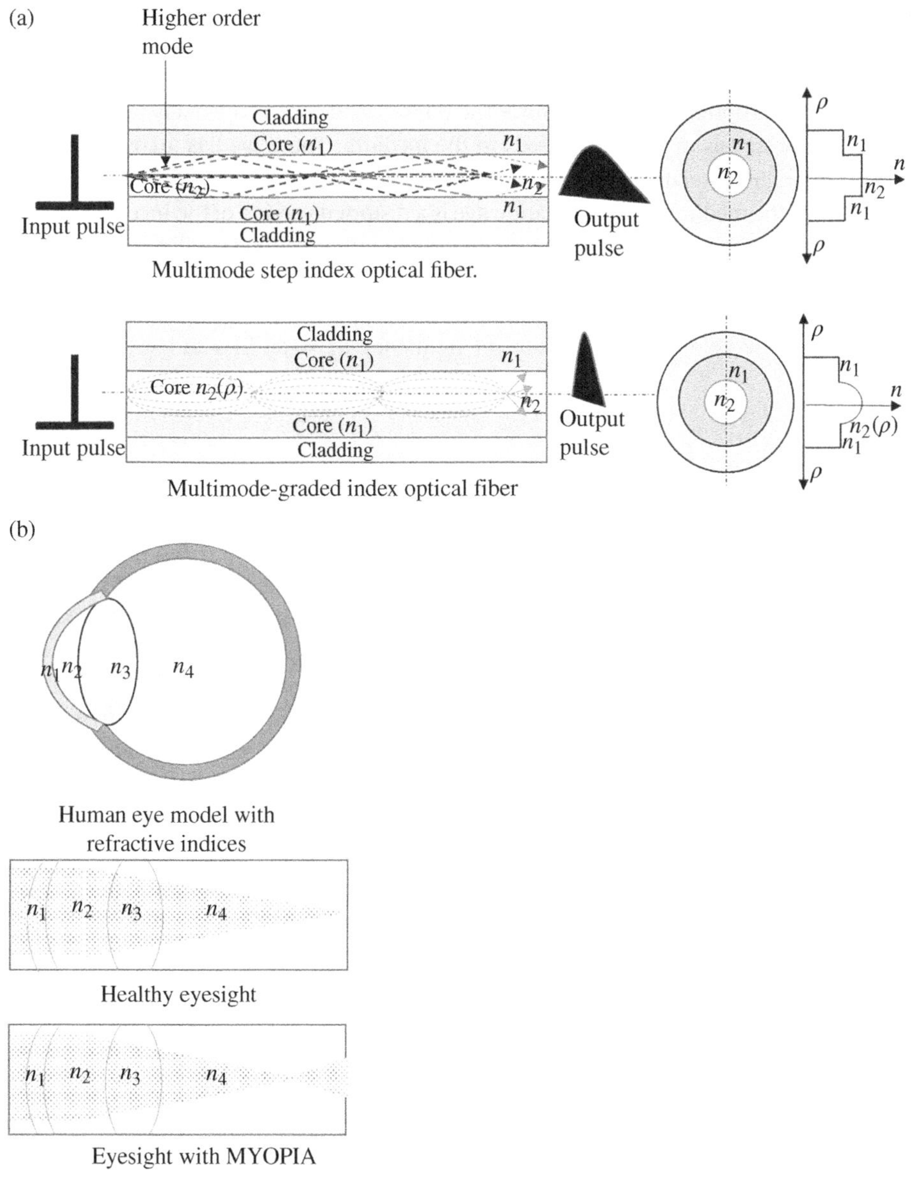

Figure 1.11 (a) Index graded optical fibers, and (b) refractive indices of various parts of a human eye and eyesight with health and diseased eyes.

changes with the frequency is called *the dispersive medium*.[21] This pulse broadening happens with the change of the relative permittivity ε_r of the medium and hence the phase velocity u_p of the wave inside the medium with frequency f. For example, the refractive index of distilled water at microwave frequency is $n = 9$. At optical frequency, such as 500 THz, the frequency of sodium light,[22]

21 The practical design of a low dispersion optical fiber consider various dispersion such as material and waveguide dispersion. Author encourages readers to investigate these topics via open sources.
22 Islam, M.A. (1969). *Electromagnetic Theory*, 348–349. Dacca, East Pakistan: EPUET.

the refractive index $n = 1.33$. This reveals that at the frequency in the optical range, the value of refractive index n does not remain the same as for the microwave frequency. In this example, there is a significant change in the relative permittivity $\varepsilon_r = n^2$ from 81 in the microwave frequency to only 1.77 in the optical frequency. Since the phase velocity u_p is inversely proportional to $\sqrt{\varepsilon_r}$ or $1/n$, the wave travels in different speeds in the medium. The wave has many harmonic components and suffers from relative displacement of the harmonic components in phase as the wave propagates along the medium. The resultant wave is a distorted version of the incoming wave at certain distance in the medium. We define this distorted version of the signal with harmonic components as *the dispersed signal.* This means the signal is scattered or spread out in space and time. This phenomenon of signal broadening is called *dispersion.* The medium as discussed above in which the phase velocity is a function of frequency $u_p = u_p(f)$ is called a *dispersive medium.*[23]

1.9.1 Summary of Wave Propagation in Lossless Medium

Now we can summarize the findings about the EM wave in lossless media as follow.

1) Phase constant beta is $\beta = \omega/u_p = 2\pi/\lambda$ (rad/m). For the last example substitution for the angular frequency ω, free space permittivity ε_0, the medium relative permittivity ε_r and the free space permeability μ_0, we get $\beta = 31.06$ (rad/m).
2) For free space phase velocity, $u_p = c = 3 \times 10^8$ (m/s).
3) The refractive index of a medium is defined as the ratio of phase velocity of a signal in free space versus the velocity in the medium: $n = u_0/u_p$ is simplified by simple mathematical manipulation to $n = \sqrt{\mu_r \varepsilon_r}$.
4) For most optical fibres which are made of dielectric media such as silica or plastic, the permeability $\mu = \mu_0$, gives rise to $n = \sqrt{\varepsilon_r}$. For the above example $n = 1.48$.
5) We use Taconic TLX0 with dielectric constant or ε_r of 2.3. As homework, you find ε_r for most used commercial substrate such as Rogers Duroid, Taconic TLX series and RF4. Example includes fire resistant FR4 $\varepsilon_r = 4.2$ is the low-cost and most-used laminate in low frequency RF while low loss Rogers Duroid, Taconic TLX laminates are used in microwave frequencies.

Review Questions 1.12: Wave Propagation in Lossy Medium

Q1. Define refractive index.
Q2. For an optical design what would be the refractive index?
Q3. Explain how the refractive index is graded in optical fiber design and why?
Q4. What is dispersive medium? How does the refractive index plays a significant role in mitigating dispersion in optical spectra?
Q5. Explain the contrasting difference in the refractive index of a medium in microwave and optical spectra.

23 Chapter 7- Wave Propagation in Dispersive Medium deals with the analyses of waves in various dispersive medium.

1.10 Time Harmonic Wave Solution

As we have defined before that the time harmonic wave equation is a function of sine and cosine functions, and therefore, is oscillatory in nature with an angular frequency $\omega = 2\pi f$ and synchronised with a time period $T = 2\pi/\omega$. The wave also varies sinusoidally as it propagates along the direction of propagation. We have defined the solution of the plane wave with the wave function $F_1(t - (z/v))$, which is a two-dimensional function with space z and time t, respectively as shown in Figure 1.12. Therefore, the second periodicity comes with respect to space with a factor kz. Where $k = 2\pi/\lambda$, where k is called *the wave number* or *phase constant* and λ is *the wavelength*. Note there is a subtle difference between the phase constant β that we derive in the above theory of plane wave and the wave number k, that usually refers to the phase constant in free space. Sometimes in textbooks, k is also replaced with β_u, the free space wave number or propagation constant and β refers to the guided wave number or phase constant.

Let us consider the one-dimensional electric field wave E_x that is propagating along $\pm z$-directions as time t progresses. Revisiting the generic form of the wave equation:

$$\frac{\partial^2 E_x}{\partial z^2} = \frac{1}{v^2}\frac{\partial^2 E_x}{\partial t^2} \tag{1.54}$$

with the propagation velocity $v = \left(1/\sqrt{\mu\varepsilon}\right)$

We already have derived the solution for the wave equation with the forward and backward travelling wave functions as follow[24]:

$$E_x = f_1\left(t - \left(\frac{z}{v}\right)\right) + f_2\left(t + \left(\frac{z}{v}\right)\right) \tag{1.55}$$

We can express the time harmonic case in the sinusoidal function of a specific angular frequency ω and phase velcoity u_p in the following form:

$$\begin{aligned} E_x(z,t) &= E_x^+(z,t) + E_x^-(z,t) \\ &= E_{x0}^+ \cos\left\{\omega\left(t - z/u_p\right) + \phi_1\right\} + E_{x0}^- \cos\left\{\omega\left(t + z/u_p\right) + \phi_2\right\} \\ &= E_{x0}^+ \cos\left\{\omega t - k_o z + \phi_1\right\} + E_{x0}^- \cos\left\{\omega t + k_o z + \phi_2\right\} \end{aligned} \tag{1.87}$$

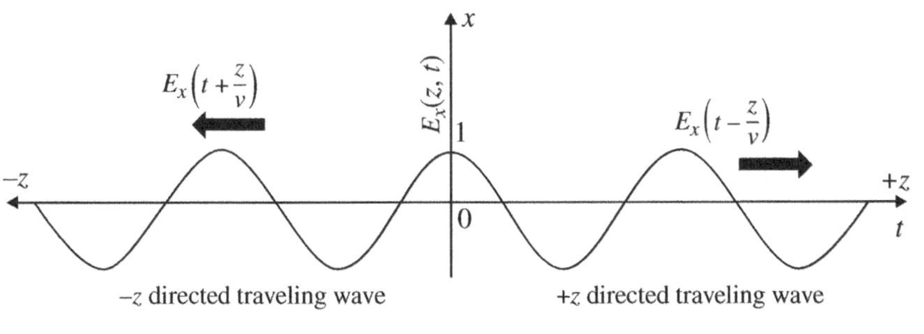

Figure 1.12 General wave solution of an x-directed field propagating in $\pm$ z-direction.

24 Similar derivation is also available in Hayt, W.H. and Buck, J.A. (2001). *Engineering Electromagnetics*, 6e. Boston, Ma, USA: McGraw Hill.

Here,

$$k_o = \omega/u_p \text{ (rad/m)} \tag{1.88}$$

is called the free space *wave number* or *phase constant* of the wave.

As discussed before, the wave number k has a very practical significance in time harmonic electromagnetic analyses. The wave number defines the number of full wave cycles in radian[25] per meter as a wave cycle periodically with 2π radians per cycle. The solution of the time harmonic electric field in the Equation (1.87) is called *the instantaneous* or *time domain* form of the field. It has two components: the first part of the right-hand side is called *the forward travelling wave* and the second part is called *the reverse travelling wave.* Now we can analyze the functions in the parenthesis of cosine functions:

$$cos\,\{\omega t \pm k_o z + \phi_i\}$$

for $i = 1, 2$.

The simplest term is the phase terms ϕ_i for $i = 1$ and 2, that represent the initial state of the signal in the forward and reverse directions, respectively. If properly referenced, the phase quantities can be made zero. For the first and second terms in the parenthesis both have units of angle and expressed in radians. The angular frequency $\omega = 2\pi f$ (rad/s) measures the phase shift of the signal in radians of a periodic signal per second. In a similar annotation, we can define k_o as *the spatial frequency* that measures the phase shift per unit length in radians per meter along the direction of propagation, in this case along z-axis. Assume the time is fixed, the wave travels a full wavelength is the distance over which the spatial phase shift is 2π (rad). At that time the spatial phase shift[26] in free space is defined as:

$$k_o z = k_o \lambda = 2\pi \text{ (rad)} \tag{1.89}$$

This understanding leads to the derivation of the wavelength in terms of the phase constant as follows:

$$\lambda = \frac{2\pi}{k_o} \text{ (m)} \tag{1.90}$$

As the distance progresses with time, the instantaneous field in Equation (1.87) above travels with the sinusoidal function in both space and time. Therefore, there are crests and troughs in the waveform. Neglecting the phase shifts $\phi_{I,2}$, which represent the initial conditions of the waves, the wave travels a full cycle of 2π from a crest to its successive one and this will continue to repeat. Therefore, we can define the spatial angular distance in radians as the integer multiple of 2π as

$$k_o z = 2\pi m \text{ (rad)} \tag{1.91}$$

Now we allow time to grow, then we can define the argument of the cosine functions as:

$$\omega t - k_o z = \omega\left(t - \frac{z}{c}\right) = 2\,m\pi \text{ (rad)} \tag{1.92a}$$

25 Note that 2π (rad) is equivalent to one cycle and a full wave length λ in meters.

26 The spatial phase shift is also called the electric length in radians of the signal.

This means that as time progresses the distance must be positively increased to make the term in the parenthesis constant. This is the definition of *the forward travelling* wave. Using similar argument, we can say that

$$\omega t + k_o z = \omega\left(t + \frac{z}{c}\right) = 2\,m\pi\ (\text{rad}) \tag{1.92b}$$

represents *the reverse travelling wave.*

Review Questions 1.13: Characteristics of electromagnetic waves
Q1. Explain the difference in wave number k and the phase constant β in the electromagnetic wave theory.
Q2. Define the instantaneous or time domain form of the field.
Q3. Define spatial frequency and its unit.

1.11 Poynting Theorem[27]

The above discussion on the fundamentals of electromagnetic laws and the electromagnetic wave theory can be derived from two aspects: (i) initially the action at a distance (force) such as the laws of Coulomb, Gauss, Biot-Savart, Ampere in electrostatic and magnetostatic cases, and Faraday in dynamic cases, and (ii) Maxwell's field theory due to the electromagnetic interaction. The legendary *continuity equation* of Maxwell had opened up the gate of the modern time varying electromagnetic wave theory. As you may recall that *the continuity equation* suggests that the time rate of diminishing charges in an enclosed surface generates the diverging current. This means the rate of the decrement of the stored charge is transformed into the diverging current. *The displacement current* concept, another legendary discovery of time varying electromagnetic field theory, is also derived from *the continuity equation* of Maxwell. Maxwell's original contribution leads to the wireless energy propagation through any medium either guided or wireless. In this section we shall discuss the energy balance equation in an enclosed surface that is derived by an English Physicist J. D. Poynting in 1883.[28] At the same time, Heaviside also reported similar theory of the energy balance equation.[29] To understand this energy balance equation, we shall first understand the energy stored and dissipated in the circuit elements as described in the circuit theory and also shown in Figure 1.13a. We shall draw an analogy of the energy stored in the circuit theory with those for the electrostatic and magnetostatic cases in a closed area which is defined by a differential volume ΔV enclosed by a closed surface area ΔS as shown in Figure 1.13b. Finally, we shall consider the dynamic electromagnetic cases and shall derive the energy balance equations for the time varying electromagnetic fields at any arbitrary time rate of changes.

27 Similar derivation is also available in Islam, M.A. (1969). *Electromagnetic Theory*. Dacca, East Pakistan: EPUET.
28 Poynting, J.D. (1883). On the transfer of energy in the electromagnetic field. *Phil. Trans.*, 174: 343 (quoted in Sadiku, M. N.O. (2001). *Elements of Electromagnetics*, 3e, 363. Oxford University Press.)
29 Islam, M.A. (1969). *Electromagnetic Theory*. Dacca, East Pakistan: EPUET.

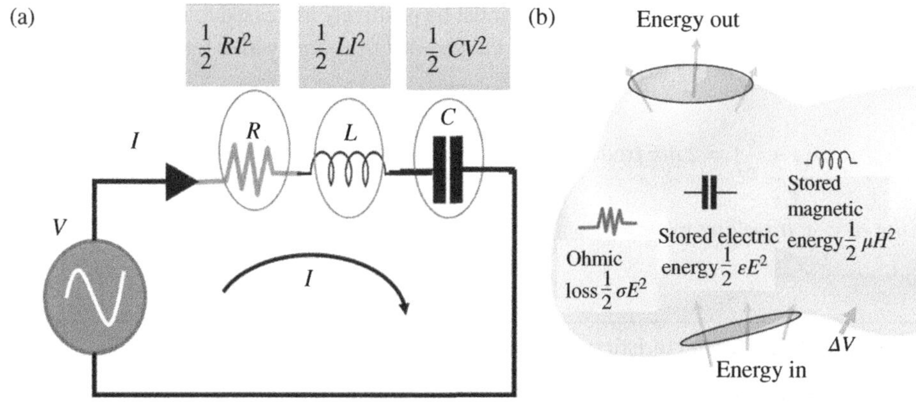

Figure 1.13 (a) An RLC AC circuit with impressed voltage *V*, and (b) equivalent Poynting theorem is a power balance equation for EM fields.

In guided electromagnetics, electromagnetic energy is transported from one point to another point via a transmission line.[30] We are more used to with the voltage V (equivalent to the electric field $\boldsymbol{E}$) across the terminal and the current I (equivalent to the magnetic field $\boldsymbol{H}$) through the circuit elements. Contrary to the guided energy case, based on Maxwell's field wave theory, the electromagnetic energy is also propagated in free space as the unguided wave. However, in static and low frequency cases, we can assume that the energy is stored in separate cases of electric and magnetic energies in the respective fields in a medium with their respective finite constitutive parameters, permittivity ε and permeability μ. For a lossless medium the conductivity and Ohmic loss can be neglected.

First, we start with the AC circuit theory of the reactive elements with inductance L and capacitance C. In the circuit theory, the magnetic energy stored in the circuit element is defined as $(1/2)LI^2$ where, I is the current. Likewise, the electric energy stored in a capacitor is defined as $(1/2)CV^2$ where V is the voltage across the capacitor. If the transmission line and the circuit elements are lossy then the real power loss is represented by a resistor R. The time–average power dissipated as heat is defined as $(1/2)RI^2$.

Now if we draw the analogy of the static electric and magnetic cases[31] with the above circuit theory, the electric and magnetic energies stored in a medium with permittivity ε and permeability μ can be defined as $(1/2)\varepsilon E^2$ and $(1/2)\mu H^2$, respectively. If the medium is lossy then we can also derive the real power loss as heat dissipation as $(1/2)\sigma E^2$, where σ is the conductivity of the medium. The above definitions of the energy equations are based on the static and low frequency dynamic cases and is perceived that the energy accumulation is happening in a slowly building up form like a summation of the consecutive static states.[32]

---

30 The transmission line is perceived as the distributive circuit elements with combinations of *R*, *L*, and *C*. We shall derive the wave equations based on the circuit elements using Telegraphist's equations in Transmission Line theory chapter later on.

31 We reserve the term electromagnetic for the time varying case and here we define the static cases in separate definitions "Static electric and magnetic cases" instead of defining "static electromagnetic case."

32 Revisit 3.9 Electric Energy and Joule's Law; Section 3.11 Electrostatic Potential Energy and Section 3.19 Magnetic Energy and Inductance.

The objective of the above analogy is to get an understanding whether such energy accumulation as the stored electromagnetic energy is valid in any arbitrary rate in the time varying cases. To validate the theory, first we take a small volume ΔV within which we would like to evaluate the energy state as shown in Figure 1.13b. Assume that the total enclosed surface area of the volume is ΔS. Here we assume that the enclosed surface ΔS with a volume ΔV can accept energy as incident electromagnetic fields denoted as "Energy in," and also energy flowing out as "Energy out."[33] In between, there are some electromagnetic interaction based on the field theory of Maxwell, in the forms of stored electric energy (symbolized as capacitor) and magnetic energy (symbolized as inductor), and the real power dissipation (symbolized as resistor). This means in between the "energy in" and "energy out" there is some energy that can be stored in the form of electric and magnetic energies as shown with the symbols of capacitance and inductance, respectively, plus some real power loss as heat in the resistance.

Based on *the principle of the conservation of energy*, any increase of energy within the volume ΔV must be accompanied by the equal amount of energy flowing into the volume via the enclosed surface ΔS. Likewise, any decrease of energy within the volume ΔV must be accompanied by the equal amount of energy flowing out of the volume via the enclosed surface ΔS. First consider the energy flowing out of the surface per unit volume "per unit time" as a vector $\boldsymbol{P}$ (W/m^2). Also, for simplicity assume that the medium is loss less $\sigma = 0$. Then there must be diminishing of energy in the volume and we can define the energy balance with the following equation:

$$\int \boldsymbol{P} \cdot d\boldsymbol{S} = -\int_v \frac{\partial}{\partial t}\left(\frac{1}{2}\varepsilon E^2 + \frac{1}{2}\mu H^2\right) dV \text{ (W)} \tag{1.93}$$

Here the unit of $\boldsymbol{P}$ is in (W/m^2) and $(1/2)\varepsilon E^2$ and $(1/2)\mu H^2$ are in (J/m^2). The above equation is the energy balance equation in the integral form for lossless case. Now considering the differential form of the equation that can be written using the divergence theorem on the enclosed surface as follows:

$$\int_v (\nabla \cdot \boldsymbol{P}) dV = -\int_v \frac{\partial}{\partial t}\left(\frac{1}{2}\varepsilon E^2 + \frac{1}{2}\mu H^2\right) dV \text{ (W)} \tag{1.94a}$$

or

$$\nabla \cdot \boldsymbol{P} = -\frac{\partial}{\partial t}\left(\frac{1}{2}\varepsilon E^2 + \frac{1}{2}\mu H^2\right) (\text{W/m}^2) \tag{1.94b}$$

Here, $\boldsymbol{P}$ is power density per unit area and is called *the Poynting vector*, after the name of J. D. Poynting.

Now consider the rapidly time varying field case in which the electric and magnetic fields are present simultaneously and the fields are governed by Maxwell's equations. Recalling from Equations (1.4)–(1.7). The time varying electromagnetic field equations for any linear, isotropic, homogeneous medium are:

$$\nabla \cdot \boldsymbol{E} = \frac{\rho_v}{\varepsilon} \tag{1.4}$$

$$\nabla \cdot \boldsymbol{H} = 0 \tag{1.5}$$

33 Note that the figure shows the representations of area of in and out of energy. Energy can be in from any point/area and out from any point/area depending on the configuration of an electromagnetic system. As for example, for an antenna, the input energy is provided via its input point and the output energy is the radiation from its radiation aperture or length.

$$\nabla \times \boldsymbol{E} = -\mu \frac{\partial \boldsymbol{H}}{\partial t} \tag{1.6}$$

$$\nabla \times \boldsymbol{H} = \boldsymbol{J}_c + \varepsilon \frac{\partial \boldsymbol{E}}{\partial t} = \sigma \boldsymbol{E} + \varepsilon \frac{\partial \boldsymbol{E}}{\partial t} \tag{1.7}$$

We observe that for a lossy medium with finite conductivity σ, we define Ohm's law $\boldsymbol{J} = \sigma \boldsymbol{E}$ that represents the Ohmic or real power loss. We drop subscript c for conduction current for simplicity. Therefore, $\boldsymbol{J}.\boldsymbol{E} = \sigma E^2$ represents the real power loss as heat dissipation. To develop the generic energy balance equation with all energy stored and losses we start with the dot product of $\boldsymbol{E}$ and Ampere's law $\nabla \times \boldsymbol{H} = \boldsymbol{J} + \varepsilon(\partial \boldsymbol{E}/\partial t)$:

$$\boldsymbol{E} \cdot (\nabla \times \boldsymbol{H}) = \boldsymbol{E} \cdot \left(\boldsymbol{J} + \varepsilon \frac{\partial \boldsymbol{E}}{\partial t}\right) = \boldsymbol{E} \cdot \boldsymbol{J} + \varepsilon \boldsymbol{E} \cdot \frac{\partial \boldsymbol{E}}{\partial t} \tag{1.95}$$

Likewise, if we take a dot product between Equation (1.6) and $\boldsymbol{H}$, we obtain

$$\boldsymbol{H} \cdot (\nabla \times \boldsymbol{E}) = -\mu \boldsymbol{H} \cdot \frac{\partial \boldsymbol{H}}{\partial t} \tag{1.96}$$

Now subtracting Equation (1.95) from (1.96) we obtain:

$$\boldsymbol{H} \cdot (\nabla \times \boldsymbol{E}) - \boldsymbol{E} \cdot (\nabla \times \boldsymbol{H}) = -\mu \boldsymbol{H} \cdot \frac{\partial \boldsymbol{H}}{\partial t} - \boldsymbol{E} \cdot \boldsymbol{J} - \varepsilon \boldsymbol{E} \cdot \frac{\partial \boldsymbol{E}}{\partial t} \tag{1.97}$$

Using the vector identity

$$\nabla \cdot (\boldsymbol{E} \times \boldsymbol{H}) = \boldsymbol{H} \cdot (\nabla \times \boldsymbol{E}) - \boldsymbol{E} \cdot (\nabla \times \boldsymbol{H}) \tag{1.98}$$

we obtain

$$\nabla \cdot (\boldsymbol{E} \times \boldsymbol{H}) = -\left(\mu \boldsymbol{H} \cdot \frac{\partial \boldsymbol{H}}{\partial t} + \varepsilon \boldsymbol{E} \cdot \frac{\partial \boldsymbol{E}}{\partial t} + \boldsymbol{E} \cdot \boldsymbol{J}\right) \tag{1.99}$$

Assuming the medium is linear, homogeneous, isotropic, and does not change with time, we can simplify the energy equations into the following forms:

$$\mu \boldsymbol{H} \cdot \frac{\partial \boldsymbol{H}}{\partial t} = \frac{1}{2} \frac{\partial(\mu \boldsymbol{H} \cdot \boldsymbol{H})}{\partial t} = \frac{1}{2} \frac{\partial}{\partial t}(\mu H^2) \tag{1.100}$$

$$\varepsilon \boldsymbol{E} \cdot \frac{\partial \boldsymbol{E}}{\partial t} = \frac{1}{2} \frac{\partial(\varepsilon \boldsymbol{E} \cdot \boldsymbol{E})}{\partial t} = \frac{1}{2} \frac{\partial}{\partial t}(\varepsilon E^2) \tag{1.101}$$

Substitution for Equations (1.100) and (1.101) into Equation (1.99) we obtain:

$$\nabla \cdot (\boldsymbol{E} \times \boldsymbol{H}) = -\frac{\partial}{\partial t}\left(\frac{1}{2}\varepsilon E^2 + \frac{1}{2}\mu H^2\right) - \boldsymbol{E} \cdot \boldsymbol{J} \tag{1.102}$$

This is the differential from of the energy balance equation. The integral from of the energy balance equation can be obtained by integrating the above equation over a volume enclosed by the closed surface area S and using the divergence theorem as follows:

$$\int_V \nabla \cdot (\boldsymbol{E} \times \boldsymbol{H}) dV = \oint_S (\boldsymbol{E} \times \boldsymbol{H}) \cdot d\boldsymbol{S} = -\int_V \left[\frac{\partial}{\partial t}\left(\frac{1}{2}\varepsilon E^2 + \frac{1}{2}\mu H^2\right) - \boldsymbol{E} \cdot \boldsymbol{J}\right] dV \tag{1.103}$$

$$\int_V \nabla.(\boldsymbol{E} \times \boldsymbol{H}) dV = \oint_S (\boldsymbol{E} \times \boldsymbol{H}) \cdot d\boldsymbol{S} = -\int_V \left[\frac{\partial}{\partial t}\left(\frac{1}{2}\varepsilon E^2 + \frac{1}{2}\mu H^2\right) - \boldsymbol{E} \cdot \boldsymbol{J}\right] dV$$

or,

$$\oint_S \boldsymbol{P}\cdot d\boldsymbol{S} = -\int_V \left[\frac{\partial}{\partial t}\left(\frac{1}{2}\varepsilon E^2 + \frac{1}{2}\mu H^2\right) - \boldsymbol{E}\cdot\boldsymbol{J}\right] dV \tag{1.104}$$

Here

$$\boldsymbol{P} = \boldsymbol{E}\times\boldsymbol{H}\ \ (\mathrm{W/m^2}) \tag{1.105}$$

is used. Equation (1.104) is referred to as *the Poynting theorem*. As stated above, the equation was derived by J. D. Poynting in 1883 and also by Heaviside at about the same time. From the examination of the left side of the above equation we can easily state that the diverging energy density term $\boldsymbol{P} = \boldsymbol{E}\times\boldsymbol{H}\,(\mathrm{W/m^2})$ on an enclosed surface S equates with the decrement of the time rate of change of the electric and magnetic energies subtracted by the real power dissipation in the volume. By *the law of conservation of energy*, we can say that the decrement of energy must be equal to the power (time rate of energy) flowing out of the volume through the enclosed surface. The power flow density vector is designated as Equation (1.105). The vector $\boldsymbol{P}$ is called *the Poynting vector*.

If we compare Maxwell's *current continuity equation* with the above *Poynting theorem* Equation (1.104), we can draw nice similarities between them. For *the continuity equation*, the negative time rate of change of the stored charge $(-(\partial Q/\partial t))$ in a volume must be equal to the diverging current density $\nabla.\boldsymbol{J}$. For *the Poynting theorem*, the negative rate of change of the stored energy minus the Ohmic loss must equate with the diverging power flow density vector $\nabla\cdot\boldsymbol{P} = \nabla\cdot(\boldsymbol{E}\times\boldsymbol{H})$ where, the power flow density vector or *the Poynting vector* is defined as: $\boldsymbol{P} = \boldsymbol{E}\times\boldsymbol{H}$ in $(\mathrm{W/m^2})$. In the following sections, we shall examine a few special cases of Poynting theorem.

Review Questions 1.14: Poynting Theorem

Q1. Draw a mud map to reflect how the energy balance equation evolved from Maxwell's current continuity equation.

Q2. Explain the difference between static and low frequency and time varying energy balance cases.

Q3. Derive the analogy between the circuit theory and the electromagnetic energy balance theory.

Q4. What is the equivalence of the circuit theory with the electromagnetic energy balance theory in the static and low frequency cases?

Q5. Derive the Poynting vector equations from the electric and magnetic potential energy theorems for the static case.

Q6. What is the relevance between the transmission line and an enclosed volume with electromagnetic phenomenon?

Q7. What is the difference between the concepts of the energy balance in the circuit theory and the theory developed by Poynting using Maxwell's equations?

Q8. Derive Poynting theorem from Maxwell's equations for the time varying case.

Q9. Relate Poynting theorem with Maxwell's current continuity equation and interpret each elements of the Poynting theorem.

1.12 Static Poynting Theorem

In static electromagnetic case such as DC or low frequency AC, there is no rate of change of the electromagnetic energy within the volume of investigation. Therefore, the Poynting theorem reduces to:

$$\int_S \boldsymbol{P} \cdot d\boldsymbol{S} = -\int_V \boldsymbol{E} \cdot \boldsymbol{J} dV \tag{1.106}$$

Assume the volume contains no source. Thus, in static case the net influx of power within the system is fully dissipated as Julian heat or power loss. In the following, an example is given for a conductor.

1.12.1 Poynting Theorem for a Wire

Figure 1.14 shows a wire with radius a and carrying a DC current I. The conductivity of the wire is σ and length is L. We investigate the Poynting vector and the total power loss due to the finite conductivity. We shall also define the real power loss in terms of the circuit parameter R.

Since the wire is a circularly symmetric configuration, we employ cylindrical coordinate system (ρ, ϕ, z). From our understanding of the DC current I and resistance R, we can assume the current I is uniformly distributed over the cross-sectional area as shown in Figure 1.14. The current density is:

$$\boldsymbol{J} = \frac{I}{\pi a^2}\boldsymbol{a}_z \ (\text{A/m}^2) \tag{1.107}$$

Applying Ohm's law the electric field is:

$$\boldsymbol{E} = \frac{\boldsymbol{J}}{\sigma} = \frac{I}{\sigma \pi a^2}\boldsymbol{a}_z \ (\text{A/m}) \tag{1.108}$$

Applying Ampere's law, the magnetic field on the surface of the wire is:

$$\boldsymbol{H} = \frac{I}{2\pi a}\boldsymbol{a}_\phi \ (\text{A/m}) \tag{1.109}$$

The Poynting vector is the product of the electric and magnetic fields:

$$\boldsymbol{P} = \boldsymbol{E} \times \boldsymbol{H} = \frac{I}{\sigma \pi a^2}\boldsymbol{a}_z \times \frac{I}{2\pi a}\boldsymbol{a}_\phi = -\frac{I^2 \sigma}{2\pi^2 a^3}\boldsymbol{a}_\rho \ (\text{W/m}^2) \tag{1.110}$$

Which is directed inside the wire along the radial direction, as shown in Figure 1.14.

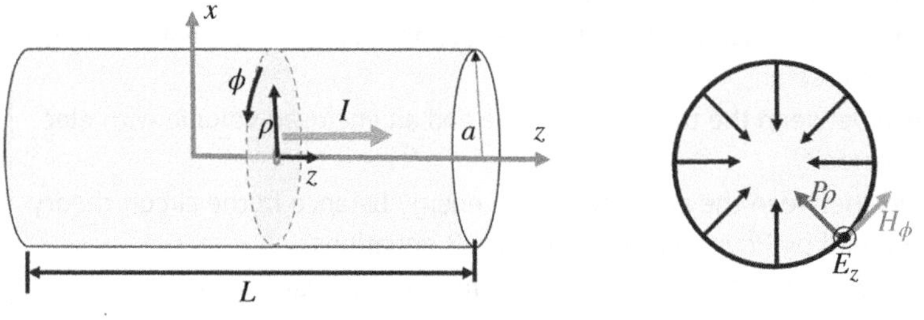

Figure 1.14 A wire with radius a which is carrying a DC current I. The conductivity of the wire is σ and length is L. We calculate the Poynting vector and the total power loss due to the finite conductivity σ. We shall also define the real power loss in term of the circuit parameter R.

Now to verify the Poynting theorem we integrate the Poynting vector $\boldsymbol{P}$ over the entire volume or the entire enclosed surface which is $2\pi aL$. Let us do the surface integration. In the integration we can ignore the two vertical side surfaces at two ends of the wire of length L due to the fact that the normal of the surfaces is along $\pm z$ whereas the Poynting vector $\boldsymbol{P}$ is along the radial direction, and therefore, the contributions of the integration of the two surfaces will be zero. Therefore, the integration of ρ over the cylindrical surface arc is:

$$\oint_S \boldsymbol{P}\cdot d\boldsymbol{S} = \oint_S -\frac{I^2}{2\sigma\pi^2a^3}(\boldsymbol{a}_\rho\cdot\boldsymbol{a}_\rho)\rho d\phi d\rho = -\frac{I^2}{2\sigma\pi^2a^3}\times 2\pi aL = -I^2\left(\frac{1}{\sigma}\times\frac{L}{\pi a^2}\right) = -I^2R \tag{1.111}$$

where, the resistance is defined as:

$$R = \frac{1}{\sigma}\times\frac{L}{\pi a^2}(\Omega). \tag{1.112}$$

If we draw the analogy of *the current continuity equation,* we can tell that the –ve sign of the Poynting vector represents the power flowing out of the surface.

1.13 Energy Balance Equation in the Presence of a Generator: In-Flux and Out-Flow of Power

So far, we did not consider any source or generator that impressed energy inside the volume. Let us start with the term $\boldsymbol{E.J}$ that represents the Ohmic power loss per unit volume. As shown in the above figure now assume that the in-flux of energy generates an impressed electric field $\boldsymbol{E}'$.[34] Also assume that the conductivity of the medium is σ. Then the total current density $\boldsymbol{J}$ generated by both the impressed electric field $\boldsymbol{E}'$ and the electric field $\boldsymbol{E}$ that is generated by the current and charges within the volume is:

$$\boldsymbol{J} = \sigma(\boldsymbol{E}+\boldsymbol{E}') \tag{1.113}$$

This leads to the definition of the original electric field $\boldsymbol{E}$ in terms of the current density and the impressed electric field as

$$\boldsymbol{E} = \frac{\boldsymbol{J}}{\sigma} - \boldsymbol{E}' \tag{1.114}$$

Therefore, the volume integration of $\boldsymbol{E.J}$ leads to the following expression:

$$\int_V(\boldsymbol{E}\cdot\boldsymbol{J})dV = \int_V\frac{J^2}{\sigma}dV - \int_V(\boldsymbol{E}'\cdot\boldsymbol{J})dV \tag{1.115}$$

Now examining the above Equation (1.115) we can easily deduce that the first term in the right-hand side which represents the energy dissipation as heat which is an irreversible process. The second term represents the impressed energy flowing inside the system. The –ve sign represents some works need to be done by the external source to impress the electric field $\boldsymbol{E}'$.[35] This supplied power

34 The prime sign is usually reserved to represented quantities and coordinate systems related to source.
35 Recall the definition of electric potential $V = -\int \boldsymbol{E}\cdot d\boldsymbol{L}$ with the negative sign, it indicates work is needed to bring a unit charge in the influence of the electric field.

is balanced by the increase in the stored electromagnetic energy and heat loss in the system. The rest is flowing out of the surface as $\oint_S \boldsymbol{P} \cdot d\boldsymbol{S}$. Let us prove the above statement with the following mathematical manipulation.

$$\oint_S \boldsymbol{P} \cdot d\boldsymbol{S} = -\int_V \left[\frac{\partial}{\partial t}\left(\frac{1}{2}\varepsilon E^2 + \frac{1}{2}\mu H^2\right)\right] dV - \int_V \frac{J^2}{\sigma} dV + \int_V (\boldsymbol{E}' \cdot \boldsymbol{J}) dV \tag{1.116}$$

or,

$$\int_V (\boldsymbol{E}' \cdot \boldsymbol{J}) dV = \oint_S \boldsymbol{P} \cdot d\boldsymbol{S} + \int_V \left[\frac{\partial}{\partial t}\left(\frac{1}{2}\varepsilon E^2 + \frac{1}{2}\mu H^2\right)\right] dV + \int_V \frac{J^2}{\sigma} dV \tag{1.117}$$

Examination of the above Equation (1.117) reveals that the left-hand side term $\int_V (\boldsymbol{E}' \cdot \boldsymbol{J}) dV$ represents the in-flux energy in the system from a generator as shown in Figure 1.15. The first term on the right-hand side $\oint_S \boldsymbol{P} \cdot d\boldsymbol{S}$ represents the energy flowing out of the volume through the enclosed surface S. The Poynting vector $\boldsymbol{P}$ represents energy flow at a point in the system. The second term on the right-hand side $\int_V [(\partial/\partial t)(((1/2)\varepsilon E^2) + ((1/2)\mu H^2))] dV$ represents the rate of change of stored electromagnetic energy in any arbitrary time rate of change in fast varying and/or quasi-static state. The last term $\int_V (J^2/\sigma) dV$ is the irreversible Ohmic loss due to finite conductivity of the system.

A good analogy of the energy balance equation as developed via the Poynting theorem is an antenna as shown in Figure 1.4 in the beginning of the chapter. A source is added to the antenna. The antenna has stored reactive electromagnetic energy which changes with the operating frequency, some power is dissipated as heat due to the finite conductivity of the antenna material, and the rest of the power is radiated as the Poynting vector as $\boldsymbol{P} = \boldsymbol{E} \times \boldsymbol{H}$ (W/m^2).

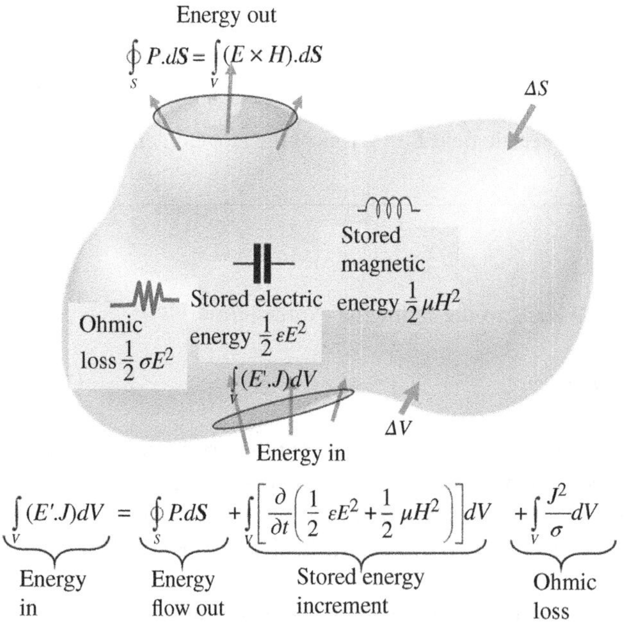

Figure 1.15 Depiction of energy balance Poynting theorem.

Review Questions 1.15: Poynting Theorem

Q1. Explain the difference of the energy balance equation without and with a generator.
Q2. Explain the difference of the Poynting theorem in static and dynamic cases in the light of Maxwell's equations.

1.14 Time Harmonic Poynting Vector

So far, we have discussed the generic time varying field quantities and developed the Poynting Theorem. However, when describing plane wave propagation, it is customary to express the field quantities in their complex phasor vector or time harmonic forms. As we know that the time domain or the instantaneous expressions of these complex field quantities are the real part of the product of the phasor vector quantity and $e^{j\omega t}$. The instantaneous expression of the field quantities that we can measure in the oscilloscope in the sine and cosine waveforms. In a lossless, linear, isotropic, and homogeneous medium, a positive z-directed electric and magnetic field can be expressed as follows:

$$E_x = \mathbf{a_x} E_0^+ \cos(\omega t - \beta z) \tag{1.118}$$

$$H_y = \mathbf{a_y} H_0^+ \cos(\omega t - \beta z) = \mathbf{a_y} \frac{E_0^+}{\eta} \cos(\omega t - \beta z) \tag{1.119}$$

where η is the intrinsic impedance of the medium.

The instantaneous Poynting vector is expressed as:

$$P(z, t) = \mathbf{E} \times \mathbf{H} = \left[\mathbf{a_x} E_0^+ \cos(\omega t - \beta z)\right] \times \left[\mathbf{a_y} \frac{E_0^+}{\eta} \cos(\omega t - \beta z)\right] = \mathbf{a_z} \frac{\left[E_0^+\right]^2}{\eta} \cos^2(\omega t - \beta z) \tag{1.120}$$

Using the trigonometric identity of $\cos^2\theta = (1/2)(1 + \cos 2\theta)$ we obtain the instantaneous expression for the Poynting vector as:

$$\mathbf{P}(z, t) = \mathbf{a_z} \frac{\left[E_0^+\right]^2}{2\eta} [1 + \cos 2(\omega t - \beta z)] \tag{1.121}$$

From the examination of the above expression (1.121), we can see that the time domain expression for the Poynting vector has two parts: (i) the first DC component with amplitude $P_{DC}(z, t) = \mathbf{a_z}\left(\left[E_0^+\right]^2/2\eta\right)$, and (ii) the second AC component with double the frequency $P_{AC}(z, t) = \mathbf{a_z} \frac{\left[E_0^+\right]^2}{2\eta} \cos 2(\omega t - \beta z)$.

If we integrate the above AC component over the cycle (2π) then we obtain zero contribution from the AC component.

We are more interested in the time average quantity as this represents the real power loss as heat. The time average power is the real power loss in all resistive elements in the network over a time period or a cycle. The time average power is defined over a time period T or a cycle (2π) as

$$\mathbf{P_{ave}} = \frac{1}{T} \int_0^T \mathbf{P}(x, y, z, t) dt \tag{1.122}$$

or,

$$\boldsymbol{P_{ave}} = \frac{1}{2\pi}\int_0^{2\pi} \boldsymbol{P}(x, y, z, t)d(\omega t) \tag{1.123}$$

Now substitution for the value of $\boldsymbol{P}(z, t) = \boldsymbol{a_z}\frac{[E_0^+]^2}{2\eta}[1 + \cos 2(\omega t - \beta z)]$ in the above Equation (1.123) we obtain:

$$\begin{aligned}\boldsymbol{P_{ave}} &= \boldsymbol{a_z}\frac{1}{2\pi}\int_0^{2\pi} P(x, y, z, t)d(\omega t) = \boldsymbol{a_z}\frac{1}{2\pi}\int_0^{2\pi}\frac{[E_0^+]^2}{2\eta}[1 + \cos 2(\omega t - \beta z)]d(\omega t)\\ &= \boldsymbol{a_z}\frac{1}{2\pi} \times \frac{[E_0^+]^2}{2\eta}\left\{\int_0^{2\pi} d(\omega t) + \int_0^{2\pi} \cos 2(\omega t - \beta z)d(\omega t)\right\} = \boldsymbol{a_z}\frac{[E_0^+]^2}{2\eta}\end{aligned} \tag{1.124}$$

$$\therefore \boldsymbol{P_{ave}} = \boldsymbol{a_z}\frac{[E_0^+]^2}{2\eta} \tag{1.125}$$

The above expression can be deduced from their phasor vector expression using the following expression:

$$\boldsymbol{P_{ave}} = \frac{1}{2} Re\left[\boldsymbol{E_s} \times \boldsymbol{H_s^*}\right] = \frac{1}{2} Re\left[\boldsymbol{E_s^*} \times \boldsymbol{H_s}\right] \tag{1.126}$$

$$\nabla \times \boldsymbol{E} = -\frac{\partial \boldsymbol{B}}{\partial t} = -\mu\frac{\partial \boldsymbol{H}}{\partial t}$$

The Phasor expressions of the electric and magnetic fields are:

$$\boldsymbol{E_{xs}} = \boldsymbol{a_x}E_0^+ e^{-j\beta z} \tag{1.127}$$

$$\boldsymbol{H_y} = \boldsymbol{a_y}H_0^+ e^{j\beta z} = \boldsymbol{a_y}\frac{E_0^+}{\eta}e^{-j\beta z} \tag{1.128}$$

Then the time average Poynting vector is defined[36] as:

$$\boldsymbol{P_{ave}} = \frac{1}{2} Re\left[\boldsymbol{E_s} \times \boldsymbol{H_s^*}\right] = \frac{1}{2} Re\left[\boldsymbol{E_s^*} \times \boldsymbol{H_s}\right] = \frac{1}{2} Re\left[\boldsymbol{a_x}E_0^+ e^{-j\beta z} \times \boldsymbol{a_y}\frac{E_0^+}{\eta}e^{j\beta z}\right] \tag{1.129}$$

$$\therefore \boldsymbol{P_{ave}} = \boldsymbol{a_z}\frac{[\boldsymbol{E_0^+}]^2}{2\eta} \tag{1.130}$$

Observation from the above procedure of taking the real part of the electric and the conjugate of the magnetic fields or vice versa of the phasor expressions of the electric and magnetic fields we obtain the same results of the procedure followed in the integration over the full cycle of 2π. Therefore, we can experience the benefit of using the phasor vector calculation in time harmonic field analysis.[37]

If the medium is lossy then we need to consider the loss term $e^{-\alpha z}$ in the above calculation. The phase difference between the field quantities due to the loss is also needed to be considered.[38] The instantaneous expressions for the field quantities are:

$$E_x = \boldsymbol{a_x}E_0^+ e^{-\alpha z}\cos(\omega t - \beta z) \tag{1.131}$$

36 Usually we take complex conjugate of the magnetic field.

37 Helmholtz devised the expressions of phasor vector method of calculation of time harmonic fields using Euler theorem.

38 Recall the intrinsic impedance becomes complex for a lossy medium. Therefore, the phase difference between the electric and magnetic fields must be considered in the calculation.

$$\boldsymbol{H_y} = \boldsymbol{a_y} H_0^+ e^{-\alpha z} \cos(\omega t - \beta z - \theta) = \boldsymbol{a_y} e^{-\alpha z} \frac{E_0^+}{\eta} \cos(\omega t - \beta z - \theta) \tag{1.132}$$

where the phase angle θ represents the initial phase state of the magnetic field compared with the electric field due to the loss factor. The phasor expression of the above field quantities are:

$$\boldsymbol{E_{xs}} = \boldsymbol{a_x} E_0^+ e^{-\alpha z} e^{-j\beta z} \tag{1.133}$$

$$\boldsymbol{H_{ys}} = \boldsymbol{a_y} e^{-\alpha z} \frac{E_0^+}{\eta} e^{-j\beta z} e^{-j\theta} \tag{1.134}$$

Using the above procedure, the time average Poynting vector for the lossy medium is:

$$\boldsymbol{P_{ave}} = \frac{1}{2} Re \left[\boldsymbol{a_x} E_0^+ e^{-\alpha z} e^{-j\beta z} \times \boldsymbol{a_y} e^{-\alpha z} \frac{E_0^+}{\eta} e^{j\beta z} e^{j\theta} \right] \tag{1.135}$$

$$\therefore \boldsymbol{P_{ave}} = \boldsymbol{a_z} \frac{\left[E_o^+\right]^2}{2\eta} e^{-2\alpha z} \cos\theta \tag{1.136}$$

The above expression uses Euler theorem: $Re[e^{-j\theta}] = Re[\cos\theta - j\sin\theta] = \cos\theta$

Review Questions 1.16: Poynting Vector and Time Average Power

Q1. Derive the Poynting vector for a lossy medium using the integration method and compared with the phasor vector method of calculation.

Q2. Explain the benefit of using the phasor vector notations of the field quantities in time harmonic electromagnetic analysis

Q3. Explain why there is a phase difference in the field quantities for a lossy medium.

Q4. Define the time average Poynting vector and its SI unit.

Q5. Explain why we are more interested in the time average values instead of the instantaneous values of the field quantities and Poynting vector.

Example 1.7: Poynting Theorem[39]
A y-polarized electromagnetic field wave is propagating along $+z$ direction. The operating frequency is 6 GHz. The time average power density of the wave is 20 (mW/m^2). Calculate the electric and magnetic fields in their instantaneous form using the above theorem. Assume that the medium is lossless.

39 Adapted from Johnk, C.T.A. (1988). *Engineering Electromagnetic Fields and Waves*, 2e. NY, USA: Wiley.

Answer:

Since the electric field is propagating along $+z$ direction with y-polarization, the generic expressions for the electric and magnetic fields are:

$$\boldsymbol{E_s} = \boldsymbol{a_y} E_0^+ e^{-j\beta z} \tag{1.137}$$

$$\boldsymbol{H_s} = -\boldsymbol{a_x} e^{-\alpha z} \frac{E_0^+}{\eta} e^{-j\beta z} \tag{1.138}$$

where the wave impedance $\eta = 120\pi$ (Ω)

The amplitude of the fields can be calculated using the time average power density P_{ave} as follow:

$$P_{ave} = \frac{\left[E_0^+\right]^2}{2\eta} = 20 \times 10^{-3} \left(\mathrm{W/m^2}\right)$$

Therefore, the peak amplitude of the electric field is:

$$E_o^+ = \sqrt{(20 \times 10^{-3} \times 2 \times 120\pi)} = 3.88\ (\mathrm{V/m})$$

The amplitude of the magnetic field is:

$$H_o^+ = \frac{E_0^+}{\eta} = \frac{3.88}{120\pi} = 0.01\ (\mathrm{A/m})$$

The angular frequency is: $\omega = 2\pi f = 2\pi \times 6 \times 10^9 = 37.7 \times 10^9$ (rad/s)

The Phase constant of the field wave is:

$$\beta = \frac{2\pi}{\lambda} = \frac{2\pi f}{c} = \frac{2\pi \times 6 \times 10^9}{3 \times 10^8} = 125.66\ (\mathrm{rad/m})$$

The expressions for the instantaneous electric and magnetic fields are:

$$\boldsymbol{E}(t) = 3.88\ \cos\left(37.7 \times 10^9 t - 125.66z\right)\boldsymbol{a_y}\ (\mathrm{V/m})$$

$$\boldsymbol{H}(t) = -0.01\ \cos\left(37.7 \times 10^9 t - 125.66z\right)\boldsymbol{a_x}\ (\mathrm{A/m})$$

Example 1.8: Poynting Theorem

A time varying incident wave is defined by

$$E_x(t) = \boldsymbol{a_x} E_0^+ \cos(\omega t - \beta z)\ H_y = \boldsymbol{a_y} H_o^+ \cos(\omega t - \beta z) = \boldsymbol{a_y} \frac{E_0^+}{\eta} \cos(\omega t - \beta z)$$

Assume that the wave is emanating out of the aperture of a rectangular horn antenna with dimensions $a = 10$ cm, $b = 5$ cm as shown in Figure 1.16. Calculate the total power flux entering a rectangular region of the same dimensions of the aperture of the horn antenna. Also verify the Poynting theorem that $\oint_S \boldsymbol{P} \cdot d\boldsymbol{S} = -\int_V [(\partial/\partial t)((1/2)\varepsilon E^2 + (1/2)\mu H^2)]dV$ assume the system is lossless $\sigma = 0$. The increase in the reactive power is a phenomenon of the near field of the antenna.

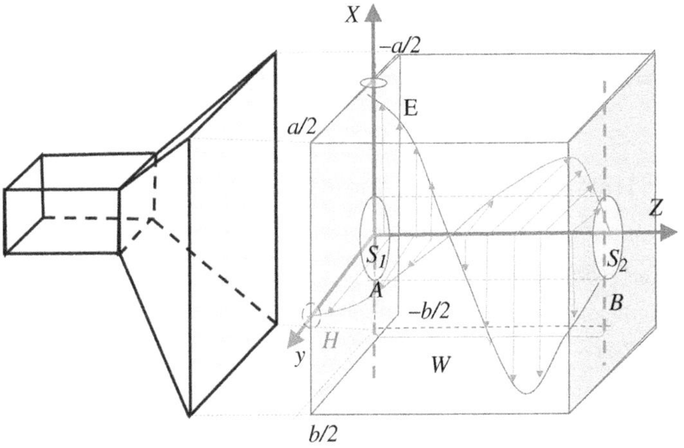

Figure 1.16 The Poynting theorem of near field of a horn antenna with dimensions *ab*.

Answer:

The contribution for surface S_1 is:

$$P_1(t) = -\int_{S_1} \boldsymbol{P}\cdot d\boldsymbol{S}\,\boldsymbol{a}_z = -\int_{y=-\frac{b}{2}}^{\frac{b}{2}}\int_{x=\frac{a}{2}}^{\frac{a}{2}}\left[\boldsymbol{a}_z\frac{\left[E_0^+\right]^2}{\eta}\{\cos^2(\omega t-\beta z)\}\right]_{z=0}$$
$$\cdot(-\boldsymbol{a}_z dxdy) = \frac{\left[E_0^+\right]^2}{\eta}ab\cos^2(\omega t) \tag{1.139}$$

The contribution from surface S_2 is:

$$P_2(t) = -\int_{S_2} \boldsymbol{P}\cdot d\boldsymbol{S}\,\boldsymbol{a}_z = -\int_{y=-\frac{b}{2}}^{\frac{b}{2}}\int_{x=\frac{a}{2}}^{\frac{a}{2}}\left[\boldsymbol{a}_z\frac{\left[E_0^+\right]^2}{\eta}\{\cos^2(\omega t-\beta z)\}\right]_{z=W}$$
$$\cdot(\boldsymbol{a}_z dxdy) = \frac{\left[E_0^+\right]^2}{\eta}ab\cos^2(\omega t-\beta W) \tag{1.140}$$

Total net power flux entering the closed surface S is:

$$P(t) = P_1(t) + P_2(t) = \frac{\left[E_0^+\right]^2}{2\eta}ab\{\cos(2\omega t) - \cos 2(\omega t-\beta W)\} \tag{1.141}$$

where we use: $\cos^2\theta = \frac{1}{2}(1+\cos 2\theta)$.

Now we shall prove $-\oint_{S_2} \boldsymbol{P}\cdot d\boldsymbol{S} = \int_V\left[\frac{\partial}{\partial t}\left(\frac{1}{2}\varepsilon E^2 + \frac{1}{2}\mu H^2\right)\right]dV$

The right-hand side of the Equation (1.141) is:

$$\int_V\left[\frac{\partial}{\partial t}\left(\frac{1}{2}\varepsilon E^2 + \frac{1}{2}\mu H^2\right)\right]dV = \frac{\partial}{\partial t}\left[\frac{\varepsilon\left[E_0^+\right]^2}{4}\int_{z=0}^{W}\int_{y=-b/2}^{b/2}\int_{x=-a/2}^{a/2}\{1+\cos 2(\omega t-\beta z)\}dxdydz\right]$$
$$+\frac{\partial}{\partial t}\left[\frac{\mu\left[E_0^+\right]^2}{4\eta^2}\int_{z=0}^{W}\int_{y=-b/2}^{b/2}\int_{x=-a/2}^{a/2}\{1+\cos 2(\omega t-\beta z)\}dxdydz\right] \tag{1.142}$$

To simplify the calculation first we take the time derivate of the function[40] inside. Thus, the unity term becomes zero and $cos\{2(\omega t-\beta z)\}$ becomes $-2\omega sin2\{(\omega t-\beta z)\}$. Also, we use the definition $\eta=\sqrt{\mu/\varepsilon}$ so that $\mu/\eta^2=\varepsilon$ and $\omega\varepsilon/\beta=\eta^{-1}$. Substitutions of these quantities into the above equation yields:

$$\begin{aligned} P(t) &= \int_V \left[\frac{\partial}{\partial t}\left(\frac{1}{2}\varepsilon E^2+\frac{1}{2}\mu H^2\right)\right]dV = \frac{\varepsilon\left[E_0^+\right]^2}{2}\int_{z=0}^{W} -2\omega\sin 2(\omega t-\beta z)dz \\ &= \frac{\omega\varepsilon\left[E_0^+\right]^2 ab}{2\beta}\left[-\cos 2(\omega t-\beta z)\right]_{z=0}^{W} \\ \therefore P(t) &= \frac{\left[E_0^+\right]^2 ab}{2\eta}\{\cos 2\,\omega t-\cos 2(\omega t-\beta W)\} \end{aligned} \tag{1.143}$$

Substitution for $E_0^+=10$ (V/m), $\eta=377$ (Ω). Operating frequency $f=6$ GHz, and $W=1$ m, we get the following closed form expression for the net power flux entering region is:

$$\begin{aligned} P(t) &= \frac{\left[E_0^+\right]^2}{2\eta}ab\{\cos(2\omega t)-\cos 2(\omega t-\beta W)\} \\ P(t) &= \frac{10^2\,(\text{V/m})^2}{2\times 377}(0.10\,(\text{m})\times 0.05\,(\text{m})) \\ &\times\left\{\cos\left(2\pi\times 6\times 10^9\times t\right)-\cos 2\left[2\pi\times 6\times 10^9\times t-\frac{2\pi\times 6\times 10^9}{3\times 10^8}\times 1\,(\text{m})\right]\right\} \\ &= 6.63\times 10^{-4}\times\left\{\cos\left(3.77\times 10^{10}\times t\right)-\cos\left[7.54\times 10^{10}\times t-251.3)\right]\right\}\,(\text{W}) \end{aligned}$$

Exercise 1.6: One Dimensional Wave Equation

Q1. Repeat the problem for finite conductivity of the medium σ.

1.15 Problems

Uniform Plane Wave Theory

P1.1 A half wavelength dipole is operating at 2.45 GHz Wi-Fi frequency band. Calculate the far field distance at which we consider the emanating out wave from the dipole is a plane wave.

P1.2 Is $v(t)=V_0\sin\omega t$ a wave equation? If so what is the amplitude, frequency of operation, direction of propagation of the voltage wave? To complete the answer, write the complete equation of the voltage wave.

P1.3 In connection to P1.1, if the radiated far electric field has an amplitude $E_0=20$ mV/m that oscillates along x-axis, propagates along z-direction and the intial phase state is 0°, write the expression of the electric field with appropriate unit. Is the field a time harmonic field?

---

40 Here we assume that the volume is not changing with neither time t nor the medium's constitutive parameters ε and μ.

P1.4 Using Maxwell's equation calculate the far magnetic field from P1.3 for the dipole antenna.

P1.5 For the complete vector time-domain expression for the electric field:

$$\boldsymbol{E}(z, t) = E_0^+ e^{-\alpha z} \cos(\omega t - \beta z)\boldsymbol{a_x} + E_0^- e^{+\alpha z} \cos(\omega t + \beta z)\boldsymbol{a_x} \text{ (V/m)}$$

what are the amplitude, phase, frequency of operation, direction of propagation of the electric field wave? Which ones are the forward travelling and reverse travelling waves' amplitudes?

P1.6 An electric field is defined as $\boldsymbol{E}(z, t) = 15e^{-10z} \cos(\omega t - \beta z)\boldsymbol{a_x} - 20e^{-\alpha z} \cos(\omega t - \beta z)\boldsymbol{a_y}$ (V/m). Determine the amplitude and direction of the field at $t = 0$ and $z = \lambda/2$.

P1.7 We know that curling electric and magnetic fields generate time varying magnetic and electric fields respectively. The fields support propagation of electromagnetic wave. Explain the justification against your belief.

P1.8 Draw a conceptual map (flow chart) to derive the general wave equations for the electric and magnetic fields, and finally, show that the time-domain field expressions for both fields are similar in form.

P1.9 A uniform plane wave which is defined with an electric field as $\boldsymbol{E}(z, t) = 250 \sin(2\pi \times 10^3 t - \beta z)\boldsymbol{a_y}$ (V/m). Calculate the magnetic field if the medium of propagation is free space.

P1.10 Write the complete time domain expression of the magnetic field for the following parameters: the peak amplitude, $H_0 = 20$ (A/m), the phase constant, $\beta = 37, 7$ (rad/m),the direction of propagation is along $-z$ direction and the field vibrates along $-\boldsymbol{a_y}$ at its initial half cycle. Derive the time domain expression for the electric field from the magnetic field. Identify the frequency operation that has commercial significance.

Propagation in Dielectric and Lossy Media

P1.11 Explain how food is cooked in a microwave oven in the perspective of wave propagation in lossy dielectric as shown in Figure P1.11.

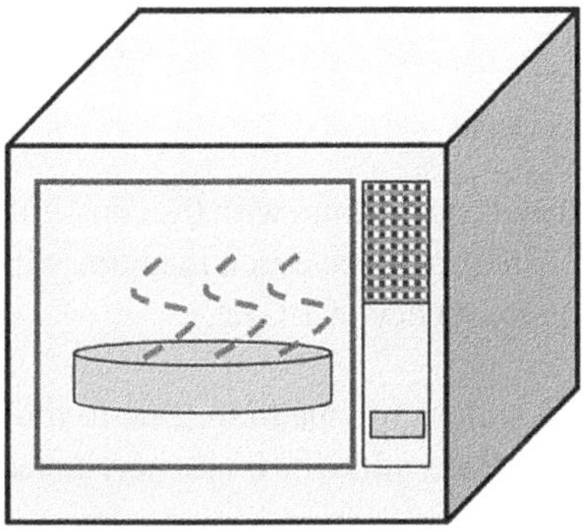

Figure P1.11 A microwave oven launches signal at 2.45 GHz from a high power microwave source called magnetron.

P1.12 What are the real power losses in the lossy dielectric materials such as foods as shown in the figure above?

1) How does the finite conductivity contribute to the loss in the dielectric when an EM wave propagates?
2) What is the other factor that contributes to the loss in the dielectric media?
3) Write the expression of loss tangent of a loss dielectric.
4) Find the data sheets for a few microwave laminates and study the dielectric constant in terms of ε', ε'' and $tan\delta$ (Some manufactures of microwave laminates are Rogers; Taconic; etc.)

P1.13 Using general expressions for the propagation constant: $\gamma = \alpha + j\beta$; where

$$\alpha = \omega\sqrt{\frac{\mu\varepsilon}{2}\left(\sqrt{1+\left(\frac{\sigma}{\omega\varepsilon}\right)^2}-1\right)};\beta = \omega\sqrt{\frac{\mu\varepsilon}{2}\left(\sqrt{1+\left(\frac{\sigma}{\omega\varepsilon}\right)^2}+1\right)}$$

1) Find the attenuation constant α and the propagation constant β in:
i) low-loss medium with $(\sigma/\omega\varepsilon) < < 1$;
ii) a good dielectric?
iii) a good conductor?

P1.14 A high intensity Electromagnetic bomb (electric bomb) may cripple a nation within a moment and could be as dangerous as atomic bomb, but without any human casualty. Design an electric bomb shield to protect the computers of a strategic information management center of Melbourne with attenuation 100 (dB/m). Figure P1.14 illustrates the attenuation of wave propagation in the conducting shield. Hints: 1 (Np/m) = 8.686 (dB/m).

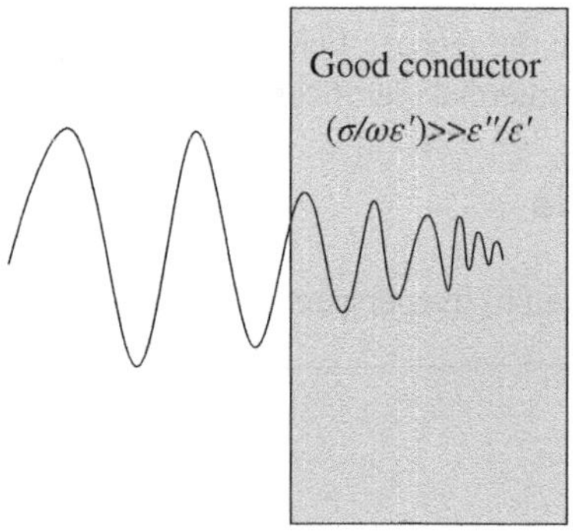

Figure P1.14 Propagation in good conductor as a shield.

P1.15 A plane wave incident from free space upon a dielectric medium with the constitutive parameters: μ_r, ε_r, σ. Calculate the propagation constant of a wave in a medium with $\mu_r = 1$. $\varepsilon_r = 10.2$, $\sigma = 24 \times 10^{-12}$ (S/m) at an operating frequency of 1 GHz.

P1.16 A plane wave is propagating from free space to a dielectric medium. Due to the medium change, the wavelength reduces from 5 to 2.5 mm. Determine the frequency, dielectric constant, and phase velocity of the medium.

P1.17 From a mobile tower an x-polarised uniform plane wave is launched to a partially conducting wall with $\varepsilon_r = 10.2$, $\mu_r = 1000$, and $\sigma = 0.4$ (S/m). The wave is propagating along z-direction with an operating frequency of 1800 MHz. Calculate attenuation constant α, phase constant, β, wave impedance η, phase velocity u_p of the wave inside the medium. If the electric field has a peak amplitude of 230 (V/m), derive the expression for both the electric and magnetic fields.

P1.18 A dual band GSM smart phone operates in 900 and 1800 MHz. The phone is protected with an aluminium cage so that the electromagnetic compatibility is maintained for underwater leakage of signal to and from the phone circuitry. Find the skin depth δ, propagation constant γ, phase velocity u_p, and the thickness of the aluminum cage so that less than 1% signal can penetrate the cage.

P1.19 If the fields from a GMS 1800 MHz mobile base station in P1.9 propagate through a thick wall of a nearby building with the following constitutive parameters, $\mu_r = 1$, $\varepsilon_r = 9$, $\sigma = 0.2$ (nS/m). Determine the propagation constant of the wave in the wall. Also determine if the wall is conductor or a dielectric in nature. If the thickness of the wall is 1 m, calculate the attenuation in the wall in dB.

P1.20 An x-polarised free space plane wave with amplitude 125 (mV/m) travelling along $+z$-direction from a GMS dual band mobile tower is normally incident on the aluminum cage of the smart phone in P1.18, calculate the wave impedance, the electric field, and the magnetic field propagating inside the aluminum cage for both bands. Assume cosine reference of the electric field and the cage is at reference position $z = 0$.

P1.21 Using the definition of attenuation constant in (Np/m) for a plane wave in lossy medium, show that 1 Np is 8.686 dB.

P1.22 If the wall shows high conductivity $\sigma = 20 \times 10^{-4}$ (S/m) due to steel frames inside the wall, but other constitutive parameters remain the same for P1.19, calculate the frequencies at which the wall is assumed to be a perfect dielectric. For a perfect dielectric use the threshold $(\sigma/\omega\varepsilon) \leq 1\%$. What is the attenuation constant at these frequencies? Does the attenuation depend on the frequency?

P1.23 The property of a material can be characterized with the attenuation profile with frequency. This can be used to detect adulteration in substances such as milk and powder for uniformity. If the amplitude of the incident electric field just at the edge of sample of substance is 20 (V/m) and the constitutive parameters of the substance is $\mu_r = 1$, $\varepsilon_r = 22$ and $\sigma = 0.58$ (S/m). Calculate the amplitude of the electric field for a 25 cm thick sample for frequencies (a) 800, (b) 1200, and (c) 1600 MHz. Draw a plot for amplitude versus frequency.

P1.24 A uniform plane wave with an electric field that vibrates along x-axis and propagates along z-direction has a maximum amplitude of 10 (mV/m) at $t = 0$ and $z = 0$. The medium of propagation is a lossless dielectric medium of dielectric constant of 10.2. The operating frequency is 850 MHz. Calculate the followings:

i) the instantaneous expression of the electric field for any t and z.
ii) the instantaneous expression of the magnetic field for any t and z.
iii) the time at which the electric field is the maximum for $t > 0$.

P1.25 Repeat P1.24 for which the maximum of the electric field is at $t = 0$, $z = 0.25$. (Hints: first calculate the initial phase ϕ).

P1.26 A radar hits a target as shown Figure P1.26. Explain in what condition the radar receive the signal. How do the stealth bombers avoid the radar?

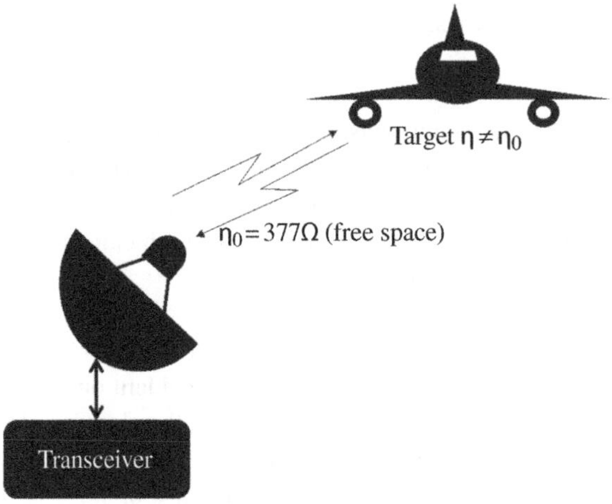

Figure P1.26 A radar targets a plane with different wave impedance than that for the free space.

P1.27 What is the intrinsic impedance of the EM wave in a lossy dielectric with a finite conductivity σ, permittivity ε, and permeability μ?

P1.28 For a propagation medium that has conductivity $\sigma = 0.02$ (S/m), dielectric constant $\varepsilon_r = 4$, and relative permeability $\mu_r = 1$. The operating frequency of a signal is 100 MHz. Calculate the propagation constant γ, attenuation constant α, phase constant β, and the wave impedance η.

P1.29 What is the phase velocity of an EM wave in a medium with the relative permittivity $\varepsilon_r = 2.2$, and the relative permeability $\mu_r = 1$?

P1.30 Show that the attenuation constant α and the propagation constant β of a good conductor are the same. Calculate the skin depth of a good conductor for which the signal attenuates to e^{-1} of its original value. And find the thickness at which the signal attenuates less than 1%.

P1.31 Show that the wave impedance of a good conductor is $\eta = \sqrt{(\omega\mu/\sigma)}\angle 45°$.

P1.32 A mobile phone cage is made of aluminum sheet to stop leakage of electromagnetic wave for electromagnetic compatibility. The incident electric field is $\boldsymbol{E_i}(z, t) = 20e^{-\alpha z}\cos(\omega t - \beta z)\boldsymbol{a_x}$ (V/m), the operating frequency is 540 MHz. The aluminum conductivity is $\sigma = 3.69 \times 10^7$ (S/m) and $\mu_r = 1$. Calculate the skin depth of aluminum at 450 MHz and find thickness of the aluminum shield so that less than 1% of the incident wave can leak through the shield.

P1.33 The phasor vector expression of a composite uniform plane wave is given by the electric field: $\boldsymbol{E}_s(x, z) = 123e^{j(23x + 15z)}\boldsymbol{a}_y$ (V/m). Find the frequency f and wavelength λ and write the instantaneous expression for the electric and the associated magnetic fields.

Poynting Theorem

P1.34 Draw a conceptual map (flow chart) of the Poynting Theorem. (Hints: take divergence of Poynting vector: $\boldsymbol{E} \times \boldsymbol{H}$). Describe the physical meaning of the Poynting Theorem.

P1.35 Energy can be transmitted wirelessly from one point in space to another point. Most recent example is the microwave power transmission from the outer orbit to the earth. About $3\text{–}9 \times 10^9$ W are expected to be collected via the rectifying antennas as the renewable energy. Read more in Wikipedia: Microwave power transmission.

P1.36 If an incident electric field is 2 (V/m) with x-polarization and a magnetic field is 2 (A/m) with y-polarization hit an antenna; what is the average power intensity at the aperture of the antenna in (W/m^2)? What is the direction of the power intensity? [Hints: $\boldsymbol{P}_{ave} = (1/2)\, Re\,(\boldsymbol{E} \times \boldsymbol{H}^*)$]

P1.37 A circular patch antenna of radius 150 mm is receiving a plane wave with the instantaneous electric field $\boldsymbol{E}(z, t) = 250\, cos\,(\omega t - \beta z)\, \boldsymbol{a}_x$ (V/m). Find the time average power incident upon the aperture of the patch antenna. Note the circular patch antenna has the diameter for the fundamental mode is very close to half of the wavelength.

P1.38 A time harmonic electromagnetic wave with electric field amplitude of 125 (V/m) from a mobile tower is launched in a highly foggy weather with relative permittivity of 9 and loss tangent of 0.2. Calculate the time average power dissipation in the atmosphere in (W/m^3). The operating frequencies are 900 and 1800 MHz for GSM mobile bands in Australia.

P1.39 An x-polarized electromagnetic field wave is propagating along $+z$ direction from a mobile tower with an operating frequency of 900 MHz. The time average power density of the wave is 30 (mW/cm^2). Calculate the electric and magnetic fields in their instantaneous form using the Poynting theorem.

P1.40 A square patch antenna is operating at 2.45 GHz wi-fi frequency band. The patch is resonant at half the wavelength of the operating frequency and designed on a microwave laminate of dielectric constant $\varepsilon_r = 4.2$. (a) Determine the patch dimension in mm. (b) If the launched average power density from the patch antenna is 50 (W/m^2), calculate the total power flux entering a square region of the same dimensions and length of 1 m of the radiating aperture of the patch antenna.

P1.41 A circular patch antenna radiates spherically symmetric electromagnetic fields. If the patch antenna radiated field is defined in the spherical coordinates as follows: $\boldsymbol{E}(r, \theta, t) = \frac{E_0}{r}\, cos^{1.2}\theta\, cos(\omega t - \beta r)\boldsymbol{a}_\theta$ (V/m), where $E_0 = 250$ (mV/m), calculate the radiated magnetic field and the time average power crossing the hemispherical shell of the radius $r = 1\ km$, $0 \leq \theta \leq (\pi/2)$. Also calculate the power density if no leakage in the bottom side of the antenna as shown in the Figure P1.41.

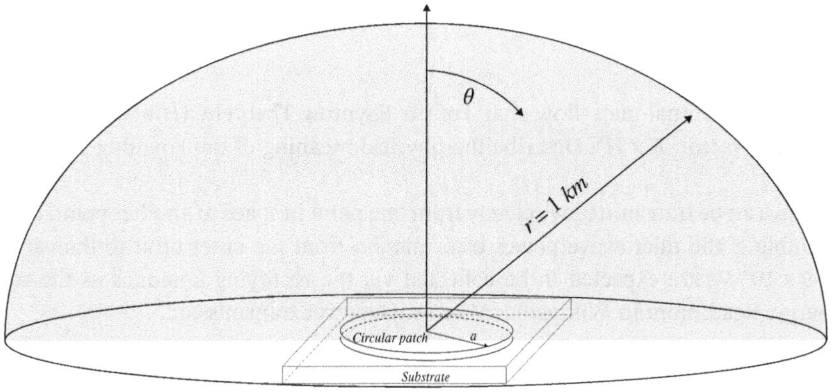

Figure P1.41 Circular patch antenna radiation mechanism.

P1.42 A uniform plane wave which is defined with $\boldsymbol{E}(z, t) = 125\, sin\,(\omega t - \beta z)\, \boldsymbol{a}_x$ (V/m) is propagating in free space. (a) Calculate the time average power density in (W/m^2). (b) Calculate the power crossing a rectangular area of sides 25 mm by 15 mm normal to the direction of power propagation.

2

Wave Propagation in Homogeneous, Nondispersive Lossy Media

2.1 Introduction

This chapter presents the electromagnetic wave propagation in homogenous nondispersive but lossy media. Both good dielectric and good conductor media are studied in this chapter followed by their composite medial for modern applications. For the good dielectric media, the effective permittivity and loss tangent are defined and the practical measurement procedure is described. For good conductors, the one-dimensional current wave equation, impedance, and power loss are analyzed. Finally, various applications are presented.

With the advent of low power low-cost wireless electronic gadgets, the use of low cost nonconventional material media for RF, microwave and millimeter wave communications is becoming prevalent. However, the low cost nonconventional materials are usually lossy and incur losses as the wave propagates in the materials. As for example, the modern chipless RFID tags that do not require any application-specific integrated circuit (ASIC) are printed on low cost composite and multilayered packaging materials in the ultra-wideband (UWB) RF, microwave and millimeter wave frequency bands. The chipless RFID is a passive printed electronic circuit on a paper or polymer packaging materials. First, on this low cost but lossy laminates, there are high demands for high integration of electronics from a few hundred MHz to 60 GHz and beyond. Second, the low-cost composite of such nonconventional dielectric materials and metallic ink paste layers in the interleaved configuration are in high demand for emerging chipless RFID technology, displays, frequency selective surfaces, microwave, and millimeter wave absorbers. Third, with popularity of the low-cost additive manufacturing using 3D printers, low-cost printing processes such as inkjet, thermal fusion, gravure, flexography, and screen printing as well as the nano-fabrication processes, the conventional bulky metallic antennas and waveguides are replaced with light weight plastics and polymer substitute materials with thin layers of metallic paints. All these state-of-the-art manufacturing facilities can satisfy the current demands of low-cost light weight electronics products in wide varieties of applications for wide range of frequency bands. Additionally, to the above exciting applications, high power radio and television transmission and even the extremely low frequency power transmission and distribution, and microwave wireless power transmission requires development of circuits on such new composite materials.

It is imperative to understand electromagnetic wave propagations in such low-cost, lossy, nonconventional composite media that were not studied with the classical electromagnetic theories before. Fourth, the new additive manufacturing such as low cost on-demand 3D printing and nano-manufacturing facilities have opened up new opportunities and infinite possibility of developing new wireless media and exotic circuit design in those frequency bands. All the current

Fields and Waves in Electromagnetic Communications, First Edition. Nemai Chandra Karmakar.

Companion website: www.wiley.com/go/fieldsandwavesinelectromagneticcommunications

demands force the RF, microwave, and millimeter wave design engineers to examine the classical electromagnetic phenomena in the new design and manufacturing environments. Understanding electromagnetic wave propagation in such complex media is paramount for efficient design, rectification of faults, and mitigation of interferences of signals. To address the current demands, the classical electromagnetic theories in lossy media is examined in details in this chapter. In Chapter 3, we shall examine wave propagation in various dispersive media and in Chapter 4, we shall examine wave propagation in more complex and practical dispersive media. As for example, optical fiber is being one of the state-of-the-art transmission lines with tera-byte of data transmission capacity, in electromagnetic wave propagation in optical fibers, we shall consider design with dispersion and analyze the fields in the dispersive media. Therefore, the theory presented in this chapter has tremendous practical significance to address the modern problems in the light of the classical electromagnetic theory.

In reality, no medium is lossless, as for example, even the expensive microwave laminates such as the commercially available popular dielectric substrates, Taconic and Duroid have losses. When electromagnetic waves propagate through these media they encounter propagation losses. Whereas the less expensive substrate, fire resistant FR4 laminate, which is commonly used in low-cost low frequency electronic PCBs, has relatively higher losses. The loss of dielectric is measured in terms of *loss tangent* of the laminate. It is a standard for commercial laminates that indicates the degree of lossy components of the materials. We shall define *the loss tangent* of a medium in this section. The *loss tangent* is defined as the subtended angle between the displacement current $\boldsymbol{J_d} = j\omega\varepsilon\boldsymbol{E}$ and conduction current $\boldsymbol{J_c} = \sigma\boldsymbol{E}$ as shown in Figure 2.1 The loss of signal during propagation in the dielectric medium is due to the finite conductivity σ and arises due to an external applied electric field $\boldsymbol{E}$ in the medium, and the damping effect of the signal due to the polarizability and dipole moments of the atomic structures of the dielectric materials. For low-loss dielectrics $(\sigma/\omega\varepsilon) << 1$, this is logical because conductivity σ is a very small value in nonconducting media. For air it is zero, for glass it is in the order of 10^{-12} (S/m). Assuming glass is a low-loss dielectric medium, you may

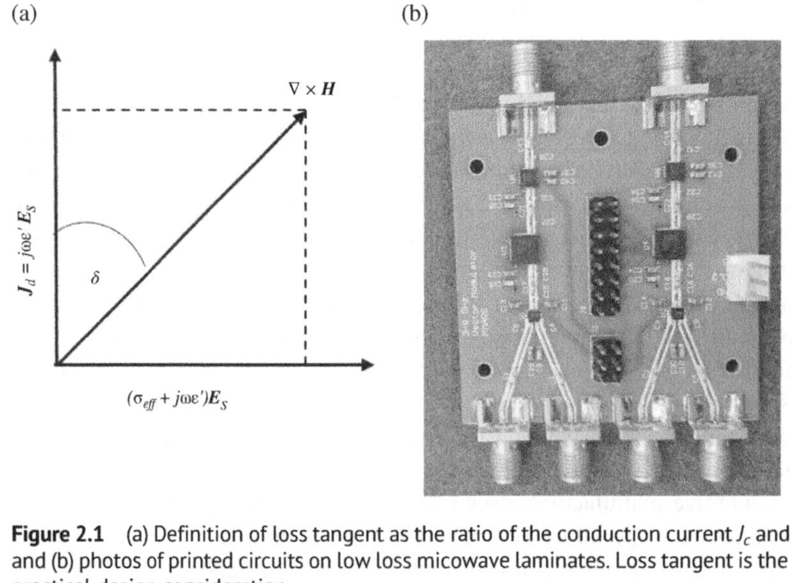

Figure 2.1 (a) Definition of loss tangent as the ratio of the conduction current J_c and displacement current J_d, and (b) photos of printed circuits on low loss micowave laminates. Loss tangent is the important parameter for practical design consideration.

easily imagine the value of σ for other good dielectric materials. The popular commercial dielectric laminates for microwave and millimeter wave frequency bands such as Taconic and Duroid are ceramic and PTFE materials, respectively. The conductivity is in the order of 10^{-15} (S/m) or less.

In Chapter 1, while defining the refractive index n and dispersion, we did not consider the loss factor of the medium. The loss comes from the finite conductivity ($\sigma \neq 0$) of the medium. To understand the electromagnetic wave signal's propagation behaviors in various media, we need to know how the attenuation constant α (Np/m) and the phase constant β (rad/m) change with frequency. All that we need to know is the behaviour of the electromagnetic wave propagation in various non-dispersive media. In this section, we are basically examining the conditions of $(\sigma/\omega\varepsilon)$ for various media at different frequency bands. Based on the condition of $(\sigma/\omega\varepsilon)$, we divide the media of propagation into two main useful categories: (i) good dielectric $(\sigma/\omega\varepsilon) \ll 1$, and (ii) good conductor $(\sigma/\omega\varepsilon) \gg 1$. First, we examine the behaviour of a good dielectric in details. We shall also learn how to measure the constitutive parameters of unknown material media. Next, we shall examine good conducting media in the light of $(\sigma/\omega\varepsilon)$ vs frequency. We shall examine the practical measurement procedure of the resistivity of conducting media. We shall examine the wave propagation in composite medium made of both dielectric and conducting media, and examine their characteristics in the light of current waves. Finally, we shall conclude the chapter with the salient features of wave propagation in various lossy media as stated.

2.2 Wave Propagation in Lossy Media

In this section, we shall examine the effects of the finite conductivity on the frequency of the propagation of plane waves in a medium. We have derived the generic expressions of the attenuation and phase constants of a lossy medium, respectively, as:

$$\alpha = \omega\sqrt{\frac{\mu\varepsilon}{2}\left[\sqrt{1+(\sigma/\omega\varepsilon)^2}-1\right]}\left(\frac{\text{Np}}{\text{m}}\right) \tag{2.1}$$

$$\beta = \omega\sqrt{\frac{\mu\varepsilon}{2}\left[\sqrt{1+(\sigma/\omega\varepsilon)^2}+1\right]}\left(\frac{\text{rad}}{\text{m}}\right) \tag{2.2}$$

As can be seen in the above expressions, both α and β are functions of $(\sigma/\omega\varepsilon)$. Since the conductivity σ is not zero for a lossy medium, therefore, there is a conduction current $\boldsymbol{J}_c = \sigma\boldsymbol{E}$, always flows in the medium due to the existence of an external applied electric field $\boldsymbol{E}$. This represents some Ohmic loss in the medium. We can define the effective conductivity σ_{eff} of the medium using Ampere's law in the time harmonic field expression:

$$\nabla \times \boldsymbol{H}_s = \sigma\boldsymbol{E}_s + j\omega\varepsilon\boldsymbol{E}_s = j\omega\left(\sigma + \frac{\sigma}{j\omega}\right)\boldsymbol{E}_s = j\omega\varepsilon_c\boldsymbol{E}_s \tag{2.3}$$

where the complex permittivity of the lossy medium is:

$$\varepsilon_c = \varepsilon' - j\varepsilon'' = \varepsilon - j\frac{\sigma}{\omega} \tag{2.4}$$

The remaining three Maxwell's equations are not included, hence they remain unchanged with the materials' property variations due to the change of conductivity σ.

The atomic point of view explanation of the complex permittivity is as follow[1]: When an external time varying electric field $\boldsymbol{E}$ propagates inside a material body, the displacement of bound charges happens. This means the change of polarization happens. As the frequency increases there is a damping factor that prevents the change of displacement and polarization. This damping factor is represented by the "susceptibility" hence the complex permittivity. There is always ohmic loss irrespective of the types of materials such as metals, semiconductors, and dielectric due to the finite conductivity of any material. Therefore, treating such media as the electromagnetic media, both damping and ohmic losses are included in the imaginary part of the permittivity ε''. Note that both real ε' and imaginary ε'' parts of the permittivity ε may be functions of frequency. We also define *the effective conductivity* that represents all losses, and therefore, can be written as:

$$\sigma_{eff} = \sigma + \omega\varepsilon'' \tag{2.5}$$

so that we can express Ampere's current law in the following form

$$\nabla \times \boldsymbol{H_s} = \sigma_{eff}\boldsymbol{E_s} + j\omega\varepsilon'\boldsymbol{E_s} = \left(\sigma_{eff} + j\omega\varepsilon'\right)\boldsymbol{E_s} = \boldsymbol{J_c} + \boldsymbol{J_d} \tag{2.6}$$

The ratio of the new displacement current $\boldsymbol{J_d} = j\omega\varepsilon'\boldsymbol{E_s}$ and the conduction current $\boldsymbol{J_c} = \sigma_{eff}\boldsymbol{E_s}$ is defined as the loss tangent of the medium[2]:

$$tan\,\delta = \frac{\sigma + \omega\varepsilon''}{\omega\varepsilon'} = \frac{\sigma_{eff}}{\omega\varepsilon'} \tag{2.7}$$

The loss tangent *tanδ* of a commercial laminate is a gold standard for substrate selection for microwave and millimeter wave frequency design. Readers are encouraged to visit the websites of Rogers and Taconic to find the loss tangents of various microwave and millimeter wave commercial laminates. Figure 2.1a illustrates the definition of loss tangent in the phasor diagram of the Ampere's circuital law with the ratio of conduction current J_c vs displacement currents J_d of a material. Figure 2.1b shows the photo of microwave printed circuit boards on low lossy dielectric materials. As stated before, the loss tangent of commercial microwave laminates in one of the most important design parameters of practical microwave circuit design. As for example if a 2.4 GHz 4×4-element array antenna is designed on a RF4 laminate with loss tangent in the order of 10^{-2} yield only 6 dBi gain. Whereas the same antenna designed on expensive laminates such as Taconic TLX0 yields about 14 dBi gain. With the lossy materials the insertion loss (forward transmission loss) in the feed network of the array as well as the real power wasted as the surface waves in the laminates and the ohmic losses due to finite conductivity significantly reduce the gain which represents the desired real power loss by radiation.

For the good conductor the term $(\sigma/\omega\varepsilon) >> 1$. In this case the conduction current $\boldsymbol{J_c} = \sigma_{eff}\boldsymbol{E_s}$ is much larger than the displacement current $\boldsymbol{J_d} = j\omega\varepsilon'\boldsymbol{E_s}$ hence very large value of *the loss tangent*. This is true in the case of conducting media where the conductivities of the metals are in the order of 10^7 (S/m). For example, the conductivity of copper is 5.8×10^7 (S/m). For the list of conductivity of metals readers are refer to the Wikipedia article[3] titled: *Electrical resistivity and conductivity*. The permittivity of the conductor can be assumed to be the same order of free space or ordinary dielectric in the order of 10^{-11} (F/m).[4] Therefore, the conduction and displacement currents to be in the same order, the frequency of operation should be in the order of $f = (\sigma/2\pi\varepsilon) = (10^7/2\pi\times10^{-11}) \cong 10^{16}$ Hz. This frequency of operation is much too high a

1 D. K. Cheng, *Fundamentals of Engineering Electromagnetics*, Addison Wesley, Reading, MA, USA, 1993, p 287.
2 There are slight variations of the definition of loss tangent and effective conductivity in different textbooks.
3 https://en.wikipedia.org/wiki/Electrical_resistivity_and_conductivity (accessed 12 December 2017).
4 For vacuum or air, the permittivity is 8.854×10^{-12} (F/m).

frequency compared that what we use in microwave, millimeter wave and light wave frequencies[5] and any frequency in the range will be used in unforeseeable future in guided and wireless communications. Therefore, we can perceive from the above discussion that for RF, microwave, millimeter wave, and light wave propagation in good conductors,[6] the conduction current J_c is much larger that the displacement current J_d. The condition is therefore $J_c \gg J_d$. For example, consider the bulk silver conductor with conductivity 6.2×10^7 (S/m) that we can use for fully printed chipless RFID tags. At the center frequency of 6 GHz of the fully printable chipless RFID tag, the loss tangent is calculated as

$$\frac{\sigma}{\omega\varepsilon} = \frac{6.2 \times 10^7 \left(\frac{\mathrm{S}}{\mathrm{m}}\right)}{2\pi \times 6 \times 10^9 \left(\frac{1}{\mathrm{s}}\right) \times \frac{10^{-9}}{36\pi}\left(\frac{\mathrm{F}}{\mathrm{m}}\right)} = 18.6 \times 10^7 >> 1$$

Therefore, for $(\sigma/\omega\varepsilon) >> 1$, the attenuation constant and phase constant for a good conductor reduces to:

$$\alpha = \beta = \omega\sqrt{\frac{\mu\varepsilon}{2}\left[\sqrt{1 + (\sigma/\omega\varepsilon)^2} \pm 1\right]} = \omega\sqrt{\frac{\mu\varepsilon}{2} \times (\sigma/\omega\varepsilon)} = \sqrt{\frac{\omega\mu\sigma}{2}} = \sqrt{\pi f \mu\sigma} \tag{2.8}$$

The skin depth is defined as the depth of penetration at which signal reduces to $e^z = e^1$ of its original value at the surface of the conductor. Therefore, *the skin depth* is defined as:

$$\delta = 1/\alpha = 1/\sqrt{\pi f \mu\sigma} \ (\mathrm{m}) \tag{2.9}$$

For good nonmagnetic conductor we assume $\varepsilon_r \cong 1$, hence $\mu = \mu_0\mu_r = 4\pi \times 10^{-7} \times \mu_r$ (H/m) and $\varepsilon = \varepsilon_0\varepsilon_r \cong 8.854 \times 10^{-12} \times \varepsilon_r$ (F/m). We obtain the expression for the phase constant with the above assumptions:

$$\beta = \alpha = \sqrt{\pi f \mu\sigma} = 1.987 \times 10^{-3}\sqrt{f\mu_r\sigma}$$

And the phase velocity

$$u_p = \omega/\beta \rightarrow \sqrt{\frac{f}{\mu_r\sigma}} \tag{2.10}$$

From the examination of the above definitions of the attenuation constant, the phase constant and the phase velocity, we can draw the following conclusions:

All three important propagation characteristic parameters α, β, and u_p are functions of frequency, f, relative permeability, μ_r and the conductivity, σ.

1) Any increase in any of these quantities, results in the increase of the attenuation constant α and phase constant β of the electromagnetic signals in the conductor.
2) The phase velocity u_p increases with the frequency f, but apparently decreases with the relative permeability μ_r, and conductivity σ. In a complex wave comprising many harmonics, the electromagnetic waves in higher frequency harmonics propagate with higher phase velocity than

5 Microwave wave frequencies range from $3 \cdot 10^8$ to $3 \cdot 10^{10}$ Hz, millimeter wave frequency ranges from $3 \cdot 10^{10}$ to $3 \cdot 10^{11}$ Hz, and light wave in the range of 10^{14} Hz.
6 In superconductor we assume zero DC resistance and $\sigma \rightarrow \infty$ (S/m). This condition of conductivity for certain materials can be achieved in cryogenic temperature. Read interesting Wikipedia article on *superconductivity* in https://en.wikipedia.org/wiki/Superconductivity, (accessed 12 December 2017).

that for the lower frequency harmonic waves. Due to the frequency dependence of the phase velocity u_p at different harmonics, signal distortion and pulse broadening occur in metallic waveguides and optical fibers. This dispersion is called *modal dispersion. The modal dispersion* creates huge design challenge in guided structures especially in optical fibers. In optical fibers, this dispersion is also called *waveguide dispersion*. The modal dispersion analysis is an integral part of practical optical fiber design.

2.3 Good Dielectric Medium

As stated above, the understanding of the electromagnetic propagation in good dielectric is of special significance in high speed microwave and millimeter wave circuit design and optical fibers. Through the above example of an antenna design on a good dielectric (e.g. Taconic TLX0) and a lossy dielectric laminate (e.g. FR4) is well perceived, it is always desirable to get a low-loss dielectric medium to design efficient active and passive microwave and mm-wave circuits. With the advent of fifth generation (5G) wireless communications and beyond 5G wireless communications in hundreds of GHz frequency bands, the demand of the electromagnetic field analysis in good dielectric materials is becoming paramount in modern era. As stated above, the dielectric loss factor is measured in terms of the loss tangent of the material medium. Usually, the materials with loss tangent in the order of 10^{-3} to 10^{-5} is desirable in microwave and millimeter wave frequency integrated circuit design. Commercial substrates such as Rogers and Taconic demonstrate quite stable low loss and dissipation factors over a wide frequency range. But these materials are very expensive. A low-cost material such as FR4 has comparatively high loss tangent in the order of 10^{-2} hence are not suitable for microwave and millimeter wave circuit design. Proper analysis of electromagnetic behavior, a composite of RF4 and a foam[7] layer can improve the situation and make very low-cost, efficient circuit design. Therefore, the electromagnetic field analysis in novel composite materials becomes very popular especially in the light of additive manufacturing such as 3D printing and non-fabrication processes.

Good dielectrics are defined as the electromagnetic media where the loss tangent $(\sigma/\omega\varepsilon) << 1$. In this case the displacement current is much larger than the conduction current. To find the approximate values of and for the low-loss dielectric material with the condition $(\sigma/\omega\varepsilon) << 1$, we can start with the generic expressions that are simplified using simple mathematical series expansion and above assumption as follows:

$$\alpha = \omega\sqrt{\frac{\mu\varepsilon}{2}\left[\sqrt{1 + (\sigma/\omega\varepsilon)^2} - 1\right]} \tag{2.11}$$

Using the series expansion method $(1 + x)^n \cong 1 + nx$ for $x << 1$ and taking only the first two terms, we obtain

$$(1 + x)^{1/2} = 1 \pm \frac{1}{2}x - \frac{1}{2}.\frac{1}{4}x^2 \pm \frac{1}{2}.\frac{1}{4}.\frac{3}{6}x^3 \quad \text{for } x \; << 1 \tag{2.12}$$

Substituting $x = (\sigma/\omega\varepsilon)^2 << 1$ into (2.12) and taking only the first two terms due to the fact that the values of x^2, x^3, etc. are negligible we obtain

---

7 Even a cushion foam laminate is made of closed air cell with polystyrene and has very low loss tangent and dielectric constant close to one.

$$\alpha \cong \omega\sqrt{\frac{\mu\varepsilon}{2}\left[1+\frac{1}{2}\left(\frac{\sigma}{\omega\varepsilon}\right)^2-1\right]} \tag{2.13}$$

Therefore, the most simplified form of expression for the attenuation constant α for the good dielectric medium is defined as:

$$\alpha = \frac{\sigma}{2}\sqrt{\frac{\mu}{\varepsilon}} \tag{2.14}$$

Likewise, we can calculate the phase constant:

$$\beta \cong \omega\sqrt{\mu\varepsilon} \tag{2.15}$$

Exercise 2.1: Propagation in Good Conductor
Q1. Show that the propagation constant for a good dielectric $\beta \cong \omega\sqrt{\mu\varepsilon}$.

You may observe that the attenuation constant α is now frequency independent due to the assumption of $(\sigma/\omega\varepsilon)^2 << 1$. This assumption of good dielectric can be perceived in two conditions[8]: (i) the conductivity σ of a good dielectric is extremely small in the order of 10^{-14} (S/m).[9] This is equivalent to almost perfect dielectric, and (ii) the conductivity σ of a good dielectric is not too large such as in sea water, but the frequency is very high so that we can fulfil the condition $J_d >> J_c$ at that frequency. From the result we can conclude that in a given medium, the attenuation increases with frequency, but if we keep on increasing the frequency then the attenuation factor α does not increase indefinitely, but this factor approach to plateau. Glass is a good dielectric and the loss tangent significantly is low and constant over the wide frequency band.

2.3.1 Wave Impedance of Good Dielectric

The wave impedance of any propagation medium is defined as the ratio of the orthogonal electric field and magnetic fields. For an x-polarized one-dimensional electromagnetic field wave, the wave impedance in a medium is defined in terms of its constitutive parameters as follows:

$$\eta = \frac{E_x^+(z,t)}{H_y^+(z,t)} = -\frac{E_x^-(z,t)}{H_y^-(z,t)} = \frac{1}{\mu u_p} = \sqrt{\frac{\mu}{\varepsilon}}\ (\Omega) \tag{2.16}$$

The general solutions of the time harmonic electromagnetic field expressions can be written as:

$$E_{xs} = E_0^+ e^{-\gamma z} + E_0^- e^{\gamma z} \tag{2.17}$$

$$H_{ys} = H_0^+ e^{-\gamma z} + H_0^- e^{\gamma z} = \frac{E_0^+}{\eta} e^{-\gamma z} - \frac{E_0^-}{\eta} e^{\gamma z} \tag{2.18}$$

where the symbols have their usual meanings.

Using Maxwell's equation, first we derive the respective x- and y-components of the time harmonic electric and the magnetic fields and then we can derive the wave impedance as follow:

$$\nabla \times \boldsymbol{E_s} = -j\omega\mu\boldsymbol{H_s} \tag{2.19}$$

8 M. A. Islam, *Electromagnetic Theory*, Dacca, East Pakistan: EPUET, 1969 p. 350.
9 Volume resistivity minimum 10^6 MΩ-cm, IPC (2014). *Specification for Base Materials for High Speed/High Frequency Applications*, IPC-4103A Amendment 1, 2014 – January, IPC, Bannockburn, Illinois, USA.

Expanding the curl of $\boldsymbol{E}_s$ in the rectangular coordinate system, we can express the above equation as follows:

$$\begin{vmatrix} \boldsymbol{a}_x & \boldsymbol{a}_y & \boldsymbol{a}_z \\ \dfrac{\partial}{\partial x} & \dfrac{\partial}{\partial y} & \dfrac{\partial}{\partial z} \\ E_0^+ e^{-\gamma z} + E_0^- e^{-\gamma z} & 0 & 0 \end{vmatrix} = \boldsymbol{a}_y\left(-\gamma E_0^+ e^{-\gamma z} + \gamma E_0^- e^{\gamma z}\right) = -j\omega\mu \boldsymbol{H} \tag{2.20}$$

We can see for the one-dimensional field wave with the x-polarized electric field, only the y-directed magnetic field exists:

$$H_y = \frac{\gamma}{j\omega\mu} E_0^+ e^{-\gamma z} - \frac{\gamma}{j\omega\mu} E_0^- e^{\gamma z} \tag{2.21}$$

Substituting for E_x and H_y from the Equations (2.17) and (2.18) in (2.21) we obtain,

$$\eta = \frac{j\omega\mu}{\gamma} = \frac{j\omega\mu}{\sqrt{j\omega\mu(\sigma + j\omega\varepsilon)}} = \sqrt{\frac{j\omega\mu}{\sigma + j\omega\varepsilon}} \tag{2.22}$$

Using the assumption of good dielectric $(\sigma/\omega\varepsilon) \ll 1$, we obtain,

$$\eta = \sqrt{\frac{j\omega\mu}{j\omega\varepsilon}} = \sqrt{\frac{\mu}{\varepsilon}} = \frac{\eta_0}{\sqrt{\varepsilon_r}} \tag{2.23}$$

where $\eta_0 = 120\pi$ (Ω), is the free space wave impedance. For the above deduction of the wave impedance for a good dielectric medium, we can conclude that the wave impedance for a good conductor is a real and positive value.

We can also generalize the above observation of the phase angles between the electric and magnetic fields using the phase angle of the complex propagation constant γ[10] as follows:

$$\tan\gamma = \frac{\alpha}{\beta} = \left(\frac{\sqrt{1 + \left(\dfrac{\sigma}{\omega\varepsilon}\right)^2} - 1}{\sqrt{1 + \left(\dfrac{\sigma}{\omega\varepsilon}\right)^2} + 1}\right)^{\frac{1}{2}} = 0 \tag{2.24}$$

From the above equation of the angle of γ, we can conclude that for perfect dielectric, $\sigma = 0$, and the angle is zero.

In the above derivation we consider the real value of the dielectric constant ε. Now we shall consider the complex permittivity value of the material and shall show that with the loss there is a phase angle between the attenuation constant and the phase constant.

2.4 Low-Loss Dielectric Medium

In the above analysis of the good dielectric we did not consider the complex permittivity $\varepsilon = \varepsilon' - j\varepsilon''$. For lossy dielectric media we need to consider the complex permittivity. In this section we consider a low-loss dielectric as a good but imperfect insulator with a nonzero equivalent conductance $\sigma_{eff} = \sigma + \omega\varepsilon''$.[11] The condition for a low-loss dielectric is defined as $\varepsilon'' \ll \varepsilon'$ in which case $(\sigma/\omega\varepsilon'') \ll 1$ is still valid. Using the condition, the propagation constant is defined as:

---

10 M.A. Islam, *Electromagnetic Theory*, EPUT, Dacca, East Pakistan: EPUET, 1969, p. 350.

11 Similar derivation is available in D. K. Cheng, *Fundamentals of Engineering Electromagnetics*, Addison Wesley, NY, USA, 1993.

$$\gamma = \alpha + j\beta \cong j\omega\sqrt{\mu\varepsilon'}\left\{1 - j\frac{\varepsilon'}{2\varepsilon''} + \frac{1}{8}\left(\frac{\varepsilon'}{\varepsilon''}\right)^2\right\} \tag{2.25}$$

Separating the real and imaginary parts, respectively, we obtain the expression of the attenuation constant α and the phase constant β as follows:

$$\alpha = Re(\gamma) \cong \frac{\omega\varepsilon'}{2}\sqrt{\frac{\mu}{\varepsilon'}}\left(\frac{\text{Np}}{\text{m}}\right) \tag{2.26}$$

$$\beta = Im(\gamma) \cong \omega\sqrt{\mu\varepsilon'}\left\{1 + \frac{1}{8}\left(\frac{\varepsilon'}{\varepsilon''}\right)^2\right\}\left(\frac{\text{rad}}{\text{m}}\right) \tag{2.27}$$

The wave impedance for the good dielectric can be deduced using the assumption $(\sigma/\omega\varepsilon) << 1$ with the complex permittivity value as follows.[12] Using the assumption of good dielectric $(\sigma/\omega\varepsilon) \ll 1$, as follows,

$$\eta = \frac{j\omega\mu}{\gamma} = \sqrt{\frac{j\omega\mu}{\sigma + j\omega\varepsilon}} = \sqrt{\frac{j\omega\mu}{j\omega\varepsilon\left(\frac{\sigma}{j\omega\varepsilon} + 1\right)}} = \sqrt{\mu/\varepsilon}\left(1 + \frac{\sigma}{j\omega\varepsilon}\right)^{-1/2} \tag{2.28}$$

Substitution for the complex permittivity: $\varepsilon_c = \varepsilon' - j\varepsilon'' = \varepsilon - j(\sigma/\omega)$ and using binomial expansion we obtain:

$$\eta = \sqrt{\mu/\varepsilon'}\left\{1 - j\frac{\varepsilon''}{\varepsilon'}\right\}^{-1/2} \cong \sqrt{\mu/\varepsilon'}\left\{1 + j\frac{\varepsilon''}{2\varepsilon'} + \frac{1}{8}\left(\frac{\varepsilon''}{\varepsilon'}\right)^2\right\}^{-1/2} \tag{2.29}$$

Since "$\varepsilon' \gg \varepsilon''$," we can deduce the wave impedance as:

$$\eta = \frac{E_x}{H_y} = \sqrt{\mu/\varepsilon'}\left(1 + j\frac{\varepsilon''}{2\varepsilon'}\right)(\Omega) \tag{2.30}$$

The wave impedance η is complex meaning that the electric and magnetic fields are not in phase in a lossy dielectric. The phase velocity is obtained as the ratio of ω/β as follows:

$$u_p = \frac{\omega}{\beta} \cong \frac{1}{\sqrt{\mu\varepsilon'}}\left\{1 - \frac{1}{8}\left(\frac{\varepsilon''}{\varepsilon'}\right)\right\}\left(\frac{\text{m}}{\text{s}}\right) \tag{2.31}$$

Comparing the lossless case and lossy case of the dielectric constant of a good dielectric we can deduce that the exact phase velocity is slightly lower than that for the lossless medium.

Comparing the derivations above considering the real value of permittivity ε and complex permittivity $\varepsilon_c = \varepsilon' - j\varepsilon'' = \varepsilon - j(\sigma/\omega)$ with the condition $(\sigma/\omega\varepsilon) << 1$ and finite conductivity σ, we can summarize the findings as shown in the following Table 2.1.

Figure 2.2 illustrates the variation of the loss tangent for the three lossy media glass, a good dielectric, a lossy the dielectric and copper, a good conductor and iron, a poor conductor with frequency. As can be seen the loss tangent for glass remains constant with a negligible value over the frequency band from 10^2 to 10^{12} Hz. Glass is a good dielectric medium for which $\varepsilon' \gg \varepsilon''$. On the other hand, being a good conductor, the loss tangent of copper reduces steadily with the frequency. Same for the iron with less loss tangent. For both copper and iron the loss tangent

12 D. K. Cheng, *Fundamentals of Engineering Electromagnetics*, Addison Wesley, NY, USA, 1993.

Table 2.1 Characteristic parameters of good and low-loss dielectric media.

	Good dielectric	Low-loss dielectric
Effective conductivity	$\sigma_{eff} = \sigma + \omega\varepsilon''$ (S/m)	$\sigma_{eff} = \sigma + \omega\varepsilon''$ (S/m)
Loss tangent	$tan\delta = (\sigma_{eff}/\omega\varepsilon')$	$tan\delta = (\sigma_{eff}/\omega\varepsilon')$
Attenuation constant	$\alpha = (\sigma/2)\sqrt{(\mu/\varepsilon)}$ (Np/m)	$\alpha = \omega\varepsilon'/2\sqrt{(\mu/\varepsilon')}$ (Np/m)
Phase constant	$\beta \cong \omega\sqrt{\mu\varepsilon}$ (rad/m)	$\beta \cong \omega\sqrt{\mu\varepsilon'}\{1 + (1/8)(\varepsilon'/\varepsilon\prime\prime)^2\}$ (rad/m)
Wave impedance	$\eta = \sqrt{(\mu/\varepsilon)} = (\eta_0/\sqrt{\varepsilon_r})$ (Ω)	$\eta = \sqrt{\mu/\varepsilon'}(1 + j(\varepsilon''/2\varepsilon'))$ (Ω)
Phase velocity	$u_p \cong (1/\sqrt{\mu\varepsilon'})\{1 - (1/8)(\varepsilon''/\varepsilon')\}$ (m/s)	$u_p \cong (1/\sqrt{\mu\varepsilon'})\{1 - (1/8)(\varepsilon''/\varepsilon')\}$ (m/s)

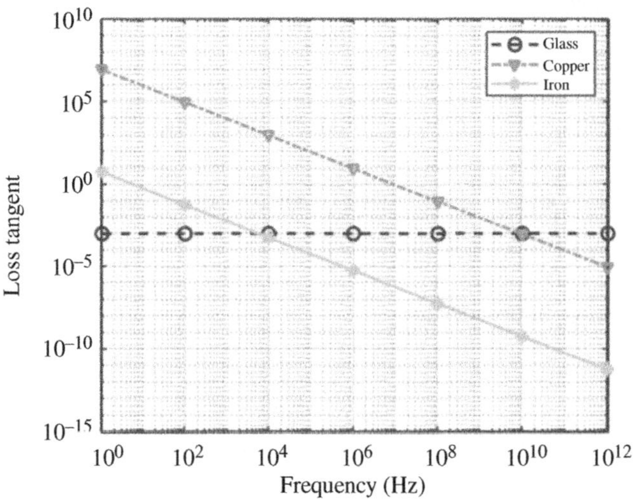

Figure 2.2 Loss tangent vs frequency of glass, copper, and iron: Copper: $\sigma = 5.8 \times 10^7$ (S/m), $\varepsilon_r' = 1$, $\varepsilon_r'' = 0$, glass: $\sigma = 10 \times 10^{-12}$ (S/m), $\varepsilon_r' = 10$, $\varepsilon_r'' = 0.01$, and iron: $\sigma = 10^7$ (S/m), $\varepsilon_r' = 1$, $\varepsilon_r'' = 0$.

changes linearly in the logarithmic scale. As a matter of fact, the loss tangent of a dielectric medium is a measurable quantity and most often measured preciously with a vector network analyzer. In the following section we present the practical measurement procedure of the dielectric constant, attenuation constant, and loss tangent of any liquid and solid medium.

Review Questions 2.1: Medium Properties and Loss Tangent

Q1. Define polarizability and polarization loss of lossy dielectric in the light of atomic model.
Q2. Define modal dispersion of materials.
Q3. What is the effective conductivity of a lossy dielectric?
Q4. Explain why the permittivity of a lossy dielectric is complex. Where do the losses come from?

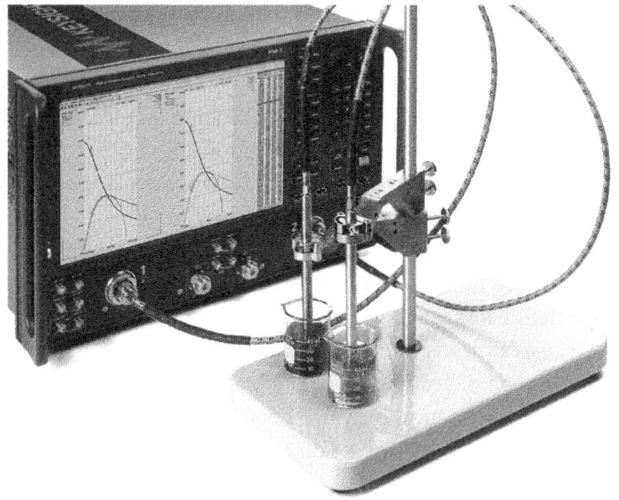

Figure 2.3 Keysight N150 A dielectric probe kit 10 MHz to 150 GHz. *Source:* Reproduced with permission from Keysight Technologies.

2.4.1 Measurement Procedure of Relative Permittivity and Loss Tangent

For designing RF/microwave/millimeter wave passive and active circuits on any unknown dielectric laminate, it is significant to measure the relative permittivity and loss tangent of the laminate. Note that the RF/microwave/millimeter laminates are medium of wave propagation at the designated frequency or a band of frequencies. Usually the commercial datasheets of those laminates do not provide sufficient and essential data for microwave design. Hence, invariably for any new design, it is essential to measure these two parameters for accurate design. The measurements of relative permittivity and loss tangent are performed in RF/microwave frequencies. The theory of measurement is based on microwave resonant structures and waveguides. Hence, it needs the knowledge of microwave engineering and an expensive equipment such as a vector network analyzer, and measurement probes. Figure 2.3 shows the photograph of Keysight technologies Inc.'s Keysight N1501A, dielectric probe kit for measurement of dielectric constant and loss tangent from 10 MHz to 50 GHz. The instrument can measure dielectric constants and loss tangent of many types of solid and liquid materials. The photograph shows two samples of liquid materials. Readers are encouraged to read more about the probe and its measurement procedure via the Keysight N150A link.

Finally, the practical measurement procedure of dielectric constant and loss tangent is presented. In the next section, the theory of wave propagation in good conductors is presented. In the modern era of wireless communications, the low cost printed electronics with advanced manufacturing process, are becoming prevalent. Therefore, the study of wave propagation in good conductors is significant to underpin the efficient design of low-cost electronic circuits in RF/microwave and millimeter domain.

2.4.2 Summary of Lossy Dielectric Materials

1) Loss tangent is a very significant concept for high speed electronic circuit design on low-loss dielectric medium. A microwave design engineer carefully selects microwave laminated with very low loss tangent. As stated above during the explanation of the atomic structure of the dielectric medium, the dielectric loss comes from the finite conductivity of the medium, and polarization loss or a combination of both. In dielectric, the molecules create dipole moments. When an electric field excitation is applied, the dipole moments try to align themselves along the field

lines and that process slows down and attenuates the wave propagation. This phenomenon is called *polarizability* of the dielectric. And the associated damping loss is called polarization loss.[13]

2) In lossy dielectric, the permittivity becomes complex. The expression for complex permittivity is: $\varepsilon_c = \varepsilon' - j\varepsilon'' = \varepsilon - j(\sigma/\omega)$. Substitution for the complex permittivity in Ampere's law we get: $\nabla \times \boldsymbol{H_s} = (\sigma_{eff} + j\omega\varepsilon')\boldsymbol{E_s} = \boldsymbol{J_c} + \boldsymbol{J_d}$. Now we replace σ with σ_{eff} for complex permittivity calculation of lossy dielectric.
3) Low loss materials have a small but finite attenuation α. For delicate microwave frequency design, we must consider it for accurate results.
4) We define loss tangent $tan\delta$, where δ is the subtended angle by which the displacement current $\boldsymbol{J_d} = j\omega\varepsilon'\boldsymbol{E_s}$ leads the total current density $\boldsymbol{J}$. See Figure 2.1 for this definition.

   For good dielectric σ is negligible and loss tangent $tan\delta$ is simply the ratio of ε'' and ε''.
5) Finally, the practical measurement procedures of the dielectric constant ε_r, attenuation constant α and loss tangent $tan\delta$ of dielectric media have been presented.

Review Questions 2.2: Lossy Dielectric Materials

Q1. Define loss tangent of a dielectric medium.

Q2. Research tanδ for a few commercial substrates such as Rogers Duroid, Taconic TLX8 and FR4.

Q3. Can you identify relations between the loss tangent, and conductivity with the price of these materials?

Q4. In good conductor $(\sigma/\omega\varepsilon)^2 >> 1$ and in good dielectric $(\sigma/\omega\varepsilon)^2 << 1$. Can you draw an inference how the frequency f, conductivity σ, permittivity ε, and permeability μ play their roles in the respective medium?

Q5. What are the values of conductivities in a good dielectric and a good conductor, respectively.

In the next section, the theory of the wave propagation in good conductors is presented. As stated above, in the modern era of wireless communications, the low cost printed electronics with advanced manufacturing processes are becoming prevalent. Therefore, the study of the wave propagation in good conductors is significant to underpin the efficient design of low-cost electronics in RF/microwave/millimeter wave domains.

2.5 Wave Propagation in Good Conducting Medium

In microwave and millimeter wave frequencies, even in low frequency electronics, conductor plays significant role in two aspects: (i) good conductors are used as the guiding structures of the electromagnetic signals, and (ii) good conductors are used for shielding and grounding to suppress electromagnetic interferences. In the first application, we use the knowledge of transmission line theory to design the conducting tracks for efficient transfer of electromagnetic energy from one point of the medium to another point. In this aspect, we learn the efficient transfer of electromagnetic energy

13 The atomic model of lossy dispersive media will be examined in details in Chapter 3: Waves in Dispersive Media.

via the transmission line and matching theories. In guiding the signal, we need high conductivity, certain thickness larger than a few skin depths and least roughness of the conductor. The first one is the matter of study in this section. The thickness is related to *the skin depth* of the conductor that defines the depth of penetration of signal in the skin of the conductor at which the signal attenuates less than 38% of its amplitude at the surface. See (2.9) for the definition of the skin depth. As we know the signal exponentially decays inside the conductor with their relaxation time. The depth of the metallic skin in which the signal reduces to e^{-1} of its original value is called *the skin depth.* Skin depth is a function of permeability, μ, conductivity σ, and the frequency, f. It is inversely proportional to the square root of the frequency f. Therefore, it is a blessing that with the advent of low cost and efficient microwave and millimeter wave generators, we can envision designing many circuits in those frequencies with the advanced printing processes. With the advent of nanotechnology and nano-fabrication, and 3D additive printing processes, it is possible to fabricate microwave/mm-wave printed electronics[14] in recent years. In this design, *the skin depth* plays a major role in efficient transfer of signal via the conducting truck. In nano-fabrication process, we can print a conducting track that guide microwave and mm-wave signals in the order of a fraction of a *skin depth.* However, if the track thickness is less than the skin depth then the signal is not strongly guided, as a result, it leaks into the laminate and air. This results in huge attenuation of guided signal and interferences in the nearby circuits. We shall examine the significance of *the skin depth* in this section. The same concept is also applicable to low cost shielding for electromagnetic compatibility (EMC). With the advent of 3D printing using additive manufacturing technology, nowadays complex mechanical structures can be built with polymers. We can coat the polymers, which are dielectric, with a thin layer of nanoparticle based conductive paint. In this regard *the skin depth* also plays a major role to shield the circuit against unwanted interference. Figure 2.4a illustrates the photograph of a flexographic printed chipless RFID tags on a polymer substrate. Figure 2.4b illustrates the measured UWB microwave response from 2–8 GHz for 300 μm and 400 μm slot width of the tag. Figure 2.4c shows the thickness measurement of printed tag with conductor thickness of 0.361 and 0.730 μm.

For the good conductor the term $(\sigma/\omega\varepsilon) >> 1$. In this case the conduction current $J_c = \sigma_{eff} E_s$ is much larger than the displacement current $J_d = j\omega\varepsilon E_s$ hence very large value of *the loss tangent.* This is true in the case of conducting media where the conductivity of the metals is in the order of 10^7 (S/m). For example, the conductivity of copper is 5.8×10^7 (S/m). For the list of conductivity of metals, readers are referred to the Wikipedia article[15] titled: *Electrical resistivity and conductivity.* The permittivity of the conductor can be assumed to be the same order of free space or ordinary dielectric media in the order of 10^{-11} (F/m).[16] Therefore, the conduction and displacement currents to be in the same order, the frequency of operation should be in the order of $f = (\sigma/2\pi\varepsilon) = (10^7/2\pi \times 10^{-11}) \cong 10^{16}$ Hz. This frequency of operation is much too high a frequency compared to what we use in microwave, millimeter wave, and light wave frequencies[17] and any frequency in the range

14 The author has been working on the fully printable chipless radio frequency identification (RFID) tags in microwave and millimeter wave frequencies using nanoparticle silver ink and low-cost laminates such as paper and plastic. Some of the printing processes such as screen printing, flexography and gravure may not yield the conductor thickness to match *the skin depth* at those frequency bands. In those cases, the metallic transmission line printed by the conducting ink cannot strongly guide the microwave and millimeter wave electromagnetic signals. As a consequence, leakages of the electromagnetic signals from the printed metallic ink track into the laminates and air results in huge signal loss and poor performance of the circuits.

15 https://en.wikipedia.org/wiki/Electrical_resistivity_and_conductivity (accessed 12 December 2017).

16 For vacuum or air, the permittivity is 8.854×10^{-12} (F/m).

17 Microwave wave frequencies range from $3{\cdot}10^8$ to $3{\cdot}10^{10}$ Hz, millimeter wave frequency ranges from $3{\cdot}10^{10}$ to $3{\cdot}10^{11}$ Hz, and light wave in the range of 10^{14} Hz.

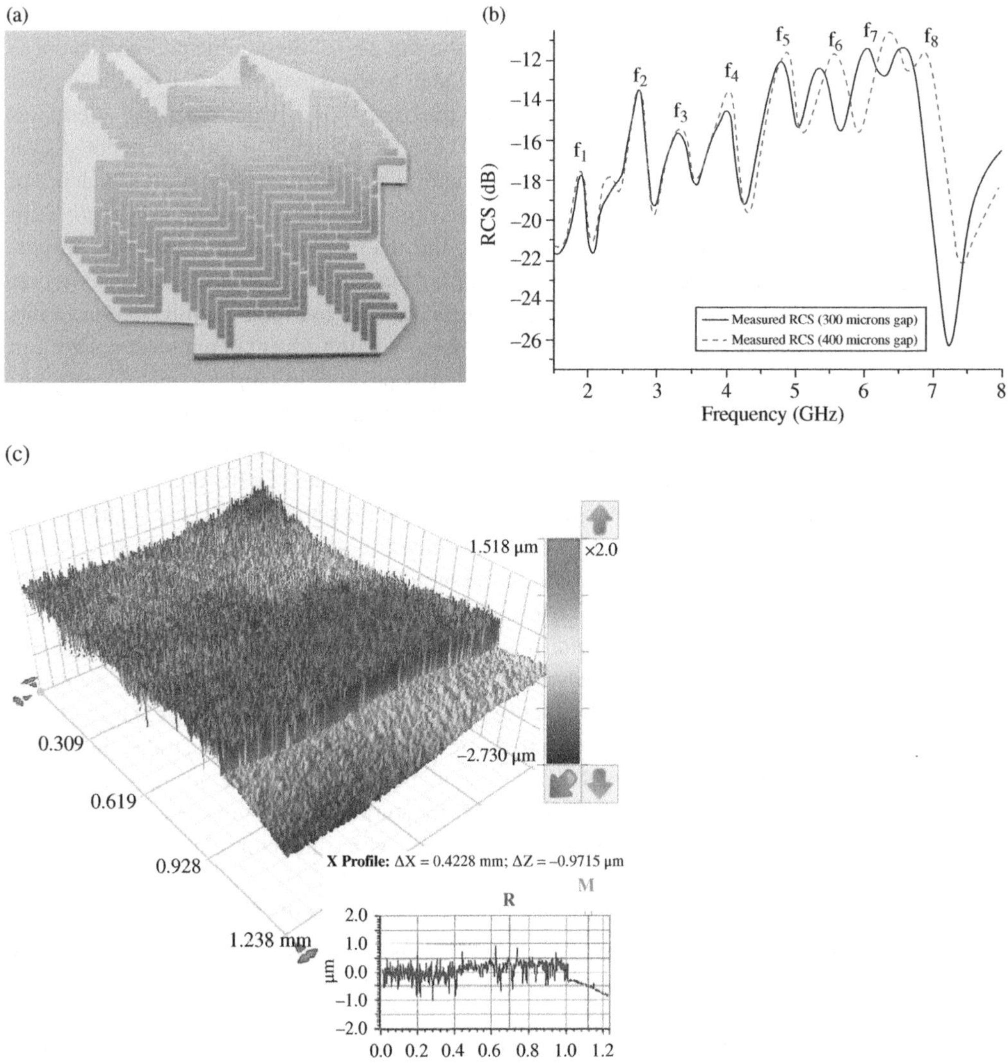

Figure 2.4 (a) Printed L-slot multi-resonator as the chipless RFID tag, (b) measured ultra-wideband microwave response for 300 and 400 μm slot width of tag, and (c) thickness measurement of printed tag with conductor thickness of 0.361 and 0.730 μm. *Source:* Reproduced with permission from Sika Shrestha.

will be used in unforeseeable future in guided and wireless communications. Therefore, we can perceive from the above discussion that for RF, microwave, millimeter wave, and light wave propagation in a good conductor,[18] the conduction current $\boldsymbol{J_c} >> \boldsymbol{J_d}$.

---

18 In superconductor we assume zero DC resistance and $\sigma \rightarrow \infty$ (S/m). This condition of conductivity for certain materials can be achieved in cryogenic temperature. Read interesting Wikipedia article on *superconductivity* in https://en.wikipedia.org/wiki/Superconductivity, (accessed 12 December 2017).

Example 2.1: Good Conductor
Consider the bulk silver conductor with conductivity 6.2×10^7 (S/m) that we can use for fully printed chipless RFID tags as shown in Figure 2.4. At the center frequency of 6 GHz, of the fully printable chipless RFID tag, calculate the loss tangent?

Answer:

$$tan\,\delta = \frac{\sigma}{\omega\varepsilon} = \frac{6.2\times10^7\left(\frac{\mathrm{S}}{\mathrm{m}}\right)}{2\pi\times6\times10^9\left(\frac{1}{s}\right)\times\frac{10^{-9}}{36\pi}\left(\frac{\mathrm{F}}{\mathrm{m}}\right)} = 18.6\times10^7 >> 1$$

Exercise 2.2: Good Conductor
Q1. Repeat Example 2.1 for copper with $\sigma = 5.8\times10^7$ (S/m).

Therefore, for $(\sigma/\omega\varepsilon) >> 1$, the attenuation constant α and phase constant β reduces to:

$$\alpha = \beta = \omega\sqrt{\frac{\mu\varepsilon}{2}\left[\sqrt{1+(\sigma/\omega\varepsilon)^2}\pm1\right]} = \omega\sqrt{\frac{\mu\varepsilon}{2}\times(\sigma/\omega\varepsilon)} = \sqrt{\frac{\omega\mu\sigma}{2}} = \sqrt{\pi f\mu\sigma} \tag{2.32}$$

This means for a good conductor, the attenuation and propagation constant are the same and equal to $\sqrt{\pi f\mu\sigma}$, where the symbols have their usual meanings.

As defined earlier, the skin depth at which the depth from the surface of the electromagnetic signal reduces the signal intensity to $e^{-\alpha\delta} = e^{-1}$ of its original value is called *the skin depth*. Therefore, *the skin depth* δ in (cm) is defined as:

$$\delta = 1/\alpha = 1/\sqrt{\pi f\mu\sigma} \tag{2.33}$$

where, the symbols have their usual meaning.

For a good nonmagnetic conductor, we assume $\varepsilon_r \cong 1$, $\mu_r = 1$ and $\varepsilon = \varepsilon_0\varepsilon_r \cong 8.854\times10^{-12}$ (F/m), hence $\mu = \mu_0\mu_r = 4\pi\times10^{-7}\mu_r$ (H/m). We obtain the expression for the phase constant with the above assumptions:

$$\beta = \alpha = \sqrt{\pi f\mu\sigma} = 1.987\times10^{-3}\sqrt{f\mu_r\sigma}$$

And the phase velocity is defined as:

$$u_p = \frac{\omega}{\beta} = 2\sqrt{\frac{\pi f}{\mu\sigma}} \rightarrow \sqrt{\frac{f}{\mu\sigma}} \tag{2.34}$$

From the examination of the above definitions of the attenuation constant α, phase constant β and the phase velocity u_p we can draw the following conclusions:

1) All three important propagation characteristic parameters α, β, and u_p are functions of frequency, *f*, relative permeability, μ_r and the conductivity, σ.

2) Any increase in any of these quantities, results in the increase of the attenuation and phase constants of the electromagnetic signals in the conductor.
3) The phase velocity u_p increases with the frequency f but apparently decreases with the relative permeability μ_r, and conductivity σ. In a complex wave comprising many harmonics, the electromagnetic waves in higher frequency harmonics propagates with higher velocity than that for the lower frequency harmonic waves. Due to the frequency dependence of the phase velocity u_p at different harmonics, signal distortion and pulse broadening occur in metallic waveguides and optical fibers. This dispersion is called *modal dispersion*. The modal dispersion creates huge design challenges in guided structures especially in optical fibers. In optical fibers, this dispersion is also called *waveguide dispersion*. These theory of *the waveguide dispersion* of optical fibers are important for practical design challenges.

For nonmagnetic conductors such as silver, copper, and gold, $\mu_r = 1$ hence the skin depth and the attenuation and propagation constants are independent of μ_r.

$$\beta = \alpha = 1.987 \times 10^{-3}\sqrt{f\sigma}$$

Review Questions 2.3: Wave Propagation in Good Conducting Media

Q1. Explain the significance of the study of skin depth in modern printed electronic circuits.
Q2. Calculate the number of skin depths of the flexographic printed chipless RFID in Figure 2.4. Do you think, it can guide the electromagnetic signal well?
Q3. Explain the limit of frequency at which we can assume the displacement current is much less than the conduction current. What does this condition physically mean to you?
Q4. Explain the dispersion related to the phase velocity of the electromagnetic signal in propagation medium.
Q5. Define electric wall. What is the second name of a perfect conductor and why?

Exercise 2.3: Propagation and Attenuation in Popular Good Conductors
Q1. Calculate the propagation and attenuation constant of silver, copper and gold at 6 GHz. (Hints: $\sigma_{silver} = 6.3 \times 10^7$ (S/m); $\sigma_{copper} = 5.8 \times 10^7$ (S/m); $\sigma_{glod} = 4.1 \times 10^7$ (S/m))

Answer:
$\beta_{sliver} = \alpha_{sliver} = 1.33 \times 10^6$ {(rad/m); (Np/m)}; $\beta_{copper} = \alpha_{copper} = 1.17 \times 10^6$ {(rad/m); (Np/m)}; $\beta_{gold} = \alpha_{gold} = 985.52 \times 10^3$ {(rad/m); (Np/m)}

2.6 Wave Impedance in Good Conductors

To derive the wave impedance in good conductors, now let us link the electric and magnetic fields and examine their characteristics in good conductor. The ratio of the orthogonal components of the propagating electric and magnetic fields is defined as the intrinsic wave impedance of the medium. For a good conductor;

$$\eta = \frac{E_x^+(z,t)}{H_y^+(z,t)} = -\frac{E_x^-(z,t)}{H_y^-(z,t)} = \frac{1}{\mu u_p} = \sqrt{\frac{\mu}{\varepsilon}}\ (\Omega) \tag{2.35}$$

The general solutions of the time harmonic field expression can be written as:

$$E_{xs} = E_0^+ e^{-\gamma z} + E_0^- e^{\gamma z} \tag{2.36}$$

$$H_{ys} = H_0^+ e^{-\gamma z} + H_0^- e^{\gamma z} = \frac{E_0^+}{\eta} e^{-\gamma z} - \frac{E_0^-}{\eta} e^{\gamma z} \tag{2.37}$$

Using Maxwell's equation, we can derive the wave impedance as follow:

$$\nabla \times \boldsymbol{E_s} = -j\omega\mu \boldsymbol{H_s} \tag{2.38}$$

Expanding the curl of $\boldsymbol{E_s}$ in the rectangular coordinate system, we can express the above equation as follows:

$$\begin{vmatrix} \boldsymbol{a_x} & \boldsymbol{a_y} & a_z \\ \frac{\partial}{\partial x} & \frac{\partial}{\partial y} & \frac{\partial}{\partial z} \\ E_0^+ e^{-\gamma z} + E_0^- e^{-\gamma z} & 0 & 0 \end{vmatrix} = \boldsymbol{a_y}\left(-\gamma E_0^+ e^{-\gamma z} + \gamma E_0^- e^{\gamma z}\right) = -j\omega\mu \boldsymbol{H} \tag{2.39}$$

The y-directed component of $\boldsymbol{H}$ is defined as:

$$H_y = \frac{\gamma}{j\omega\mu} E_0^+ e^{-\gamma z} - \frac{\gamma}{j\omega\mu} E_0^- e^{\gamma z} \tag{2.40}$$

Therefore, the intrinsic wave impedance of a good conductor is:

$$\eta = \frac{j\omega\mu}{\gamma} = \frac{j\omega\mu}{\sqrt{j\omega\mu(\sigma + j\omega\varepsilon)}} = \sqrt{\frac{j\omega\mu}{\sigma + j\omega\varepsilon}} \tag{2.41}$$

Using the assumption of good conductor $(\sigma/\omega\varepsilon) >> 1$, we obtain,

$$\eta = \sqrt{\frac{j\omega\mu}{\sigma}} = \sqrt{\frac{\omega\mu}{\sigma}} e^{j45^\circ} \tag{2.42}$$

Defining $\sqrt{j} = 1e^{j45^\circ}$, we obtain the above relationship. For the above deduction, we can conclude that the magnetic field lags the electric field by 45° in a good conductor.

We can also generalize the above observation of the phase angles between the electric and magnetic fields using the phase angle of the complex propagation constant γ[19] as follows:

$$\tan\gamma = \frac{\alpha}{\beta} = \left(\frac{\sqrt{1 + (\sigma/\omega\varepsilon)^2} - 1}{\sqrt{1 + (\sigma/\omega\varepsilon)^2} + 1}\right)^{\frac{1}{2}} \tag{2.43}$$

From the above equation of the angle of γ, we can conclude that for the perfect dielectric $\sigma = 0$, and the angle is zero. When the medium is very conductive, we assume $(\sigma/\omega\varepsilon) >> 1$, hence $\tan\gamma \to 1$. Therefore, in metals the angle of γ becomes 45° and in good conductors, the electric and magnetic field vectors are 45° out of phase.

19 M.A. Islam, *Electromagnetic Theory*, EPUT, Dacca, East Pakistan: EPUET, 1969, p. 350.

Example 2.2: Wave Impedance of Good Conductor
Calculate wave impedance of copper where $\mu_r = 1$, $\varepsilon_r = 1$, $\sigma = 5.8 \times 10^7$ (S/m) at 1.2 GHz.

Solution:

$$\eta = \sqrt{\frac{j\omega\mu}{\sigma + j\omega\varepsilon}} = \sqrt{\frac{j\,2\pi\;\;1.2\;\;10^9 \times 4\pi \times 10^{-7}}{5.810^7 + j\,2\pi \times 1.2 \times 10^9 \times 8.85 \times 10^{-12}}}$$

$$\eta = \sqrt{\frac{9465.2j}{5.810^7 + 66.6 \times 10^{-3}j}} = \frac{\sqrt{9465.2\angle 90^\circ}}{\sqrt{5.8 \times 10^7 \angle 0^\circ}} = 0.012\angle 45^\circ\ (\Omega).$$

2.6.1 Practical Applications: Geophysics

Practical applications of the theory of propagation of electromagnetic wave in good conductors can be found in areas other than the electrical/electronic/telecommunications engineering. For example,[20] in classification of hard rocks in groundwater geophysics, a very low frequency resistance measurement is performed with a very low frequency resistance (VLF-R) meter. If the phase difference between the electric and magnetic fields is 45° or more then the rock is classified as conductive, and if it is less than 45° then it is classified as the resistive surface. Figure 2.5 shows the working principle of the meter. As can be seen, the electromagnetic signal incidents via the current probes, and the voltage is measured with the potential electrodes and voltmeter. Based on the geophysical compositions of the soil, the signal strength is different due to the variation of the impedance. Thus, such measurements are used for mining, explorations, and geophysics. We shall discuss the relationship between electric and magnetic fields in the interface of different media in the transmission and reflection theory in the later chapters.

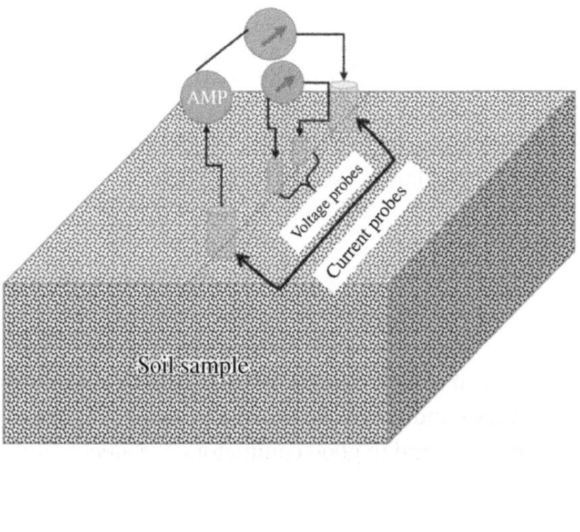

Figure 2.5 VLF-R meter method to measure the profile of underground earth layers and water profile.

20 Prabhat Chandra, *Groundwater Geophysics in Hard Rock*, CRC Press, 2015, p. 224.

Example 2.3: Attenuation, Propagation and Wave Impedance of Good Conductor
As for the example of the fully printable chipless RFID with a bulk silver track shown in Figure 2.5, calculate the attenuation and phase constants, wave impedance and phase velocity at 6 GHz. Assume for silver $\mu_r = 1$.

$$\alpha = \beta = \sqrt{\pi f \mu \sigma} = \sqrt{\pi \times 6 \times 10^9 (\text{Hz}) \times 4\pi \times 10^{-7} \left(\frac{\text{H}}{\text{m}}\right) \times 6.2 \times 10^7 \left(\frac{\text{S}}{\text{m}}\right)} = 1.212 \times 10^6 \,(\text{m}^{-1})$$

$$\eta = \sqrt{\frac{2\pi \times 6 \times 10^9 \times 4\pi \times 10^{-7}}{6.2 \times 10^7}} e^{j45^\circ} = 27.6\angle 45^\circ \text{ m}\Omega$$

If we compare this characteristic parameter of bulk sliver at 6 GHz with those for the signal propagation in air, we have significant contrast as follows:
For air, $\mu_r = 1$ and $\sigma = 0$. Hence $\alpha = 0$.
The propagation constant

$$\beta = \frac{2\pi \times 6 \times 10^9}{3 \times 10^8} = 40\pi \left(\text{rad/m}^{-1}\right)$$

The wave impedance is $\eta = \sqrt{(\mu_o/\varepsilon_0)} = \sqrt{(4\pi \times 10^{-7}/8.854 \times 10^{-12})} = 120\pi \,(\Omega)$

Comparing the good conductor with the air, we can deduce that the conductor appears to be almost short circuit to the wave. For a perfect conductor, $\alpha = \beta \to \infty$ with infinite σ. The intrinsic wave impedance $\eta = 0\,(\Omega)$, offers perfect short circuit to the electromagnetic wave. A perfect conductor is also called *the electric wall* in the microwave nomenclature. Such a nomenclature is used to define the boundary conditions of the field wave in the analysis of microwave resonators, waveguides, and microstrip patch antennas. A perfect conductor also offers an equipotential surface as the tangential electric field E_t on the conductor is zero.
In the following section, we shall derive the current wave equation in highly conductive media followed by the current distribution in a flat conductor.

2.7 Current Wave Equation in High Conductivity Materials

From the above discussion we have made the following assumptions for a good conductor[21]:

1) There is no charge density ρ_v within the conductor. We also examined during the development of Maxwell's current continuity equation and relaxation time calculation that the charge inside the conductor, if any, surfaces out very rapidly with the very short relaxation time $\tau = \varepsilon/\sigma$.
2) The conduction current density $\boldsymbol{J_c} = \sigma \boldsymbol{E_s}$ is much larger that the displacement current density $\boldsymbol{J_d} = j\omega\varepsilon \boldsymbol{E_s}$ due to the fact that $\sigma >> \omega\varepsilon$.

Based on the above assumptions for good conductors, we can use Maxwell's equation in the simplified form:

$$\nabla \times \boldsymbol{H} = \sigma \boldsymbol{E} \tag{2.44}$$

21 Similar derivation is also available in M. A. Islam, *Electromagnetic Theory*, Dacca, East Pakistan: EPUET, 1969.

Taking curl on both sides of the above equation we obtain,

$$\nabla \times \nabla \times \boldsymbol{H} = \sigma \nabla \times \boldsymbol{E} \tag{2.45}$$

Using the vector identity

$$\nabla \times \nabla \times \boldsymbol{H} = \nabla(\nabla . \boldsymbol{H}) - \nabla^2 \boldsymbol{H} \tag{2.46}$$

And Faraday's law

$$\nabla \times \boldsymbol{E} = -\mu \frac{\partial \boldsymbol{H}}{\partial t} \tag{2.47}$$

We obtain:

$$\nabla(\nabla \cdot \boldsymbol{H}) - \nabla^2 \boldsymbol{H} = -\sigma\mu \frac{\partial \boldsymbol{H}}{\partial t} \tag{2.48}$$

Since $\nabla .\ \boldsymbol{H} = 0$, we obtain:

$$\nabla^2 \boldsymbol{H} = \sigma\mu \frac{\partial \boldsymbol{H}}{\partial t} \tag{2.49}$$

Similar derivation for the field wave equation for the electric field, we obtain:

$$\nabla^2 \boldsymbol{E} = \sigma\mu \frac{\partial \boldsymbol{E}}{\partial t} \tag{2.50}$$

Using the definition of conduction current $\boldsymbol{J_c} = \sigma \boldsymbol{E_s}$ and substitution for the current in the above equation we can also get the wave equation for current inside the conductor:

$$\nabla^2 \boldsymbol{J} = \sigma\mu \frac{\partial \boldsymbol{J}}{\partial t} \tag{2.51}$$

Using the harmonic time varying field expressions, we can write the above three equations in the following form:

$$\nabla^2 \boldsymbol{H} = j\omega\sigma\mu \boldsymbol{H} \tag{2.52}$$

$$\nabla^2 \boldsymbol{E} = j\omega\sigma\mu \boldsymbol{E} \tag{2.53}$$

$$\nabla^2 \boldsymbol{J} = j\omega\sigma\mu \boldsymbol{J} \tag{2.54}$$

In the following sections, we shall examine the current distributions inside conductors for a few specialized cases that are frequently used in the industrial applications.

Exercise 2.4: Current Wave Equations

1) Derive $\nabla^2 \boldsymbol{E} = \sigma\mu(\partial \boldsymbol{E}/\partial t)$
2) Derive $\nabla^2 \boldsymbol{J} = \sigma\mu(\partial \boldsymbol{J}/\partial t)$
3) Derive $\nabla^2 \boldsymbol{H} = j\omega\sigma\mu \boldsymbol{H}$; $\nabla^2 \boldsymbol{E} = j\omega\sigma\mu \boldsymbol{E}$ and $\nabla^2 \boldsymbol{J} = j\omega\sigma\mu \boldsymbol{J}$

2.7.1 Current in a Conducting Sheet

Figure 2.6 illustrates a semi-infinite rectangular metallic block on a rectangular coordinate system. The block is extended along all x, y,and z directions infinitely. Assume the surface current density J_s is along z-axis, and for simplicity we assume no variation along y-axis, this means, $\partial/\partial y = 0$. Now we shall solve the one-dimensional current wave equation for this scenario, and deduce a few very important concepts about the current inside such a conductive medium. The one-dimensional current equation is deduced from the above three-dimensional current equation as:

$$\frac{d^2J_z}{dx^2} = j\omega\mu\sigma J_z = u^2J_z \quad (2.55)$$

[22]

$$u = \sqrt{j\omega\mu\sigma} \quad (2.56)$$

We can simplify the above expression using complex mathematical manipulation:

$$u = \sqrt{j}\sqrt{\omega\mu\sigma} = \frac{1+j}{\sqrt{2}}\left(\sqrt{2\pi f\mu\sigma}\right) = \frac{1+j}{\delta} \quad (2.57)$$

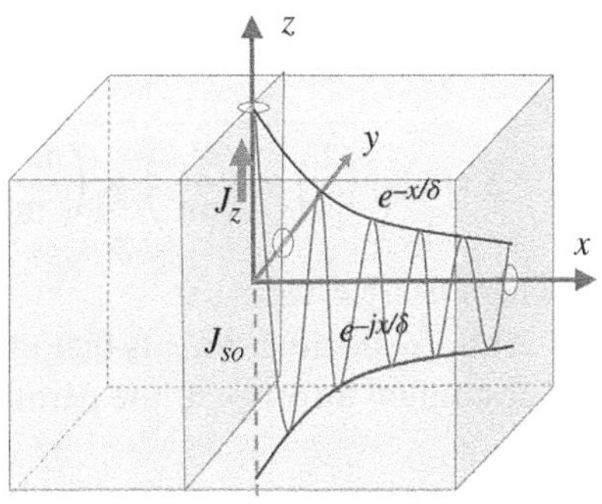

Figure 2.6 A current carrying conducting medium with surface current density J_z (A/m) in rectangular coordinate system. Current density at $x = 0$ is J_{so}.

where, the skin depth is defined as: $\delta = 1/\sqrt{\pi f\mu\sigma}$.

The complete solution to the one-dimensional current wave equation: $(d^2J_z/dx^2) = u^2J_z$ is then:

$$J_z = c_1e^{-ux} + c_2e^{ux} \quad (2.58)$$

where c_1 and c_2 are arbitrary constants. We assume that the current must diminish to zero as x approaches infinity. And using $J_z = 0$ at $x = \infty$, we obtain:

$$\lim_{x\to\infty} J_z = c_1 0 + c_2e^{ux} = 0; \text{leads to } c_2 = 0. \quad (2.59)$$

Therefore, we obtain the pertinent solution of J_z as:

$$J_z = c_1e^{-ux} = J_{s0}e^{-x/\delta}e^{-jx/\delta} \quad (2.60)$$

where $c_1 = J_{so}$ is the surface current density at $x = 0$ on yz-plane as shown in Figure 2.6. This figure illustrates the solution of the one-dimensional current wave equation in the conducting medium. We can infer the following salient points from the deduction:

1) The current has two components: (i) an exponentially decaying component $e^{-(x/\delta)}$, the current decays exponentially with the distance x, and (ii) an oscillatory component $e^{-j(x/\delta)}$, that oscillates with the factor j/δ with distance x.
2) If the factor δ increases at a fixed distance x, then the current also decrease exponentially. As can be seen and also, we have defined δ, *the skin depth* is a function of the frequency f, conductivity σ and permeability μ, and inversely proportional to the square root of the product of the three parameters $\delta = 1/\sqrt{\pi f\mu\sigma}$. Therefore, as the frequency or any of the constitutive parameters μ or σ, increases, the skin depth reduces as a consequence, the surface current J_z inside the conductor increases.

22 The current amplitude is varying along z-axis but propagating along x-axis.

Now we can examine the unit of the skin depth δ through the units of the three parameters as follows:

$$\delta = 1/\sqrt{\left(\frac{1}{\text{s}}\right) \times \left(\frac{\Omega.\text{s}}{\text{m}}\right) \times \left(\frac{mho}{\text{m}}\right)} = \text{m}$$

where $henry = \Omega.\text{s}$

This examination reveals that the unit of δ is length (m).

3) Substitution for $x = \delta$ in the pertinent solution of J_z we obtain:

$$J_z|_{x=\delta} = J_{s0}e^{-1}e^{-j} = J_{s0}e^{-1}\angle -90°$$

This means the amplitude of $|J_z|_{x=\delta}$ reduces to $J_{s0}e^{-1}$ which is 36.9% of the original value of J_{s0} at $x = 0$. Now we are well positioned to provide the scientific definition of *the skin depth* which is also called *the depth of penetration*. The current penetration depth at which the current reduces to $1/e$ of its initial value at the surface of the conductor is called *the skin depth*.

At sufficiently high frequency the current only resides at the surface of the conductor with very little penetration. However, at this little penetration, the signal gets attenuated if the depth of the conductor is less than *the skin depth*. In modern multilayer integrated circuits design, nano-fabrication, and conducting ink printing process we need at least a few skin depth to avoid such signal attenuation and leakage. Following numerical example illustrates the issue. The thumb rule in microwave frequency is about five time the skin depth of a conductive layer is safe for proper guiding of the signal and least attenuation.[23] In high power waveguide and transmission line design for radio and TV transmitters or even microwave frequency radar communications, outer structure is made of aluminum for light weight and low-cost fabrication but inner layer is made of higher conductivity metal such as copper, silver, and gold with five skin depth thickness. Electroplating processes are used to develop the thin inner layer on the base metal, for low loss propagation of signal. Also, higher conductivity metal such as copper has better heat conductivity thus quickly can dissipate heat as well.

Now examining the phase of the current we observe that the current lags from its phase of the surface current at $x = \delta$ it lags by 90° from its original phase at the surface of the conductor.

2.7.2 *Skin Effect* and Internal Impedance

From the above examination of the current wave equation, we have derived the following salient characteristics of the current in a good conductor[24]:

1) As the frequency is getting higher and higher compared to low frequency and decreases the current density $\boldsymbol{J_z}$ tends to concentrate on the surface of the conductor.
2) As the conductivity of the medium goes high, the current penetration inside the conductive medium is less and less.

Therefore, the characteristics of the current wave propagation in a conductive medium where $\boldsymbol{J_c} = \sigma \boldsymbol{E_s}$ is the dominant component, we experience two degrees of freedom in designing circuits. The first consideration is the frequency of operation f and the second is the conductivity σ of the medium. Understanding these two considerations is paramount for efficient circuit design.

23 S. Wentworth, *Electromagnetics with Engineering Applications*, Wiley, NY, USA, 2005.
24 Similar derivation is also available in M. A. Islam, *Electromagnetic Theory*, Dacca, East Pakistan: EPUET, 1969.

Review Questions 2.4: Current Wave Equation and Skin Depth

Q1. For an electric wall which component of the electric field vector is zero? Explain why it is called an equipotential surface. Justify your answer in the light of the atomic model of the conductor.
Q2. Provide the scientific definition of skin depth. What is the second name of skin depth?
Q3. What is the skin depth of a conducting medium? What is its unit? What else is it called?
Q4. Explain the significance of skin depth in the light of signal propagation in a conductive medium.
Q5. Explain the gold standard of electroplating of high-power RF/microwave passive design.
Q6. Give the relevance of skin depth in frequency of operation.
Q7. What significance you can find in modern low-cost electronics about the skin depth?
Q8. What is the thumb rule of skin depth in practical microwave circuit designs?
Q9. What is the phase difference of current at the skin depth from its original signal on the surface?
Q10. What is *skin effect*? Explain *skin effect* of the outer jacket of a coaxial cable.
Q11. Explain the effect of the skin depth in the light of the internal resistance and internal inductance. Which one is the most prominent phenomenon of high frequency?

If we connect a current source to an infinitely extending solid conducting tube then the current will flow at the outer skin of the conductor. For a hollow conducting tube of infinite extent, then the current will reside only on the outer surface of the conductor if we connect the source at the outer side. Likewise, if we connect the source to the bottom surface, the current will reside only in the bottom surface. This phenomenon of the current only flowing at the skin of the conductor is called *the skin effect. The skin effect* has two consequences:

i) As the area of current penetration in the medium is less, the effective internal resistance R_s is large and
ii) *the skin effect* has also slight reduction of the internal inductance L_s to certain extent.

The increase of the internal resistance R_s is due to the reduced effective cross-sectional area of the current flow within the conductor. The reduction of the internal inductance is due to the fact that in general, the penetration of any field in the conductor is less. Therefore, the total magnetic flux ϕ_m at high frequency is small compared to a dielectric medium. Hence the inductance, which is defined as the ratio of total magnetic flux over the total current, $L_S = (\phi_m/I)$ is less in some extent at high frequency. Therefore, the increase in internal resistance R_s in the conductor is the most predominant phenomenon at high frequency. This resistance is proportional to the real power loss in the conductor and the lost power is usually defined as $P = I^2R_s$ where I is the total current in the conductor. Note that the resistance we define with the concept of *the skin effect*, is different than the Ohmic resistance that we encounter in low frequency and DC cases, because R_s is not same across the cross section of the conductor. The fact is that due to *the skin effect*, $R_s >> R_{dc}$.[25]

Quantitatively we can define the effective impedance as:

$$Z = \frac{V}{I} = R_S + jX_S \tag{2.61}$$

25 There are two types of resistances: AC and DC resistances. For a conductor of length l, radius a and conductivity σ, the DC resistance is $R_{dc} = (l/\sigma\pi a^2)$ and AC resistance is $R_{ac} = (l/2\pi a\delta\, \sigma)$ where, $\delta = \left(1/\sqrt{\pi f \mu\sigma}\right)$, the sink depth.

where, V = the effective RMS voltage drop across the conductor

I = the effective current through the conductor
R_s = the internal resistance
X_s = the internal inductive reactance = $j\omega L_s$

Here, the effective resistance is the real part of the impedance Z. In order to calculate the impedance, we need to calculate the total current through and voltage across the conducting block. Since the conducing block, which was considered in our previous examination of the surface current, was a semi-infinite conducting block, the current is going to be infinite. The total current is defined as the integration of the surface current density across the cross-sectional area of the conductor: $I = \int_s \boldsymbol{J_s} \cdot d\boldsymbol{S}$. To simplify the calculation let us consider a unit width (along y-axis). The total current across the semi-infinite conducting block can be defined by taking the integration of the complete solution of the current density as follows:

$$I_z = \int_0^\infty J_z dx = \int_0^\infty J_{s0} e^{(1+j)(x/\delta)} dx = \frac{J_{s0}\delta}{1+j} = \frac{(1-j)J_{s0}\delta}{2} \tag{2.62}$$

The current is complex with equal real and imaginary parts and with a phase lag of 45°. We shall use the current expression to calculate *the resistivity* and *sheet resistance* of the conductor. We are more interested in the resistance because it represents the real power loss in the conductor as heat dissipation in the conductor. Also, the sheet resistance is an important parameter to examine the performance of many modern high frequency electronic circuits.

As shown in Figure 2.7, take a slice of conductor of unit width W, assume that a voltage is applied across the surface V_{BA}. Look at the microscopic examination of the internal current and the voltage drop across the conductive slab. Take a circular cross-sectional area on both sides of the slab with areas $S_1 = S_2$. Assume that due to the applied voltage across the circular tube, the current is flowing uniformly through the cross-sectional area. The electric field on the surface is uniform due to the flow of uniform current and is defined as:

$$E_s = \frac{\boldsymbol{J_s}}{\sigma}\left(\frac{\text{V}}{\text{m}}\right). \tag{2.63}$$

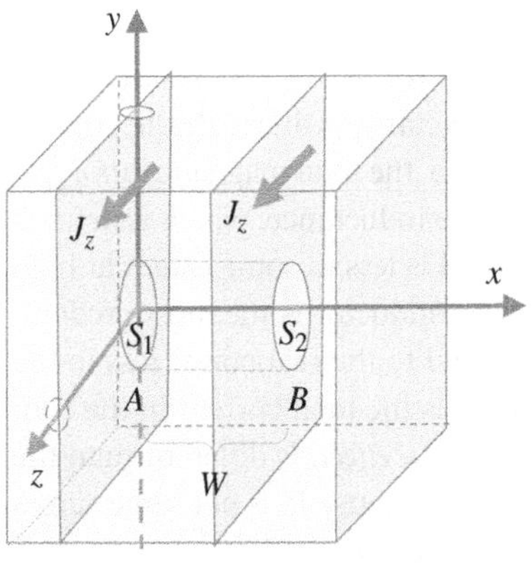

Figure 2.7 A solid circular conductor with two surfaces S_1 and S_2 at locations A and B in which the uniform current leaving the conductor. Both surfaces S_1 and S_2 are equipotential surfaces of different voltage drops due to the finite conductivity

The voltage drops across the width W due to the uniform field distribution E_s across the cross-sectional area:

$$V_{BA} = -\int_A^B E_s\, dl = E_s W = \frac{J_s}{\sigma} W \tag{2.64}$$

The effective impedance per unit length and unit width is defined as:

$$Z = \frac{V_{BA}/W}{I_z} = \frac{J_s}{\sigma} \div \frac{J_s\delta}{1+j} = \frac{1+j}{\sigma\delta} \tag{2.65}$$

The real part of the effective impedance is *the surface resistance*:

$$R_s = \frac{1}{\sigma\delta} = \sqrt{\frac{\pi f \mu}{\sigma}} \tag{2.66}$$

Now examine the imaginary part which is *the effective inductive reactance*:

$$X = \omega L_S = Im(Z) = \frac{1}{\sigma\delta} = R_S \tag{2.67}$$

This reveals that the real and imaginary parts of the effective impedance of a conductor is equal and the phase angle of the impedance is 45° at all frequencies.

The resistance of a plane conductor of unit length and unit width is called *the surface resistivity* of the conductor. For any other length and width, the effective resistance is defined by the product of the effective areas:

$$R = R_S \frac{L}{W} (\Omega - \square) \ (Ohm - square). \tag{2.68}$$

Review Questions 2.5: Sheet Resistance

Q1. Show that $R_S = L_S = 1/\sigma\delta$

Q2. Define sheet Resistance. What is unit of sheet resistance?

Q3. Find sheet resistance of few popular conductors from the internet.

Q4. Define surface resistance of a planar conductor.

Example 2.4: Sheet Resistance and Impedance of Good Conductors
Calculate the sheet resistance and impedance of copper $\sigma = 5.8 \times 10^7$ (S/m) at 6 GHz frequency.

Answer:

$$\delta = 1/\sqrt{\pi f \mu \sigma}$$

$$\delta = 1/\sqrt{\pi \times 6 \times 10^9 \times 4\pi \ \times 10^{-7} \times 5.8 \times 10^7} = 853 \times 10^{-9} \text{ m}$$

$$R_S = \frac{1}{\sigma\delta} = \frac{1}{5.8 \times 10^7 \times 853 \times 10^{-9}} = 0.02\,\Omega = X$$

$$Z = R_s + j\ X = 0.02 + j\,0.02\ (\Omega)$$

Comments: Impedance is too low at 6 GHz.

Exercise 2.5: Sheet Resistance
Q1. Repeat above example for silver. $\sigma = 6.2 \times 10^7$ (S/m)

2.7.3 Sheet Resistance

In the earlier section, we have derived the general solution of an AC impedance of a good conductor. There we found that the internal resistance $R_s = X_s = (1/\sigma\delta)\angle 45°\ (\Omega)$, where R_s is called *the skin effect resistance*. We also define the effective resistance: $R_s = R_{skin} \times (L/W)$. There is a standard term used for the unit as $(\Omega - \square)$ called (Ohm-square) to generalize the definition for unit width and unit length. In this section, we examine the high frequency effect of the skin resistance and its industrial applications.

Hypothesis 1: Examination of R_{sheet} at high frequency to calculate *the skin effect* resistance.

As shown in Figure 2.7, the sheet resistance with the proper choice of the length and width of the conductor, the ratio is defined as

$$R_{sheet} = \frac{1}{\sigma t}\left(\frac{\Omega}{\square}\right) \text{ (per unit thickness)} \tag{2.69}$$

The total resistance is defined for any width and length ratio as:

$$R_{total} = R_{sheet}\left(\frac{\Omega}{\square}\right) \times \left(\frac{L}{W}\right)(\square) = \frac{1}{\sigma t} \times \left(\frac{L}{W}\right)(\Omega) \tag{2.70}$$

Example 2.5: Sheet Resistance and Impedance of Good Conductors

Calculate the effective resistance of a silver sheet of width $w = 10$ cm and length $L = 100$ cm at 6 GHz.

Answer:

$$R = R_s\frac{L}{W} = \sqrt{\frac{\pi f \mu}{\sigma}} \times \frac{L}{W} = \sqrt{\frac{\pi \times 6 \times 10^9 \times 4\pi \ \times 10^{-7}}{6.2 \times 10^7}} \times \frac{100}{10} = 0.195\,(Ohm - square).$$

Example 2.6: Sheet and Surface Resistances of Good Conductors

Calculate R_s, R, R_{sheet}, and R_{total} for a gold plate of thickness 2 mm. $L = 100$ cm and $W = 10$ cm *at* 6 GHz.

Hint: $\sigma = 4.11 \times 10^7$ (S/m)

i) $R_s = \dfrac{1}{\sigma\delta} = \sqrt{\dfrac{\pi f \mu}{\sigma}} = \sqrt{\dfrac{\pi \times 6 \times 10^9 \times 4\pi \ \times 10^{-7} \times 5.8 \times 10^7}{4.11 \times 10^7}} = 0.024\,\Omega$

ii) $R_{eff} = R_s\dfrac{L}{W} = 0.024 \times \dfrac{100}{10} = 0.24\,(Ohm - square).$

iii) $R_{sheet} = \dfrac{1}{\sigma t} = \dfrac{1}{4.11 \times 10^7 \times 10^{-3}} = 24.33 \times 10^{-6}\,(\Omega)$

iv) $R_{total} = \dfrac{1}{\sigma t} \times \left(\dfrac{L}{W}\right) = 2.433 \times 10^{-5} = 243.3\,(\mu\Omega)$

Review Questions 2.6: Sheet Resistance

Q1. Explain difference between surface resistance, effective resistance, sheet resistance, and total resistance.

2.7.4 High Frequency Effect

See Figure 2.8. A printed circuit on a laminate is shown in a magnified state.[26] Assume that the sheet of the printed electronics is semi-infinite in extent. The width and length of the sheet conductor is W and L, respectively. A high frequency voltage is applied across the sheet with the rms value of V_{rms}. The current J_s is flowing along z-axis with an exponentially decaying function as shown in the figure. See the configuration of current closely, we can see that the conductor thickness is along x-axis, the current is flowing along z-axis due to the applied voltage V_{rms}. Assume that

26 This type of printed circuit has huge application in printed electronics at high frequencies such as RF, microwave, and mm-wave antennas and passive design.

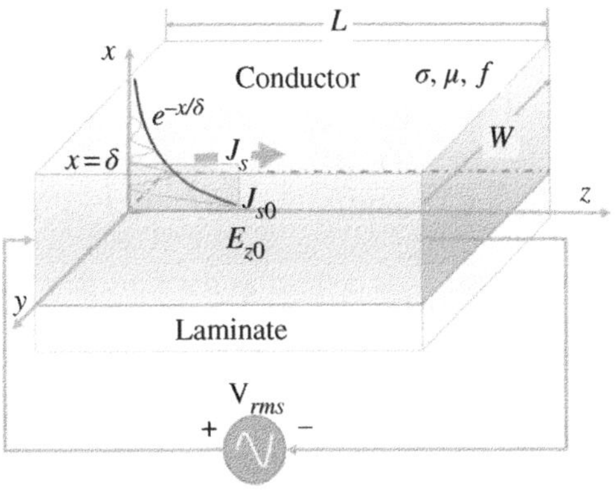

Figure 2.8 A sheet conductor on a PCB laminate, of width *W*, and length *L* along *z*-axis is extended a few skin depths along *x*-axis is excited with a voltage source of *rms* value V_{rms}.

the initial value of the *z*-directed current is J_{so}. This current is exponentially decaying as it penetrates in deeper in the conductor thickness along *x*-axis with the rate of $e^{-(x/\delta)}$. As the frequency goes up, the penetration is less. Now we are in a position to calculate the current using the applied external electric field due to the potential V_{rms} using Ohm's law as follows:

1) The external applied electric field, which is polarized along *z*-direction, penetrates inside the conducting sheet along *x*-direction. Note that the electric field also decays exponentially with a factor $e^{-x/\delta}$ as it penetrates inside the conductor thickness:

$$E_z = E_{z0}e^{-x/\delta} \tag{2.71}$$

The current density due to the applied electric field is in the same polarization of the field and can be expressed with the aid of Ohm's law as follow:

$$J_{sz} = \sigma E_{z0}e^{-x/\delta} \tag{2.72}$$

We assume that the conductor is semi-infinite along *x*-axis and the width of the conductor is *W* along *y*-axis and length *L* along *z*-axis. Therefore, we are in a position to calculate the current penetration through the conducting slab by integrating the current density of the above expression as follows:

$$I = \int J_z d\mathbf{S} = \int_{x=0}^{\infty}\int_{y=0}^{W} \sigma E_{zo}e^{-x/\delta}dydx = \frac{\sigma W E_{zo}}{(-1/\delta)}e^{-x/\delta}\bigg|_{x=0}^{\infty} = \sigma\delta W E_{zo} \tag{2.73}$$

Substitution for δ from (2.33) yields:

$$I = WE_{zo}\sqrt{\frac{\sigma}{\pi\mu f}}\ (\text{A}) \tag{2.74}$$

Examining the current density profile J_{sz} in the above illustration we can say that the current is an exponentially decaying function along the thickness of the conducting sheet. For the above derivation, we can draw an equivalent current profile assuming that the current is constant with a fixed amplitude over the extent of *the skin depth*.

Now we are in a position to recalculate *the skin effect* resistance in the light of high frequency phenomenon. Take the total length *L* of the conductor across which the applied voltage is V_{rms}.

$$V = E_{z0}L\ (\mathrm{V}) \tag{2.75}$$

The total resistance of the slab of length L and width W can be obtained from the above two important calculations of current and voltage respectively.

$$R = \frac{V}{I} = \frac{E_{z0}L}{\sigma\delta W E_{zo}} = \left(\frac{1}{\sigma\delta}\right)\cdot\left(\frac{L}{W}\right) = R_{skin}\left(\frac{L}{W}\right) \tag{2.76}$$

where, $R_{skin} = \dfrac{1}{\sigma\delta}$ is defined as *the skin effect resistance.*[27]

Review Questions 2.7: High Frequency Effect
Q1. Explain the assumption made in calculation of current inside a conducting sheet.
Q2. What is *skin effect* resistance?
Q3. What is high frequency approximation?
Q4. Explain significance of skin depth in the light of current penetration in good conductors.

Example 2.7: High Frequency Effect
An *rms* voltage of 15 *V* is applied across a conductor coated laminate of sufficient thickness. The area of conductor is 100 cm by 10 cm, the conductivity is $\sigma = 5.8 \times 10^7$ (S/m), and the operating frequency is 500 MHz. Calculate the total current penetrating the conductor.

Solution:

As $V = E_{z0}L$

$$E_{z0} = \sqrt{2}\frac{V}{L} = \sqrt{2}\frac{15}{1m} = 21.2\,\mathrm{V/m}$$

$$\mathrm{I} = 0.1(\mathrm{m}) \times 15\left(\frac{\mathrm{V}}{\mathrm{m}}\right) \times \sqrt{2} \times \sqrt{\frac{5.8 \times 10^7}{\pi \times 5 \times 10^8 \times 4\pi \times 10^{-7}}} = 363.63\,\mathrm{A}.$$

Example 2.8: *Skin Effect* and Total Resistance
Calculate the total resistance R and *skin effect* resistance of the conductor coated laminate.

Solution:

$$R_{skin} = \frac{1}{\sigma\delta} = \sqrt{\frac{\pi f \mu}{\sigma}} = \sqrt{\frac{\pi \times 5 \times 10^8 \times 4\pi \times 10^{-7}}{5.8 \times 10^7}} = 0.006\,\Omega$$

$$R = R_{skin}\left(\frac{L}{W}\right) = 0.006 \times \frac{100}{10} = 0.06\,\Omega.$$

Hypothesis 2: Industrial applications of skin depth.

In the above theory, we consider a semi-infinite block of conductor and examine the effect of *the skin effect* resistance that decays exponentially with thickness. However, in reality it is possible to have a devise with semi-infinite thickness. Understanding of the realistic value of the conductor

27 Adopted from M. Wentworth, *Fundamentals of Electromagnetics with Engineering Applications*, Wiley 2005.

thickness that provides substantial reduction of signal penetration meaning diminishing current depth has an important implication in industry. This reduces weight, cost, and complexity of the circuit. As for example, classic bulky metallic waveguides, coaxial cables, coaxial type expensive connectors and many printed electronics at microwave and millimeter wave frequencies, can exploit the theory of *skin effect* resistance. Those devices can be constructed with cheaper and lower loss conductive materials as the outer shells and then can be coated with high conductivity materials. For metallic waveguide the structure can be made of brass and a thin silver coating of a few skin depth improves the overall performance. In microwave and mm-wave frequency connectors, a few skin depth thicknesses of gold plating are used for the modern microwave and mm-wave printed circuits. As for example, the fully printed chipless RFID tag printed of a polymer, a few skin depths of silver ink printed layer can be used for efficient guiding of electromagnetic signals. Both silver and gold are excellent conductors with the conductivity in the range of 6.30×10^7 and 4.10×10^7 (S/m), respectively. The skin depth of these conductors is so small at microwave and millimeter wave frequencies that the penetration depth is realistically very small and in micron levels. With the advent of 3D printing, many ambitious projects come into reality. Many complex microwave and mm-wave devices of sizable complexity can be made simple with the 3D printing. As for example, the coatings of silver and gold nanoparticle ink on plastics housing for light weight devices with the additive manufacturing. This is the modern drives for many wireless communications engineering manufacturers to make the antennas, waveguides, circuit housing light weight and cheap without sacrificing the microwave and mm-wave electrical performance. In this section, we shall examine the realistic thickness of these conductive coating that make the microwave and millimeter wave devices and circuits very cheap. Assume a finite conductive sheet of thickness t and width W is used and we would like to examine the effectiveness of the shielding in reference to the resistance of the semi-infinite thick conductor.

The current of finite conductive layer of thickness $x = t$ and width $y = W$, can be calculated as:

$$I_t = \int J_z d\mathbf{S} = \int_{x=0}^{t}\int_{y=0}^{W} \sigma E_{zo} e^{-x/\delta} dy dx = \frac{\sigma W E_{zo}}{(-1/\delta)} e^{-x/\delta}\Big|_{x=0}^{t} = \sigma\delta W E_{zo}\left(1 - e^{-t/\delta}\right) \quad (2.77)$$

Now if we take the ratio of the two currents, we can find the realistic thickness of the conductor in which the current diminishes to zero:

$$\frac{I_t}{I} = \left(1 - e^{-t/\delta}\right) \quad (2.78)$$

We plot the function in Figure 2.9 and find that after 5δ the ratio approaches to 1. Therefore, in industrial applications, a high conductivity material can be coated with the thickness of 5δ at a specific frequency is used.

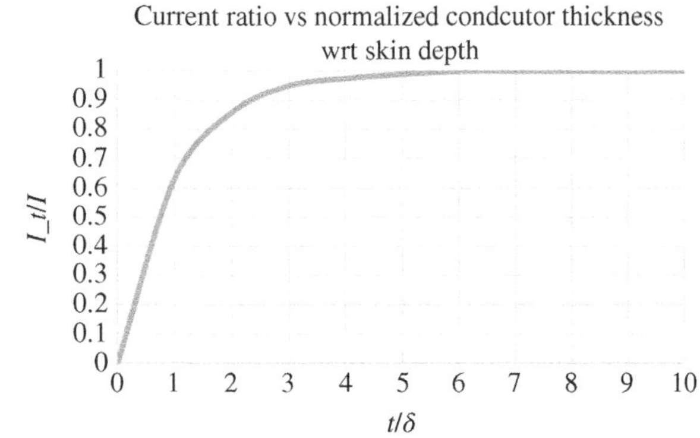

Figure 2.9 Calculation of penetration depth of the electric current in terms of skin depth of a sheet conductor to calculate the realistic thickness at which the current becomes negligibly small. Equation $\frac{I_t}{I} = \left(1 - e^{-t/\delta}\right)$ is used to plot the graph.

Exercise 2.6: Chipless RFID and Printed Electronics

Q1. Calculate the practical thickness of a silver ink printed track on a paper laminate at 6 GHz chipless RFID application, where σ of ink = 6.3 $\times 10^7$ (S/m)

Review Questions 2.8: Industrial Applications of Skin Depth
Q1. Explain industrial application of skin depth.   **Q2.** What is the realistic thickness of higher conductive material on a low conductive material without sacrificing the electrical performance?

Example 2.9: High Frequency Effect 1

A metallic waveguide is made of aluminum on the strickled material. Gold is plated on the top of aluminum. Calculate the realistic thickness of gold plating at 500 MHz.

Solution:

$$T = 5\sqrt{\frac{1}{\pi f \mu \sigma}} = 5\sqrt{\frac{1}{\pi \times 5 \times 10^8 \times 4\pi \ \times 10^{-7}}} = 17.55\ \mu\text{m}.$$

Example 2.10: High Frequency Effect 2

Calculate total current for the above example where $V_{rms} = 15\ V$ and $W = 10$ cm.

Solution:

From previous calculation, $I = 363.63$ A

$$I_t = I\left(1 - e^{-t/\delta}\right) = 363.63\left(1 - e^{-\frac{5}{\delta}}\right) \ = 361.2\,\text{A}$$

2.8 Sheet Resistance of a Wire and a Coaxial Line

Circularly symmetric conductors are of special interests in high frequency electronics. Examples include bonding wires for the legs and vias between layers in a high frequency IC, coaxial cables, circular waveguides, rotary joints, and circular horn antennas. However, the solutions of wave equations in circular cylindrical coordinate system involves solutions of the independent variables in terms of Bessel's functions. The wave equations in cylindrical symmetric boundary value problem such as round wires, coaxial cable, circular waveguides, and circular patch antennas have solutions in terms of Bessel's function. The readers can refer to any advanced mathematics book for the solution. This makes the calculation of circularly symmetric configurations bit complex. However, in this section we shall summarize the analyses of wire, hollow tubes (circular waveguides), and coaxial cables. Figure 2.10a shows the round wire of radius a and (b) shows the half cross section of a coaxial cable of inner radius a and outer radius b. First, we shall examine the current distribution of a circular wire and then a coaxial cable. The penetration depth of the electric current in terms of skin depth of a sheet conductor is to calculate the realistic thickness at which the current becomes negligibly small. Equation $\left(\frac{I_t}{I}\right) = \left(1 - e^{-t/\delta}\right)$ is used to plot the curve.

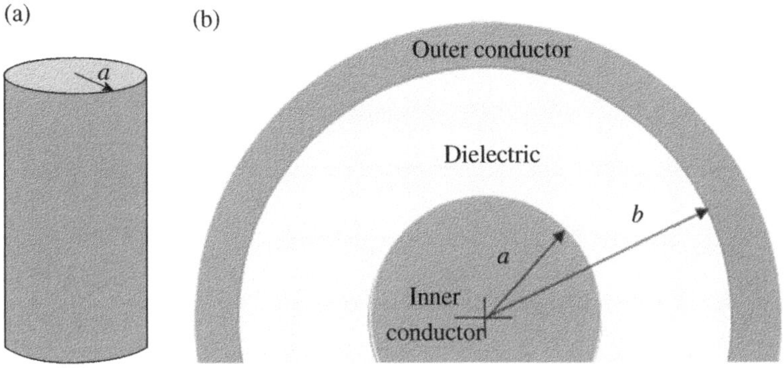

Figure 2.10 (a) Round wire of radius *a*, and (b) half cross-section of coaxial cable with the inner conductor radius a and outer conductor radius *b*.

2.9 Current Distribution on a Wire

Assume that current is flowing along z-axis and there is no variation along the axial direction $z\,(\partial/\partial z = 0)$ and azimuth angle φ $((\partial/\partial\phi) = 0)$.[28] Only variation is along the radial distance ρ. With this assumption the one-dimensional current wave equation in cylindrical coordinate is:

$$\frac{d^2J_z}{d\rho^2} + \frac{1}{\rho}\frac{dJ_z}{d\rho} = j\omega\mu\sigma J_z \tag{2.79}$$

Assume,

$$\xi = j^{-1/2}\sqrt{\omega\mu\sigma} = \frac{1}{\sqrt{j}}\sqrt{\frac{2}{\delta}} \tag{2.80}$$

Then,

$$\frac{d^2J_z}{d\rho^2} + \frac{1}{\rho}\frac{dJ_z}{d\rho} + \xi^2 J_z = 0 \tag{2.81}$$

The above equation is a Bessel equation. Assume the current is finite at $t = 0$, the solution of J_z inside the conductor is in the form of Bessel's function

$$J_z = AJ_o(\xi\rho) \tag{2.82}$$

where A is the amplitude of J_z, and $J_o()$ is the zeroth order Bessel function. Assume the current density at the surface $\rho = a$ is J_a, then the value of A is:

$$A = \frac{J_a}{J_o(\xi a)} \tag{2.83}$$

Substituting for A in the earlier equation we obtain

$$J_z = J_a\frac{J_o(\xi\rho)}{J_o(\xi a)} \tag{2.84}$$

28 A similar derivation is available in S. Ramo and J.R. Whinnery, *Fields and Waves in Modern Radio*, Wiley 2e 1953. M.A. Islam, *Electromagnetic Theory*, Dacca, East Pakistan: EPUET, 1969.

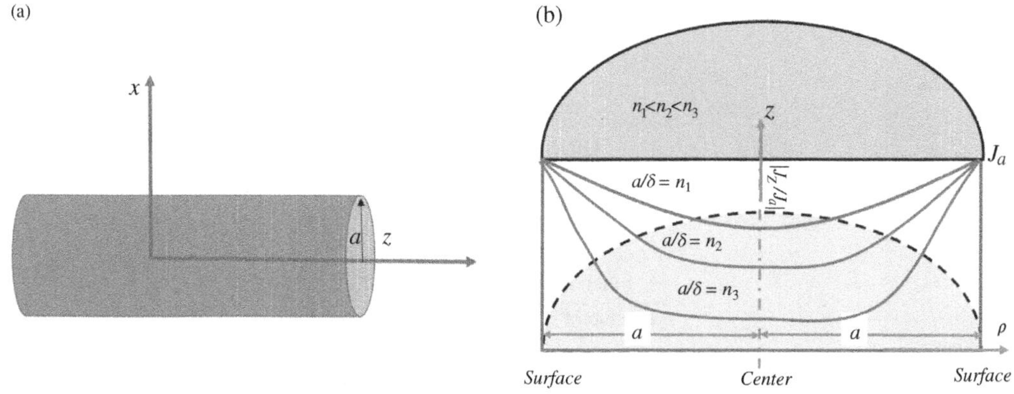

Figure 2.11 (a) A wire of radius a, and (b) current distribution across the wire. Calculation of penetration depth of the electric current in terms of normalized radius (a/δ) of a conducting wire. $J_z = J_a\left(\left(Ber\left(\sqrt{2}\rho/\delta\right) + jBei\left(\sqrt{2}\rho/\delta\right)\right) / \left(Ber\left(\sqrt{2}a/\delta\right) + jBei\left(\sqrt{2}a/\delta\right)\right)\right)$ is used to plot the curves.

Here, ξ *is* a complex quantity hence we need to break Bessel's function into its real and imaginary components. The solution can be rewritten as

$$J_z = J_a \frac{Ber(\sqrt{2}\rho/\delta) + jBei(\sqrt{2}\rho/\delta)}{Ber(\sqrt{2}a/\delta) + jBei(\sqrt{2}a/\delta)} \tag{2.85}$$

where,

$$Ber(x) = \left(J_0\left(j^{-1/2}x\right)\right) \text{ and } Bei\,(x) = \left(J_0\left(j^{-1/2}x\right)\right) \tag{2.86}$$

Figure 2.11b illustrates the normalized current density J_z/J_a versus the radial distance ρ for different values of normalized penetration depth a/δ. As can be seen from the figure, as the frequency f increases hence δ decreases, the normalized current J_z/J_a decreases in the middle of the conductor. This means that at very high frequency, the current only concentrates on the surface of the wire.

2.9.1 Rayleigh Approximation of Finite Conductor Thickness

From the above examination of the current profile inside a circular wire, we can conclude that at a very high frequency, the effective AC resistance R_{ac} of a very large conductor which has thickness many times *the skin depth* δ, may be approximated by the DC resistance R_{dc} of a superficial layer of the conductor of thickness equal to *the skin depth* δ. We use both R_{ac} and R_{dc} resistances in calculations of the equivalent circuit model of a wire in RF of electromagnetic interference theory. This approximation is called *Rayleigh approximation*.[29] Let us examine the statement of the approximation for a conducting wire. For a conductor of length of l meter and thickness of δ and conductivity σ, the DC resistance per unit length $R_{dc} = \frac{1}{\sigma\delta}$. If we consider δ to be *the skin depth*, then according to the Rayleigh approximation

$$R_{dc} = R_s = \frac{1}{\sigma\delta} \tag{2.87}$$

29 M.A. Islam, *Electromagnetic Theory*, Dacca, East Pakistan: EPUET, 1969 p. 322.

This surface resistance R_s at high frequency was calculated using the derivation of the internal impedance of a good conductor. Thus, for a semi-infinite conductor, the effective AC resistance R_{ac} is equal to the DC resistance R_{dc} of a superficial layer of conductor thickness δ. For the finite thickness of the conductor, we have already calculated that the current does not extend beyond a few skin depths before it diminishes to zero in the case of planar conductor as shown in Figure 2.11. Therefore, if the conductor thickness is many times of the skin depth δ at a particular frequency then the semi-infinite conductor can be approximated as a conductor of finite thickness in the point of view for the depth of penetration of the electromagnetic signal at high frequency. Now we consider the calculation of internal impedance of a wire based on the current wave solution above.

Review Questions 2.9: Rayleigh Approximation

Q1. What is Rayleigh approximation? Explain Rayleigh's approximation in the light of depth of penetration of the electromagnetic signal at high frequency.

Q2. Draw the profile of current distribution and comment on high frequency effect in relation to a conducting wire.

Example 2.11: Rayleigh Approximation of Surface Resistance
Calculate DC resistance, AC resistance and R_s using Rayleigh approximation for conductor of length $l = 3$ cm,radius $a = 0.3$ mm,conductivity $\sigma = 5.8 \times 10^7$ (S/m) and frequency $f = 500$ MHz.

i) First we calculate the skin depth:

$$\delta = \frac{1}{\sqrt{\pi f \mu \sigma}} = \frac{1}{\sqrt{\pi \times 5 \times 10^8 \times 4\pi \times 10^{-7} \times 5.8 \times 10^7}} = 2.955\ \mu\text{m}$$

Using conventional method of uniform current over the wire cross-section:

$$R_{dc} = \frac{l}{\pi a^2} = \frac{3 \times 10^{-2}}{\pi \times (0.3 \times 10^{-3})^2 \times 5.8 \times 10^7} = 183\ \text{m}\Omega$$

$$R_{ac} = \frac{l}{2\pi a \delta} = \frac{3 \times 10^{-2}}{2\pi \times 0.3 \times 10^{-3} \times 2.955 \times 10^{-6} \times 5.8 \times 10^7} = 92.86\ \text{m}\Omega$$

Hence, $R_{av} \gg R_{dc}$.

ii) Using Rayleigh's approximation:

Total resistance for length l : $R_{dc} = R_{ac} = R_s = \dfrac{l}{\sigma\delta} = \dfrac{0.03}{5.8 \times 10^7 \times 2.955 \times 10^{-6}} = 175\ \mu\Omega$

As can be seen, R_s using Rayleigh approximation is quite low for a good conducting wire.

2.9.2 Internal Impedance of a Round Wire

Understanding the internal impedance at high frequency of a wire leads to the definition of the AC resistance and its relationship with its DC counterpart. Usually for a good conductor $R_{ac} >> R_{dc}$. This condition $R_{ac} >> R_{dc}$ is also evident from the numerical example above. We shall apply this relationship in the derivation of the internal impedance of a round wire when we shall study the electromagnetic interference (EMI) and electromagnetic compatibility (EMC). In EMI/EMC, we need to consider all delicate issues of the high frequency effects on the passive circuit components.

In high frequency AC, the passive components exhibit spurious effects, which are also called high frequency parasitic effect and nonideal behaviors of passive components at RF. For accurate prediction of their behavior we need to consider the electromagnetic analysis. As for example, for a simple wire with finite conductivity at DC, we may consider the wire to be a short circuit. At high frequency such as at RF and microwave frequency, the wire can be a combination of an AC resistance which is much larger that it's DC counterpart plus the inductance associated with the induced magnetic field inside the wire due to the flow of current at RF and microwave frequency.

In calculating the internal impedance of a round conductor of radius a we first attempt to calculate the total current from the previous solution for the current density J_z. We assume the same conditions as before, and define the total current I_z as:

$$I_z = \int_S \boldsymbol{J} \cdot d\boldsymbol{S} = \int_{\rho=0}^{a} J_z 2\pi\rho d\rho \tag{2.88}$$

Substitution for $J_z = J_a \frac{J_o(\xi\rho)}{J_o(\xi a)}$ in the above equation yields

$$I_z = \frac{2\pi J_a}{J_o(\xi a)} \int_{\rho=0}^{a} J_o(\xi\rho)\rho d\rho \tag{2.89}$$

Using the integration identity $\int_0^{\alpha} J_o(x)xdx = \frac{x}{\alpha} J_1(\alpha x)$, we obtain the total current as:

$$I_z = \frac{2\pi J_a}{J_o(\xi a)} \times \frac{a}{\xi} J_1(\xi a) \tag{2.90}$$

where $J_1(\xi a)$ is the first-order Bessel's function of (ξa).

The internal impedance per unit length is defined as the voltage over current per unit length. The electric field is the voltage per unit length. Therefore, the internal impedance per unit length is:

$$Z = \frac{V}{I_z} = \frac{E_S}{I_z} = \frac{J_a}{\sigma} \div \left(\frac{2\pi J_a}{J_o(\xi a)} \times \frac{a}{\xi} J_1(\xi a) \right) = \frac{\xi J_o(\xi a)}{2\pi a\sigma J_1(\xi a)} \tag{2.91}$$

where,

$$\xi = j^{-1/2}\sqrt{\omega\mu\sigma} = \frac{1}{\sqrt{j}}\sqrt{\frac{2}{\delta}} \tag{2.92}$$

We can simplify the above expression using the identity: $J_1(\xi a) = -J_0'(\xi a)$ and the following calculation of the derivative of Bessel's function:

$$j^{-1/2} J_0' \nu \left(j^{-1/2} \right) = \frac{d}{d\nu} \{Ber(\nu) + jBei(\nu)\} = Ber'(\nu) + Bei'(\nu) \tag{2.93}$$

Here,

$$\xi a = j^{-1/2} \left(\sqrt{2} a/\delta \right) = j^{-1/2} \nu \tag{2.94}$$

where,

$$\nu = \sqrt{2}(a/\delta) \tag{2.95}$$

Therefore, we obtain the internal impedance per unit length as:

$$Z = R_{eff} + jX = \frac{\sqrt{2}}{\delta} \times \frac{jBer(\nu) - Bei(\nu)}{2\pi a\sigma \times \{Ber'(\nu) + jBei'(\nu)\}} \tag{2.96}$$

Separating the real and imaginary parts after rationalization we obtain,

$$R_{eff} = \frac{1}{\sqrt{2}\pi a\sigma\delta} \times \frac{Ber(\nu)Bei'(\nu) - Ber'(\nu)Bei(\nu)}{\{Ber'(\nu)\}^2 + \{Bei'(\nu)\}^2} \tag{2.97}$$

$$X = \omega L_i = \frac{1}{\sqrt{2}\pi a\sigma\delta} \times \frac{Ber(\nu)Bei'(\nu) + Ber'(\nu)Bei(\nu)}{\{Ber'(\nu)\}^2 + \{Bei'(\nu)\}^2} \tag{2.98}$$

For $(a/\delta) >> 1$ we can deduce the above two expressions[30] as follows:

$$R_{eff} \cong \frac{1}{2\pi a\sigma\delta} \text{ and } L_i = \frac{\mu\delta}{4\pi a} \tag{2.99}$$

We shall be using the above two expressions in the AC internal resistance and inductance calculations of a conducting wire in the EMI/EMC theory.

Example 2.12: Internal Impedance at High Frequency
Calculate the internal impedance per unit length of the wire in the previous example.

$$R_{eff} = R_{ac}/l = \frac{92.85 \times 10^{-3}}{3 \times 10^{-2}} = 3.095\,\Omega/\text{m}$$

$$L_i = \frac{4\pi \times 10^{-7} \times 2.955\,\mu}{4\pi \times (0.3 \times 10^{-3})} = 0.985\,\text{H/m}$$

Internal impedance per unit length at 500 MHz:

$$Z = R_{eff} + jX = 3.095 + j \times 2\pi \times 5 \times 10^8 = 3.095 + j3.1\,(\Omega/\text{m}).$$

2.10 Low Frequency Approximation

Using the series expansion of Bessel's functions and after complex mathematical manipulations and simplifications,[31] we obtain the low frequency impedance for the wire:

$$Z = R + j\omega L = \frac{1 + \frac{j\omega\mu\sigma a^2}{8}}{\pi a^2 \sigma} \tag{2.100}$$

Separating the real and imaginary parts we obtain per unit resistance and inductance as follows;

$$R = \frac{1}{\pi a^2 \sigma}, L = \frac{\mu}{8\pi} \tag{2.101}$$

You may easily recognize the resistance value is the DC resistance. Likewise, the inductance is also the DC inductance. However, the external inductance is much larger than the internal inductance.

30 S. M. Wentworth, *Fundamentals of Electromagnetics with Engineering Applications*, Wiley 2005, pp. 459–460.
31 Detailed derivation can be obtained from M.A. Islam, *Electromagnetic Theory*, Dacca, East Pakistan: EPUET, 1969 pp. 323–324.

Review Questions 2.10: Low Frequency Approximation
Q1. Explain the contrast between Rayleigh approximation and low frequency approximation.
Q2. Explain why the external inductance is much larger than the internal inductance.
Q3. Explain the significance of R_{ac}, R_{dc}, L_i of wire in EMC/EMI.

Exercise 2.7: Internal Impedance

Q1. Repeat previous example for low frequency approximation.

2.11 *Skin-Effect* Resistance and Inductance Ratios

The effective internal resistance at high frequency AC is called *the skin effect resistance*. The ratio of the AC and DC resistances of a wire R_{ac}/R_{dc} is known as *the skin-effect resistance ratio*. If we use the above approximated values of the respective resistances then we can deduce *the skin-effect resistance ratio* as follows:

$$R_{ac}/R_{dc} = \frac{\pi a^2 \sigma}{2\pi a \sigma \delta} = \frac{a}{2\delta} \tag{2.102}$$

Likewise, with the approximated values of L_{ac} above, we can also calculate *the skin effect inductance ratio* is as follows:

$$L_{ac}/L_{dc} = \frac{\mu\delta}{4\pi a} \times \frac{8\pi}{\mu} = \frac{2\delta}{a} \tag{2.103}$$

When the frequency is very large the skin depth δ is very small resulting in very large ratio of a/δ. According to the Rayleigh approximation as discussed, above the resistance ratio is <1% for $a/\delta > 50$ and for $a/\delta > 50$ the error is less than 10%.[32] For frequency is moderately high so that $4 < a/\delta < 50$, we can use Rayleigh approximation to provide less than 1% error as follows:

$$R_{ac}/R_{dc} = \frac{1}{2(\delta/a) - (\delta/a)^2} \tag{2.104}$$

This approximation is obtained using the assumption that the AC current is confined with the outer layer of thickness δ equal to the skin depth.

Exercise 2.8: Internal Impedance and *Skin Effect*

Q1. Consider a copper wire operating at microwave frequency of 10 GHz. If the same conductor is used for AC power line at 60 Hz, calculate above the skin depth for both cases and justify if we need solid copper wires to carry signals. The conductivity of copper is 5.8×10^7 (S/m).

Q2. A copper wire of radius $a = 2$ mm and length 15 m carrying a current at 10 GHz, calculate the AC and DC resistance and *the skin effect resistance ratio*.

Q3. Repeat the problem Q2 to calculate *the skin effect inductance ratio*.

Q4. A nanoparticle silver ink has the permeability $\mu = 4\pi \times 10^{-7}$ (H/m) and $\sigma = 5.8 \times 10^6$ (S/m). It is used to print a chipless RFID tag that works at 6 GHz. The tag circuit has been printed on a circular tube of paper of diameter 10 mm and length 10 cm. Calculate the followings: (i) skin depth, (ii) the AC and DC resistances, (iii) the DC and AC inductances, and (iv) *the skin effect* resistance and inductance ratio.

32 M. A. Islam, *Electromagnetic Theory*, Dacca, East Pakistan: EPUET, 1969.

Review Questions 2.11: *Skin Effect* Resistance and Inductance Ratios

Q1. What is *the skin effect* resistance?
Q2. What is *the skin effect* resistance ratio?
Q3. Define *skin effect* inductance ratio.

2.12 Impedance of a Circular Tube and Coaxial Cable

In the previous section, we have examined the internal current distribution of a solid wire. We see that the current penetration reduces with the frequency. In this section we shall examine the surface current of a metallic tube. This configuration has special significance as many practical high-power high frequency communications systems use metallic tubes. Examples include circular waveguides, coaxial cable, circular horn antennas and rotary joints. Even the theory can be extended to the optical fiber. Due to its circular symmetric configuration it has many potential applications. Figure 2.11 depicts the metallic tube with inner radius a and outer radius b. The tube is extended along z-axis and the radial regions are separated in *Region* 1 (outer radius b metal) and *Region* 2 (inner radius a air) and are extended along ρ-axis. Assume that the material of the tube has a conductivity σ. Two conditions exist: (i) if the conductor thickness is large compared to the skin depth than according to the theory of the depth of penetration, it does not matter if the conductor is solid or hollow. In this regard, the internal impedance of the metallic tube is the same as for the solid conductor, and (ii) if the thickness of the metallic tube is comparable to the depth of penetration then in the formulation of the internal impedance, we need to consider the finite thickness of the conductor. Based on the above observation, we divide the tube into *Region* 1, the metallic region and *Region* 2 is the air or dielectric region. As we did for the wire, we now start with the wave equation for z-directed current in *Region* 1 as follow:

$$\frac{d^2 J_{z1}}{d\rho^2} + \frac{1}{\rho}\frac{dJ_{z1}}{d\rho} + \xi^2 J_{z1} = 0 \tag{2.105}$$

where

$$\xi = j^{-1/2}\sqrt{\omega\mu\sigma} = \frac{1}{\sqrt{j}}\sqrt{\frac{2}{\delta}} \tag{2.106}$$

The boundary conditions of the current in the tube are somewhat different than those for the current in the wire. Here the current is confined within the metallic *Region* 1 extending from $a << b$. Therefore, the solution of the z-directed current in *Region* 1 is somewhat different than that for the case of the current in the wire. The solution is:

$$J_{z1} = AJ_o(\xi\rho) + BY_o(\xi\rho) \tag{2.107}$$

where, A and B are the coefficients and $J_0()$ and $Y_o()$ are the zeroth-order Bessel's functions of the first and second kinds, respectively. Figure 2.12 illustrates the graphical solutions of both functions. It can easily be visualized that due to the zero current in *Region* 2, we choose $Y_o()$.

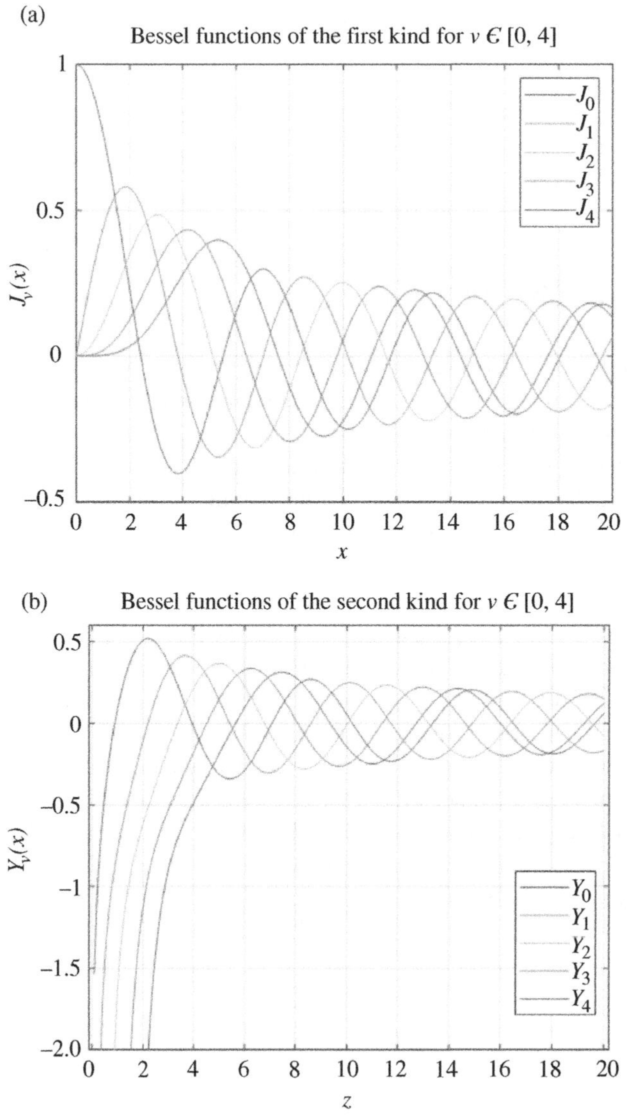

Figure 2.12 Graphical solutions of Bessel's functions of (a) first kind, and (b) second kind.

The procedure to calculate the exact solutions is presented in Figure 2.13, the procedure is as follows:

1) Write current wave equation as above in the cylindrical coordinate system. This is the Bessel's equation. With the direction of the current flow along a particular direction, in this case z-axis, assume the one-dimensional current, in this case J_z.
2) Based on the geometry of the metallic tube where the metallization is bounded by the condition from the extent of the metallic boundaries, $a << b$, we can predict that there is no current in the region from $0 < \rho < a$. Therefore, we can predict a solution with the combination of Bessel's function of the first kind due to the solid region and Bessel's function of the second kind due

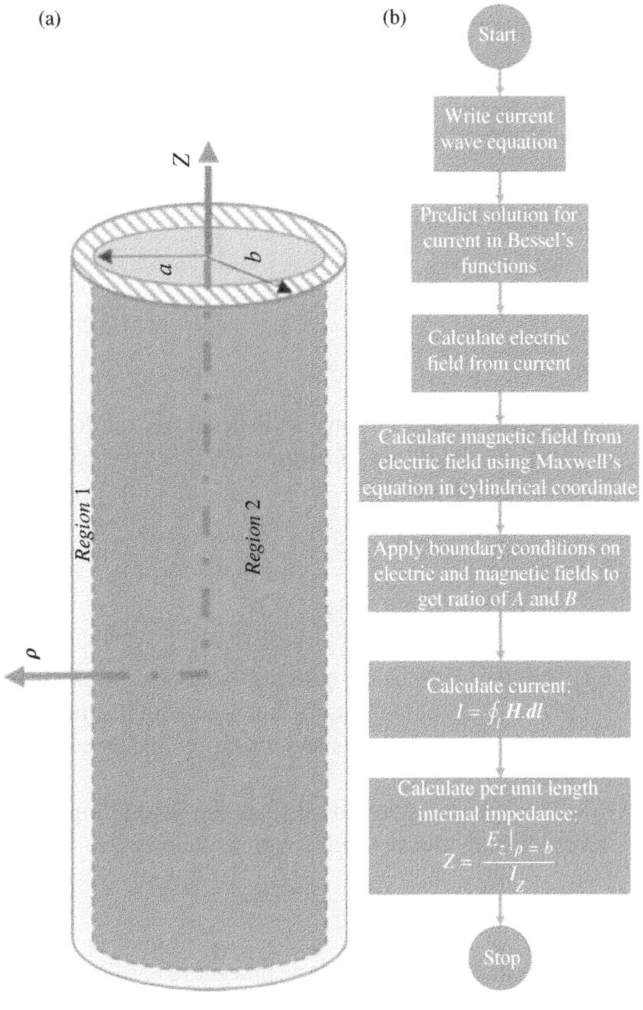

Figure 2.13 Calculation of penetration depth of the electric current in terms of normalized radius (a/δ) of a conducting wire is used to plot the curves. *Source:* Adopted from Islam, M.A. (1969). *Electromagnetic Theory*, Dacca, East Pakistan: EPUET. (a) Shows the conducting tube with the cylindrical coordinate system, and (b) shows the flow chart of the analytical procedure.

to the dielectric region where the current is indefinite. Therefore, the current becomes the combination of $J_0()$ and $Y_o()$ in the following form:

$$J_{z1} = AJ_o(\xi\rho) + BY_o(\xi\rho) \tag{2.108}$$

3) Next aim is to find the unknown coefficients A and B in terms of the known constitutive parameters and the operating frequency f via the calculations of the electric and the magnetic fields and applications of appropriate boundary conditions over the calculated fields. The steps are as follows:

   i) First, we calculate the electric field E from the one-dimension current using Ohm's law: $E = \dfrac{J_z}{\sigma}$. Now the field expression becomes a function with two unknown coefficients A and B.

ii) Calculate the magnetic field H from the electric field E using Faraday's law in cylindrical coordinate:

$$\nabla \times \boldsymbol{E} = j\omega\mu \boldsymbol{H} \tag{2.109}$$

Expanding the curl of $\boldsymbol{E}$ in the cylindrical coordinate system, we can express the above equation as follows:

$$\begin{vmatrix} \boldsymbol{a}_\rho & \boldsymbol{a}_\phi & \boldsymbol{a}_z \\ \dfrac{\partial}{\partial \rho} & \dfrac{\partial}{\partial \phi} & \dfrac{\partial}{\partial z} \\ 0 & 0 & E_z \end{vmatrix} = \boldsymbol{a}_\phi \frac{\partial E_z}{\partial \rho} = -j\omega\mu \boldsymbol{H} \text{ or, } H_\phi = -\frac{1}{j\omega\mu}\frac{\partial E_z}{\partial \rho} \tag{2.110}$$

iii) Both the electric and magnetic fields are continuous at the interface of the two regions at $\rho = a$. Therefore, we can write the following equations to satisfy the boundary conditions at $\rho = a$ as follows:

$$E_{z1} = E_{z2} \textit{ for } \rho = a \tag{2.111}$$

$$H_{\phi 1} = H_{\phi 2} \textit{ for } \rho = a \tag{2.112}$$

iv) From the above field continuity equations, we obtain the ratio of the unknown coefficients B/A in terms of the Bessel's functions with the independent variables which are the known constitutive parameters and operating frequency.

4) Now we are in a position to calculate the closed form expression for the internal impedance of the conducting tube per unit length as follows:

$$Z = \frac{E_{z1}|_{\rho = a}}{I_z} \tag{2.113}$$

Now let us implement the generic procedure to calculate the internal impedance for the specific case of the metallic tube of the internal and external radii a and b, respectively.

1) The electric field in region $a < \rho < b$ is:

$$E_{z1} = J_{z1}/\sigma = \frac{A}{\sigma}J_o(\xi\rho) + \frac{B}{\sigma}Y_o(\xi\rho) \tag{2.114}$$

2) The magnetic field is:

$$E_{z1} = \frac{J_{z1}}{\sigma} = \frac{A}{\sigma}J_o(\xi\rho) + \frac{B}{\sigma}Y_o(\xi\rho) \tag{2.115}$$

$$H_{\phi 1} = -\frac{1}{j\omega\mu}\frac{\partial E_{z1}}{\partial \rho} = \frac{\xi}{j\omega\mu\sigma}\{AJ_0'(\xi\rho) + BY_0'(\xi\rho)\} \tag{2.116}$$

Here, we use:

$$\xi^2 = -j\omega\mu\sigma \text{ that leads to } \frac{\xi}{j\omega\mu\sigma} = -\frac{1}{\xi}. \tag{2.117}$$

3) Now the applications of the boundary conditions at the interface of the two regions $\rho = a$: Assume the current density at $\rho = a$ is:

$$J_b = AJ_o(\xi a) + BY_oY_o(\xi a), \tag{2.118}$$

This leads to the closed form expression for A as follows:

$$A = \frac{J_b}{\{J_o(\xi a) + BY_o(\xi a)\}} \tag{2.119}$$

The current in the region $\rho < a$ in the hollow conductor *Region* 2 is zero. Therefore, the magnetic field H_2, which is enclosed inside *Region* 2, must be zero. Now applying the continuity of the magnetic fields at the interface of *Regions* 1 and 2 we get,

$$H_{\phi 1} = H_{\phi 2} = 0 \, at \, \rho = a \tag{2.120}$$

This yield

$$AJ_0'(\xi a) + BY_0'(\xi a) = 0 \tag{2.121}$$

The ratio of the unknown coefficients is:

$$\frac{B}{A} = -\frac{J_0'(\xi a)}{Y_0'(\xi a)} \tag{2.122}$$

So far, we have defined the unknown coefficients using the boundary conditions at $\rho = b$ and $\rho = a$ respectively. Now we are in a position to calculate the current and the internal impedance per unit length respectively.

4) The total current is: $I = \oint \boldsymbol{H} . \boldsymbol{dl}$. Performing the integration over the outer periphery of radius $\rho = b$ we obtain

$$I_z = \oint H_{\phi 1} \, d\rho = 2\pi b H_{\phi 1}|_{\rho = b} \tag{2.123}$$

Substituting for H_1 from the above equation, we obtain

$$I_z = \frac{2\pi b}{-\xi} A \left\{ J_0'(\xi b) + \frac{J_0'(\xi a)}{Y_0'(\xi a)} Y_0'(\xi b) \right\} \tag{2.124}$$

5) Using $-\xi = j^{-1/2}\sqrt{\omega\mu\sigma} = (1/\sqrt{j})\sqrt{(2/\delta)}$ and the electric field expression at $\rho = b$ and after some mathematical manipulations, we obtain the internal impedance per unit length as follows:

$$Z = \frac{E_{z1}|_{\rho = b}}{I_z} = \frac{j^{-1/2}}{\sqrt{2}\pi b \sigma \delta} \left\{ \frac{J_o(\xi b)Y_0'(\xi a) - J_0'(\xi a)Y_o(\xi b)}{J_0'(\xi b)Y_0'(\xi a) - J_0'(\xi a)Y_0'(\xi b)} \right\}. \tag{2.125}$$

In the derivation, we have used –ve sign of ξ to make the physical sense that the impedance of such a passive device must be positive. A MathWorks MATLAB code can be written for the expression for the internal impedance to calculate the internal impedance per unit length.

A few exceptions are as follows[33]:

1) If we connect the source in the inner side of the tube at $\rho = a$ then we simply need to interchange a and b in the above expression.
2) If the wall thickness $(b - a)$ is much less than the radius of the tube then we can use Rayleigh's approximation and calculate the internal impedance by considering the flat surface. Then we

33 M.A. Islam, *Electromagnetic Theory*, Dacca, East Pakistan: EPUET, 1969.

can use the theory of the coated conductor assuming that $R_{s2} \rightarrow \infty$ meaning $\sigma_2 \rightarrow 0$. In this case, the closed formed expression of the internal impedance can be written as[34]:

$$Z = R_s \times \left[\frac{\sinh \nu + \sin \nu}{\cosh \nu - \cos \nu} + j \frac{\sinh \nu - \sin \nu}{\cosh \nu - \cos \nu} \right] \tag{2.126}$$

where, $\nu = 2(b - a)\delta$.

Therefore, the tube with small conductor wall thickness we can calculate the AC resistance and the reactance per unit length as follows:

$$R_{ac} \cong \frac{R_s}{2\pi\rho} \times \left[\frac{\sinh \nu + \sin \nu}{\cosh \nu - \cos \nu} \right] \text{ and } \omega L_i \cong \frac{R_s}{2\pi\rho} \times \left[\frac{\sinh \nu - \sin \nu}{\cosh \nu - \cos \nu} \right] \tag{2.127}$$

3) If the source connected to the outer side of the conducting tube then $\rho = b$ and for inner side $\rho = a$. The other parameters are:

$$R_s = 1/(\delta\sigma) = \sqrt{\pi f \mu / \sigma}; \tag{2.128}$$

$$\nu = \frac{2(b - a)}{\delta} = \frac{2(b - a)}{\sqrt{\pi f \mu \sigma}} \tag{2.129}$$

For very high frequency $\left[\frac{\sinh \nu \pm \sin \nu}{\cosh \nu \pm \cos \nu} \right]$ reduces to 1 so that the expressions for R_{ac} and L_i can be approximated as follows:

$$R_{ac} \cong \omega L_i \cong \frac{R_s}{2\pi\rho} \tag{2.130}$$

Exercise 2.9: AC Resistance of Conducting Tube

Q1. Calculate the R_{ac} resistance of a conducting tube of radius 10 cm at 10 MHz.

2.13 Impedance of a Coaxial Cable

Consider the coaxial cable of length L as shown in Figure 2.14a has the radius of the solid inner conductor c, and the inner and outer radii of outer conducing tube are b and a, respectively. Assuming both conductors have conductivity, σ using Rayleigh approximation, we calculate the internal resistance R_a of the outer conductor of radius a, the internal resistance of the inner conductor R_b and the total resistance R_{total}. We assume the total resistance is a series combination of the two resistances and $a >> b$. The dielectric material that fills the tube has a relative permittivity of ε_r and the relative permeability of μ_r.

With Rayleigh approximation, the surface currents reside at the outer side of the inner conductor and inner side of the outer conductor. Since $a >> b$, we can assume the depth of penetration δ in both inner and outer conductors as shown in Figure 2.14b. Now we calculate the resistances.

The internal resistance of the outer conductor is:

$$R_a = \frac{L}{2\pi a \sigma \delta} \tag{2.131}$$

34 M. A. Islam, *Electromagnetic Theory*, Dacca, East Pakistan: EPUET, 1969.

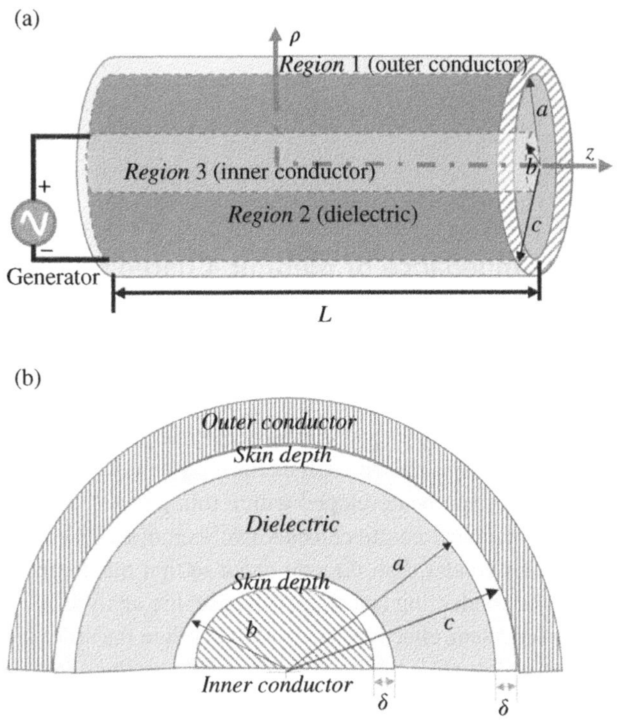

Figure 2.14 (a) Configuration of a coaxial cable with the radius of the solid inner conductor b, inner and outer radii of outer conducing tube are a and c, respectively and connect with a generator with lead at the inner solid conductor and inner side of the outer conductor, and (b) the cross section of (a).

where, the skin depth is defined as:

$$\delta = \frac{1}{\sqrt{\pi f \mu \sigma}} \tag{2.132}$$

The internal resistance of the inner conductor is:

$$R_b = \frac{L}{2\pi b \sigma \delta} \tag{2.133}$$

The total resistance is:

$$R_{total} = R_a + R_b = \frac{L}{2\pi \sigma \delta}\left(\frac{1}{a} + \frac{1}{b}\right) \tag{2.134}$$

Example 2.13: Total Resistance of Coaxial Cable
Calculate the total resistance for a coaxial cable of the inner conductor of 1 mm diameter and outer conductor of 3 mm. The conductivity of the conductor is 5.8×10^7 (S/m) and permeability of $4\pi \times 10^{-7}$ (H/m). The operating frequency of 10 GHz. The length of the cable is 1 m.

Answer:
First, we calculate the skin depth at 10 GHz using the above given parameters.
The skin depth is:

$$\delta = \frac{1}{\sqrt{\pi f \mu \sigma}} = \frac{1}{\sqrt{\pi \times 10^{10}(\text{Hz}) \times 4\pi \times 10^{-7}(\text{H/m}) \times 5.8 \times 10^{7}(\text{S/m})}} = 6.6 \times 10^{-7}\ (\text{m})$$

The total internal resistance:

$$R_{total} = \frac{L}{2\pi\sigma\delta}\left(\frac{1}{a}+\frac{1}{b}\right) = \frac{1\,(\text{m})}{2\pi \times 5.8 \times 10^{7}\left(\frac{\text{S}}{\text{m}}\right) \times 6.6 \times 10^{-7}\,(\text{m})}\left(\frac{10^{3}}{0.5}+\frac{10^{3}}{1.5}\right)(\text{m}^{-1}) = 11\,\Omega.$$

2.14 Impedance of Metallic-Coated Conductors and Laminates

Different conductor coated layered media are very common in microwave circuits. Examples include the tin plating on copper wires, copper plating on iron or iron alloys and electron tube leads, silver plating in waveguides and resonant cavities,[35] and most recently nanoparticle silver ink printed chipless RFID tags on packaging materials. With the advent of 3D printing, very low-cost microwave passive design such as horn antennas, waveguides, rotary joints, and low-cost microwave systems can be developed with a thin layer of spray painting on rigid dielectric materials.[36] In these examples, we can consider two scenarios: (i) when the coated conducting layers are made many times greater than *the skin depth* so that the depth of electromagnetic signal penetration is within the conducting layer. In such case, the analysis is simplified with Rayleigh approximation that considers the effective penetration depth in the order of a few *skin depth*,[37] and (ii) if the coated conducting layer is less than *the skin depth* then the signal penetration inside the laminates occurs. The consequence is the huge loss of signal due to penetration loss at it propagates along the conductive guiding medium. Because in this case, as the microwave frequency current flows through conducting layers it also leaks through the dielectric substrate layer. Figure 2.15 illustrates the configuration of the conducting ink coated laminate with thickness t_1, and conductivity σ_1 for the conductive layer, and t_2 and σ_2 for the laminates. In this case, we assume $t_2 >> t_1$ and $\sigma_1 >> \sigma_2$. We assume that a z-directed current J_z is propagating along x-axis. For simplicity of calculation and

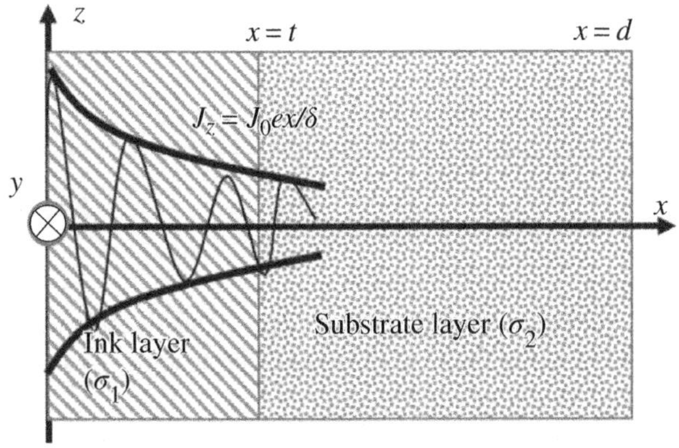

Figure 2.15 Conducting ink coated dielectric layer with x-directed current $J_z = J_o e^{-x/\delta}$ is propagating along x-direction, conducting ink layers of thickness $x = t$ and dielectric layer thickness $x = d$.

35 M.A. Islam, *Electromagnetic Theory*, Dacca, East Pakistan: EPUET, 1969.

36 https://www.optomec.com/printed-electronics/aerosol-jet-core-applications/printed-antennas/ (accessed 21 December 2017).

37 Usually with five times *the skin depth* brings the conduction current to zero in a good conductor.

also for legitimate gauss we assume that the multilayered medium is extended infinitely along both y- and z-directions.

For electromagnetic wave propagation in a good conductor we have the following assumptions:

1) There is no charge density ρ_v within the conductor.
2) From the examinations of Maxwell's current continuity equation and relaxation time, we assume that the charge inside the conductor, if any, surfaces out very rapidly with the very short relaxation time $\tau = \varepsilon/\sigma$. Where the symbols have their usual meanings.
3) With the applied z-directed electric field, the conduction current density $\boldsymbol{J}_c = \sigma \boldsymbol{E}_s$ is much larger than the displacement current density $\boldsymbol{J}_d = j\omega\varepsilon\boldsymbol{E}_s$ due to the fact that $\sigma \gg \omega\varepsilon$. This is a valid assumption even up to THz light wave frequencies.

Based on the above assumptions for good conductors, we can write Maxwell's equation in the simplified form:

$$\nabla \times \boldsymbol{H} = \sigma \boldsymbol{E} \tag{2.135}$$

Taking curl on both sides of the above equation we obtain,

$$\nabla \times \nabla \times \boldsymbol{H} = \sigma \nabla \times \boldsymbol{E} \tag{2.136}$$

From expansion of left-hand side and substitution for Faraday's law we get the wave equation for the magnetic field:

$$\nabla(\nabla \cdot \boldsymbol{H}) - \nabla^2 \boldsymbol{H} = -\sigma\mu \frac{\partial \boldsymbol{H}}{\partial t} \tag{2.137}$$

In source free condition, we simplify the above equation

$$\nabla^2 \boldsymbol{H} = \sigma\mu \frac{\partial \boldsymbol{H}}{\partial t} \tag{2.138}$$

Similar derivation of other Maxwell's equation we can obtain the wave equation for the electric field:

$$\nabla^2 \boldsymbol{E} = \sigma\mu \frac{\partial \boldsymbol{E}}{\partial t} \tag{2.139}$$

Using the definition of conduction current $\boldsymbol{J}_c = \sigma \boldsymbol{E}_s$ and substitution for the current in the above equation we can also get the wave equation for the current inside the conductor:

$$\nabla^2 \boldsymbol{J} = \sigma\mu \frac{\partial \boldsymbol{J}}{\partial t} \tag{2.140}$$

For convenience, the subscript c is dropped from the current. In this case, we assume is both spatial and time invariant. Using the harmonic time varying field expressions, we can write the wave equation in the following form:

$$\nabla^2 \boldsymbol{J} - j\omega\sigma\mu \boldsymbol{J} = 0 \tag{2.141}$$

In the following section, we shall solve the one-dimensional current wave equation to analyze the behavior of the current wave. From there, we derive the wave impedance and the reflection coefficient.

2.15 1D Current Wave Equation in Multilayered Media

As shown in Figure 2.15, a semi-infinite rectangular metallic block resides on a rectangular coordinate system. The block is extended along *y*-, and *z*-directions infinitely. Assume the surface current density J_z is along *z*-axis, and for simplicity we assume no variation along *y*- and *z*-axes, this means, $(\partial/\partial y) = (\partial/\partial z) = 0$. Now we shall solve for the 1D current wave equation for the current, and deduce a few very important concepts about the current inside such a multilayer medium. The 1D current equation is deduced from the above 3D current equation as follow [5]:

$$\frac{d^2 J_z}{dx^2} = j\omega\mu\sigma J_z = u^2 J_z{}^{38} \tag{2.142}$$

where,

$$u = \sqrt{j\omega\mu\sigma} \tag{2.143}$$

We can simplify the above expression using complex mathematical manipulation:

$$u = \sqrt{j}\sqrt{\omega\mu\sigma} = \frac{1+j}{\sqrt{2}}\left(\sqrt{2\pi f\mu\sigma}\right) = \frac{1+j}{\delta} \tag{2.144}$$

where, $\delta = 1/\sqrt{\pi f\mu\sigma}$ is *the skin depth* or the depth of penetration in the medium. The complete solution to the 1D current wave equation:

$$\frac{d^2 J_z}{dx^2} = u^2 J_z \tag{2.145}$$

is then [5]:

$$J_z = c_1 e^{-ux} = J_{s0} e^{-\left(\frac{x}{\delta}\right)} e^{-j\left(\frac{x}{\delta}\right)} \tag{2.146}$$

where $c_1 = J_{s0}$ is the surface current density at $x = 0$ on *yz*-plane as shown in Figure 2.15. The figure also illustrates the solution of the exponentially decaying 1D current wave in any medium with finite conductivity. We can infer the following salient points from the deduction:

1) The current has two components: (i) an exponentially decaying component $e^{-x/\delta}$, the current decays exponentially with the propagation distance *x*, and (ii) an oscillatory component $e^{-jx/\delta}$, that oscillates with the factor (j/δ) with distance *x*.
2) The factor δ increases at a fixed distance *x*, then the current also decrease exponentially. As can be seen and also, we have defined δ, *the skin depth* is a function of the frequency *f*, conductivity σ and permeability μ, and inversely proportional to the square root of the product of the three parameters ($\delta = 1/\sqrt{\pi f\mu\sigma}$). Therefore, as the frequency increase, the skin depth reduces.

In the next section, we shall solve for the current in such a multilayered complex medium using the electromagnetic field boundary conditions and Maxwell's equations.

---

38 The current amplitude is varying along *z*-axis but propagating along *x*-axis.

2.16 Boundary Conditions and Exact Solution of Surface Current of a Multilayered Medium

A possible and most generic solution for the surface current inside the lossy medium of finite conductivity is defined with the hyperbolic *sinh*() and *cosh* () functions for medium 1 and exponential decaying function in medium 2[39] as follows [5]:

$$J_{z1} = A\,sinh(u_1x) + B\,cosh(u_1x) \;\; \text{for } 0 < x < t \tag{2.147}$$

$$J_{z2} = Ce^{-u_2x} \;\; \text{for } t < x < \infty \tag{2.148}$$

In medium 2, we may assume the thickness of the second medium 2 is quite large $d >> t$ and can be extended to infinity so that the current diminishes to zero as it propagates from $x = t$ to $x = \infty$. Here, $u_1{}^2 = j\omega\mu_1\sigma_1$ and $u_2{}^2 = j\omega\mu_2\sigma_2$. Note that in medium 1, the current is an oscillatory decaying function and in medium 2 it is an oscillatory function as u_2 is a complex quantity. Therefore, it is a valid guess that the leakage current wave propagates inside the laminate. To determine the values of A, B, and C we assume a known current density J_o at $x = 0$, and the continuous tangential electric and magnetic fields at the interface of the two media at $x = t$. Furthermore, we also use Maxwell's equations to define the relationship between the electric and magnetic fields as follows [5]:

$$E_z = J_z/\sigma \;\; \text{and} \;\; \nabla \times \boldsymbol{E} = j\omega\mu\boldsymbol{H} \tag{2.149}$$

In this case, we assume the electric field E has only z-directed component E_z, and there are no variations of the current along y- and z-directions. Simple mathematical manipulation on Faraday's law in the rectangular coordinate system yields:

$$\begin{vmatrix} \boldsymbol{a_x} & \boldsymbol{a_y} & \boldsymbol{a_z} \\ \frac{\partial}{\partial x} & \frac{\partial}{\partial y} & \frac{\partial}{\partial z} \\ 0 & 0 & E_z \end{vmatrix} = -\boldsymbol{a_y}\frac{dE_z}{dx} = -j\omega\mu\boldsymbol{H} \tag{2.150}$$

Equating individual field components yield:

$$H_y = \frac{1}{j\omega\mu}\frac{dE_z}{dx} \tag{2.151}$$

Combining the general solution and the above relation we obtain:

$$J_{z1} = \sigma_1 E_{z1} = A\,sinh\,(u_1x) + B\,cosh\,(\,u_1x) \tag{2.152}$$

$$J_{z2} = \sigma_2 E_{z2} = -Ce^{u_2x} \tag{2.153}$$

$$u_1 H_{y1} = A\;sinh\;(u_1x) + B\;cosh\;(\,u_1x) \tag{2.154}$$

$$u_2 H_{y2} = -Ce^{u_2x} \tag{2.155}$$

At $x = t$, the tangential electric and magnetic fields are continuous. Therefore, $E_{z1} = E_{z2}$ and $H_{y1} = H_{y2}$. Applying these boundary conditions, we can write $(x = t)$ [5]:

$$H_{y2} = -Ce^{u_2x}/u_2 = H_{y1} \tag{2.156}$$

---

39 Assuming medium 2 is lossy and extended infinitely so that the current diminishes in the end as it starts propagating in the medium. We shall be using such assumption in optical/dielectric waveguides and optical fibers in which the fields in the second medium decays exponentially due to the internal reflection of signals. We shall examine the internal reflection of signal in the oblique incident cases in later chapter.

$$Ce^{-u_2 t} = \left(\frac{\sigma_2}{\sigma_1}\right)[A\ sinh\ (u_1 t) + B\ cosh\ (u_1 t)] \tag{2.157}$$

$$Ce^{-u_2 t} = -(u_2/u_1)[A\ sinh\ (u_1 t) + B\ cosh\ (u_1 t)] \tag{2.158}$$

or,

$$\frac{C}{u_2}e^{-u_2 t} = -(1/u_1)[A\ sinh\ (u_1 t) + B\ cosh\ (u_1 t)] \tag{2.159}$$

From the above equation we can develop the relation of unknown coefficient A and B, that paves the way for the pertinent solution of the current.

$$\frac{B}{A} = -\frac{sinh\,(u_1 t) + (u_2\sigma_1/u_1\sigma_2)\,cosh\,(u_1 t)}{cosh\,(u_1 t) + (u_2\sigma_1/u_1\sigma_2)\,sinh\,(u_1 t)} \tag{2.160}$$

Now we can calculate the total current inside the multilayered medium for unit width along y-axis:

$$I_z = \int_{y=0}^{1} dy \int_{x=0}^{t} J_{z1}dx + \int_{y=0}^{1} dy \int_{x=t}^{\infty} J_{z1}dx \tag{2.161}$$

$$I_z = \frac{A}{u_1}cosh\,(u_1 t) + \frac{B}{u_1}sinh\,(u_1 t) - \frac{A}{u_1} + \frac{C}{u_2}e^{-u_2 t} = -\frac{A}{u_1} \tag{2.162}$$

Substitution for $(C/u_2)e^{-u_2 t} = -(A/u_1)\,cosh\ (u_1 t) + (B/u_1)\,sinh\ (u_1 t)$ in the equation yields the total current in a closed form expression as follow:

$$I_z = -\frac{A}{u_1} \tag{2.163}$$

Here, we assume $d >> t$ and tends to infinity. We can also obtain the same results using Ampere's law in integral form by integrating over a loop of unit width along y-axis and the loop extending to infinity along x-axis. The impedance per unit width ($y = 1$) is:

$$Z = \frac{E_{z1}|_{x=0}}{I_z} = \frac{u_1 B}{\sigma_1 A} = \frac{u_1}{\sigma_1} \times \frac{sinh\,(u_1 t) + (u_2\sigma_1/u_1\sigma_2)\,cosh\,(u_1 t)}{cosh\,(u_1 t) + (u_2\sigma_1/u_1\sigma_2)\,sinh\,(u_1 t)} \tag{2.164}$$

Here, we deduce $u_i = (1 - j/\delta_i)$ and $R_{si} = (1/\delta_i\sigma_i)$ for $i = 1, 2$. Therefore, the complex internal impedance per unit width is [5]:

$$Z = (1 + j)R_{s1} \times \frac{sinh\,(u_1 t) + (R_{s2}/R_{s1})\,cosh\,(u_1 t)}{cosh\,(u_1 t) + (R_{s2}/R_{s1})\,sinh\,(u_1 t)} \tag{2.165}$$

The values of R_{s1} and R_{s2} can be obtained from the measurement of sheet resistance measuring probe, thickness t can be obtained from the measurement of thickness of the silver ink printed layer, and the conductivity σ_1 and σ_2, permeability μ_1 and μ_2 can be obtained from the datasheet. From the specific band of operating frequency f, we can calculate the impedance vs frequency over the band. The reflected and transmitted signals from the multilayer medium can be obtained from the following transmission and reflection coefficient expressions:

$$\Gamma = \frac{Z - Z_0}{Z + Z_0}, \text{and } \tau = 1 + \Gamma \tag{2.166}$$

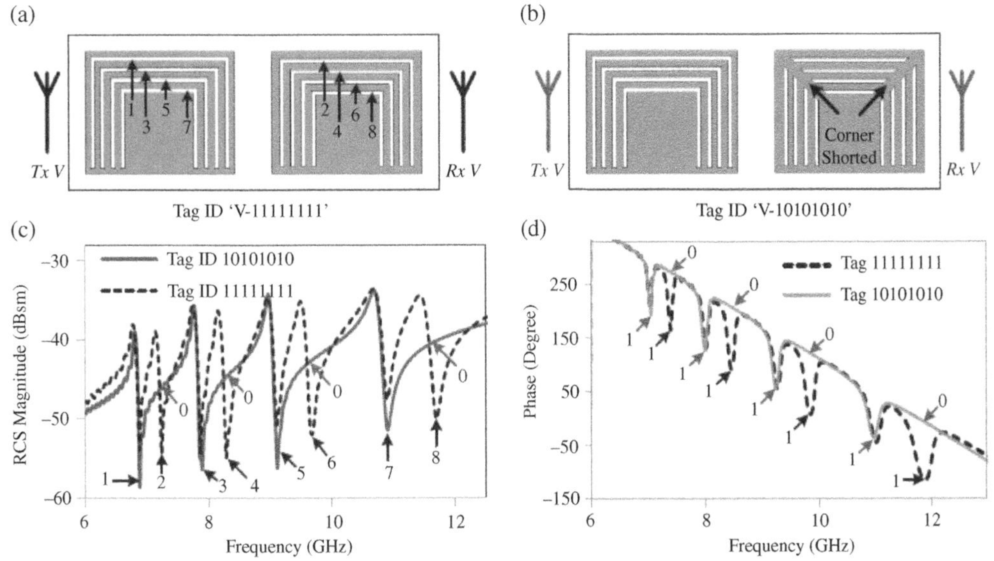

Figure 2.16 (a) 8-bits all 1s, (b) 8-bits 1 and 0, (c) response of (a), and (d) response of (b). *Source:* Islam, M.A. (2012). Compact Printable Chipless RFID Systems. PhD thesis. Monash University.

Here $Z_o = 377\ \Omega$, the free space intrinsic impedance. The frequency response from the printed chipless RFID tag can be determined by calculating the reflected power vs frequency from the tag at a particular distance in the form of $|\Gamma|^2$. The leakage through the tag can also be determined by $1 - |\Gamma|^2$. The second method could be the conversion of the reflection coefficient into the scattering parameters in radar cross section (RCS) in dBsm vs frequency and plots the frequency response of the tag. Figure 2.16 shows the scattering parameters vs frequency in the form of RCS of an 8-bit U-slot chipless RFID tag for different bit combinations from 6 – 12 GHz frequency band.[40] Distinct resonant frequency nulls represent the frequency signature of the chipless RFID.

2.17 Design of Multi-Bit Chipless RFID Tags

A multi-bit tag is designed on a paper and a polymer substrate using CST Microwave Studio. Figure 2.17 shows the square patch resonator-based multi-bit chipless RFID tag. The patch is half wavelength resonator hence each one has its distinct frequency response.

So far, we have examined the currents in lossy conducting media and multilayered media comprehensively. In the following section, we shall examine the power loss in good conducting media.

Review Questions 2.12: Wave Propagation in Good Conducting Media

Q1. What are assumptions that we can be made to simplify the field analysis in good conductors?

Q2. Using Maxwell's Curl equation of *E*, show that $\nabla^2 \mathbf{E} = \sigma\mu(\partial \mathbf{E}/\partial t)$ for a good conductor.

40 Islam, M.A. (2012). Compact Printable Chipless RFID Systems. PhD thesis. Monash University.

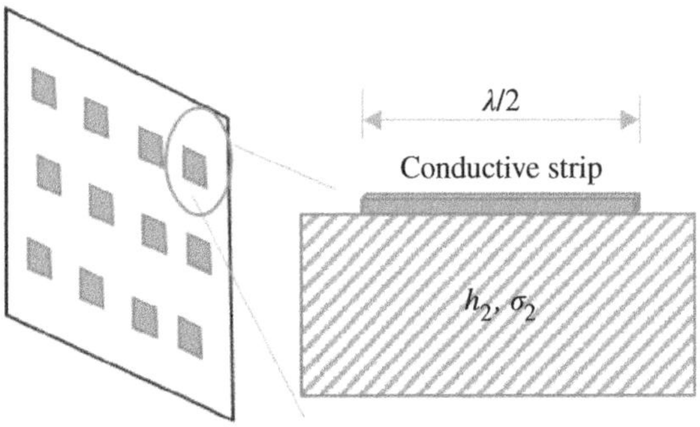

Figure 2.17 Multi-bit chipless tag with square patch resonators. Inset picture shows magnified view of single patch configuration.

2.18 Power Loss in Good Conductor

Calculation of power loss in good conductors has special significance in both very high-power transmission of high frequency waves as well as low power transmission in microwave printed circuits.[41] The finite conductivity of the conductor imposes many restrictions in power budgets in wireless power transmission. As for example, in radio and TV transmitters many kilowatts of power are transmitted by the coaxial cables, filter banks, and antenna assemblies. Due to the finite conductivity, the devices create huge real power loss as the dissipation of heat.[42] The depth of penetration plays a significant role in the internal impedance as current flows on the surface of the conductor. In contrast, in the modern printed electronics in microwave frequency band finite conductivity also causes losses as it propagates in the conducting layers. In the following we enumerate the conductor loss at high frequencies. Based on Rayleigh's approximation, we can approximate the circular shape conductor as the planar conductor with finite depth and width. Consider the situation of a conductor of such a case where we assume the conductor has a unit width and extended infinitely along z-axis as shown in the figure.

We assume the current is flowing along z-axis and there is no variation along y-axis. The z-directed current destiny is deduced as:

$$J_z = c_1 e^{-ux} = J_{s0} e^{-x/\delta} e^{-jx/\delta} \tag{2.167}$$

where $\gamma = u = (1 + j)/\delta$.

The total current per unit width is also deduced as:

$$I_z = \frac{J_{s0}\delta}{1 + j} \tag{2.168}$$

where J_{so} is the surface current at $x = 0$. We also have deduced the surface resistivity as the resistance per unit length per unit width as

$$R_s = \frac{1}{\sigma\delta} = \sqrt{\frac{\pi f \mu}{\sigma}} \tag{2.169}$$

41 Similar deduction is also available in M. A. Islam, *Electromagnetic Theory*, Dacca, East Pakistan: EPUET, 1969.
42 Analysis of heat distribution in high power passive microwave circuit and cooling system have significant practical implications. To reduce loss due to reflections, the circuits are made precisely so that the reflection coefficient is close to zero between different blocks and cable assemblers. Both liquid and fan cooling are used to dissipate heat to avoid dielectric breakdown and catastrophe.

We also have defined the internal impedance per unit length per unit width as:

$$Z = R_s + jX_s = \frac{1+j}{\sigma\delta} \tag{2.170}$$

From the above definitions of the current, impedance and the relation of the surface current and the electric field in terms of the conductivity σ, we can reduce the total current as:

$$I_z = \frac{J_{s0}\delta}{1+j} = \frac{J_S}{\sigma Z} = \frac{E_S}{Z} \tag{2.171}$$

where, E_s is the electric field at the surface of the conductor. Next, we express the current with the magnetic field on the conductor using Ampere's circuital law as follow:

$$I_t = \oint \boldsymbol{H} \cdot \boldsymbol{dl} \tag{2.172}$$

Now for Figure 2.18 above with the closed contour *abcd*, we shall perform the closed contour integration of the above Ampere's circuit law. Looking closely, you may easily comprehend that the points *b* and *c* are at infinity where the current diminishes. This has been shown with dotted lines on the arms *ab* and *cd*. Since *bc* is located at infinity the field $\boldsymbol{H}$ is also zero at this point. Therefore, the only line integration that we need to perform is along *da*. The distance between *da* is unity. Therefore, performing the integration yields:

$$I_t = \oint \boldsymbol{H} \cdot \boldsymbol{dl} = \int_d^a H_y\, \boldsymbol{a_y} \cdot dy\, \boldsymbol{a_y} = -H_y|_{x=0} \tag{2.173}$$

Second, the surface integration of $\boldsymbol{J}$ is simply total current along z-axis I_z. Therefore, for both examinations of line and surface integration, we have deduced the total current per unit width as:

$$I_t = I_z = \int_d^a H_y\, \boldsymbol{a_y} \cdot dy\, \boldsymbol{a_y} = -H_y|_{x=0} \tag{2.174}$$

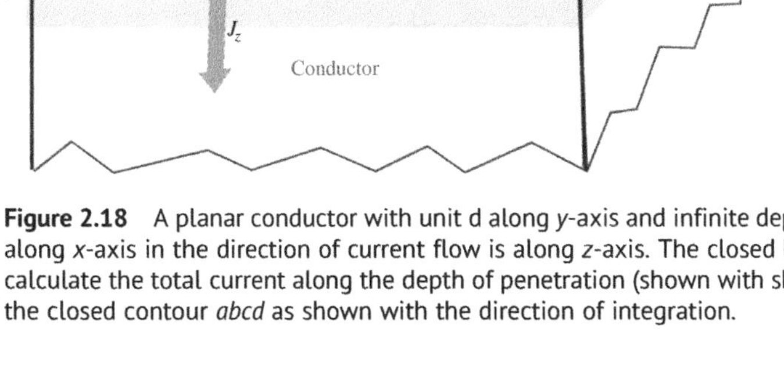

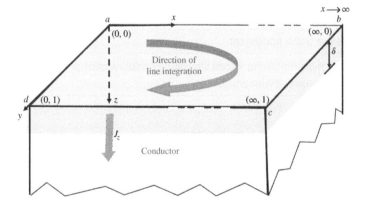

Figure 2.18 A planar conductor with unit d along *y*-axis and infinite depth along *z*-axis. It extends infinitely along *x*-axis in the direction of current flow is along *z*-axis. The closed loop integration is performed to calculate the total current along the depth of penetration (shown with skin depth δ) $I_z = \oint \boldsymbol{H.dl} = \int_S \boldsymbol{J.dS}$ along the closed contour *abcd* as shown with the direction of integration.

Since the current is flowing along z-axis, we can easily deduce the power flow along the depth of penetration if we know the fields. Assume that the electric field is an x-polarized field E_x. and the magnetic field is along y-axis H_y. These two conditions ensure the power propagation along z-axis. Now we are in a position to calculate the power loss per unit area by simply using Poynting theorem as follows:

$$P_{L-ave} = \frac{1}{2} Re\left(\mathbf{E} \times \mathbf{H}^*\right) = \frac{1}{2} Re\left(E_x H_y^*\right) = \frac{1}{2} Re\left(Z I_z I_z^*\right) = \frac{1}{2} Z|I_z|^2 \tag{2.175}$$

Since the current per unit width is the tangential magnetic field H_t on the surface and $Re(Z) = R_s$ then we can approximate the above equation as: follows:

$$P_{L-ave} = \frac{1}{2} R_s |H_t|^2 \tag{2.176}$$

Remarks

In the above approximation we assume infinite conductivity for guessing H_t is the tangential magnetic field on the surface. This is a legitimate guess due the fact that the internal resistance Z is much smaller compared to the intrinsic impedance or air or vacuum $\eta = 377\ (\Omega)$.

Example 2.14: Power Loss in Conductors

Calculate the surface resistance of copper at 10 GHz and compare the results with the intrinsic impedance of air. $\sigma_{cu} = 5.8 \times 10^7$ (S/m) and permeability is: $\mu = 4\pi \times 10^{-7}$ (H/m). If the total current is 10 (mA) calculate the real power loss per unit area.

Answer:

The skin depth of copper

$$\delta = \frac{1}{\sqrt{\pi f \mu \sigma}} = 6.61 \times 10^{-7}(\text{m})$$

$$R_s = \frac{1}{\sigma \delta} = \frac{1}{\sqrt{5.8 \times 10^7 \left(\frac{\text{S}}{\text{m}}\right) \times 6.61 \times 10^{-7}(\text{m})}} = 0.026\ (\Omega)$$

$$P_L = \frac{1}{2} R_s I_z^2 = \frac{1}{2} \times 0.026 \times \left(10^{-2}\right)^2 = 1.3 \times 10^{-6}\ \left(\text{W/m}^2\right)$$

Review Questions 2.13: Power Loss in Good Conductor
Q1. Explain how infinite extend of material along x-axis simplifies calculation of power.
Q2. Explain the significance of microwave engineering.

2.19 Practical Measurement of Sheet Resistance

Practical measurements of sheet resistance have industrial significances in semiconductor industries and modern printed electronics such as digital displays, frequency selective surfaces and chipless RFID tags.[43] These measurements are usually done in DC or low frequency AC. With the

43 Excerpt from "Sheet Resistance" Wikipedia Source: en.wikipedia.org/wiki/Sheet_resistance (accessed 28 December 2017).

emergence of new technologies at microwave and mm-wave printed devices, the significance of the sheet resistance at *DC* and its microwave and mm-wave transmissions and reflections are becoming too important to avoid. Usual sheet resistance of less than (0.1 Ω/□) gives acceptable results for transmitted and reflected signals from a printed circuit. Sheet resistance is a measure of the resistance of thin films that are normally deposited in semiconductor doping, metal deposition, resistive paste printing, and glass coating. For quality assurance and characterization of uniformity in conducting and semi-conductive coatings and materials, sheet resistance measurement has become a standard practice in industries and research institutions. The applications areas are inline process control of metal, transport conductive oxide (TCO), conductive nanomaterials and nanoparticle ink on laminates and architectural glass, wafers, flat panel displays, polymers foils, organic light emitting diodes (OLEDs), ceramics, etc. Measurements are usually done to measure the sheet resistance at DC or low frequency AC. There is no direct measurement procedure to measure the AC sheet resistance in microwave and mm-wave frequencies. With the advent of modern microwave and mm-wave printed electronics, as for example, UWB fully printable chipless RFID tags at microwave and mm-wave frequency band up to 60 GHz and above, the understanding of measurement of sheet resistance has become very important. Although the sheet resistance measurement is performed in DC and low frequency range using standard four probe or four terminal resistance measuring device, the measurement at DC can provide some correlation between the low frequency and microwave and mm-wave frequency measurements. In microwave and mm-wave frequency bands the microwave transmission and reflection performance of the thin film printed device is measured in terms of scattering parameter (s-parameters) vs frequency or the quality factor of the device at the design frequency. Therefore, it is very hard to develop a direct correlation between the sheet resistance measured in low frequency or DC. In this section first, the sheet resistance is defined from the fundamental electromagnetic principle and then in light of measurement perspective as showin in Figure 2.19. We shall also use these concepts of AC resistance in EMI/EMC theory.

There are two methods used in sheet resistance measurements: (i) four-point probe as shown in the following Figure 2.20a, and (ii) noncontact eddy current-based testing. Sheet resistance measuring four-point probe is a direct contact measuring method as shown in Figure 2.20. *Sheet resistance* is a standard measurement procedure for comparing the electrical properties of devices and components of various shapes. As shown in the figure, the four-point probe for sheet resistance measurement is a two-dimensional measurement system. Basically, it measures the resistance of the thin film assuming that there is a uniformity of the thickness of the conductive coating and the current is flowing along the

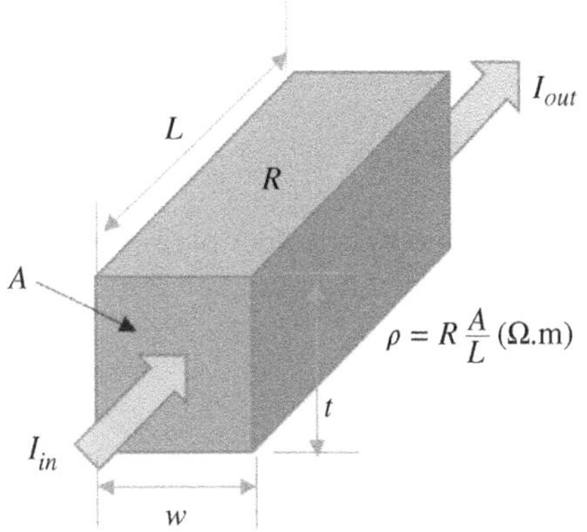

Figure 2.19 A block of resistive material with surface area A and length L, total resistance R and carrying a current I along its length. The surface resistivity is defined as the total resistance time area A over the length L to make the resistivity an intrinsic quantity irrespective of the shape and size. This relationship tells that the resistance increases with the length of the material, but not the resistivity.

plane of the sheet.[44] To understand the sheet resistance in physical sense first we define the *resistivity* in terms of the applied electric field E, the conductivity σ (S/m) and the inverse of the conductivity is *the resistivity* ρ (Ω.m). Then, we shall develop the relation between the *resistivity* ρ with respect to the total resistance, R and the geometrical configuration of the conductive layer. All we know about resistance R is that it is the aggregated impedance that it offers to the flow of electricity (current). The resistance varies with length, shape, and size of a conductor or device. Therefore, we need to define a nonvariant universal or an intrinsic quantity for the conductor or device that does not depends on the shape, size and length. This quantity is called *the resistivity*. As shown in the figure below, the current flows through a rectangular block of surface area A and length l, two terminals are added to flow the current. Assume that the applied electric potential creates an applied electric field across the material block. Using Ohm's law and assuming the current density J that is flowing through the material surface is[45]:

$$E = \rho J \text{ (V/m)} \tag{2.177}$$

where J is the volume current density (A/m^2) and ρ is the proportionality constant (Ω.m). We can also define Ohm's law as follows:

$$J = \sigma \boldsymbol{E} = \frac{I}{A} \text{ (A/m}^2\text{)} \tag{2.178}$$

$$E = \frac{I}{\sigma A} = \frac{\rho I}{A} \text{ (V/m)} \tag{2.179}$$

where, σ is the conductivity and inverse of the resistivity ρ and has unit of (S/m). A is the total surface area in m^2 and I is the total current in (A).

Now we can define the applied voltage which is the product of the electric field and the length,

$$V = EL = \frac{I\rho L}{A} = RI \text{ (V)} \tag{2.180}$$

We always measure the total resistance R using a two-probe multimeter.

$$\rho = R\frac{A}{L} \text{ (Ω.m)} \tag{2.181}$$

Now take the example of a block of resistive martials with area $A = W \times t$, where W is the width of the surface and t is the thickness. Let us define resistance of the conductor of length L as shown in Figure 2.19.

$$\rho = R\frac{A}{L} \text{ (Ω.m)} \tag{2.182}$$

Here, ρ is the surface resistivity (Ω-m), L is the length, W is the width and t is the thickness of the conductor sheet. Thus, the above discussion supports the definition of *resistivity* as the intrinsic property irrespective of the shape and size of the device. However, the resistance R can be different depends of the shape and size of the device and R increases with the dimensions.

44 In this regard our comprehensive study of *depth of penetration*, and *skin effect* resistance is the most appropriate exercise to understand the concept of the sheet current and resistance.

45 M.B. Heaney, Electrical conductivity and resistivity, In *Electrical Measurement, Signal Processing, and Displays*, Eds. J. G. Webster, CRC Press, 2003, 7–1.

Now dividing *the resistivity* with thickness t, the resistance can be defined in terms of R_S, *the sheet resistance* as follows:

$$R = \left(\frac{\rho}{t}\right) \times \left(\frac{L}{W}\right) = R_S \times \left(\frac{L}{W}\right) \tag{2.183}$$

In the above definition, $R_S = \rho/t\,(\Omega/\square)$ is used. The unit is called "Ohms per square." *The sheet resistance* of a conductor is reserved for defining *the resistivity* of a uniform sheet thickness whereas *the resistivity* is defined as the volume resistance meaning the volume resistance per meter length.

2.19.1 Measurement of Sheet Resistance

Now we discuss the practical measurement procedure of the sheet resistance in industry and research arenas. The four-point probe sensor has two separate pairs of electrodes for voltage sensing and current measurements as shown in Figure 2.20a. As you know the voltage meter provides infinite resistance at the contact points of its electrodes whereas the current meter offers zero resistance at the contact points of its electrodes. This provides more accurate measurement than a 2-probe multimeter that we usually use. Thus, the separate voltage and current probes eliminates the lead and contact resistance from the measurement and provide precise measurement of low resistance values. Such precise measurements of the sheet resistance of the printed circuits at microwave and mm-wave frequency provides a sound understanding of transmission and reflections of signals through those circuits. Lord Kelvin invented the Kelvin Bridge to measure very low resistance using the four terminal probes in 1861.

Therefore, it is also called Kelvin sensing. The operating principle is as follows: When the four probes are used on a sample of conducting sheet, the current is supplied to current probes 1 and 4 that resides outside of the voltage probes. The current generates voltage drop across the impedance to be measured according to Ohm's law $V = IR$. The voltage electrodes are applied adjacent to the target impedance so that the measurement does not include the extraneous impedance adjacent to the current leads. Since the voltage lead has infinite impedance it does not include the lead impedance and makes the measurement accurate. Figure 2.20(b, c) show the photographs of a four-point sheet resistance measuring instrument.

Review Questions 2.14: Measurement of Sheet Resistance

Q1. Define surface resistivity ρ and relate it with conductivity.

Q2. What is fundamental difference between total resistance R and surface resistivity ρ?

Q3. How do you measure the sheet resistance of a conductor?

Q4. Explain why measurement of sheet resistance of a conducting laminate is important in modern arenas.

Q5. Define resistivity and sheet resistance. Explain their differences.

Q6. Explain the difference of a 4-probe sheet resistance meter and a simple 2-probe multimeters.

Q7. When Lord Kelvin invented his Kelvin Bridge?

Q8. Explain the characteristics of a voltage sensing probe. Explain how does it make accurate measurements.

Q9. Who invented the 4-probe meter?

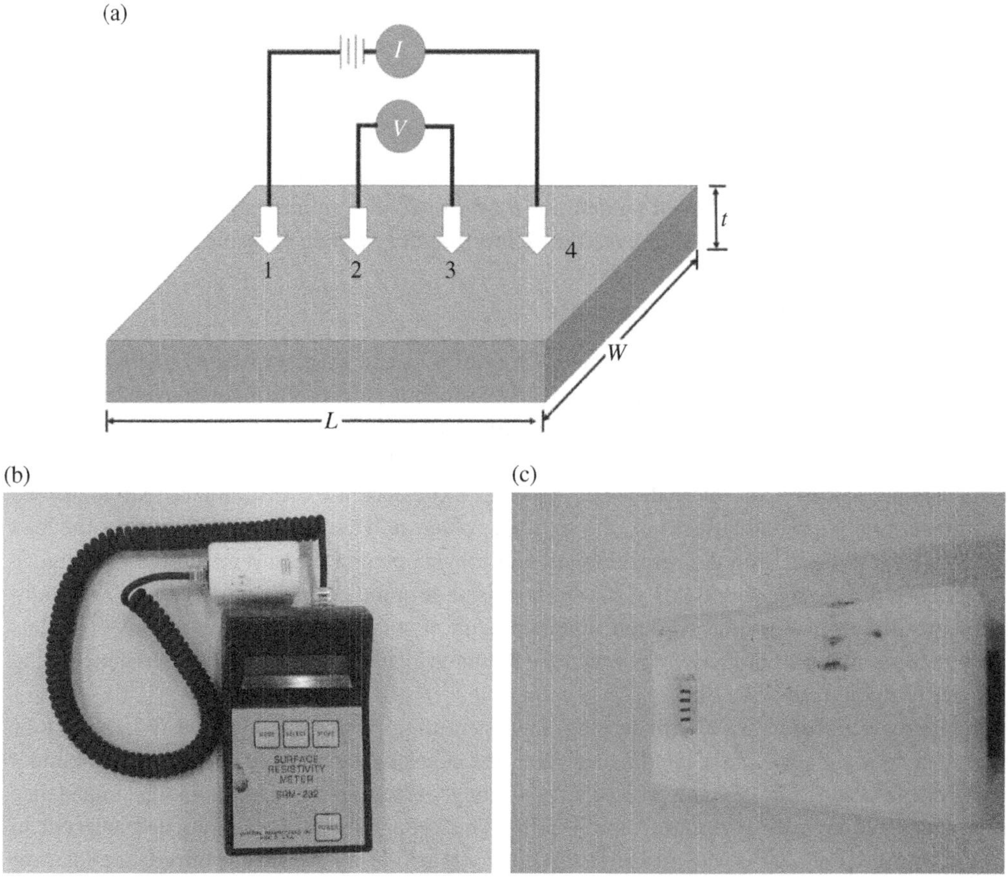

Figure 2.20 (a) Sheet resistance measurement procedure, (b) surface resistivity meter SRM-232 with 4-point probe head, and (c) 4-point probes.

2.19.1.1 Sheet Resistance Meter

Figure 2.20b and inset (Figure 2.20c) shows Guardian Manufacturing Inc, SRM-232[46,47,48] Sheet resistance meter. We use this instrument to measure the sheet-resistance in Ohms-per-square (Ω/□) of our chipless RFID that is developed with silver nanoparticle ink on a polymer laminate. This hand-held device uses 4-point probe to perform the measurement and can measure from 0.00 up to 9.90 (Ω/□) with a resolution of 0.04 (Ω/□). It operates on one 9-volt battery. The resistance measurement is performed by placing the probe head shown in Figure 2.20c on top of the printed chipless tag. A calibrated constant current-source of 4.53 mA through the two outer tips of the 4-point probe is passed through the printed chipless tag circuit and the voltage across the two inner probe tips is measured. The sheet resistance is calculated using this voltage and the current and is

---

46 http://www.guardianmfg.com/media/SRM_10_operators_guide-3-02.pdf (accessed 29 December 2017).

47 Manufacturing, G. SRM Probe Head. http://four-point-probes.com/srm-probe.pdf

48 G. M. Bridge technology. Sheet Resistance Meter, Model SRM – 232, Operator's Guide. http://www.bridgetec.com/srm_guide.pdf

displayed on the 16-digit LCD display. The functional block diagram of SRM-232 is similar to Figure 2.20a. Figure 2.21 shows how the microwave performance in terms of reflected signal or the RCS and the thickness of the conducting printed ink layer influence the spreading of bandwidth (lowering Q-factor). Sheet resistance and skin depth have profound effect in the microwave performance. However, there is no direct correlation that can be established through DC measurement of the sheet resistance as stated above.

Rigorous analyses of the electromagnetic signal with full wave electromagnetic solver such as Microwave Studio can predict such advanced level microwave and millimeter wave performances. Figure 2.21 shows the extracted data from CST Microwave Studio simulation results.

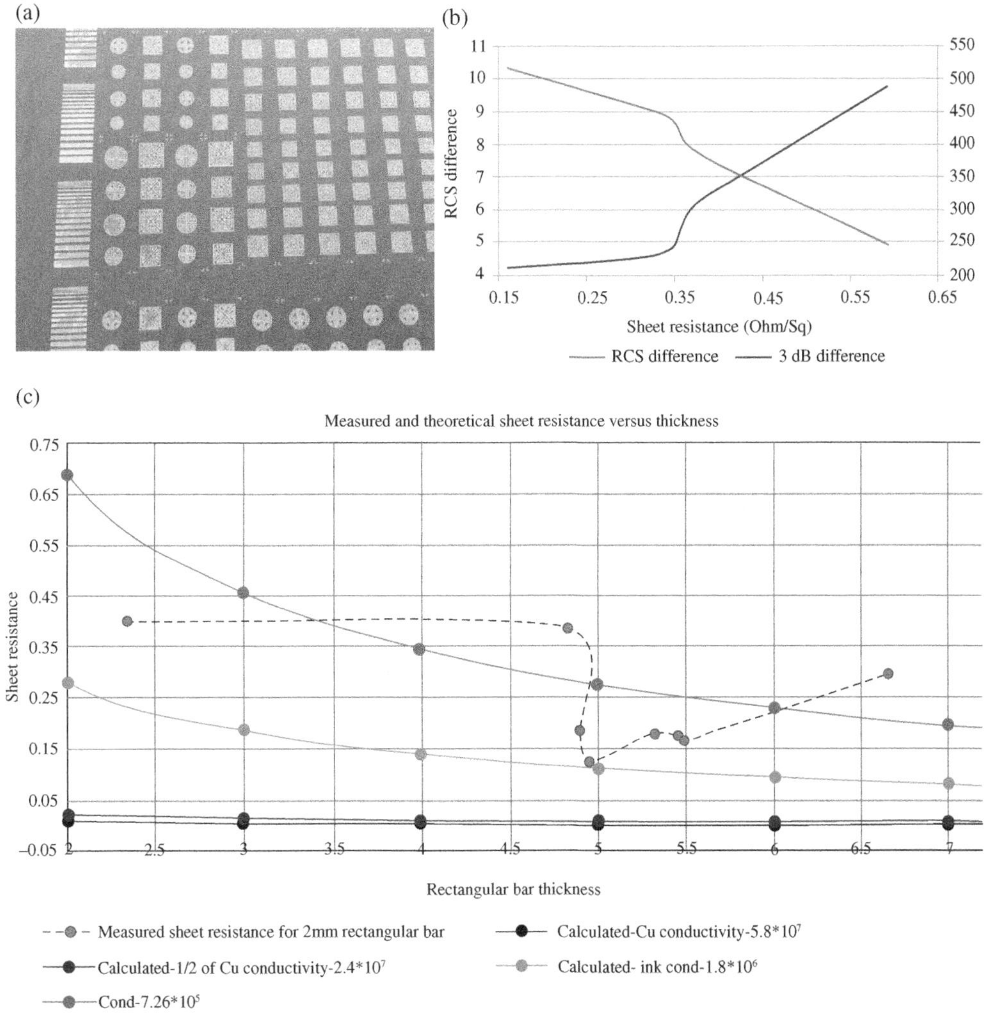

Figure 2.21 (a) Different printed chipless RFID, the left horizontal bars are used for sheet resistance measurements, (b) microwave performance for difference sheet resistance value, and (c) measured and theoretical sheet resistance versus thickness of chipless RFID (ink conductivity varies between 1.8×10^6 and 7.26×10^5 (S/m). *Source:* Sika Shrestha.

2.20 Summary of Propagation in Conducting Media

From Sections 2.8–2.18 we have examined the electromagnetic wave propagation in the conducting medium. Frist, we started with the example of a chipless RFID printed circuit on a laminate to understand the practical significance of the theory. Then we have derived the current wave equation in the conducting medium. We have defined the wave impedance in good conductors. We have examined the current wave equations in a wire, a coaxial cable, and a metallic coated conductor and a dielectric laminate. For circularly symmetric configurations such as wires and coaxial cables, we have used Bessel's functions to solve for the current in those media. We observed that the current surfaces out to the outer perimeter as the frequency increases. We also comprehensively study the skin depth and *skin effect* of the conducting media. We examined mathematically that if the conducting layer thickness is made five times the skin depth, the current wave is guided properly and efficient circuit can be designed. In industrial applications, low-cost conductor such as aluminum and copper are used as the structural support and coated with higher conducting media such as silver and gold with thickness of five times the skin depths at the operating frequency for efficient guiding of the electromagnetic signals. Examples include metallic waveguides, coaxial cables, and antennas. We also have calculated the surface impedance of such metallic coated structures. We have examined Rayleigh approximation for high frequency. Rayleigh hypotheses that the surface current confines within the skin depth of the surface of the medium. We also examined the high frequency approximation in the light of current wave equations. We have experience that the penetration of the signal in the conducting medium decreases exponentially with the frequency. For metallic coated dielectric media, we have examined one-dimensional wave equation and derived the current wave solutions. In the analysis of the electromagnetic wave propagation in the conducting media, *the resistivity* and *the surface resistance* are the performance parameters for benchmarking. However, there is no direct correlation between the two quantities and the electromagnetic propagation performance in microwave and millimeter wave frequencies. We can get only a good indication that with the surface resistance also called the sheet resistance of the conducting medium the attenuation of the signal propagation increases. We have examined the microwave performance in the light of the scattering parameters versus frequency and developed design curves for the special cases of printed electronics in microwave UWB frequencies. We have concluded that as the conductivity is high enough close to the bulk conductor, the guiding of the electromagnetic signal through the medium at microwave and millimeter wave frequencies is least attenuated. We have also given a practical example of a chipless RFID tag design and its performance characteristics in the light of the wave propagation in the conductor and dielectric composite medium. Finally, we have demonstrated how the sheet resistance of a conducting sheet can be measured with a sheet resistance meter. Such measurement has huge practical significance in industrial applications.

2.21 Chapter Remarks

This chapter presents wave propagation in lossy but nondispersive dielectric and conducting media. In reality, every medium is lossy and we have to deal with the loss. Knowing the wave phenomena in such lossy media has many practical significances for efficient circuit design in wide bands of frequencies. In commercial applications, we usually use two types of media for electromagnetic wave propagation: (i) a good dielectric usually used as the laminates of the printed circuits and filling media in waveguides, coaxial cables and many sorts of passive design, and (ii) a good conductor for guiding the electromagnetic signal. This chapter first examines the propagation of the electromagnetic wave in good dielectric media. In the course of the development of the theory of wave propagation in the medium, we have defined the propagation constant α and attenuation constant β and use the approximation $(\sigma/\omega\varepsilon) \ll 1$. Based on the approximation we have simplified the propagation and attenuation constants and derived the field wave equations and the wave impedance. We also examine how the characteristic parameters of various lossy dielectric media change with frequency. As for example, for sea water as a lossy dielectric medium, the loss tangent changes significantly with frequency. Finally, we experienced the practical measurement procedure of unknown solid and liquid dielectric media using a commercially available VNA and probe.

A good conductor holds the condition $(\sigma/\omega\varepsilon) \gg 1$. We have examined the attenuation and propagation constants, phase velocity and wave impedance using the condition of the good conductor. We have given two practical applications of the measurements of these parameters for low cost printed electronics such as a chipless RFID and geophysical measurements for mining applications. We have derived the one-dimensional current wave equations in various media and examined a few specific cases of industrial significance. The skin depth is an important parameter for signal guiding in the conducting media. In this aspect, we have examined Rayleigh approximation and high frequency effect. Finally, we have presented the measurement procedure of the sheet resistance with a sheet resistance meter. This chapter has huge practical significance as the new technology demands analyzing wave propagation in such complex media and advanced manufacturing technology augment such development in new industrial applications.

2.22 Problems

P2.1 Explain how the medium is classified in three categories: (i) good dielectric, (ii) good conductor, and (iii) lossy medium.

P2.2 Explain the difference between dispersive and nondispersive media. Define modal dispersion.

P2.3 Explain the significance of the skin depth and *skin effect* in industrial applications. What is the influence of the skin depth and *skin effect* for printed electronics in RF, microwave, and millimeter wave frequencies?

Good and Low-loss Dielectric Media

P2.4 Explain the loss tangent of a dielectric laminate and its impact in microwave circuits.

P2.5 The datasheet of Roger's corporation high frequency laminate RT/Duriod 5880 reveals the following constitutive parameters at 10 GHz: (i) $\varepsilon_r = 2.20$, $tan\delta = 0.0009$. Calculate the imaginary part of the dielectric constant and the effective conductivity. Write the complete expression of the complex dielectric constant. If an electric field of amplitude 1 (V/m) propagates through the medium calculate the power loss in (W/m^2) in the medium. What is the bulk conductivity of the medium? Justify if RT/Duroid is a good dielectric.

P2.6 Repeat P2.5 for 1 GHz and show if there is any contrasting difference in results.

P2.7 Calculate the attenuation and propagation constants, wave impedance and phase velocity of RT/Duriod 5880 at 10 GHz.

P2.8 Commercially available fire resistance laminate FR4 has the following parameters. Frequency 1 GHz, dielectric constant 4.4 and loss tangent 0.03. Is it a good or a low-loss dielectric?

P2.9 Assume FR4 is a low loss conductor calculate the attenuation and propagation constants, wave impedance and phase velocity at 1 GHz. Does the complex nature of the wave impedance of FR4 laminate make any sense?

P2.10 Describe the experimental measurement procedure of dielectric constant and loss tangent of a medium.

Good Conducting Medium

P2.11 A silver ink paste nanoparticle ink has the effective conductivity of 3.2×10^6 (S/m). Calculate the loss tangent of the ink at UWB center frequency of 6 GHz. Assume that the permittivity of the ink is similar magnitude of bulk silver. Compare the value with a good dielectric like RT/Durioid 5880.

P2.12 Calculate the conduction and displacement current for an applied electric field of 500 (mV/m) at 6 GHz for the sliver ink in P2.11.

P2.13 Calculate the propagation and attenuation constants, the skin depth and phase velocity of the silver ink in P2.11. Assume the silver ink is made of nonmagnetic nano particles. Calculate the time average dissipated power density with an applied electric field of 500 (mV/m).

P2.14 Calculate the wave impedance and the phase angle between the electric and the magnetic fields for the sliver ink in P2.11.

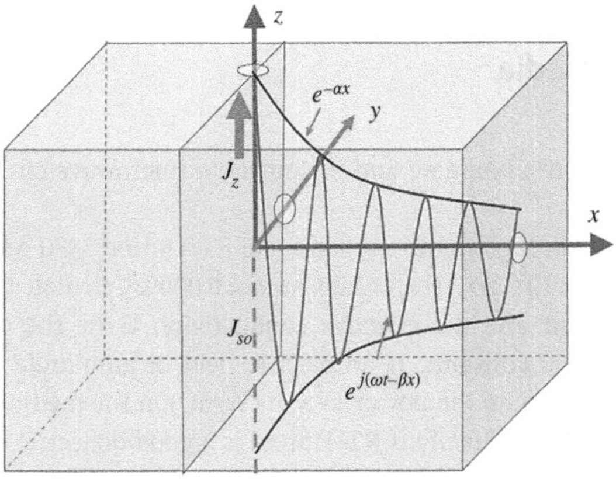

Figure P2.22 Current density profile in the conductor.

P2.15 An AC current density $J_z = 1.0e^{-\alpha x}e^{-j(\omega t - \beta x)}$ (A/m^2) with the operating frequency of 540 MHz at the surface of the sliver ink in P2.11 located at $x > 0$ as shown in Figure 6.22. Write the time domain expression of the current density. Examine the attenuation of the wave inside the conducting paste.

P2.16 Calculate the total current across the semi-infinite cross section of the block shown in Figure P2.22.

P2.17 Calculate the sheet resistance and impedance of the conducting ink in P2.11 at 6 GHz.

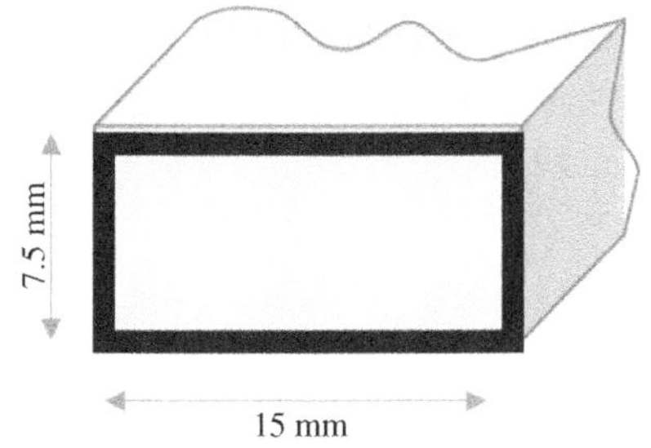

Figure P2.23 Metalic rectangular waveguide.

P2.18 A rectangular waveguide (Figure P2.23) is made of copper and operating at 12 GHz band. The inner surface is plated with gold. Calculation the acceptable thickness of the gold plating. Calculate the surface resistance, effective resistance, sheet resistance, and total resistance of the gold plating. The width of the waveguide opening is 15 mm by 7.5 mm by 20 cm. Examine the difference of each resistance and make inference. Hints: gold bulk conductivity 44.2×10^6 (S/m).

P2.19 Printed electronics is designed for electronic identification cards. It is a conductor silver ink coated laminate of sufficient thickness. An *rms* voltage of 12 *V* is applied across it. The area of conductor is 100 mm by 10 mm, the conductivity is $\sigma = 5.1 \times 10^5$ (S/m), and the operating frequency is 4.5 GHz. Calculate the total current penetrating the conductor. Estimate *the skin effect* resistance and the total resistance of the printed electronics.

P2.20 Explain the high frequency approximation for wave propagation in conducting medium.

P2.21 A conducting wire is used as a via between multiple layers of a wireless communication circuit say for the transceiver circuit of a smartphone. Calculate DC resistance, and AC resistance using (i) conventional method, and (ii) Rayleigh approximation for a conducting wire of length $l = 3$ cm, radius $a = 0.3$ mm, $\sigma = 5.8 \times 10^7$ (S/m) and frequency $f = 2450$ MHz.

P2.22 It is important to consider the parasitic effect at high frequency for passive devices such as via wires, coaxial cables, resistors, inductors, and capacitors. Explain the parasitic effect of a round wire.

P2.23 Calculate internal impedance for a wire of length $l = 5$ mm, radius $a = 0.3$ mm, conductivity $\sigma = 5.8 \times 10^7$ (S/m) and frequency $f = 2.5$ GHz.

P2.24 Calculate the total resistance for a coaxial cable of the inner conductor of 0.5 mm diameter and inner radius of the outer conductor of 3 mm. The conductivity of the conductor is 5.8×10^7 (S/m) and permeability of $4\pi \times 10^{-7}$ (H/m). The operating frequency of 18 GHz. The length of the cable is 20 cm.

P2.25 Calculate the surface resistance of a chipless RFID printed with silver nanoparticle ink at 8 GHz and compare the results with the intrinsic impedance of air and calculate the reflection coefficient. $\sigma_{silver} = 3.2 \times 10^6$ (S/m) and permeability is: $\mu = 4\pi \times 10^{-7}$ (H/m). If the total current is 50 (mA) calculate the real power loss per unit area.

P2.26 A two-probe multimeter is used to measure a resistive element of $W \times t \times L = 1 \times 1 \times 10$ mm. The reading is 10 Ω. Calculate the resistivity and sheet resistance. Comment on the microwave performance of such a resistive surface.

3

Uniform Plane Wave in Dispersive Media

3.1 Introduction

This chapter presents many physical perspectives of the uniform plane wave. First, it presents the complete one-dimensional field wave solution in a generic lossy medium with the attenuation constant α in (Np/m) and the phase constant β (rad/m). Then the physical phenomena of the field wave are described in its characteristic parameters such as the phase and group velocities, the wave impedance, and polarization. While observing a group velocity of a band of frequencies, the dispersion of the analogue signal, broadening of spectrum, digital pulses, and digital signal processing in the presentive of the fundamental electromagnetic wave theory are thoroughly analyzed. It is interesting to observe that the digital world is not out of bound of the analogue electromagnetic world. This is the virtue of the Fourier series expansion that converts any signal into a series combination of discrete analogue sinusoids. Then the chapter analyses the polarization in more depth with the help of trigonometric theory. Finally, the chapter concludes with various electromagnetic field waves. This chapter has laid the foundation for more advanced topic of wave propagation in dispersive media as presented in the next chapter.

In Chapters 1 and 2, we have examined the generic wave properties of plane waves. There we have provided the fundamental concept of the electromagnetic wave propagation, comprehensively studied how to derive the generic time varying wave equations from Maxwell's equations, time harmonic field wave equations, the medium properties in the light of the electromagnetic wave propagations, and finally the Poynting theorem. We also studied the time harmonic wave equations and understood some basic definitions. In Chapter 2, we studied the wave propagation in lossy media, specifically good dielectric and conducting media. In this chapter, we shall specifically examine the time harmonic plane wave equations and their comprehensive solutions. We shall also examine dispersion and the polarization of the plane waves. Although the polarization of an electromagnetic wave is an intrinsic property, we have discussed this behavior of the wave in this chapter because the nature of polarization is defined in terms of the trigonometric functions. Therefore, we discuss this topic with its physical interpretation in comprehensive details. Then in the next chapter we shall examine the incident electromagnetic fields impinging at the interface of multiple media and their interactions at the discontinuities using appropriate boundary conditions.

Figure 3.1 depicts the complete solutions of an electric field wave of a uniform plane wave which is directed along x-axis (x-polarized field wave) and propagating in both $\pm z$ directions in a lossy medium. As can be seen a dipole antenna element is the source of such a field wave on the xz-plane that is radiating in a lossy medium. The positive z-directed field is called the "forward travelling wave" and the negative z-directed field is called the "reverse travelling wave." We shall learn about

Fields and Waves in Electromagnetic Communications, First Edition. Nemai Chandra Karmakar.

Companion website: www.wiley.com/go/fieldsandwavesinelectromagneticcommunications

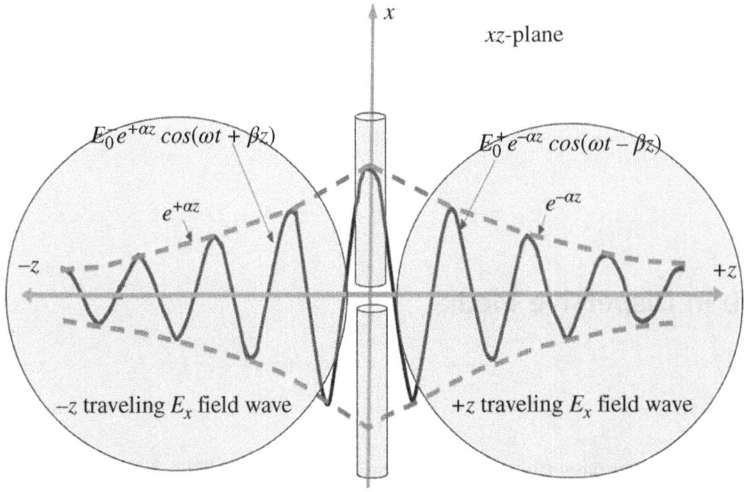

Figure 3.1 Concept of a uniform plane wave emanating out from a dipole antenna with forward and reverse travelling waves in a loss medium with loss factor α in (Np/m) and β in (rad/m). Figure shows the complete solutions that will be derived in details in Section 3.2.1 1D wave solution.

the complete solution of this plane wave in this chapter. This represents an one-dimensional homogeneous wave equation. A homogeneous wave equation is the one which is not associated with any source such as charge density ρ_v and current density J.[1] We shall solve the one-dimensional wave equation and find its general solution. Then we shall discuss in more details the wave characteristic parameters such as phase velocity, group velocity, the phase constant, and intrinsic impedance. We have already defined the parameters for the generic wave equations except the group velocity in the last chapter. During the discussion of the group velocity we shall define dispersion of the medium. Then we shall examine the dispersion of impulses in digital communications and signal processing methods. In this chapter, all the generic parameters will be derived from the solution of the time harmonic wave equations. Then we shall discuss the polarizations of EM waves and their significance in wireless communications and many other applications.

3.2 One-Dimensional Wave Equation

In Chapter 2, we have examined the 3D plane wave equation in both generic and time harmonic cases. There we also dealt with the one-dimensional wave equations. In the one-dimensional wave equation, the partial double derivatives of x, y, and z in the three-dimensional wave equation become the simple double derivatives of x, y, and z of the field vectors. Here, we deal with a one-dimensional wave equation that is propagating along z-axis hence only the field wave equation with the variation of d^2E_x/dz^2 is solved. Here, E_x is the x-directed field propagates along z-axis.

As stated in the previous chapter, for the solution of the field wave equation, there is no easy way to solve the wave vectors which are the electric and magnetic fields. The first step of the procedure is

---

1 Poisson's equation deals with the wave equation with the source and the equation is called a nonhomogeneous wave equation.

to derive the scalar differential equations of the field waves from which some vector function may be constructed. Unfortunately, three-dimensional wave equation cannot be solved using three-dimensional scalar wave equations using *separation of variable method*. Can you explain why? Recall *the separation of variable method* that we did use in deriving field solutions in the previous chapter.

Now we shall work on one-dimensional *x*-directed electric field to get into more details of the wave solutions. Similar exercise applies to *y*- and *z*-component wave fields. We shall work on *x*-component wave field which is travelling along *z*-axis. It is your homework to do similar exercise to solve wave equations for *y* and *z* components travelling in other directions say, *z* and *x*, respectively.

Review Questions 3.1: Concept of Uniform Plane Wave

Q1. Write the *x*-oriented electric field that is propagating along +ve *z*-direction.

Q2. Explain why the field solutions are not possible from using simple separation of variable method.

Q3. Explain why the *x*-directed dipole antenna has a *x*-polarized field wave.

Q4. Explain why the electric field distribution of the fundamental mode field on the aperture of a rectangular waveguide is a half sine wave.

Q5. In which direction the wave propagated in both the cases of the dipole antenna and the rectangular waveguide? Explain why.

See the two examples of an *x*-axis oriented (Figure 3.2a) dipole antenna with its radiation pattern and (Figure 3.2b) a rectangular waveguide with sides *a* and *b*. As shown in Figure 3.2a dipole antenna is oriented along the direction of *x*-axis so the electric field wave can vary along *x*-axis and propagate along *z*-axis. The second example is a rectangular waveguide where the electric field is oriented along the *x*-axis in the rectangular aperture with the half sine variation[2] and in both cases the electromagnetic waves propagate along positive *z*-axis.

Let us write the one-dimensional wave electric field which is oriented or polarized along *x*-axis and propagates along *z*-axis.

$$\boldsymbol{E_x}(z) = \boldsymbol{a_x} E_x(z) \tag{3.1}$$

Our exercise is to find the solution for the scalar function $E_x(z)$. The subscript *x* indicates that the field's amplitude varies along *x*-axis but propagates along *z*-axis.

Therefore, the one-dimensional wave becomes

$$\frac{d^2 E_x}{dz^2} - \gamma^2 E_x = 0 \tag{3.2}$$

where γ is called the propagation constant. $\gamma = \alpha + j\beta$ is a complex quantity for a lossy medium. Here, α is called the attenuation constant and β is called the phase constant.

2 The boundary conditions of the fundamental modes field wave inside the waveguide forces the field wave to be half sine wave. Recall the boundary conditions for the electric and magnetic fields at the interface of two media.

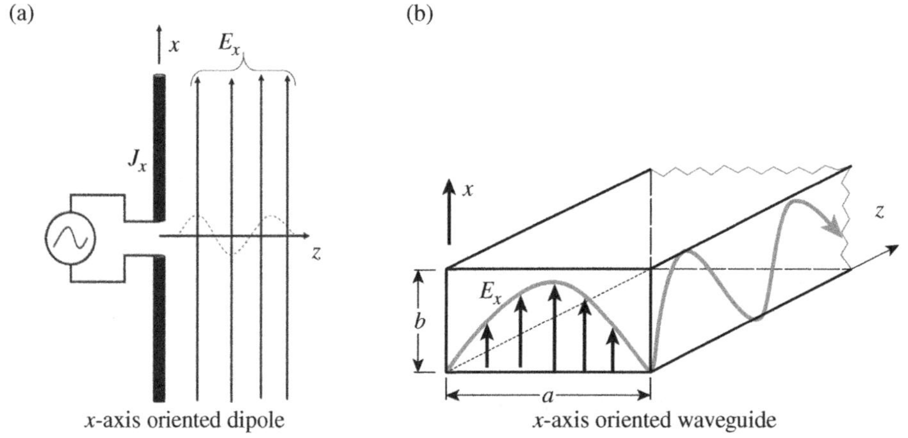

Figure 3.2 The electric field orientations (a) on a dipole antenna with its plane wave radiation pattern, and (b) in the aperture of a rectangular waveguide with sides *a* and *b*.

Recall the one-dimensional wave equations in the previous chapter. Here, we use only *x*-directed electric field. It is easier and practical to generate and also compute one dimensional field wave equation like this using Maxwell's equation. The generation of such one-dimensional field wave is shown via the dipole antenna system. In the system, a generator is connected to the dipole antenna with a terminal pair.[3] The antenna is orientated along the *x*-axis hence the current is along the orientation of the dipole antenna which is *x*-directed. Thus, it generated the *x*-directed electric field. Finally, the above one-dimensional field wave equation is a second-order linear homogeneous differential equation. Homogenous means there is no source present and therefore the right-hand side of the equation is equal to zero.

Review Questions 3.2: Plane Wave Propagation

Q1. Can you differentiate the homogeneous wave with respect to a wave if ρ_v volume charge density present in free space?

Q2. Write the expression of the one-dimensional field wave equations that has different polarizations say *y* and *z*, and propagates along other directions, say *x* and *y*?

Q3. Write down the homogeneous field wave equation for the other two field polarizations and directions of propagation.

Q4. What is the unit of the propagation constant? Define the attenuation and phase constants.

Q5. Explain when the propagation constant can have only one component say a real number and an imaginary number.

By applying simple solution of the second-order differential equation (3.2) above, the possible solution for the electric field is:

$$E_x(z) = Ae^{az} \tag{3.3}$$

3 The terminal pair is called a *port* in microwave engineering nomenclature.

where both A and a are coefficients. Here A represents the peak amplitude of the field wave and a is the exponent factor. Substitution for (3.3) into (3.2) yields,

$$a^2E_x(z) - \gamma^2E_x(z) = 0 \quad (3.4)$$

$$E_x(z)\left(a^2 - \gamma^2\right) = 0 \quad (3.5)$$

The first factor, the field $E_x(z)$ is finite and cannot be zero. Therefore, the second part yields $\gamma = \pm a$.

Therefore, the general solutions for the x polarized electric field is:

$$E_x(z) = Ae^{\pm \gamma z} \quad (3.6)$$

We shall be conveniently using this expression to find the exact solutions for $E_x(z, t)$.

Remember that so far, we worked on the phasor fields with $e^{j\omega t}$ understood, and therefore, for simplicity the exponential term is ignored in the above calculation. Once we want to compute the time varying instantaneous fields, we just multiply the field solution with $e^{i\omega t}$ term and then take the real part of the product.

Now we can write more conveniently, the general solution of $\boldsymbol{E}_x(z)$ by substituting $A = E_0^+$ for forward travelling wave amplitude of $\boldsymbol{E}_x$, $e^{-\gamma z}$ is the exponent for the forward travelling wave, this yields $E_0^+ e^{-\gamma z}$–the solution for the forward travelling wave. Likewise, substitution for $A = E_0^-$ for the reverse travelling wave amplitude of $\boldsymbol{E}_x$, $e^{+\gamma z}$ as the exponent for the reverse travelling wave, this yields $E_0^- e^{+\gamma z}$ as the solution for reverse travelling wave. Combining the two solutions and taking the real part we can obtain the time varying instantaneous field solutions of the electric field that is depicted in Figure 3.3 as:

$$Re\left[E_x(z)e^{j\omega t}\right] = Re\left[E_0^+ e^{-(\alpha + j\beta)z}e^{j\omega t} + E_0^- e^{(\alpha + j\beta)z}e^{j\omega t}\right] \quad (3.7)$$

Now we look for time varying instantaneous electric field expression. This is the most convenient expression to deal with time varying instantaneous expression for circuit analysis.

Substitution for $\gamma = \alpha + j\beta$, multiplied with $e^{j\omega t}$ and taking the real parts yields the complete solution for the electric field which is the most logical expression for one dimensional x-polarized wave travelling in both positive and negative z-directions. This procedure is illustrated in Figure 3.3 with complete solution and Figure 3.4 with the complete graphical representation of the field wave.

Here, we explain every details of the components of the complete solution of the one-dimensional x-directed electric field. Look at the illustration and the equation of the complete solution of the $\boldsymbol{E}(z, t)$ field.

Review Questions 3.3: Concepts of Plane Wave

Q1. Explain how you know that the electric field is a wave. And if so what is the direction of the field wave? Can you answer this looking at the expression and the illustration of $\boldsymbol{E}(z, t)$?

Q2. Derive the field wave solution for the magnetic field. Assume the magnetic field is y-directed and propagating along z-axis.

Figure 3.3 The complete solution for the instantaneous time varying electric field.

$$E(z,t) = \underbrace{\boldsymbol{a}_x E_0^+ e^{-\alpha z} \underbrace{\cos(\omega t - \beta z)}_{+z \text{ directed phase}}}_{+z \text{ directed field component}} + \underbrace{\boldsymbol{a}_x E_0^- e^{+\alpha z} \underbrace{\cos(\omega t - \beta z)}_{-z \text{ directed phase}}}_{-z \text{ directed field component}}$$

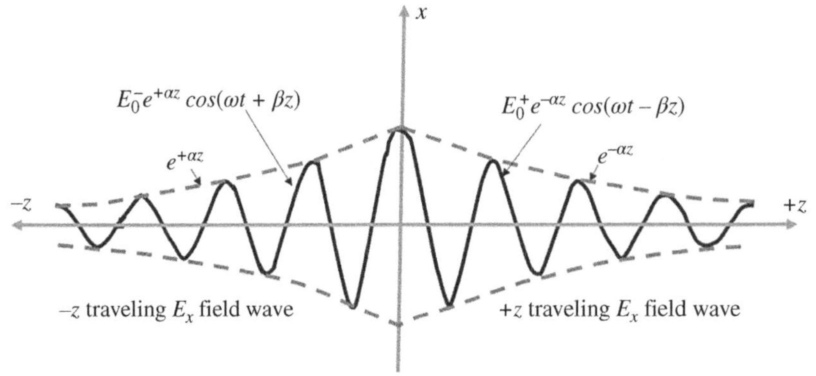

Figure 3.4 Graphical solution of one-dimensional *x*-polarized electric field with propagation constant α and the phase constant β.

3.2.1 Field Solutions in Different Forms

You may notice that we have dealt with various forms of the same field quantity. Here we summarize the three forms of field expressions. The generic or exponential form is:

$$\boldsymbol{E}(z, t) = \boldsymbol{a_x} E_0^+ e^{j\omega t - \gamma z} \tag{3.8}$$

$$\boldsymbol{H}(z, t) = \boldsymbol{a_y} H_0^+ e^{j\omega t - \gamma z} = \boldsymbol{a_y} \frac{E_0^+}{\eta} e^{j\omega t - \gamma z} \tag{3.9}$$

Notice the exponents here we use $e^{j\omega t - \gamma z}$. Explain if it is a time harmonic field? If so why?

The phasor form with $e^{j\omega t}$ understood, this means we exclude $e^{j\omega t}$ in the expression of the phase form of the field quantities. This exclusion of $e^{j\omega t}$ enormously helps mathematical manipulations of the complex vector field quantities. The expressions of the phasor form of the electric and magnetic field quantities are:

$$\boldsymbol{E_s}(z) = \boldsymbol{a_x} E_0^+ e^{-\gamma z} \tag{3.10}$$

$$\boldsymbol{H_s}(z) = \boldsymbol{a_y} H_0^+ e^{-\gamma z} = \boldsymbol{a_y} \frac{E_0^+}{\eta} e^{-\gamma z} \tag{3.11}$$

Finally, the most physically perceived form is the time varying instantaneous form of the field quantities which are associated with the cosine function. On a oscilloscope screen, we see the time varying instantaneous voltage and current waveforms. To obtain the instantaneous form of the field quantities from the phasor form we use Euler's theorem $e^{j\omega t} = \cos(\omega t) + j\sin(\omega t)$. We take the product of the phasor form of the field quantity and $e^{j\omega t}$, and then take only the real part of the product to get the time domain expression of the field quantities as follows:

$$\boldsymbol{E}(z, t) = \boldsymbol{a_x}\ Re\ \left\{E_0^+ e^{-\gamma z} e^{j\omega t}\right\} = \boldsymbol{a_x}\ E_0^+ e^{-\alpha z} \cos(\omega t - \beta z) \tag{3.12}$$

Here, we use $\gamma = \alpha + j\beta$

Likewise, we can calculate the instantaneous from of the time harmonic magnetic field vector as:

$$\boldsymbol{H}(z, t) = \boldsymbol{a_y}\ Re\ \left\{H_0^+ e^{\gamma z} e^{j\omega t}\right\} = \boldsymbol{a_y}\ H_0^+ e^{-\alpha z} \cos(\omega t - \beta z) = \boldsymbol{a_y} \frac{E_0^+}{\eta} e^{-\alpha z} \cos(\omega t - \beta z) \tag{3.13}$$

We shall use conveniently all forms of expressions during the analysis of the electromagnetic fields in specific problems. Finally, the time average power flow density vector is defined as:

$$\boldsymbol{P}_{ave} = Re\{\boldsymbol{E}(z,\ t)) \times \boldsymbol{H}^*(z,\ t)\} = Re\left\{\boldsymbol{a_x} E_0^+\, e^{-\alpha z} e^{j\omega t - \beta z} \times \boldsymbol{a_y} \frac{E_0^+}{\eta} e^{-\alpha z} e^{-(j\omega t - \beta z - j\theta)}\right\} \tag{3.14}$$

or,

$$\boldsymbol{P}_{ave} = \boldsymbol{a_z} \frac{|E_0^+|^2}{\eta} e^{-2\alpha z} cos\theta\ (\mathrm{W/m^2}) \tag{3.15}$$

In this expression a phase difference θ is added between the electric and magnetic fields.

Attenuation: The uniform plane wave signal suffers from huge loss as it propagates through the lossy medium. This loss of power is defined as attenuation of the signal power and is usually expressed in dB scale. The attenuation factor is defined as the ratio of the original power and the propagated power at certain distance in logarithmic scale. Mathematically the attenuation in dB is a positive quantity.

$$A_{dB} = 10\,log\,\frac{P_1}{P_2}\ (\mathrm{dB}) \tag{3.16}$$

where P_1 is the initial time average power density and P_2 is the transmitted power density.

For lossy dielectric medium, the propagation loss is so huge that we hardly get any signal after some distance. A good example is sea water. Fire Resistant FR4 laminate is also a lossy material at microwave frequencies due to its finite conductivity and higher loss tangent compared to the expensive microwave laminates such as Rogers Duriod and Taconic TLX. The downside on the cost point of view the low loss materials are very expensive.

Remarks

Observe the sign change in the reverse field ratio. Explain why negative sign in the form of the reversed field amplitudes? Because η cannot be negative in a lossless, linear, homogeneous, and isotropic medium.

Example 3.1: One-dimensional Plane Wave Field Solution

An x-polarized uniform plane wave with amplitude 10 (V/m) is propagating along z-direction. The wave is operating at 1 GHz and is propagating in a medium with the following constitutive parameters: $\varepsilon_r = 9$, $\mu_r = 1$ and $\sigma = 0.1$ (S/m).

a) Calculate the propagation constant γ (1/m), attenuation constant α (Np/m), phase constant β (rad/m), and the wave impedance η (Ω).

b) From results in part (a) calculate the forward travelling wave vector electric and magnetic fields, and the time average Poynting vector of the wave.

c) If the uniform plane wave travels 1 m from the transmitter in the loss medium calculation the total attention in dB.

Answer:

a) First calculate the propagation constant. Substitution for the values given above, the propagation constant is:

$$\gamma = \alpha + j\beta = \sqrt{j\omega\mu(\sigma + j\omega\varepsilon)} = \sqrt{j \times 2\pi \times 10^9 \times 4\pi \times 10^{-7}\left(0.1 + j \times 2\pi \times 10^9 \times \frac{10^{-9}}{36\pi} \times 9\right)}$$

$$= \sqrt{j7896(0.1 + j0.5)} = 6.25 + j63.1\left(\mathrm{m}^{-1}\right)$$

Therefore,

$$\alpha = 6.25\ (\mathrm{Np/m});\ \beta = 63.1\ (\mathrm{rad/m})$$

The impedance is: $\eta = 124\angle 12.3°\Omega$.

b) The electric field is x-polarized and is propagating along z-direction:

$$\boldsymbol{E_x}(z, t) = \boldsymbol{a_x} 10e^{-6.25z} \cos\left(2 \times \pi \times 10^9 t - 61.3z\right)\ \ (\mathrm{V/m})$$

The magnetic field is: $\boldsymbol{H}_y(z, t) = \boldsymbol{a_y}\dfrac{\boldsymbol{E_x}(z, t)}{\eta} = \boldsymbol{a_y}\dfrac{10}{124\angle 12.3°}e^{-6.25z} \cos\left(2 \times \pi \times 10^9 t - 61.3z\right)$ or,

$\boldsymbol{H}(z, t) = \boldsymbol{a_y} 0.081e^{-6.25z} \cos\left(2 \times \pi \times 10^9 t - 61.3z - 12.3°\right)\ \ (\mathrm{A/m})$

The time average power density or Poynting vector is:

$$P_{ave} = Re\{\boldsymbol{E} \times \boldsymbol{H}^*\}$$

$$= Re\left\{\boldsymbol{a_x} 10e^{-6.25z} e^{j2 \times \pi \times 10^9 t - 61.3z}\right.$$

$$\left. \times \boldsymbol{a_y} 0.081e^{-6.25z} e^{-(j2 \times \pi \times 10^9 t - 61.3z - j12.3°)}\right\}$$

or, $P_{ave} = \boldsymbol{a_z} 0.81e^{-12.5z} \cos(12.3°) = \boldsymbol{a_z} 0.78e^{-12.5z}\left(\mathrm{W/m^2}\right)$

c) The time average power density at $z = 10$ km,

$$\boldsymbol{P}_{ave} = 0.78e^{-12.5\,(\mathrm{Np/m}) \times 10\,(\mathrm{m})} = 2.91\left(\mathrm{\mu W/m^2}\right)$$

The attenuation of the signal after 1 m of distance is:

$$A_{dB} = 10 \log \frac{P_1}{P_2} = 10 \log \frac{0.78}{2.91 \times 10^{-6}} = 54.29\ (\mathrm{dB})$$

Exercise 3.1: One-Dimensional Plane Wave Field Solution
Q1. Redo the problem for a lossless medium.

3.2.2 Wave Motion

As stated in the discussion of the generic wave equation, *wave motion* means a disturbance in which a constant *phase front* of a wave is propagating with a uniform speed. Each surface of a constant phase is called a *phase front*. That we have shown in our colored surface plot of the uniform plane wave from a horn antenna in Figure 1.8. For each wave, there is a characteristic distance between two successive wavefronts. This distance is called wavelength λ. As we showed before, depending on the wave front there is a name of the wave. As for example, if the phase front is plane meaning the amplitude and phase of the wave quantity are constant at the plane then it is called a *uniform plane wave*. Which is the present case for the wave that we have just derived. If the phase front consists of successive concentric spheres it is called *spherical wave*, as for example in the near field case of an antenna as illustrated for the dipole antenna in Figure 1.4a. Likewise, if the phase front is cylindrical it is called a *cylindrical wave*. Understanding of the phase front where the phase of the wave is constant is very important because it provides the fundamental characteristics of the electromagnetic field waves.

Now take the phase function inside the parenthesis of *cos*, which is $(\omega t - \beta z)$, a constant. Here it has two components. The first one is ωt in radians, which represents the time harmonic or oscillating nature of the wave. The second term βz is also in radians, and represents the electrical length of the wave in spatial coordinate. This is also a phase. The resultant of the two components which is the phase of the wave front must be constant in both time and spatial coordinates, respectively. Since the angular frequency ω and phase constant or wave number β are fixed for a wave with its operating frequency, the only two orthogonal and independent variables are t and z-the variables for the independent time and spatial coordinates, respectively. As time t increases, z must increase to keep the phase of the phase front constant. The figure below illustrates the physical concept of plane wave motion.

Take an example of the car. If you want to measure exactly the distance travelled by a car at a particular distance then we need to find a point on the car and multiplied by two variables, time t that the car travelled for and the velocity v of the car at which the car has travelled. These two quantities are needed to get the exact distance R that you can travel by the car at that particular velocity and time as shown in Figure 3.5a. Likewise, for the plane wave shown in Figure 3.5b, we can fix a point on the waveform of the electric field and measure the distance travelled by the wave at a particular time t and at a particular velocity u_p. The velocity of the wave motion is called the phase velocity u_p of the wave. It is the velocity at which the wave travels along a particular direction in this case the z-direction. The dynamic position of the wave as it travels with time can also be shown in electrical length βz in radians along z-axis. Normally we set time $t = 0$ to observe the physical distance the wave travels in terms of the electrical length βz in radian at it travels from a

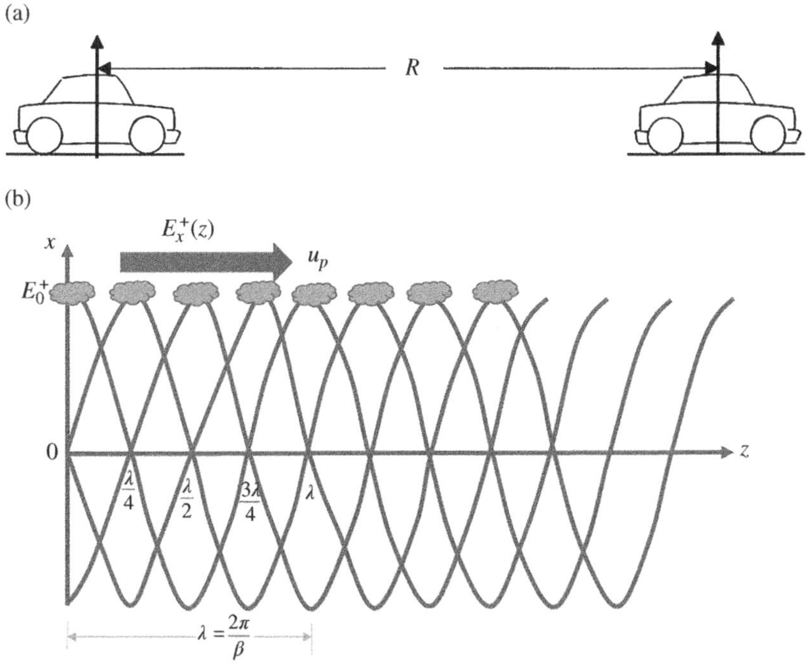

Figure 3.5 The physical meaning of such a wave motion concept.

physical point in space to another point. Likewise, we measure the time phase ωt in radians we also set $z = 0$ as the reference plane. In both cases, the time harmonic wave is represented with the sine/cosine wave function as depicted in Figure 3.5.

Now came back to the constant phase concept in the parentheses of the cosine function. As the time t increases, z must increase positively to keep the phase constant. This wave solution is called the positive z-directed or *the forward travelling wave.* The forward travelling wave is the wave from $z > 0$ to infinity in the envelop $e^{-\alpha z}$, it travels along $+z$-direction as the wave attenuates with the factor $e^{-\alpha z}$. Likewise, for a lossy medium the reverse travelling wave is the wave from $z < 0$ to ∞ in the envelop $e^{\alpha z}$, it travels along $-z$-direction as the wave attenuates with the factor $e^{\alpha z}$. The positive $e^{\alpha z}$ is also making the wave attenuated as in this $z < 0$ region, meaning z is increasing negatively as we progress toward left hand side of $z = 0$. The interior wave function with $cos(\omega t + \beta z)$ with $(\omega t + \beta z) =$ constant represents a fixed reference point on the wavefront that is travelling in the negative z-direction. With time, the distance z must be negatively increasing to keep the phase constant. This is called the negative z-directed or *the reverse travelling wave.* Therefore, what we have learned here is that the complete solution of a one-dimensional plane wave has two components. One is the forward travelling wave with −ve sign in front of the electrical length βz and the reverse travelling wave with +ve sign in front of the electrical length βz. If the wave travels along x and y directions, respectively, we shall consider the phase terms $(\omega t - \beta x)$ and $(\omega t - \beta y)$, respectively, for *the forward travelling waves*, and $(\omega t + \beta x)$ and $(\omega t + \beta y)$, respectively, for *the reverse travelling waves*. Understanding of the forward travelling wave is easy as we can easily perceive a radiated field emanating out in a particular direction from an antenna. Another example is the wave travelling along the transmission line from the generator. A reflected wave from a reflector is an example of the reverse travelling wave. As for the antenna example if the antenna is not properly matched with the source impedance then the signal from the generator as it comes to the antenna input terminals gets reflected. For the terminated transmission line, the signal gets reflected if the load in not properly matched. These reflected signals are called the reverse travelling wave. For the examples, the amplitude of the reflected or the reverse travelling waveform as illustrated in the figure, will not be the same. This depends on the amplitude and phase of the reflected wave signal. The ratio of the reflected and the incident travelling wave is called *the reflection coefficient*. The reflection coefficient varies depending on the types of the reflecting reference plane. As for example, if we put a perfect conducting reflector in front of the antenna then the incident signal completely reflects back from the perfect conducting reflector and in this case the reflected signal is with equal amplitude and out of phase with respect to the incident wave. We exploit such reflection to enhance the antenna gain in the forward directions as in the parabolic reflector antennas, and a grounded reflector at the quarter wavelength distance of a dipole antenna. For the terminated transmission line a short circuit and an open circuit also creates total reflection. For the short circuit load of a transmission line the reflection coefficient is unity with a 180° phase reversal with respect of the incident wave. For the open circuit, the reflection coefficient is unity with a 0° phase (also called in-phase). We shall be discussing the theory of reflection in the next couple of chapters. In the next chapters we shall utilize this concept of the forward and reverse travelling waves in normal and oblique incidences at the interface of two media. In all cases of the plane wave propagation we always use right hand rule to understand the directions of the field vectors. Remember, in all cases we perceive the incident wave meaning *the forward travelling wave* and the reflected wave meaning *the reverse travelling wave.* Now we shall calculate the velocity of the wave front which is called the phase velocity u_p.

Note that so far, we discussed the time harmonic electric field wave equation. We can also develop the solution following the similar procedure for the time harmonic magnetic field wave equation.

Review Questions 3.4: Analysis of Plane Wave

Q1. Explain why the reverse travelling wave is hard to perceive physically. Give an example of the reverse travelling wave.

Q2. Write the complete solutions for the wave travelling along the positive x-and y-directions when the fields are oriented along z-direction.

Q3. Look at the bottom plot of the complete solution in expression in Figure 3.4 top. Can you identify, the decaying factor associated with the attenuation constant α and the oscillating component with the phase constant β? Can you write a math essay about the complete solution for the expression?

3.2.3 Phase Velocity

In the above discussion we consider a plane wave at a fixed frequency that is propagating in a homogeneous, linear, and isotropic medium with certain loss factor due to which the envelop of the wave attenuates with the exponential decaying factor $e^{-\alpha z}$. We consider the velocity of a single frequency to be a pure sine wave at the phase velocity of wave motion u_p. The velocity of the single wave is defined as the phase velocity assuming that a pure and unperturbed signal is propagating with a fixed velocity. From the elementary physics, the velocity is defined as the derivative of the distance travelled z with respect to time t (dz/dt). The velocity also gives rise to a relationship with the angular frequency and the phase constant as $u_p = \omega/\beta$ where the phase constant β is ω/u_p. In the free space lossless case, the free space velocity is defined as:

$$c_o = \frac{1}{\sqrt{\mu_o \varepsilon_o}} = 3 \times 10^8 \text{ (m/s)}.$$

The refractive index of a dielectric medium is defined by

$$n = \sqrt{\frac{\mu\varepsilon}{\mu_o\varepsilon_o}} = \sqrt{\frac{\mu_o\mu_r\varepsilon_o\varepsilon_r}{\mu_o\varepsilon_o}} = \sqrt{\varepsilon_r} \tag{3.17}$$

For the most optical materials μ_r is 1, leading to $n = \sqrt{\varepsilon_r}$. We have discussed the refractive index in greater details in the preceding chapter.

3.3 Dispersion of Media and Group Velocity

In ideal terms, we assume that the phase velocity of the signal at different frequencies do not change with their frequencies of operation in the propagating medium. Such medium is called *nondispersive* medium. In a *nondispersive* medium, the signal does not get distorted. If the medium is *nondispersive* then signals of various frequencies propagate with the same velocity. The medium is called *dispersive* if the phase velocity u_p of the wave in the medium depends on frequency meaning $u_p(\omega)$. Even if the signal with a single frequency propagates in a dispersive medium then this dispersion phenomenon results in several harmonics components those suffer relative displacements in phase in the direction of propagation of the components.[4] The resultant wave at a distant point is not an exact replica of the incoming wave as shown in Figure 1.11a.[5] We say that the original wave is "dispersed" meaning scattered or spread out in space and time and the phenomenon is known as *dispersion*. The medium in which the phase velocity, u_p is a function of the frequency it is known as a *dispersive* medium. We deal with material and waveguide[6] dispersions during optical fiber design.

When the medium becomes dispersive in nature or when the signal contains multiple frequencies[7] then the signal propagates at groups that are also called as *wave packets*. In this case, the velocity of the composite wave or groups reduces compared to that for the phase velocity of the single wave. The velocity of *the wave packet* is called *the group velocity*. Here, we are detailing the difference between *the phase* and *the group velocities* and define the group velocity in various scenarios as stated above in details.

Remarks

The phase velocity is meaningful when we consider a single frequency wave in a *nondispersive medium*. However, in the absence of dispersion meaning when the refractive index of the medium is not a function of the frequency, the field waves at all the frequencies travel at the same velocity. Thus, all electromagnetic waves including the light waves travel in vacuum or air at the same velocity of 3×10^8 m/s.

Consider a *wave packet* with multiple frequency components are travelling in a certain direction. If all the components of the wave constituting the group travels with the same velocity then apparently the group as a whole travel with the same velocity as any of the individual waves. The dispersion can happen in two ways: (i) if all the waves are very close by frequencies and the mechanism of superposition happens such as in the amplitude modulation of two frequency (tone) signals then the resultant modulated signal is slower than the individual signal waves. Here, the medium does not play any significant role in the dispersion due to the fact that the modulation process of the signal in the medium, by its inherent nature, slows down the modulated signal compared to the individual signals, and (ii) if the medium is dispersive then the different frequency signals experience different constitutive parameters due to the frequency dependence of the medium.[8] Here we shall examine the first case of the dispersion and define *the group velocity*. Then, we amalgamate the first phenomenon with the medium property and its dispersion. Let us have an examination of the two tone[9] signals to define the group velocity followed by the relationship between the group and phase velocities in terms of the dispersion of the medium.

---

4 The decomposition of the white light into ROYGB lights via a prism is a good example of dispersion. In this case different color lights get different angular dispersion. The term dispersion implies the separation of distinguishable components of a signal. Hayt, W.H. and Buck, J.A. (2006). *Engineering Electromagnetics*, 7e, 467. Boston, USA: McGraw Hill.

5 Pulse broadening happens when the signal passes through a dispersive medium. In optical fiber design, materials dispersion is the main cause of the pulse broadening.

6 Waveguide dispersion is an intrinsic phenomenon inside the waveguide due to the boundary conditions of the waveguide. Therefore, it is different from the material dispersion that we deal within the optical fiber.

7 Such wave with multiple frequencies is called *wave packet*. In modern ultra-wideband (UWB) communication systems, we encounter such wave packets in many folds.

8 The dependence of the loss characteristics or attenuations of the atmosphere at different frequency bands is a significant area of research. In satellite communications and optical fiber design, we take the advantages of the frequency/wavelength dependent attenuation profile for efficient communications.

9 Frequency is also defined as tone in acoustic and electromagnetic alike.

Figure 3.6 illustrates a two-tone signal with time periods T_1 and T_2 at the operating frequencies f_1 and f_2, respectively. The angular frequencies of the two-tone signals are defined as:

$$\omega_1 = 2\pi f_1 = \frac{2\pi}{T_1} \text{ (rad/s)} \tag{3.18}$$

$$\omega_2 = 2\pi f_2 = \frac{2\pi}{T_2} \text{ (rad/s)} \tag{3.19}$$

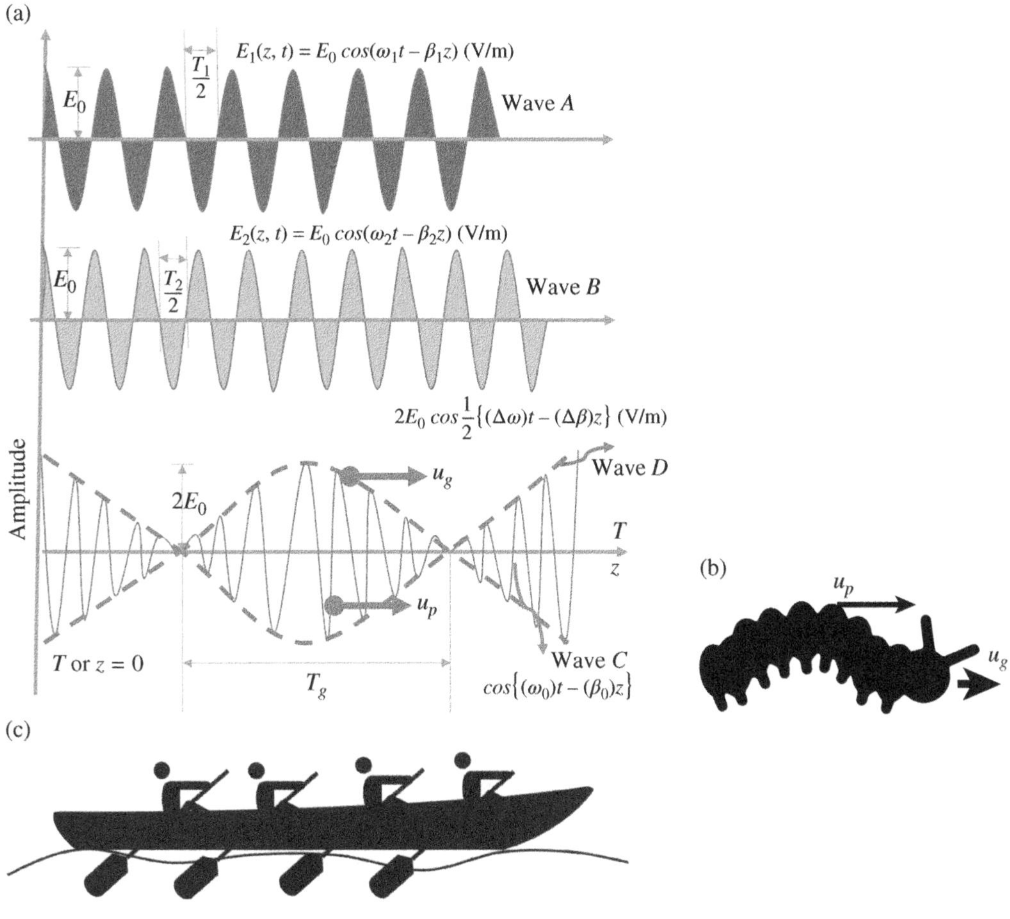

Figure 3.6 (a) Amplitude modulation of two independent frequency waves with time periods T_1 (Wave A) and T_2 (Wave A) due to in-phase superposition into a single wave with envelop (Wave D) and composite (Wave C). The group velocity, u_g, is the velocity of the envelop and the phase velocity, u_p, is the velocity of the composite Wave C, (b) Caterpillar locomotion with an analogy of phase velocity u_p at which the humps on its back is moving and the group velocity u_g at which the actual movement of the caterpillar, and (c) rowing in a group is a good example of the phase and group velocity difference. The rower paddles move much faster than the speed of the canoe.

If we consider the phase constants β_1 and β_2 (rad/m), which are associated with the spatial variation with the coordinate system say along the z-axis (direction), then we can also define these two-phase constants as follows:

$$\beta_1 = \frac{\omega_1}{c} = \frac{2\pi}{\lambda_1} \text{ (rad/m)} \tag{3.20}$$

$$\beta_2 = \frac{\omega_2}{c} = \frac{2\pi}{\lambda_2} \text{ (rad/m)} \tag{3.21}$$

where c is the speed of the electromagnetic wave in the medium, λ_1 and λ_2 are the wavelengths of the signals at the two frequencies f_1 and f_2, respectively.

For simplicity we consider the two waves are travelling in the positive z-direction and they have the same amplitude E_0 (V/m). Then we can write that the one-dimensional time domain field wave equations as:

$$E_1(z, t) = E_0 \cos(\omega_1 t - \beta_1 z) \text{ (V/m)} \tag{3.22}$$

$$E_2(z, t) = E_0 \cos(\omega_2 t - \beta_2 z) \text{ (V/m)} \tag{3.23}$$

These two field waves are denoted as Wave A and Wave B, respectively, in Figure 3.6a. The resultant amplitude modulated wave is the direct superposition of the two signals through a linear channel as follows:

$$E_{Total} = E_1(z, t) + E_2(z, t) = E_0\{\cos(\omega_1 t - \beta_1 z) + \cos(\omega_2 t - \beta_2 z)\} \text{ (V/m)} \tag{3.24}$$

Using the trigonometric identity $\cos A + \cos B = 2\cos[(B - A)/2] \times \cos[(A + B)/2]$ we obtain

$$E_{Total} = 2E_0\left[\cos\frac{1}{2}\{(\omega_1 - \omega_2)t - (\beta_1 - \beta_2)z\} \times \cos\frac{1}{2}\{(\omega_1 + \omega_2)t - (\beta_1 + \beta_2)z\}\right] \text{ (V/m)} \tag{3.25}$$

$$E_{Total} = 2E_0 \cos(\Delta\omega t - \Delta\beta z) \times \cos(\omega_c t - \beta_c z) \text{ (V/m)} \tag{3.26}$$

where $\omega_c = (\omega_1 + \omega_2)/2$; $\beta_c = (\beta_1 + \beta_2)/2$; $\Delta\omega = (\omega_1 - \omega_2)/2$ and $\Delta\beta = (\beta_1 - \beta_2)/2$ are used.

Here we assume $\omega_1 > \omega_2$ and $\beta_1 > \beta_2$. If we decompose the above expression for the total electric field wave E_{Total} in (3.26), then apparently, we can define two separate waves as defined by *Wave C* – the normalized modulated wave with the career frequency ω_0 and the phase constant β_0 and *Wave D – the modulating beat frequency* wave with much lower frequency $\Delta\omega = (\omega_1 - \omega_2)/2$ and the phase constant $\Delta\beta = (\beta_1 - \beta_2)/2$. Therefore, we can write mathematically,

$$E_{Total} = \text{Wave } C \times \text{Wave } D = \text{Normalised carrier wave} \times \text{Beat frequency wave} \tag{3.27}$$

where, the carrier wave or the modulator wave is

$$\text{Wave } C = \cos(\omega_c t - \beta_c z) \tag{3.28}$$

and the beat frequency wave is

$$\text{Wave } D = 2E_0 \cos(\Delta\omega t - \Delta\beta z) \tag{3.29}$$

From the examination of the expressions (3.28) and (3.29), we can deduce that *Wave C* is the normalized *carrier frequency* wave or *the modulated signal wave* with the mean frequency ω_c

and the mean phase constant β_c. *Wave D* is *the beat frequency wave*[10] which is oscillating at much lower beat frequency $\Delta\omega = ((\omega_1 - \omega_2)/2)$. This wave is also called *the modulating signal* envelop with an *envelope* with the peak amplitude twice that for the original signals meaning $\pm 2E_0$. If we can draw the analogy of *the beat* frequency electromagnetic wave with that in acoustic case then we can realize that the *beat frequency wave* is originated from the two-tone signals of close by much higher frequencies ω_1 and ω_2. They generate a much lower *beat frequency signal* with double the amplitude. The electronic process of modifying a sinusoidally varying high frequency *signal wave packet* to produce a slowly varying envelop superimposed on a high frequency wave is called *modulation.* In the current example, the amplitude of the resultant wave is modulated, therefore, the process is called *the amplitude modulation.*[11] As we can see that there are two components of the resultant wave: (i) one is the high frequency component also called *the carrier frequency wave* with the average angular frequency, $\omega_c = (\omega_1 + \omega_2)/2$. The envelop of the modulation has *the beat frequency*, which is much lower than the career frequency and is defined as the mean of the difference of the two original signal frequencies: $\Delta\omega = (\omega_1 - \omega_2)/2$. The resultant *carrier wave* has a wavelength which can be derived from the average phase constant or wave number as follows:

$$\beta_c = \frac{\beta_1 + \beta_2}{2} = \pi\left[\frac{1}{\lambda_1} + \frac{1}{\lambda_2}\right] = \frac{2\pi}{\lambda_c} \tag{3.30}$$

Here, $\beta_1 = 2\pi/\lambda_1$ and $\beta_2 = 2\pi/\lambda_2$ are used.

Therefore, the resultant wavelength of the carrier wave is:

$$\lambda_c = \frac{2\lambda_1\lambda_2}{\lambda_1 + \lambda_2} \tag{3.31}$$

The amplitude of the resultant wave is modulated to form as indicated by the dotted envelopes in Figure 3.6a. The phase constant of the envelop or the beat frequency wave is defined as

$$\Delta\beta = \frac{\beta_1 - \beta_2}{2} = \pi\left[\frac{1}{\lambda_1} - \frac{1}{\lambda_2}\right] = \frac{2\pi}{\Delta\lambda} \tag{3.32}$$

Therefore, the resultant wavelength of the envelop is:

$$\Delta\lambda = \frac{2\lambda_1\lambda_2}{\lambda_2 - \lambda_1} \tag{3.33}$$

The average velocity of the individual waves that also constitutes the velocity of the carrier wave is:

$$u_p = \frac{\omega_c}{\beta_c} = \frac{\omega_1 + \omega_2}{\beta_1 + \beta_2} = \frac{c}{\lambda_c} \tag{3.34}$$

where, λ_c is the carrier wave's wavelength as defined in equation (3.31). Therefore, the resultant wavelength of the carrier wave is:

$$\lambda_c = \frac{2\lambda_1\lambda_2}{\lambda_1 + \lambda_2} \tag{3.31}$$

10 Read *beat frequency waves* and listen to different beat frequencies in Wikipedia (https://en.wikipedia.org/wiki/Beat_(acoustics)#Binaural_beats) (accessed 07 January 2018).

11 You may recall that we produce such amplitude modulation so easily by passing two-tone signals via an active device such as bipolar junction transistor modulator.

The velocity of the group or the envelop is:

$$u_g = \frac{\Delta\omega}{\Delta\beta} = \frac{\omega_1 - \omega_2}{\beta_1 - \beta_2} \tag{3.35}$$

For negligible differences in frequencies of the original signals we can deduce the group velocity of the carrier or modulated signal as

$$u_g = \lim_{\Delta\to 0} \frac{\Delta\omega}{\Delta\beta} = \frac{d\omega}{d\beta} \tag{3.36}$$

Review Questions 3.5: Dispersion of Media and Group Velocity

Q1. Define dispersion and differentiate dispersion in single tone and multitone signals.
Q2. What is a dispersive media?
Q3. Define phase and group velocities and explain their differences.
Q4. What is a wave packet?
Q5. Define amplitude modulation.
Q6. Define modulation in electronic process. What type of modulation happens in the beat frequency envelope?
Q7. If the difference between the two frequencies f_1 and f_2 is negligibly small compared to their original values show that the group velocity is deduced as $u_g = d\omega/d\beta$.
Q8. Explain dispersion of white light in a prism.
Q9. What is a nondispersive medium?
Q10. Explain the consequence of dispersion? what are the various types of dispersions?

Figure 3.7 illustrates the dispersion diagram that shows the relationship between the phase constants β_1 and β_2 for the individual single tone signals of frequencies ω_1 and ω_2, respectively. They generate the superposed carrier frequency ω_c, which has the phase constant β_c. The light line is defined as the straight line drawn as the inverse of the slope of curve in the nondispersive medium.

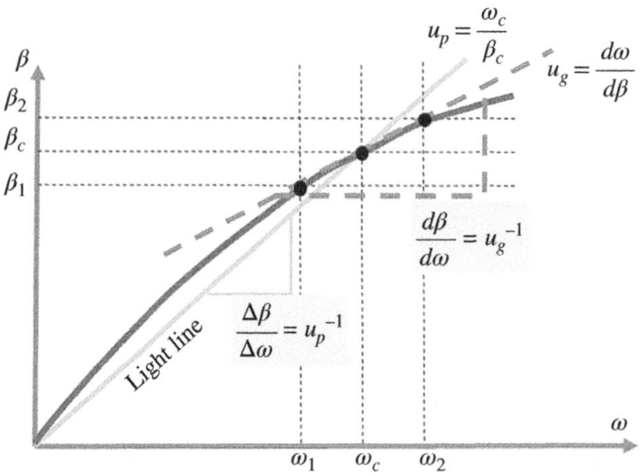

Figure 3.7 Dispersion diagram showing the relationship between the phase constants β_1 and β_2 for the individual single tone frequencies ω_1 and ω_2, respectively that generate the superposed carrier frequency ω_c, which has the phase constant β_c. The light line is defined as the straight line drawn as the inverse of the slope of curve in the nondispersive medium. In that medium, the phase velocity is defined as $u_p = \omega_c/\beta_c$ where the line light is bent in the dispersive medium and the equivalent two-tone signal has the carrier frequency which is defined as $\omega_c = (\omega_1 + \omega_2)/2$.

In that medium the phase velocity is defined as $u_p = \omega_c/\beta_c$. The light line is bent in the dispersive medium and the equivalent modulated signal of the two-tone signals has the carrier frequency which is defined as $\omega_c = (\omega_1 + \omega_2)/2$. To examine the relationship between the phase and the group velocities let us consider that the difference between the two angular frequencies is too small so that we can define the difference between the two signals as

$$d\omega = \underset{\Delta\omega \to 0}{limit} \Delta\omega = \omega_1 - \omega_2 \tag{3.37}$$

where, the difference in angular frequencies, $d\omega$ is infinitesimally small.

Therefore, we can likewise define the difference of the respective wave numbers (phase constants) as $d\beta = \lim_{\Delta\beta \to 0} \Delta\beta = \beta_1 - \beta_2$. Also, we have already defined the carrier frequency and the phase constant, as $\omega_c = (\omega_1 + \omega_2)/2$ and $\beta_c = (\beta_1 + \beta_2)/2$, respectively. From these two definitions and the graphical representation above we can define the phase and the group velocities as follows:

$$u_p = \frac{\omega_c}{\beta_c} \tag{3.38}$$

and

$$u_g = \frac{d\omega}{d\beta} \tag{3.39}$$

Therefore, we can define the angular frequency of the carrier wave signal as:

$$\omega_c = u_p \beta_c \tag{3.40}$$

Taking the differential form of the above expression we have

$$d\omega_c = u_p d\beta_c + \beta_c du_p \tag{3.41}$$

From rearranging the above expression with the generalization of the equation by dropping the subscript c for the carrier frequency and the phase constant, we obtain the relationship between the phase and group velocities as follows:

$$u_g = \frac{d\omega}{d\beta} = u_p + \beta\frac{du_p}{d\beta} \tag{3.42}$$

Note for the dispersive medium the phase constant β is a function of the frequency ω hence $\beta = \beta(\omega)$ is understood. In the above equation, we shall develop the relationship between $\beta/d\beta$ with the wavelength using the definition of $\beta\lambda = 2\pi$. Taking the differential form of $\beta\lambda = 2\pi$ we obtain

$$\beta d\lambda + \lambda d\beta = 0 \tag{3.43}$$

or,

$$\frac{\beta}{d\beta} = -\frac{\lambda}{d\lambda} \tag{3.44}$$

Substituting for the above relation into (3.42) we obtain:

$$u_g = u_p - \lambda\frac{du_p}{d\lambda} \tag{3.45}$$

This is an important relationship between the phase and group velocities. Here, λ is the actual wavelength of the carrier frequency in the medium and is given by

$$\lambda_c = \frac{2\lambda_1\lambda_2}{\lambda_1 + \lambda_2} \tag{3.46}$$

Here, the velocity u_g represents the velocity of the slowly varying envelop with the normalized wave function $cos(\Delta\omega t - \Delta\beta z)$ and is called *the group velocity*.

Remarks

From the above expression of the group velocity in terms of the phase velocity and the wavelength we can deduce the following conclusions[12]:

1) A dispersive medium is one in which the phase velocity is a function of the frequency and hence the free space wavelength λ.
2) The medium is called *normal dispersive medium* if the change of the phase velocity with wavelength is positive meaning $du_p/d\lambda > 0$. For such a medium $u_p > u_g$.
3) The medium is called *anomalously dispersive medium* if the change of the phase velocity with wavelength is negative meaning $du_p/d\lambda < 0$. For such a medium $u_g > u_p$.

Now we take the constant phase term in the parenthesis of the slowly varying envelop $cos(\Delta\omega t - \Delta\beta z)$ so that we can define the phase velocity of the group as follows:

$$\lim_{\Delta\omega \to 0} (\Delta\omega t - \Delta\beta z) = t d\omega - z d\beta \tag{3.47}$$

Therefore, we can define the phase velocity of the group using the fundamental definition of velocity as distance over time as:

$$u_g = \frac{dz}{dt} = \frac{d\omega}{d\beta} \tag{3.48}$$

We have already defined the above expression of the group velocity, but in a little bit different perspective.

Another convenient expression for the group velocity can be obtained from the relationship

$$\omega = u_p \beta \tag{3.49}$$

Taking the differential from of the above definition we obtain

$$d\omega = u_p d\beta + \beta du_p \tag{3.50}$$

or,

$$1 = u_p \frac{d\beta}{d\omega} + \beta \frac{du_p}{d\omega}, \tag{3.51}$$

or,

$$u_p \frac{d\beta}{d\omega} = 1 - \beta \frac{du_p}{d\omega} \tag{3.52}$$

12 J. D. Kraus and D. A. Fleisch, *Electromagnetics with Applications*, 5e McGraw Hill, Boston, MA, USA, 1999, p. 204–205.

or,

$$\frac{u_p}{u_g} = 1 - \left(\frac{\omega}{u_p}\right) \times \left(\frac{du_p}{d\omega}\right) \tag{3.53}$$

or,

$$u_g = \frac{u_p}{1 - \left(\frac{\omega}{u_p}\right) \times \left(\frac{du_p}{d\omega}\right)} \tag{3.54}$$

The above expression is another important deduction for the group velocity of a wave packet.

Remarks
From the above deduction, we can conclude the following points:

1) The phase velocity is equal to the group velocity if the phase velocity does not change with frequency meaning $du_p/d\omega = 0$. This means the medium is nondispersive.
2) The medium is called *normal dispersive medium* if the change of the phase velocity with frequency is positive meaning $du_p/d\omega > 0$. For such a medium $u_p > u_g$.
3) The medium is called *anomalously dispersive medium* if the change of the phase velocity with wavelength is negative meaning $du_p/d\omega > 0$. For such a medium $u_g > u_p$.

The definitions of the normal and anomalous dispersive media are arbitrary. The significance can be realized with contain media as for example in plasma and in conductor, respectively.

Remarks: Unearthing Cornerstone of Phase and Group Velocities
The experimental determination of the velocity of a wave necessarily involves the group velocity and not the phase velocity. Because there is no way to measure the movement of an individual wave in a group. Lord Rayleigh was the first to point out this phenomenon of the electromagnetic wave in connection with the measurement of the velocity of light. However, in a dispersionless medium the phase velocity and the group velocity are the same because the refractive index does not change with frequency and all frequency components in the group travels at the same velocity. From the above deduction we have seen that the group velocity can be larger than the phase velocity if the medium is anomalously dispersive. Under *the normal dispersion, the group velocity* is less than *the phase velocity*. The fact is that *the group or phase velocity* sometimes can be greater than the speed of light in vacuum. This is a contradiction with Einstein's theory of relativity that suggest that the maximum attainable velocity is c, the speed of light in vacuum. We may approach c but never exceed it. Even in a Cyclotron or a Betatron, it has not been possible to attain a velocity greater than c. We must point out that the relativity postulate applies to the velocity at which the energy can be transmitted. In a medium where the dispersion is normal, and only moderately strong, a wave packet travels in a considerable distance with negligible diffusion. In such a case the velocity of the energy flow is approximately equal to the group velocity. We can see from the above derivation and explanation that the refractive index $n(\omega)$ or the phase constant $\beta(\omega)$ increases so that the slope of $\beta(\omega)$ is positive. In this case, *the phase velocity* increases with frequency and the medium is *normally dispersive.*

For an *anomalously dispersive medium* the case is opposite, the refractive index $n(\omega)$ or the phase constant $\beta(\omega)$ decreases so that the slope of $\beta(\omega)$ is negative. In this case the phase velocity decreases with frequency and the group velocity is larger than the phase velocity. In certain cases, as in the neighborhood of the absorption band due to the existence of conductivity in the medium, for example, the term $du_p/d\lambda$, is so negative that the group velocity may be larger without bound and even negative. In such case, our concept of the group velocity has very little physical significance.

The practical difficulty of the concept of the velocity of energy propagation in an anomalously dispersive medium is due to the fact that the wave packet or pulse becomes highly distorted and because of the attenuation, the amplitude of the wave reduces to almost zero within a very short distance so that our definition of group velocity no longer possesses any physical significance. The problem, corresponding to a plane wave that starts through an anomalously dispersive medium has been treated by Summerfield[13] and Brillouin. Their conclusion is that the velocity of energy propagation never exceeds the velocity of light in vacuum. In the treatment of the waveguides and the obliquely incident cases of the plane wave at the interface of media, we shall come across the group velocities, which are greater than the speed of light. These examples do not contradict with the principle of relativity as no material object moves with these velocities. It is the velocity of a fictitious point on the wave. A couple of trivial examples are shown in Figure 3.8. First one is a sea way launches on the shore of a sea. The second one is the contact point of intersection of the straight edges of the blades of a pair of scissors. In the first example, when the wave was launched on the sea shore, the velocity is larger than the actual wave inside the sea. The shore wave moves much faster than the wave in the sea that carries the real energy. This is the fictitious contact point of the water and the shore wave that move very fast. The second example is the infinitely long blade of a pair of scissors. Opening and closing of the blades with moderate velocity results in the contact point of the intersection of the blades moves very fast may be as high as or even greater than c. However, as pointed out above, in these two examples, the velocity does not refer to the motion of a material object or that of energy propagation and hence is not governed by the relativity postulates.

The illustrations of the two examples where the group velocity can be larger than the phase velocity are shown in Figure 3.8a,b.

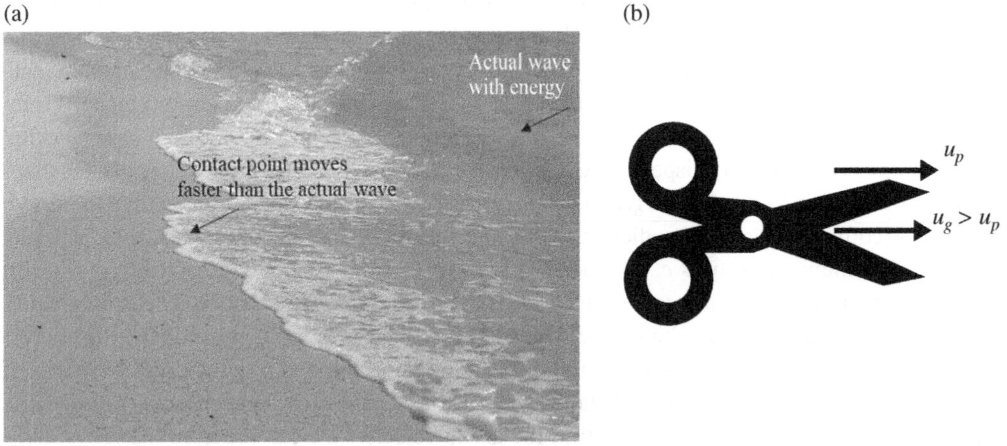

Figure 3.8 Examples of anomalous dispersion where the group velocity $u_g > u_p$ and may even exceed the velocity of light c (a) fictitious contact point of wave at sea shore (*Source:* Maki Akiyama/Wikimedia Commons/ CC BY-SA 3.0), and (b) contact point of blades of a pair of scissors.

13 A. Sommerfeld, *Optics*, Sec. 22 Academic Press, 1954 and L. Brillouin, *Wave Propagation and Group Velocity* 1946 Springer in M.A. Islam, *Electromagnetic Theory*, EPUET, Dacca, East Pakistan 1969 p. 401.

3.4 Dispersion in Digital Signal Processing and Information Theory

In this section, we shall discuss two topics: (i) the group velocity of the signal that carries information, and (ii) the pulse broadening due to the dispersion.

3.4.1 Group Velocity in Information Theory

In the preceding section, we have examined the two simple sinusoidally varying electromagnetic wave when their frequency difference is very small. We have found that the resultant wave of the two waves is the product of the modulated carrier wave and the modulating envelop. We also assume that the signal is propagating in the positive z-direction for infinite distance and time. In reality the information carrying signal are finite in time length and arbitrary in the waveform. *Therefore, in information theory and digital signal processing perspectives, the simple time harmonic electromagnetic wave that is periodic in nature and propagates indefinitely in time and space cannot be a signal.* In information theory, we are more interested with the information carrying pulses that are generated by "turning on" and "off" the oscillator for short intervals with time period T, were T is infinitesimally small. The turning on sequence is defined as digit "1" and the turning off sequence is defined as digit "0." We are familiar with the Fourier transformation of a periodic signal, and Fourier series expansion of an arbitrary waveform in information theory. To represent such pulses, we use a series expansion of sine and cosine functions. Therefore, with the help of Fourier theory we can define such an arbitrary pulse with the summation of single frequency waves without the beginning or end. Let us define any arbitrary function $\psi(z, t)$ satisfying certain conditions using Fourier integral theory[14] as:

$$\psi(z, t) = \int_{-\infty}^{\infty} A(\beta) e^{j(\omega t - \beta z)} d\beta \tag{3.55}$$

where, $A(\beta)$ is the amplitude function that represent the pulse. Since we are referring to a pulse of very short duration, therefore, we can define the pulse amplitude as:

$$A(\beta) = \begin{cases} A_o \text{ for } \beta_c - \delta\beta \leq \beta \leq \beta_c + \delta\beta \\ 0 \text{ otherwise} \end{cases} \tag{3.56}$$

where $\delta\beta \to 0$. This means we are referring to a pulse which extends within a very narrow band of frequencies.

Therefore, as stated before, the concept of group velocity is applicable to a pulse which extends within a very narrow bandwidth. Using the condition of the narrow band pulse we define the above function as:

$$\psi(z, t) = \int_{\beta_c - \delta\beta}^{\beta_c + \delta\beta} A(\beta) e^{j(\omega t - \beta z)} d\beta \tag{3.57}$$

The above equation is the definition of a wave packet. For a dispersive medium, β is a function of the frequency ω. Applying the conditions as stated above for the infinitesimal bandwidth signal, we

14 Similar derivation is also available in M.A. Islam, *Electromagnetic Theory*, EPUET, Dacca, East Pakistan, 1969, pp. 399–400.

can expand the frequency ω around the carrier frequency wave number β_c using Taylor series expansion[15]:

$$\omega(\beta) \simeq \omega(\beta_c) + (\beta - \beta_c)\left(\frac{d\omega}{d\beta}\right)_{\beta = \beta_c} \tag{3.58}$$

Replacing $\omega(\beta)$ with the phase factor $(\omega t - \beta z)$ we can define the phase factor in Taylor series expansion as follows:

$$\omega t - \beta z \simeq (\omega_c t - \beta_c z) + (\beta - \beta_c)\left[\left(\frac{d(\omega t - \beta z)}{d\beta}\right)_{\beta = \beta_c}\right] = (\omega_c t - \beta_c z) + (\beta - \beta_c)\left[t\left(\frac{d\omega}{d\beta}\right)_{\beta = \beta_c} - z\right] \tag{3.59}$$

Substitution for the phase factor in the above integral definition of the arbitrary pulse function $\psi(z, t)$ we obtain:

$$\psi(z, t) = \int_{\beta_c - \delta\beta}^{\beta_c + \delta\beta} A(\beta) e^{j\left\{(\omega_c t - \beta_c z) + (\beta - \beta_c)\left[t\left(\frac{d\omega}{d\beta}\right)_{\beta = \beta_c} - z\right]\right\}} d\beta = \psi_0 e^{j(\omega_c t - \beta_c z)} \tag{3.60}$$

Here, ψ_0 is the mean amplitude of the function $\psi(z, t)$. The phase factor of the mean amplitude ψ_0 is:

$$(\beta - \beta_c)\left[t\left(\frac{d\omega}{d\beta}\right)_{\beta = \beta_c} - z\right] = \text{constant} \tag{3.61}$$

The velocity of the wave function is defined as:

$$u_g = \frac{dz}{dt} = \left(\frac{d\omega}{d\beta}\right)_{\beta = \beta_c} \tag{3.62}$$

From the above exercise we find that the group velocity of an arbitrary wave function $\psi(z, t)$ is the same as that obtained from the combination of the two single tone sinusoidally varying wave functions as defined in the very beginning of the analysis.

Remarks

From the above analysis of the arbitrary pulse as used in information theory and digital signal processing, we can deduce the following inferences:

1) A true information carrying signal has finite extents in both time and space. Such signals are in the forms of pulses with very narrow bandwidth of frequency spectrum. Therefore, we can use the narrowband group velocity definition of the electromagnetic theory exactly in the same expression as for the information theory and digital signal processing.
2) The concept of group velocity in the electromagnetic theory is physically significant only when the dispersion is small. For large dispersion, we cannot use the definition as expanded using the Fourier theory because the range $\beta_c - \delta\beta \leq \beta \leq \beta_c + \delta\beta$ is no longer small and we cannot use the definition of group velocity that has just been derived above.
3) For wideband signal, the definition is not more valid as the group velocity changes with the large variation. The signal gets distorted and hence it is not possible to define a single group velocity of such large varying signal.
4) For large dispersion, we may subdivide the spectrum into small ones as we do in the discrete and fast Fourier transformations and calculate the individual group velocities of the segmented spectrum.

15 See also the $\omega\beta$ diagram in Figure 3.7 above for the explanation of the group velocity around β_c.

Review Questions 3.6: Dispersion in Digital Signal Processing

Q1. Define a *signal* in the information theory and digital signal processing perspectives.
Q2. Explain the difference between sinusoidally varying electromagnetic wave and an arbitrary wave function that represents a pulse.
Q3. Write the expression of an arbitrary signal $\psi(z, t)$. In Fourier series expansion.
Q4. Derive $u_g = dz/dt = (d\omega/d\beta)_{\beta=\beta_c}$ from the beginning of the definition of $\psi(z,t) = \int_{-\infty}^{\infty} A(\beta)e^{j(\omega t-\beta z)}d\beta$ using Fourier integral theory of an arbitrary function.
Q5. Explain what happens if the dispersion is wide.

3.4.2 Pulse Broadening in Dispersive Medium

In digital communications systems, the carrier frequency wave is modulated with an information carrying impulse before it is transmitted. The presence of the pulse is represented as binary digit "1" and absence as "0." When the modulated information carrying signal passes through a dispersive medium, the pulse gets distorted and broadened. This phenomenon is called *dispersion*.[16] In this section, we shall examine the pulse broadening of a Gaussian pulse which is the most common type of information carrying pulse. The time domain expression of a Gaussian pulse is[17]:

$$\psi(0, t) = \psi_0 e^{\frac{1}{2}\left(\frac{t}{T}\right)^2} e^{-j\omega_c t} \tag{3.63}$$

Here, ψ_0 is the initial reference amplitude of $\psi(z, t)$ at $t = z = 0$, ω_c = the carrier frequency, and T *is the characteristic half width* at which the amplitude $\psi_0(z, t)$ reduces to $1/e$ of its original value of ψ_0.

Taking the Fourier transform of the time domain Gaussian function above we obtain the frequency domain expression of the function as follows:

$$\psi(0, \omega) = \frac{1}{2\pi}\int_{-\infty}^{\infty} \psi_0 e^{\frac{1}{2}\left(\frac{t}{T}\right)^2} e^{-j\omega_c t} e^{j\omega t} dt = \frac{1}{2\pi}\int_{-\infty}^{\infty} \psi_0 e^{\frac{1}{2}\left(\frac{t}{T}\right)^2} e^{j(\omega-\omega_c)t} dt = \frac{T\psi_0}{\sqrt{2\pi}} e^{-\frac{1}{2}T^2(\omega-\omega_c)^2} \tag{3.64}$$

Figure 3.9 illustrates the normalized spectral density of the Gaussian pulse that peaks at the carrier frequency ω_c and the half width bandwidth at which the spectral density falls by $1/e$ of its peak amplitude. At this location the frequency is designated as ω_1 and ω_2 respectively. Where we define the half bandwidth $\Delta\omega = \omega_2 - \omega_c$. The ω–β diagram[18] in Figure 3.9b that has three different velocities (slopes) at the location of ω_1, ω_2, and ω_c, which are defined as:

$$u_{g-\omega_1} = \left.\frac{d\omega}{d\beta}\right|_{\omega=\omega_1}, \tag{3.65}$$

$$u_{g-\omega_c} = \left.\frac{d\omega}{d\beta}\right|_{\omega=\omega_c} \tag{3.66}$$

16 In optical fiber, the pulse broadening happens due to the materials and waveguide dispersions. The unit of the dispersion factor is ps/nm-km, where the unit in "ps" represents the extent of pulse broadening, the unit "nm" represents the wavelength of visible light spectrum and unit km represents the length of fiber.
17 Similar derivation can also be obtained in Hayt, W.H. and Buck, J.A *Engineering Electromagnetics*, 7e, 471–475. McGraw Hill.
18 This diagram is inverse of the β–ω diagram as depicted in the previous topic to define the group velocity without taking inverse.

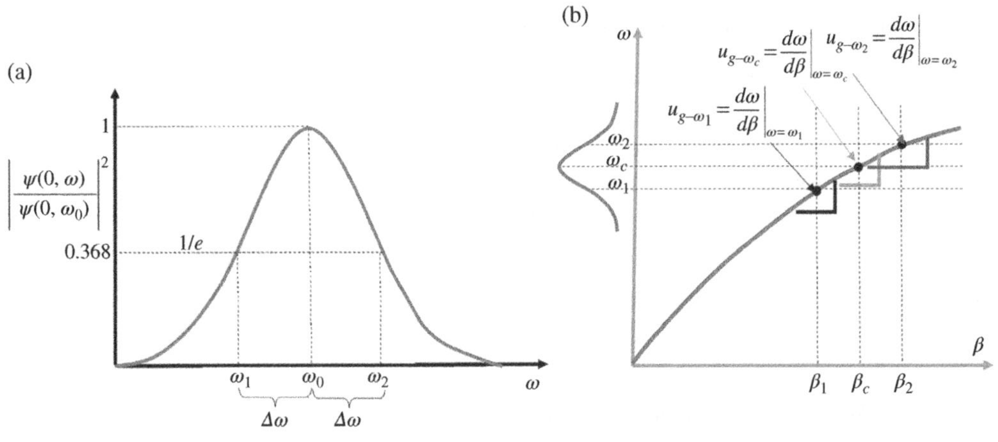

Figure 3.9 (a) Normalized spectral density of Gaussian pulse showing the half width bandwidth $2\Delta\omega = \omega_2 - \omega_1$ centered at carrier frequency ω_c, and (b) group velocities at the three frequency locations ω_1, ω_2 and ω_c.

and

$$u_{g-\omega_2} = \left.\frac{d\omega}{d\beta}\right|_{\omega=\omega_2} \tag{3.67}$$

Due to the different group velocities at the three frequency locations the signal with different frequency components arrive at different times. This phenomenon is called pulse broadening. The pulse broadening of different frequency components happens in time, not in space. Assume that the pulse travels a distance z at a time t. Then the time the signals at frequencies ω_2 and ω_c take are defined as:

$$t_2 = \frac{z}{u_{g\_\omega_2}} \tag{3.68}$$

and,

$$t_c = \frac{z}{u_{g\_\omega_c}} \tag{3.69}$$

If we define the pulse broadening in time that can be quantified with respect to the peak amplitude at the carrier frequency ω_c, then in this instance, the pulse broadening factor in time is defined as:

$$\Delta\tau = \frac{z}{u_{g\_\omega_2}} - \frac{z}{u_{g\_\omega_c}} = z\left(\left.\frac{d\omega}{d\beta}\right|_{\omega=\omega_2} - \left.\frac{d\omega}{d\beta}\right|_{\omega=\omega_c}\right) \tag{3.70}$$

Like before, we need to do some mathematical manipulation to get the closed form expression of the delay spread of the pulse using Taylor series expansion as follows:

$$\beta(\omega) \simeq \beta(\omega_c) + (\omega-\omega_c)\left.\frac{d\beta}{d\omega}\right|_{\omega=\omega_c} + \frac{1}{2}(\omega-\omega_c)^2\left.\frac{d^2\beta}{d\omega^2}\right|_{\omega=\omega_c} \tag{3.71}$$

From the examination of the Taylor series expansion of $\beta(\omega)$ above due to its infinitesimal variation around ω_c we can easily deduce that if we consider the first two terms then that represents the straight line meaning no dispersion.[19] This is called *linear variation*.

Linear variation: $\beta(\omega) \simeq \beta(\omega_c) + (\omega - \omega_c)\,(d\beta/d\omega)|_{\omega = \omega_c}$ follows a variation in the form of a line, $y = mx + c$ where, c is a constant and equivalent to $\beta(\omega_c)$ and the gradient $m = (d\beta/d\omega)|_{\omega = \omega_c}$.

The third term $\left((d^2\beta)/d\omega^2\right)|_{\omega = \omega_c}$ represents the curvature meaning the dispersion. This is called *quadratic variation*. This term is called *dispersion parameter*. The unit is time^2/distance.

Quadratic variation: $1/2(\omega - \omega_c)^2\left((d^2\beta)/d\omega^2\right)|_{\omega = \omega_c}$; where the double derivate represents a curvature derivative of the group velocity around the carrier frequency ω_c.

Usually in fiber optics, the pulse width is measured in picoseconds and the distance in kilometers. Therefore, the unit for *the dispersion parameters* is (ps^2/km).

Since $\beta(\omega_c)$, $(d\beta/d\omega)|_{\omega = \omega_c}$ and $\left(d^2\beta/d\omega^2\right)|_{\omega = \omega_c}$ are constants, we can take the derivation of $\beta(\omega)$ to find the closed formed expression of the delay spread:

$$\frac{d\beta}{d\omega} = \left.\frac{d\beta}{d\omega}\right|_{\omega = \omega_c} + (\omega - \omega_c)\left.\frac{d^2\beta}{d\omega^2}\right|_{\omega = \omega_c} \tag{3.72}$$

Therefore, we can define $d\beta/d\omega$ at $\omega = \omega_2$ and ω_c to calculate the delay spread in its closed formed expression.

$$\left.\frac{d\beta}{d\omega}\right|_{\omega = \omega_2} = \left.\frac{d\beta}{d\omega}\right|_{\omega = \omega_c} + (\omega_2 - \omega_c)\left.\frac{d^2\beta}{d\omega^2}\right|_{\omega = \omega_c} \tag{3.73}$$

Substituting for the above equation in the original expression for the delay spread we obtain:

$$\begin{aligned}\Delta\tau &= z\left(\left.\frac{d\omega}{d\beta}\right|_{\omega=\omega_2} - \left.\frac{d\omega}{d\beta}\right|_{\omega=\omega_c}\right) = z\left(\left.\frac{d\beta}{d\omega}\right|_{\omega=\omega_c} + (\omega_2 - \omega_c)\left.\frac{d^2\beta}{d\omega^2}\right|_{\omega=\omega_c} - \left.\frac{d\beta}{d\omega}\right|_{\omega=\omega_c}\right)\\ &= z\Delta\omega\left.\frac{d^2\beta}{d\omega^2}\right|_{\omega=\omega_c} = \left.\frac{z}{T}\frac{d^2\beta}{d\omega^2}\right|_{\omega=\omega_c}\end{aligned} \tag{3.74}$$

where, $T = 1/\Delta\omega$ is used.

Remarks

Depending of the width of the input pulse, there are two cases of considerations for pulse broadening $\Delta\tau$:

Case 1: If the transmitted pulse is shorter and smaller compared to the pulse spread $\Delta\tau$, then after the travel distance z at the end of the transmission, the pulse spread will be considered as $\Delta\tau$:

$$\Delta\tau_{Short} = \frac{z}{T} \times \left.\frac{d^2\beta}{d\omega^2}\right|_{\omega = \omega_c} \tag{3.75}$$

---

19 The first two terms represent a straight line with the form of $y = mx + c$ where m is the slope $(d\beta/d\omega)$ and the constant c is defined as the value of β at ω_c: $\beta(\omega_c)$. Such straight line of β–ω diagram is referred to as the light line that represent the nondispersive medium. This basically defines the refractive index n of the medium.

Case 2: If the transmitted pulse is comparable to the pulse spread $\Delta\tau$, then after the travel distance z at the end of the transmission, the pulse spread can be found by taking the convolution of the initial Gaussian pulse with half width T and the Gaussian pulse which has the width $\Delta\tau$. In this case, the pulse width after transmission is:

$$\Delta\tau_{wide} = \sqrt{T^2 + (\Delta\tau)^2} \tag{3.76}$$

The amplitude is also reduced in the spread pulse after transmission[20] as follows:

$$|\psi_0|_{max} = \sqrt[4]{\frac{T^4}{T^4 + (\Delta\tau)^4}} \tag{3.77}$$

Example 3.2: Dispersion in Optical Fiber
An optical fiber has a dispersion parameter of 14 ps^2/km. A Gaussian input signal with a half period of $T = 8$ ps is passed through the optical fiber of length 20 km. Determine the output pulse width and normalized amplitude.

Answer:
Given parameters $T = 8$ ps, $(d^2\beta/d\omega^2)|_{\omega=\omega_c} = 14$ ps^2/km, length $z = 20$ km.

The pulse spread is: $\Delta\tau_{Short} = (z/T) \times (d^2\beta/d\omega^2)|_{\omega=\omega_c} = (20\ (\text{km})/8\ (\text{ps})) \times 14\ (\text{ps}^2/\text{km}) = 35$ ps

The output pulse width: $\Delta\tau_{wide} = \sqrt{8^2 + (35)^2} = 35.9$ ps

The normalized output pulse amplitude is:

$$|\psi_0|_{max} = \sqrt[4]{\frac{8^4}{8^4 + (35)^4}} = 0.228$$

The effect of pulse spreading and amplitude reduction as explained above is defined as *the group velocity dispersion* or *chromatic dispersion*. Figure 3.10 illustrates the amplitude reduction and pulse spread.

The chromatic dispersion, which is associated with the pulse broadening, also suffers from chirping of the carrier frequency signal. A chirped signal has an instantaneous frequency that changes linearly with time. A chirped signal can be represented as

$$e^{j\omega_c t} \Rightarrow e^{j\left(\omega_c t + \frac{1}{2}t^2\frac{d\omega}{dt}|_{\omega=\omega_c}\right)} \tag{3.78}$$

The phase term is represented by $\theta(t) = \omega_c t + (1/2)t^2(d\omega/dt)|_{\omega=\omega_c}$

The instantaneous frequency is defined as the derivative of $\theta(t)$:

$$\frac{d\theta(t)}{dt} = \omega_c + \left.\frac{d\omega}{dt}\right|_{\omega=\omega_c} \times t \tag{3.79}$$

---

20 The step is called quadrature approximation. The detailed convolution of the two function is given in detail in Orfanidis, *Electromagnetic Waves and Antennas*, http://www.ece.rutgers.edu/~orfanidi/ewa/ch03.pdf (accessed 09 January 2018). pp. 95–104.

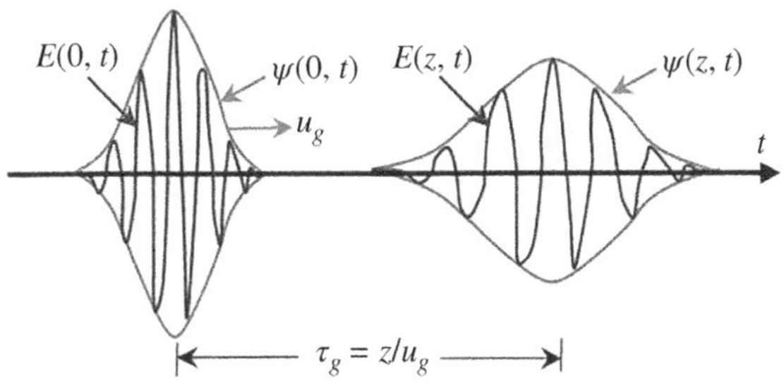

Figure 3.10 Pulse spread and amplitude reduction of envelop after transmission through the dispersive medium. Here $E(0, t)$ is the modulated carrier frequency signal and $\psi(0, t)$ is the Gaussian pulse envelop. After chromatic dispersion, due to travelling a distance z with the group velocity u_g, the broadened pulse envelop is $\psi(z, t)$ and the carrier frequency is $E(z, t)$ as illustrated. *Source:* Picture is adopted from Orfanidis, Electromagnetic Waves and Antennas, http://www.ece.rutgers.edu/~orfanidi/ewa/ch03.pdf (accessed 09 January 2018). p. 100 Figure 3.6.1.

where, *the chirping parameter* $(d\omega/dt)|_{\omega=\omega_c}$ represents the linear variation of the frequency with time around ω_c. *The chirping parameter* can be positive or negative resulting in increase or decrease in frequency. The chirped Gaussian pulse is obtained by modulating the chirped sinusoid by a Gaussian envelop as depicted in the above figure. The chirping is a natural phenomenon of the pulse broadening as the pulse broadens out, different spectral components propagate at different group velocities. Let us consider the illustration above, and define the group delay as:

$$\tau_g = \frac{z}{u_g} = z\frac{d\beta}{d\omega} = z\frac{d}{d\omega}\left\{\beta(\omega_c) + (\omega-\omega_c)\frac{d\beta}{d\omega}|_{\omega=\omega_c}\right\} = \left\{\frac{d\beta}{d\omega}|_{\omega=\omega_c} + (\omega-\omega_c)\frac{d^2\beta}{d\omega^2}|_{\omega=\omega_c}\right\}z \tag{3.80}$$

The Equation (3.80) is in the form of $y = mx + c$, a line. Where $c = (d\beta/d\omega)|_{\omega=\omega_c}$ = constant, $m = \left(d^2\beta/d\omega^2\right)|_{\omega=\omega_c}$ = constant slope. The only independent variable in the right-hand side is the frequency ω. Therefore, examination of the above equation reveals that the group delay is a linear function of the frequency. If m is positive then the frequency increases linearly with time and the lower frequency leads the higher frequency. Contrary to this if m is negative then the lower frequency lags the higher frequency. In optical fibers, the waveguide dispersion can be negative value. This means the lower order mode frequency lags the higher order mode frequency.

3.5 Wave Impedance of Uniform Plane Wave

So far, we have dealt with the one-dimensional wave equations and their solutions in time harmonic condition. In the examination of the wave velocities in the free space and in the dispersive medium we have covered a fair bit of theories in the topic of group velocity. We also have dealt with the pulse broadening effect due to the variation of the group velocities in the dispersive medium. In this section, we now examine the wave impedance for the cases for the completeness of the time harmonic field theory. First, we shall work on the solution for the time harmonic magnetic field from the known electric field solution that we have derived above. The relation between the time varying magnetic and electric fields is derived from Faraday's law. Therefore, we shall calculate the magnetic field vector using the known electric field vector and imposed the solution in Faraday's law. Once the complete solution of the electric and the magnetic fields in the time varying cases are obtained then we shall determine the wave impedance η. However, it is noteworthy to mention that the wave impedance of the time harmonic fields remains the same as for the arbitrary time carrying magnetic case that we have derived earlier. This is due to the fact that the impedance is the ratio of the electric and magnetic fields. And eventually the impedance becomes a function of the constitutive parameters ε, μ, σ, and the operating frequency f.

Using Faraday's law, we can calculate the magnetic field from the electric field as shown here:

$$\begin{aligned} \nabla \times \boldsymbol{E} &= -j\omega\mu\boldsymbol{H} \\ \boldsymbol{H} &= \frac{j}{\omega\mu}(\nabla \times \boldsymbol{E}) = \boldsymbol{a}_y \frac{j}{\omega\mu}\frac{\partial \boldsymbol{E}_x}{\partial z}. \\ \boldsymbol{H} &= \frac{j}{\omega\mu}(\nabla \times \boldsymbol{E}) = \boldsymbol{a}_y \frac{j}{\omega\mu}\frac{\partial \boldsymbol{E}_x}{\partial z} \end{aligned} \tag{3.81}$$

Substitutions for $E_x(z, t)$ in the above equation and with simple curl manipulation in determinant form[21] results in:

$$\boldsymbol{H} = \frac{j}{\omega\mu}(\nabla \times \boldsymbol{E}) = \boldsymbol{a}_y \frac{j}{\omega\mu}\frac{\partial \boldsymbol{E}_x}{\partial z} \tag{3.82}$$

Now substitution for the complete solutions of $E_x(z, t)$ yields

$$\boldsymbol{H}(z, t) = \boldsymbol{a}_y \frac{\gamma}{j\omega\mu}\left(E_0^+ e^{-\gamma z} - E_0^- e^{+\gamma z}\right) \tag{3.83}$$

or,

$$\boldsymbol{H}(z, t) = \boldsymbol{a}_y\left(H_0^+ e^{-\gamma z} + H_0^- e^{+\gamma z}\right) \tag{3.84}$$

Observe the sign changes of the reverse travelling wave of the magnetic field components in Equation (3.83). Can you explain why? In the reverse wave, the magnetic field (equivalent to current direction) has changed its direction.

The ratio of the respective electric and magnetic field components with the appropriate signs defines the intrinsic wave impedance of the medium.

$$\eta = \frac{E_0^+}{H_0^+} = -\frac{E_0^-}{H_0^-} = \frac{j\omega\mu}{\gamma} = \sqrt{\frac{j\omega\mu}{\sigma + j\omega\varepsilon}} \tag{3.85}$$

The above was also derived for the lossy medium in the previous chapter.

3.6 Polarization of Wave Fields

Figure 3.11 depicts classification of polarizations of wave fields. We have already defined the polarization of the electromagnetic field waves in the very beginning of the textbook in Chapter 1 during the introduction to the plane wave in induction topics. We have also defined polarization of the uniform plane wave and solve x-polarized electromagnetic fields. You have seen the two examples – the orientations of the propagating electric field vector of a dipole antenna, and the electric field distribution of the guided wave on the aperture of the rectangular waveguide in the beginning of the chapter. In both cases, we have seen that the orientation of the electric field vector with time

21 We have derived the magnetic field using the rectangular coordinate system of curl operation in the previous chapter. There we obtain only magnetic field component due to the electric field component.

is along the x-axis. This means that the polarization of the electromagnetic field is x-directed. The orientation of the electric field vector with time is called *polarization* of the electromagnetic field. To understand the polarization in physical sense, in this case we assume the travel distance is frozen at a wave plane and the tip of the electric field vector is tracing a locus and the geometrical shape of the locus, defines the type of polarization.[22] Such conceptual definition is illustrated in Figure 3.12. What are the polarizations of the wave for both example? In both cases the polarizations of the waves are linearly polarized in x-orientation. This means they are called *x-polarized waves*. Since the fields in both cases traces a line, they are called *linearly polarized* (LP) wave and both the devices are called the *LP devices*. Again, the antenna is vertically oriented with respect to the ground plane, therefore, the antenna is called *vertically polarized* antenna. In both cases, the tips of the electric field vectors meaning the amplitude variation of the electric field vectors are always along the vertical direction with respect to the earth. If the antenna is rotated at 90° and made horizontally oriented with the ground then the electric field is also oriented along the horizontal direction all the time and the antenna is called *the horizontally polarized antenna*. Therefore, we define the devices[23] based on the polarization of the field waves that the devices launch or convey. As for example, LP and circularly polarized (CP) antennas launch (radiate) and also receive the linearly and circularly polarized signals, respectively. In the cases of both the antenna and the waveguide are LP and the direction of propagation is z. If the receiving and the transmitting antennas or the components in a system are in the same polarization then they are called *the polarization matched* devices. The polarization match components receive and transmit signals without any excess loss due to the polarization mismatch.[24] If any polarization mismatch happens between the coupling components in a system then the polarization mismatch results in the excess loss in the system. This loss is quantified as *the polarization loss factor*. *The polarization loss factor* is defined as:

$$PLF = cos^2\theta \tag{3.86}$$

where θ is the angular spread between the field vectors. Assume that a transmitting LP antenna has a radiated electric field has zero-degree angle with respect to its LP receiving counterpart. This means the *PLF* is 1 and there is no excess loss due to the polarization mismatch in the reception. However, if the two LP antennas are oriented 90° with respect to each other then, the polarization loss factor is zero. This means that ideally no signal communication happens in between the two antennas. Therefore, understanding the polarization of the signal is not only very significant for the fundamental physical point of view of the characteristic of the electromagnetic wave signals but also equally important in the practical design aspects of the wireless communications channels and their link budget calculations. In the next section, we shall study the polarization of plane wave in more detail to fulfill both objectives.

The uniform plane wave refers to that type of electromagnetic waves in which there is no component of both electric and magnetic fields in the direction of the wave propagation. Since the electric and magnetic fields are orthogonal to each other in a uniform plane wave and also to the

---

22 In optics, some instances of the orientation of the magnetic field vector are also considered as polarization. It is important to point out that the polarization of the electromagnetic fields that we are presenting here is very different than the polarization of the dipole moments in dielectric and other material bodies. The definition of the polarizability is a fully different concept that is linked with the dipole moments and the force acting between different polarity charge bodies.

23 In this case, the dipole antenna and waveguides are considered. This can be extended to any device such as optical fibers, resonators, etc.

24 Note that there are path losses due to the intensity reduction with distance in guided media such as transmission lines and waveguides. There are also mismatch losses when the devices and components are not impedance matched. We shall examine such losses in the development of wave incident and reflection at the discontinuity of two media and during interfacing of devices of different impedances.

direction of wave propagation, the wave is also called *transverse electromagnetic* (TEM) waves. If the boundary conditions of the device or the component that carries the electromagnetic waves are such that they can support waves with the electric and magnetic field components are in the direction of the wave propagation then the wave is no more TEM waves. A good example is a metallic waveguide. In a TEM wave, both longitudinal electric and magnetic field components are zero if we assume the direction of wave propagation is along z-axis. Depending on the direction of the electric and magnetic field vectors with respect to the direction of the wave propagation, there are other two different types of wave fields. If the electric field is traverse meaning no component of the electric field is along the direction of wave propagation then the wave is called *transverse electric* (TE) wave. As for example, the electric field is transverse to the direction of propagation, say z-direction, then the z-directed field component $E_z = 0$, but E_x, E_y, and all components of the magnetic field viz H_x, H_y, and H_z are not zero. If the magnetic field is traverse meaning no component of the magnetic field is along the direction of wave propagation then the wave is called *transverse magnetic* (TM) wave. As for example, the magnetic field is transverse to the direction of propagation, say z-direction, then the z-directed magnetic field component $H_z = 0$, but H_x and H_y are not zero. Likewise, all electric field components – E_x, E_y, and E_z – are not zero. Usually waveguides cannot support TEM waves due to their boundary conditions that they imposed on the propagating field vectors.[25] In waveguide nomenclature, we define the field types as *modes*. As for example, we call TE and TM modes. Besides the three main propagation modes TEM, TE, and TM of the electromagnetic waves, we also define a mode that is called *hybrid mode*. When there are nonzero electric and magnetic field components in the direction of propagation we called the mode as the *hybrid mode*. We shall discuss the mode fields in the waveguide theory in more details.[26]

In the above discussion, we have defined the various types of the field waves based on their directions of amplitude variations – tips of the electric and magnetic field vectors with respect to the direction of wave propagation. However, the concept of *wave polarization* is somewhat different than the various *modes* of field waves. Figure 3.11 illustrates the classification of various types of *polarizations* which will be covered in the following subsections. The polarization of the field

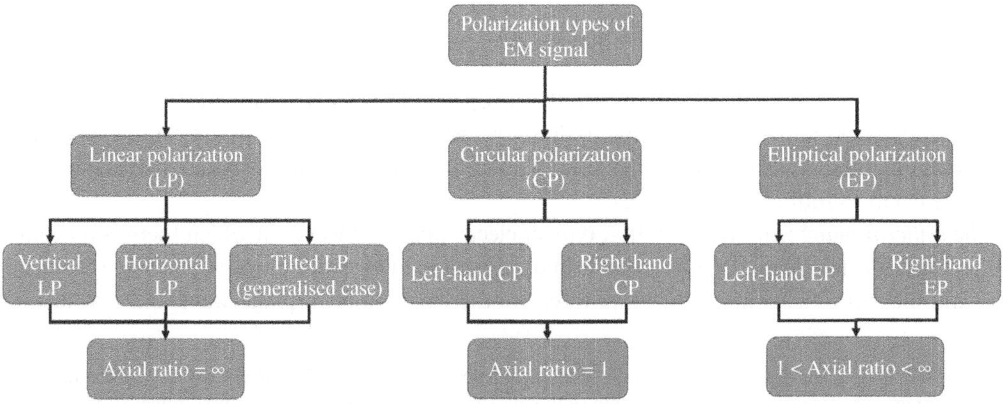

Figure 3.11 Classification of polarizations of electromagnetic uniform plane wave.

25 In waveguide field analysis, we shall use this boundary conditions and definition of the field wave modes conveniently to find the solutions of the fields inside the waveguide. As for example, we shall find the solution for E_z for the TM mode field first and then find the solutions for the other field components, later.

26 Besides the three main propagation modes in waveguides we also define a mode is called *hybrid modes* when there are non-zero electric and magnetic field components in the direction of propagation. We shall discuss the mode fields in the waveguide theory.

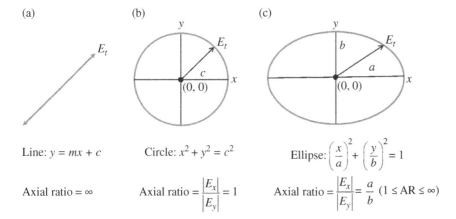

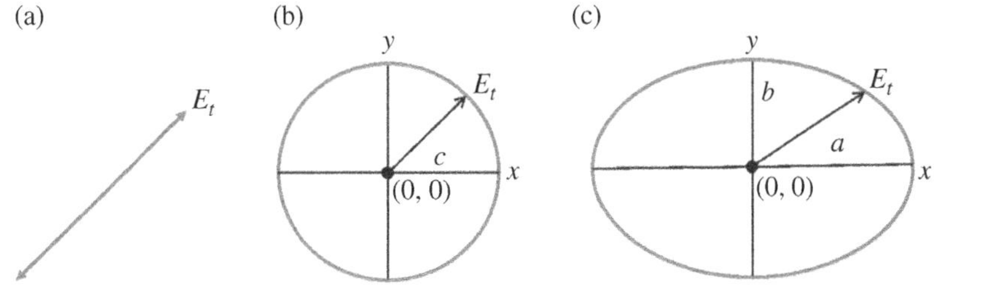

Line: $y = mx + c$ | Circle: $x^2 + y^2 = c^2$ | Ellipse: $\left(\frac{x}{a}\right)^2 + \left(\frac{y}{b}\right)^2 = 1$

Axial ratio $= \infty$ | Axial ratio $= \frac{|E_x|}{|E_y|} = 1$ | Axial ratio $= \frac{|E_x|}{|E_y|} = \frac{a}{b}$ $(1 \le \text{AR} \le \infty)$

Figure 3.12 Different traces of resultant field vector E_t and their field components and resulting axial ratios: (a) linear, (b) circular, and (c) elliptical polarization.

wave is solely determined with the orientation of the electric field wave vector with time as it propagates in the medium. In this case, we freeze the distance and place a polarization screen (wavefront) as shown in Figure 3.13. Based on the orientation of the electric field with time, we define three types of polarizations – linear, circular, and elliptical. In general, all field waves in any arbitrary orientations can be decomposed into two orthogonal field components. Based on the amplitude and the phase differences of the two orthogonal field components, we define the different polarizations of the wave fields. As for shown in Figure 3.12a the wave field has only a field component which is transverse to the direction of the wave propagation then we call the wave as the linearly polarized wave. The linearly polarized wave can also be made with two orthogonal field components with different amplitudes but equi (0° phase difference) or (180° phase difference) out of phase. In this case, the resultant field still traces a line but with an angle depending on the amplitudes of the two orthogonal contributing field components. As shown in Figure 3.12b a wave is called circularly polarized wave if the two orthogonal field components, that make the resultant circularly polarized field wave E_t, have equal amplitude and $\pm 90°$ phase difference between them. If the individual contributing field components are $\pm 90°$ out of phase with respect to each other they, individually, are called the orthogonally polarized waves. We can create such orthogonally polarized discrete field components with an orthogonally polarized antenna feeding the antenna with 90° spatial and electrical phase differences. We shall discuss orthogonally polarized antennas in the later chapters of the book. Finally, about the circular polarization, depending on the lagging and leading phase differences between the orthogonal, individual field components, the resultant circularly polarized field wave has handedness – left- and right-hand circular polarizations. We shall cover these aspects in more details with illustration in the following subsection. Finally, about the classification of the polarizations of the field waves, the last one is *the elliptically polarized* wave as shown in Figure 3.12c. Any wave in general is *elliptically polarized, the linear* and *circular polarizations* are the special cases of *the elliptically polarized* waves. The definitions of the former two types have explained the special cases. An *elliptically polarized* resultant wave can have the two individual contributing field components with any amplitudes and phase differences except the conditions for the linearly and circularly polarized waves. This means, the wave components can have any amplitudes and any phase differences except 0, 90, and 180°. Finally, a critical parameter about the polarization of the field waves is *the axial ratio.* The definition of an *axial ratio* is better perceived if we visualize the configuration of an ellipse of its major arm representing one field component, say E_x field, and the minor arm representing the other orthogonal field component say, E_y field. *The axial ratio* is defined as the ratio of the two field components. Therefore, we are in a position to define the *axial ratio* (AR) of the three types of polarizations based on the geometrical shapes of the resultant field vectors' tips which trace the loci. Figure 3.12 illustrates the three scenarios of LP, CP, and EP, respectively. As can be seen, for the linearly polarized wave, the field traces a line hence the axial ratio is infinity. For the circularly polarized wave, the axial ratio is one and for the elliptically polarized wave the axial ratio ranges from one to infinity ($1 \leq AR \leq \infty$). In the following subsection, we are discussing the three different types of polarizations of the field waves in details.

3.6.1 Linearly Polarized Waves

In the preceding sections of the chapter we have derived one-dimensional wave solutions for the uniform plane wave. In this case we assume that the electric field orientated along x-axis, the magnetic field along y-axis and the direction of power propagation is along z-axis. During the discussion of the three-dimensional wave equations in the preceding chapter, we have decomposed the wave fields into three sets of waves (x, y, z) to simplify the analysis of the electromagnetic fields so that the individual sets of field wave solutions can be vectorially synthesized to get the exact solution of the resultant field waves. Therefore, we can deduce that the field wave can be oriented at any direction and can propagate also at any direction.[27] We have also studied the group velocity of a group of waves in significant details in Section 3.6. We have seen that although the group of waves propagates in the same directions, the field vectors may orient in any arbitrary direction, with arbitrary amplitude and phase, even though they are lying transverse to the direction of propagation. In this case, the wave is called *unpolarized.* However, we are more interested in the defined direction of polarization of the wave for the simplicity of the analysis of the waves. Here we always stick to the orientation of the electric field vector as the exact definition of polarization of the electromagnetic wave.

See the Figure 3.13 for the definition of a linearly polarized wave. The wave is described by the equation:

$$\boldsymbol{E}(z, t) = \boldsymbol{a}_x E_0^+ \cos(\omega t - \beta z) \tag{3.87}$$

The one-dimensional vector field wave equation of the instantaneous electric field is a function of both distance z and time t. In this case, the electric field is oriented[28] (directed) along x-axis and propagates along z-axis. Now we know the field wave is travelling along $+z$-direction as we have a $-$ve sign in front of spatial phase factor β_z.[29] The polarization of a field wave is defined with the orientation of the electric field vector with respect to time.[30] Polarization describe the path taken by

27 Though in the uniform plane wave we assume that the wave propagates along the orthogonal directions of the electric and magnetic fields, the uniform plane wave, later on we shall analyse the fields in the waveguide and shall find the fields can be oriented in the direction of the propagation. The uniform plane wave is called the transverse electromagnetic wave, or the TEM wave in short. In waveguide, the fields are excited in such a way that due to the boundary conditions of the field waves inside the waveguide, either the electric field or the magnetic field only is transverse to the direction of the wave propagation. In these cases, they are called transverse electric field (TE) or transverse magnetic (TM) field cases. Thus, in general, waves are classified into three groups: TEM, TE, and TM. There are combinations also possible in the complex vector analysis of the fields in complex devices where the boundary conditions of the field waves in the devices are loosely defined. In such cases they are called hybrid TEM, hybrid TE and hybrid TM waves.

28 The direction of the field orientation means the amplitude of the time harmonics (sinusoidally varying) field varies only in that direction.

29 The spatial phase factor βz is also called *the electrical length*. We shall be using this nomenclature in the analysis of the transmission lines.

30 You must consider the associated magnetic field and its direction for the electromagnetic wave. The direction of the magnetic field can be calculated once both the orientation of the electric field and the direction of power wave propagation are known. For example, the generic definition of the magnetic field calculations for the known electric field and power wave direction is: $\boldsymbol{H} = (1/\eta)\boldsymbol{a}_p \times \boldsymbol{E}$ where, η is the wave impedance, $\boldsymbol{a}_p$ is the unit vector of the power wave and $\boldsymbol{E}$ is the electric field vector.

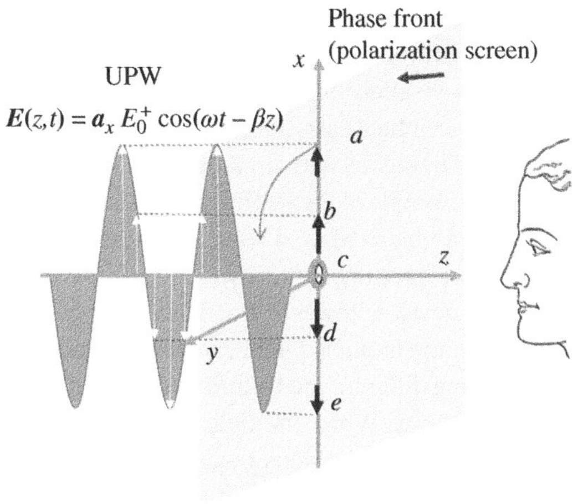

Figure 3.13 A polarized uniform plane wave $E(z, t) = \mathbf{a}_x E_0^+ \cos(\omega t - \beta z)$ traces a line along x-axis at $z = 0$ phase front at different instances of time during one cycle. A person looking toward the direction of propagation in this case z-axis sees a trace of line $a \to b \to c \to d \to e$ in the half cycle and then as opposite direction $e \to d \to c \to b \to a$ and then continue to going on and on the same linear trace the wave continues propagating. Here, the right-hand rule is applied for the coordinate system as shown with curvy arrow and the direction of propagation is along z-axis. This is equivalent to curl your four fingers toward your palm and the thumb points out toward the direction of propagation, which is in this case z-axis.

the tip of the electric field intensity vector on a fixed orthogonal plane, which is basically the phase front. In this case, $z = 0$ plane. This is plotted against time. If you see the phase front on an oscilloscope you will see that the wave vector is tracing a line with time. This is called linear polarization. In this case, x-polarized vector field. See the illustration. A person is looking at the phase front (screen) at $z = 0$ location. The field is launching as a one-dimensional wave with the tip of the electric field is oriented along x-axis (vertical axis) and propagating the field along the direction of propagation z-axis. As time progresses the electric field vector tip is at locations $a \to b \to c \to d \to e$ in the first half cycle and then $e \to d \to c \to b \to a$ in the second half cycle and comes back to the original position at "a." The person who looks at the phase front (screen) can only see a trace of line that moves back and forth in between points a and e.

You may notice that in the above example we only consider lossless condition without the exponential attenuation factor $e^{-\alpha z}$. We also consider the forward travelling wave only. The attenuation factor is common to all field components and does not affect the polarization. Since the field wave does not change its orientation and direction of the propagation with the loss factor[31] the polarization of the field wave remains the same irrespective of the loss properties of the media as long as the medium is considered homogeneous, linear, and isotropic. Likewise, we also consider the positively z-directed travelling wave. Since the general solution of a field wave has both positive and negative z-directed travelling waves, the direction of propagation will not affect the polarization of the wave as long as the medium is linear, homogeneous, and isotropic. Therefore, we can omit the attenuation factor $e^{-\alpha z}$ in our analysis.

Look at the animation of the tip of the vector field. As you can see in the 2D plane at $z = 0$ (xy-plane) the electric field vector tipping along x-direction is going downward and upward as it travels along the sine wave envelop and the direction of propagation. Now assume that you are looking at the wave, you will see a line going downward and upward as the time progresses meaning the wave propagates with time. The field vector is tracing a line all the time. Now imagine a

31 We consider $e^{\pm\alpha z}$ based on the direction of the field wave, however, the direction of propagation does not affect the field components, because in the reverse travelling wave the attenuation factor is the same as for the forward travelling wave due to the fact that for reverse travelling wave the distance z is increasing negatively hence brings the same result as for the forward travelling wave.

wave with an operating frequency of 1 GHz,[32] it will travel up and down so fast that you will see only a line on the phase plane. It is the linearly polarized wave in short LP wave.[33]

In the above analysis and the definition of the polarization of a field wave we only consider the one direction at field variation, which in this case, is along *x*-direction. This kind of field wave can be created and launched with an antenna if the configuration of the antenna is specifically fixed to do so. For the example of the vertically polarized dipole antenna in Figure 3.2a, we can easily perceive a well-defined polarized field wave, likewise the case for the rectangular waveguide in Figure 3.2b. Although we have the authority to launch the specific polarized waves via specifically oriented and configured devices, however, the field can be of any angle on the polarization screen.[34] This is due to the fact that the launched field can be reflected, refracted, and diffracted at discontinuities as it propagates along different media and obstacles that we always encounter in our wireless mobile phone communications. When the field waves bounce back and forth by some discontinuities and are transmitted through various media, the direction of the field vector may change its orientation.[35] Therefore, in all cases we do not have the authority to assume that in the electromagnetic system, the field wave may always be in the vertical or horizontal direction. Also, in mobile applications, the orientation of the antenna cannot be fixed and is varying continuously all the time. Therefore, we must consider a generic case of any arbitrary orientation of the electric field vector on the polarization screen. Still in this case the field wave is a linearly polarized one meaning it traces on a line on the polarization screen. In this case, the electric field must be specifically a superposition of two independent, linearly polarized, and orthogonal field components, say E_x and E_y. We use these superimposed waves in satellite communications, RFIDs and many other exciting applications so that if we miss one component then we can work with another component of the signal and the signal is not totally lost. Now we are going to analyze such arbitrary oriented electric field which is linearly polarized but creates an angle with *x*-axis as shown in Figure 3.14.

(a)

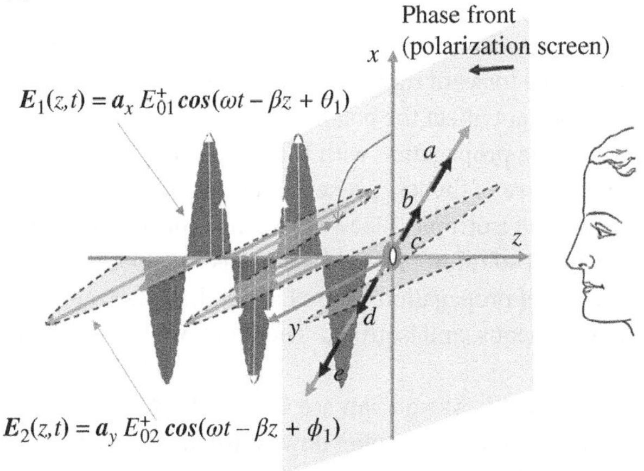

Figure 3.14 Two linearly polarized in equi-phase ($\phi x - \phi y = 0°$) or out of phase ($\phi x - \phi y = 180°$) but with different amplitudes can also form a linearly polarized signal with a tilted linear trace that creates an angle τ with *x*-axis. (a) Generic case, and (b) specific case with different amplitudes and equi- or out of phase.

32 In the GSM mobile frequency band region.

33 On an oscilloscope screen you can see a linear trace of a sign wave when you select *xy*-plot.

34 Such change of orientation or angular modulation is called refraction is electromagnetic nomenclature.

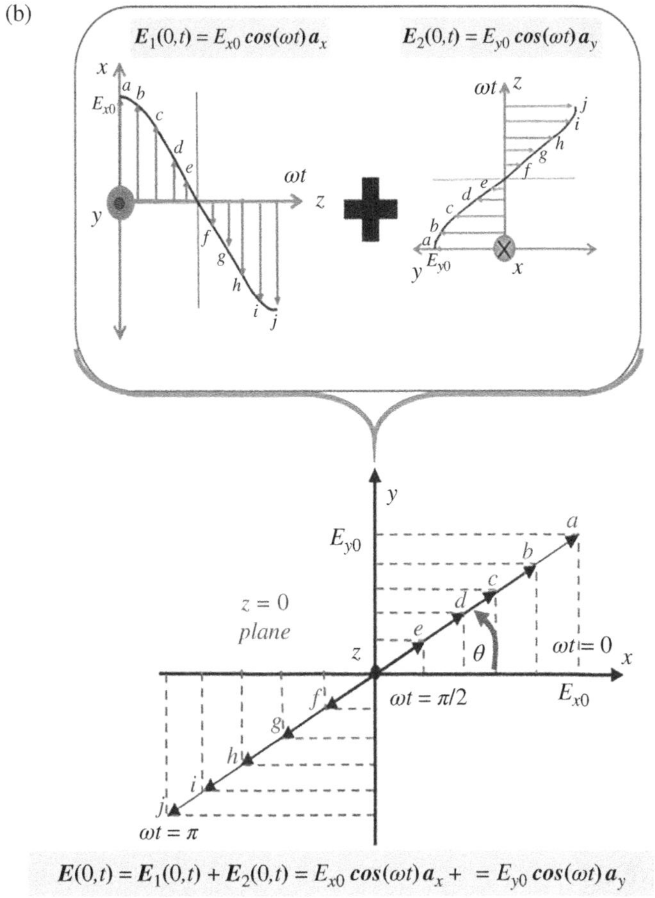

Figure 3.14 (Continued)

Review Questions 3.7: Concept of Wave Polarization

Q1. Explain why we do not consider the attenuation factor $e^{-\alpha z}$ in our analysis of the polarization of the field wave.

Q2. Does the polarization change with the direction of the field wave?

Q3. Explain why we cannot keep the polarization of the field wave fixed in mobile applications.

Q4. Explain the source of arbitrary polarization of the linearly polarized field wave.

Q5. Explain why we do not consider the magnetic field in the analysis of the field wave polarization.

Q6. How do you differentiate the field wave if the wave is arbitrary linearly polarized with respect to the well-defined vertically or horizontally polarized wave?

Q7. Explain why an arbitrary linearly polarized field wave is a superposition of the two perpendicular linearly polarized field waves.

From the above discussion we understand that in general cases, the electric field vector in a plane (polarization screen) is the superposition of two linearly independent orthogonal field waves as shown in Figure 3.14. For simplicity we consider only the forward travelling wave and omit the loss term $e^{-\alpha z}$ in our analysis. The two fields have arbitrary phase angle and different amplitudes.

The instantaneous or time domain field expression of the arbitrary-oriented fields are defined as:

$$\boldsymbol{E}_1(z, t) = \boldsymbol{a_x} E_{x0} \cos(\omega t - \beta z + \phi_x) \tag{3.88}$$

and

$$\boldsymbol{E}_2(z, t) = \boldsymbol{a_y} E_{0y} \cos(\omega t - \beta z + \phi_y) \tag{3.89}$$

The resultant field waves that we basically see in the polarization screen is the superposition of the two individual field components

$$\boldsymbol{E}(z, t) = E_1(0, t) + E_2(0, t) = E_{x0} \cos(\omega t - \beta z + \phi_x)\boldsymbol{a_x} + E_{y0} \cos(\omega t - \beta z + \phi_y)\boldsymbol{a_y} \tag{3.90}$$

Look at the expression of this superimposed UPW with x and y components with their respective phases ϕ_x and ϕ_y. As can be seen now, both are travelling in $+z$ direction with a phase constant β. The term βz represents the spatial phase variation along the direction of propagation. For simplicity, let us begin with ignoring the phases of the two field components meaning $\phi_x - \phi_y = 0°$ or $\pm 180°$. At $z = 0$ plane, in the superimposed wave is defined as:

$$\boldsymbol{E}(0, t) = \boldsymbol{E_1}(0, t) + \boldsymbol{E_2}(0, t) = E_{x0} \cos(\omega t)\boldsymbol{a_x} \pm E_{y0} \cos(\omega t)\boldsymbol{a_y} \tag{3.91}$$

If we take the ratio of the scalar fields E_1 and E_2, then we obtain:

$$\frac{E_2(0, t)}{E_1(0, t)} = (-1)^m \frac{E_{y0}}{E_{x0}} \tag{3.92}$$

or,

$$E_2(0, t) = (-1)^m \left(\frac{E_{y0}}{E_{x0}}\right) \times E_1(0, t) \tag{3.93}$$

The above equation is equivalent to an equation of a line with the general expression: $y = mx$ where, m is the slope. The line passes through the origin with a slope of m. Therefore, the above field equation represents a linearly polarized signal of which the tip of the electric field vector traces a line of slope

$$(-1)^m \left(\frac{E_{y0}}{E_{x0}}\right) \tag{3.94}$$

with $m = 0, 1, 2,...$ etc.

The locus of the tip of the electric field vector of $\boldsymbol{E}(0, t)$ on the xy-plane (at $z = 0$) traces a straight line passing through the origin with an angle θ with the x-axis given by

$$\theta = \tan^{-1}\left(\frac{E_2(0, t)}{E_1(0, t)}\right) = \tan^{-1}\left((-1)^m \frac{E_{y0}}{E_{x0}}\right) \tag{3.95}$$

where $m = 0, \pm 1, \pm 2, \pm 3, ...$

The above conditions constitute the definition of *the linear polarization*. We can examine that even though the amplitudes of the field components E_{y0} and E_{x0} change with time at an angular frequency ω, they are making a constant angle θ with the x-axis.

Look at the superimposed field expression that contains only ωt term in the parenthesis meaning only oscillating components. Figure 3.14b shows the plot of the two signals at $z = 0$ which are defined as

$$\boldsymbol{E}_1(0, t) = E_{x0} \cos(\omega t)\boldsymbol{a_x} \tag{3.96}$$

and,

$$\boldsymbol{E}_2(0, t) = E_{y0} \cos(\omega t)\boldsymbol{a_y} \tag{3.97}$$

They are pure sinusoidal wave with their respective peak amplitudes E_{x0} and E_{y0}, respectively.[35] We have taken ten instances of the orthogonal plane waves from $a \rightarrow b \rightarrow c \rightarrow d \rightarrow e \rightarrow f \rightarrow g \rightarrow h \rightarrow i \rightarrow j$ as the time progresses and then reversing their direction as the waves cover the next half cycle on the trace. The tip of the resultant electric field vector[36] tip traces the line, which in this case is an inclined line that makes an angle τ with x-axis, here τ is called the "tilt angle."[37] Using a software tool such as MATLAB, you may plot the resultant field trace for the cases when the phase differences are not zero or out of phase. We shall study such cases for circular and elliptical polarizations in the next sections. In summary, linear polarization occurs when $\phi_y - \phi_x = 0°$ or $\pm 180°$.

Let us briefly summarize the linear polarization.

1) In conclusion, the linear polarization occurs when the phase difference between the two field components are $0°$ or $\pm 180°$.
2) The tilt angle τ is the angle that is created by the trace of the resultant electric field vector with the x-axis.
3) See the cases for both 1D and 2D fields and their polarizations in Figure 3.14. The former one is a vertically polarized field wave and the latter one is a titled linearly polarized field wave with tilt angle τ.

Such a "tilted linearly polarized signal" with many such discrete tilt angles of EM wave may emanate out from a polarization diversity chipless RFID tag to increase the data capacity. This technique a changing angle of signal polarization is called *polarization diversity*. *Polarization diversity* is conveniently used in wireless communications to increase the data capacity as well.[38] Finally, the linearly polarized waves are used in AM radio broadcasting and TV transmission. While AM radio transmits and receives signal in the vertical polarization, the TV broadcasting transmitter sends signal in horizontally linearly polarized waves. You may notice olden day the car aerials are vertically polarized. The TV broadcasting antennas[39] are horizontally polarized and you may notice that the home TV antennas are placed horizontally with respect to the earth to receive the TV broadcasting signal. Can you differentiate the signal received between AM and FM radios? Usually with the same volume of your radio receiver, an AM radio is louder compared to a FM radio. Can you explain why it is so? The FM radio broadcast signal in circular polarization which is the next topic of the section. We shall discuss this phenomenon of different polarized signals in the new section of the circular and elliptical polarizations.

---

35 For the case of $\phi_x - \phi_y = 180°$, the y-field components reverse its direction and the effect on the inclined trace will be just the opposite movement on the same inclined linear trace.

36 See the resultant field is vectorially added with their respective components E_{xo} and E_{yo} at the specific time intervals in 20 steps to obtain the trace of the inclined resultant field as in Figure 3.14b.

37 See YouTube videos for such an inclined linearly polarised wave: https://www.youtube.com/watch?v=Fu-aYnRkUgg (accessed 12 January 2018).

38 N.C. Karmakar, R. Koswatta, P. Kalansuriya and E Rubayet, *Chipless RFID Reader Architecture*, Artech House Publishing, 2013.

39 Home TV antennas are called Yagi-Uda antennas after the two Japanese inventors, Yagi and Uda. It has a folded dipole in the middle of the antenna as the active element and sandwiched by the reflector dipole in the end and a series of director dipoles in the front. Japanese are very proud of the adopted TV antenna worldwide which are visible on the roof top of every household.

Review Questions 3.8: Linear Polarization

Q1. Explain how a titled linearly polarized signal is formed.
Q2. Explain what happen for $\phi_y - \phi_x = \pm 180°$. Look at the diagrams about, shift one of the field vectors by 180° and draw conclusion.
Q3. Discuss different applications of linearly polarized signal.
Q4. What types of polarizations are used for AM Radio and FM radio broadcasting and TV broadcasting?

Example 3.3: Linear Polarization
An arbitrary polarized signal is defined as:

$E(z, t) = \boldsymbol{a}_x E_{xo} \sin(\omega t - \beta z) + \boldsymbol{a}_y E_{yo} \sin(\omega t - \beta z + \phi_y)$. Explain the conditions for the wave to be defined as the linearly polarized wave. Assume $E_{x0} = 10\,(\text{V/m})$, $E_{y0} = 15\,(\text{V/m})$, and $\phi_y = 0°$, calculate the angle at which the signal will make with x-axis.

Answer:
To be a linearly polarized wave $\phi_y = 0, \pm 180°$. The angle that the signal create with the x-axis is:

$$\theta = tan^{-1}\left(\frac{15}{10}\right) = 56.3°$$

3.6.2 Circularly Polarized Waves

Next is the case for circular polarization in short CP waves. The general wave equation of the decomposed x- and y-directed field components are:

$$\boldsymbol{E}_1(z, t) = \boldsymbol{a}_x E_{x0} \cos(\omega t - \beta z + \phi_x) \tag{3.98}$$

and

$$\boldsymbol{E}_2(z, t) = \boldsymbol{a}_y E_{y0} \cos\left(\omega t - \beta z + \phi_y\right) \tag{3.99}$$

In the case for the circularly polarized field wave, the following conditions must be met:

Condition 1: $|E_{x0}| = |E_{y0}|$ (the amplitudes of the two orthogonal field components must be equal)
Condition 2: $\Delta\phi = (\phi_y - \phi_x) = m(\pi/2)$; $m = \pm 1, \pm 3, \pm 5$.

This condition means that for *the circular polarization*, the amplitudes of the two orthogonal (x- and y-directed electric fields) field components are equal and the phase difference between the vectors must be $\pm 90°$, $\pm 270°$, and $\pm 450°$, etc. Using the above conditions, we can deduce the two field components as follows:

$$\boldsymbol{E}_1(z, t) = E_0 \cos(\omega t - \beta z + \phi_x)\boldsymbol{a}_x \tag{3.100}$$

$$\boldsymbol{E}_2(z, t) = E_0 \cos\left(\omega t - \beta z + \phi_x - m\frac{\pi}{2}\right)\boldsymbol{a}_y = E_0 \sin(\omega t - \beta z + \phi_x)\boldsymbol{a}_y \tag{3.101}$$

From the above two field components we can easily deduce that the addition of the square of the normalized scalar field functions is a circle:

$$\left[\frac{E_1(z, t)}{E_o}\right]^2 + \left[\frac{E_2(z, t)}{E_o}\right]^2 = 1 \tag{3.102}$$

The above equation is equivalent to $(x/a)^2 + (y/a)^2 = 1$, which is the equation of a circle with radius a. Therefore, the composite field of the circularly polarized field wave traces a circle with amplitude $|E_0|$. This is shown in the perspective view of the circularly polarized wave propagating in the z-direction in Figure 3.15.

The resultant field vector is obtained by vectorially adding the two contributing/decomposed equal amplitude and orthogonal field components with the phase difference of $m(\pi/2)$ at $z = 0$ plane. For positive $m = 1, 3, 5$, etc. the x-directed field leads the y-directed field by 90°. The instantaneous resultant field is defined as:

$$\boldsymbol{E}(z, t) = E_0\left[\cos(\omega t - \beta z + \phi_x)\boldsymbol{a_x} - E_{y0}\sin(\omega t - \beta z + \phi_x)\boldsymbol{a_y}\right] \tag{3.103}$$

The angle between the two instantaneous vector field components determine *the handedness* of the circularly polarized field:

$$\tan\theta = \frac{|E_2|}{|E_1|} = -\tan(\omega t - \beta z + \phi_x) \tag{3.104}$$

or,

$$\theta = -(\omega t - \beta z + \phi_x) \tag{3.105}$$

In this case, the resultant field is right-hand circular polarized (*RHCP*) signal. For the left-handed circular polarized (*LHCP*) signal, the phase angle is:

$$\theta = \omega t - \beta z + \phi_x \tag{3.106}$$

(a)

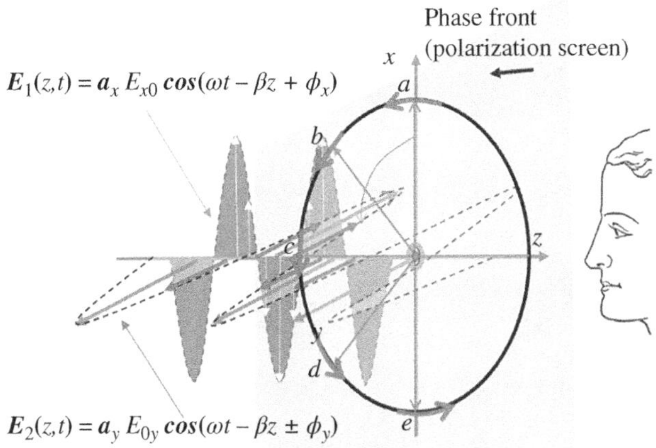

Condition 1: $|E_{x0}| = |E_{y0}|$

Condition 2: $\Delta\phi = (\phi_x - \phi_y) = m\frac{\pi}{2}$; $m = \pm 1, \pm 3, \pm 5 \ldots$

Figure 3.15 Circularly polarized wave is formed with the two orthogonally polarized field vectors that has same amplitude and phase difference by ±90°. (a) The wave traces a circle when a person sees the wavefront at $z = 0$, (b) 3D rendering of such wave propagation, and (c) analysis of two $\pi/2$ phase difference sinusoids that create a circular trace. The measure of the effectiveness of the circularly polarized signal is defined as the axis ratio which is linked with the definition of the elliptically polarized wave. The ratio of the major over minor axes is called axial ratio.

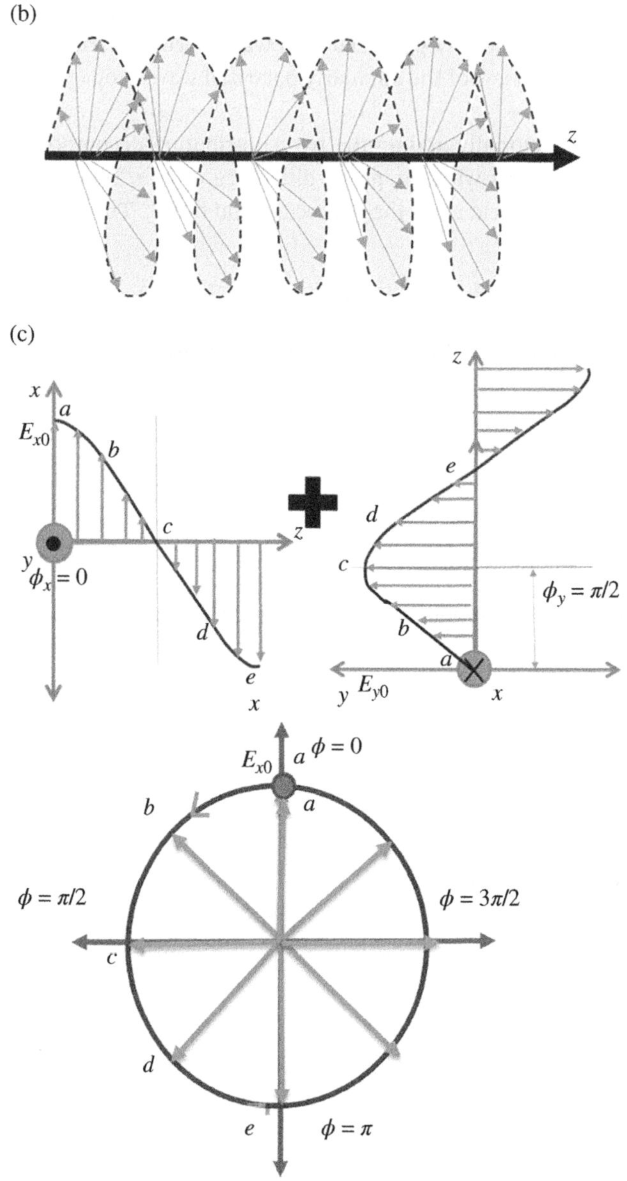

Figure 3.15 (Continued)

In this case, the y-directed field components lead the x-directed field component by 90°. We can explain the handedness of the circular polarization mathematically as follows: For the right-handed case,

$$\boldsymbol{E_1}(0, t) = E_0 \cos(\omega t)\boldsymbol{a_x} \text{ and } \boldsymbol{E_2}(0, t) = -E_0 \sin(\omega t)\boldsymbol{a_y} \tag{3.107}$$

$$\therefore \boldsymbol{E}(0, t) = E_1(0, t) + E_2(0, t) = E_0\left[\cos(\omega t)\boldsymbol{a_x} - \sin(\omega t)\boldsymbol{a_y}\right] \tag{3.108}$$

At $\omega t = 0$, we have

$$\boldsymbol{E}_1(0, 0) = E_0\boldsymbol{a}_x \text{ and } \boldsymbol{E}_2(0, 0) = 0. \text{ In this case, } \boldsymbol{E}(0, 0) = \boldsymbol{E}_1(0, 0) + \boldsymbol{E}_2(0, 0) = E_0\boldsymbol{a}_x \tag{3.109}$$

This instant of time at $t = 0$ and its equivalent time phase of $\omega t = 0$, is shown as point "*a*" on the polarization screen. At $\omega t = \pi/2$,

$$\boldsymbol{E}_1(0, t) = 0\boldsymbol{a}_x \text{ and } \boldsymbol{E}_2(z, t) = E_0\boldsymbol{a}_y \tag{3.110}$$

$$\therefore \boldsymbol{E}(0, t) = E_1(0, t) + E_2(0, t) = E_0\boldsymbol{a}_y \tag{3.111}$$

This instantaneous electric field vector is shown as point "*c*" on the polarization screen. This way we can plot the field vector's tip with increasing time instances *t* (points) of the instantaneous field with its angular frequency ω on the polarization screen.

Figure 3.15b illustrate the 3D perspective view of the individual components of the field wave with their unique conditions and the trace of the resultant field at $z = 0$ plane for different instances of time.

As shown in Figure 3.15c, if you vectorially sum up the two sinusoidally varying *x*- and *y*-directed field components in the gradual steps from $a \rightarrow b \rightarrow c \rightarrow d \rightarrow e$ ($\omega t = 0, \pi/4, \pi/2, 3\pi/4$, and π) then you observe the tip of the resultant field vector is tracing a circle either in clockwise or counter clockwise depending on the polarity of the differential phase. In the current example the *x*-directed electric field leads the *y*-directed electric field by the odd multiple of 90° and the resultant field vector's tip is tracing a circle counter-clockwise as the time progresses. This means if you turn a screw with your right hand and the field propagates toward you along the *z*-axis the resultant field traces a gradual circle counter clockwise as the time progresses.[40] This handedness is an inherent phenomenon of a circularly polarized wave. In this case, the circularly polarized wave is called *right-hand circularly polarized* (*RHCP*) wave. If the phase difference changes its polarity meaning the difference phase of the two field components are −ve odd multiple of −90° then the resultant electric field's vector tip traces a circle clockwise. This is equivalent to as if you turn a screw with your left hand, and the field propagates toward you along the *z*-axis, the resultant field traces a gradual clockwise circle as the time progresses. This type of wave is called *left-hand circular polarization* (*LHCP*) wave. In wireless communications the handedness has special significance. As they create *polarization diversity* meaning the polarization mismatch between the handedness. Therefore, theoretically you can only receive the same *handed signal* if the transmitting and receiving antennas or the coupling components in the communications system are in the same *handedness*. Contrary to this if the transmitting and receiving antennas or the coupling components in the communications system are in the opposite *handedness* then you will not receive any signal and communication stops. These phenomena of communication disruption and effective communication can be explained with the polarization loss factor that we have defined earlier. This means the subtended angle between the two opposite handed signals are the odd multiple of ±90° and therefore the *PLF* is zero.

Finally, we can write the resultant vector field in the complex from as follows:

$$\text{Right} - \text{hand circular polarization: } E = E_1 + jE_2 \tag{3.112}$$

$$\text{Left hand circular polarization: } E = E_1 - jE_2 \tag{3.113}$$

40 Note that the travel distance of the wave front is also progresses, but we freeze the wave front to see the trace of the resultant electric field vector on the fixed wave front located at $z = 0$.

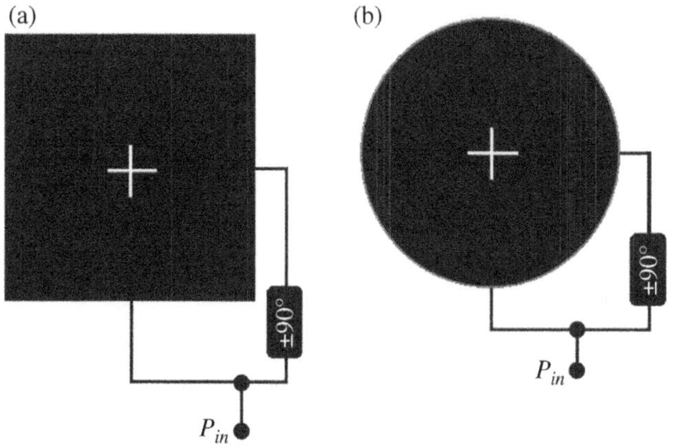

Figure 3.16 Mechanism to generate circularly polarized signals using (a) square, and (b) circular patch antennas with two orthogonal feeds with 90° delay line in one of the branches of the power divider that connects the input port (P_{in}) with the patch antenna. This feed mechanism generates degenerate modes meaning field vectors with equal amplitude with 90° phase difference. Depending on the polarity of the phase difference we can create right or left handedness of the circularly polarized signal.

3.6.2.1 Practical Design of Circularly Polarized Wave

The mechanism to create a circularly polarized signal of a particular handedness is shown in Figure 3.16. A square and a circular shaped microstrip patch antennas are fed at two orthogonally spatial feed positions with a phase difference of ±90° or the odd multiple of it. For compact design $m = 1$ is selected so that the phase difference is ±90°.[41] The compact design has also another advantage besides the compactness. The shorter length of the delay line that creates the ±90° phase difference between the two orthogonal feed ports[42] has minimum loss. With equal power division from the input port, the two orthogonal ports ensures equal amplitudes of the orthogonal electric field vectors that means $E_{x0} = E_{yo}$. This generates circularly polarized electromagnetic field wave that propagates with the trace of a circle as it progresses in space. Figure 3.15b shows the 3D rendering of the electromagnetic field waves as its progresses in both time and space.

In a circularly polarized wave, you will receive same power all the time irrespective of the orientation of the receiving antenna. The most beautiful aspect of circular polarization is its orientation independency. We shall discuss this beautiful feature of the CP wave in antennas as well as optical fiber applications.

Can you identify the direction of the arrow tip on the circle? It is going downward the plane in first half cycle and then upward the plane in second half cycle, counter clockwise. If you look at the vector tip it rotates counter clockwise. This rotation direction gives us *the handedness* of CP waves.

See the handedness of the polarization. A wave is *RHCP* if the thumb of your right-hand points in the direction of propagation, the fingers curl in the direction that the vector traces with time on the polarization plot, in this case a circle. In this case, $\phi_y - \phi_x = -90°$. See the graphical plots and tell what the handedness of the CP wave is. It is *RHCP.*

If the trace goes in other direction meaning $\phi_y - \phi_x = +90°$ then the wave is left-hand CP (*LHCP*). Handedness is also applied to elliptical polarization that will be discussed next. Also see the video as stated before.

The axial ratio is the ratio of the vertical length to the horizontal length of the field lines. The ratio is one for the circle. The axial ratio is 1 for a CP wave. *The axial ratio bandwidth* is one of the most prominent design parameters for a CP antenna for satellite communications. In industry standard, it is the bandwidth over which the axial ratio drops by 3 dB.

---

41 Phase differences of 90° is created with a microstrip transmission line of a quarter wavelength.

42 Note that in microwave engineering nomenclature, the port is a terminal pair and usually the terminals are match to a 50 Ω impedance. One terminal carries the signal and other terminal is the return path of the current to the ground.

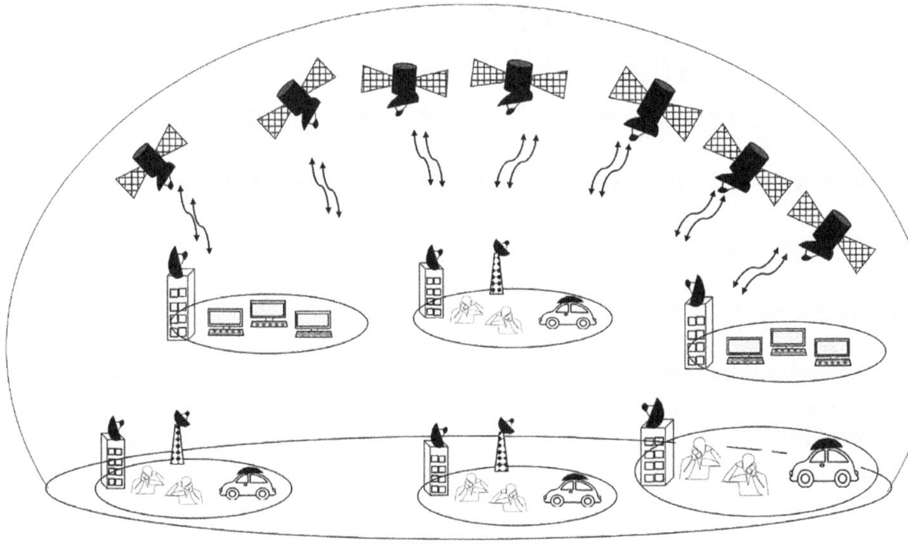

Figure 3.17 The illustrations depict the applications of CP waves in modern wireless communications. Optus Mobilesat/GPS use *RHCP* wave for communications. In outer space, ionized layer called Clark belt that rotates EM field's direction. Many emerging wireless communications will use CP wave, e.g. 5G communications, mm-wave chipless RFID is using CP wave to provide short distance but orientation insensitive communications.

3.6.2.2 Applications of CP Waves

Circular polarization is used in FM radio and satellite communications. Figure 3.17 shows the futuristic radio communications using circularly polarized waves. The modern and the futuristic wireless communications systems exploit the orientation independent characteristic of the field waves for efficient communications. In the satellite communications paths between the ground terminals and the satellite transponders,[43] an ionized layer exists in upper atmospheric layer which is called ionosphere. Ionized gas molecules fill up the atmospheric layer. Scientists say that the color the of ionized gas is blue that is why the sky on a clear day looks blue. This ionized layer is also called *Clark belt* after the name of Arthur C. Clark, a British origin Sri Lankan Scientist. In 1945, he proposed satellite communications. Due to the rotation of EM wave when it passes through the Clark belt the signal changes its orientation due to the interaction with the ionosphere. CP EM wave is usually used for satellite communications.[44] The author has designed L-band Optus Mobilesat antenna at parts of his PhD studies and industry R&D projects. In RFID we also use CP wave so that the RFID scanner can read the tag at any orientation. How wonderful this is! The hind side of CP wireless communications is that you may lose 50% power if you receive the CP signal with a linearly polarized antenna. Explain why the half power reduction seeing the trace diagram. Answer the following review questions in relation to CP EM wave.

43 Transponder is the short form of transmitter and responder. It is an on-board microwave transceiver circuit with a big dish antenna. In modern satellites, low cost conformal microstrip patch antennas replaces the bulky dish antenna. Although the wind drug of such antenna is zero in outer space, it is always convenient to use conformal low profile low cost microstrip patch antenna with equivalent performance of that for a dish antenna.

44 Otherwise, the ground antenna needs to adjust manually to receive the best signal.

Review Questions 3.9: Circularly Polarized Wave

Q1. What is axial ratio? What is the axial ratio bandwidth for a CP wave?
Q2. Why satellite communications use CP EM wave in transmission and reception?
Q3. Why RFID technology use CP EM waves?
Q4. Research wireless technologies where CP EM wave are the backbone.
Q5. Derive $[(E_1(z, t))/E_o]^2 + [(E_2(z, t))/E_o]^2 = 1$

3.6.3 Elliptical Polarization

In the above sections we have comprehensively cover the theories of *the linear* and *circular* polarizations of the uniform plane wave. There, we have concluded that a *linear polarized wave* may have both *x*- and *y*-components with phase difference 0° or ±180°. For *the circular polarization*, the amplitudes of the two orthogonal field components have the same amplitudes and the difference in their phase angles is ±90°. The circular polarization also has handedness such as *LHCP* and *RHCP*. These are the special cases of wave polarizations. The generic polarization can have the phase difference in the two field components at any angle except the special cases of 0°, ±90°, or ±180°. The amplitudes can have the same or different. This generic polarization that does not satisfy the conditions for *the linear* and *circular* polarizations is called *the elliptical polarization*. In other words, in general the polarization of the uniform plane wave is *elliptical* and *the linear* and *circular* polarizations of the wave are the special conditions of *the elliptical polarization*. In this section we are going to examine this hypothesis from the very basic definition of the two orthogonal field components. The instantaneous expressions of the two orthogonal field components are:

$$\mathbf{E}_1(z, t) = \boldsymbol{a}_x E_{x0} \cos(\omega t - \beta z + \phi_x) \tag{3.114}$$

and

$$\mathbf{E}_2(z, t) = \boldsymbol{a}_y E_{y0} \cos(\omega t - \beta z + \phi_y) \tag{3.115}$$

where the symbols have their usual meanings. Our objective is to show that the magnitude of the resultant field $|\mathbf{E}(z, t)| = \sqrt{\{\mathbf{E}_1(0, t)\}^2 + \{\mathbf{E}_2(0, t)\}^2}$ traces an ellipse. The normalized field components of the individual orthogonal field components are:

$$E_{xn} = \frac{E_1(z, t)}{E_{x0}} = \cos(\omega t - \beta z + \phi_x) \text{ and } E_{yn} = \frac{E_2(z, t)}{E_{0y}} = \cos(\omega t - \beta z + \phi_y) \tag{3.116}$$

Here, the phase angles ϕ_x and ϕ_y represent their initial conditions of the two field waves, respectively. Let the phase angle of the wave propagation along *z*-axis with time $\theta = \omega t - \beta z$, which is the variable part in the phase components with the two independent variables *t* and *z*. Then the normalized field components can be defined as:

$$E_{xn} = \cos(\theta + \phi_x) \text{ and } E_{yn} = \cos(\theta + \phi_y) \tag{3.117}$$

Now multiplying the *x*-directed normalized field component E_{xn} with $\sin\phi_y$ and the y-directed normalized field component E_{yn} with $\sin\phi_x$ and then subtracting them we obtain:

$$E_{xn}\sin\phi_y = \cos(\theta + \phi_x)\sin\phi_y = \cos\theta\cos\phi_x\sin\phi_y - \sin\theta\cancel{\sin\phi_x}\sin\phi_y \tag{3.118}$$

$$E_{yn}\sin\phi_x = \cos(\theta + \phi_y)\sin\phi_x = \cos\theta\cos\phi_y\sin\phi_x - \sin\theta\cancel{\sin\phi_y}\sin\phi_x \tag{3.119}$$

$$- \qquad\qquad\qquad - \qquad\qquad\qquad +$$

$$E_{xn}\sin\phi_y - E_{yn}\sin\phi_x = \cos\theta\left(\cos\phi_x\sin\phi_y - \cos\phi_y\sin\phi_x\right) = \cos\theta\sin\left(\phi_y - \phi_x\right) = \cos\theta\sin\delta \tag{3.120}$$

where, the phase difference between the two signal components $\delta = \Delta\phi = \phi_y - \phi_x$ is used.

Likewise multiplying the x-directed normalized field component E_{xn} with $\cos\phi_y$ and the y-directed normalized field component E_{yn} with $\cos\phi_x$ and then subtracting them we obtain:

$$E_{xn}\cos\phi_y = \cos(\theta + \phi_x)\cos\phi_y = \cos\theta\cancel{\cos\phi_x}\cos\phi_y - \sin\theta\sin\phi_x\cos\phi_y \tag{3.121}$$

$$E_{yn}\cos\phi_x = \cos(\theta + \phi_y)\cos\phi_x = \cos\theta\cancel{\cos\phi_y}\cos\phi_x - \sin\theta\sin\phi_y\cos\phi_x \tag{3.122}$$

$$- \qquad\qquad\qquad - \qquad\qquad\qquad +$$

$$E_{xn}\cos\phi_y - E_{yn}\cos\phi_x = \sin\theta\left(\sin\phi_y\cos\phi_x - \sin\phi_x\cos\phi_y\right) = \sin\theta\sin\left(\phi_y - \phi_x\right) = \sin\theta\sin\delta \tag{3.123}$$

Therefore, we have obtained two orthogonal equations as follows:

$$E_{xn}\sin\phi_y - E_{yn}\sin\phi_x = \cos\theta\sin\delta \tag{3.124}$$

$$E_{xn}\cos\phi_y - E_{yn}\cos\phi_x = \sin\theta\sin\delta \tag{3.125}$$

Now we square the above two equations and then add them together:

$$\left(E_{xn}\sin\phi_y - E_{yn}\sin\phi_x\right)^2 + \left(E_{xn}\cos\phi_y - E_{yn}\cos\phi_x\right)^2 = \cos^2\theta\sin^2\delta + \sin^2\theta\sin^2\delta \tag{3.126}$$

or,

$$E_{xn}^2\left(\sin^2\phi_y + \cos^2\phi_y\right) + E_{yn}^2\left(\sin^2\phi_x + \cos^2\phi_x\right) - 2E_{xn}E_{yn}\left(\sin\phi_x\sin\phi_y + \cos\phi_x\cos\phi_y\right) = \sin^2\delta\left(\sin^2\theta + \cos^2\theta\right) \tag{3.127}$$

or,

$$E_{xn}^2 + E_{yn}^2 - 2E_{xn}E_{yn}\cos\left(\phi_y - \phi_x\right) = \sin^2\delta \tag{3.128}$$

or,

$$E_{xn}^2 + E_{yn}^2 - 2E_{xn}E_{yn}\cos\delta - \sin^2\delta = 0 \tag{3.129}$$

Note that in the above equations, $E_{xn} = (E_1(z,t)/E_{x0})$, $E_{yn} = (E_2(z,t)/E_{0y})$ and $\delta = \phi_y - \phi_x$ are used. Now we can write a general *conic equation* as follows:

$$AE_1^2 + BE_1E_2 + CE_2^2 + D = 0 \tag{3.130}$$

where, $A = 1/E_{x0}^2$, $B = 2\cos\delta/E_{x0}E_{y0}$, $C = 1/E_{y0}^2$ and $D = -\sin^2\delta$ are used.

The generic equation of a *conic intersection* with the independent variables (x, y) can be written as:

$$ax^2 + bxy + cy^2 = d \tag{3.131}$$

that represents a *conic section* created by the curve generated with the intersection of a plane and a cone at different angles as shown in the following Figure 3.18. As can be seen in the figure, the

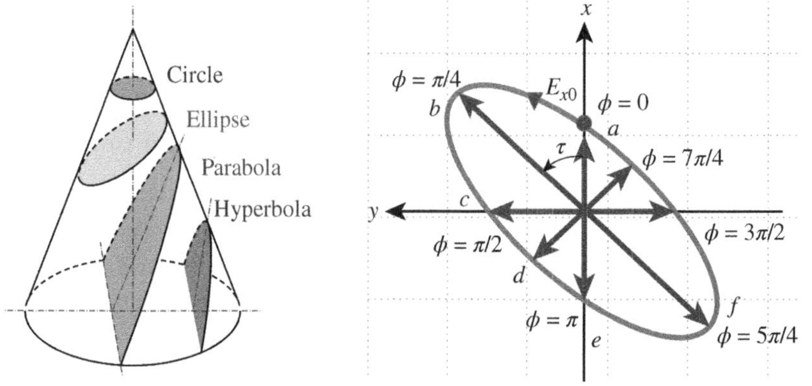

Figure 3.18 Intersections of a plane on the surface of a cone trace circle, ellipse, parabola and hyperbola as the angle of the intersection increases. *Source:* https://en.wikipedia.org/wiki/Conic_section (accessed 14 January 2017).

contour that the plane traces on the surface of a cone has different configurations depending of the angle of the intersection. If the angle between the plane and the cone is zero the intersection traces a circle and with the increase of angle we obtain an ellipse, a parabola and a hyperbola. The discriminant of the above equation determines the different configurations. The conditions are:

$$b^2 - 4ac < 0(-\text{ve}) \text{ is an ellipse}$$
$$b^2 - 4ac = 0 \text{ is a parabola}$$
$$b^2 - 4ac > 0(+\text{ve}) \text{ is a hyperbola}$$

In the generic conic equation, the independent variables are:

$$x = E_1, \tag{3.132}$$

$$y = E_2, \tag{3.133}$$

$$a = \frac{1}{E_{x0}^2}, \tag{3.134}$$

$$b = \frac{2\cos\delta}{E_{x0}E_{y0}}, \tag{3.135}$$

$$c = \frac{1}{E_{y0}^2} \tag{3.136}$$

and

$$d = -\sin^2\delta \tag{3.137}$$

Substitutions for the coefficients we obtain the discriminant as:

$$b^2 - 4ac = \frac{4\cos^2\delta}{E_{x0}^2E_{y0}^2} - \frac{4}{E_{x0}^2E_{y0}^2} = -\frac{4\sin^2\delta}{E_{x0}^2E_{y0}^2} \tag{3.138}$$

This is a negative quantity as long as $\sin\delta \neq 0$ and $\delta = \Delta\phi = \phi_y - \phi_x$ is a real quantity. Therefore, it can be concluded that the orthogonal field components E_1 and E_2 with arbitrary amplitudes and phases trace an ellipse. In other words, the locus of the vector $\boldsymbol{E}(0, t)$ having components E_1

and E_2 is an ellipse on the xy-plane ($z = 0$) provided that $\sin \delta \neq 0$. Therefore, the electromagnetic field wave that satisfies the above relationship in between the field components is called an *elliptically polarized wave*. An ellipse in a rectangular coordinate system is defined as:

$$\left(\frac{x}{a}\right)^2 + \left(\frac{y}{b}\right)^2 = 1 \tag{3.139}$$

where, a is the major axis and b is the minor axis. Now we are going to prove that the field components can satisfy the above equation. If $\cos\delta = 0$ then $\delta = (\pi/2)$. With this condition the above equation reduces to:

$$E_{xn}^2 + E_{yn}^2 = 1 \tag{3.140}$$

Substitutions for $E_{xn} = E_1(z, t)/E_{x0}$, $E_{yn} = E_2(z, t)/E_{0y}$, we obtain:

$$\left(\frac{E_1(z, t)}{E_{x0}}\right)^2 + \left(\frac{E_2(z, t)}{E_{0y}}0\right)^2 = 1 \tag{3.141}$$

This means we have an ellipse having its center is at the origin of the xy-plane with major axis of length $2E_{x0}$ and the minor axis of length $2E_{y0}$. For $\cos\delta \neq 0$, the ellipse has its major and minor axes align them at a tilt angle from OA and OB direction. The conditions for the *handedness* depend on the phase difference $\delta = \phi_y - \phi_x$. The conditions are:

$\sin\delta = \sin(\phi_y - \phi_x) > 0$ for *RHCP* or clockwise rotation
$\sin\delta = \sin(\phi_y - \phi_x) < 0$ for *LHCP* or counter clockwise rotation

In general, the curve traces at a given position (polarization screen) on the xy-plane in general is a titled ellipse as shown in the following Figure 3.19. The ratio of the major axis to the minor axis is referred to as *the axial ratio* in short *AR* and is defined as:

$$AR = \frac{\text{major axis}}{\text{minor axis}} = \frac{OA}{OB} \tag{3.142}$$

where,

$$OA = \sqrt{\left[\frac{1}{2}\left(E_{xo}^2 + E_{y0}^2\right) + \sqrt{E_{xo}^4 + E_{y0}^4 + 2E_{xo}^2 E_{y0}^2 \cos(2\delta)}\right]},$$

$$OB = \sqrt{\left[\frac{1}{2}\left(E_{xo}^2 + E_{y0}^2\right) - \sqrt{E_{xo}^4 + E_{y0}^4 + 2E_{xo}^2 E_{y0}^2 \cos(2\delta)}\right]}$$

Axial ratio is always a positive quantity within the limit of $1 < AR < \infty$.[45] The tilt angle is defined as:

$$\tau = \frac{1}{2}\tan^{-1}\left[\frac{2E_{xo}E_{y0}}{E_{xo}^2 - E_{y0}^2}\cos(\delta)\right] \tag{3.143}$$

When the ellipse is aligned with the principal x and y axes, the tilt angle is defined as: $\tau = n\pi/2$ for $n = 0, 1, 2, 3...$ the major or minor axis are equal to E_{xo} or E_{yo} and the axis ratio is equal to $AR = E_{xo}/E_{y0}$ or E_{yo}/E_{x0} depending on the alignment of the field components with the major axis. The major axis of the ellipse is the numerator and the other one will be the denominator in the above equation.

45 C. A. Balanis, *Antenna Theory Design and Analysis*, Wiley, 1997, Section 2.12 Polarisation, pp 64–68.

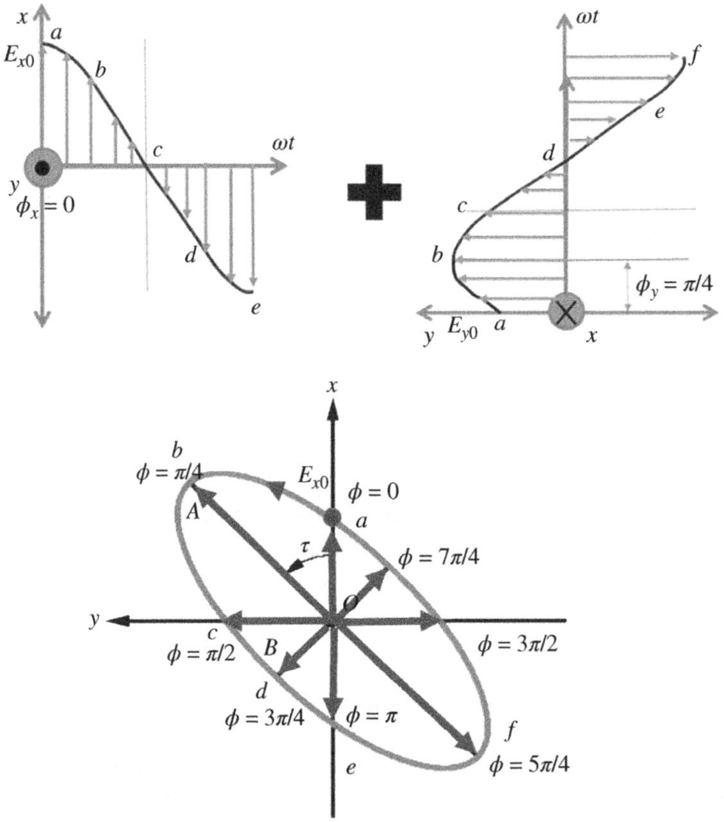

Figure 3.19 Creation of elliptically polarized wave with two orthogonal components of 45° phase difference.

Therefore, the electromagnetic field wave that satisfies the following conditions[46] is called an *elliptically polarized wave.*

$E_{x0} \neq E_{y0}$

$$\delta = \Delta\phi = \phi_y - \phi_x = \left\{ \begin{array}{l} \left(\frac{1}{2} + 2n\right)\pi \text{ for } RHCP \\ -\left(\frac{1}{2} + 2n\right)\pi \; LHCP \end{array} \right\} \text{ for } n = 0, 1, 2, 3 \text{ etc.,}$$

or, $E_{x0} = E_{y0}$

$$\delta = \Delta\phi = \phi_y - \phi_x \neq \left\{ \begin{array}{l} \frac{n\pi}{2} > 0 \text{ for } RHCP \\ \frac{n\pi}{2} < 0 \; LHCP \end{array} \right\} \text{ for } n = 0, 1, 2, 3 \text{ etc.,}$$

So far, we have examined the hypothesis of an *elliptically polarized wave* which is generated with the above conditions. Now we are going to interpret the phenomenon of the elliptical wave in the

46 C. A. Balanis, *Antenna Theory Design and Analysis*, Wiley, 1997, Section 2.12 Polarisation, pp 64–68.

graphical solutions of two arbitrary orthogonal sinusoidal field waves. For simplicity we assume the phase difference between the E_x and E_y field components is $\delta = \Delta\phi = \phi_y - \phi_x = \frac{\pi}{4} = 45°$ as shown in the diagram. The expressions for the individual and the superimposed electric field components are given in the following equations:

$$x-\text{directed field component: } \boldsymbol{E}_1(z, t) = E_{x0}\cos(\omega t)\boldsymbol{a_x} \tag{3.144}$$

$$y-\text{directed field component: } \boldsymbol{E}_2(z, t) = E_{y0}\cos\left(\omega t + \frac{\pi}{4}\right)\boldsymbol{a_y} \tag{3.145}$$

$$\text{The composite field: } \boldsymbol{E}(z, t) = E_{x0}\cos(\omega t)\boldsymbol{a_x} + E_{y0}\cos\left(\omega t + \frac{\pi}{4}\right)\boldsymbol{a_y} \tag{3.146}$$

As can be seen as the time progresses the tip of the resultant electric field vector

$$|\,E(z, t)\,| = \sqrt{\{E_1(0, t)\}^2 + \{E_2(0, t)\}^2} \tag{3.147}$$

at $z = 0$ traces an ellipse.

We have taken six instances of the orthogonal plane waves from points *a→b→c→d→e→f* as shown in Figure 3.19b. As the time progresses the time phase $\phi = \omega t = 0$ (*a*), $(\pi/4)$ (*b*), $(\pi/2)$(*c*), $(3\pi/4)$ (*d*), π (*e*), $(4\pi/4)$(*f*) progresses tracing the ellipse with the major axis *OA* and the minor axis *OB* and the tilt angle τ. The tip of the resultant electric field vector tip traces the line, which is in this case an inclined line that makes an angle τ with *x*-axis, here τ is called the "tilt angle."[47] Using MATLAB, you may plot the trace of the resultant field for the cases when the phase difference is not zero or out of phase. In summary, *elliptical polarization* occurs when $\delta = \phi_y - \phi_x \neq 0$, ± 90 or $\pm 180°$. Let us briefly summaries the elliptical polarization as we did for the linear and circular polarizations.

1) *The elliptical polarization* is the generic polarization of electromagnetic field waves. *The linear* and *circular* polarizations are the special cases for *the elliptical polarization. The elliptical polarization* occurs when the phase difference between the two field components can take any value except 0, ± 90 or $\pm 180°$ and/or the amplitudes of the two orthogonal field components are not the same.
2) The tilt angle τ is the angle that is created by the trace of the resultant electric field vector with the *x*-axis. The title angle is defined by the subtended angle of the major axis *OA* with *x*-axis and is defined in terms of the amplitude and the phase difference between the two orthogonal field components. For $\cos\delta = 0$ or $\delta = (\pi/2)$, the major axis and the minor axis align with the two principal axes and the tilt angle $\tau = 0$. For other cases the tilt angle is defined by the following expression:

$$\tau = \frac{1}{2}\tan^{-1}\left[\frac{2E_{xo}E_{y0}}{E_{xo}^2 - E_{y0}^2}\cos(\delta)\right] \tag{3.148}$$

3) When the ellipse is aligned with the principal axes then the axial ratio is defined by:

$$AR = \frac{E_{xo}}{E_{y0}} \text{ or } \frac{E_{yo}}{E_{x0}} \tag{3.149}$$

---

47 See YouTube videos for electromagnetic wave polarization: https://www.youtube.com/watch?v=BycPkRIutqg (accessed 15 January 2018).

4) *The handedness* of *the elliptically polarized* wave is determined with its difference in the phase angle δ. The conditions are:

$sin\delta = sin(\phi_y - \phi_x) > 0$ for *RHCP* or clockwise rotation if you look at the polarization screen from the direction of the wave propagation.[48]

$sin\delta = sin(\phi_y - \phi_x) < 0$ for *LHCP* or counter clockwise rotation if you look at the polarization screen from the direction of the wave propagation.

Review Questions 3.10: Elliptical Polarization
Q1. If the major axis is 3 time larger than minor axis, what is the axial ratio of the elliptically polarized signal.
Q2. What is the axial ratio of a LP signal?

Look at the trace of the tip of the resultant vector. In the animation it traces an ellipse with major axis and a minor axis. The tilt angle of the polarization is $\tau = 45°$ here. The tilt angle as before the angle between the major axis and the x-axis. And *the axial ratio* is the ratio of major to minor axes.

If the major axis is 3 time larger than minor axis, what is the axial ratio of the elliptically polarized signal? Answer is 3 or $20\ log_{10}\ 3 = 9.54$ dB. What is the axial ratio of a LP signal? Infinity. Therefore, *AR* can vary from 1 to ∞. *The handedness* is also applied here. Identify the handedness of the elliptical polarization.

3.6.4 Polarization Loss Factor and Polarization Efficiency

We have stated in the very beginning of the introduction of the polarization loss factor due to the polarization mismatch between the two signals. In general, a signal is launched in free space with a particularly polarized antenna and also received with a receiving antenna which is also polarized in a particular direction. Likewise, in a guided configuration such as in a waveguide and a transmission line, this polarization match between different sections of the guiding structure is also critical to avoid signal loss due to polarization mismatched.

3.6.4.1 Polarization Loss Factor

This loss due to the polarization mismatch is quantified with a factor is called *polarization loss factor*. In this section, we shall define *the polarization loss factor* using the field configurations. As shown in Figure 3.20 assume that an antenna is transmitting an electric field is defined as:

$$\boldsymbol{E}_1 = \boldsymbol{a}_1 E_1 \tag{3.150}$$

48 In previous figures, we showed you are looking toward the direction of wave propagation on the polarization screen. Therefore, the definition of handedness in terms of clockwise and counter clockwise is opposite. Many book have defined this way.

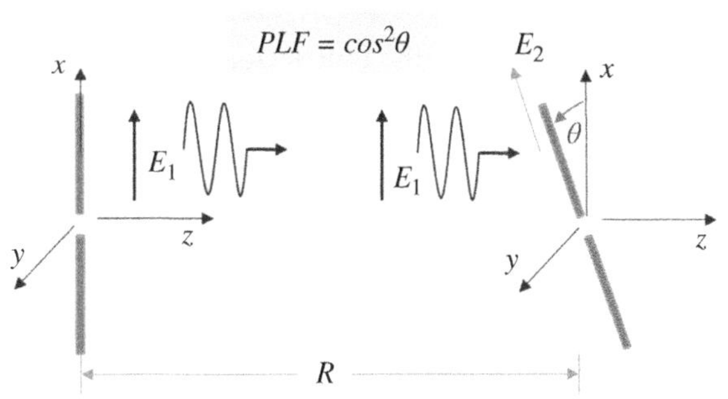

Figure 3.20 Illustration of polarization loss factor with two dipole antenna system rotated at an angle θ with respect to each other.

where $\boldsymbol{a}_1$ is the unit vector of the incident field wave and E_1 is the scalar time harmonic field. Assume a second antenna is designed and/or oriented in such a way that it radiates the electric field as follows:

$$\boldsymbol{E}_2 = \boldsymbol{a}_2 E_2 \tag{3.151}$$

If we use the second antenna as the receiving antenna then the antenna will receive the maximum power if the antennas are polarization matched. This means the condition of maximum receive power is:

$$cos\,\theta = \boldsymbol{a}_1.\boldsymbol{a}_1 = cos\,0 = 1 \tag{3.152}$$

The angle θ between the two signal vectors is called *the polarization angle*. Since we are interested how much power is lost due to the polarization mismatch, the polarization loss factor is defined as:

$$PLF = |\boldsymbol{a}_1.\boldsymbol{a}_1|^2 = cos^2\theta \tag{3.153}$$

PLF is a dimensionless quantity (ratio) and can vary from 1 to ∞. The sources of polarization mismatch are mainly for two reasons: (i) if the sources of the electromagnetic signals are not at the same polarization or (ii) the antennas that we use for transmission and reception are not at the same polarization and/or orientation. We shall do the following example to clarify the theory.

Example 3.4: Polarization Loss Factor 1

Assume that a linearly polarized transmitting dipole antenna is oriented along x-axis and is launching an incoming electric field intensity signal as follows:

$$E_1 = a_x E_{10}\,cos\,(\omega t - \beta z + \theta_1)$$

The receiving antenna is designed for receiving a dual polarized signal in orthogonal planes[49] of which we can define the field vector as:

$$E_2 = \left(\boldsymbol{a}_x + j\boldsymbol{a}_y\right) E_2\,cos\,(\omega t - \beta z + \theta_2)$$

Calculate the polarization loss factor.

49 For satellite communications we frequently use such orthogonally polarized antennas one polarization for transmission and one for reception. Also, in chipless RFID we use orthogonally polarized antennas to exploit the polarization diversity due to the fact that the cross polar signal is immune to noises. Therefore, same antenna can be used to transmit the signal in one polarization and receive the returned echo in the other polarization from the chipless tag.

Answer:
For the incident wave, the unit vector is:

$$\boldsymbol{a}_1 = \boldsymbol{a}_x$$

For the receiving antenna signal the unit vector is:

$$\boldsymbol{a}_2 = \frac{1}{\sqrt{2}}\left(\boldsymbol{a}_x + \boldsymbol{a}_y\right)$$

Therefore, the polarization loss factor is:

$$PLF = |\boldsymbol{a}_1.\boldsymbol{a}_2|^2 = \left|\frac{1}{\sqrt{2}}\left(\boldsymbol{a}_x.\boldsymbol{a}_x + \boldsymbol{a}_x.\boldsymbol{a}_y\right)\right|^2 = \frac{1}{2}$$

Usually polarization loss factor is defined in decibel unit as follows:

$$PLF = 10\,log\,(PLF)\ \ (\mathrm{dB})$$

Therefore, in the current example, $PLF = 10\,log\,(1/2) = -3$ (dB)

Example 3.5: Polarization Loss Factor 2
Assume that a linearly polarized transmitting dipole antenna is oriented along y-axis and is launching an incoming electric field intensity signal that is propagating along z-axis as follows:

$$\boldsymbol{E}_1 = \boldsymbol{a}_y E_{10}\cos\left(\omega t - \beta z + \theta_1\right)$$

The receiving antenna is designed for receiving a right hand circularly polarized signal in orthogonal planes[50] of which we can define the field vector as:

$$\boldsymbol{E}_2 = \left(\boldsymbol{a}_x - j\boldsymbol{a}_y\right)E_{20}\cos\left(\omega t - \beta z + \theta_2\right)$$

Calculate the polarization loss factor.

Answer:
Foe the incident wave, the unit vector is:

$$\boldsymbol{a}_1 = \boldsymbol{a}_y$$

For the receiving antenna signal the unit vector is:

$$\boldsymbol{a}_2 = \frac{1}{\sqrt{2}}\left(\boldsymbol{a}_x - j\boldsymbol{a}_y\right)$$

50 For mobile satellite communications we frequently use circularly polarized antennas so that signal can be received from any orientation. As stated above, due to Clark belt in the ionosphere the signal rotates in arbitrary angle. Using a circularly polarized antenna we can receive the signal in any orientation as the signal is always rotating in clockwise/counter clockwise direction seeing from the direction of wave propagation opposite to the diagram shown above. This convention of seeing the field from the direction of propagation match with the IEEE definition of polarization. The concept is derived from the radiation property of a helix antenna. The winding direction of a helix represents the handedness of the radiated circularly polarized signal. Therefore, if you look behind the antenna toward the direction of propagation from which the signal is propagated, it will give you the handedness.

Therefore, the polarization loss factor is:

$$PLF = |\boldsymbol{a}_1.\boldsymbol{a}_2|^2 = \left|\frac{1}{\sqrt{2}}\left(\boldsymbol{a}_y.\boldsymbol{a}_x - j\boldsymbol{a}_y.\boldsymbol{a}_y\right)\right|^2 = \frac{1}{2}$$

Or, $PLF = 10\,log\,(1/2) = -3$ (dB) Still half power will be received by the receiving antenna.

Two most commonly used linearly polarized antennas are a horn antenna and a dipole antenna. The horn antenna is called an aperture antenna due to its rectangular aperture that is accountable for radiation. Usually the length of the wider side is accountable for the defined resonant frequency. The dipole antenna is a wire antenna with a defined resonant length. Figure 3.21 illustrates various antenna configurations and their polarization loss factor for a linearly polarized incoming/incident uniform plane wave at different incident angles. As can be seen in the following figure, four different types of antennas are considered that generate various types of polarizations. The linearly polarized horn antenna has a specific half sine wave field distribution along x-direction due to

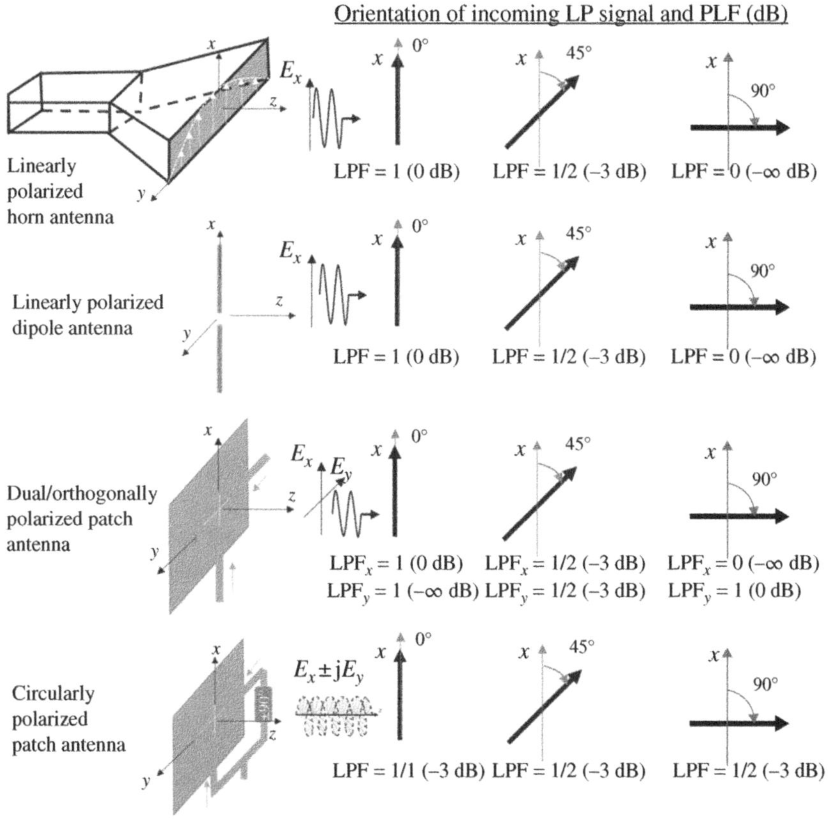

Figure 3.21 Orientations of incoming linearly polarized signals and the polarization loss factors for various antenna configurations.

its boundary conditions.[51] The horn antenna launches or receives only x-directed linearly polarized incoming signal with the maximum intensity meaning the polarization loss factor is 1. As the incoming linearly polarized signal changes its orientation the angle between the antenna's field orientation and the incoming signal increases. Thus the polarization loss factor also reduces. At $\pm 90°$ the polarization loss factor will be zero. The relevant decibel values of the polarization loss factor are also shown alongside its ratio values. For the dipole antenna the polarization loss factor is the same as for the horn antenna. For the dual/orthogonally polarized patch antenna the linearly polarized incoming signal is not totally lost as it happened for the cases of the $\pm 90°$ incoming field's orientation. In this case if the signal lost ($PLF = 0$) for one polarity, the signal is received in the fullest strength ($PLF = 1$) to another polarity. This phenomenon of signal isolation between the two polarities is called *polarization diversity*. Using a dual orthogonal polarized antenna, we can exploit *the polarization diversity* to enhance channel capacity to double as well as to suppress noise. The circularly polarized antenna always receives the linearly polarized signal irrespective of the orientation of the incoming signal. However, the polarization loss factor is always ½ (−3 dB) meaning half power is lost in the reception process with the gain of no total power lost due to the change of the orientation of the incoming signal. This circular polarization is used in mobile communications where orientation of the incoming signal is always changing especially for the mobile satellite communications.

3.6.4.2 Polarization Efficiency

We cannot warrant that the polarization of the incoming signal will be matched with that for the receiving antenna all the time. Besides *the polarization loss factor*, the degree of the polarization mismatch viz the signal loss due to the polarization mismatch between the incoming signal and the receiving antenna we define another figure of merit called *polarization efficiency* (PE). PE describes the polarization characteristic of an antenna. Like *the antenna efficiency*,[52] *the polarization efficiency* represents the power loss due to the polarization mismatch and the degree of polarization mismatch that it offers to an incoming signal. It is the power ratio of the power received by the antenna from an arbitrarily polarized incoming signal to the power the antenna would normally receive if the antenna is polarization matched for the maximum received power with the same incoming signal of same power flux density. Eventually *the polarization loss factor* and *the polarization efficiency* represent the degree of the polarization mismatch. We examine *the polarization efficiency* of antennas in more detail in antenna theory.

3.7 Specific Topics on Polarizations of Uniform Plane Wave

So far, we have comprehensively studied the three types of polarizations. In all cases we only analyzed the electric field vectors as the generic definition of the polarization meaning the orientation of the electric field vector with respect to time is defined as the polarization. However, in a uniform

---

51 We usually encounter the half sign wave field distribution on the rectangular apertures in the rectangular waveguide theory.

52 *Antenna efficiency* is defined as the ratio of the antenna's effective aperture area, A_e to its actual physical aperture area, A. *Antenna effective area* is an abstract concept that is also related to the polarization loss factor. It is a ratio between the power in watts that the antenna actually captures from an incoming signal to the power flux density (W/m^2) that impinges on the antenna aperture from the incoming signal. Therefore, the antenna effective area has unit in m^2. *Antenna radiation efficiency* refers to the ratio of the real power loss due to radiation and the real power losses in the antenna incurs due to the finite conductivity of the antenna materials, reflection loss at the antenna input port, the loss of signal due to surface wave that propagates in the dielectric of a patch antenna and a waveguide antenna, and any other substantial losses that affect the overall gain of the antenna. In this regard, *polarization efficiency* is a part of *the antenna efficiency*.

plane wave the magnetic field is associated with the electric field all the time. Also, we would like to know the power, the wave conveys from one point in time and space to another point. Also, the polarization is affected by the anisotropy of the medium in which the signal propagates. In the above analysis we always assume that the medium is homogeneous, linear, isotropic, and lossless. In the beginning we also mentioned that the polarization would not be affected by the loss factor of the medium. However, we shall have a close look at the Poynting vector calculation of differently polarized field waves in both lossless and lossy media. We have defined the relationship between the three polarizations and have told that the elliptical polarization is the general polarization of signal waves, and the linear and circular polarizations are the specific cases of the elliptical polarization. We shall develop new relationships in between these three polarizations. Can we create elliptical polarization from the two circularly polarized waves of opposite handedness? In this section, we are going to discuss these special features of the theory of polarizations.

3.7.1 Magnetic Field in Plane Wave with Generic Polarization

In uniform plane wave propagation, the magnetic field is always associated with the electric field. The two fields are linked with the wave impedance, η. The wave impedance is real for the lossless medium and complex in the lossy medium. Therefore, irrespective of the medium property we always take the definition of the magnetic field in terms of the electric field vector and the direction of wave propagation as:

$$\boldsymbol{H} = \frac{1}{\eta}\boldsymbol{a_p} \times \boldsymbol{E} \tag{3.154}$$

where $\boldsymbol{a}_p$ is the unit vector along the direction of propagation. So far, we have analyzed mainly the field solutions for the one-dimensional wave equation. However, when we consider the polarization of the electromagnetic fields, depending on the type of polarization we deal with more than one-dimensional field. Although the resultant field is always one-dimensional for the linearly polarized field, they are decomposed into two orthogonal field components. That raises the complexity of the magnetic field calculation. For the case of an elliptically polarized signal as defined above, the electric field is decomposed into x- and y-directed field vectors as follows:

$$E(z, t) = \boldsymbol{a_x}E_x(z, t) + \boldsymbol{a_y}E_x(z, t)\ ^{53} \tag{3.155}$$

where,

$$E_x = E_{x0}\cos(\omega t - \beta z + \phi_x) \text{ and } E_y = E_{y0}\cos(\omega t - \beta z + \phi_y).$$

Using the above relationship between the electric and magnetic fields, from the known electric field, we derive the magnetic field as

$$\boldsymbol{H} = \frac{1}{\eta}\boldsymbol{a}_p \times \boldsymbol{E} = \frac{1}{\eta}\left(\boldsymbol{a_z} \times \boldsymbol{a_x}E_x + \boldsymbol{a_z} \times \boldsymbol{a_y}E_y\right) = \frac{1}{\eta}\left(E_x\boldsymbol{a_y} - \boldsymbol{a_x}E_y\right) \tag{3.156}$$

Substitution for the above time domain (instantaneous) expressions for E_x and E_y we obtain the expression for the time domain (instantaneous) magnetic field expression

$$\boldsymbol{H} = \frac{1}{\eta}\left[-E_{y0}\cos\left(\omega t - \beta z + \phi_y\right)\boldsymbol{a_x} + E_{x0}\cos\left(\omega t - \beta z + \phi_x\right)a_y\right] \tag{3.157}$$

---

53 Note in the previous example we use subscript 1 and 2 instead of x and y, respectively to represent the two field components.

Therefore, the individual instantaneous magnetic field components are:

$$H_x = \frac{1}{\eta} - E_{y0}\cos\left(\omega t - \beta z + \phi_y\right) \text{ and } H_y = \frac{1}{\eta}E_{x0}\cos\left(\omega t - \beta z + \phi_x\right) \tag{3.158}$$

Therefore, we can conclude that the calculation of the magnetic fields for a particular polarization of the uniform plane waves is derived from the universal relationship of the wave impedance η and the known orthogonal electric field vector $\boldsymbol{E}(z, t)$ and the direction of wave propagation with unit vector $\boldsymbol{a_p}$. If we closely look at the derived individual x- and y-directed field components then we can perceive that they individually are linearly polarized waves. However, when the two linearly polarized waves are with unequal amplitude and arbitrary phase difference then in their superimposed form they produce an elliptically polarized wave.[54] The handedness comes from the difference in the phase angle and the polarity as defined above.

The amplitude of the resultant magnetic field is:

$$H(z, t) = \sqrt{\left(\frac{-E_y(z, t)}{\eta}\right)^2 + \left(\frac{E_x(z, t)}{\eta}\right)^2} = \frac{E(z, t)}{\eta} \tag{3.159}$$

Therefore, the relationship between the scalar field components holds with the wave impedance in every instance of time and every point in space.

3.7.2 Poynting Vector Calculation in Different Polarizations of Electromagnetic Fields

Like the universal relationship between the electric and magnetic field vectors for any arbitrary polarization, Poynting vector can also be derived from the known electric and magnetic field components. The instantaneous Poynting vector expression is calculated as:

$$\boldsymbol{P} = \boldsymbol{E} \times \boldsymbol{H} \tag{3.160}$$

Therefore, using the above example for the elliptically polarized electromagnetic field wave we can calculate the instantaneous expression for the Poynting vector by simple substitution of the two field components in the above equation.

$$\begin{aligned}\boldsymbol{P} = \boldsymbol{E} \times \boldsymbol{H} = &\left\{E_{x0}\cos(\omega t - \beta z + \phi_x)\boldsymbol{a_x} + E_{y0}\cos\left(\omega t - \beta z + \phi_y\right)\boldsymbol{a_y}\right\} \\ &\times \frac{1}{\eta}\left\{-E_{y0}\cos\left(\omega t - \beta z + \phi_y\right)\boldsymbol{a_x} + E_{x0}\cos(\omega t - \beta z + \phi_x)\right\}\boldsymbol{a_y}\end{aligned} \tag{3.161}$$

or,

$$\boldsymbol{P} = \frac{1}{\eta}\left\{E_{x0}^2\cos^2(\omega t - \beta z + \phi_x) + E_{y0}^2\cos^2\left(\omega t - \beta z + \phi_y\right)\right\}\boldsymbol{a_z} \tag{3.162}$$

The time average Poynting vector can also be calculated from the above expression by considering the phasor vector form of the two field vectors.

$$\boldsymbol{E_s} = E_{x0}e^{-j\beta z}e^{j\phi_x}\,\boldsymbol{\alpha_x} + E_{y0}e^{-j\beta z}e^{j\phi_y}\,\boldsymbol{\alpha_y} \tag{3.163}$$

$$\boldsymbol{H_s} = \frac{1}{\eta}\left[-E_{y0}e^{-j\beta z}e^{j\phi_y}\,\boldsymbol{\alpha_x} + E_{x0}e^{-j\beta z}e^{j\phi_x}\,\boldsymbol{\alpha_y}\right] \tag{3.164}$$

---

54 We shall also show that when two opposite polarity signals of circularly polarized waves are vectorially added together then they create an elliptically polarized wave.

$$\boldsymbol{P}_{\boldsymbol{ave}} = \frac{1}{2} Re\left[\boldsymbol{E}_{\boldsymbol{s}} \times \boldsymbol{H}_{\boldsymbol{S}}^{*}\right] = \frac{1}{2}\left[E_{x0}e^{-j\beta z}e^{j\emptyset_x}\boldsymbol{\alpha_x} + E_{y0}e^{-j\beta z}e^{j\emptyset_y}\boldsymbol{\alpha_y}\right] \times \frac{1}{\eta}\left[-E_{y0}e^{-j\beta z}e^{j\emptyset_y}\boldsymbol{\alpha_x} + E_{x0}e^{-j\beta z}e^{j\emptyset_x}\boldsymbol{\alpha_y}\right] \tag{3.165}$$

$$\boldsymbol{P}_{\boldsymbol{ave}} = \frac{1}{2\eta}\left[E_{x0}^2 + E_{y0}^2\right] \cos\left(\phi_y - \phi_x\right)\boldsymbol{a_z} = \frac{|E|^2}{2\eta} \cos\left(\phi_y - \phi_x\right)\boldsymbol{a_z} \tag{3.166}$$

where $\boldsymbol{E} = \boldsymbol{a}_x E_x + \boldsymbol{a}_y E_y$ is used.

From the above derivation we can see that the time average Poynting vector is a real quantity. From the above derivation we can conclude that the time average Poynting vector for any polarized wave reduces the same results as for the linearly polarized wave. Let us now verify the time average Poynting vector for a circularly polarized wave. The instantaneous vector forms of the two field vectors in the circular polarization are:

$$\boldsymbol{E}_{RHCP} = \boldsymbol{E_x} + j\boldsymbol{E_y} = E_{01}\cos(\omega t - \beta z + \phi_x)\boldsymbol{a_x} + E_{01}\sin(\omega t - \beta z + \phi_x)\boldsymbol{a_y} \tag{3.167}$$

$$\begin{aligned}\boldsymbol{H}_{RHCP} &= \frac{1}{\eta}\boldsymbol{a_p} \times \boldsymbol{E}_{RHCP} = \frac{1}{\eta}\boldsymbol{a_z} \times \left[E_{01}\cos(\omega t - \beta z + \phi_x)\boldsymbol{a_x} + E_{01}\sin(\omega t - \beta z + \phi_x)\boldsymbol{a_y}\right] \\ &= \frac{1}{\eta}\left[-E_{01}\sin(\omega t - \beta z + \phi_x)\boldsymbol{a_x} + E_{01}\cos(\omega t - \beta z + \phi_x)\boldsymbol{a_y}\right]\end{aligned} \tag{3.168}$$

The phasor vector forms of the two field vectors in the circular polarization are:

$$\boldsymbol{E}_s = E_{x0}e^{-j\beta z}e^{j\emptyset_x}\boldsymbol{\alpha_x} + E_{yo}e^{-j\beta z}e^{j\left(\emptyset_x - \frac{\pi}{2}\right)}\boldsymbol{\alpha_y} \tag{3.169}$$

$$\boldsymbol{H}_s = \frac{1}{\eta}\left[-E_{y0}e^{-j\beta z}e^{j\left(\emptyset_x - \frac{\pi}{2}\right)}\boldsymbol{\alpha_x} + E_{x0}e^{-j\beta z}e^{j\emptyset_x}\boldsymbol{\alpha_y}\right] \tag{3.170}$$

$$\boldsymbol{P}_{ave} = \frac{1}{2} Re\left[\boldsymbol{E}_s \times \boldsymbol{H}_s^{*}\right] = \frac{1}{2}\left[E_{10}e^{-j\beta z}e^{j\emptyset_x}\boldsymbol{\alpha_x} + E_{10}e^{-j\beta z}e^{j\left(\emptyset_x - \frac{\pi}{2}\right)}\boldsymbol{\alpha_y}\right] \times \frac{1}{\eta}\left[-E_{10}e^{j\beta z}e^{-j\left(\emptyset_x - \frac{\pi}{2}\right)}\boldsymbol{\alpha_x} + E_{10}e^{j\beta z}e^{-j\emptyset_x}\boldsymbol{\alpha_y}\right] \tag{3.171}$$

$$\boldsymbol{P}_{ave} = \frac{1}{2\eta} Re\left[E_{10}^2 + E_{10}^2\right]\boldsymbol{\alpha_z} \tag{3.172}$$

$$\boldsymbol{P}_{ave} = \frac{1}{\eta} E_{10}^2\boldsymbol{\alpha}_z \tag{3.173}$$

Here. $E_{x0} = E_{yo} = E_{10}$ and $\phi_y - \phi_x = \pi/2$ are used.

Now we can define the Poynting vector for the three cases of polarizations of the field waves:

1) For the case of an elliptically polarized signal with equal amplitude $E_{x0} = E_{y0}$ and arbitrary phase difference $\delta = \phi_y - \phi_x \neq 0, \pm 90$ and $\pm 180°$ we can define

$$\boldsymbol{P} = \frac{1}{\eta}E_{x0}^2\left\{\cos^2(\omega t - \beta z + \phi_x) + \cos^2\left(\omega t - \beta z + \phi_y\right)\right\}\boldsymbol{a_z} \tag{3.174}$$

2) For the case of a linearly polarized signal with unequal amplitudes $E_{x0} \neq E_{y0}$ and phase difference: $\delta = \phi_y - \phi_x = 0$ and $\pm 180°$, we can define

$$\boldsymbol{P} = \frac{1}{\eta}\left\{E_{x0}^2\cos^2(\omega t - \beta z) \pm E_{y0}^2\cos^2(\omega t - \beta z)\right\}\boldsymbol{a_z} \tag{3.175}$$

3) For the case of a linearly polarized signal with only one field component E_{x0} or E_{y0} we can define

$$\boldsymbol{P} = \frac{1}{\eta}\{E_{x0}^2 \cos^2(\omega t - \beta z)\}\boldsymbol{a_z} \tag{3.176}$$

4) For the case of an elliptically polarized signal with equal amplitude $E_{x0} = E_{y0}$ and arbitrary phase difference $\delta = \phi_y - \phi_x = \pm 90°$, we can define

$$\boldsymbol{P} = \frac{1}{\eta}E_{x0}^2\{\cos^2(\omega t - \beta z) + \sin^2(\omega t - \beta z)\}\boldsymbol{a_z} = \frac{1}{\eta}E_{x0}^2\boldsymbol{a_z} \tag{3.177}$$

3.7.3 Elliptically Polarized Wave from Two Unequal Cross-Polar Circularly Polarized Wave

The elliptically polarized wave can also be created not only by superimposition of two linearly polarized waves with unequal amplitudes and/or an arbitrary phase difference between the two orthogonal signal components. We can also create elliptically polarized wave from two unequal cross-polar meaning counter rotating circularly polarized waves. The generic expressions for the two orthogonal field components of two counter rotating circularly polarized waves are defined as:

$$\boldsymbol{E}_x(z, t) = E_{x0}\cos(\omega t - \beta z + \phi_x)\boldsymbol{a_x} = E_{01}\cos(\omega t - \beta z + \phi_x)\boldsymbol{a_x} \tag{3.178}$$

$$\boldsymbol{E}_y(z, t) = E_{y0}\cos\left(\omega t - \beta z + \phi_x \pm m\frac{\pi}{2}\right)\boldsymbol{a_y} = \mp E_{01}\sin(\omega t - \beta z + \phi_x)\boldsymbol{a_y} \tag{3.179}$$

Therefore, the two unequal counter-rotating circularly polarized waves can be written as:

$$\boldsymbol{E}_{RHCP} = \boldsymbol{E_x} + j\boldsymbol{E_y} = E_{01}\cos(\omega t - \beta z + \phi_x)\boldsymbol{a_x} \pm E_{01}\sin(\omega t - \beta z + \phi_x)\boldsymbol{a_y}\ (RHCP) \tag{3.180}$$

$$\boldsymbol{E}_{LHCP} = \boldsymbol{E_x} - j\boldsymbol{E_y} = E_{02}\cos(\omega t - \beta z + \phi_x)\boldsymbol{a_x} \mp E_{02}\sin(\omega t - \beta z + \phi_x)\boldsymbol{a_y}(LHCP) \tag{3.181}$$

Adding these two fields we obtain:

$$\begin{aligned}\boldsymbol{E} = \boldsymbol{E}_{RHCP} + \boldsymbol{E}_{LHCP} &= E_{01}\cos(\omega t - \beta z)\boldsymbol{a_x} \pm E_{01}\sin(\omega t - \beta z)\boldsymbol{a_y}\\ &+ E_{02}\cos(\omega t - \beta z)\boldsymbol{a_x} \pm E_{02}\sin(\omega t - \beta z)\boldsymbol{a_y}\end{aligned} \tag{3.182}$$

or,

$$\boldsymbol{E} = [\{E_{01} \pm E_{02}\}\cos(\omega t - \beta z)]\boldsymbol{a_x} \pm [\{E_{01} \mp E_{02}\}\sin(\omega t - \beta z)]\boldsymbol{a_y} \tag{3.183}$$

where $E_{01} \pm E_{02}$ and $E_{01} \mp E_{02}$ are different in amplitudes hence constitute an elliptically polarized wave. Next, we can also do the same exercise to solve for the magnetic fields. This will be exactly the same calculation except the relationship holds with the wave impedance as per the above example for the calculation of the magnetic fields from the electric fields.

3.7.4 Effect of Medium Characteristics on Polarization-Anisotropic Medium

For simplicity, we considered the medium is linear, homogeneous and isotropic and even lossless in the above analyses. In an isotropic medium the atomic and molecular polarization $\boldsymbol{P}$ per unit volume[55] depends on the direction of the applied field $\boldsymbol{E}$. Therefore, polarization of the electromagnetic field waves suffers from the anisotropy of the medium. In the anisotropic medium the crystal structure of the medium lacks spherical symmetry.[56] Therefore, the polarization $\boldsymbol{P}$ produced in one direction (say x) may be greater than when the field $\boldsymbol{E}$ is applied in y-or z-direction. This is because of the greater ease of polarization $\boldsymbol{P}$ in the x axis. In such a case, we must have different displacement densities $\boldsymbol{D} = \varepsilon \boldsymbol{E}$ in different direction in the medium due to the fact that the medium exhibits different constitutive parameters ε in different axes. The dielectric constant is expressed in the principal dielectric constant in the following matrix form:

$$\varepsilon = \begin{bmatrix} \varepsilon_{xx} & & \\ & \varepsilon_{yy} & \\ & & \varepsilon_{zz} \end{bmatrix} \tag{3.184}$$

where, $\varepsilon_{xx} = \varepsilon_{rx}\varepsilon_0 \neq \varepsilon_{yy} = \varepsilon_{ry}\varepsilon_0 \neq \varepsilon_{zz} = \varepsilon_{rz}\varepsilon_0$ is the principal dielectric constants in x-, y-, and z-axes, respectively. The symbols have their usual meanings.

Therefore, an anisotropic medium exhibits different phase constant $\beta = \omega\sqrt{\varepsilon\mu}$ in different directions to the wave hence the polarization suffers. The relative permittivities of the principal planes are defined as $\varepsilon_{rx} = 2.25$ and $\varepsilon_{ry} = 2.8$ in a crystalline dielectric medium. In this section we shall investigate the effect of the anisotropic medium when a plane wave signal with a defined polarity impinges on the medium. First case, we assume a linearly polarized incoming signal with the electric field intensity $\boldsymbol{E}$ impinges on the anisotropic medium as follows:

$$\boldsymbol{E} = \boldsymbol{a_x}E_{10}\cos(\omega t - \beta z) + \boldsymbol{a_y}E_{20}\cos\left(\omega t - \beta z + \frac{\pi}{4}\right) \tag{3.185}$$

In this case, we assume that the phase difference between the orthogonal signal components is 45°. Since the medium is anisotropic, the phase constants will be different in the principal x- and y-directions:

$$\beta_x = \omega_0\sqrt{\varepsilon_{rx}\varepsilon_0\mu_0} = \frac{2\pi}{\lambda_0}\sqrt{\varepsilon_{rx}} \tag{3.186}$$

and,

$$\beta_y = \omega_0\sqrt{\varepsilon_{ry}\varepsilon_0\mu_0} = \frac{2\pi}{\lambda_0}\sqrt{\varepsilon_{ry}} \tag{3.187}$$

Here $\mu_r = 1$ is assumed for the dielectric medium. This means that the two principal components of the electric field travels in two different velocities $u_x = 1/\sqrt{\varepsilon_{rx}\varepsilon_0\mu_0}$ and $u_y = 1/\sqrt{\varepsilon_{ry}\varepsilon_0\mu_0}$ along z-axis. Therefore, the phase difference between the two components increases as they travel along z-axis with time.[57] In optics, this phenomenon of light propagation in an anisotropic medium is

55 This polarization is due to the asymmetric nature of the atoms and molecules in a crystal and is different from the wave polarization that we are studying here. This polarization is due to the relative displacement of electron cloud due to the applied field $\boldsymbol{E}$ and the direction of the applied electric field $\boldsymbol{E}$ that produces different damping effect in the medium. The molecular polarization is defined as $\boldsymbol{P} = \varepsilon_0\chi_e\mathbf{E}$,where ε_0 is the permittivity of the medium and χ_e is the electric susceptibility of the medium.

56 Similar discussion is also available in R. E. Collin, *Foundations for Microwave Engineering*, McGraw Hill, NY, USA, 2e, 1992. Section 2.2 Constitutive Relations, pp. 23–28.

57 We usually study this phenomenon of travelling two orthogonal polarized signals in two different velocities in optical fiber dispersion. This phenomenon is called *birefringence*. Read more details about birefringence in https://en.wikipedia.org/wiki/Birefringence (accessed 17 January 2018).

called *birefringence*. Therefore, the relative phase difference between the two field components at a distance d is defined as:

$$\delta = \phi_y - \phi_x = -(\beta_y - \beta_x)d = -\frac{2\pi d}{\lambda_0}\left(\sqrt{\varepsilon_{ry}} - \sqrt{\varepsilon_{rx}}\right) \tag{3.188}$$

In the above equation, we use negative sine for the phase delay as it propagates along z-direction on distance d. Here $\delta > 0$ for $\varepsilon_{rx} > \varepsilon_{ry}$ and $\delta < 0$ for $\varepsilon_{ry} > \varepsilon_{rx}$.

Now if we assume that the amplitude of the two principal components have equal amplitude and at a minimum distance d_1 at which $\delta = (\pi/2)$, by the definition the outcome of the signal at the distance:

$$\delta = \frac{\pi}{2} = \frac{2\pi d_1}{\lambda_0}\left(\sqrt{\varepsilon_{rx}} - \sqrt{\varepsilon_{ry}}\right) \text{ or, } d_1 = \frac{1}{\left(\sqrt{\varepsilon_{rx}} - \sqrt{\varepsilon_{ry}}\right)} \times \left(\frac{\lambda_0}{4}\right) \tag{3.189}$$

For $\delta = +90°$, the handedness of *the circularly polarized wave* is left hand circular polarization. This type of anisotropic medium of a specific unequal principal plane permittivity that changes the polarization of incoming signal's polarization is called *a polarizer*. *The polarizer* has substantial applications in microwave engineering, optics, and many other applications. As for example, in satellite communications, we can use waveguide polarizer to change the polarization so that the signal can be transmitted in one polarization and received in another polarization. The cross polar components of a signal are immune to the noise which is amplified with the co-polar components. By extracting the cross polar component, we can suppress and filter out noises. In mm-wave chipless RFID tag design we have been using such principle of polarizer to send the signal in one polarization and receive the information carrying echo from the tag in the orthogonal polarization. This way we can enhance the detection capacity of the tag by using this polarization diversity in the transmission and reception. In optics, we also used crystal pieces cut at the length d_1 to make such optical polarizer. Figure 3.22 shows a chipless RFID that exploits such principle.

Now we shall study the case for the length $d_2 = 2d_1$, this means twice the length of the anisotropic medium and find the output of the signal as it passes through the medium. In this case:

$$\delta = \pi \text{ and } d_2 = 2d_1 = \frac{1}{\left(\sqrt{\varepsilon_{rx}} - \sqrt{\varepsilon_{ry}}\right)} \times \left(\frac{\lambda_0}{2}\right) \tag{3.190}$$

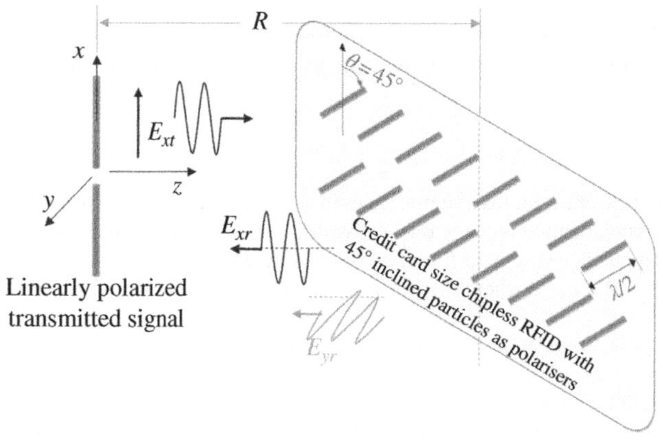

Figure 3.22 A mm-wave chipless RFID with 45° inclined polarizers provide the returned echoes in both co- and cross-polar components. The co-polar component E_{xr} contains the noise of surrounding environment. The cross polar component E_{yr} contains only the data hence enhance the detection capacity of the tag.

For a signal with phase difference $\delta = \phi_y - \phi_x = \pi$ with any amplitudes of the two orthogonal components represents a linearly polarized signal. The input and the output superposed signals, respectively, can be expressed as:

$$\boldsymbol{E}_1 = (\boldsymbol{a}_x + \boldsymbol{a}_y)E_{10}\cos(\omega t - \beta z) \tag{3.191}$$

and,

$$\boldsymbol{E}_2 = (\boldsymbol{a}_x - \boldsymbol{a}_y)E_{10}\cos(\omega t - \beta z) \tag{3.192}$$

From the expression we can deduce that the output signal of the anisotropic medium changes its direction compared to the incoming signal. If we vectorially draw each orthogonal component, we can see that the output signal is shifted to its direction by 90° compared to the input signal. Thus, if we used an anisotropic medium of length d_2 that generates the phase difference between the signals by $\delta = 180°$ then we can rotate the field by 90°.

Example 3.6: Polarization in Anisotropic Medium
An anisotropic crystal has $\varepsilon_{rx} = 4.2$ and $\varepsilon_{ry} = 4.0$, calculate the lengths d_1 and d_2 for such an anisotropic medium that provides *LHCP* and linear polarizations, respectively. The operating frequency is 60 GHz.

Solution:
The operating wavelength at 60 GHz is:

$$\lambda_0 = \frac{3 \times 10^8\,(\text{m/s})}{60 \times 10^9\,(1/\text{s})} = 5\,\text{mm}.$$

$$d_1 = \frac{1}{\left(\sqrt{4.2} - \sqrt{4}\right)} \times \left(\frac{5\,\text{mm}}{4}\right) = 25.31\,(\text{mm})$$

$$d_2 = 2d_1 = 50.62\,(\text{mm})$$

Therefore, an anisotropic crystal with lengths of 25.31 and 50.62 mm can generate an *LHCP* and 90° rotated LP signals, respectively, at its output end.

3.8 Chapter Remarks

- One-dimensional, x-polarized electric field was physically defined in the contexts of a dipole and a waveguide. These types of physical examples provide enhanced understanding of the abstract nature of EM field waves and their applications in reality.
- We considered homogenous one-dimensional plane wave equation meaning no free charge and current are present in space. A wave equation is homogenous if the right-hand side is zero. This reduces the computational complexity in greater degree.
- We found general solutions of one-dimensional plane wave equation. It has $\pm z$-directed (forward and reverse) wave components. The forward wave is easy to comprehend as the incident wave is along $+z$ direction. The reverse wave is an example of the reflected wave from a reflector.
- We explained complete wave solutions with a diagram for lossy medium. The envelop diminishes exponentially with distance with a factor $e^{\pm \alpha z}$.
- We explained wave motion and derived wave velocity, propagation, phase, and attenuation constants. These constants define the characteristics of the wave.
- We derived intrinsic wave impedance and magnetic field solutions from the electric field. The intrinsic wave impedance is complex for finite conductivity medium or lossy medium. For free space it is 377 Ω.

- We examined the dispersion of field waves in dispersive media in details. We comprehend that the digital impulses and signal processing methods are integral parts of the electromagnetic theory.
- Wave polarizations were explained in details. Elliptical polarization is *the general case for all polarizations* (analogy to curvilinear coordinate). The special cases are linear polarization with phase difference of two field components is zero or $\pm 180°$.
- Finally, we also did a worked out exercise, on power flow density and EM field calculations.
- Finally, we examined the polarization in anisotropic media and defined polarizer.

3.9 Problems

Propagation in Medium

P3.1 Name a few devices that generate one dimensional x-polarized electric field waves. Is a magnetic field associated with the electric field? If yes, what is the polarization of the magnetic field? Explain how does the magnetic field generate.

P3.2 Derive an expression for a one-dimensional wave equation for a magnetic field of amplitude H_0 that varies along y-axis and propagates along x-axis.

P3.3 Show that the general solution for the y-directed magnetic field that propagates along x-axis is derived as: $\boldsymbol{H}(x, t) = \boldsymbol{a}_y H_0^+ e^{-\alpha x} \cos(\omega t - \beta x) + \boldsymbol{a}_y H_0^- e^{\alpha x} \cos(\omega t + \beta x)$ (A/m).

P3.4 Write the phasor and exponential expression of the magnetic field of P3.3.

P3.5 Using the right-hand rule derive the complete solution for the associated electric field of P3.3.

P3.6 Draw the diagram of the complete solution of the electric field in P3.5 and explain each component of the wave in the diagram.

P3.7 Show that the phase velocity is: $u_p = (\omega/\beta)$.

P3.8 A y-polarized uniform plane wave with amplitude 12 (V/m) is propagating along z-direction. The wave is operating at 900 MHz and is propagating in a medium with the following constitutive parameters: $\varepsilon_r = 4.2$, $\mu_r = 1$ and $\sigma = 0.01$ (S/m).

a) Calculate the propagation constant γ, attenuation constant α (Np/m), phase constant β (rad/m) and the wave impedance η (Ω).
b) From results in part (a) calculate the forward travelling wave vector electric and magnetic fields, and the time average Poynting vector of the wave.
c) If the uniform plane wave travels 1 (m) from the transmitter in the lossy medium calculation the total attention in dB.

Dispersion, Pulse Broadening, Group, and Phase Velocities

P3.9 Determine the phase constant, phase and group velocities of a dispersive medium in which the refractive index varies linearly with a defined bandwidth: $n(\omega) = 2(\omega/\omega_0)$. Where, $n(\omega_0) = n_0 = 2$ at the operating frequency ω_0.

P3.10 For a frequency dependent medium with a nonlinear function refractive index as: $(\omega) = c_1\omega^2 + c_2\omega^{-2} + c_3$. Derive the phase constant, the phase and group velocities of the dispersive medium.

P3.11 If the operating wavelength of an optical waveguide is 1550 nm that uses the dispersive medium of P3.9, determine the operating frequency, the operating angular frequency ω_0, the phase constant, the phase and group velocities at that wavelength.

P3.12 If the operating frequency of an optical waveguide is 250 THz that uses the dispersive medium of P3.10, with the constants $c_1 = 5\times10^{-7}$ ((ps)/rad)^2, $c_2 = -2\times10^{-5}$ ((rad)/ps)^2, $c_3 = 0.3463$, determine the refractive index at the operating frequency, the operating angular frequency ω_0, the phase constant, the phase and group velocities at that wavelength.

P3.13 An optical fiber has a dispersion parameter of 5 ((ps)/km^2). A Gaussian input signal with a half period of $T = 2$ ps is passed through the optical fibre of length $z = 30$ km. Determine the output pulse width and normalized amplitude.

P3.14 For silica the refractive index varies with wavelength which is approximated as $n(\lambda) = a_1 + a_2\lambda^2 + a_3\lambda^{-2}$ where a's is constant. Show that the dispersion is zero at the wavelength given by $\lambda_{ZD} = -((3a_3)/a_2)$.

P3.15 For P3.14 if the material is a pure silica that is used in optical fiber $a_1 = 1.45084$, $a_2 = -0.00343$ μm^{-2}, and $a_3 = 0.00292$ μm^2, calculate the zero dispersion wavelength λ_{ZD}.

P3.16 Determine the length of the optical fiber in P3.13 if the output pulse width is time the input short transform pulse of $T = 5$ ps.

Polarization

P3.17 A wave is defined with the electric field: $\boldsymbol{E}(z, t) = 125\,sin\,(\omega t - \beta z)\boldsymbol{a_x} + 125\,cos\,(\omega t - \beta z)\boldsymbol{a_y}$ (V/m). Plot the composite field at $z = 0$ and $\omega t = 0, 90°, 180°, 270°, 360°$ and show that the tip of the electric field vector traces a circle. Identify the handedness of the wave.

P3.18 A uniform plane wave propagates in a dielectric medium along z-direction is defined with the electric field: $\boldsymbol{E}(z, t) = 125\,cos\,(1.131\times10^{10}t - 12\pi z)\boldsymbol{a_x} + 250\,cos\,(1.131\times10^{10}t - 12\pi z)$ $\boldsymbol{a_y}$ (V/m). (a) Determine the polarization and tilt angle of the plane wave. Calculate the followings: (b) the frequency of operation and the wavelength of the plane wave; (b) the dielectric constant of the propagation medium; and (c) the instantaneous magnetic field.

P3.19 A uniform plane wave propagates in a dielectric medium along z-direction is defined with the electric field: $\boldsymbol{E}(z, t) = 123\,cos\,(2\pi\times10^9 t - 20\pi z)\boldsymbol{a_x} - 146\,sin\,(2\pi\times10^9 t - 20\pi z)\boldsymbol{a_y}$ (V/m). (a) Determine the polarization, handedness, axial ratio and tilt angle of the plane wave and (b) the instantaneous magnetic field.

P3.20 The magnetic field of a uniform plane wave is given $\boldsymbol{H}(z, t) = 240\cos(1.131 \times 10^{10}t - 143z)\,\boldsymbol{a}_x + 480\cos(1.131 \times 10^{10}t - 143z - 75°)\boldsymbol{a}_y$ (mA/m), (a) derive the instantaneous electric field and (b) determine the polarization and handedness of the wave.

Polarization Loss Factor

P3.21 A z-directed travelling wave from a linearly polarized antenna creates a tilt angle of 35°. The wave impinges on an antenna whose polarization vector is given by: $\boldsymbol{a}_p^2 = \dfrac{4\boldsymbol{a}_x + j3\boldsymbol{a}_y}{5}$. Calculate the polarization loss factor *PLF* in both ratio and dB.

P3.22 An elliptically polarized antenna has a radiated field with major and minor axes ratio 3 : 2. The field is expressed as: $\boldsymbol{E}(z, t) = 15(3\boldsymbol{a}_x + j2\boldsymbol{a}_y)\cos(\omega t - \beta z)$ (V/m). A receiving antenna is linearly polarized along x-axis. Calculate the *PLF*.

P3.23 Repeat P3.22 if the receiving antenna is y-polarized calculate the *PLF*.

P3.24 An elliptically polarized antenna transmits a signal which is defined as: $\boldsymbol{E}(z, t) = 10(3\boldsymbol{a}_x + j\boldsymbol{a}_y)\cos(\omega t + \beta z)$ (V/m); the receiving antenna is a circularly polarized antenna. Calculate the polarization loss factor if: (a) the receiving antenna is a *RHCP*, and (b) the receiving antenna is an *LHCP*.

P3.25 Two elliptically polarized waves are defined as: $\boldsymbol{E}_1(z, t) = \boldsymbol{a}_x 1.3\cos(\omega t - \beta z) + \boldsymbol{a}_y 3.2\cos(\omega t - \beta z + (\pi/2))$ (V/m), and $\boldsymbol{E}_2(z, t) = \boldsymbol{a}_x 7.3\cos(\omega t - \beta z) + \boldsymbol{a}_y 4.2\cos(\omega t - \beta z - (\pi/2))$ (V/m), calculate the resultant electric field, the axial ratio of the resultant field and the handedness of the resultant field.

Poynting Vector in Different Polarizations

P3.26 A 1.8 GHz GSM elliptically polarized signal has two components $E_{x0} = E_{y0} = 125$ (V/m) that impinges on an antenna with an effective aperture area of 2.5 m^2. The phase difference between the two signals $\delta = \phi_y - \phi_x = 67°$. Calculate the time domain expression for Poynting vector. Determine the maximum power in watts delivered by the antenna to the receiver.

P3.27 Repeat P3.26 for $E_{x0} = 2E_{y0}$ and the phase difference is −180°.

P3.28 Repeat P3.26 if the signal has only y-directed field component.

P3.29 Repeat P3.26 for $E_{x0} = 2E_{y0}$ and the phase difference is 90°.

Anisotropic Medium

P3.30 An anisotropic crystal has $\varepsilon_{rx} = 10.2$ and $\varepsilon_{ry} = 4.2$, calculate the lengths d_1 and d_2 for such anisotropic medium that provides *LHCP* and linear polarizations, respectively. The operating frequency is 18 GHz.

4

Wave Propagation in Dispersive Media

4.1 Introduction

The frequency dependence of the characteristic parameter of the medium of electromagnetic wave propagation is called dispersion. Dispersion of propagation media causes significant design challenges of electromagnetic devices and systems. Therefore, understanding the dispersion effects in propagation media is important for efficient devices and system design. This chapter presents dispersion phenomena in solid, liquid, and gaseous media with the classical electron theory of matters, the mechanics model and the macroscopic model. The theory also answers why circular polarization signal is essential for satellite communications. Finally, the theory of anisotropic dielectric media is presented.

So far, we have examined the wave propagation in various homogeneous and nondispersive but lossy[1] media. We have given special attention to the conductive medium and surface resistivity as a function of frequency due to their industrial significance and technological innovations. We have developed the sound concept of *the skin effect* for a good conductor, and the loss tangent for a good dielectric.[2] However, for all these investigations, we assume that the conductivity of the medium is invariant with the operating frequency *f*. And based on the assumption we also have examined the performance of a microwave printed electronics, and found that the conductivity has a profound effect on the performance and the quality factor of microwave frequency electronics. In this chapter, we shall learn the electron theory of dispersion of electromagnetic waves in different conducting and semiconducting media. As shown in Figure 4.1, the media include dielectric media such as gas, solid, and liquid, conducting media such as metal, special media such as plasma and ionosphere. This understanding has a profound impact on the modern electronic circuit design for emerging applications.

The chapter is organized as follows: First, we shall discuss the dispersion of general materials at microwave and mm-wave frequencies. In this regard, we shall cover the dispersion phenomena in

1 In previous study, we have considered the loss factor comes from the finite conductivity of the medium. In this section, the dispersive nature of the medium will be investigated using the classical electron theory of materials using Maxwell's equations so that we can demonstrate the interaction of charged particles and the electromagnetic waves.

2 Basically, we define conductors with conductivity parameters and dielectric with loss tangent. Although both types of materials are having finite conductivity and the loss tangent is partly due to the finite conductivity of the medium σ. In this section, we shall study the macroscopic permittivity parameter of the medium in the light of the microscopic view of electron theory of a single electron and their macroscopic effect with the combination of many electrons in per unit volume. This microscopic view will provide the fundamental mechanism of dispersion and the frequency dependence of the constitutive parameters, ε_r. Any dispersive medium is represented by the relative permittivity or the refractive index which is $n = \sqrt{\varepsilon_r}$.

Fields and Waves in Electromagnetic Communications, First Edition. Nemai Chandra Karmakar.

Companion website: www.wiley.com/go/fieldsandwavesinelectromagneticcommunications

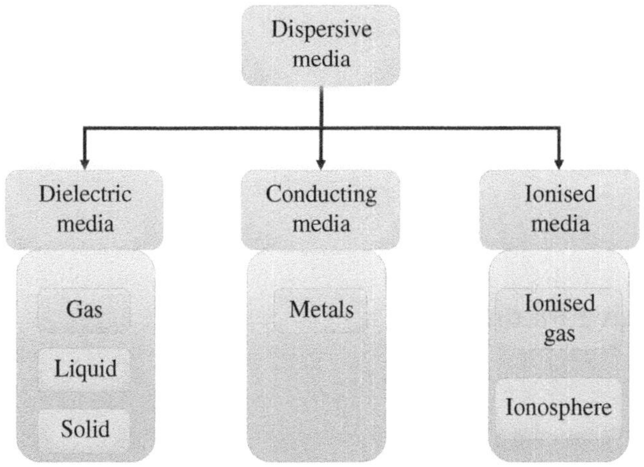

Figure 4.1 Classification of dispersive medium.

dielectric materials and metals. We shall understand the profound impact of frequency of operation and the frequency dependence of wave propagation in different media. We shall use classical electron theory to examine the behavior of the discrete charged particles in the influence of steady and time varying electromagnetic fields. Then we shall study the propagation of the electromagnetic waves in the plasma and ionosphere. In modern satellite communications, the impact of the wave propagation in plasma is highly significant. It is noteworthy to mention that besides the study of wave propagation in telecommunications media, the classical theories of electrons and its link to wave propagation in dispersive media, also help develop solid foundations for electromagnetic wave interaction in particle physics, quantum mechanics, and linear accelerators.[3]

4.2 Dispersion in Materials

The electromagnetic wave always encounters materials either it is in unbounded propagation in air, vacuum and dielectric media,[4] or in a guided propagation in a circuit. The properties of materials are defined with the constitutive parameters ε, μ, and σ, and associated electromagnetic wave's intrinsic parameters n, η, γ, α, and β. Where the symbols have their usual meanings. So far, we have assumed the medium to be homogeneous, linear, isotropic, and lossless. However, in reality, due to the loss factor and anomalies in the structures of the atoms and molecules of the medium, they exhibit dispersive nature. The frequency dependence of the refractive index n and the propagation constant β is defined as *dispersion*. What causes the dispersion in the propagation medium? Why analysis of the dispersion of electromagnetic wave is so important? You will realize the significance

3 An experienced RF engineer understands the practical electromagnetics better than most physicists. When electromagnetic wave interacting with the charged particles such as an electron beam, the RF/microwave engineers usually look into the problem with Maxwell's equations perspective. Therefore, an RF/microwave engineer does not use any simplified guess. As for example in a linear accelerator, an electron beam bunches back and forth and passes through a periodic structure of cavities (with external RF fed to it) is nothing but another form of cavity excitation to an RF engineer. The beam loads the cavity and given the usually wide RF spectrum of the beam, certain resonances are excited and can cause problems in terms of multi-modal dispersion. It is an area where the expertise of the physicists and RF engineers overlap. Therefore, when RF engineers work together with the particle and quantum physicists, there is a synergy and a powerful teamwork to solve the complex problems.

4 We shall also study the wave propagation in a magnetic medium such as ferrite medium in a separate section.

of dispersion of wave in dispersive media when we shall move on section to section and topic to the topic of the applications of the electromagnetic theory.[5] We shall discuss these dispersion phenomena for the guided passive structures such as transmission lines,[6] metallic waveguides, and specifically, in optical fibers. The dispersion covers a big part of the design, because it concerns the pulse that is transmitted via the waveguide and the optical fiber and does get distorted,[7] broadened, and delayed significantly in the other end of the reception. Most modern electromagnetic and microwave design software tools have embedded routines to include the dispersion of the propagation medium. Therefore, the knowledge of dispersion in a medium is imperative for accurate design of devices and systems in modern microwave engineering, optical communications, and antenna technology. Also, the proper design of a wireless propagation channel needs knowledge of dispersion. Ionosphere is a very good example; through the upper atmospheric layers satellite and interstellar communications occur. This proper knowledge of propagation via those media is very important for every component and system design of such technologies.[8] In particle physics, there are huge applications of dispersion. This is an important platform in which both quantum physicists and electromagnetic theorists and microwave engineers can work together. Semiconductor industry also uses microwave for proper control of doping concentration of device fabrication. Here is also the electron theory of materials and dispersion is an important consideration.

Dispersion of propagation media is a complex subject which is usually dealt with complex quantum mechanics and the classical electron theory of matters. In telecommunications, the propagation medium is considered to be dielectric in nature. For example, air is a dielectric medium and its constituents are made of varieties of atoms and molecules, mainly of oxygen, nitrogen, and carbon dioxide.[9] This is about an unbounded medium in which the wireless channel exists. On the other hand, in circuits, we make use of some specific compound such as polytetrafluoroethylene (PTFE), ceramic and similar substrate materials to make the microwave printed circuit boards (PCBs) as the guided propagation medium. To understand the nature of *dispersion* of materials without going through the complex quantum theory of matters, we shall study those dielectric materials first from their classical electron theory perspective where the charged particles such as electrons, ions, and molecules interact with applied electric and magnetic fields.[10] In this regard, we shall use our basic understanding of electron theory of matters that we have learned in basic physics and chemistry. The beautiful aspect of the electromagnetic wave theory of *dispersion* of materials is that with the simple knowledge of the constructions of atoms and molecules of the propagation medium, and their polarities of electric charges, applying simple electromagnetics principles, and some basic theory of mechanics that we have learned in preliminary physics, we can predict the phenomenon of *dispersion* in material media, in which the wave propagates, surprisingly accurately. This understanding not only helps precious design of microwave and mm-wave passive and active circuits,

5 The material and waveguide dispersion have profound impact in passive and active device design such as waveguides and optical systems design.

6 Even in two-wire transmission lines, coaxial cables and modern microwave and mm-wave frequency microstrip transmission lines, slotted lines, and coplanar waveguides, have dielectric substrates as the media of propagation. These dielectric materials are lossy, nonhomogeneous, an-isotropic, and nonlinear in reality. Therefore, precious design of these transmission lines also need to consider the dispersion phenomena and appropriate corrections compared to the ideal design exercise.

7 We also discuss the dispersion and pulse broadening in the definitions of the group velocity and group delay.

8 The author designed a mobile satellite communication antenna that considers the wave propagation through atmospheric layers.

9 The moisture contents (H_2O) has profound impact in loss and dispersion and a big part of terrestrial propagation studies of electromagnetic waves.

10 Both steady state and time varying field waves are considered in the study of dispersion of materials.

and optical systems but also help the scientists enormously in their experiments of particle physics, and understanding of complex characteristics of the discipline. In this section, we shall study the *dispersion* phenomenon of the dielectric materials, firstly in gaseous medium and then in liquid and solid. In the next section, we shall study *the dispersion* of metals.

Review Questions 4.1: Dispersion in Materials

Q1. Define dispersion.
Q2. Explain the cause of dispersion in propagating media.
Q3. Explain why analysis of dispersion is so significant in modern applications?
Q4. Explain various significant applications of dispersion theory in modern industries?
Q5. Explain how dispersion problems are handled in advanced electromagnetics?
Q6. Explain how the simple electron theory helps scientists and microwave engineers solve complex dispersion problems accurately?

4.3 Classical Electron Theory and Dispersion in Material Media

The classical electron theory of matter deals with the motion of free and bound charge particles, usually electrons, under the influence of electromagnetic fields. The moving free electron under the influence of the electromagnetic field constitutes a current. However, the bound electrons may undergo oscillatory motion around its mean equilibrium position. This oscillation at a given frequency for a particular matter is called *the natural frequency*. Matters are formed with interacting atoms and molecules. In a classical atomic model, the electrons with elementary charge of 1.602×10^{-19} C is revolving around a nucleus. The main constituent of the nucleus is the proton which carries a positive charge. The charges of the electron and proton are perceived to be distributed in a volume of radius in the order of 10^{-15} m. The current is produced by the moving charge particles. This means a charged particle undergoes motion with a velocity $\boldsymbol{v}$ due to the applied electric field $\boldsymbol{E}$ with the associated potential difference V. If the charge density is ρ and is moving with a velocity $\boldsymbol{v}$, then the current density is the product of the charge density and velocity: $\boldsymbol{J} = \rho\boldsymbol{v}$. The electromagnetic interaction of the microscopic charge particles of matters is given by Maxwell–Lorentz equations:

Faraday's law:

$$\nabla \times \boldsymbol{E} = -\frac{\partial \boldsymbol{B}}{\partial t} \tag{4.1}$$

Gauss' law for magnetic case:

$$\nabla \cdot \boldsymbol{B} = 0; \tag{4.2}$$

Ampere's circuital law:

$$\nabla \times \boldsymbol{H} = \rho\boldsymbol{v} + \frac{\partial \boldsymbol{D}}{\partial t}; \tag{4.3}$$

Gauss law for electric case:

$$\nabla \cdot \boldsymbol{D} = \rho; \tag{4.4}$$

Constitutive equation for electric case:

$$D = \varepsilon_0 E; \tag{4.5}$$

Constitutive equation for magnetic case:

$$B = \mu_0 H; \tag{4.6}$$

Lorentz force law, here $\boldsymbol{F}$ volume density of force:

$$F = \rho (E + v \times B); \tag{4.7}$$

The contrasting difference between the electron theory and the electromagnetic theory, in this case, is that in the former one, we assume the interacting particles are in vacuum ($\mu = \mu_0$ and $\varepsilon = \varepsilon_0$) whereas in the latter case, it can be in any material media.[11] In the following, we shall examine the interaction of the charge particles in the presence of the electromagnetic fields using the classical electron theory.

Review Questions 4.2: Classical Electron Theory

Q1. Define the natural frequency of a matter.
Q2. Explain the atomic structure and its interaction with the electromagnetic field.
Q3. Explain the difference between the electron theory of matter and the electromagnetic theory of matter.

4.4 Discrete Charged Particles in Static Electromagnetic Fields

To understand the dispersion of materials, the first step is to understand the mechanics of a discrete charge particle when it undergoes motion and then oscillation in the electromagnetic field. To simplify our understanding of the mechanics we first consider the electric field and then the magnetic field. In the next section, we shall consider both fields simultaneously. Figure 4.2 illustrates the mechanics of a charged particle initially at rest, then it gets the velocity due to the potential difference V across a parallel plate capacitor and then the moving charge comes under the influence of a static magnetic field and starts revolving around the magnetic field. Therefore, we start by considering that the charge is at a stationary position (in rest). Then under the influence of the applied static electric field $\boldsymbol{E}$ due to the potential difference V_{BA} at two points A and B in space, the charged particle undergoes motion with a fixed velocity $\boldsymbol{v}$. Then, we consider the moving charge is under the influence of a static magnetic field $\boldsymbol{B}$. Once the moving charge[12]

11 In this regard, we can find *oneness* of the universe with our planetary system and the atomic model of a matter. As the sun is the center of gravity of our planetary system around which all planets are revolving in distinctive orbits in vacuum (space), the nucleus is the center of gravity and all electrons are orbiting in their own trajectories in space. This is the Newtonian particle theory of electron which is still valid. Bhor's distributive model of electron cloud is appropriate for the electromagnetic theory where the charged particles are strongly bonded with each other (electrons and nucleus) with Coulomb's force.

12 A moving charge is the current density. According to the Biot–Savart law, two current elements exert forces to each other. In this case the Lorentz force of magnetic case $\boldsymbol{F_e} = \rho \boldsymbol{v} \times \boldsymbol{B}$ prevails. Due to the cross product of $\boldsymbol{v}$ and $\boldsymbol{B}$, the normal component of the velocity $\boldsymbol{v}$ acts on the magnetic field and causes the rotation of the charged participle. Based on the polarity of the charged particle the spin has handedness – right-hand rotation for positive charge and left-hand rotation for negative charge.

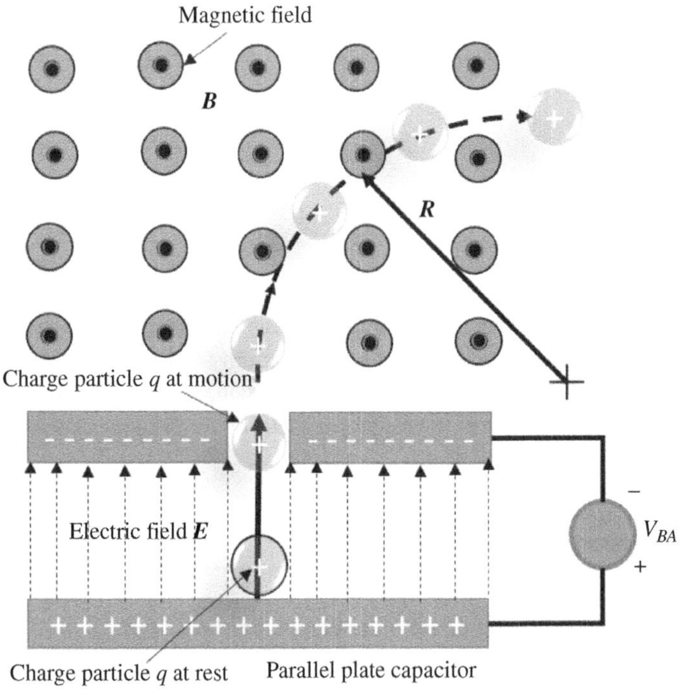

Figure 4.2 Illustration of a charge particle under the influence of electromagnetic field created with a parallel plate capacitor and a current carrying coil (not shown).

comes under the influence of the magnetic field, it undergoes circular motion around the magnetic field. Figure 4.2 illustrates the scenario that is created with a capacitor for the electric case and then a current carrying coil to create the magnetic field (coil is not shown here). As can be seen, a potential difference V is applied to a parallel plate capacitor between the two terminals A and B. A is at zero potential such that $V_{BA} = V$. The associated applied electric field is $\boldsymbol{E}$ and the field lines of the field conversion is from the negative to the positive plates of the capacitor. Assume a charge particle is at rest at point A.

Once the voltage is applied at the terminals A and B of the capacitor, due to the impressed electric field $\boldsymbol{E}$, the charged particle gets acceleration and moves from A to B. Assume the velocity at point A is v_A (zero as it was initially at rest) and the final velocity at point B is v_B. The charge particle has q coulombs of charge. According to the theory of Lorentz force for the electrical case when a charged particle with charge q coulombs is under the influence of a static electric field, $\boldsymbol{E}$ the force acting on it is:

$$\boldsymbol{F_e} = q\boldsymbol{E} \tag{4.8}$$

Assume the charge is moving with a fixed velocity $\boldsymbol{v}$. Due to the influence of the static electric field $\boldsymbol{E}$ it undergoes an acceleration $\boldsymbol{a}$. Assume that the mass of the charged particle is m. Then according to the theory of mechanics, we can write the force equation and equate it with the above equation as follows:

$$\boldsymbol{F_e} = q\boldsymbol{E} = m\boldsymbol{a} \tag{4.9}$$

The work done in moving the charge particle from its initial state to its final state due to the potential difference V_{BA}, where the charge was at rest in its initial position A and moved to final point B is given by:

$$\frac{1}{2}m\left(v_A^2 - v_B^2\right) = qV_{BA} \tag{4.10}$$

Here, we assume $v_A = 0$ as the charge was at rest. The final velocity $v = v_B$ is defined as:

$$v = \sqrt{\frac{2eV}{m}} = 5.93 \times 10^5 \sqrt{V} \tag{4.11}$$

where, $V_{BA} = V$, $q = e$ electron charge, and $v = v_B$ the final velocity is used. From the above analysis, we can deduce that under the influence of the electric field and the associated potential difference V, we can calculate the velocity of the charged particle. For a static field, the velocity is constant. Now we shall examine the influence of the static magnetic field $\boldsymbol{B}$ on the moving charge particle.

If the charged particle moves under the influence of the static magnetic field, the force experienced by the particle can be expressed in terms of the Lorentz force again:

$$\boldsymbol{F_m} = q\boldsymbol{v} \times \boldsymbol{B} \tag{4.12}$$

The velocity can be in any arbitrary direction with respect to $\boldsymbol{B}$. Therefore, we decompose the velocity vector into two parts, $v_{\parallel}$ = velocity component parallel to $\boldsymbol{B}$, and $v_{\perp}$ = velocity component perpendicular to $\boldsymbol{B}$. Now we examine the circulation of the charged particle around $\boldsymbol{B}$ based on the classical theory of mechanics. The force can also be decomposed based on the two component velocities. However, due to the cross product of the magnetic force, the force acting on the charged particle is obviously the normal component[13]:

$$F_{\perp} = qv_{\perp}B \tag{4.13}$$

For a given condition, the force is constant if q, $v_{\perp}$ and B are constant. The force is also perpendicular to both the velocity and the magnetic field. From the classical theory of mechanics, we know that this perpendicular direction of the force with respect to the constant velocity and the fixed direction of the magnetic field results in spinning or circular motion of the charged particle. If r_B is the radius of the circle then the centrifugal force[14] is defined as

$$F_{\perp} = \frac{mv_{\perp}^2}{r_B} \tag{4.14}$$

Equating the above two equations, we obtain *the radius of gyration*

$$r_B = \frac{mv_{\perp}}{qB} \text{ (m)} \tag{4.15}$$

13 For simplicity of calculation we now consider only the scalar components of the vector quantities $\boldsymbol{F}$, $\boldsymbol{v}$, and $\boldsymbol{B}$.
14 Refresh the definition and theory of centrifugal and centripetal forces via Wikipedia https://en.wikipedia.org/wiki/Centrifugal_force and https://en.wikipedia.org/wiki/Centripetal_force, respectively, (accessed 13 February 2018).

In plasma physics, the radius is called *Larmor radius*.[15] The angular frequency is called *Larmor frequency* and is defined as:

$$\omega_B = \frac{v_\perp}{r_B} = \frac{qB}{m}\ \text{(rad/s)} \tag{4.16}$$

ω_B is also called *cyclotron resonance frequency*. *Cyclotron resonance* describes the interaction of the external force exerted on the charged particle moving in a circular path under the influence of a magnetic field. It is interesting to note that the frequency is independent of the radius, velocity, and hence the kinetic energy.[16] All particles with the same charge to mass ratio (q/m) rotates around the magnetic field vector $\boldsymbol{B}$ with the same frequency ω_B.

The rotational frequency is given by:

$$f_B = \frac{v_\perp}{2\pi r_B} = \frac{qB}{2\pi m}\ \text{(Hz)} \tag{4.17}$$

The time period of the circular motion is given by

$$T_B = \frac{1}{f_B} = \frac{2\pi m}{qB}\ \text{(s)} \tag{4.18}$$

In the following, we do some review questions and examples to support the basic concept of the theory.

Review Questions 4.3: Charge Particle in Magnetic Field

Q1. Explain the interaction of a charge under the influence of the applied electromagnetic field.
Q2. Show the velocity of a charged particle under the influence of voltage *V*.
Q3. Define the radius of gyrotron. What is the other name of the radius?
Q4. Define the cyclotron frequency and cyclotron resonance.
Q5. Define the rotational frequency and time period of a charged particle.
Q6. Explain why *q/m* is so significant?

Example 4.1: Cyclotron Frequency
Calculate the cyclotron frequency for an electron under the influence of 2.2 T steady magnetic field. Electron mass 9.109×10^{-31} kg and electric charge 1.6×10^{-19} C.

Answer:

$$f_B = \frac{1.6 \times 10^{-19} \times 2.2}{2\pi \times 9.109 \times 10^{-31}} \approx 61.5\,\text{GHz}$$

Exercise 4.1: Cyclotron Frequency
Q1. Recalculate the cyclotron frequency for 1.2 T and mass of the charged particle 1.7×10^{-27} kg and charge 1.6×10^{-19} C.

Answer:
18 MHz.

---

15 Read definition of *Larmor radius* in https://de.wikipedia.org/wiki/Larmor-Radius (accessed 13 February 2018).
16 https://en.wikipedia.org/wiki/Cyclotron_resonance (accessed 13 February 2018).

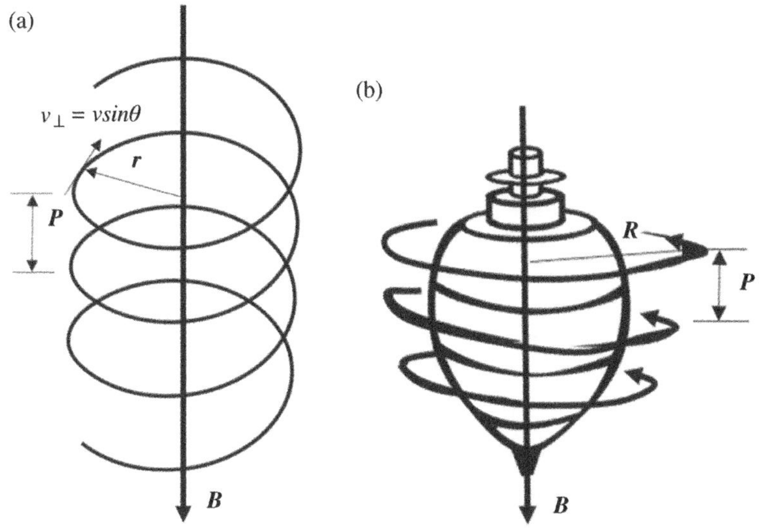

Figure 4.3 Trajectory of charged particles under the influence of steady magnetic field (a) same velocity, and (b) different velocities. *Source:* Adapted from https://physics.stackexchange.com/questions/311836/formula-of-the-radius-of-the-circular-path-of-a-charged-particle-in-a-uniform-ma/311847.

Examining the definition of radius r_B we can see that it is directly proportional to the velocity $v_\perp$ provided that the same charge to mass ratio (q/m)[17] is maintained and the particle rotates around the magnetic field vector $\boldsymbol{B}$ with the same frequency ω_B. Therefore, the particle with a larger velocity will travel along circle of larger radius, but both slow- and fast-moving particles will complete one revolution at the same time. This can also happen with nonuniform magnetic fields, in which case the magnetic field can be assumed to be slowly varying in amplitude with time. In both cases, the resultant motion gives a trajectory in the shape of a helix. Figure 4.3 shows the spinning charged particle along the magnetic field and its analogy with a revolving top. Here $v_{||}$, the velocity component parallel to $\boldsymbol{B}$, will not contribute in the resultant motion due to the Lorentz force $\boldsymbol{F_m} = q\boldsymbol{v} \times \boldsymbol{B}$. However, the velocity component parallel to $\boldsymbol{B}$ continues to travel along the direction of $\boldsymbol{B}$. Thus, the resultant velocity, is the trajectory of a spring, as shown in Figure 4.3a. The positive charge moves in the direction of the right-hand screw and the negative charge in the left-hand screw. For the generalized case, the axis of the helix is parallel to the magnetic field line, as shown in Figure 4.3b.[18] The pitch of the helix is defined as:

$$P = v_B T = \frac{2\pi m v_B}{qB}\ (m) \tag{4.19}$$

Figure 4.4 illustrates the cyclotron of positive charge under the influence of the static magnetic field $\boldsymbol{B}$. Also, see the YouTube video on the motion in a magnetic field via the link.[19]

17 The charge to mass ratio is called *specific charge* first calculated by J. J. Thomson in 1897 https://en.wikipedia.org/wiki/Mass-to-charge_ratio (accessed 18 February 2018).
18 In the next section the relations between vectors are examined using the theory of mechanics.
19 https://www.youtube.com/watch?v=3NMQtVnd0-s and https://www.youtube.com/watch?v=84ZTzeCfswg (accessed 18 February 2018).

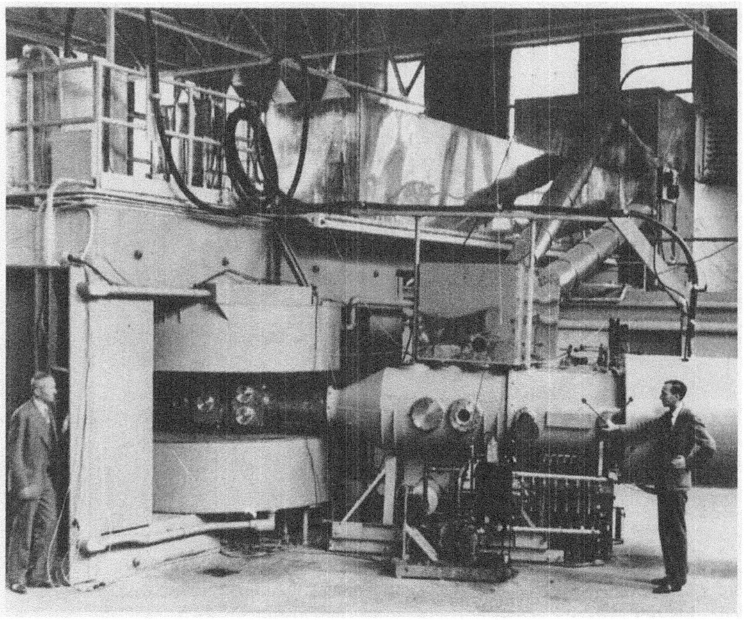

Figure 4.4 Cyclotron. *Source:* U.S. Department of Energy/Wikimedia Commons/Public domain.

Example 4.2: Proton in Magnetic Field

A proton is accelerated from rest through a potential difference of 2400 V in the capacitor. It is then injected into a uniform magnetic field $\boldsymbol{B}$. of 0.5 T.[20]

i) Find the radius of its orbit. Ignore the effect due to gravity
ii) Find the speed of the particle upon exiting the capacitor, use conservation of energy between the kinetic energy and electromagnetic energy.

Answer:

i) $v = 5.93 \times 10^5 \times \sqrt{2400} = 29.051 \times 10^6 (\text{m/s})$

ii) The radius of gyratron $r = \dfrac{1.76 \times 10^{-27} \times 29.051 \times 10^6}{1.6 \times 10^{-19} \times 0.5} = 63.91\ (\text{cm})$

Practical Application[21]

"E.O. Lawrence and M.S. Livington first developed cyclotron device for imparting very high energy to positive ions. A **cyclotron** is a type of *particle accelerator* invented by *Ernest O. Lawrence* in 1934 in which *charged particles* accelerate outward from the center along a spiral path. The particles are held to a spiral trajectory by a static magnetic field and accelerated by a rapidly varying (*radio*

20 https://www.slideshare.net/abhinaypotlabathini/magnetic-39152188 (accessed 13 February 2018).

21 Adapted from https://en.wikipedia.org/wiki/Cyclotron (accessed 13 February 2018).

frequency) electric field. Lawrence was awarded the 1939 *Nobel prize in physics* for this invention. Cyclotrons were the most powerful particle accelerator technology until the 1950s when they were superseded by *the synchrotron*, which are still used to produce particle beams in physics and *nuclear medicine*. The largest single-magnet cyclotron was the 4.67 m (184 in) *synchrocyclotron* built between 1940 and 1946 by Lawrence at *the University of California at Berkeley*, which could accelerate protons to 730 *MeV*. The largest cyclotron is the 17.1 m (56 ft) multimagnet *TRIUMF* accelerator at *the University of British Columbia* in *Vancouver, British Columbia* which can produce 500 MeV protons. Over 1200 cyclotrons are used in nuclear medicine worldwide for the production of radionuclides." Wikipedia *https://en.wikipedia.org/wiki/Cyclotron* (accessed 13 February 2018).

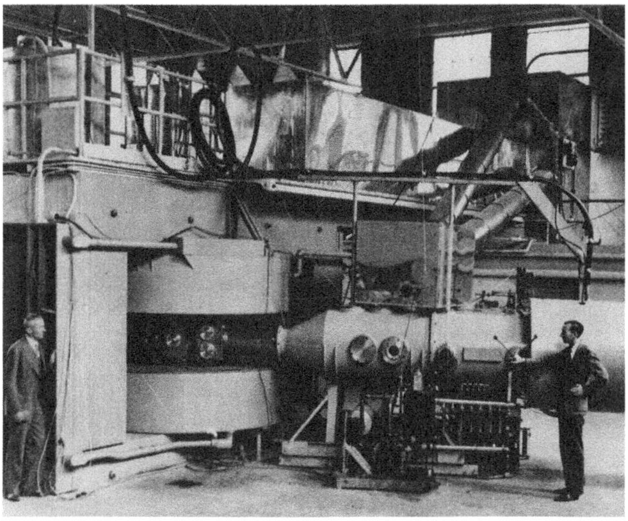

"Lawrence's 60-inch cyclotron, with magnetic poles 60 inches (5 feet, 1.5 meters) in diameter, at the University of California Lawrence Radiation Laboratory, Berkeley, in August, 1939, the most powerful accelerator in the world at the time. Glenn T. Seaborg and Edwin M. McMillan (right) used it to discover plutonium, neptunium, and many other transuranic elements and isotopes, for which they received the 1951 Nobel Prize in chemistry. The cyclotron's huge magnet is at left, with the flat accelerating chamber between its poles in the center. The beamline which analyzed the particles is at right." *https://en.wikipedia.org/wiki/Cyclotron* (accessed 13 February 2018).

Review Questions 4.4: Discrete Charged Particles in Magnetic Field

Q1. Explain what happens to a charged particle under the influence of a steady and time varying magnetic fields?

Q2. Explain the handedness of revolution of the charged particle based on its polarity.

Q3. Explain the significances of cyclotron and synchrotron.

4.5 Classical Mechanics Model of Matters

In the previous section, we have examined the spinning charged particle under the influence of the magnetic field. In that aspect we have defined *the cyclotron frequency,* which is *the natural frequency* of a matter.[22] In this section, we use the mechanics model to understand the relationship of different vectors and converge to the same results as obtained from the electron theory of the charged particle and electromagnetic interaction. Figure 4.5a illustrates the 2D

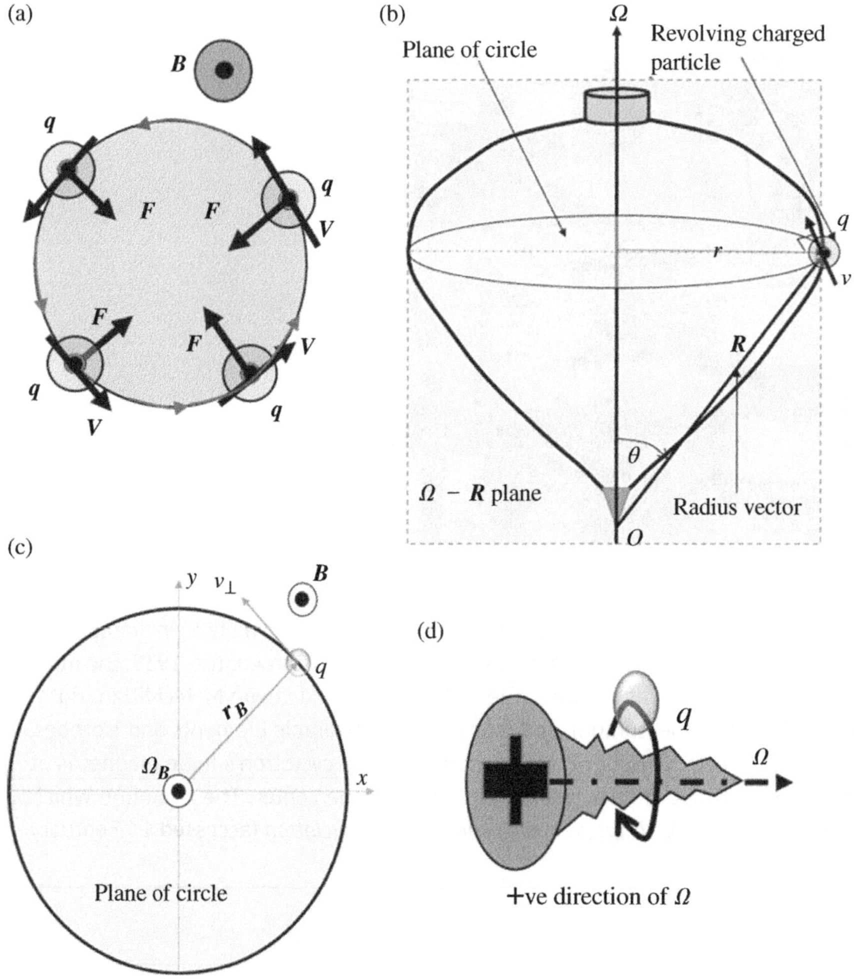

Figure 4.5 (a) Revolving charged particle q under the influence of $\boldsymbol{B}$, (b) mechanical model of a revolving charge particle in presence of $\boldsymbol{B}$, and (c) systemization of vectors on the place of circle, and (d) direction of Ω that follows right hand screw rotation.

---

22 Each charged particle has its natural frequency and distinctive from each other like the atomic structure of individual atoms of a particular matter. Materials behave very differently near its natural frequency and goes from resonances.

aspect of the revolving charged particle of charge q under the influence of the steady magnetic field $\boldsymbol{B}$. Due to the interaction between the field and the charged particle, the force is acting toward the center. And the velocity is always perpendicular to the force, the magnetic field and tangent to the locus of rotation of radius r. The cross section of the spinning charged particle keeps the center of gravity of the magnetic field at the center with origin O,[23] as shown in Figure 4.5b. Consider the plane of circle with radius r, the charge particle revolves around the circle with velocity $\boldsymbol{v}$. *The radius vector* from the origin to the charge particle is $\boldsymbol{R}$[24] and makes an angle θ with the vector Ω. The velocity vector is perpendicular to both the radius r and the Ω-R plane. Based on the classical mechanics, we define *the linear momentum* of the charged particle with mass m and velocity $\boldsymbol{v}$ as:

$$\boldsymbol{p} = m\boldsymbol{v}\,(\mathrm{kg-m/s}) \tag{4.20}$$

To understand the spinning of the charged particle around the axis of the magnetic field, we need to consider its angular momentum.[25] Angular momentum represents the charged particle's rotational inertia and rotational velocity about a particular axis. As shown in Figure 4.5c, the particle's trajectory lies in a single plane (plane of circle), the angular momentum can be considered as rotational analog of the linear momentum. Thus, we can write the angular momentum $\boldsymbol{L}$ as:

$$\boldsymbol{L} = \boldsymbol{R} \times \boldsymbol{p} = \boldsymbol{R} \times m\boldsymbol{v} \tag{4.21}$$

where $\boldsymbol{R}$ is the radius vector from the origin to the center of the particle. In relation to the angular momentum we also define an angular velocity vector $\boldsymbol{\Omega} = \omega \boldsymbol{a_u}$ as shown in Figure 4.5d. Where ω is the ampltude of the angular velocity and $\boldsymbol{a_u}$ is the unit vector. The direction of $\boldsymbol{\Omega}$ is parallel to the axis of rotation as shown in the figure. The linear velocity is defined as $\boldsymbol{v} = \omega \boldsymbol{r} = \omega \boldsymbol{R} sin\theta$. This tells us the perpendicular (cross product) relationship between the angular velocity $\boldsymbol{\Omega}$ and the radius vector $\boldsymbol{R}$. This means the linear velocity vector $\boldsymbol{v}$, $\boldsymbol{R}$, and Ω form a right-hand system of vectors and can be written as:

$$\boldsymbol{v} = \boldsymbol{\Omega} \times \boldsymbol{R} \tag{4.22}$$

Therefore, the velocity vector has one component $\boldsymbol{v}_\perp$ that is always perpendicular to $\Omega - R$ plane as shown in Figures 4.5b, c. Now transferring the analysis to the plane of circle and representing all relevant components with subscript "B," we designate the rotational gyration vector $\boldsymbol{\Omega_B}$ then we can write;

$$\boldsymbol{v} = \boldsymbol{v}_\perp = \boldsymbol{\Omega_B} \times \boldsymbol{r_B} \tag{4.23}$$

where $\boldsymbol{r_B}$ is *the radius of gyration* and likewise, the rotation vector $\boldsymbol{\Omega}$ becomes $\boldsymbol{\Omega_B}$ and the amplitude ω becomes ω_B. We assume the system is under the influence of a uniform magnetic field so that $\boldsymbol{\Omega_B}$

23 Selecting the position of the origin is arbitrary and should not influence the interaction.

24 Note $Rsin\theta = r$ here from the geometry of rotation.

25 Read details about angular momentum in https://en.wikipedia.org/wiki/Angular_momentum (accessed 15 February 2018).

is also uniform. In this instance, the time derivative of the vector $\boldsymbol{\Omega_B}$ is zero. Therefore, the acceleration vector is defined as:

$$\boldsymbol{a} = \frac{d\boldsymbol{v}}{dt} = \frac{d\boldsymbol{v}_\perp}{dt} = \frac{\boldsymbol{d}}{\boldsymbol{dt}}(\boldsymbol{\Omega_B} \times \boldsymbol{r_B}) = \boldsymbol{\Omega_B} \times \frac{d\boldsymbol{r_B}}{dt} = \boldsymbol{\Omega_B} \times \boldsymbol{v}_\perp \tag{4.24}$$

However, $\boldsymbol{v}_\perp = d\boldsymbol{r_B}/dt = \boldsymbol{\Omega_B} \times \boldsymbol{r_B}$; therefore, we can define acceleration as follows:

$$\boldsymbol{a} = \boldsymbol{\Omega_B} \times (\boldsymbol{\Omega_B} \times \boldsymbol{r_B}) = \boldsymbol{\Omega_B}(\boldsymbol{\Omega_B} \cdot \boldsymbol{r_B}) - \boldsymbol{\Omega_B^2 r_B} \tag{4.25}$$

Since the orthogonality of the two vectors $\boldsymbol{\Omega_B}$ and $\boldsymbol{r_B}$, $\boldsymbol{\Omega_B}.\ \boldsymbol{r_B} = 0$. Therefore, the acceleration vector is reduced to:

$$\boldsymbol{a} = \frac{d\boldsymbol{v}}{dt} = -\boldsymbol{\Omega_B^2 r_B}\left(\frac{\mathrm{m}}{\mathrm{s}^2}\right) \tag{4.26}$$

This is an interesting derivation that links the acceleration of the rational charge particle with its angular frequency ω_B and the radius vector $\boldsymbol{r_B}$. Now we consider Lorentz force equation:

$$\boldsymbol{F} = m\boldsymbol{a} = m\frac{d\boldsymbol{v}}{dt} = -m(\boldsymbol{\Omega_B} \times \boldsymbol{v}) = q(\boldsymbol{v} \times \boldsymbol{B}) \tag{4.27}$$

Since there is an orthogonal relationship existing between $\boldsymbol{\Omega_B}$ and $\boldsymbol{v}$, and $\boldsymbol{v}$ and $\boldsymbol{B}$, therefore we can express the following scalar relation between the angular velocity vector $\boldsymbol{\Omega_B}$ and $\boldsymbol{B}$ as follows:

$$\boldsymbol{\Omega_B} = \boldsymbol{\omega_B} = -\frac{q\boldsymbol{B}}{m} \tag{4.28}$$

where we derived in the previous section $|\Omega_B|$ is ω_B, *the cyclotron frequency.*

A few remarks can be made from the above physical analysis of the circular motion of a charged particle under the steady magnetic field:

1) The mechanics model provides a much clearer understanding of the microscopic vector system and the relative position of different vector components of the system.
2) The mechanics system is understood in terms of linear velocity, the linear momentum and then the angular momentum and the radius vector. These three vector quantities form the right-hand vector system from there we have derived a relationship of the circulation of the charged particle with respect to the applied magnetic field.
3) Due to the orthogonal relationship, the velocity that contributes to the spinning on the charged particle is the orthogonal velocity vector. This simplifies the derivation of the acceleration of the charged particle and the acting force on the particle.
4) From the right-hand vector relation between the linear and angular velocities and the radius vector, we are also able to calculate the acceleration in terms of the amplitudes of the rotation vector squared and the velocity vector ($\omega_B^2 r_B$). The sign has some physical significance which tells us that the force acting on the particle is centrifugal or centripetal force depending on the sign of the charge and the direction of rotation.
5) Finally, we have derived the same *cyclotron frequency* using the Lorentz force of the mechanics model as we did use the electronic model of the charged particle under the influence of electromagnetic fields.

Review Questions 4.5: Mechanics Model of Charged Particles

Q1. Discuss the mechanics model of a charged particle q under the influence of a magnetic field.

Q2. What is radius of gyration?

Q3. From the mechanics model, show that the Lorentz force $\boldsymbol{F} = q\,(\boldsymbol{v} \times \boldsymbol{B})$.

Q4. Explain the difference between electronic and mechanics model of a spinning charge under the influence of the magnetic field.

4.6 Motion of Charged Particle in Steady Electric and Magnetic Fields

So far, we have described the motion of the charged particle under the influence of the electric field $\boldsymbol{E}$ and the magnetic field $\boldsymbol{B}$ separately. We have found that the charged particle under rest gets momentum with the influence of the applied electric field and then starts travelling with a uniform velocity $\boldsymbol{v}$. Once the particle comes under the influence of the magnetic field, it starts revolving with a helical trajectory around the magnetic field. In this section, we shall examine the motion of the very charged particle under steady electric and magnetic fields simultaneously, as shown in Figure 4.6. As can be seen, a charged particle of charge $+q$ undergoes a revolution under the influence of simultaneous steady electric and magnetic fields. The forces acting on the charge is the electric force $\boldsymbol{F_e}$ and the magnetic force $\boldsymbol{F_m}$. The orthogonal relationship between the motion of the charge particle, and the electric and magnetic fields are shown in the figure. This is the most simplistic case that is demonstrated here. In general, the fields are not always orthogonal to each other and likewise, the force exerted on the charge and its resultant motion.

Therefore, the motion of a charged particle of charge q under the influence of both the electric and the magnetic fields has varieties of manifestations, starting from a straight-line motion to the cycloid and other complex motion.[26] While the force of the magnetic field is responsible for the circular motion, the force of the electric field tends to bring changes in both direction and magnitude depending on the direction of the charge particle's velocity with respect to the applied electric field. Figure 4.6 illustrates the "crossed fields" in which case the electric and the magnetic fields are perpendicular to each other. The charged particle under the influence of the crossed field has special significance and application in electromagnetic measurements and cyclotrons, synchrotron, etc.

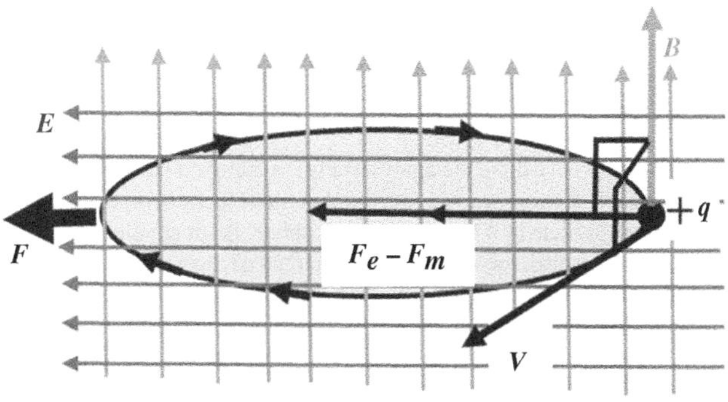

Figure 4.6 A charged particle $+q$ under the influence of simultaneous steady electric and magnetic fields. *Source:* Figure adopted from http://scientificsentence.net/Equations/Electrostatics/index.php?key=yes&Integer=motion_charges (accessed 17 February 2018).

26 S. K. Singh, *Motion of a Charged Particle in Electric and Magnetic Fields*, OpenStax-CNX module: m31547, pp.1–12 http://creativecommons.org/licenses/by/3.0/ (accessed 19 February 2018).

As stated above, the relative orientation of $\boldsymbol{v}$, $\boldsymbol{E}$, and $\boldsymbol{B}$ are too broad and can have many possible manifestations. Based on the preceding examination of the relative position and orientation of these vectors, we can consider the three simplified cases: (i) the velocity $\boldsymbol{v}$ is parallel to $\boldsymbol{E}$, (ii) the velocity $\boldsymbol{v}$ is parallel to $\boldsymbol{B}$, and (iii) the velocity $\boldsymbol{v}$ is perpendicular to $\boldsymbol{B}$. Using these three conditions of the orientations of the velocity vector with respect to the electric and the magnetic fields, we can analyze the velocity vector and its effect into three contexts: (i) $\boldsymbol{v}_1$ is the velocity that is the result due to the interaction of the charge particle q with the magnetic field alone,[27] and (ii) $\boldsymbol{v}_2$ is the velocity that is the result due to the interaction of the charge particle q with both the electric field and the magnetic field together. Here, we are considering the specific orientation of the electric and the magnetic fields and assume that they can have any orientation between themselves and with respect to the velocity of the charged particle. This assumption is made to make the analysis general and can be justified for any specific cases later on. Since we are considering the DC case for the electromagnetic field, both $\boldsymbol{E}$ and $\boldsymbol{B}$ are time independent. Therefore, we can consider that $\boldsymbol{v}_2$ is some sort of a uniform drift velocity of the charge particle.[28] Now let us analyze the motion of the charged particle in the electromagnetic field. The velocity is defined as:

$$\boldsymbol{v} = \boldsymbol{v}_1 + \boldsymbol{v}_2 \tag{4.29}$$

The time independent condition is:

$$\frac{d\boldsymbol{v}_2}{dt} = 0 \tag{4.30}$$

Lorentz force equation due to the velocity $\boldsymbol{v}_1$ and the magnetic field $\boldsymbol{B}$ is:

$$m\frac{d\boldsymbol{v}_1}{dt} = q\left(\boldsymbol{v}_1 \times \boldsymbol{B}\right) \tag{4.31}$$

Lorentz force equation due to the composite velocity $\boldsymbol{v}$ and the electromagnetic field is:

$$m\frac{d\boldsymbol{v}}{dt} = q\left(\boldsymbol{E} + \boldsymbol{v} \times \boldsymbol{B}\right) \tag{4.32}$$

Substitution for (4.29) and (4.31) into (4.32) we obtain

$$m\frac{d\left(\boldsymbol{v}_1 + \boldsymbol{v}_2\right)}{dt} = q\{\boldsymbol{E} + \left(\boldsymbol{v}_1 + \boldsymbol{v}_2\right) \times \boldsymbol{B}\} \tag{4.33}$$

Since $(d\boldsymbol{v}_2/dt) = 0$ and $m\,(d\boldsymbol{v}_1/dt) = q\,(\boldsymbol{v}_1 \times \boldsymbol{B})$ we obtain

$$\boldsymbol{E} = -\boldsymbol{v}_2 \times \boldsymbol{B} \tag{4.34}$$

27 We already have done the analysis of the velocity due to the interaction of the charged particle under the influence of the magnetic field alone. We realized that the velocity component of $\boldsymbol{v}_1$, which is perpendicular causes the rotation of the charged particle in a circular motion about the magnetic field. The component of $\boldsymbol{v}_1$, which is parallel to the magnetic field causes the motion of the charged particle along the magnetic field. The combined effect of the motion is the trajectory of helical orbit about the magnetic field. In this section, we examine the effect of the velocity $\boldsymbol{v}_2$, which is the velocity of the charged particle that undergoes the simultaneous influence of both the electric and magnetic fields.
28 *The drift velocity* of a charged carrier is the average velocity that the particle attains in a material under the influence of an applied electric field. Since the velocity is assumed to be due to the interaction of the charged particle with both the electric and magnetic fields, we can identify some component of the velocity that are truly *the drift velocity* and this specifically refers to *the electric drift velocity. The drift velocity* is perpendicular to both the electric and the magnetic fields.

To understand the properties of the velocity $\boldsymbol{v}_2$ in more details, let us consider the product of $\boldsymbol{E} \times \boldsymbol{B}$ as follows:

$$\boldsymbol{E} \times \boldsymbol{B} = (-\boldsymbol{v}_2 \times \boldsymbol{B}) \times \boldsymbol{B} \tag{4.35}$$

Now using the vector identity $(\boldsymbol{A} \times \boldsymbol{B}) \times \boldsymbol{C} = (\boldsymbol{A}.\ \boldsymbol{C})\boldsymbol{B} - (\boldsymbol{B}.\ \boldsymbol{C})\boldsymbol{A}$ we obtain the following expression for the right-hand side of (4.35),

$$(-\boldsymbol{v}_2 \cdot \boldsymbol{B})\boldsymbol{B} + (\boldsymbol{B} \cdot \boldsymbol{B})\boldsymbol{v}_2 = -v_{EB}B^2\boldsymbol{a_B} + B^2\boldsymbol{v}_2 \tag{4.36}$$

where, v_{EB} is the magnitude of the velocity component of v_2 along $\boldsymbol{B}$, and $\boldsymbol{a_B}$ is the unit vector of $\boldsymbol{B}$. Therefore, equating (4.35) and (4.36), we obtain the exact solution for the velocity $\boldsymbol{v}_2$ as follows:

$$\boldsymbol{v}_2 = \boldsymbol{v}_2' + \boldsymbol{v}_2'' = \frac{\boldsymbol{E} \times \boldsymbol{B}}{B^2} + v_{EB}\boldsymbol{a_B} \tag{4.37}$$

From the above analysis, we can draw the following remarks for the charged particle that undergoes motion due to the influence of the simultaneous electric field and the magnetic field:

1) The velocity vector that is governed by simultaneous electric and magnetic fields, has two components: $\boldsymbol{v}_2'$ *and* $\boldsymbol{v}_2''$. The first one $\boldsymbol{v}_2' = (\boldsymbol{E} \times \boldsymbol{B}/B^2)$ is perpendicular to both the electric and the magnetic fields ($\boldsymbol{v}_2' \perp \boldsymbol{E} \perp \boldsymbol{B}$). The velocity is called *the electric drift velocity*. The drift velocity is a function of the electromagnetic fields only and independent of the charge q of the particle. The instantaneous center of the rotation of the particle is called *guiding* center.
2) The second component $\boldsymbol{v}_2'' = (v_{EB}\boldsymbol{a_B})$ is parallel to the applied magnetic field $\boldsymbol{B}$ ($\boldsymbol{v}_2' \parallel \boldsymbol{B}$).
3) Judging the above manifestations of the velocity $\boldsymbol{v}_2$; we can argue that when the velocity is perpendicular to both $\boldsymbol{E}$ and $\boldsymbol{B}$ then the component $\boldsymbol{v}_2'' = (v_{EB}\boldsymbol{a_B})$ must vanish.

Figure 4.7 illustrates the various manifestations of the motion of the charged particle under the influence of the electromagnetic field:

Let us do an example of the motions of charged particles under the influence of electromagnetic fields.

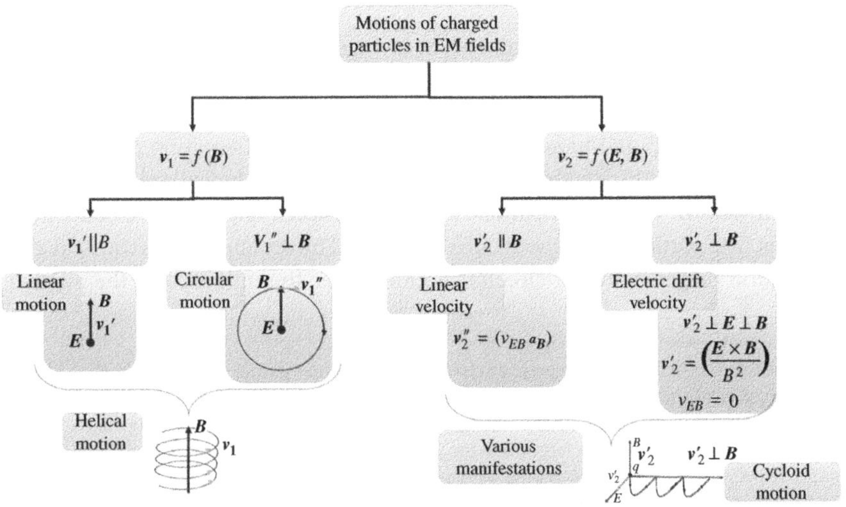

Figure 4.7 Various manifestations of charge particle under the influence of electromagnetic fields.

Example 4.3: Motions of Charged Particles in Electromagnetic Fields
Consider the electric field $\boldsymbol{E}$, magnetic field $\boldsymbol{B}$, and velocity vector $\boldsymbol{v}$ of a charged particle of charge q. The three vectors are orthogonal to each other. Calculate the amplitude of the velocity vector so that the net force on the particle is zero.[29]

Solution:
The Lorentz force on the charged particle is:

$$\boldsymbol{F} = q\boldsymbol{E} + q(\boldsymbol{v} \times \boldsymbol{B})$$

Let the three vectors be defined as: $\boldsymbol{E} = E\boldsymbol{a_x} : \boldsymbol{B} = B\boldsymbol{a_y}$ and $\boldsymbol{v} = v\boldsymbol{a_z}$
The force is then $\boldsymbol{F} = \boldsymbol{F_e} + \boldsymbol{F_m} = qE\boldsymbol{a_x} + q\,(v\boldsymbol{a_z} \times B\boldsymbol{a_y}) = q\,(E - vB)\boldsymbol{a_x}$

From the above calculation we realize that the net force on the charged particle is the vector subtraction of the electric and magnetic forces and acts along x-direction, which is the direction of the electric field vector. If the electric force $F_e > F_m$ then the net force acts along the same direction of the electric field. If the magnetic force $F_m > F_e$ then the net force $\boldsymbol{F}$ acts along the opposite direction of the electric field.

If the electric field, magnetic field, and velocity are such that the net force is zero, then we have

$$E - vB = 0, \text{ or, } v = \frac{E}{B}$$

This tells us that though the net force on the particle is zero, still the charged particle moves with its original direction with a velocity which is the ratio of the electric and the magnetic fields. There is no deviation of the amplitude and direction of the particle and can be separated. This technique is called velocity selector or filter. Therefore, if we know the amplitude of the electric and the magnetic fields then we can measure the velocity of the charged particle. This technique is used to understand the electromagnetic waves, and in various measurement systems and applications such as cyclotron as shown in the above example.

4.7 Theory of Cyclotron

Cyclotron was first invented by E. O. Lawrence and M. S. Livingston in 1932. It is used to impart very high energies in the order of mega electron volts to positive ions, as shown in Figure 4.8.[30] The cyclotron has two D-shape regions also called dees[31] separated by a uniform gap g. In the gap, there is a uniform electric field (Felming's left hand rule) established with an alternating potential difference supplied by a high frequency oscillator. The magnetic field $\boldsymbol{B}$ is perpendicular to the surface D and the applied electric field $\boldsymbol{E}$. The charge particle leaves a dee accelerated by the electric field across the gap of two dees. The particle gains speed once accelerated by the applied electric field and enters the other dee with a bigger radius of circular motion under the influence of the applied

29 https://www.youtube.com/watch?v=53w2GJFF3Cs.

30 It is not encouraged to use the electric field to accelerate electrons. Being a lighter particle with mass in the order of $9.10938356 \times 10^{-31}$ kg, the velocity of the acceleration electron reaches close to the velocity of light. Based on the theory of relativity the gain of mass of an accelerated particle is defined as $m = m_0/\{1-(v/c)^2\}$, where, v is the velocity, c is the speed of light and m_0 is the mass at rest. Therefore, when the particle's velocity approaches the speed of light the mass goes infinitely large. However, this issue is overcome with synchronous cyclotron also called synchrotron by adjusting the oscillator frequency with the change of the mass. (See https://www.youtube.com/watch?v=--9CHTtTdV0 for more explanation (accessed 19 February 2018)).

31 The dee is made of hollow, evacuated metallic chamber.

magnetic field. The particle returns to the gap with a larger radius. By this time the polarity of the gap has changed due to the alternating nature of the oscillators' potential difference between the gap. The direction of the electric field is reversed in this case. The particle is accelerated and collected at the other edge of the dee. The particle gains a larger radius. Thus, with the acceleration due to the alternating electric field from the oscillator, and the force[32] acting on the moving charge particle on the Dees, the spiraling circulation of the charged particle in the dee region happens, the particle gains the maximum radius in which end the accelerated charge particle is collected with high energy in the order of mega eV.

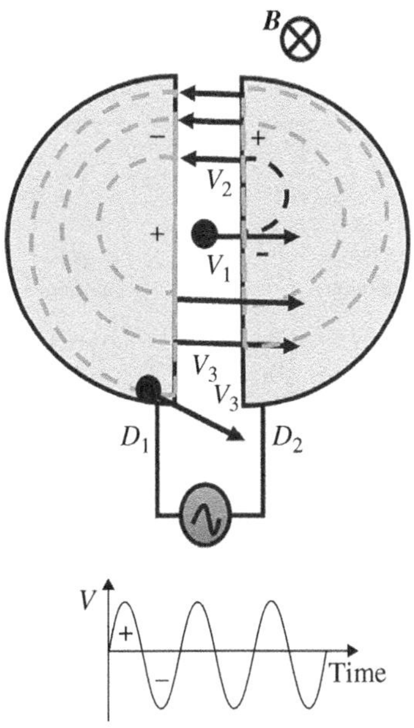

Figure 4.8 Working principle of basic cyclotron.

The oscillator synchronizes the frequency of the applied alternating potential with the same angular frequency of the charged particle's cyclotron frequency so that no phase mismatch happens. The magnetic field cannot accelerate but only make the circular motion of the charge particle in the dee region. The radius of the rotation is: $R = (mv/qB)$. The time $t = (\pi m/qB)$ that rotates the particle and comes to the edge of the dee. The time period of one circulation is: $T = 2\pi m/(qB)$. Once the particle reaches the edge of the dee, the particle gets an acceleration with increased velocity in the gap and:

$R_{max} = (mv_{max}/qB)$, the frequency $f_c = (qB/2\pi m)$ is the cyclotron frequency. The maximum speed is: $v_{max} = (R_{max}qB/m)$; the maximum kinetic energy is: $K_{max} = (1/2)mv_m^2 = (1/2)m (R_{max}qB/m)^2 = \left((R_{max}qB)^2/2m\right)$.

The maximum of this kinetic energy is limited when the speed of the particle approaches that of light. The relativistic effects make the frequency ω depend on the speed. In a synchrocyclotron, to overcome this effect, we change the frequency of the oscillator to synchronize, at each cycle, the two frequencies. Thus, the synchrotron is evolved.

As stated above, the applications of cyclotron are: (i) synthesizing of new materials or substances, (ii) creating radioactive isotopes for medical treatments, and (iii) bombardment of nuclei.[33]

Example 4.4: Theory of Cyclotron

For example, an accelerated proton $+e$ emerging from a cyclotron of radius $R = 50.00$ cm with a magnetic field of magnitude $B = 1.5$ T has[34,35]:

- The cyclotron frequency: $\omega = (e/m)B = ((1.6 \times 10^{-19})/(1.67\ x\ 10^{-27})) \times 1.5 = 1.5 \times 10^8$ (rad/s)
- The speed: $v = \omega R = 1.5 \times 10^8$ (rad/s) $\times$ 0.5 (m) $= 4.5 \times 10^7$ (m/s)
- The kinetic energy: $KE = (1/2)\ m\ v^2 = (1/2)\ (1.67 \times 10^{-27}) \times (4.5 \times 10^7)^2 = 16.9 \times 10^{-13}$ (J)
- Converting *Joules* to eV: $KE = ((16.9 \times 10^{-13})/(16 \times 10^{-19})) = 10.57 \times 10^6$ eV $\cong$ 11 MeV.

32 The force acting on the charged particle follows Fleming's left hand rule where the force acting on the charge particle follows: $\boldsymbol{F} = q\boldsymbol{v} \times \boldsymbol{B}$.
33 https://www.youtube.com/watch?v=98-ruTyK8uE.
34 http://scientificsentence.net/Equations/Electrostatics/index.php?key=yes&Integer=motion_charges (accessed 17 February 2018).
35 https://www.youtube.com/watch?v=98-ruTyK8uE. This is a Hindi version of detailed lecture on cyclotron (accessed 19 February 2018).

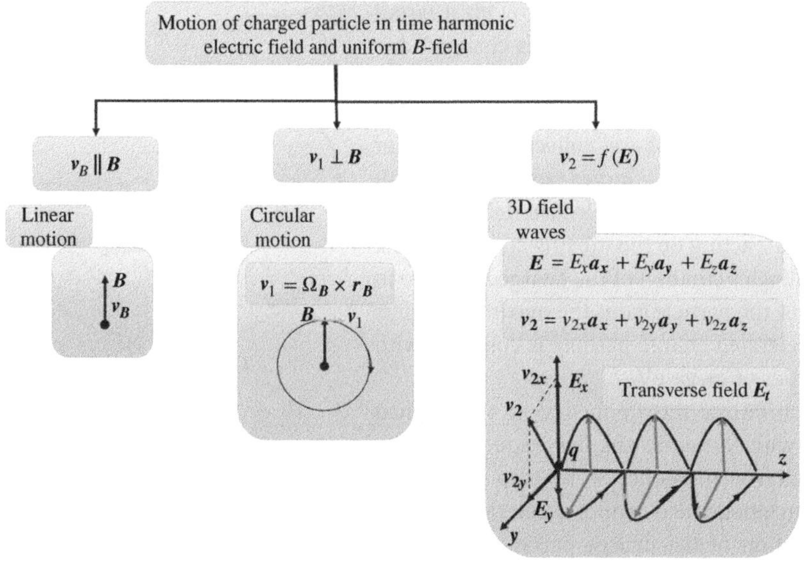

Figure 4.9 Decomposed velocities of charged particle in time harmonic electric field and uniform magnetic fields.

4.8 Analysis of Charged Particle in Time Harmonic Electric Field and Uniform Magnetic Field

In the preceding section we have examined the working principle of cyclotron, in a cyclotron an alternating electric field in the order of 10^4 Hz (operating at high frequency (HF)) is used. Strong uniform magnetic field is used for circulation of the charged particle in the dee region. We also discussed important applications of cyclotrons and synchrotrons in synthesizing new materials, analyzing materials, developing isotopes for biomedical applications and bombardments of charged particle for high energy emission. As the concept of the interaction of the charged particles is highly useful to understand the dispersive natures of the material media that carries electromagnetic waves for communications.[36] Therefore, it is significant to analysis the motion of a charged particle under the influence of the time harmonic electric field and a uniform magnetic field as shown in Figure 4.9. Let us start the analysis of the motion of the charged particle under the time varying electric and uniform magnetic fields.

Let the time harmonic electric field expression is:

$$\boldsymbol{E}(t) = \boldsymbol{E}e^{j\omega t} \tag{4.38}$$

The electric field can have three components in x, y and z-directions.

The uniform (DC) magnetic field expression is:

$$\boldsymbol{B} = B\boldsymbol{a_u} \tag{4.39}$$

where, $\boldsymbol{a_u}$ is the unit vector of the uniform magnetic field. In this case, we usually create the magnetic field with some sort of permanent magnet with its north and south poles along the plane of rotation of the charged particle, and we may consider the field in one dimension. For simplicity, we can consider $\boldsymbol{a_u} = \boldsymbol{a_z}$ without loss of any generality of the configuration of the vector system. In these field conditions, the motion of the charge particle under the influence of the electromagnetic field can be expressed as:

---

36 We shall study such dispersions of material media of electromagnetic waves in the rest of the chapter. The studies are vital to understand dispersions and precious design of microwave, mm-wave and optical devices.

$$m\frac{d\boldsymbol{v}}{dt} = q(\boldsymbol{E} + \boldsymbol{v} \times \boldsymbol{B}) \tag{4.40}$$

where the symbols have their usual meanings. Similar to the analysis of the motion of the charged particle under simultaneous steady electric and magnetic fields, we can consider that the velocity vector has three components:

$$\boldsymbol{v} = \boldsymbol{v_B} + \boldsymbol{v_1} + \boldsymbol{v_2} \tag{4.41}$$

where, (i) $\boldsymbol{v_B}$ is the uniform velocity along the steady magnetic field $\boldsymbol{B}$. ($\boldsymbol{v_B} || \boldsymbol{B}$), (ii) $\boldsymbol{v_1}$ is the cyclotron motion and is perpendicular to $\boldsymbol{B}$. As we have derived in the previous section $\boldsymbol{v_1} = \boldsymbol{\Omega_B} \times \boldsymbol{r_B}$, where $\boldsymbol{\Omega_B} = -q\boldsymbol{B}/m$ is the circular velocity vector and $\boldsymbol{r_B}$ is the radius vector as explained in the previous section. As we have examined that $\boldsymbol{\Omega_B} \| \boldsymbol{B}$, and (iii) the third component, $\boldsymbol{v_2}$, is the velocity vector, which is solely governed by the time varying electric field $\boldsymbol{E}$. Since $\boldsymbol{E}(t) = \boldsymbol{E}e^{j\omega t}$ therefore, we can also assume $\boldsymbol{v_2}(t) = \boldsymbol{v_2}e^{j\omega t}$. Likewise, $\boldsymbol{v_2}$ can have x-, y-, and z-directed components. The figure illustrates the velocity vectors under the influence of simultaneous electromagnetic fields. For the time harmonic (sinusoids) $\boldsymbol{E}$ and $\boldsymbol{v_2}$ we only depict the transverse components. This is a legitimate guess for most electromagnetic cases, we are more interested in the transverse field components, which are orthogonal to the direction of the magnetic field and the propagation direction.

Now we can consolidate the conditions of the velocities and go ahead with the analysis of the equation of motion as above.

Conditions:	$\boldsymbol{v_B} \times \boldsymbol{B} = 0$; as $v_B \| B$
	$\boldsymbol{\Omega_B} = -(q\boldsymbol{B}/m)$; the angular velocity vectors
	$(d\boldsymbol{\Omega_B}/dt) = 0$; as the field $\boldsymbol{B}$ is assumed uniform
	$(d\boldsymbol{v_B}/dt) = 0$; likewise, the condition for the uniform magnetic field
	$(d\boldsymbol{r_B}/dt) = \boldsymbol{v_1}$; the velocity of cyclotron
	$\boldsymbol{v_1} = \boldsymbol{\Omega_B} \times \boldsymbol{r_B}$; the definition of cyclotron velocity
	$(d\boldsymbol{v_2}/dt) = \boldsymbol{v_2}e^{j\omega t}$; the time harmonic nature of the velocity governed by the electric field

Impositions of the above conditions in the equation of motion of the charged particle we can write:

$$m\,\boldsymbol{\Omega_B} \times \boldsymbol{v_1} + j\omega m\boldsymbol{v_2} = q\{\boldsymbol{E} + (\boldsymbol{v_1} + \boldsymbol{v_2}) \times \boldsymbol{B}\} \tag{4.42}$$

Substituting for $\boldsymbol{\Omega_B} = -(q\boldsymbol{B}/m)$ in the first left-hand side term of the above equation, and equating for $m\ \boldsymbol{\Omega_B} \times \boldsymbol{v_1} = q(v_1 \times \boldsymbol{B})$, we obtain an equation for the velocity vector $\boldsymbol{v_2}$ in terms of $\boldsymbol{E}$ as follows:

$$(-j\omega + \boldsymbol{\Omega_B} \times)\boldsymbol{v_2} = -\frac{q\boldsymbol{E}}{m} \tag{4.43}$$

Now we are in a position to solve for $\boldsymbol{v_2}$ in the rectangular coordinate system. Let,

$$\boldsymbol{v_2} = v_{2x}\boldsymbol{a_x} + v_{2y}\boldsymbol{a_y} + v_{2z}\boldsymbol{a_z} \tag{4.44}$$

and

$$\boldsymbol{E} = E_x\boldsymbol{a_x} + E_y\boldsymbol{a_y} + E_z\boldsymbol{a_z} \tag{4.45}$$

Assume $\boldsymbol{B}$ has only z-directed component so that $\boldsymbol{B} = B_z\boldsymbol{a_z}$. Likewise, the angular velocity vector has only z-directed components so as $\Omega_B = \omega_B\boldsymbol{a_z}$. Now we are going to analyze the components of the equation of motion in the x-, y-, z-coordinate system one by one as follows.

$$\boldsymbol{\Omega_B} \times v_2 = \begin{bmatrix} \boldsymbol{a_x} & \boldsymbol{a_y} & \boldsymbol{a_z} \\ 0 & 0 & \omega_B \\ v_{2x} & v_{2y} & v_{2z} \end{bmatrix} = \boldsymbol{a_x}(-v_{2y}\omega_B) + \boldsymbol{a_y}(v_{2x}\omega_B) \tag{4.46}$$

Substitution of the above equation in (4.43), we obtain,

$$(-j\omega v_{2x} - v_{2y}\omega_B)\boldsymbol{a_x} + (-j\omega v_{2y} + v_{2x}\omega_B)\boldsymbol{a_y} + (-j\omega v_{2x})\boldsymbol{a_z} = -\frac{q}{m}(E_x\boldsymbol{a_x} + E_y\boldsymbol{a_y} + E_z\boldsymbol{a_z}) \tag{4.47}$$

The solutions for the velocity components are:

$$v_{2x} = \frac{\left(j\omega\frac{q}{m}E_x - \frac{q\omega_B}{m}E_y\right)}{-\omega^2 + \omega_B^2} = -j\frac{q}{m\omega}\left[\frac{\omega^2}{\omega^2 - \omega_B^2}E_x + \frac{j\omega\omega_B}{\omega^2 - \omega_B^2}E_y\right] \tag{4.48}$$

Likewise,

$$v_{2y} = -j\frac{q}{m\omega}\left[\frac{j\omega\omega_B}{\omega^2 - \omega_B^2}E_x + \frac{\omega^2}{\omega^2 - \omega_B^2}E_y\right] \tag{4.49}$$

$$v_{2Z} = -j\frac{q}{m\omega}E_z \tag{4.50}$$

Let us assume: $\xi^2 = (\omega^2/\omega^2 - \omega_B^2)$ and $\tau^2 = (\omega\omega_B/\omega^2 - \omega_B^2)$, then we can write the solution in a matrix form as follows:

$$\begin{bmatrix} v_{2x} \\ v_{2y} \\ v_{2z} \end{bmatrix} = -j\frac{q}{m\omega}\begin{bmatrix} \xi^2 & j\tau^2 & 0 \\ -j\tau^2 & \xi^2 & 0 \\ 0 & 0 & 1 \end{bmatrix}\begin{bmatrix} E_x \\ E_y \\ E_z \end{bmatrix} \tag{4.51}$$

We can make a few remarks from the above analysis as follows:

1) There is always a special interest in many application areas where in most cases we consider that the electric field is perpendicular to the magnetic field.[37] For such cases, the dot product of the transverse electric field and the magnetic field is zero: $\boldsymbol{E_T}.\ \boldsymbol{B} = 0$. In this case, $\boldsymbol{E_T} = E_x\boldsymbol{a_x} + E_y\boldsymbol{a_y}$. The velocity will also be transverse to the direction of $\boldsymbol{B}$ and has two components: $\boldsymbol{v_{2T}} = v_{2x}\boldsymbol{a_x} + v_{2y}\boldsymbol{a_y}$. This means, $\boldsymbol{v_{2T}}$ will remain entirely transverse to the magnetic field and follow the force equation with Fleming's left-hand rule: $\boldsymbol{F} = q\boldsymbol{v_2} \times \boldsymbol{B}$. In this case, $\boldsymbol{v_B}$ and $\boldsymbol{v_1}$ will remain unaffected and continue to follow $\boldsymbol{B}$ and spiral around $\boldsymbol{B}$, respectively.
2) We also see that near $\omega \approx \omega_B$ the amplitude of the velocity gets a very high valuc. If we draw special attention to the theory of cyclotron, we have seen that the charged particle gets accelerated at large values every time it crosses the gap between the two dees under the influence of the alternating HF electric field of angular frequency of ω. We also have learned that light particles such as electrons are not suitable for use as they quickly get large velocities. In the later section, we shall study the damping effect of the material media due to the collisions of charged particles in the medium volume and that controls the singularities at the resonant frequency $\omega \approx \omega_B$.
3) For a single component electric field E_x, the velocity vector $\boldsymbol{v_2}$ still has two transverse components v_{2x} and v_{2y}, and attains large values near $\omega \approx \omega_B$. In this case, $\boldsymbol{v_B}$ and $\boldsymbol{v_1}$ will still remain unaffected and continue to follow $\boldsymbol{B}$ and spiral around $\boldsymbol{B}$, respectively with its original velocity. In such casc, the original conditions hold:

$$\boldsymbol{v_1} = \boldsymbol{\Omega_B} \times \boldsymbol{r_B};\ \boldsymbol{\Omega_B} = -\frac{q\boldsymbol{B}}{\boldsymbol{m}};\boldsymbol{v_1} = \frac{d\boldsymbol{r_B}}{dt}\ \boldsymbol{v_B}\|\boldsymbol{B}\ \text{and}\ \boldsymbol{\Omega_B}\|\boldsymbol{B}. \tag{4.52}$$

In the next sections, we shall cover the dispersion of materials in the light of the classical electron theory of matters.

37 Even if we examine the example of the cyclotron, the magnetic field is perpendicular to the electric field.

Review Questions 4.6: Charged Particles in Time Harmonic Fields

Q1. Explain the significance of a charged particle in the presence of a time harmonic electric field.

Q2. Draw a conceptual map of the charged particle under the influence of time harmonic electric field and steady magnetic field.

4.9 Dispersion in Gaseous Media

The media of the electromagnetic wave propagation are broadly classified as *dielectric* and *conducting* materials. The dielectrics or insulators are those sorts of materials that do not carry appreciable electric current under the influence of the applied electric field of appreciable magnitude. Although no appreciable electric current conduction occurs in the presence of the impressed electric field in the dielectric, some molecular changes occur that impede the normal flow of electromagnetic waves through the dielectric. This abnormal behavior of the dielectric medium can be explained via the microscopic model of the atoms and molecules. In normal condition without the presence of an impressed electric field, the molecules are statistically electrically neutral and this phenomenon is called *electrically equilibrium state*. In this electrically equilibrium condition, the epicenter of the positively charged nucleus of the atoms and molecules are in the same position of the epicenter of the negatively charged electron cloud,[38] as shown in Figure 4.10.

When an electric field is applied in a dielectric medium, there is a microscopic displacement between the two epicenters of the positively and negatively charged particles occurs. Note that the classical electron theory of matters is a dynamic system. In that dynamic system, the charged particles are in continuous thermal vibration and the charge particles also emit electromagnetic radiations. This dynamic system is represented by the natural oscillation of the particles at its *natural frequency*. Figure 4.10a depicts the statically electrically neutral charged particles where the epicenters of the two charge clouds are in the same position. Figure 4.10b shows the minute displacement $\boldsymbol{R}$ due to the applied electric field $\boldsymbol{E}$. Note that the restoration force between the particles is strong and thereby, the displacement acts against the applied electric field. This restoration force can be represented by the spring force according to Hooke's law. Besides the restoration force that impedes the deflection of the two epicenters of the opposite charge particles, there is also a damping force that dampens the natural oscillation so that the particles do not go infinite oscillation and reach to a break down position. The restoration or the spring force is governed by Hooke's law and is proportional to the displacement $\boldsymbol{R}$ and related with a constant $\aleph$. The damping force is proportional to the velocity meaning the rate of change of the displacement $\boldsymbol{v} = (d\boldsymbol{R}/dt)$ and the proportionality constant is γ which is called *the damping factor*. The displacement of the epicenters of the two oppositely charged particles in a system of molecules is also defined by the polarization vector $\boldsymbol{p}$, which is defined as the product of the electron charge and the displacement vector $\boldsymbol{p} = -e\boldsymbol{R}$. For a system of charge particles with N number of electrons per volume, the polarization

38 In some materials there is a natural misalignment of the epicenters of the positive and negative charge particles. These types of materials are called *polar molecules*. Water (H_2O) is a good example of a bipolar material. The electromagnetic characteristics of *the polar molecules* is beyond the scope of the study and we shall assume that the dielectric materials which are in equilibrium state in the absence of the applied electric field.

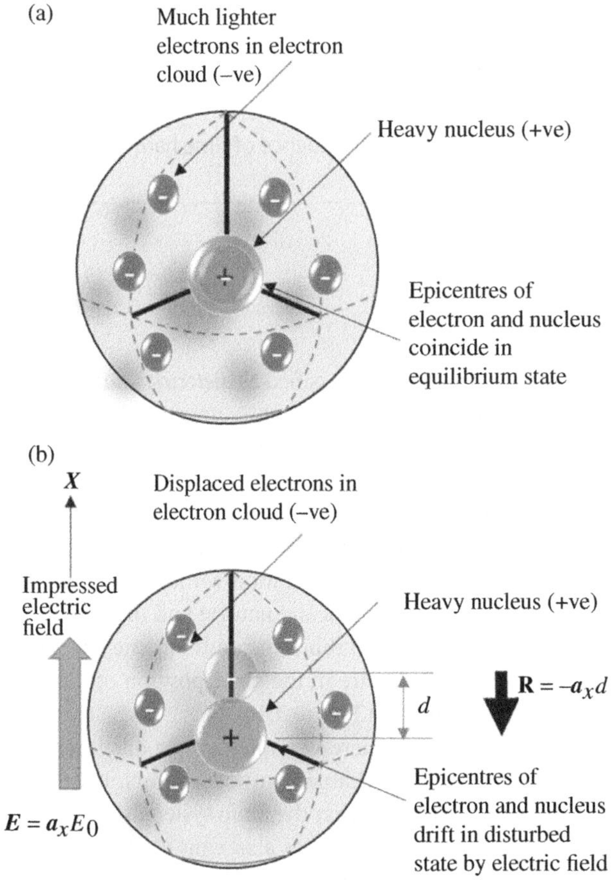

Figure 4.10 Classical electron models of atom and molecules (a) in equilibrium state, and (b) with applied electric field.

of the system is defined as $\boldsymbol{P} = -Ne\boldsymbol{R}$.[39] It is assumed that each electron in the dynamic microscopic molecular system is acting according to the Lorentz force: $\boldsymbol{F} = -e(\boldsymbol{E} + \boldsymbol{v} \times \boldsymbol{B})$. We assume the velocity of the charged particle, which is the rate of change of the microscopic displacement $\boldsymbol{R}$, is much smaller than the velocity of light in free space $c = 3 \times 10^8$ m/s. Also, we can assume that in the contribution of the magnetic field in the associated electromagnetic field is many orders less than that of the electric field.[40] In these approximations, we can simplify Lorentz force equation using only the contribution from the electric field to: $\boldsymbol{F} = -e\boldsymbol{E}$. Considering all these factors of the dynamically varying microscopic charge particle system and the electromagnetic interaction, we can write an equation of motion as follows:

$$m\frac{d^2\boldsymbol{R}}{dt^2} + m\gamma\frac{d\boldsymbol{R}}{dt} + \aleph\boldsymbol{R} = -e\boldsymbol{E} \tag{4.53}$$

39 We have discussed the polarization of dielectric materials in Chapter 3.

40 This approximation of the contribution of magnetic field of the electromagnetic wave is more profound in metals and ionized media. In the current investigation, the assumption of velocity less than the velocity of light suffices the requirement to simplify the analysis.

Consider the time harmonic case with the applied electric field $\boldsymbol{E} = \boldsymbol{E_0} e^{j\omega t}$ and the displacement vector $\boldsymbol{R} = \boldsymbol{R_0} e^{j\omega t}$ we can write the motion equation above as:

$$\left(-m\omega^2 + jm\gamma\omega + \aleph\right)\boldsymbol{R} = -e\boldsymbol{E} \tag{4.54}$$

or

$$\boldsymbol{R} = -\frac{e\boldsymbol{E}}{m\left(\omega_0^2 - \omega^2 + j\omega\gamma\right)} \tag{4.55}$$

where, $\omega_0^2 = (\aleph/m)$; ω_0 is called *the natural frequency* of the medium.

Now consider the molecular dipole moment $\boldsymbol{p_i}$ for individual ith electron

$$\boldsymbol{p_i} = -e\boldsymbol{R} = \alpha'\varepsilon_0\boldsymbol{E_0} \tag{4.56}$$

where α' is called *the molecular susceptibility* or *polarizability* of the medium. The unit is m^3. For the N electrons per unit volume, the average dipole moment per unit volume is defined as:

$$\boldsymbol{P} = \sum_i \boldsymbol{p_i} = \sum_i Ne\boldsymbol{R_i} = -Ne\boldsymbol{R_{ave}} = N\alpha'\varepsilon_0\boldsymbol{E_0} \tag{4.57}$$

Substituting for $\boldsymbol{R}$ from the above equation in $\boldsymbol{P}$ we obtain the expression for *the polarizability* of the medium as follows:

$$\alpha' = \frac{e^2}{m\varepsilon_0(\omega_0^2 - \omega^2 + j\omega\gamma)}\left(\mathrm{m}^3\right) \tag{4.58}$$

Now invoking the classical electromagnetic theory of the displacement vector:

$$\boldsymbol{D} = \varepsilon\boldsymbol{E} = \varepsilon_0\boldsymbol{E} + \boldsymbol{P} = (1 + N\alpha')\varepsilon_0\boldsymbol{E} = \varepsilon_0\varepsilon_r\boldsymbol{E} \tag{4.59}$$

Here we define the relative permittivity

$$\varepsilon_r = 1 + N\alpha' = 1 + \frac{Ne^2}{m\varepsilon_0(\omega_0^2 - \omega^2 + j\omega\gamma)} = 1 + \frac{a^2}{(\omega_0^2 - \omega^2 + j\omega\gamma)}$$

where we define *the plasma frequency* of the dielectric as: $a = \omega_p = \sqrt{(Ne^2/m\varepsilon_0)}$. From the examination of the expression of the relative permittivity ε_r,we can see that it is a complex and frequency dependent quantity. We can write the complex relative permittivity as:

$$\varepsilon_r = \varepsilon' - j\varepsilon'' = 1 + \frac{a^2}{(\omega_0^2 - \omega^2 + j\omega\gamma)} \tag{4.60}$$

The refractive index for the nonmagnetic and dielectric materials is defined as:

$n = \sqrt{\varepsilon_r} = n_r - jn_i$ is a complex number. We can separate the real and imaginary parts of n as follows:

$$n^2 = n_r^2 - n_i^2 - 2jn_rn_i = \varepsilon_r = 1 + \frac{a^2}{(\omega_0^2 - \omega^2 + j\omega\gamma)} \tag{4.61}$$

Separating the real and the imaginary parts of n we obtain $n_r^2 - n_i^2 = 1 + \left(\left(a^2\left(\omega_0^2 - \omega^2\right)\right)/\left(\omega_0^2 - \omega^2 + \omega^2\gamma^2\right)\right)$; and $2n_rn_i = \left(a^2\omega\gamma/\left(\omega_0^2 - \omega^2 + \omega^2\gamma^2\right)\right)$. As we know from the lossy dielectric, the real part of the refractive index n is responsible for dispersion and the imaginary part is responsible for absorption of the electromagnetic signal. To find the closed form expression for the two quantities we need to go through rigorous mathematical manipulations. Best is to write a MATLAB code to understand their behavior in terms of the operating frequencies. After rigorous mathematical manipulation we have obtained the closed form expressions for n_r and n_i as follows:

$$n_r^2 = -\frac{1+E^2}{2} \times \left[1 \mp D^2\sqrt{1+\frac{C^2}{D^2(1+E^2)^2}}\right] \tag{4.62}$$

and

$$n_i^2 = -\frac{1+E^2}{2} \times \left[1 \mp D^2\sqrt{1+\frac{C^2}{D^2(1+E^2)^2}}\right] - \frac{D^2+E^2}{D^2} \tag{4.63}$$

where,

$$C^2 = a^2\omega\gamma; D^2 = \omega_0^2 - \omega^2 + \omega^2\gamma^2; \text{and } E^2 = a^2\left(\omega_0^2 - \omega^2\right) \tag{4.64}$$

Since n_r and n_i are real positive numbers we shall select the positive solutions for them.[41] A closed form approximation for the real and imaginary parts of the refractive index is derived as follows:

$$n_r \approx 1 + \frac{a^2\left(\omega_0^2 - \omega^2\right)}{2\left\{\left(\omega_0^2 - \omega^2\right)^2 + \left(\omega^2\gamma^2\right)\right\}} \tag{4.65}$$

and

$$n_i \approx \frac{a^2\omega\gamma}{2\left\{\left(\omega_0^2 - \omega^2\right)^2 + \left(\omega^2\gamma^2\right)\right\}} \tag{4.66}$$

A few remarks can be made from the above analysis:

1) The real part of the refractive index is accountable for the dispersion and the imaginary part is for damping. We also learned from the loss tangent calculation that the finite conductivity of the materials makes the dielectric constant complex and provides the loss term of the materials.
2) The dispersion caused by the real part of the refractive index is called *anomalous dispersion*. This is caused by the finite conductivity of the medium.
3) From the above expressions of n_r and n_i we see that the refractive index is a function of frequency and resonance in nature. n_r is $(1+a^2/2)$ at DC, and then increases with frequency initially, at resonance $(\omega \cong \omega_0)$ it gets zero, and then becomes negative. At a very large frequency it tends to be one. n_i peaks at $(\omega \cong \omega_0)$ and then damps out as the frequency increase.

Exercise 4.2: Refractive Index

Q1. Derive the closed form solutions for n_r and n_i.

(Hints: use the solution of quadratic equation; $ax^2 + bx + c = 0$ is $x = -(b/2a) \pm \left(\sqrt{b^2 - 4ac}/2a\right)$

41 Negative refractive index is present in metamaterials. Metamaterials are artificially created materials that are used in various emerging applications in microwave passive and active design. For more details read *Metamaterials* in Wikipedia https://en.wikipedia.org/wiki/Metamaterial (accessed 03 February 2018).

Exercise 4.3: Dispersion in Gaseous Media

Q1. A material with the following parameters:

No of electrons per unit volume $N = 2.418 \times 10^{22}$ (m^{-3}); $e = 1.602 \times 10^{-19}$C; $m = 9.109 \times 10^{-31}$kg; $\gamma = 10^{22}$ (s^{-1});

i) Calculate the natural frequency ω_0; Hints: use $\aleph = Ne^2/\varepsilon_0$, where $\varepsilon_0 = 8.854 \times 10^{-12}$ F/m
ii) Calculate a.
iii) Calculate ε_r and plot $\varepsilon_r vs\ f$.

From the above analysis of the relative permittivity, we can deduce the following special cases:

1) Case 1: DC case when $\omega = 0$.
2) Case 2: low frequency case when $\omega < \omega_0$
3) Case 3: in the vicinity of $\omega \cong \omega_0$
4) Case four: high frequency case when $\omega > \omega_0$

We shall also examine two cases of the damping factor: no damping $\gamma = 0$ and with damping $\gamma \neq 0$.

Case 1: DC case when $\omega = 0$ and no damping.

In this case, the relative permittivity becomes:

$\varepsilon_r = 1 + \left(Ne^2/m\varepsilon_0\omega_0^2\right)$; this is a constant. The DC refractive index without damping is $n^2 = (\mu_r\varepsilon_r) = 1 + \left(Ne^2/m\varepsilon_0\omega_0^2\right)$; we assume $\mu_r \cong 1$. Figure 4.11 illustrates the DC and low frequency dispersion of the medium without damping.

Case 2: low frequency case when $\omega < \omega_0$, $\varepsilon_r = 1 + N\alpha' = 1 + \left(Ne^2/m\varepsilon_0\left(\omega_0^2 - \omega^2\right)\right)$ and $\varepsilon_r = 1 + \left(a^2/\left(\omega_0^2 - \omega^2 + j\omega\gamma\right)\right)$ without and with damping, respectively. As the frequency increases, the relative permittivity does not remain constant, but increases with the frequency until it reaches to a singularity for no damping case at $\omega \approx \omega_0$. With damping, we get the peak at $\omega \approx \omega_0$, but without singularity.

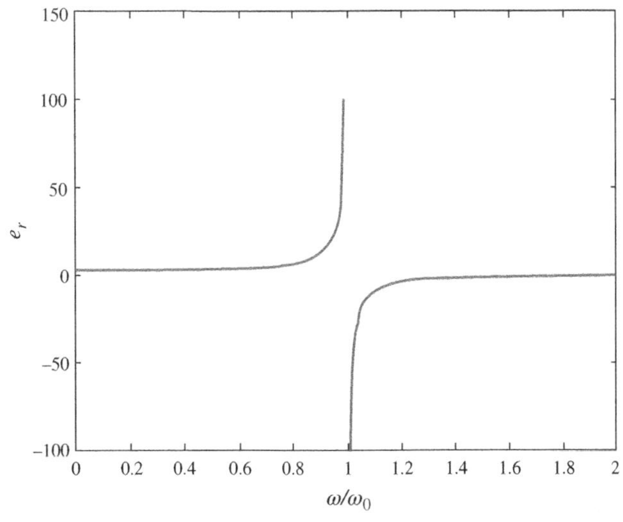

Figure 4.11 Dispersion curve of materials without damping shows the singularity at the normalized frequency $\omega = \omega_0$. In a DC the relative permittivity is equal to $\varepsilon_r = 1 + \left(Ne^2/m\varepsilon_0\left(\omega_0^2\right)\right)$. The natural frequency has effect in the dielectric materials.

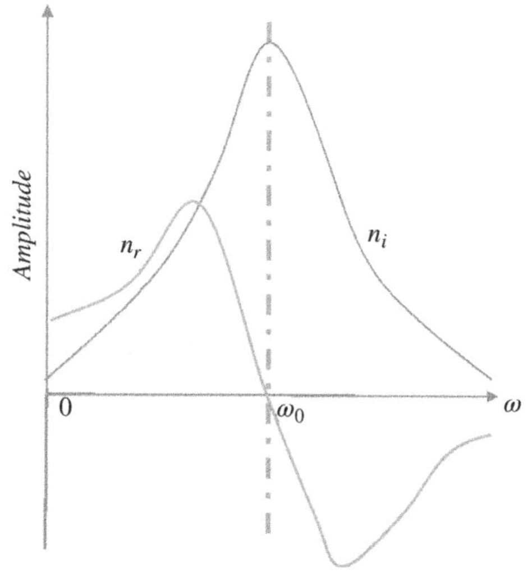

Figure 4.12 Dispersion curve for material with damping effect.

Case 3: In the vicinity of $\omega \cong \omega_0$. In this case, the refractive index becomes infinity without damping. In reality, this does not happen due to the intrinsic damping factor γ that causes the dampening of the peak amplitude to a finite value. Figure 4.12 shows the plot of n^2 vs ω for both cases. Looking at the curve, we can easily perceive that the nature of resonance of the refractive index is like that for a parallel RLC resonant circuit. Here the imaginary part n_i represents the loss term R, the resistance. The loss term peaks near the resonant frequency. The real part of the refractive index is like a reactive element $\pm jx$. Depending on the frequency the real part of the refractive index takes the form of inductive reactance when $\omega < \omega_0$ and at resonance it goes to zero, for $\omega > \omega_0$ it becomes capacitive with the negative values. Bandwidth of the resonant frequency is proportional to the damping factor. What does the real part of the refractive index represent? It represents the oscillatory nature of the wave propagation without the loss term and is very similar to $j\beta$ term of the propagation constant.

The above examination of the dispersion of the medium considers single atom and molecules and there is no interaction between them. This happens in gaseous medium where the electrons are dispersedly located and do not interact with each other. However, in liquids and solids, the molecules are packed heavily together. Therefore, we need to consider liquid and solid medium with a different aspect of analysis due to their macroscopic construction. We shall use the above understanding of dispersion and add some extra features of molecular structures and their interaction with the electromagnetic field waves.

4.10 Dispersion in Liquid and Solid Media

As stated above, the generic dispersion theory has been presented where we assume the electron concentration is moderate and the particles do not interact with each other. However, in liquid and solid, the situation is different where the interatomic forces are so strong that the simple impressed field could not deviate their original shape. However, the polarizations in liquid and solid can be classified into three types, as depicted in Figure 4.13.

As shown in Figure 4.14, in *the electronic polarization*, the epicenter of the electron cloud is displaced due to the impressed electric field. However, the difference between the previous analysis of the generic displacement of the medium with the one in solid and liquid is that there is a localized electric field is called E_{local} is acting on the polarized atoms and molecules instead of the impressed electric field. Therefore, we shall develop a relationship between the impressed electric field and the local electric field that is accountable for the displacement. In *the ionic polarization*, the polarization is similar in nature. The only difference is the charge carriers are the positive and negative ions depending on the particular medium. The dipolar polarization is also called *the orientational polarization*. In some specific materials, permanent electric dipoles are randomly oriented in the absence

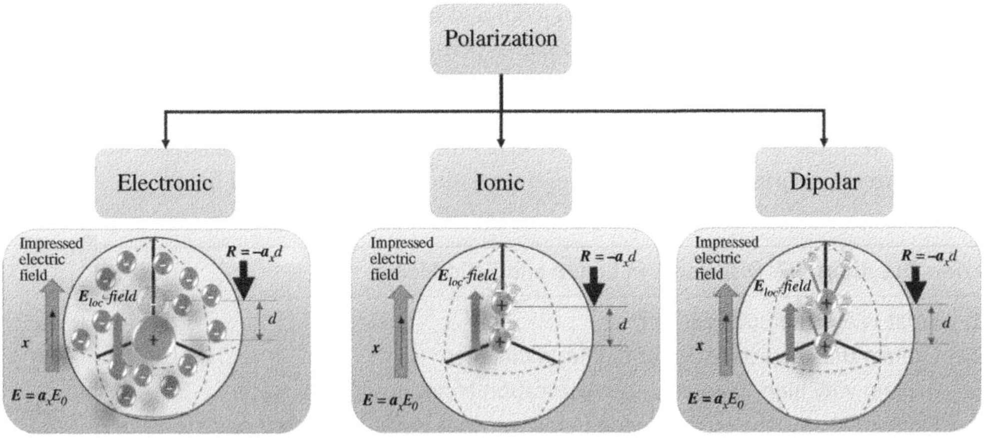

Figure 4.13 Types of polarizations in liquid and solid.

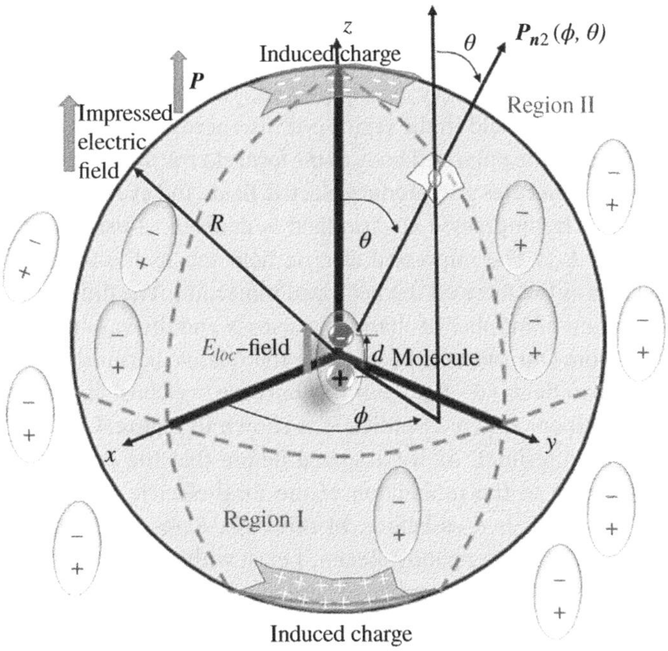

Figure 4.14 A polarized molecule in the influence of a local field E_{loc} in Region I of radius R. We assume there are such polarized molecules in Region I but the net contribution is zero. In Region II ($r > R$), the regional field and polarization are $\boldsymbol{E}$ and $\boldsymbol{P}$, respectively. There is some representative induced charge due to the contribution of the molecules in Regions I and II. At the observation point, the normalized polarization in Region II is P_{n2} creates an angle θ with P and z-axis.

of the impressed electric field. Once the electric field is applied the permanent dipole moments try to orient themselves along the impressed electric field. A few examples of bipolar materials are water (H_2O), hydrochloric acid (HCl), and carbon dioxide (CO_2). *The dipole molecules* can be assumed as the combination of the positive and negative ions with the coulomb's force acting between them. Usually, *bipolar molecules* resonate at a much lower frequency than that of the other materials. Water is an interesting *bipolar material* and behaves very uniquely at electromagnetic wave propagation. At low frequencies, the relative permittivity of water is 81. At microwave frequencies, the relative permittivity varies in between 10 and 20 with a high loss tangent. This characteristic of water bars microwave propagation in water. That is why submarine communications happened in very low frequencies in the lower kHz range. In the dispersion analysis, we consider only *the electronic polarization* due to its prevalent nature of most solid and liquid.

Review Questions 4.7: Dispersion in Liquid and Solid Media
Q1. What is electric polarization?
Q2. Explain the characteristics of bipolar molecules.
Q3. Explain why high frequency communication is not possible in water?
Q4. Explain what is the local electric field and what is the impressed electric field?

4.11 Ionic Dispersion in Liquid and Solid Media

As stated above, for liquid and solid we need to consider the electronic polarization and Coulomb's force effect of adjacent molecules which are very closely packed. Therefore, in addition to the force acting due to the impressed electric field in the region, we also need to add the force created by the interaction of the molecules of the polarized body as *the local electric field*. Because, in dense materials, the nearby atoms and molecules also produce electric fields and exert force on the molecule/atom under investigation. The impressed electric field is denoted as usual as $\boldsymbol{E}$, whereas *the local field* is represented by $\boldsymbol{E}_{loc}$. The impressed electric field acts as the force on the region and the local electric field acts as the force on the polarized molecules. The difference between the two arises due to the variation of number of charge particles N and the varying gap exists between the molecules in the region. Our motto is to calculate the relative permittivity of the medium from the difference of the two fields ($\boldsymbol{E}-\boldsymbol{E}_{loc}$). For this purpose, we consider a particular molecule at the center of an imaginary sphere of radius R as shown in Figure 4.14. The radius of the sphere is small but not infinitesimal. As we assumed before that the epicenters of the positive and negative charges is due to the interaction of the local electric field, $\boldsymbol{E}_{loc}$. The displacement between the charge centers is d as before. In this case, $R >> d$. Therefore, we divide the complete system of charges with two regions: Region I exist within the imaginary sphere of radius R. and the region outside the imaginary sphere where the radial distance $r > R$ is Region II. We assume that there are plenty of such charged molecules as the polarized body inside and outside the imaginary sphere in Regions I and II, respectively. In Region II, we assume that the impressed electric field is $\boldsymbol{E}$ and is acting along z-axis for simplicity of analysis. The resulting polarization vector due to the impressed electric field in $\boldsymbol{P}$ is also acting along z-axis. Here, we assume $\boldsymbol{P}$ is the contribution of the molecules outside the sphere. The normal directed charge polarization is $\boldsymbol{P}_{n2}(\varphi, \theta)$ creates an angle θ with z-axis as shown in Figure 4.14. Due to the system of charges of the polarized body, we also assume that there are resulting induced charge accumulations on the top and bottom of the imaginary sphere. For the time being, we also assume that the contribution of the molecules inside the sphere has no effect on the central polarized molecule.[42]

42 We shall prove the zero effect of the inside molecules on the central molecule as an example later on.

Now we calculate the difference ($\boldsymbol{E_{loc}}-\boldsymbol{E}$) from the impressed field $\boldsymbol{E}$ due to $\boldsymbol{P}$ outside the sphere. Assume that the polarization charge density is ρ_p. Therefore, we can derive the polarization $\boldsymbol{P}$ using divergence theory:

$$\int_v \nabla \cdot \boldsymbol{P} dv = \int_v \rho_p dv \tag{4.67}$$

This yield:

$$\nabla \cdot \boldsymbol{P} = \rho_p \tag{4.68}$$

The result is exactly similar to the point form of the divergence theory of $\nabla . \boldsymbol{D} = \rho_v$. This means we can assume a parallel concept of the polarization charge density of the charge molecule due to the impressed field, $\boldsymbol{E}$ outside the sphere. Now apply the boundary condition to find a relation between the local and the impressed fields as follows:

$$P_{n1} - P_{n2} = \rho_{sp} \tag{4.69}$$

where, P_{n1} and P_{n2} are the normal polarization vectors inside and outside of the sphere, respectively, and ρ_{sp} is the polarization surface charge density. Since we have already assumed that the net contribution of the inside molecules on the central charged molecule is zero, this argument leads to

$$-P_{n2} = \rho_{sp} \tag{4.70}$$

Therefore, using the system configuration of the macroscopic polarized body as illustrated above, we can obtain the expression of polarization inside the surface P_{n2} in terms of P as follows:

$$P_{n2} = \boldsymbol{P} \cdot (-\boldsymbol{a_r}) = -P\boldsymbol{a_z} \cdot \boldsymbol{a_r} = -Pcos\theta = \rho_{sp} \tag{4.71}$$

We assume a differential surface area on which the different charge exists. We can define the differential charge from the polarization and then we can calculate the z-directed electric field.

$$dq = \rho_{sp} d\boldsymbol{S} = -Pcos\theta d\boldsymbol{S} \tag{4.72}$$

The resulting differential field that acts on the central molecule due to the Coulomb force can be obtained from the above differential charge as follows:

$$d\boldsymbol{E} = -\frac{dq\boldsymbol{a_r}}{4\pi\varepsilon_0 R^2} = \frac{Pcos\theta d\boldsymbol{S}\boldsymbol{a_r}}{4\pi\varepsilon_0 R^2} \tag{4.73}$$

The scalar expression for the z-directed differential electric field is:

$$dE_z = \frac{Pcos\theta d\boldsymbol{S}}{4\pi\varepsilon_0 R^2} \tag{4.74}$$

Now we can calculate the total z-directed electric field acting on the central molecule by integrating the above differential electric field over the spherical surface area as follows:

$$\begin{aligned} E_z &= \int_s dE_z = \frac{P}{4\pi\varepsilon_0}\int_s \frac{cos\theta d\boldsymbol{S}}{R^2} = \frac{P}{4\pi\varepsilon_0}\int_0^{2\pi} d\phi \int_0^{\pi} cos^2\theta sin\theta d\theta \\ &= \frac{P}{4\pi\varepsilon_0} \times 2\pi \times \frac{2}{3} = \frac{P}{3\varepsilon_0} \end{aligned} \tag{4.75}$$

This is the electric field on the polarized molecule at the center of the sphere due to the polarization $\boldsymbol{P}$:

$$E_z = \frac{P}{3\varepsilon_0} \tag{4.76}$$

Therefore, it can be said that the local field that is acting on the molecule is the summation of the impressed field plus the contribution from the other polarized molecules, and can be defined as using the above derivation:

$$\boldsymbol{E}_{loc} = \boldsymbol{E} + \frac{\boldsymbol{P}}{3\varepsilon_0} \tag{4.77}$$

With this net acting local field on the molecule, we can now find the polarization $\boldsymbol{P}$ from the derivation for gaseous medium as follows:

$$\boldsymbol{P} = N\alpha'\varepsilon_0\boldsymbol{E}_{loc} = N\alpha'\varepsilon_0\left(\boldsymbol{E} + \frac{\boldsymbol{P}}{3\varepsilon_0}\right) \tag{4.78}$$

However, Maxwell's equation says $\boldsymbol{D} = \varepsilon_0\boldsymbol{E} + \boldsymbol{P} = \varepsilon_0\varepsilon_r E$. This leads to

$$\boldsymbol{P} = \varepsilon_0(\varepsilon_r - 1)\boldsymbol{E} \tag{4.79}$$

Substituting the above expression in the local field we have

$$\boldsymbol{E} + \frac{\boldsymbol{P}}{3\varepsilon_0} = \boldsymbol{E} + \frac{\varepsilon_0(\varepsilon_r - 1)}{3\varepsilon_0}\boldsymbol{E} = \frac{(\varepsilon_r + 2)}{3}\boldsymbol{E} \tag{4.80}$$

First substitution for (4.78) into (4.80) we obtain

$$\boldsymbol{P} = N\alpha'\varepsilon_0 \times \frac{(\varepsilon_r + 2)}{3}\boldsymbol{E} \tag{4.81}$$

Then from Equations (4.79) and (4.81) we obtain

$$\boldsymbol{P} = \varepsilon_0(\varepsilon_r - 1)\boldsymbol{E} = N\alpha'\varepsilon_0 \times \frac{(\varepsilon_r + 2)}{3}\boldsymbol{E} \tag{4.82}$$

This leads to the *Clausius–Mosotti formula* as follows:

$$\frac{(\varepsilon_r - 1)}{(\varepsilon_r + 2)} = \frac{1}{3}N\alpha' \tag{4.83}$$

The formula tells us that the ratio $((\varepsilon_r - 1)/(\varepsilon_r + 2))$ is directly proportional to the density of the dielectric. This is a unique relationship between the microscopic quantities N and α', and the macroscopic parameter of the material—the dielectric constant ε_r. In modern technologies, it is very common that the propagation medium is composed of multiple materials species. In this realm, the formula is derived as the superposition of all constituents as follows[43]:

$$\frac{(\varepsilon_r - 1)}{(\varepsilon_r + 2)} = \sum_i \frac{1}{3}N_i\alpha_i' \tag{4.84}$$

43 *Clausius–Mossotti relation* https://en.wikipedia.org/wiki/Clausius%E2%80%93Mossotti_relation (accessed 11 February 2018) also read article P. Melman and R. Davies, Application of the Clausius-Mossotti equation to dispersion calculations in optical fibers, *Journal of Lightwave Technology* (Volume: **3**, Issue: 5, 1985, pp. 1123–1124. *Application of the Clausius-Mossotti Equation to Dispersion Calculations* in. https://www.researchgate.net/publication/3241365_Application_of_the_Clausius-Mossotti_Equation_to_Dispersion_Calculations_in_Optical_Fibers (accessed 11 February 2018).

where i is the integer representing the number of constituents. *Clausius–Mossotti equation* is heavily used in the calculation of the dispersion properties of optical fibers, particularly for graded fiber optics.

It is also customary to calculate the refractive index of the material media using the above formulation. This relationship is called *Lorentz–Lorenz relation*.[44]

$$\frac{(n^2-1)}{(n^2+2)} = \frac{1}{3}N\alpha' \tag{4.85}$$

Since N is proportional to the density ρ of the material and α' being a constant at a particular frequency, we can say that $((n^2-1)/(n^2+2))$ is directly proportional to the density of the material. Therefore, we can write the relation for a given frequency as:

$$\frac{(n^2-1)}{(n^2+2)} = \frac{K}{\rho} \tag{4.86}$$

where, K is the proportionality constant. The constant K is called *the atomic refractivity* of the material. Both *Clausius–Mossotti equation* and *Lorentz–Lorenz relation* are used conveniently for calculations of material dispersions in propagation medium. In the above derivations, the following assumptions are used[45]:

1) As illustrated above, the electric fields of the molecules are the fields created by the dipole moment of charge q.
2) We assumed the perfect symmetry in the construction of the polarized molecules.
3) The dipole moments of all molecules are assumed to be identical for the simplicity of calculation.

The above assumptions are arguably valid for isotropic, homogeneous, and linear medium, although some deviations can be observed in an actual situation. In reality, the complex permittivity is not dependent on the electronic displacement or polarization. The other factors such as ionic, dipolar, and atomic polarization also have effects on the dispersion though each polarization type exhibits distinct frequency dependence as shown in the following figure. The real and imaginary parts of the complex dielectric constant ε_r' and ε_r'' are related to each other with *Kramer–Kronig relation*. This is equivalent to the complex impedance, which is a function of frequency and the real and imaginary parts are related to each other. In the vicinity of $\omega \approx \omega_0$, the real part peaks R and the imaginary part changes over its role from inductive to capacitive (at $\omega \approx \omega_0$, $jX_r = jX_c$).

Figure 4.15 shows the representative frequency response of the dielectric medium with the contribution of different polarization types. As can be seen, the frequency responses have many resonances from low frequency in kHz to the ultraviolet in PHz region. The width of the resonance is determined by the damping factor as illustrated in the electronic polarization. The frequency dependence of the complex dielectric constant $\varepsilon' - j\varepsilon''$ are used conveniently in many practical applications.[46] For example, board bipolar resonance in UHF and microwave frequency region is responsible for microwave heating. Modern microwave heating has many domestic and industrial applications. A resonance in water molecule in the frequency range of 24 GHz (1.25 cm

44 H. A. Lorentz of Leyden, The Netherlands derived the relationship in 1880 and L. Lorenz of Copenhagen, Denmark in 1881 (in M.A. Islam, *Electromagnetic Theory*, EPUET, Dacca, East Pakistan: EPUET, 1969, pp. 385–386) also read article: *Lorentz-Lorenz relation*, http://www.tau.ac.il/~tsirel/dump/Static/knowino.org/wiki /Lorentz-Lorenz_relation.html (accessed 11 February 2018).
45 M.A. Islam, *Electromagnetic Theory*, EPUET, Dacca, East Pakistan: EPUET, 1969, p. 385.
46 U. S. Inan and A.S. Inan, *Electromagnetic Waves*, Prentice Hall, Upper Saddle River, NJ, USA, 2000.

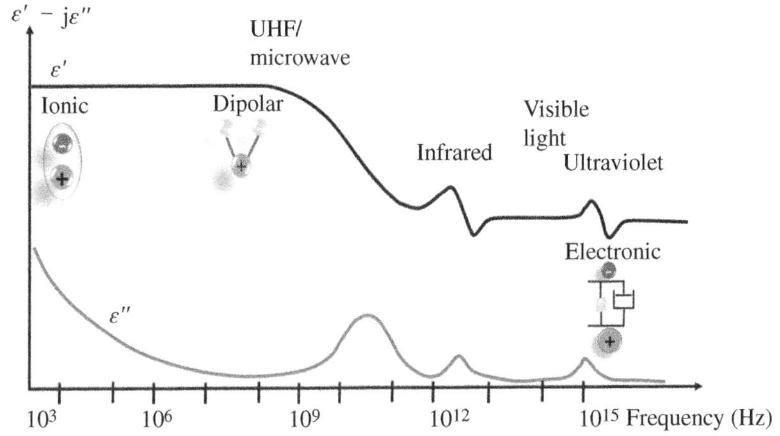

Figure 4.15 A dielectric permittivity spectrum over a wide range of frequencies. ε′ and ε″ denote the real and the imaginary part of the permittivity, respectively. Various processes are labeled on the image: ionic and dipolar relaxation, and atomic and electronic resonances at higher energies.

wavelength) limits long range radar applications in microwave and mm-wave frequencies.[47] Oxygen molecules resonate at 60 GHz frequency band and create huge challenge of long-range communications at this frequency band. Recent unlicensed frequency band of 60 GHz find potentials in many short, medium, and long-range communications systems. A good example is the chipless RFID as electromagnetic barcodes at 60 GHz using imaging techniques finds short range potential applications in pervasive data transmission and receptions for identifications, tracking and tracing of objects, personnel, and goods. Ozone and nitrogen in high altitude keep most ultraviolet energy of sun from penetrating to a lower altitude. We shall study the dispersion of ionosphere in the next section and understand the activity of the ultraviolet ray to create ionized layer at a high altitude. Understanding the broader aspect of the electromagnetic wave propagation in dispersive medium has a profound impact on optical fiber design, and many active and passive devices designs in modern microwave/mm-wave and visible light communications.

4.12 Dispersion in Metals

4.12.1 Significance of Dispersion in Metals in Mixed Signal Electronics?

In the modern information and communication (ICT) era, the high-speed data transmission via very low cost printed electronics is getting huge momentum. The demands push the electronics engineers understand the electromagnetic theory in greater importance than ever before. The electronic engineers are pushed to design low cost printed electronics in microwave and mm-wave even in sub-mm-wave domains over hundreds of GHz. It is a blessing that *the skin effect* becomes less significant as *the skin depth* δ of the conductors reduces with the frequency of operation. This is true about the bulk conductors. However, understanding the conductivity of the nanoparticle conducting inks[48] in microwave and mm-wave frequencies and their ultra-wide band behavior

47 Water and oxygen absorptions are also dominant in modern emerging 60 GHz mm-wave and visible light frequencies. Contamination of water molecules also creates high attenuation in optical fibers.

48 The nanoparticle conducting ink is formed with the nanoparticles of bulk conductors such as silver and copper submersed in glue which is more dielectric in nature such as resin.

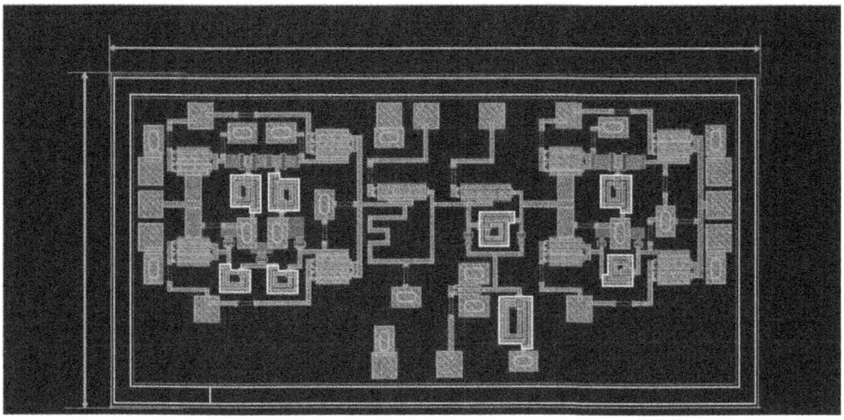

Figure 4.16 The layout of an active phase shifter at Ku-band has both biasing as well as microwave frequency active and passive circuits. *Source:* Trinh, K.N. (2020). Phase shifters for multi-band soil moisture radiometer phased array antennas. PhD thesis. Monash University, printed with permission.

becomes more and more important for the emerging ICT technologies. Although *the skin effect* is less significant at high microwave and mm-wave frequencies, the cost reduction efforts force to use thin conducting ink layer and also the existing printing methods do support thin printing processes. This *skin effect* becomes highly significant in modern printed electronics at microwave and mm-wave frequencies. Figure 4.16 shows the photograph of a microscopic layout of a mixed signal integrated circuit as an active phase shifter at Ku-band. As can be seen from the complexity and broad disciplines of circuits from DC to microwave frequency AC, the modern engineers need to learn all aspects of microwave and mm-wave theories as we are discussing here.

Some excellent examples include transparent Mobile phones, and fully printable chipless RFID in microwave and mm-wave frequencies.[49] Such circuits can be made transparent with the transparent, highly conducting inks. In this section, we shall study the wave propagation in metal and its dispersion, meaning the frequency dependency of the conductivity of the metals. As we have studied the surface resistivity R_s of a bulk conductor is inversely proportional to its conductivity σ as $R_s = (1/\sigma\delta)$, where the conductivity σ is assumed to be frequency independent. Here, we shall study the dispersion of the electromagnetic signal in the metallic medium and try to understand the upper limit of the conductivity in terms of the frequency of operation. This is very vital for modern electronics design engineers who are dealing with the cutting-edge mixed signal integrated circuit design.[50] This understanding will help them to efficiently design printed electronics at any

49 The author has been doing research on the fully printable chipless RFID for more than a decade.

50 A modern-day mixed signal circuit encompasses a variety of integrated circuit design areas and also use different substrate materials to cater for digital design, analogue electronics and microwave and mm-wave circuit and systems. Good example is the modern smartphone. Digital circuits such as microprocessors including multi-bit digital to analogy and analogue to digital converters in a very dense and fast processing package, analogue circuits such as sensors, communications, and power supply and control circuits, RF meaning microwave and mm-wave transceivers, on-chip antennas, memory (dense/vertical), and microelectromechanical system (MEMS) circuits. Microwave and mm-wave IC design have push electronics design engineers to work on GaAs and GaN high electron mobility field effect transistor (HEMT) design. These are the cutting edge technologies used in current and emerging fifth generation (5G) wireless communications to address the high data transmission demands for internet of thing (IoT) and internet of everything (IoE) markets. The key vendors for mixed signal IC design software and hardware testing facilities are Teradyne, Keysight, and Texas Instruments. Read more about mixed signal circuits via Wikipedia *Mixed-signal integrated circuit* https://en.wikipedia.org/wiki/Mixed-signal_integrated_circuit, (accessed 26 January 2018).

frequency from DC to sub-mm-wave AC. From there we shall derive the upper limit of the frequency of operation where we can consider the conductivity σ is frequency independent and above which we shall consider the dispersion effect of the metal.

4.12.2 What Are Metals Made of: The Classical Electron Theory and Electromagnetic Wave Interaction?

According to the classical electron theory every matter in the universe is made of an atom with specific electron counts in orbital structures. The electrons revolve around the heavy nucleus made of protons. We perceived that electrons are negatively charged particles and the protons are positively charged particles. Multiple atoms coupled with the outer orbital electrons in a specified fashion to make a molecule. The electrons are the fundamental unit charge with the elementary charge $e = 1.602 \times 10^{-19}$ C. We also know that for metals the electrons in the outer orbits are loosely coupled and can easily be detached from its outer orbit[51] and become free to conduct current under the influence of the applied electric field.

According the electronic theory it is also perceived that the metals are crystalline in structure as shown in Figure 4.17. This means that they form three-dimensional crystalline lattice of atoms and molecules. Although the outer electrons are loosely bonded and free to detach from the atoms for current conduction, the interatomic force is extremely strong to resist any deformation of the physical shape.[52] The outer electrons under the influence of the applied electric fields, which are liable for current conduction are called *free electrons*. The electrons move freely thoroughly the crystalline lattices of the metal under the influence of the applied electric field. Thus, under the influence of the applied electromagnetic fields we can assume a piece of metal is a crystalline lattice structure with the heavy ions tightly bonded to the crystalline lattices and a count of free electrons, which are free to move through the lattice structure of the metal, conduct current. Without any external force, the atoms move randomly, collide each other and with ions and molecules and, become statically neutral, hence do not conduct any current. Because of these collisions and encounters and the constant

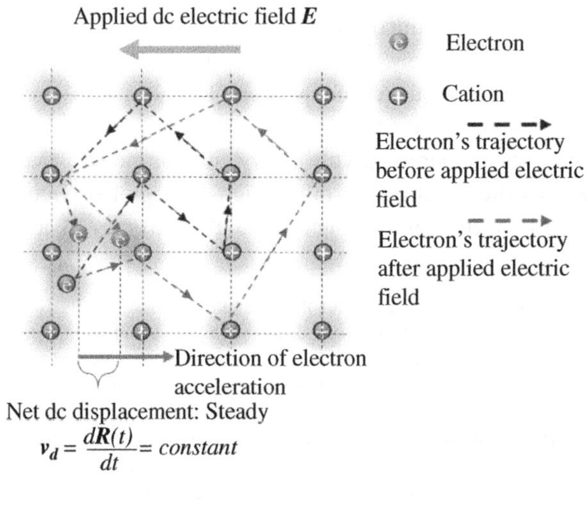

Figure 4.17 The Drude model approximates the metal to a lattice of cations through which delocalized electrons flow.

51 The energy required to move a charge particle (electron) from its orbit is called Fermi energy after the name of the Italian-American physicist Enrico Fermi.

52 We create alloy with extreme heat and pressure of two metals, even in the alloy the characteristics of the base metals remain intact. In olden days permanent magnets were also created with such extreme heat and pressure aligning the iron with the earth's poles.

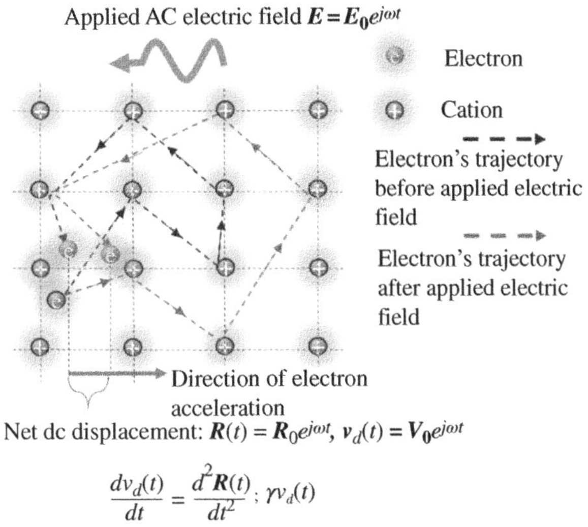

Figure 4.18 The Drude model approximates the metal to a lattice of cations through which delocalized electrons flow.

thermal vibrations of ions, the drift velocity dampens out. In 1900, a German physicist Paul Karl Ludwig Drude developed the powerful model of the thermal, electrical, and optical properties of matter.[53] The damping factor is proportional to the drift velocity v_d and also related to the conductivity of the metal. In this investigation we shall define the conductivity in terms of the atomic model and operating frequency of the wave.

When an external electric field $\boldsymbol{E}$ is applied to the free electrons of the metal which is statistically neutral and in equilibrium state initially, the displacement of the free electron is represented by a spatial displacement vector $\boldsymbol{R}$ also called net drift or displacement as shown in Figure 4.18. As the electric field line generates from the positive charge and ends at the negative charge, the net drift and associated acceleration is opposite to the direction of the electric field. In this case, we assume a volume of elements N in which the free electrons collide with the ions in the lattice and give up an *average momentum* mv_d at a *mean free time* of collision T_m. In this case, we can define the average drift velocity as:

$$\boldsymbol{v}_d = \frac{1}{N}\sum_i \boldsymbol{v}_i \tag{4.87}$$

Here, m is the mass of the free electron. The average rate of the momentum an electron releases to the ion lattice is (mv_d/T_m), the SI unit is kilogram meters per second (kg·m/s). Momentum is the mass in motion and defined by Newton's second law of motion—a body's rate of change in momentum is equal to the net force acting on it.[54] Therefore, the electron and electromagnetic wave interaction can be defined with the help of Lorentz force as follows:

$$\frac{m\boldsymbol{v}_d}{T_m} = -e\boldsymbol{E} \tag{4.88}$$

53 See Wikipedia for details about Paul Karl Ludwig Drude via https://en.wikipedia.org/wiki/Paul_Drude, (accessed 26 January 2018). Also read *atomic vibration*, *Free electron model* and *elastic wave* in Wikipedia and internet. A nice depiction of Drude mode and ion and free electron interaction can also be found via *Semiconductor theory-Background, Electron Conduction transport* via http://www.chm.bris.ac.uk/webprojects2000/igrant/theory.html (accessed 26 January 2018).
54 *Momentum*, https://en.wikipedia.org/wiki/Momentum (accessed 27 January 2018).

Table 4.1 Electron mobilities and conductivities for common metals at room temperature.

Metal	Electron mobility μ_e m^2/(V.s)	Conductivity σ (S/m)
Copper	3.2×10^{-3}	5.8×10^7
Aluminum	1.4×10^{-3}	3.54×10^7
Silver	5.2×10^{-3}	6.17×10^7
Gold	4.26×10^{-3}	4.10×10^7

Now we can define the drift velocity which is the time rate of change of the average displacement vector $\boldsymbol{R}$.

$$\boldsymbol{v}_d = \frac{d\boldsymbol{R}}{dt} = -\frac{eT_m}{m}\boldsymbol{E} = -\mu_e\boldsymbol{E} \tag{4.89}$$

where, $\mu_e$is a positive number and called *the electron mobility*, μ_e represents how quickly an electron moves when pulled by the applied electric field and is defined as:

$$\mu_e = \frac{eT_m}{m} \tag{4.90}$$

The unit of *the electron mobility* is m^2/(V·s). From the linear relationship we can say that *the high electron mobility* means long *mean free time*. Table 4.1 shows the electron mobilities and conductivities for common metals at room temperature[55]:

The conduction currents are the result of the drift velocity of the charge carriers[56] under the influence of the applied electric field and directly proportional to the applied electric field. Therefore, definition of the current density is defined by: $\boldsymbol{J} = -Ne\boldsymbol{v_d}$, where, N is electron density per unit volume, e is the charge of the free electron. Therefore, the current is defined as:

$$\boldsymbol{J} = -Ne\boldsymbol{v}_d = -Ne\frac{d\boldsymbol{R}}{dt} = -\frac{Ne^2\,T_m}{m}\boldsymbol{E} = \sigma\boldsymbol{E} \tag{4.91}$$

The above definition of the current is the point form of *Ohm's law* which defines the current in terms of the conductivity of the medium time the impressed electric field. The unit of the current density is (A/m^2). Here, N denotes the free electron density, the number of free electrons per m^3 and σ is the conductivity of the metal medium (S/m). Therefore, the conductivity is defined as

$$\sigma = \frac{Ne^2\,T_m}{m} = \rho_v\mu_e \tag{4.92}$$

where we can define the volume carrier charge density as $\rho_v = Ne$ (C/m^3). Let us do an example to clarify the theory for the time varying case for the drift velocity of the metallic dispersion.

Example 4.5: Metallic Dispersion

Find the mean free time, and the conductivity of a medium for its electron mobility 3.2×10^{-3} m^2/(V.s). Assume the mass density of copper is 8.93×10^3 kg/m^3 and its atomic mass is 63.55 g/mol. Hints: Avogadro constant: 6.022×10^{23} mol^{-1}.

---

55 D. K. Cheng, *Fundamentals of Engineering Electromagnetics*, Addison-Wesley, Reading, USA, 1993, pp 153–154.
56 In this study of metal dispersion, the charge carrier is *the free electron*. For semiconductor we also consider the drift velocity of *holes*, in that case the charge carrier is *the hole*.

Answer:

The atomic number of electrons per unit volume

$$N = 8.93 \times 10^3 \left(\frac{\text{kg}}{\text{m}^3}\right) \times \frac{6.022 \times 10^{23}\left(\frac{1}{\text{mol}}\right)}{63.55 \times 10^3\left(\frac{\text{kg}}{\text{mol}}\right)} = 8.462 \times 10^{22}\left(\text{m}^{-3}\right)$$

$$T_m = \frac{\mu_e m}{e} = 3.2 \times 10^{-3}\frac{\text{m}^2}{(\text{V}\cdot\text{s})} \times \frac{9.109 \times 10^{-31}\,\text{kg}}{1.602 \times 10^{-19}\,\text{C}} = 1.82 \times 10^{-14}\,\text{s}$$

$$\sigma = Ne\mu_e = 8.462 \times 10^{22}\left(\text{m}^{-3}\right) \times 1.602 \times 10^{-19}(\text{C}) \times 3.2 \times 10^{-3}\left(\frac{\text{m}^2}{(\text{V}\cdot\text{s})}\right) = 43.3\left(\frac{\text{S}}{\text{m}}\right)$$

This shows rather low conductivity compared to the metal which are in the order of 10^7 (S/m).

Example 4.6: Electron Drift Velocity of Conductor

The current density of a copper rod is measured at 10 (A/m^2). $\sigma = 5.8 \times 10^7$ (S/m). Calculate the electric field intensity and electron drift velocity v_d.

Answer:

$$\text{The electric field intensity}: E = \frac{J}{\sigma} = \frac{10\left(\frac{\text{A}}{\text{m}^2}\right)}{5.8 \times 10^7\left(\frac{\text{S}}{\text{m}}\right)} = 1.724 \times 10^{-7}\left(\frac{\text{V}}{\text{m}}\right)$$

$$\text{The drift velocity } v_d = \mu_e E = 3.2 \times 10^{-3}\left(\frac{\text{m}^2}{(\text{V}\cdot\text{s})}\right) \times 1.724 \times 10^{-7}\left(\frac{\text{V}}{\text{m}}\right) = 5.52 \times 10^{-10}\left(\frac{\text{m}}{\text{s}}\right)$$

The above case is for the steady electric field and velocity means for the DC case. In the following we shall examine the influence of the applied AC electric field.

The applied AC field is a function of time as it is oscillatory and the velocity is also function of time and undergoes damping with a *damping factor* $\gamma = (1/T_m)$. The motion equation in terms of velocity $\mathbf{v}_d$ is defined as:

$$m\frac{d\mathbf{v_d}}{dt} + m\gamma\mathbf{v_d} = -e\mathbf{E} \tag{4.93}$$

For time harmonic field with $E = E_0\,e^{j\omega t}$ and $v_d = v_0 e^{j\omega t}$, we have the solution for the velocity as:

$$mj\omega\,\mathbf{v_d} + m\gamma\mathbf{v_d} = -e\mathbf{E} \tag{4.94}$$

or,

$$\mathbf{v_d} = -\frac{e\mathbf{E}}{m\,(j\omega + \gamma)} \tag{4.95}$$

The current density is:

$$\boldsymbol{J} = \frac{Ne^2\,\boldsymbol{E}}{m\,(j\omega + \gamma)} = \sigma\boldsymbol{E} \tag{4.96}$$

The conductivity of the metal at the operating frequency ω is defined as:

$$\sigma = \frac{Ne^2}{m\,(j\omega + \gamma)} \tag{4.97}$$

The conductivity is a complex quantity and can be expressed in the real and imaginary components:

$$\sigma = \sigma_r - j\sigma_i \tag{4.98}$$

where,

$$\sigma_r = \frac{Ne^2\gamma}{m(\omega^2 + \gamma^2)}; \quad \sigma_i = \frac{Ne^2\omega}{m(\omega^2 + \gamma^2)} \tag{4.99}$$

As can be seen from the expression of σ that it is a function of frequency ω. For sufficiently low frequency the conductivity reduces to

$$\sigma_{dc} = \frac{Ne^2}{m(j\omega + \gamma)} = \frac{Ne^2}{m\gamma} = \frac{Ne^2 T_m}{m} \tag{4.100}$$

Exactly the same expression for the DC conductivity. The frequency dependence of the phase constant can be expressed as:

$$\gamma_{prop} = jk = \sqrt{\omega^2\mu\varepsilon - j\omega\mu\sigma} = \beta - j\alpha \tag{4.101}$$

where the symbols have their usual meanings. For most metals, the permittivity $\varepsilon = \varepsilon_0$ and for the mon-magnetic metal the permeability $\mu = \mu_0$. Therefore, we can find the solution for the frequency dependant expression for the phase constant as:

$$k^2 = \omega^2\mu_0\varepsilon_0\left(1 - \frac{j\sigma}{\omega\varepsilon_0}\right) = \beta^2 - \alpha^2 - j2\alpha\beta \tag{4.102}$$

Substitution for σ from the above expression we obtain:

$$\beta^2 - \alpha^2 - j2\alpha\beta = \left(\frac{\omega}{c}\right)^2 \times \left(1 - \frac{jNe^2}{\omega\varepsilon_0 m(\gamma + j\omega)}\right) \tag{4.103}$$

Separating the real and imaginary parts we obtain:

$$\beta^2 - \alpha^2 = \left(\frac{\omega}{c}\right)^2 \times \left(1 - \frac{Ne^2}{\varepsilon_0 m(\gamma^2 + \omega^2)}\right) \tag{4.104}$$

$$2\alpha\beta = \left(\frac{\omega}{c}\right)^2 \times \left(\frac{Ne^2}{\omega\varepsilon_0 m(\gamma^2 + \omega^2)}\right) \tag{4.105}$$

There are two unknowns α and β and two equations; therefore, we can have solutions of α and β with some mathematical manipulations.[57] Let $A = (Ne^2/\varepsilon_0 m\omega(\gamma^2 + \omega^2))$ and $B = (\omega/c)^2$. Then we have,

$$\beta^2 - \alpha^2 = B(1 - A\omega) \tag{4.106}$$

$$2\alpha\beta = AB \tag{4.107}$$

Substitution for $\beta = (AB/2\alpha)$ into the above equation leads to:

$$\beta^2 - \alpha^2 = \left(\frac{AB}{2\alpha}\right)^2 - \alpha^2 = B(1 - A\omega) \tag{4.108}$$

57 Taking only the positive real parts of the solutions.

Simple manipulation leads to a quadratic equation in the form:

$$aX^2 + bX + c = 0 \tag{4.109}$$

where, $X = \alpha^2$, $a = 1$, $b = 4B(1 - \omega A)$ and $c = -(AB/2)^2$. The solution for the quadratic equation is: $X = \left(\left(-a \pm \sqrt{b^2 - 4ac}\right)/2a\right) = -(1/2) \pm (\omega/c)\sqrt{(1 - (Ne^2/\varepsilon_0 m(\gamma^2 + \omega^2)))}$; this leads to the closed form solutions for α and β as follows:

$$\alpha = \sqrt{-\frac{1}{2} \mp 2A\sqrt{1 - \left(\frac{1}{A} - \omega\right)^2}} \tag{4.110}$$

$$\beta = \frac{AB}{2\sqrt{-\frac{1}{2} \mp 2A\sqrt{1 - \left(\frac{1}{A} - \omega\right)^2}}} \tag{4.111}$$

A MATLAB code can be written to find the dispersion phenomenon ($\alpha - \beta$ vs ω diagram) of the metal. A few remarks can be made from the above derivations[58]:

1) For the AC case, the conductivity of metal is a function of frequency ω. For sufficiently low frequency $\omega \ll \gamma = (1/T_m)$, the AC conductivity is exactly similar to the DC case. In 1903, Hagen and Rubens demonstrated that the DC conductivity is suitable for all frequencies up to the significant portion of the infrared spectrum. The theory is called Hagen–Rubens relation.[59]
2) Observing the closed form solutions for the phase constant, we can say that the dispersion formula should work with a sufficient degree of accuracy up to the optical frequency.
3) If experimental results have to be in agreement with the theory, N, the number of free electrons, should be comparable with the number of atoms in the material. The free electron counts to atoms ratio should be 1 to 3[60] or 2 to 7 with elaborate experiment by Nicholson in 1911.[61]
4) Substitution for the DC conductivity expression from above into the generic conductivity then we have:

$$\sigma = \frac{\sigma_{dc}}{1 + \frac{j\omega}{\gamma}} \tag{4.112}$$

The above expression is derived by Drude and describes the AC conductivity as a function of frequency. Figure 4.19 illustrates the applications of electromagnetic spectrum based of frequency dependent characteristic of AC conductance. From the observation of the above expression for the AC conductivity which is a function of frequency we can draw a few important conclusions[62] with illustrations:

1) At low frequencies including microwave and mm-wave and up to the infrared region ($\omega \ll \sigma/\varepsilon$), the free electron term dominates ($\gamma \gg \omega$) and we can approximate $\sigma = \sigma_{dc}$.
2) In optical regions ($\omega \cong \sigma/\varepsilon$), both real and imaginary terms agree well with the experimental results. In this region, the conductivity varies to a large extent with the frequency.

58 M.A. Islam, *Electromagnetic Theory*, EPUET, Dacca, East Pakistan: EPUET 1969.
59 *Hagen–Rubens relation*, https://en.wikipedia.org/wiki/Hagen%E2%80%93Rubens_relation (accessed 29 January 2018).
60 A. Schuster, *An introduction to the Theory of Optics*, Edward Arnold, London 1904.
61 M.A. Islam, *Electromagnetic Theory*, EPUET, Dacca, East Pakistan: EPUET 1969.
62 PHYS370-Advanced Electromagnetism, Part 3L Electromagnetic Wave in Conducting Media, Liverpool University, UK, pp. 27–32.

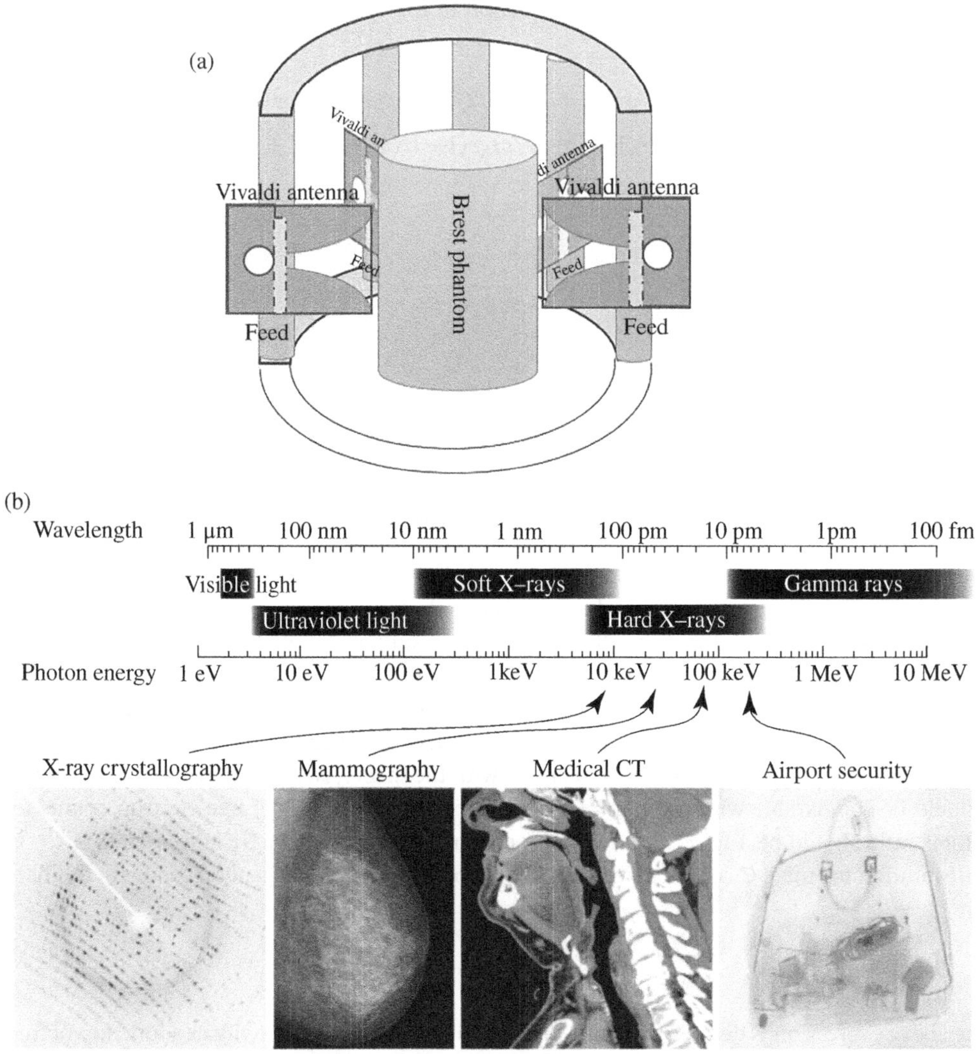

Figure 4.19 (a) Microwave impinging of metal at UWB band form 3-10.6 GHz with experimental set up and (b) top: X-rays are part of the electromagnetic spectrum, with wavelengths shorter than UV light. Different applications use different parts of the X-ray spectrum by Ulflund is licensed to CC BY-SA 3.0) left: Each dot, called a reflection, in this diffraction pattern forms from the constructive interference of scattered X-rays passing through a crystal. The data can be used to determine the crystalline structure. *Source:* Ulflund/ Wikimedia Commons/CC BY-SA 3.0.

3) In high frequencies, such as X- and γ-rays, ($\omega \gg (\sigma/\varepsilon)$ and $\omega \gg \gamma$), the free electron term is small and the material behaves like a dielectric. The electromagnetic wave becomes transparent in the metal as shown in the X-ray films in the bottom of Figure 4.19b.

In the above analysis, we have not considered the effect of specific heat in the conductivity calculation and dispersion in metals. Definitely, more free electronics will take part in conduction if heat is considered and we need to consider far greater number of free electrons than that needed to agree with the experimental values of the specific heats. It may be justified to perceive that the free electrons take part in conduction current and do not affect the property of the specific heats of metals. In many modern Multiphysics software tools such as COMSOL and CST Microwave Studio consider the thermal effects in electromagnetic design of microwave circuits. This is beyond the scope of the textbook and is not discussed further.

Review Questions 4.8: Dispersions in Metals

Q1. Explain Drude model of metal.
Q2. Explain significance of dispersion of metal in modern applications.
Q3. How a metal behaves in different frequency bands up to X-rays?

Example 4.7: Dispersion in Metal
Using Drude model (4.112), calculate the σ_{ac} at (i) 500 MHz, (ii) 60 GHz, and (iii) 1 THz for Example 5.6.

Answer:

$$\sigma_{dc} = \frac{Ne^2 T_m}{m} = \frac{2.418 \times 10^{22} \times (1.602 \times 10^{-19})^2 \times 1.82 \times 10^{-14}}{9.109 \times 10^{-31}} = 216.8 \left(\frac{\text{S}}{\text{m}}\right)$$

i) At 500 MHz, $\omega = 2\pi \times 5 \times 10^8 = 3.14 \times 10^9 (\text{rad/m})$

$$\sigma = \frac{\sigma_{dc}}{1 + \frac{j\omega}{\gamma}} = \frac{216.8}{1 + j\frac{3.14 \times 10^9}{5.49 \times 10^{13}}} \cong 216.8 \left(\frac{\text{S}}{\text{m}}\right)$$

ii) At 60 GHz, $\omega = 2\pi \times 6 \times 10^{10} = 3.77 \times 10^{11} (\text{rad/m})$

$$\sigma = \frac{\sigma_{dc}}{1 + \frac{j\omega}{\gamma}} = \frac{216.8}{1 + j\frac{3.77 \times 10^{11}}{5.49 \times 10^{13}}} \cong 216.8 \left(\frac{\text{S}}{\text{m}}\right)$$

iii) At 1 THz, $\omega = 2\pi \times 1 \times 10^{12} = 6.28 \times 10^{12} (\text{rad/m})$

$$\sigma = \frac{\sigma_{dc}}{1 + \frac{j\omega}{\gamma}} = \frac{216.8}{1 + j\frac{6.28 \times 10^{12}}{5.49 \times 10^{13}}} \cong 216.8 \left(\frac{\text{S}}{\text{m}}\right)$$

Therefore, it can be concluded that the σ_{ac} is equal to σ_{dc} the frequency up to 1 THz.

4.13 Waves Propagation in Plasma

An ionized gas is called *plasma*. The volume of ionized gas may be neutral electrically, but when an electric field is applied, the behavior of plasma on the applied electric field is very different than that in a dielectric or a conductor. In the outer atmospheric layer of the earth, a continuous bombardment of ultraviolet ray creates plasma in the ionosphere. However, the activity is not a regular phenomenon and the plasmonic contents in the ionosphere change with time and season and sudden natural activities such as solar storms. In the universe, 99% matters are made of plasma such as sun, stars, and cosmic objects. The term *plasma* was introduced by Langmuir and Tong in 1929.[63] Space age and satellite communications need understanding the radio wave propagation in the plasma. For example, the propagation of radio wave via the satellites to the earth's ground stations and vice versa and the signal loss due to the anomalies in atmospheric layer have made the studies of radio wave propagation through plasma significant. Figure 4.20 shows an example of radio communications via the satellite and the effect of the ionosphere on the radio waves. An erroneous GPS signal is received due to the anomalies of the ionospheric layer in the atmosphere. Interplanetary communications between the space vehicles and space stations are also going through the plasma medium. The author developed smart antennas for Australian Optus mobile satellite communications to establish an effective communications channel between the mobile base station and the geostationary earth orbit (GEO) satellites that is installed at approximately 37 000 km above the earth. The ionosphere layer is called *Clerk belt* after the name of Arthur C. Clerk who envisioned the GEO satellite communication in 1945.[64] Since the ionosphere is made of plasma,[65] when the signal from the satellite to the ground station is launched, and vice versa, due to the plasma concentration in the ionosphere the orientation of the electric field vector randomly changes. To avoid the mismatch of polarizations between the signals transmitted and received, we use *circularly polarized signals* so that the antennas can receive the signal in any orientation. As for example, the current GPS system uses right hand circularly polarized signal. Therefore, the antenna design required right hand circularly polarized feed network. We covered the theory of the polarization of plane waves in the last chapter.[66] Besides the satellite communications, semiconductor integrated circuit fabrication also deals with the ionized vapours of doped materials that are precisely controlled by the electromagnetic signals at microwave frequencies. Therefore, understanding the interaction between the plasma and the applied electric field covers many disciplines. Plasma is called the fourth state of the matter because its characteristics are unique and different than solid, liquid, and gas in many senses. In early stages of electromagnetism, electric discharge through low pressure gas was experimented for the plasma studies. In moisty days, the humming noises from the high-tension power lines due to the breakdown of air molecules are examples of plasmonic activities. The author also did research on the development of probes to identify *the partial discharge* which is also a plasmonic activity in the high-tension power equipment. The partial discharge is an arbitrary and very short duration electric impulse of random natures that is generated by the faulty high-power equipment due to embryonic faults during manufacturing, and wear and tear. They emit broad spectrum of microwave signal power that is detectable and preventive measures can be taken to avoid regional black outs. Lightning strikes in sky in stormy days are also plasmatic activities. The unique characteristic of plasma is that it possesses *the high pass characteristics* of radio wave signals. *The cut-off frequency* below which the signal cannot pass is called *the plasma frequency*. In the next few sections, we shall examine the field theory of plasma and understand the unique propagation characteristics of plasma.

---

63 Tonks, Lewi; Langmuir, Irving (1929). Oscillations in ionized gases, *Physical Review*. **33**: 195–210.

64 http://lakdiva.org/clarke/1945ww/ (accessed 20 January 2018).

65 As the plasma in the ionosphere is continuously going through storms, called SED (storm-enhanced density), the communications signals get disturbed. This is called ionosphere threat by the authors below.

66 Tsujii, T., Fujiwara, T., and Kubota, T. (2012). Novel navigation systems ensuring the safety of satellite navigation. http://www.aero.jaxa.jp/eng/publication/magazine/apgnews/2012_no24/apn2012no24_01.html (accessed 20 January 2018).

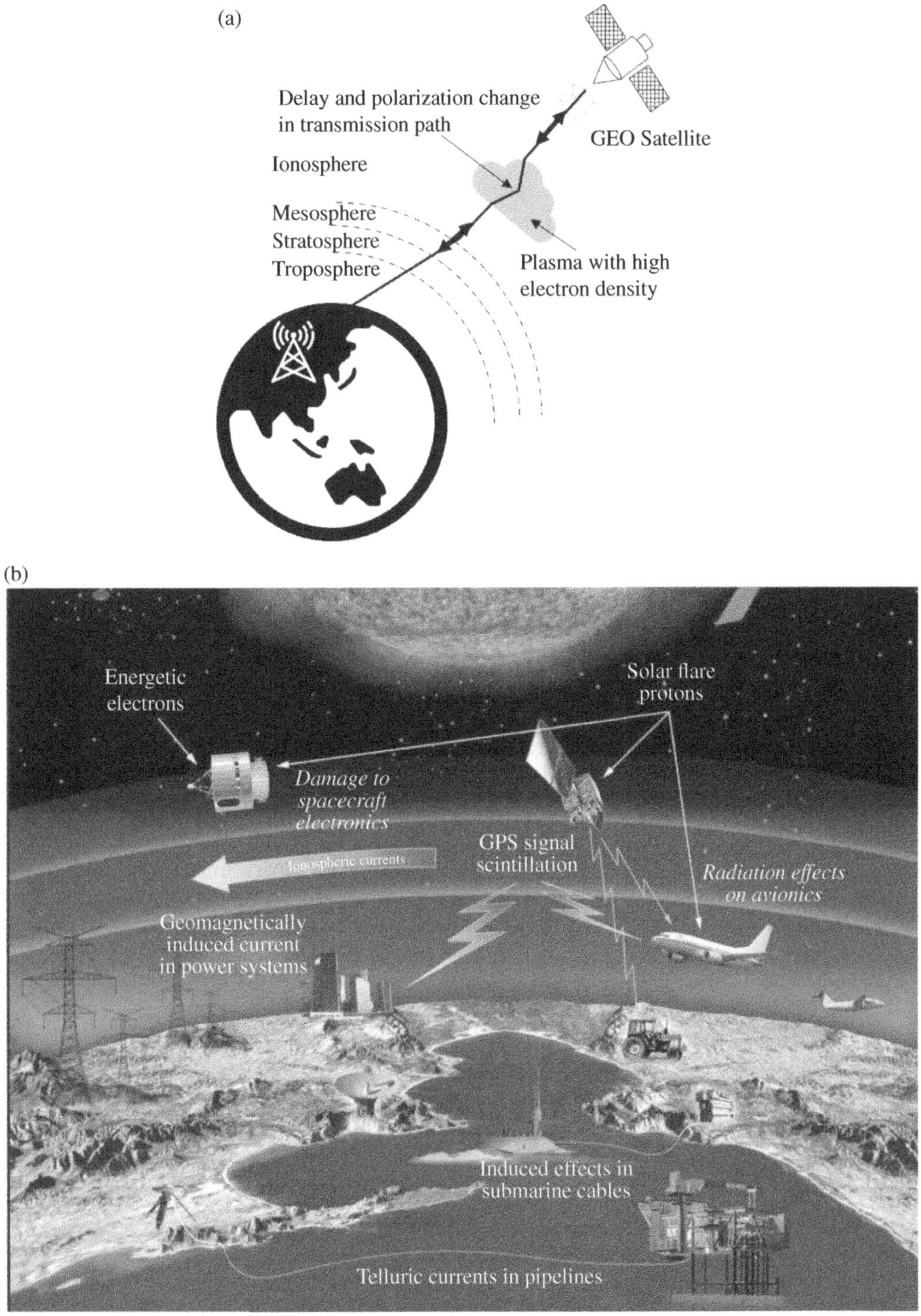

Figure 4.20 (a) Satellite signal interrupted by ionosphere creating hazard in navigation and anomalies of atmospheric layer of earth causing radio signal delay, and (b) solar storm that causes strong interaction with earth's magnetic fields and creates shockwaves that also interrupts the radio communications. *Source:* NASA.

The objective of the study leads to an understanding about the uniform plane wave propagation through an ionized medium such as ionosphere. That understanding also touches the following disciplines[67]:

1) Knowledge about the celestial bodies, interstellar, and interplanetary space
2) Harnessing natural energy from cosmic sources
3) Radio communications with the outer space
4) Radio astronomy
5) Controlled thermo-fusion to develop the new energy sources[68]
6) Thermo-nuclear activities in laboratory setting and many more.

Review Questions 4.9: Electromagnetic Wave Propagation in Plasma
Q1. What is plasma? What are its spectral characteristics?
Q2. Explain the significance of electromagnetic wave propagation in plasma.

4.13.1 Electromagnetic Wave Interaction with Plasma

What does happen when an electromagnetic field ($\boldsymbol{E}, \boldsymbol{B} = \mu_0\boldsymbol{H}$) starts propagating through an ionized gas?[69] As stated above that although the ionized molecules are charged particles, in ionized volume they exhibit like an electrically neutral medium. When an electromagnetic field is applied to the ionized gas medium, the equilibrium states is disturbed with the interaction between the field waves and the ionized particles and electronics. This disturbance produces momentum of the charged particle. This interaction produces a drift velocity v of the whole system of the particles. Usually the electrons are much lighter than the ionized particles, and therefore, in the analysis we assume that the free electrons are the charged particles that take part in the drift motion.[70] The electrons are in random collisions due to the drift velocity and causes the total loss of momentum. ν is defined as the collision frequency at which the total loss of momentum happens. The equation of motion of an electron of mass m, which undergoes through the drift velocity in the influence of an electromagnetic field exhibiting *Lorentz's force* on the electron, can be written as:

$$m\frac{d\boldsymbol{v}}{\partial t} + m v \nu = -e\left(\boldsymbol{E} + \boldsymbol{v} \times \boldsymbol{B}\right) \tag{4.113}$$

where, $\boldsymbol{v}$ is the drift velocity of the free electrons, ν is the effective collision frequency of the electrons,[71] m is the mass and e are the charge of the electron.[72] Also, note that the time varying volume charge density or the electron cloud is a four-dimensional system with spatial and temporal coordinates (x, y, z, t). Therefore, the total time derivative of the velocity can be written[73] as:

67 B. M. Notaros, *Electromagnetics*, Prentice Hall, Upper Saddle River, USA 2011.
68 B. M. Notaros, *Electromagnetics*, Prentice Hall, Upper Saddle River, USA 2011.
69 Similar derivation is also available in M. A. Islam, *Electromagnetic Theory*, Dacca, East Pakistan: EPUET, 1969.
70 The mass of an ionized molecule is much heavier compared to the free electrons that take part in the drift velocity.
71 Collision frequency is assumed as the sequence of time period at which the momentum of the drift electrons vanishes.
72 Mass and charge of an electron are: 9.1×10^{-31} kg and 1.6022×10^{-19} C, respectively.
73 S. Ramo, J. R. Whinnery and T. V. Duzer, *Fields and Waves in Communication Electronics*, Wiley, 3e, 1994, NY, USA, pp. 684–686.

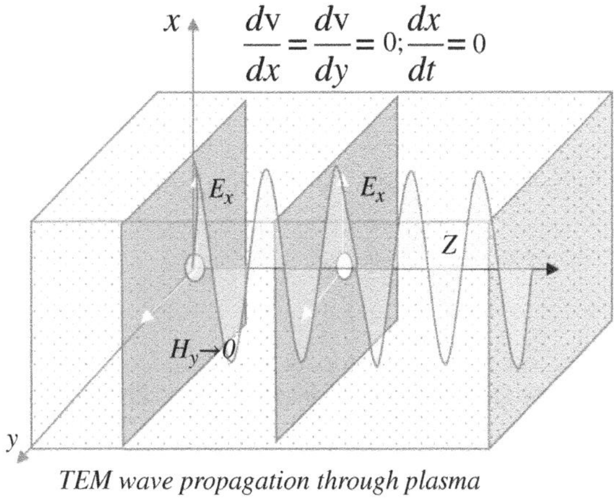

Figure 4.21 Propagation of a TEM wave in plasma medium.

$$\frac{dv\,(x, y, z, t)}{dt} = \frac{\partial v}{dt} + \frac{\partial v}{\partial x} \cdot \frac{dx}{dt} + \frac{\partial v}{\partial y} \cdot \frac{dy}{dt} + \frac{\partial v}{\partial z} \cdot \frac{dz}{dt} \tag{4.114}$$

Now consider that a TEM wave is propagating through the plasma medium, as shown in Figure 4.21. The electronic movement is also aligned with the field. Therefore, $dz/dt = 0$. Uniformity of the field in the transverse plane as shown in the figure leads to $\partial v/\partial x = \partial v/\partial y = 0$. These assumptions leaves $(dv\,(x, y, z, t)/dt) = (\partial v/dt)$.

For the time harmonic field analysis, the expressions for the electric field and the velocity vector can be represented as $\boldsymbol{E} = \boldsymbol{E}e^{j\omega t}$ and $\boldsymbol{v} = \boldsymbol{v}e^{j\omega t}$, respectively.

As we know, the movement of electron is the current. Therefore, we can define the current density $\boldsymbol{J}$ in terms of the mobile electron cloud with N number of electrons in a unit volume. Therefore, we can define the current density as:

$$\boldsymbol{J} = -eN\boldsymbol{v} \tag{4.115}$$

Also, note that the volume charge density or the electron cloud is a three-dimensional spatial system hence all vectors: $\boldsymbol{J}$, $\boldsymbol{v}$, $\boldsymbol{E}$ and $\boldsymbol{B}$ are also three-dimensional vectors with components in x, y, and z directions. Multiplying with $-eN$ on both sides of original motion equation of the free electron and substitution for the expression of $\boldsymbol{J}$ in the equation, we obtain,

$$m\frac{d\boldsymbol{J}}{\partial t} + m\nu\boldsymbol{J} = e^2N\boldsymbol{E} - e\boldsymbol{J} \times \boldsymbol{B} \tag{4.116}$$

For the time harmonic field quantities, we can replace $(\partial/\partial t) = j\omega t$ and the new equation will be

$$m\,(j\omega + \nu)\boldsymbol{J} + e\boldsymbol{J} \times \boldsymbol{B} = e^2N\boldsymbol{E} \tag{4.117}$$

The external/applied electric field can be AC or DC and have any arbitrary orientation. The magnetic field of the wave is much smaller compared to the applied electric field $\boldsymbol{E}$ and can be neglected in this instance.[74]

$$m\,(j\omega + \nu)\boldsymbol{J} = e^2N\boldsymbol{E} \tag{4.118}$$

74 The magnetic field $\boldsymbol{H}$ of the wave is linked with the electric field $\boldsymbol{E}$ through the intrinsic impedance of the plasma medium, which is much larger than the free space intrinsic impedance $\eta_0 = 120\,\Omega$. In plasma, we have infinity group velocity u_p and zero phase velocity u_g. Therefore, we can neglect the contribution of the magnetic field of the wave in the analysis.

With the above assumptions of the TEM wave, we can consider that the electron density N remains undisturbed, then we can express the current density as:

$$\boldsymbol{J} = \frac{e^2 N\boldsymbol{E}}{m\,(j\omega + \nu)} = \sigma\boldsymbol{E} \tag{4.119}$$

Where, we can define the high frequency conductivity of the medium as:

$$\sigma = \frac{e^2 N}{m\,(j\omega + \nu)} \tag{4.120}$$

Review Questions 4.10: Interaction of EM Waves with Plasma

Q1. Explain why the effect of magnetic field is neglected in the field analysis in plasmas.
Q2. Show that for an applied electric field $\boldsymbol{E}$, the current density in plasma is $\boldsymbol{J} = e^2N\boldsymbol{E}/m\,(j\omega + \nu)$.
Q3. Show that the high frequency conductance is given by $\sigma = e^2N/m\,(j\omega + \nu)$.

Now we can substitute the equation for conductivity in Faraday's law to find the complex permittivity of the plasma. The complex permittivity of the medium simplifies the field analysis of the plasma.

$$\nabla \times \boldsymbol{H} = \sigma + j\omega\varepsilon_1\boldsymbol{E} \tag{4.121}$$

where, ε_1 is the complex permittivity of the plasma medium. Substituting for σ from the above equation we obtain

$$\nabla \times \boldsymbol{H} = \frac{e^2 N}{m\,(j\omega + \nu)}\boldsymbol{E} + j\omega\varepsilon_1\boldsymbol{E} \tag{4.122}$$

or,

$$\nabla \times \boldsymbol{H} = j\omega\varepsilon_1\boldsymbol{E} + \frac{e^2 N(\nu - j\omega)}{m\,(\nu^2 + \omega^2)}\boldsymbol{E} \tag{4.123}$$

or,

$$\nabla \times \boldsymbol{H} = \left[j\omega\varepsilon_1 + \frac{Ne^2\nu}{m\,(\nu^2 + \omega^2)} - j\frac{Ne^2\omega}{m\,(\nu^2 + \omega^2)}\right]\boldsymbol{E} \tag{4.124}$$

or,

$$\nabla \times \boldsymbol{H} = j\omega\left[\left\{\varepsilon_1 - \frac{Ne^2}{m\,(\nu^2 + \omega^2)}\right\} - j\frac{Ne^2\nu}{m\omega\,(\nu^2 + \omega^2)}\right] = j\omega\varepsilon_p\boldsymbol{E} \tag{4.125}$$

where, the effective plasma complex permittivity is defined as:

$$\varepsilon_p = \varepsilon_p' - j\varepsilon_p'' = \left\{\varepsilon_1 - \frac{Ne^2}{m\,(\nu^2 + \omega^2)}\right\} - j\frac{Ne^2\nu}{m\omega\,(\nu^2 + \omega^2)} \tag{4.126}$$

where the real part of the effective plasma permittivity is:

$$\varepsilon_p' = \varepsilon_1 - \frac{Ne^2}{m\,(\nu^2 + \omega^2)} \tag{4.127}$$

And the imaginary part of the plasma permittivity is:

$$\varepsilon_p'' = \frac{Ne^2\nu}{m\omega(\nu^2 + \omega^2)} \tag{4.128}$$

For the plasma medium with ionized gas, as for example, the ionospheric layer of the earth,[75] the ion particles with very low density and negligible collision frequency ($\nu \cong 0$), we can deduce the complex permittivity assuming $\varepsilon_1 = \varepsilon_0$ as follows:

$$\varepsilon_p = \varepsilon_p' = \varepsilon_0\left[1 - \frac{Ne^2}{m\varepsilon_0\omega^2}\right] = \varepsilon_0\left(1 - \frac{\omega_p^2}{\omega^2}\right) \tag{4.129}$$

where,

$$\omega_p = \sqrt{\frac{Ne^2}{m\varepsilon_0}} \tag{4.130}$$

And the imaginary part of the plasma permittivity is:

$$\varepsilon_p'' = \frac{Ne^2\nu}{m\omega\,(\nu^2 + \omega^2)} = 0 \tag{4.131}$$

ω_p is called *the plasma frequency*. As we can see, *the plasma frequency* is independent of the operating frequency ω and only depends on the intrinsic properties of ionized molecules. "The physical meaning of *the plasma frequency* is defined as follows: As shown in Figure 4.22, if the alternate plane layers of compression and rarefaction of the electronics in a uniform positive ion background were set as initial conditions, the charge would subsequently oscillate at *the plasma frequency* because of the interaction of forces between charges and the inertial effect."[76] A few conclusions can be drawn from the expression of the effective permittivity of the plasma (4.129) as follows:

1) The wave number of the TEM wave in the plasma can be expressed as:

$$k = \omega\sqrt{\mu_0\varepsilon_p} = \omega\sqrt{\mu_0\varepsilon_0\left(1 - \frac{\omega_p^2}{\omega^2}\right)} = k_0\sqrt{1 - \frac{\omega_p^2}{\omega^2}} \tag{4.132}$$

where k_0 is the free space wave number in (rad/m). From the above expression of k there are two possible cases:

i) When the *plasma frequency* is greater than the operating frequency of the TEM wave ($\omega < \omega_p$) then ε_p becomes negative and the wave number in the plasma becomes imaginary $k = j\alpha$. The propagation factor is defined as $e^{-jkz} = e^{-\alpha z}$. This means the TEM wave attenuates exponentially with the factor $-\alpha z$ as it propagates along +ve z-direction. This means for $\omega < \omega_p$, the electromagnetic wave is reflected at the plasma boundary and the transmitted component heavily

75 Note that in the next section we shall derive the fields and complex permittivity in the ionosphere. There we consider the strong steady magnetic field of the earth. In the current field analysis, we ignore this effect, but only consider the negligible collision frequency ($\nu \cong 0$).

76 S. Ramo, J. R. Whinnery and T. V. Duzer, *Fields and Waves in Communication Electronics*, Wiley, 3e 1994, NY, USA, pp. 684–686.

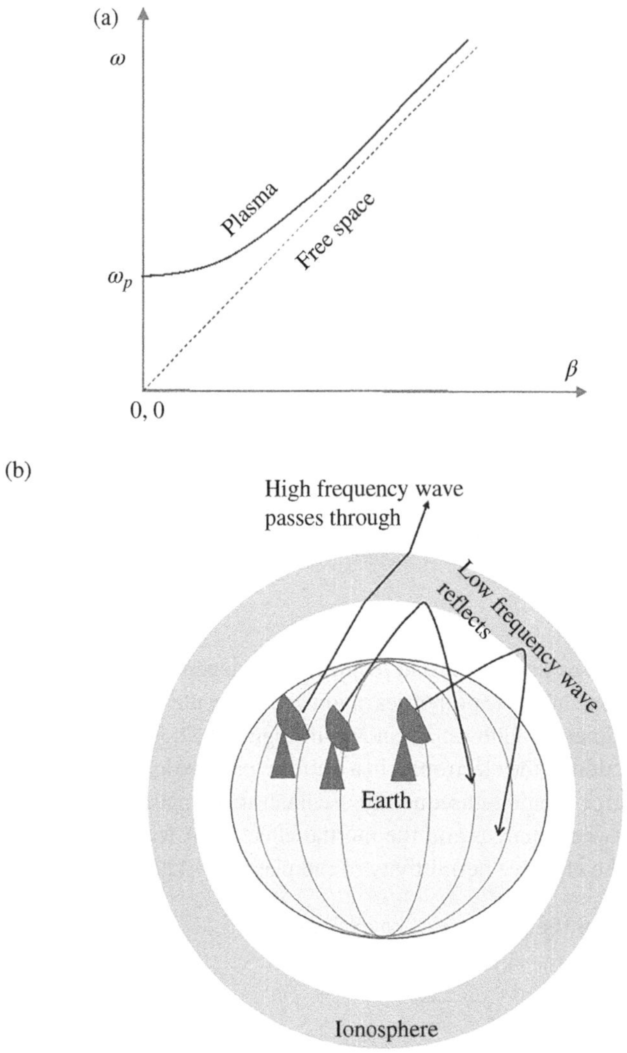

Figure 4.22 (a) $\omega - \beta$ diagram of a plane wave propagation through plasma (—) and free space (- - -), and (b) the radio wave reflection and propagation in the ionosphere of earth.

attenuates inside the plasma medium. The complete expression for the x-polarized electric field can be expressed for $\omega < \omega_p$,

$$E_x = E_{x0}^{+} e^{-\alpha z} + E_{x0}^{-} e^{+\alpha z} \tag{4.133}$$

This means a pure attenuated wave exists in the plasma medium when the operating frequency is below *the plasma frequency*.

ii) The Poynting vector of the wave in the plasma for $\omega < \omega_p$ can be calculated using the electric field expression:

$$\boldsymbol{P} = \boldsymbol{E} \times \boldsymbol{H} = \frac{E e^{-\alpha z} \boldsymbol{a_x} \times \boldsymbol{a_y} E\, e^{-\alpha z}}{\eta} = \frac{E_{x0}^2}{\eta} e^{-2\alpha z} \boldsymbol{a_z} \tag{4.134}$$

The impedance is a complex quantity and is defined as:

$$\eta = \sqrt{\frac{\mu_0}{\varepsilon_r}} = \sqrt{\frac{\mu_0}{\varepsilon_0\left(1-\frac{\omega_p^2}{\omega^2}\right)}} = \frac{\omega\mu_0}{\omega\sqrt{\mu_0\varepsilon_0\left(1-\frac{\omega_p^2}{\omega^2}\right)}} = \frac{\omega\mu_0}{k} \tag{4.135}$$

Substitution for η in the above equation yields:

$$\boldsymbol{P} = \boldsymbol{E}\times\boldsymbol{H} = \frac{Ee^{-\alpha z}\boldsymbol{a_x}\times\boldsymbol{a_y}\,E\,e^{-\alpha z}}{\eta} = \frac{E_{x0}^2}{\omega\mu_0}ke^{-2\alpha z}\boldsymbol{a_z} = (j\beta)\frac{E_{x0}^2}{\omega\mu_0}e^{-2\alpha z}\boldsymbol{a_z} \tag{4.136}$$

The average power flow is the real of $\boldsymbol{P}$ hence the power flow through the plasma is zero

$$\boldsymbol{P}_{ave} = \frac{1}{2}RE\left(\boldsymbol{E}\times\boldsymbol{H}^*\right) = RE\left[(j\beta)\frac{E_{x0}^2}{\omega\mu_0}e^{-2\alpha z}\boldsymbol{a_z}\right] = 0 \tag{4.137}$$

This means the incident power in plasma is reflected back completely. If there would have been any energy is absorbed by the plasma there must be net power flow inside the plasma.

iii) At $\omega < \omega_p$, the phase velocity is defined as $u_p = 1/\sqrt{\mu_0\varepsilon} = c_0/\sqrt{1-\left(\omega_p^2/\omega^2\right)} > 1$.[77] Therefore, the phase velocity is greater than the velocity of light in free space. This velocity becomes infinity at the plasma frequency.
iv) The group velocity is determined by the following relationship[78]:

$$u_g = \frac{dk}{d\omega} \tag{4.138}$$

However, the product of the group and phase velocities of a wave is equal to the velocity of light and is given by:

$$u_g u_p = \frac{k}{\omega}\frac{dk}{d\omega} = c_0^2 \tag{4.139}$$

Substitution for the phase velocity from the above equation leads to the expression for the group velocity as:

$$u_g = \frac{c_0^2}{u_p} = c_0\sqrt{1-\frac{\omega_p^2}{\omega^2}} \tag{4.140}$$

Therefore, the group velocity is much less than the velocity of light in free space. This velocity becomes zero at *the plasma frequency*.

v) When *the plasma frequency* is below the operating frequency of the TEM wave ($\omega > \omega_p$) then ω_p becomes positive, and the wave number k in the plasma becomes real $k = \beta$. The propagation

77 Note that this condition of the phase velocity greater than the speed of light does not violate Einstein's relativity theory. In the theory, it is told that the information cannot propagate at the speed greater than the speed of light. The phase velocity does not carry any real energy. Therefore, the derivation of the phase velocity is a valid judgment. We shall also see the similar phenomenon of the greater phase velocity in the reflection and transmission theory of the plane waves.

78 *Dielectric constant of a plasma* http://farside.ph.utexas.edu/teaching/em/lectures/node100.html (accessed 22 January 2018).

factor is defined as $e^{-jkz} = e^{-j\beta z}$. This means the TEM wave propagates with the oscillatory factor $-j\beta z$ as it propagates along +ve z-direction. This means for the condition of $\omega > \omega_p$, the electromagnetic wave continues to propagate inside the plasma medium.

vi) The complete expression for the x-polarized electric field can be expressed for $\omega > \omega_p$,

$$E_x = E_{x0}^{+} e^{-j\beta z} + E_{x0}^{-} e^{+j\beta z} \tag{4.141}$$

Figure 4.22a depicts the dispersion diagram of the plasma medium. This effect of frequency selective surface has long been used in radio wave communications. The transmitted signals from the radio transmitters, which operate below *the plasma frequency* of the ionosphere of the Earth reflect back to the earth and thus, long-range wireless communication is made possible in those frequencies.

As shown in Figure 4.22b, the ionosphere acts as a dish antenna for the low frequency radio waves. Why is the ionosphere form? Due to the continuous bombardment of the ultraviolet rays from the sun, the air molecules of[79] the outermost layer of the Earth's atmosphere, the ionosphere, becomes thinly ionized. The ion counts of the layer depend on the activities of the sun. Therefore, the free electron counts N per unit volume changes with the time of the day and also seasons.

Usually, *the plasma frequency* in the ionosphere is about 1 MHz.[80] From the examination of the theory of plane wave propagation in the plasma, we understand that the low frequency radio signals below *the plasma frequency* are totally reflected off the ionosphere. We have also derived the Poynting vector inside the ionosphere at low frequency and found that the time average real power radio signal in the plasma is zero. This observation concludes that the ionosphere acts as a reflective shield for the low frequency radio wave in the radio frequency range below 1 MHz. Due to this reflection of the incident wave on the ionosphere, the signal can travel further and can be detected over the horizon. Indeed, the signal at long wave (low frequency radio wave) bounces multiple time and can be detected all over the world. On the other hand, the signal higher than the plasma frequency such as FM radio above MHz bands pass through the ionosphere with some reflection and refraction (bent pathway as shown in the figure). Therefore, the radio band signals above *the plasma frequency*, as for example all FM bands in MHz and GHz regions pass through the ionospheric layer and cannot be detected over the horizon.[81] This is the reason for the relatively low coverage of FM radio stations. We also found that the plasma frequency is proportional to the square root of the free electron density N. As stated above, the concentration of ions and hence free electrons in the ionosphere is created by ultraviolet light from the sun. Since at dark nights the activity of the sun in the ionosphere diminishes hence the free electron density counts N also drops significantly. This causes the marked deterioration of low frequency wave propagation and reception of such radio wave signals becomes poor.

---

79 Air layer of the outer atmosphere are getting thinner as we move away from the earth. Moreover, the bombardment of strong ultraviolet ray from the sun, the thin molecules get thinly ionized. The ionized layer is called the ionosphere.

80 Learners are encouraged to do some examples on this number of electron counts and the plasma frequency based on the data given by National Aeronautics and Space Administration (NASA) USA. Read more about NASA via https://en.wikipedia.org/wiki/NASA (accessed 23 January 2018).

81 Renewed interest of electromagnetic wave reflection and propagation has developed in defense industries recently.

Review Questions 4.11: Plasma and EM Waves

Q1. Define plasma. Explain how the plasma is created in the outer layer of the earth's atmosphere. What is the name of the atmospheric layer?

Q2. Who introduced plasma in the first place and when? Search the internet to find how he defined plasma.

Q3. Why wave propagation in plasma is so important in the modern-day radio communications? What are the other fields the study touches?

Q4. How far are the geostationary earth orbiting satellites placed? Can you explain what the meaning of geostationary in the light of electromagnetic wave propagation is? What is the beam coverage of the on-board antenna with respect to the earth?

Q5. What is *Clark belt*? Who is Arthur C. Clerk? What did he envision and when?

Q6. From the internet find the annual market impact of the satellite communications. Give a few examples of satellite communications in entertainments, earth exploration and remote sensing and telecommunications.

Q7. Explain why satellite communication systems usually use the circularly polarized signals. Give a few examples of circularly polarized satellite communications systems that we use in modern time.

Q8. Explain why the plasma is called the fourth state of the matter.

Q9. The plasma has some frequency selective characteristic. Explain what this is in terms of filter nomenclature.

Q10. What is the name of the cut off frequency below which the electromagnetic wave cannot propagate through ionosphere?

Q11. Explain the objectives of electromagnetic propagation through plasma.

Q12. Explain the interaction of the electromagnetic wave and the charged particles in the plasma.

Q13. Explain why only the free electrons take part in the interaction with the electromagnetic field when a plane wave propagates through the plasma medium.

Q14. Write the motional equation of the free electrons with a drift velocity, collision frequency, mass, that undergoes through the influence of the electromagnetic fields. Explain why Lorentz force is active in the equation.

Q15. Define the current density for a free electron cloud with density N and displaced at a velocity $\mathbf{v}$. From there derive the motion equation for the current density $\mathbf{J}$: $m\,(dJ/\partial t) + m\nu\,\mathbf{J} = e^2 N\mathbf{E} - e\,\mathbf{J} \times \mathbf{B}$

Q16. Using the motion equation above show that the current density is defined as: $\mathbf{J} = e^2 N\mathbf{E}\ [m\,(j\omega + \nu)] = \sigma\mathbf{E}$

Q17. Explain why the contribution of the magnetic field $\mathbf{B}$ is ignored in the derivation of current in the above equation.

Q18. Prove that for a TEM wave the plasma medium offer a complex permittivity to the wave which offers as a high pass characteristic.

Q19. Show that the plasma frequency is: $\omega_p = \sqrt{\dfrac{Ne^2}{m\varepsilon_0}}$

Q20. Show the high frequency conductivity in the plasma medium is $\sigma = e^2N/m\,(j\omega + \nu)$

Q21. Calculate the complex permittivity of the plasma and from there show that the plasma frequency is: $\omega_p = \sqrt{Ne^2/m\varepsilon_0}$

Q22. Explain what happen to the propagation of the plane wave when the radio wave frequency is below and above *the plasma frequency*. Explain in the light of the phase constant (wave number).

Q23. Derive the expression for the *y*-polarized electric field wave in the plasma *for* $\omega < \omega_p$.

Q24. Derive the wave impedance of the plasma.

Q25. Show that the time average Poynting vector below the plasma frequency vanishes inside the plasma.

Q26. Show that the phase velocity of the electromagnetic wave inside the plasma is infinity at the plasma frequency.

Q27. Derive the group velocity and show that the group velocity is zero at the plasma frequency. Explain its physical significance.

Q28. What is the dispersion of a medium? Draw the dispersion ($\omega - \beta$) diagram from plasma showing the impact of *plasma frequency*.

Q29. Explain why the ionosphere acts as a dish antenna for the low frequency radio waves. How the radio communications over the horizon is possible?

Q30. Explain why AM radio can be heard at long distance over the horizon, but FM not.

Q31. Explain why AM radio reception becomes poor at night.

Exercise 4.4: Plasma Frequency

Q1. A plane wave is propagating through a uniform plasma medium. The radio broadcasting frequency of the waves is 477 KHz. Explain is the wave can propagate through the plasma medium.

Q2. Show that the Poynting vector of a wave below the plasma frequency in a plasma medium of the wave is zero.

Q3. In an initially neutral plasma, the electron density is *N* electrons/m^3. If due to the ultraviolet ray interaction with the ions, all electrons are displaced by a distance l_x in the *x*-direction, calculate the charge density and the induced electric field. Calculate the plasma frequency assuming the heavy ions are stationary. Assume that the induced electric field tries to restore the electrons in their original position, and the restoring force is proportional to the displacement, electron-plasma oscillations occurs. Show that the oscillation frequency is given by: $\omega_p = Ne^2/m\varepsilon_0$. Show that the equation of motion is given by $(m_e d^2 l_x/dt^2) + (Ne^2/\varepsilon_0)l_x = 0$ (Islam p. 416 Prob. 43)

Q4. In the interstellar space, one the average, there is about 1 electron and 1 proton per cubic meter. Determine the lowest frequency of the electromagnetic wave that can propagate through the medium without attenuation. Discuss the result. (Islam p. 416 Prob. 44)

Q5. The lowest frequency detectable, known as the critical frequency, is related to the density of electrons by the equation: $f = 9 \times 10^{-3} \times \sqrt{N}$ MHz. In this equation, *f* is the critical frequency and *N* is the electron density, $\sqrt{\ }$ means to take the square root of the electron density. The electron density ranges from 33300 (electrons/cm^3) (dark blue) to 249750 (electrons/cm^3) (green) to 552780 (electrons/cm^3) (red). Calculate *the plasma frequency* for the three cases.

4.14 Wave Propagation in Plasma and Satellite Communications

In satellite communications, especially in mobile satellite communications systems, we use circularly polarized waves for efficient communications. In this section, we examine why we do need circularly polarized wave for satellite communications.[82] The main outcome of the investigation is that when a linearly polarized signal wave at a frequency greater than *the plasma frequency* enters the ionosphere of the earth, the linearly polarized signal distorts and is resolved into two opposite handed circularly polarized signal with different phase velocities. [83] This is the effect when the signal has just entered into the plasma medium. What will happen to the linearly polarized signal when it is leaving the medium? This is the main motto of the examination of the propagation theory in this section. As you may recall that polarization matching is an important consideration for signal propagation to avoid excessive loss of signal. This is the main reason that we use circularly polarized signal for the mobile satellite communications system such as global position satellite system, Inmarsat maritime navigation system and Australia Optus mobile satellite communications system called Optus MobileSat, just to name a few. All these mentioned satellite communications systems use right hand circularly polarized electromagnetic wave at around 1.6 GHz, which is much higher than the typical *plasma frequency* of the ionosphere. In this section, we would like to examine the characteristics of the plane wave propagation in the presence of the earth's magnetic field as shown in Figure 4.23. As shown in the figure, we assume that a plane wave propagates along

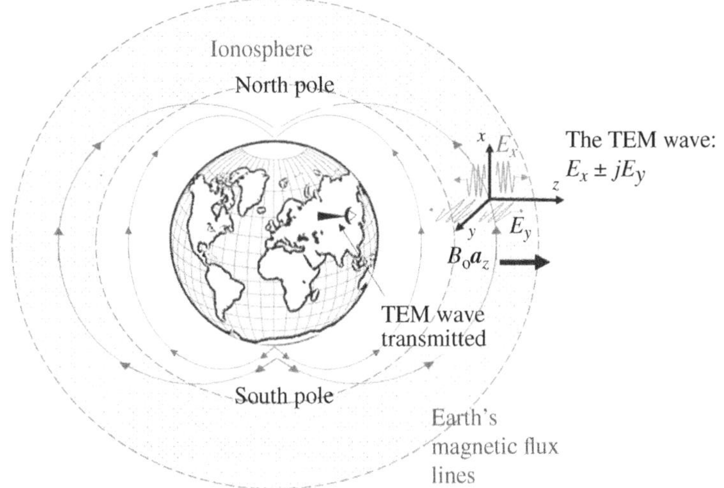

Figure 4.23 TEM wave propagation through ionosphere in the presence of the earth's static magnetic field B_o.

82 For some fixed satellite communication systems such as very small aperture terminal (VSAT) we use two orthogonal linearly polarized signals: one for transmission and one for reception. We usually use lower frequency band for reception where the atmospheric loss is the least and higher frequency band for transmission. The remote sensing satellite system for soil moisture measurements also use dual orthogonal polarized signals to discriminate the natural emissions received from the earth and measures the soil moisture contents from the received orthogonally polarized signals. For mobile satellite communications and navigation systems, the antenna position and orientation continuously change hence the best approach is to use circularly polarized signals. Although we loss −3 dB power when we receive the signal using a linearly polarized antenna in any arbitrary orientation, we can achieve assured reception and transmission of signals for any orientation. We usually use same circularly polarized antennas in both on-board and mobile base station antennas to get full reception.

83 Recall we have derived arbitrary linearly polarized signal from the opposite handed circularly polarized signal in Chapter 3.

positive z-direction. In the TEM case, the electric and the magnetic fields are transverse to the direction of propagation and have x- and y-components. As stated earlier in the derivation of the field waves in the plasma that the activity of the magnetic field associated with the plane wave is a few orders smaller than the electric field. Additionally to this, the earth's magnetic field is much stronger than the magnetic field associated with the plane wave. Therefore, we can neglect the contribution of the magnetic field of the plane wave with the charged particles in the plasma in the analysis. As depicted in the figure below, the earth's magnetic flux lines originate from the North Pole and end at the South Pole. In this case, we assume that the steady magnetic field of the earth's magnetic field is defined as

$$\boldsymbol{B} = B_o \boldsymbol{a}_z \tag{4.142}$$

Therefore, we can consider the component of the earth magnetic field, which is parallel to the direction of propagation of the uniform plane wave. The situation is exactly similar in the last analysis, except we did not consider the earth's magnetic field $\boldsymbol{B}_o$ and the second transverse component along y-direction of the electric field E_y. We assume the effective collision frequency ν is zero for the ionosphere. The complex transverse electric field can be written as

$$E = E_x + jE_y \tag{4.143}$$

In the presence of the earth's magnetic field, B_o and with zero collision frequency ν[84] we can write the equation of motion as follows:

$$m\frac{d\boldsymbol{J}}{dt} = e^2 N\boldsymbol{E} - e(\boldsymbol{J} \times \boldsymbol{B}) \tag{4.144}$$

Assuming that the current density also transverse to the direction of propagation, we can have two independent equations for the motion of current densities J_x and J_y.

$$m\frac{dJ_x}{dt} + eJ_yB_o = e^2NE_x \text{ and } m\frac{dJ_y}{dt} - eJ_xB_o = e^2NE_y \tag{4.145}$$

The above equations indicate that motion takes place on the xy-plane and the field vectors are also on the same plane. If we define the current density[85] vector $\boldsymbol{J}$ is the derivation of the displacement vector $\boldsymbol{R}$, therefore, we can write

$$\boldsymbol{J} = -eN\frac{d\boldsymbol{R}}{dt} \tag{4.146}$$

Substitution for the above equation in the motion equations for J_x and J_y above we obtain

$$-eN\left[m\frac{d^2R_x}{dt^2} + eB_o\frac{dR_y}{dt}\right] = e^2NE_x \text{ and} -eN\left[m\frac{d^2R_y}{dt^2} + eB_o\frac{dR_x}{dt}\right] = e^2NE_y \tag{4.147}$$

Since the displacement (spatial variation due to motion), R, the electric field E and the associated magnetic field H are on the xy-plane we can write the expression of the complex quantities as:

$$R = R_x + jR_y = x + jy, E = E_x + jE_y \text{ and } H = H_x + jH_y \tag{4.148}$$

84 We can assume zero collision frequency in the ionosphere as the ion concentration is very thinly concentrated in ionosphere.

85 This current density is convection in nature due to the fact that the physical movement of the charged particles in this case the free electron of mass m is happening in the medium.

Dividing the above equation by $-eNm$, assuming the time harmonic case so that $d/dt = j\omega$, and using the above relationship for the displacement vector $R = R_x + jR_y$, we obtain:

$$\frac{d^2\boldsymbol{R}}{dt^2} - j\frac{eB_o}{m}\frac{d\boldsymbol{R}}{dt} = -\frac{e}{m}\boldsymbol{E} \quad or, \frac{d^2\boldsymbol{R}}{dt^2} - j\omega_B\frac{d\boldsymbol{R}}{dt} = -\frac{e}{m}\boldsymbol{E} \tag{4.149}$$

where $\omega_B = eB_0/m$ is called *the gyration frequency*. The electrons gyrate (spin) around the static magnetic field.[86] Using Maxwell's two curl equations we can develop a system of equations as follows:

$$\nabla \times \boldsymbol{E} = \begin{bmatrix} \boldsymbol{a}_x & \boldsymbol{a}_y & \boldsymbol{a}_z \\ \frac{\partial}{\partial x} & \frac{\partial}{\partial y} & \frac{\partial}{\partial z} \\ E_x & E_y & 0 \end{bmatrix} = -\mu_0\frac{\partial \boldsymbol{H}}{\partial t} \qquad \nabla \times \boldsymbol{H} = \begin{bmatrix} \boldsymbol{a}_x & \boldsymbol{a}_y & \boldsymbol{a}_z \\ \frac{\partial}{\partial x} & \frac{\partial}{\partial y} & \frac{\partial}{\partial z} \\ H_x & H_y & 0 \end{bmatrix} = \boldsymbol{J} + \varepsilon_0\frac{\partial \boldsymbol{E}}{\partial t}$$

$$\frac{\partial E_y}{\partial z} = -\mu_0\frac{H_x}{\partial t}; \quad \frac{\partial E_x}{\partial z} = -\mu_0\frac{H_y}{\partial t}; -\frac{\partial H_y}{\partial z} = J_x + \varepsilon_0\frac{\partial E_x}{\partial t}; \frac{\partial H_x}{\partial z} = J_y + \varepsilon_0\frac{\partial E_y}{\partial t}$$

Adding the above electric field x- and y-components vectorially in both sides we obtain

$$\frac{\partial \boldsymbol{E}}{\partial z} = -\mu_0\frac{\partial \boldsymbol{H}}{\partial t}$$

Adding the magnetic field vectorially with respective components we obtain

$$\frac{\partial \boldsymbol{H}}{\partial z} = -\left(\boldsymbol{J} + \varepsilon_0\frac{\partial \boldsymbol{E}}{\partial t}\right)$$

Substitution for $\boldsymbol{J} = -eN(d\boldsymbol{R}/dt)$ we obtain

$$\frac{\partial \boldsymbol{H}}{\partial z} = jeN\frac{d\boldsymbol{R}}{dt} - j\varepsilon_0\frac{\partial \boldsymbol{E}}{\partial t}$$

We would like to obtain the solution in the time harmonic exponential form of the three vectors[87] as follows:

$$E = E_0\, e^{\mp j(\omega t - \beta z)}; H = H_o\, e^{\mp j(\omega t - \beta z)}; R = R_o\, e^{\mp j(\omega t - \beta z)} \tag{4.150}$$

Substitutions for the above solutions of the plane wave from of the field and displacement vectors, we obtain,

$$\frac{e}{m}\boldsymbol{E_0} + \left[-\omega^2 \pm \omega_B\omega\right]\boldsymbol{R_0} = 0 \tag{4.151}$$

$$\beta\boldsymbol{E_0} + j\omega\mu_0\boldsymbol{H_0} = 0 \tag{4.152}$$

$$\omega\varepsilon_0\boldsymbol{E_0} + j\beta\boldsymbol{H_0} - \omega eN\boldsymbol{R_0} = 0 \tag{4.153}$$

86 In the next section, we shall discuss the interaction of the charged particle in the presence of steady field and rotation of the free charge around the magnetic field. This phenomenon is called gyration.

87 Note we are using both $\omega t - \beta z$ because the derivatives are in both temporal and spatial domains, respectively. From there we derive all the characteristic parameters of the medium such as the plasma frequency ω_p and gyration frequency ω_B, the phase constant, the plasma dielectric constant, refractive index n and phase velocity v_p, etc.

We have three unknowns and three equations. Therefore, we can easily derive the solutions of the unknown as we did in the previous section. Using Cramer's law, we can obtain the nontrivial solutions for the vector quantities. From there, we obtain the phase constant:

$$\beta^2 = \omega^2\mu_0\varepsilon_0\left[1 - \frac{\omega_p^2}{\omega^2 \mp \dfrac{\omega_B}{\omega}}\right] \tag{4.154}$$

where, $\omega_p = (e^2N/m\varepsilon_0)$ is *the plasma frequency* of the wave and ω_B is *the gyration frequency* of the free electron. Since $\beta^2 = \omega^2\mu_0\varepsilon$ we can define the permittivity of the ionosphere;

$$\varepsilon = \frac{\beta}{\omega^2\mu_0} = \varepsilon_0\left[1 - \frac{\omega_p^2}{\omega^2 \mp \dfrac{\omega_B}{\omega}}\right] \tag{4.155}$$

The relative permittivity is then defined as:

$$\varepsilon_r = 1 - \frac{\omega_p^2}{\omega^2 \mp \dfrac{\omega_B}{\omega}} \tag{4.156}$$

This is a contrasting result compared to the case of plane wave propagation in ionized medium in the absence of the earth's magnetic field. In the ionosphere of the earth, due to the presence of the earth's strong magnetic field B_0, the free electronics gyrate around the magnetic field with a gyration frequency ω_B. This gyration effect of the electron also causes the rotation of the electric field components in two opposite directions. One rotates in the direction of the electron's gyration and one rotates in the opposite direction of the electrons.

Closely examining the phase constant β and the permittivity ε, we can see there are two solutions for both quantities. This means after entering in the ionosphere, the field rotates in two opposite directions with different phase constants and phase velocities.[88] We are now examining the two cases due to the $\pm$ sign in the expressions of β and ε individually.

Case 1: For the case of $E = E_x + jE_y$, $J = J_x + jJ_y$ we obtain the solution for $\varepsilon_{r+} = 1 - \omega_p^2/(\omega^2 - (\omega_B/\omega))$.[89] In this case electric field, vector spins in the same direction of the electron gyration. When the electric field rotates in the same direction of the spinning electrons the wave is called *extraordinary wave*. This is equivalent to the movement of a positively charged particle around the magnetic field. Figure 4.24a shows the rotating electric field keeping the magnetic field at the center of the locus. From the theory of the polarization of the electromagnetic field,[90] when the electric field rotates clockwise direction,[91] as shown in the figure, the handedness of the polarization is the right-hand circular polarization (RHCP).

---

88 Due to the different phase constants and hence the phase velocities, the linearly polarized wave rotates from its original direction. This twist of wave in ionosphere causes misalignment of signals. This is one of the main reasons that the satellite communications predominantly use circularly polarized signals.
89 For this conditions of the electric field $E = E_x + jE_y$ and current density $J = J_x + jJ_y$, the y-component leads the x-components for both the quantities by 90°. The solution of ε_0 brings negative sign.
90 Which will be discussed in the next chapter in details.
91 In defining the handedness, we need to look at the rotation toward the direction of propagation.

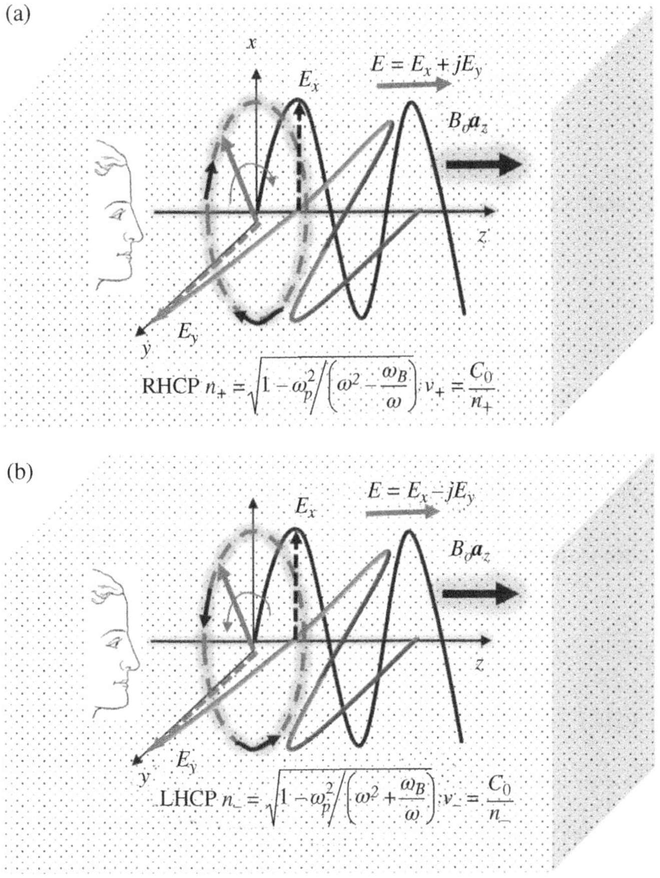

Figure 4.24 Decomposition of a linearly polarized wave into two opposite handed circular polarized wave in ionosphere: (a) RHCP, and (b) LHCP.

<u>*Case 2:*</u> For the case of $E = E_x - jE_y$, $J = J_x - jJ_y$ we obtain the solution for $\varepsilon_{r-} = 1 - \left(\omega_p^2/\omega^2 + (\omega_B/\omega)\right)$.[92] In this case, electric field vector spins in the opposite direction of the electron gyration. This is equivalent to the movement of a negatively charged particle around the magnetic field. Figure 4.24b shows the rotating electric field keeping the magnetic field at the center of the locus. From the theory of the polarization of the electromagnetic field,[93] when the electric field rotates clockwise direction[94] then we define the handedness of the polarization is the left-hand circular polarization (LHCP).

4.14.1 Refractive Indices and Phase Velocities for RHCP and LHCP Cases

The refractive index and phase velocity of a medium and an electromagnetic wave, respectively, are the two most important characteristics parameters for the field analysis, they are illustrated in Figure 4.24. Here, we shall derive both quantities from the previous results for RHCP and LHCP cases.

---

92 For this condition of the electric $E = E_x - jE_y$ and current density $J = J_x - jJ_y$, the y-component lags the x-components of the both quantities by 90°. The solution of ε_0 bring positive sign.

93 Which will be discussed in the next chapter.

94 In defining the handedness, we need to look at the rotation toward the direction of propagation.

The refractive index for the RHCP case is defined as:

$$n_+ = \sqrt{\varepsilon_{r+}} = \sqrt{1 - \frac{\omega_p^2}{\omega^2 - \frac{\omega_B}{\omega}}} \tag{4.157}$$

The refractive index for the LHCP case is defined as:

$$n_- = \sqrt{\varepsilon_{r-}} = \sqrt{1 - \frac{\omega_p^2}{\omega^2 + \frac{\omega_B}{\omega}}} \tag{4.158}$$

The phase velocity for the RHCP case is defined as:

$$v_+ = \frac{\omega}{\beta_+} = \frac{\omega}{\omega\sqrt{\mu_0\varepsilon_0\varepsilon_{r+}}} = \frac{c_0}{\sqrt{\varepsilon_{r+}}} \tag{4.159}$$

The phase velocity for the LHCP case is defined as:

$$v_- = \frac{\omega}{\beta_-} = \frac{\omega}{\omega\sqrt{\mu_0\varepsilon_0\varepsilon_{r-}}} = \frac{c_0}{\sqrt{\varepsilon_{r-}}} \tag{4.160}$$

The intrinsic wave impedance for RHCP is:

$$\eta_+ = \sqrt{\mu_0/\varepsilon_0\varepsilon_{r+}} = \frac{\eta_0}{\sqrt{1 - \frac{\omega_p^2}{\left(\omega^2 - \frac{\omega_B}{\omega}\right)}}} \tag{4.161}$$

The intrinsic wave impedance for LHCP is:

$$\eta_- = \sqrt{\mu_0/\varepsilon_0\varepsilon_{r-}} = \frac{\eta_0}{\sqrt{1 - \frac{\omega_p^2}{\left(\omega^2 + \frac{\omega_B}{\omega}\right)}}} \tag{4.162}$$

We can also derive the total impedance $\boldsymbol{E}(z)/\boldsymbol{H}(z)$ and Poynting vector $\boldsymbol{E}(z) \times \boldsymbol{H}(z)$ of the plane wave in the ionosphere as we did in the ionized medium without the presence of the earth's magnetic field. These are left as the exercise questions. Let us now recap our findings in the analysis of the wave propagation in the ionosphere as follows:

1) The ionosphere in the presence of the earth's magnetic field exhibits two different indices of refraction n_+ and n_- corresponding to the two types of circularly polarized waves: RHCP and LHCP.
2) The two wave types in the ionized medium travel with two different phase velocities v_+ and v_- respectively.
3) A linearly polarized wave ends up to two circularly polarized waves with two opposite handedness. This means when it enters the ionosphere, a linearly polarized electromagnetic wave resolves itself into two circularly polarized components of RHCP and LHCP.

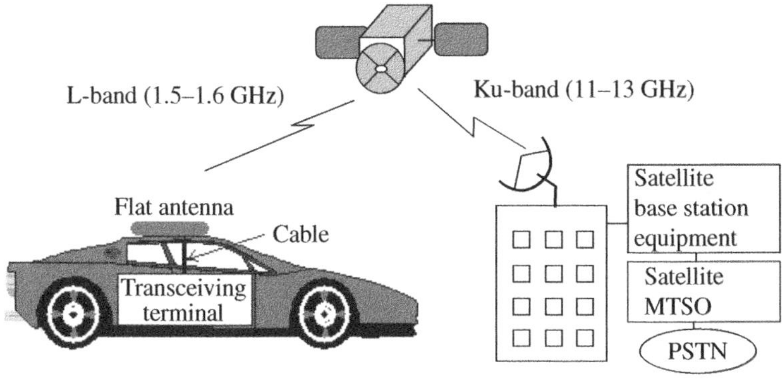

Figure 4.25 Basic configuration of an MSAT system is comprised of the space segment (the satellite), the network gateway control system and the ground segment consisting of a vehicle mounted antenna and a transceiving terminal. *Source:* Karmakar, N.C. (1999). Antennas for mobile satellite communications. PhD thesis. The University of Queensland, Australia. p. 7 (Figure 1.1).

4) This resolution provides a scientific clue as to why we prefer circularly polarized wave for satellite communications, specifically for mobile satellite communications systems as stated above. The ionosphere which is also called *Clerk Belt* causes the rotation of the linearly polarized signal. And this rotation is random as the ion concentration in the ionosphere changes with the height from the earth, the time of a day, and with different seasons due to the fluctuations of solar activities. Therefore, to avoid the random twist of the linearly polarized wave during the propagation through ionosphere, we prefer circularly polarized wave for satellite communications.[95] Figure 4.25 shows the configuration of a mobile satellite communications system where the mobile ground station is situated inside and the smart antenna on the roof of the vehicle. The antenna uses RHCP to match with the signal from the satellite. Since the signal is already rotated due to its circular polarization, it is not affected by the twist of the ionosphere and received by any orientation of the vehicle on the ground when it is on the move. This example is drawn from the PhD thesis of the author titled: *Antennas for Mobile Satellite Communications.*

In this section,[96] we shall examine how the difference in the phase constants of the two handedness of the circularly polarized waves can be used to show that a linearly polarized wave is rotated as it passes through the ionosphere. Then we shall draw the conclusions of the importance of the circularly polarized signals for satellite communications. In 1845, Faraday first observed the rotation of a linearly polarized light when it passes through a gyrotropic medium in the presence of an applied magnetic field.[97] Hence the rotation is called *Faraday rotation*. Using the two different phase constants, we can write the complex transverse field components:

$$\boldsymbol{E}_{RHCP} = \left(\boldsymbol{a}_x + j\boldsymbol{a}_y\right)\frac{E_o}{2}e^{-j\beta_+ z} \text{ and } \boldsymbol{E}_{LHCP} = \left(\boldsymbol{a}_x - j\boldsymbol{a}_y\right)\frac{E_o}{2}e^{-j\beta_- z} \tag{4.163}$$

95 The mismatch between the polarizations of the transmitted and received signals is measured with the polarization loss factor, in short *PLF*. *PLF* is defined as the square of the cosine of the subtended angle θ between the resultant field vectors of the transmitted and received signals.

96 This section is extracted from S. Ramo and J.R. Whinnery, *Fields and Waves in Modern Radio*, Wiley 2e. 1953, Section 13.14 *Faraday Rotation pp. 721—722.*

97 The ionosphere of the earth's atmosphere is a gyratropic medium due to the earth's magnetic field B_0 and the plasma concentrations in the ionosphere.

The resultant field is:

$$
\begin{aligned}
\boldsymbol{E} &= \boldsymbol{E}_{RHCP} + \boldsymbol{E}_{LHCP} = \frac{E_o}{2}\left[\boldsymbol{a_x}\left(e^{-j\beta_+ z} + e^{-j\beta_- z}\right) + j\boldsymbol{a_y}\left(e^{-j\beta_+ z} - e^{-j\beta_- z}\right)\right] \\
&= \frac{E_o}{2}e^{-j\left(\frac{\beta_+ + \beta_-}{2}\right)z}\left[\boldsymbol{a_x}\left(e^{-j\left(\frac{\beta_+ - \beta_-}{2}\right)z} + e^{j\left(\frac{\beta_+ - \beta_-}{2}\right)z}\right) + j\boldsymbol{a_y}\left(e^{-j\left(\frac{\beta_+ - \beta_-}{2}\right)z} - e^{j\left(\frac{\beta_+ - \beta_-}{2}\right)z}\right)\right] or, \\
&= E_o e^{-j\beta_{ave}}\left[\boldsymbol{a_x}\cos\theta - j\boldsymbol{a_y}\sin\theta\right]
\end{aligned}
\tag{4.164}
$$

where,

$$\beta_{ave} = \frac{\beta_+ + \beta_-}{2} \quad \text{and}\ \theta = \left(\frac{\beta_+ - \beta_-}{2}\right)z \tag{4.165}$$

From the examination of the above derivation of the composite electric field wave equation, we can draw the following remarks:

1) Observing the field vector term in the parenthesis, we find that there is no phase difference between the component fields along x- and y-directions. Therefore, the composite field wave has a fixed direction at each value of z, which is defined by the electrical length or phase angle of the wave $\theta = ((\beta_+ - \beta_-)/2)z$. The direction of the field vector relative to $z = 0$ plane is defined by:

$$tan\theta = \frac{E_x}{E_y} = tan\left[\left(\frac{\beta_+ - \beta_-}{2}\right)z\right] \tag{4.166}$$

2) Therefore, the incident linearly polarized wave rotates as it passes through the ionosphere with the propagation constant β_{ave}.
3) The rotation remains the same, irrespective of the direction of propagation either positive or negative z-direction.[98] *Faraday rotation* is strong in ferrites,[99] some magnetized plasma such as ionosphere and also in ordinary dielectrics though the angle of rotation is very small in ordinary dielectrics. Figure 4.26 illustrates the rotation of a linearly polarized signal as it passes through the plasma in the presence of the steady magnetic field.

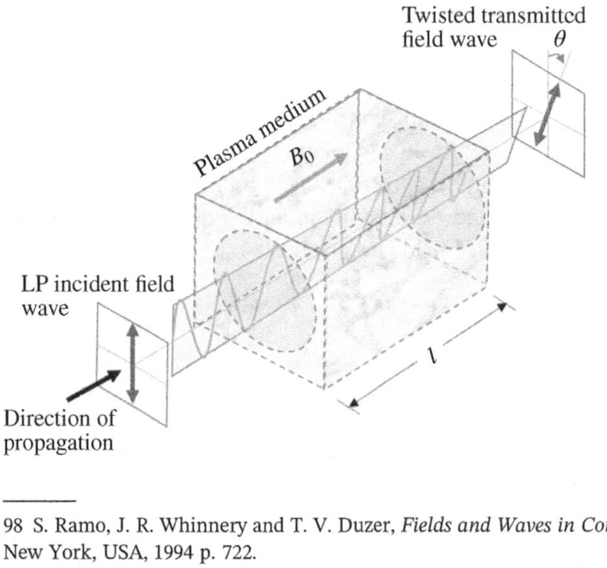

Figure 4.26 The rotation of linearly polarized signal as it passes through the ionosphere.

---

98 S. Ramo, J. R. Whinnery and T. V. Duzer, *Fields and Waves in Communications Electromagnetics*, 3rd Ed, Wiley, New York, USA, 1994 p. 722.

99 Faraday's rotation is used in many important microwave devices designs such as circulators. Using circulators, we can make the signal propagation in one directional in the system so that the transmitted signal cannot bounce back to the transmitter and hence safe guards expensive power amplifiers from damages by strong return signals.

A few important points to conclude on the issue of the use of the circularly polarized signal and its advantages in satellite communications:

1) *Faraday rotation* affects the linearly polarized signal. This has huge effects on the L, S, and C-band signals in which band the most of the existing satellite communications systems operate.
2) Due to *the Faraday rotation* of the linearly polarized signal, alignments of the transmitting and receiving antennas between the on-board satellite antennas and the ground terminal antennas becomes a huge impediment. The alignment is time consuming and costly.
3) Linearly polarized signal is highly susceptible to the fluctuating atmospheric conditions whereas the circularly polarized signal is less susceptible to these conditions.
4) Circularly polarized antennas are much easier to install as they can receive signal in any orientations
5) Due to the above advantages the circularly polarized systems offer high link reliabilities.

Review Questions 4.12: Satellite Communications and Plasma

Q1. Explain why mobile satellite communication systems use circularly polarized signals. Give a few examples of right-hand circularly polarized mobile satellite communications systems that we use in modern time. Also search for LHCP satellite systems in the Internet.

Q2. Explain what has happened when a linearly polarized signal incidents on the ionosphere of the earth.

Q3. Investigate the operating frequencies of GPS, Inmarsat, Singapore-Taiwan (ST) VSAT systems. What handedness of these systems have?

Q4. Explain the difference between the magnetic field associated with the plane wave and that for the earth.

Q5. The transverse current density vector has the motion equation defined by $m\,(d\boldsymbol{J}/dt) = e^2N\boldsymbol{E} - e\,\boldsymbol{J} \times \boldsymbol{B}$. Using the basic definition of the convection current density $\boldsymbol{J} = -\,eN\,(d\boldsymbol{R}/dt)$ show that the vector motion equation for displacement vector $\boldsymbol{R}$ is: $(d^2\boldsymbol{R}/dt^2) - j\omega_B\,(d\boldsymbol{R}/dt) = -\,(e/m)\boldsymbol{E}$. Here symbols have their usual meanings.

Q6. What is gyration frequency ω_B? Explain its physical meaning.

Q7. Using the three equations $\frac{e}{m}\boldsymbol{E}_0 + [-\omega^2 \pm \omega_B\omega]\,\boldsymbol{R}_0 = 0$; $\beta\,\boldsymbol{E}_0 + j\omega\mu_0\boldsymbol{H}_0 = 0$ and $\omega\varepsilon_0\boldsymbol{E}_0 + j\beta\boldsymbol{H}_0 - \omega eN\,\boldsymbol{R}_0 = 0$, show that the phase constant of the ionosphere is defined as:

$$\beta^2 = \omega^2\mu_0\varepsilon_0\left[1 - \frac{\omega_p^2}{\omega^2 \mp \dfrac{\omega_B}{\omega}}\right]$$

Q8. Show that the relative permittivity (dielectric constant) of the ionosphere is defined as: $\varepsilon_r = 1 - \left(\omega_p^2/\omega^2 \mp (\omega_B/\omega)\right)$.

Q9. Explain the physical significance of the two solutions of the phase constant β.

Q10. What is an *extra ordinary wave*? In which direction the wave rotates? What is the polarization of the wave?

Q11. Calculate the Poynting vector of an electromagnetic field in the ionosphere.

Q12. Explain why the ion concentration changes in ionosphere and what impact it creates in the radio communications.

Q13. Calculate the polarization loss factor of a signal with the subtended angle of $\theta = 45°$.

Q14. What is *Faraday rotation*? Explain how it affects the linearly polarized signal as it passes through the ionosphere.

Q15. Show that when a linearly polarized wave is incident on the ionosphere the transmitted wave is rotated by an angle θ. Hints: The amount of rotation depends on the two different phase constants in the ionosphere and the distance travel by the wave through the ionosphere.

Q16. Explain the advantages of using circularly polarized wave over the linearly polarized wave in satellite communications in the light of *Faraday rotation*.

Exercise 4.5: Wave Propagation in Plasma

Q1. An electromagnetic wave at 1.6 GHz is propagating through the ionosphere with the electron concentration of 3.3×10^{12} electrons/m^3. The earth's magnetic field $B_0 = 50 \times 10^{-6}$ T. Calculate the gyration frequency ω_B, the plasma frequency ω_p, the refractive indices, phase velocities and propagation constants for both RHCP and LHCP. The mass of electron $m = 9.11 \times 10^{-31}$ kg.

4.15 Waves in Dielectric Media

So far, we have discussed many complex media in which the electromagnetic signals show dispersion phenomena. However, in all cases we assume the media is homogeneous, linear, and isotropic. In this section, we shall discuss about the anisotropic nature of the dielectric medium and its modern applications in microwave engineering and antenna technology. The anisotropy of a material medium means that the macroscopic constitutive parameters ε and μ are dependent on the direction of the wave propagation. A nice illustration on anisotropy in a manmade–pavement and natural materials — cross section of a log of wood, are shown in the following Figure 4.27. Similar to

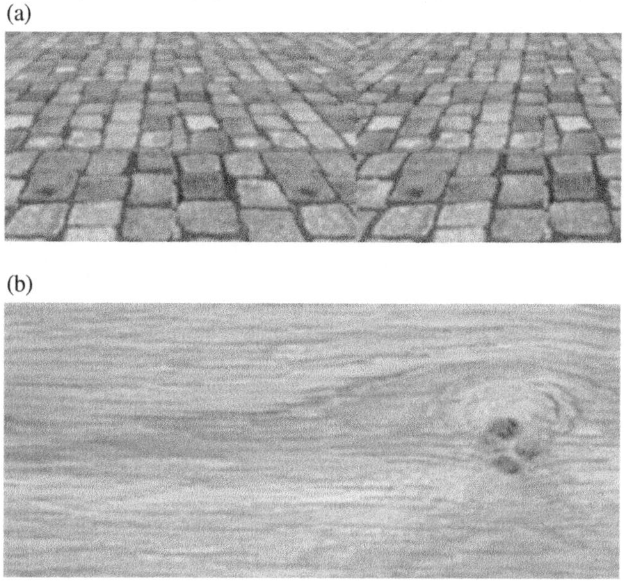

Figure 4.27 Anisotropic medium (a) manmade–pavement (An illustration of texture filtering methods showing a texture with trilinear mipmapping (left) and anisotropic texture filtering. *Source:* Thomas/ Wikimedia Commons/CC BY-SA 3.0.), and (b) wood knot in vertical section. *Source:* F pkalac/Wikimedia Commons/ CC BY-SA 3.0.

these configurations, the anisotropic medium shows different characteristics in different directions to the electromagnetic wave propagation.

Since we are not considering magnetic materials[100] in our analysis, and are considering only the dielectric medium, therefore, in the analysis, we can assume $\mu = \mu_0$. Hence our analysis considers the determination of the relative permittivity ε_r of the medium in the macroscopic level of a region of the propagation medium. In this section, we divide our analysis of the properties of materials and the progression of coverage for three fundamental concepts as shown in Figure 4.28:

i) First, we look into all dielectric and nonmagnetic materials into the perspective of the classical electron theory, and from there, we shall try to understand how the charged atoms and molecules interact with the external applied electric field, and the resulting resistance the bound charge of the dipole imposes on it. Based on this we shall draw an interesting conclusion of the different types of material media.
ii) Second, we shall look into the same very concept into *the macroscopic model* considering a region of the dielectric medium, and from the analysis, we shall draw a closed form expression of the dielectric constant (relative permittivity) as the macroscopic characteristic of the medium. In this regard, we shall define *the electric dipole moment* in a volume of the dielectric region and *the polarization vector* of *the bound charge excess* as coined by Johnk.[101] This equivalent charge density and polarization vector form the concept of the divergence nature of the equivalent charge and define the constitutive relation of the medium relative to the free space. This very concept of the divergence nature of the equivalent polarization charge density is the constituent to define the macroscopic definition of the dielectric constant of the medium.

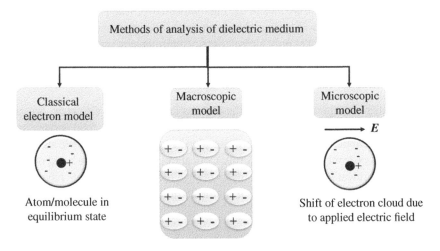

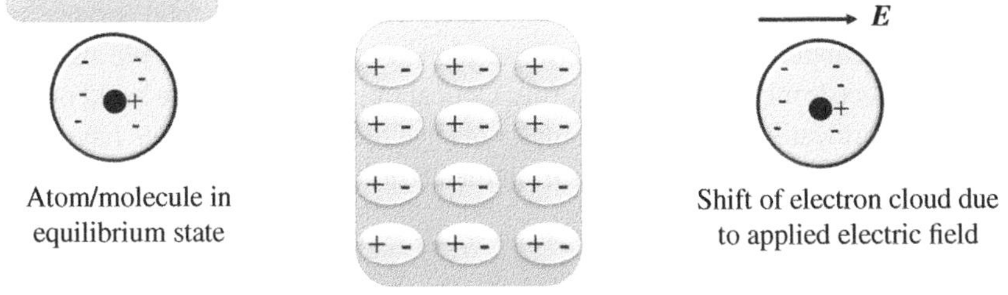

Figure 4.28 View point of analyses of dielectric medium.

100 The book mainly focuses on the passive communication devices and electromagnetic wave propagation through nonmagnetic materials, therefore, the magnetic materials are out of scope of the book.
101 C. T. A. Johnk, *Engineering Electromagnetic Fields and Waves*, 2e Wiley, NY, USA, 1988 pp. 116–119 for more elegant explanation.

iii) Third, we shall look into *the microscopic view* of the atomic and molecular models of the dipole moment due to the application of the applied electric field, and examine the electrostatic energy that it possesses in the light of the electric potential energy. From there, we shall examine the divergence nature of the equivalent electric charge density due to *the bound charge excess.* We shall land to the same conclusion as above in *the macroscopic model.*

After the analyses, we shall examine the characteristics of the anisotropic dielectric from the view point of the molecular configuration of the crystalline dielectric media. The anisotropy has many applications in modern telecommunications engineering. Many exceptional materials have shown exotic characteristics when it interacts with the electromagnetic fields, and we exploit these characteristics for many modern devices that did not exist before.[102] Example includes manmade artificial materials called metamaterials. For the big volume of the theory and special character of *the anisotropic media* we shall discussion this topic in a separate section.

In brief, the summary of the section is as follows: we look into three aspects of analyses of the relative permittivity of an anisotropic medium, and its effect on the field wave propagation. We first consider a dielectric medium in both microscopic and macroscopic views. To do so, first, we define the dielectric medium and its microscopic construction assuming the medium is homogeneous, linear, and isotropic. Then we shall try to understand the change in the microscopic construction in the medium due to the applied electric field across the medium. From there we shall develop the concept of deriving the macroscopic parameter, the relative permittivity, ε_r, first for the isotropic medium and then for the anisotropic medium.

4.15.1 Classical Electron Theory of Dielectric

Dielectric materials are insulators. When we look at the high-tension overhead power transmission lines, we see many porcelain insulators of various shapes separate the high-tension overhead conductors from the poles and cross arms. Not only for isolating the high-tension power lines from the poles, we also use dielectric materials for microwave frequency PCBs and devices. Therefore, understanding the dielectric materials and electromagnetic field wave interaction touches many disciplines.

In insulators, electric current cannot flow rather the charges localize in the dielectric. We are living in a dielectric medium such as air. All our dining plates and utensils for food serving and storage are made of dielectric materials. Dielectrics are also good insulators of heat.[103] We also communicate wirelessly via dielectric media in various forms. Optical fibers which are the backbone for long haul broadband communications are good examples of such dielectric media. We have given many examples of such propagation media starting from dielectrics, conductors, semiconductors, and plasma in this chapter.

---

102 With the advent of modern 3D printing and manufacturing tools many new artificial manmade materials are possible to make in nanometer accuracy. The theory of the negative refractive and combined left-hand and right-hand materials were there before, but it was only materialized in recent years due to advanced computer aided design and computer aided manufacturing tools.

103 Corning has high quality plates and food storage materials made of ceramic and glass. Some leads of the utensils are made of microwave transparent plastics and are called *microwave safe.* Corning developed first optical fibers using glass. It is worth visiting Corning Museum of Glass in NY, USA for their invention in glass related research. (https://www.cmog.org/, accessed 22 February 18).

What is a dielectric? A perfect dielectric is a medium in which the free charge cannot move through the atomic structure as opposed to that for the conductor. Therefore, when a charge is given to a dielectric medium, the charge remains localized in that region. According to the classical electron theory of dielectrics, they are classified as nonpolar and polar dielectrics. In a nonpolar dielectric, we can assume that the atoms of the dielectric are constructed and arranged in such a manner that the epicenters of the negative and positive charges in the microscopic body[104] coincides with each other in equilibrium. However, in polar dielectrics such as water (H_2O) and hydrochloric acid (HCl) the epicenters of the positive and the negative charges are little bit shifted to each other with a fixed distance with a strong elastic force. In this study our focus is on the non-polar dielectric medium. When an external electric field is applied, the molecules and atoms are rearranged in such a way that the epicenter of the positive and negative charges in the atoms and molecules are shifted a minute distance. Due to the strong elastic force that acts between the positive and the negative charges the Coulomb force acts between them hence an opposing electric field between the charge particles establishes. The hypothesis is that each charged atom and molecule are considered to be an electric dipole having +ve and −ve charge centers located at some distance apart, say d. Each such dipole has a dipole moment, $\boldsymbol{P} = q\boldsymbol{d}$, where q is the charge and $\boldsymbol{d}$ is the distance[105] vector between the charges. In normal condition these particles are so haphazardly distributed that the net charge of the dipoles is zero. However, when an electrostatic field is applied, the positive and negative charged centers are stretched and shift with the positive charge center a bit toward the electric field line as shown in Figure 4.29. The same very nature of the dielectric offers resistance to the electric field. That is why the electric field intensity is less in a dielectric medium than that in the free space. Based on the value of the relative permittivity of the media, they are defined as lighter and denser. In lighter medium, the electric field intensity is more than that for the denser medium

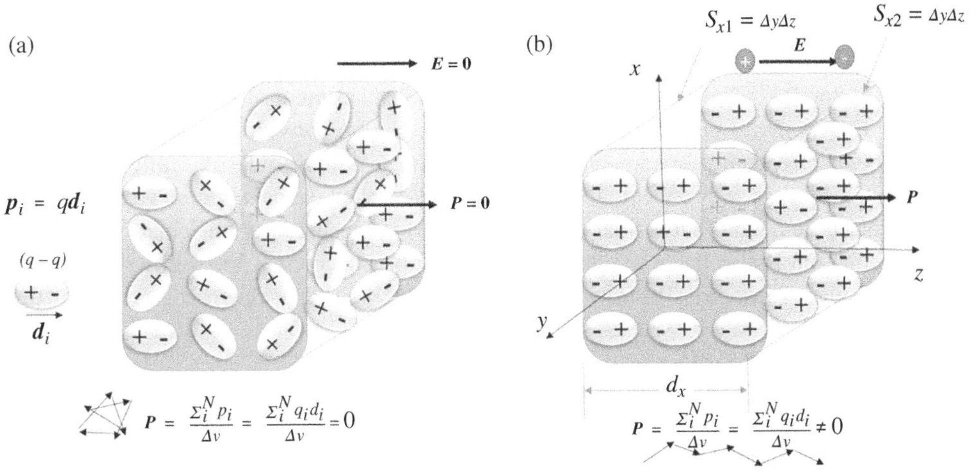

Figure 4.29 Concept of equivalent bound charge excess and polarization vector ***P*** due to the applied electric field ***E*** in a dielectric medium: (a) no field applied, and (b) field applied.

---

104 Here "body" means the charged particle that can be either atoms or molecules.

105 You may imagine how much the value of $\boldsymbol{d}$ due to the fact that the size of an atom and a molecule is in the order of Armstrong 10^{-12} m.

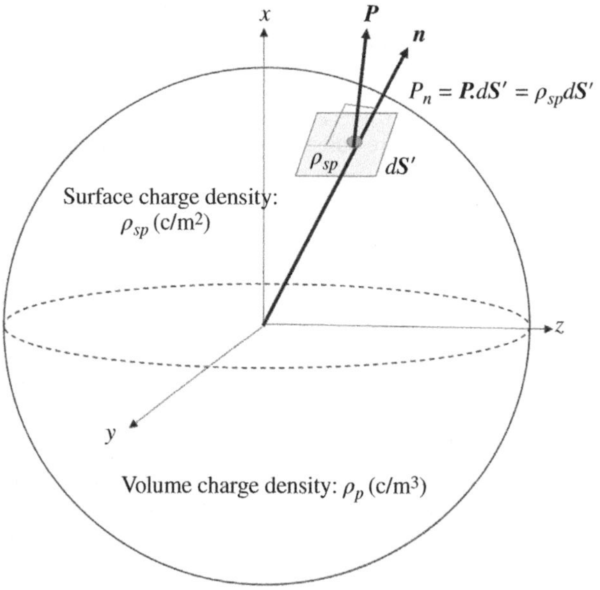

Figure 4.30 Relationships of polarization volume and surface charge densities and polarization vector of a dielectric medium.

due to the fact that the lighter medium offers less resistance to the field lines or displacement vector. A molecule which possesses such a fixed distance between the epicenters of the positive and negative charges is called electric dipole. The orientation of the dipole toward the applied electric field is called *polarized medium* or *the polarized state* of the medium.[106] What is the effect of *the polarized state* of the dielectric medium? When the medium is polarized with an applied electric field into it, the slight shift in inter-particle and molecular orientations and displaced epicenters of the positive and negative charges produces a displacement of charge as if the real charged particle does for the displacement vector ***D***.[107] The very cause that makes this displacement of charge is called *polarization vector* ***P***. The motto of the study is to develop a relation between the two quantities in the light of the displacement phenomenon of the dielectric medium under the influence of the applied electric field. First, we conceptualize a vector ***P***, also called *the polarization vector* per unit volume in a dielectric region that is linked with a new sort of localized volume charge density ρ_p with the material and is also liable to create a surface charge density ρ_{sp} due to the setup of the polarized molecules. These are illustrated in Figure 4.30. This is the macroscopic view of the

106 *The polarized state* is coined by Michael Faraday.
107 To understand *the real charge* and *the polarized charge* (also called *bound charge access* by Johnk) we need to assume that the charge is acting in free space so that we can define the constitutive relationship $\boldsymbol{D} = \varepsilon_0\boldsymbol{E}$ first. This way we can coincide our understanding of the permittivity parameter ε_0 in vacuum. Now we would like to introduce a dielectric medium in the free space when the applied field is in existence, and consider *the polarized state* of the material due to the shift of orientation of the $\pm q$ charge pairs in atoms and molecules in the region under the influence of the applied electric field. In this case, there is an excess bound charge due to the polarized charged particles. This *polarized state* is defined with a displacement vector similar to ***D*** so that we can develop another constitutive relationship for the presence of the dielectric medium in the region. We define the additional displacement vector also called *polarization vector* $\boldsymbol{P} = \varepsilon_0\chi_e\boldsymbol{E}$, where χ_e is a proportionality constant and is defined as *susceptibility* of the dielectric medium. The superposition of the two displacement vectors gives the total constitutive relationship of the dielectric region as: $\boldsymbol{D} = \varepsilon_0\,\boldsymbol{E} + \boldsymbol{P} = \varepsilon_0\,(1+\chi_e)\,\boldsymbol{E} = \varepsilon_0\varepsilon_r\,\boldsymbol{E}$. Therefore, the relative permittivity is defined as $\varepsilon_r = 1+\chi_e$. This is the macroscopic understanding of the relative permittivity of the dielectric medium.

polarization vector. The second concept is the microscopic view of the electric dipole $\boldsymbol{p}$ that set up in a dielectric molecule due to the application of the external electric field $\boldsymbol{E}$. Then we shall sum up the microscopic quantity $\boldsymbol{p}$ of the molecules to get the macroscopic quantity $\boldsymbol{P}$ for a region. In this aspect, we apply the common knowledge of the classical electron theory as shown in Figure 4.29.

4.15.2 Macroscopic View of Dielectric

Consider a system of charged body, which constitutes a dielectric medium, exists within the fictitious bounded area as shown in Figure 4.29. In the equilibrium condition, the molecules are arranged in such a way that the net field due to the haphazardly arranged bound charges are zero. Once the electric field $\boldsymbol{E}$ is applied, the dipole moment occurs and the molecules are rearranged in such a way that it exerts electric field that is opposed to the applied electric field. Due to the elastic nature of the atoms and molecules, the Coulomb force counteracts the net electric field defined by the Lorentz force and the net field is nonzero.[108] The Lorentz force due to the electric field acting on the charge particle of charge q is defined as: $\boldsymbol{F_e} = q\boldsymbol{E}$. In a dielectric, there are many such charge particle pairs that constitute the vary many dipole moments $\boldsymbol{p}$ and hence the combined effect of the polarized molecules is systematized as the polarization vector $\boldsymbol{P}$. The dipole moment created by the *i-th* such charged particle pair is defined as $\boldsymbol{p_i} = q\boldsymbol{d_i}$. The unit is coulomb-meter. Here, q is the positive charge of the charge pair $(q, -q)$. Here, $\boldsymbol{d_i}$ is the vector that represents the distance[109] of the couplet. If there is N number of such couplets per unit volume, then the net or linear vector sum of the total dipole moments within the volume is defined as *the polarization* of the dielectric medium and is defined as:

$$\boldsymbol{P} = \frac{\sum_i^N \boldsymbol{p_i}}{\Delta v} = \frac{\sum_i^N q_i \boldsymbol{d_i}}{\Delta v} \ (\text{C/m}^3) \tag{4.167}$$

Assume that all couplet has identical elementary charge q, then we can define a *bound charge excess* density ρ_p per unit volume as: $\rho_p = (Nq/\Delta v)$ and the average dipolar displacement as $\boldsymbol{d} = \left(\sum_i^N \boldsymbol{d_i}/N\right)$. Then the polarization vector of the dielectric region is defined as:

$$\boldsymbol{P} = \rho_p \boldsymbol{d} \ (\text{C/m}^3) \tag{4.168}$$

This situation can be represented as some net charge density, which is called *bound charge excess*[110] within a volume and is defined as ρ_p (C/m^3).[111] To systematize the concept of the displacement nature of the polarization vector due to *the bound charge excess*, let us consider a nonuniform polarization vector $\boldsymbol{P}$ exists in a fictitious elementary volume element $\Delta v = \Delta x \Delta y \Delta z$ with sides Δx,

---

108 If the same field is applied to the conducting medium, due to the free valance electrons in the atoms and molecules, the charged particles so rearrange that the positive and negative charges nullify the applied electric field hence the resultant field is zero across the region of investigation.

109 We can also define as the strain or deformation state of the charge couplet $(q, -q)$.

110 This term "bound charge excess" is coined by C. T. A. Johnk, *Engineering Electromagnetic Fields and Waves*, 2e., Wiley, NY, USA, 1988.This is formed due to the lengthy bonded charged molecules which are slightly shifted or deformed and create diploe moment due to the applied electric field.

111 Subscript p stands for *the polarized state* of the charged particle in the dielectric medium.

Δy, and Δz, respectively.[112] The distance between the two surfaces along x-axis is assumed to be d_x. Then assume that the net charge entering the left side of the surface $S_{x1} = \Delta y \Delta z$ is defined as:

$Q_{px1} = P'_x = (\rho d_x \Delta y \Delta z)$. As the charge passes through the differential volume along x-direction, it also collects other *bound charge excess* within the volume. Therefore, the net charge coming out of the right-hand surface S_{x2} at distance d_x is defined as:

$$Q_{px2} = P''_x \Delta y \Delta z = \left(\rho d_x + \frac{\partial(\rho d_x)}{\partial x}\Delta x\right) \times \Delta y \Delta z \tag{4.169}$$

As can be seen, due to the applied electric field from the left to right (from positive charge to the negative charge polarity of the actual/source charge), the polarized molecules are arranged in such a way that the positive charge of the dipole moment is arranged in the right hand side keeping the negative charge in the left hand side as if the negative charge is left in the volume and the positive charge is coming out of the volume and make the volume as an excess negative bound charge. Therefore, if we subtract $Q_{p1}$from Q_{p2}, we ge the net bound charge excess coming out of the surface, which is also equal to the net negative charge excess left within the volume, is defined as:

$$Q_{px1} - Q_{px2} = -\frac{\partial(\rho d_x)}{\partial x}\Delta x \Delta y \Delta z = -\frac{\partial P_x}{\partial x}\Delta x \Delta y \Delta z \tag{4.170}$$

Thereafter, we take y- and z-directed contributions of the net charge flowing out of these two surfaces and add all three contributions to get the net charge out (likewise same net negative charge excess is left within the volume) from the differential volume is:

$$\Delta Q_v = -\left(\frac{\partial P_x}{\partial x} + \frac{\partial P_y}{\partial y} + \frac{\partial P_z}{\partial z}\right)\Delta v = \rho_p \Delta v \tag{4.171}$$

Here, $\Delta Q_v = \rho_p \Delta v$, the net bound charge excess within the differential volume Δv. Dividing both sides of (4.171) with $\Delta v = \Delta x \Delta y \Delta z$ and taking limit $\Delta v \to 0$, means $\left.\begin{matrix}\Delta x\\ \Delta y\\ \Delta z\end{matrix}\right\} \to 0$, we obtain the following results[113]:

$$-\nabla \cdot \boldsymbol{P} = -\lim_{\left.\begin{matrix}\Delta x\\ \Delta y\\ \Delta z\end{matrix}\right\} \to 0} \frac{\left(\frac{\partial P_x}{\partial x} + \frac{\partial P_y}{\partial y} + \frac{\partial P_z}{\partial z}\right)}{\Delta x \Delta y \Delta z} = \rho_p \tag{4.172}$$

Due to the displacement phenomenon of the equivalent charge confined within the volume, we also assume customarily that there is displacement of the charge via a fictitious enclosed surface area and this phenomenon is defined with the surface charge density ρ_{sp}. Therefore, to comply with the diverging nature of *the polarization* vector, we can have two relations from the hypothesis. The first one is the point forms expression above derived from the very basic understanding at

112 Refer to C. T. A. Johnk, *Engineering Electromagnetic Fields and Waves*, 2e, Wiley, NY, USA, 1988 pp. 116–119 for more elegant explanation.

113 The definition of divergence of a vector $\boldsymbol{P}$ is: $\nabla \cdot \boldsymbol{P} = \lim_{\left.\begin{matrix}\Delta x\\ \Delta y\\ \Delta z\end{matrix}\right\} \to 0} \frac{\left(\frac{\partial P_x}{\partial x} + \frac{\partial P_y}{\partial y} + \frac{\partial P_z}{\partial z}\right)}{\Delta x \Delta y \Delta z}$

microscopic level. And the second one is the Gauss law of the polarization vector $\boldsymbol{P}$ in a region hence the macroscopic level:

$$\oint_s \boldsymbol{P} \cdot d\boldsymbol{S}' = \int_v (\nabla \cdot \boldsymbol{P}) dv' = -\int_v \rho_p dv' \tag{4.173}$$

The physical meaning of the negative sign on the right-hand side is already defined during the process of examining the flow of the net bound charge excess ($+\rho_p \Delta v$) flowing out of a differential volume that forms a closed surface. According to the displacement concept of $\boldsymbol{P}$, here, $-\rho_p \Delta v$, the resulting equivalent polarized charge within the volume, due to the applied electric field $\boldsymbol{E}$ that sets up *polarized state* within the dielectric region, is displaced as the surface charge density ρ_{sp} from the surface as $\oint_s \boldsymbol{P}.\, d\boldsymbol{S}' = \oint_s \rho_{sp} d\boldsymbol{S}'$. Hence the normal to the surface component of the polarization vector is equal to the surface charge density,

$$P_n = \boldsymbol{P} \cdot d\boldsymbol{S}' = \rho_{sp} d\boldsymbol{S}' \tag{4.174}$$

Here, the prime sign represents the source region. This means the bound charge excess that has been carried out by the displacement vector $\boldsymbol{P}$ out of the volume, hence the negative of the integral of ρ_p, the volume charge density, is imposed.

Applying the divergence theorem, we obtain:

$$\int_v (\nabla \cdot \boldsymbol{P})\, dv' = -\int_v \rho_p dv' \tag{4.175}$$

The differential form of the relation yields:

$$\nabla \cdot \boldsymbol{P} = -\rho_p \tag{4.176}$$

This is a significant relation between the dipole charge density ρ_p and the polarization $\boldsymbol{P}$. We shall be using conveniently this hypothesis to understand the macroscopic view of the dielectric and derive the relative permittivity of the dielectric medium. This is the contribution of *the polarized state* of the dielectric medium. To get the net contribution due to the applied electric field $\boldsymbol{E}$, we need to add this contribution with the free space case. This is equivalent to the calculation of net electric flux density coming out of the medium. We can define the constitutive relationship $D = \varepsilon_0 E$ for the free space case and add the effect of the material with it to yield the net displacement vector as:

$$\boldsymbol{D} = \varepsilon_0 \boldsymbol{E} + \boldsymbol{P} \tag{4.177}$$

where ε_0 is the free space permittivity and $\boldsymbol{E}$ is the applied electric field. We define *the polarization vector* in terms of the applied electric field which is a new constitutive relation:

$$\boldsymbol{P} = \varepsilon_0 \chi_e \boldsymbol{E} \tag{4.178}$$

Here, χ_e is a proportionality constant and is defined as *susceptibility* of the dielectric medium. χ_e is a measurable quantity and can be determined from the applied electric field and the polarization vector of the dielectric medium. Substitution for the new constitutive relation with the above displacement vector yields the total constitutive relationship of the dielectric region as:

$$\boldsymbol{D} = \varepsilon_0 \boldsymbol{E} + \boldsymbol{P} = \varepsilon_0 (1 + \chi_e) \boldsymbol{E} = \varepsilon_0 \varepsilon_r \boldsymbol{E} = \varepsilon \boldsymbol{E} \tag{4.179}$$

Therefore, the relative permittivity is defined as:

$$\varepsilon_r = 1 + \chi_e. \tag{4.180}$$

This is the macroscopic relation of the relative permittivity of the dielectric medium with respect to vacuum. Addition of χ_e with the absolute dielectric constant $\varepsilon_r = 1$ for vacuum, we can realise that the electric field reduces for the same electric flux density as for free space when the field is applied to the dielectric material. This deterrent effect on the resultant electric field in the dielectric medium is due to the Coulomb force acting on the bound couplet in a polarized molecule and their net contribution to the applied electric field.

Review Questions 4.13: Classical Electron Theory of Dielectric Materials

Q1. Discuss the very nature of the dielectric materials in terms of its classical electron theory.

Q2. Explain the difference between the dipole moment and the polarization vector. Is there any free charge involved in creating the polarization vector?

Q3. What is the susceptibility of a dielectric material? Show that $\boldsymbol{P} = \chi_e \boldsymbol{E}$.

Q4. Show that $\nabla.\ \boldsymbol{P} = -\rho_p$.

Q5. Show that $\varepsilon_r = 1 + \chi_e$

Q6. Explain the effect of insertion of a dielectric material on the applied field and displacement vector.

4.16 Microscopic View of Dielectric

In the microscopic view of the dielectric medium, we start with the electron theory of the couplet of charge $(q, -q)$ separated with a distance d, as shown in Figure 4.31.[114] First, we have two parameters for *the polarized state* of the material body as before: (i) the polarization vector $\boldsymbol{P}$ is the electric dipole moment per unit volume, and (ii) the individual dipole moment of the couplet $\boldsymbol{p} = q\boldsymbol{r}$, where $\boldsymbol{r}$ is the distance between the charges of the couplet. Assume that the charges $+q$ and $-q$ are located at spatial points A and B, respectively. The vector joining the two points is $\boldsymbol{r}$, which is directed from B to A (from the negative to the positive charges). This means the individual dipole moment of the couplet $\boldsymbol{p}$ has the same direction from A to B. As before, the polarization vector is defined by

$$\boldsymbol{P} = \rho_p \boldsymbol{r}\ (\mathrm{C/m^3}) \tag{4.181}$$

Now let us take an observation point in space $P(x, y, z)$ where we would like to calculate the potential difference due to the couplet. Here, we set the coordinate system so that $AP = R_1$, $BP = R_2$ as shown in Figure 4.31. We set the origin at the middle point of AB so that $OP = R$. Note that $R_1 >> r$ and $R_2 >> r$. Therefore, the potential difference is:

$$\phi = \frac{q}{4\pi\varepsilon_0}\left[\frac{1}{R_1} - \frac{1}{R_2}\right] = \frac{q}{4\pi\varepsilon_0} \times \frac{R_1 - R_2}{R_1 R_2} = \frac{qr\cos\theta}{4\pi\varepsilon_0 R^2} \tag{4.182}$$

114 Similar derivation is available in M. A. Islam, *Electromagnetic Theory*, Dacca, East Pakistan: EPUET, 1969.

Here, $R_1 = R_2 \approx R$ is assumed and from the geometry $R_1 - R_2 = rcos\theta$. By definition,

$$R = \sqrt{(x-x')^2 + (y-y')^2 + (z-z')^2} \tag{4.183}$$

where (x, y, z) is for the observation coordinate and (x', y', z') is for the source coordinate.

$$rcos\theta = \frac{\boldsymbol{r} \cdot \boldsymbol{R}}{R} \tag{4.184}$$

Substitution of the above relationship we obtain

$$\phi = \frac{q\,(\boldsymbol{r} \cdot \boldsymbol{R})}{4\pi\varepsilon_0 R^3} \tag{4.185}$$

However,

$$\frac{\boldsymbol{R}}{R^3} = -\nabla\left(\frac{1}{R}\right) \tag{4.186}$$

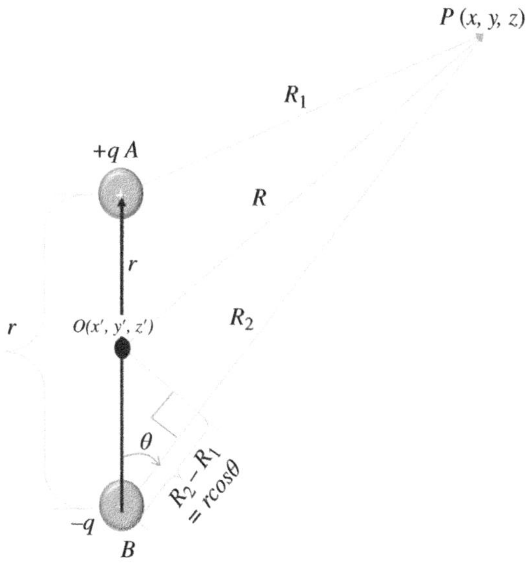

Figure 4.31 Coordinate system of couplet (q, $-q$) at the source point $O(x', y', z')$ to calculate the potential at the observation point $P(x, y, z)$.

Using the definition of gradient $\nabla = \boldsymbol{a_x}(\partial/\partial x) + \boldsymbol{a_y}(\partial/\partial y) + \boldsymbol{a_z}(\partial/\partial z)$ for the observation coordinate, $\nabla' = \boldsymbol{a_x}(\partial/\partial x') + \boldsymbol{a_y}(\partial/\partial y') + \boldsymbol{a_z}(\partial/\partial z')$ at the source point, and using the definition of $R = \sqrt{(x-x')^2 + (y-y')^2 + (z-z')^2}$, we can easily show that $\nabla(1/R) = (\boldsymbol{R}/R^3) = -\nabla'(1/R)$. Substitutions for the above all entities into (4.185), we obtain the definition of the potential in terms of the dipole moment as follows:

$$\phi = -\frac{\boldsymbol{P}}{4\pi\varepsilon_0} \cdot \nabla\left(\frac{1}{R}\right) = \frac{\boldsymbol{P}}{4\pi\varepsilon_0} \cdot \nabla'\left(\frac{1}{R}\right) \tag{4.187}$$

Now consider the aggregated dipole moment $\boldsymbol{p}$ due to the couplets within the volume and its equivalent polarization vector $\boldsymbol{P}$ per unit volume, the net potential due to the contribution of all dipole moments in the volume is defined by the following integration:

$$\phi = \frac{1}{4\pi\varepsilon_0}\int_{v'}\left[\boldsymbol{P} \cdot \nabla'\left(\frac{1}{R}\right)\right]dv' \tag{4.188}$$

The prime symbol with the divergence ∇' and v' indicates that we are performing the integration in the source coordinate system. Now we use the following identity and then perform the above integration to analyse the potential function in the light of the diverging nature of the bound charge excess.

$$\nabla' \cdot \left(\frac{\boldsymbol{P}}{R}\right) = \boldsymbol{P} \cdot \nabla'\left(\frac{1}{R}\right) + \frac{(\nabla' \cdot \boldsymbol{P})}{R} \tag{4.189}$$

Substitution for $\boldsymbol{P}.\ \nabla'(1/R) = -((\nabla'.\ \boldsymbol{P})/R) + \nabla'.\ (\boldsymbol{P}/R)$ in the above integral, we obtain

$$\phi = \frac{1}{4\pi\varepsilon_0}\left[-\int_{v'}\left[\frac{(\nabla' \cdot \boldsymbol{P})}{R}\right]dv' + \int_{v'}\left[\nabla' \cdot \left(\frac{\boldsymbol{P}}{R}\right)\right]dv'\right] \tag{4.190}$$

Using the divergence theorem for the closed surface containing the volume *dv'* we obtain:

$$\phi = \frac{1}{4\pi\varepsilon_0}\left[-\int_{v'}\left[\frac{(\nabla'\cdot \boldsymbol{P})}{R}\right]dv' + \oint_{s'}\left(\frac{\boldsymbol{P}}{R}\right)\cdot d\boldsymbol{S}'\right] \tag{4.191}$$

Now substitution for $\nabla'.\boldsymbol{P} = -\rho_p$ and $\boldsymbol{P}.\ d\boldsymbol{S}' = \rho_{sp}d\boldsymbol{S}'$ in the above equation, we obtain the following expression:

$$\phi = \frac{1}{4\pi\varepsilon_0}\left[\int_{v'}\left[\frac{\rho_p}{R}\right]dv' + \oint_{s'}\left(\frac{\rho_{sp}}{R}\right)d\boldsymbol{S}'\right] = \phi_v + \phi_s \tag{4.192}$$

If we compare the generic definition of the electric potential function due to a charge q as $\phi = (q/4\pi\varepsilon_0 R)$, then we can define two types of charges that cause the two different potential functions ϕ_v and ϕ_s due to the polarization vector $\boldsymbol{P}$. The first one is due to the $\nabla'\cdot\boldsymbol{P} = -\rho_p$ is the volume charge density due to polarization residing in the volume and the second one is due to $\boldsymbol{P}\cdot d\boldsymbol{S}' = \rho_{sp}$ the casualty effect — the emanating out surface charge density of the closed surface formed by the volume. We can also convert the above potential definition to the equivalent electric energy[115] as follows:

$$U_E = \frac{1}{2}\left[\int_{v'}\rho_p\phi_v\, dv' + \oint_{s'}\rho_{sp}\phi d\boldsymbol{S}'\right] \tag{4.193}$$

The above expression tells that the electric energy is also composed of two components — the stored energy in the volume and the emanating out of the enclosed surface energy.

Comparing the connotations of the real charge q and the relevant electric flux density or displacement vector $\boldsymbol{D}$, with that for the polarization vector $\boldsymbol{P}$, we can draw a similar analogy of the classical Gauss theory of the enclosed volume charge density that emanates out as the normal displacement vector from the enclosed surface. This concept coincides with the previous concept of the two kinds of charges – one is enclosed charge within the volume and the second one is the charge emanating out of the closed surface that encloses the volume. This represents the causality effect of the polarization of the material body due to the presence of the applied electric field. For completeness, we can invoke the definition of the displacement vector again:

$$\boldsymbol{D} = \varepsilon_0\boldsymbol{E} + \boldsymbol{P} = \varepsilon\boldsymbol{E}\ \text{where}, \boldsymbol{P} = \varepsilon_0\chi_e\boldsymbol{E}\ \text{and}\ \varepsilon_r = 1 + \chi_e. \tag{4.194}$$

Therefore, we can see that for a positive value of susceptibility χ_e, the relative permittivity is always greater than 1.[116] We can define this understanding by considering the different electric fields:

i) $\boldsymbol{E}$ is the net electric field intensity within the dielectric body
ii) $\boldsymbol{E}$ is the applied electric field due to the exterior source or charge body that is causing polarization. Assume that $\boldsymbol{E_0}$ was present before the dielectric body was introduced.
iii) $\boldsymbol{E_p}$ is the electric field intensity due to the polarized state of the charge body that is opposing $\boldsymbol{E_0}$.

Therefore, from the above relationship, we can define:

$$\boldsymbol{E} = \boldsymbol{E_0} + \boldsymbol{E_p} \tag{4.195}$$

115 S. Ramo, J.R. Whinnery and T. V. Duzer, *Fields and Waves in Modern Radio*, Wiley 2e 1994.
116 In metamaterials, the susceptibility can be negative and the permittivity value can also be negative.

We can also write:

$$\varepsilon_r \boldsymbol{E} = \boldsymbol{E} + \chi_e \boldsymbol{E} \tag{4.196}$$

or,

$$\boldsymbol{E} = \varepsilon_r \boldsymbol{E} - \chi_e \boldsymbol{E}$$

Comparing the above two equations, we can define the exterior field $\boldsymbol{E_0}$ and the resisting field due to polarization $\boldsymbol{E_p}$ in terms of the interior net field $\boldsymbol{E}$ as

i) $\boldsymbol{E} = \varepsilon_r \boldsymbol{E}$ is the applied electric field
ii) $\boldsymbol{E}_p = -\chi_e \boldsymbol{E}$ is the resisting electric field intensity due to the polarized state of the charge body

We can also define the resisting field due to the polarized sate as follows:

$$-\boldsymbol{E}_p = \boldsymbol{E} - \boldsymbol{E_0} = (\varepsilon_r - 1)\boldsymbol{E} = \left(1 - \frac{1}{\varepsilon_r}\right)\boldsymbol{E_0} \tag{4.197}$$

This relation gives us the opportunity to calculate the field due to the polarized state of the dielectric if we can know the applied field and the dielectric constant of the medium.

Review Questions 4.14: Microscopic View of Dielectric

Q1. Show that $\nabla\left(\frac{1}{R}\right) = \left(\frac{\boldsymbol{R}}{R^3}\right) = -\nabla'\left(\frac{1}{R}\right)$. Where ∇ and ∇' are for observation and source coordinate systems, respectively, and $R = \sqrt{(x-x')^2 + (y-y')^2 + (z-z')^2}$.

Q2. Calculate the field due to the polarized state for an applied field $E_0 = 10$ V/m and the dielectric constant of the medium is 3.

Q3. An atom of a dielectric material has charge pair $(q, -q)$ and creates a dipole moment $\boldsymbol{p} = q\boldsymbol{r}$ where $\boldsymbol{r}$ is the distance vector of the dipole. An external applied field is $\boldsymbol{E}$. Calculate (i) the force experience by the atom due to the field, and (ii) the potential energy of the dipole.

Answer:
(i) $\boldsymbol{F} = (\boldsymbol{p} \cdot \nabla)\boldsymbol{E}$, (ii) $U = -\boldsymbol{p}.\ \boldsymbol{E}$.

4.16.1 Waves in Anisotropic Dielectric Medium

So far, we have considered the medium is isotropic, this means, the response to the applied electric field is the same in all directions. Materials, in which the polarization vector $\boldsymbol{P}$ is linearly related to and parallel to $\boldsymbol{E}$, are called *isotropic media*.[117,118] However, there are some technically important materials such as quartz, ferrites, and metamaterials, and propagation media such as ionosphere[119] in which the response to the electric field of the materials are not uniform and the response differs from that for the isotropic media with directions and the orientation of the material medium with respect to the applied electric field. In a material medium, when the response of the electric field is different in different directions, the materials are called *anisotropic materials*. This leads to the equivalent concept that the displacement vector $\boldsymbol{D}$ is not parallel to the electric field $\boldsymbol{E}$ in the

117 R.E. Collin, *Foundations for Microwave Engineering*. McGraw Hill, 2e, 1992, pp. 23–28.
118 Nonlinearity happens for large applied fields hence for wireless communications in microwave and mm-wave frequency bands, this effect can be neglected. However, in optics, the laser sources carry high energy and also for optical fibers, due to its small size and small wavelength in micron order, the nonlinear effect of the medium is considered in the design. We shall discuss those nonlinearity effect of optical fiber in the relevant chapters.
119 Plasma is an anisotropic medium due to the application of applied magnetic field.

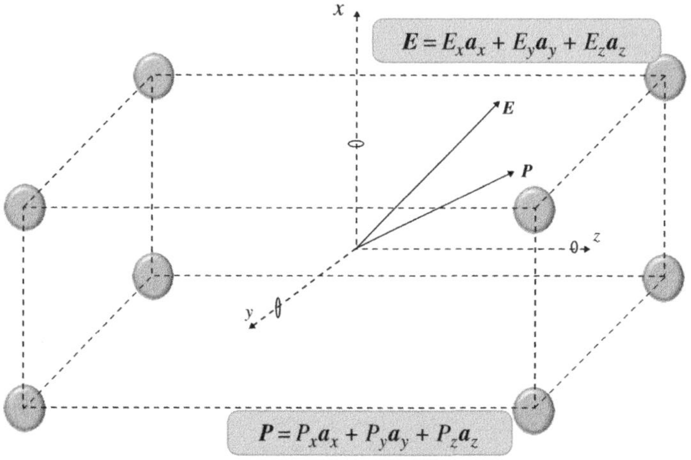

Figure 4.32 A representative noncubic crystalline lattice in the influence of the applied electric field ***E***.

medium. The anisotropy may occur in response to the electric or magnetic field.[120] However, we shall confine our discussion only for the nonmagnetic medium. Crystalline materials possess regular or ordered lattices of atoms and molecules. If the crystalline structures lack spherical symmetry due to the disparities in the interatomic spacing associated with several symmetric axes of the crystalline lattices[121] then that leads to *anisotropic media*. For example a noncubic crystal may exhibit anisotropic behavior. In those materials the polarization vector ***P*** is not parallel to ***E*** as shown in Figure 4.32. Such materials are termed as *anisotropic materials*. In these materials μ, ε, and σ differ significantly with direction. Application of anisotropic materials is increasing with the advent of new types of microwave materials. Anisotropic materials include crystals, plasmas, ferrites, various semiconductors, and manmade composite materials, also called metamaterials. These materials find many useful applications in many classical and modern microwave devices.

Assume again the crystalline lattice structure, which is a noncubic crystal, and an electric field is applied to the medium is defined as:

$$\boldsymbol{E} = \boldsymbol{a_x}E_x + \boldsymbol{a_y}E_y + \boldsymbol{a_z}E_z \tag{4.198}$$

The corresponding polarization vectors are defined as:

$$\begin{aligned} \boldsymbol{P_x} &= \varepsilon_0\chi_{11}\boldsymbol{E_x} = \varepsilon_{xx}\boldsymbol{E_x} \\ \boldsymbol{P_y} &= \varepsilon_0\chi_{22}\boldsymbol{E_y} = \varepsilon_{yy}\boldsymbol{E_y} \\ \boldsymbol{P_z} &= \varepsilon_0\chi_{33}\boldsymbol{E_z} = \varepsilon_{zz}\boldsymbol{E_z} \end{aligned} \tag{4.199}$$

In this case $\chi_{11} \neq \chi_{22} \neq \chi_{33}$. Due to the asymmetric nature of the crystalline lattice structure of the material, the polarization vector produced along *x*-axis may be greater than that produced along *y*- and *z*-axes, respectively. For example, for gypsum $\chi_{11} = 8.9$, $\chi_{22} = 4.1$ and $\chi_{33} = 4.0$.[122] In this case, the displacement vector and the electric field are not parallel and is generalized as:

$$\boldsymbol{D_i} = \varepsilon_{ij}\,\boldsymbol{E_j} \tag{4.200}$$

120 Our study is concerned only with the dielectric media therefore, the magnetic effect is neglected except the case for ionosphere.

121 C. T. A. Johnk, *Engineering Electromagnetic Fields and Waves*, 2e, Wiley, NY, USA, 1988 pp. 116–119 for more elegant explanation.

122 C. T. A. Johnk, *Engineering Electromagnetic Fields and Waves*, 2e, Wiley, NY, USA, 1988 p. 163.

where i and j are integers and ε_{ij} is the tensor of second order. Therefore, we can define the component level displacement vectors for the three principal planes as

$$D_x = \varepsilon_{xx} E_x; D_y = \varepsilon_{yy} E_y \text{ and } D_z = \varepsilon_{zz} E_z \tag{4.201}$$

where, $\varepsilon_{xx} = \varepsilon_0 \chi_{11}$, $\varepsilon_{yy} = \varepsilon_0 \chi_{22}$ and $\varepsilon_{zz} = \varepsilon_0 \chi_{33}$ and the relevant relative permittivities are: $\varepsilon_{rx} = (\varepsilon_{xx}/\varepsilon_0)$, $\varepsilon_{ry} = (\varepsilon_{yy}/\varepsilon_0)$ and $\varepsilon_{rz} = (\varepsilon_{zz}/\varepsilon_0)$. They are called *principal dielectric constants*. We can generalize the constitutive relation if the coordinate system[123] is different with respect to the crystal structure as follows:

$$\begin{aligned} D_x &= \varepsilon_{xx} E_x + \varepsilon_{xy} E_y + \varepsilon_{xz} E_z \\ D_y &= \varepsilon_{yx} E_x + \varepsilon_{yy} E_y + \varepsilon_{yz} E_z \\ D_z &= \varepsilon_{zx} E_x + \varepsilon_{zy} E_y + \varepsilon_{zz} E_z \end{aligned} \tag{4.202}$$

We can write the above system of equations in a matrix form as follows:

$$\begin{bmatrix} D_x \\ D_y \\ D_z \end{bmatrix} = \begin{bmatrix} \varepsilon_{xx} & \varepsilon_{xy} & \varepsilon_{xz} \\ \varepsilon_{yx} & \varepsilon_{yy} & \varepsilon_{yz} \\ \varepsilon_{zx} & \varepsilon_{zy} & \varepsilon_{zz} \end{bmatrix} \begin{bmatrix} E_x \\ E_y \\ E_z \end{bmatrix} \tag{4.203}$$

or,

$$[\boldsymbol{D}] = [\boldsymbol{\varepsilon}][\boldsymbol{E}] \tag{4.204}$$

From the above analysis a few remarks can be made:

1) If the anisotropic material is orientated arbitrarily with respect to the applied electric field then we expect to have all nine components of the permittivity.
2) Once the material is oriented along the principal axis of the coordinate system in which the electric field is applied, then we can have the three principal dielectric constants.
3) For physically real materials with complex permittivity, it can be shown that $\varepsilon_{ij} = \varepsilon_{ji}^*$. This means the permittivity operator is Hermitian.[124] We shall find a similar relationship for the non-diagonal entities for the gyro-tropic materials.
4) Anisotropic materials such as crystals, ferrites, plasma, varieties of semiconductors and man-made artificial composite materials called *metamaterials* exhibit many interesting phenomena and are used extensively in high frequency and microwave devices, optics and many other applications. They exhibit specific propagation properties in one chosen direction compared to the other direction. For example, metamaterials and similar anisotropic materials are used for gain enhancement of antennas, beam steering, circulators, and many high-quality factor devices with low loss and compact design.[125]

Similar to the definition of the polarization vector using the electric energy equation, we can also formalize the definition of the anisotropic media using the definition of anisotropic dielectric constant. For the isotropic medium, we define the electric energy as:

$$U_E = \frac{1}{2}\boldsymbol{D} \cdot \boldsymbol{E} \tag{4.205}$$

123 The coordinate system is in regard to the applied electric field vector $\boldsymbol{E}$.
124 S. Ramo, J. R. Whinnery and T. V. Duzer, *Fields and Waves in Communication Electronics*, Wiley, 3e, 1994, NY, USA, p. 700.
125 *Topics: Twenty-first Century Electromagnetics, Dr. Raymond C. Rumpf, rcrumpf@utep.edu* https://www.youtube.com/watch?v=RjxpRbM7-Lw (accessed 27 February 2018).

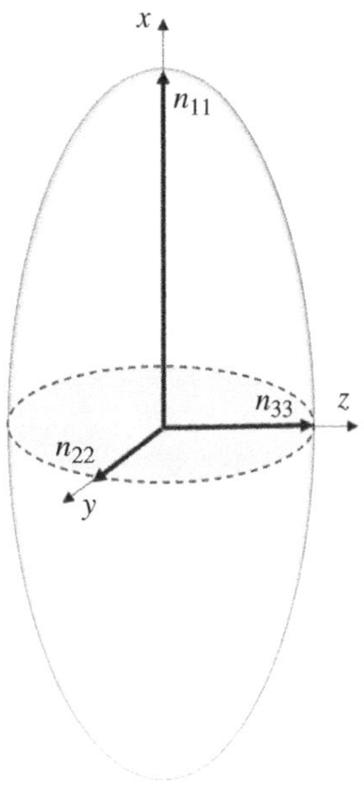

Figure 4.33 The index ellipsoid with three principal refractive indices.

Substitution for the principal constitutive parameters from (4.201), we can define:

$$U_E = \frac{1}{2}\left(\frac{D_x^2}{\varepsilon_{xx}} + \frac{D_y^2}{\varepsilon_{yy}} + \frac{D_z^2}{\varepsilon_{zz}}\right) \tag{4.206}$$

or,

$$\frac{X^2}{n_{11}^2} + \frac{Y^2}{n_{22}^2} + \frac{Z^2}{n_{33}^2} = 1 \tag{4.207}$$

where,

$$X = \frac{D_x}{\sqrt{2\varepsilon_0 U_E}}; Y = \frac{D_y}{\sqrt{2\varepsilon_0 U_E}} \text{ and } Z = \frac{D_z}{\sqrt{2\varepsilon_0 U_E}} \tag{4.208}$$

where, the principal refractive indices are defined as: $n_{11} = \sqrt{(\varepsilon_{11}/\varepsilon_0)}$; $n_{22} = \sqrt{(\varepsilon_{22}/\varepsilon_0)}$ and $n_{33} = \sqrt{(\varepsilon_{33}/\varepsilon_0)}$. In this case we assume nonmagnetic materials so that $\mu = \mu_0$. We can also define the permeability tensor μ for the magnetic materials. The above Equation (4.207) is the equation of an ellipsoid, which is called *index of ellipsoid* as shown in Figure 4.33.

As shown in the figure, the three semi axes are the principal refractive indices along three principal directions as shown in the figure. This is the general case for dielectric materials. Based on the construction of the crystalline structure of materials, they can be classified into different media such as *isotropic, uniaxial,* and *biaxial* mediums. For *the isotropic* medium, the dielectric property is independent of the direction of propagation such that $\varepsilon_{11} = \varepsilon_{22} = \varepsilon_{33}$. In this case, the ellipsoid reduces to a sphere. Examples of isotropic materials include silicon, gallium arsenide, and cadmium telluride. In *uniaxial materials,* we obtain one axis of symmetry and the ellipsoid become trigonal, tetragonal, or hexagonal. In this case, the permittivity parameters are: $\varepsilon_{11} = \varepsilon_{22} \neq \varepsilon_{33}$; one principal permittivity parameter is different from the other two.

$$[\varepsilon] = \begin{bmatrix} \varepsilon_{11} & 0 & 0 \\ 0 & \varepsilon_{11} & 0 \\ 0 & 0 & \varepsilon_{33} \end{bmatrix} \tag{4.209}$$

The unique permittivity (in this case ε_{33}) is called *extraordinary permittivity* (or *extraordinary refractive index*) and the relevant axis is called *extraordinary axis* or *optical axis.* Examples of uniaxial materials include quartz, calcite ($CaCO_3$), ruby (Al_2O), quartz SiO_2, sellaite or magnesium fluoride (MgF_2), and rutile (TiO_2). *The uniaxial materials* show birefringent properties.[126]

In *biaxial materials* all the diagonal elements of the permittivity are different such that $\varepsilon_{11} \neq \varepsilon_{22} \neq \varepsilon_{33}$ as defined as follows:

$$[\varepsilon] = \begin{bmatrix} \varepsilon_{11} & 0 & 0 \\ 0 & \varepsilon_{22} & 0 \\ 0 & 0 & \varepsilon_{33} \end{bmatrix} \tag{4.210}$$

126 *Uniaxial crystal*, https://en.wikipedia.org/wiki/Uniaxial_crystal; Double Refraction and Birefringence, http://photonicswiki.org/index.php?title=Double_Refraction_and_Birefringence (accessed 27 February 2018).

The third group is called *gyrotropic medium* that includes ferrite and magnetized plasma (ionosphere).[127,128] In these materials, the permittivity tenors take complex forms, the nondiagonal tensors with imaginary values.

$$[\varepsilon] = \begin{bmatrix} \varepsilon_{11} & j\varepsilon_{21} & 0 \\ -j\varepsilon_{21} & \varepsilon_{11} & 0 \\ 0 & 0 & \varepsilon_{33} \end{bmatrix} \tag{4.211}$$

Likewise, in ferrite, the tensor has a similar form with μ_{ij}. We shall be using the permittivity tensors in Maxwell's equations for analysing the electromagnetic wave fields in these anisotropic materials.

Review Questions 4.15: EM Waves in Anisotropic Dielectric Media

Q1. Define *the anisotropic materials* in terms of the relationships between ***E*** and ***P*** and ***E*** and ***D***.
Q2. Name a few classes of anisotropic materials that are used in microwave and optical devices.
Q3. Explain what happens if the axis of the coordinate system does not match with that for the principal axes of the crystal.

4.17 Problems

Classical Electron Theory

P4.1 Calculate the cyclotron frequency for a proton under the influence of 1.2 T steady magnetic field. Proton mass 1.67×10^{-27} kg and electric charge $+1.6 \times 10^{-19}$ C.

P4.2 A proton is accelerated from rest through a potential difference of 2100 V in a capacitor. It is then injected into a uniform magnetic field B of 1 T.
i) Find the speed of the particle upon exiting the capacitor, use conservation of energy between the kinetic energy and electromagnetic energy.
ii) Find the radius of its orbit. Ignore the effect due to gravity

P4.3 Consider the electric field 10 (V/m), magnetic field 5×10^{-6} T, and velocity vector v of a charged particle of charge 1.69×10^{-19} C. The three vectors are orthogonal to each other. Calculate the amplitude of the velocity vector so that net force on the particle is zero.

Theory of Cyclotron

P4.4 An accelerated proton $+e$ emerging from a cyclotron of radius $R = 25.00$ cm with a magnetic field of magnitude $B = 0.5$ T. Calculate (i) the cyclotron frequency, (ii) the speed, and (iii) the kinetic energy.

127 D. M. Pozar, *Microwave Engineering*, Addison Wesley, 1990, pp.53–56.
128 *Electromagnetic Properties of Materials -Part 2 Nonlinear and Anisotropic Materials, ECE5390 Special Topics: Twenty-first Century Electromagnetics, Dr. Raymond C. Rumpf, rcrumpf@utep.edu* https://www.youtube.com/watch?v=RjxpRbM7-Lw (accessed 27 February 2018).

Charge Particle in Time Harmonic Field

P4.5 Derive the expression for the displacement vector for an x-directed applied electric field of magnitude $E_0\boldsymbol{a_x}$ (V/m) with an angular frequency ω (rad/s), natural frequency ω_0 (rad/s), damping factor γ (S^{-1}) and electron charge e (C) and mass m (Kg).

Dispersion in Gaseous Medium

P4.6 A materials with the following parameters: No of electrons per unit volume $N = 2.418 \times 10^{22}$ (m^{-3}); $e = 1.602 \times 10^{-19}$ C; $m = 9.109 \times 10^{-31}$ kg; $\gamma = 10^{22}$ (s^{-1});

i) Calculate the natural frequency ω_0; Hints: use $\aleph = Ne^2/\varepsilon_0$, where $\varepsilon_0 = 5.584 \times 10^{-12}$ (F/m)
ii) Calculate the plasma frequency ω_p.
iii) Calculate ε_r at 500 MHz.

Dispersion in Metallic Medium

P4.7 Find the mean free time, and the conductivity of a metal. Its electron mobility 1.2×10^{-3} m^2/(V.s), the mass density is 2.7 (g/cm^3) and its atomic mass is 26.98 (g/mol). Hints: Avogadro constant: 6.022×10^{23} (mol^{-1}).

P4.8 The current density of an aluminum rod is measured at 1.0 (A/m^2). The conductivity and electron mobility are $\sigma = 3.8 \times 10^7$ (S/m), $\mu_e = 1.2 \times 10^{-3}$ (m^2/(V.s)), respectively. Calculate the electric field intensity and electron drift velocity v_d.

Waves Propagation in Plasma

P4.9 A plane wave is propagating through a uniform plasma medium. The radio broadcasting frequency of the wave is 900 KHz. Explain whether the wave can propagate through the plasma medium.

P4.10 Show that the Poynting vector of a wave below the plasma frequency in a plasma medium of the wave is zero.

P4.11 Calculate the wave number, the phase velocity and the group velocity of GPS signal at 1.54 GHz through the ionosphere for its plasma frequency is approximately 1 MHz.

P4.12 In an initially neutral plasma, the electron density is $N = 10^3$ electrons/cm^3. Calculate the plasma frequency assuming the heavy ions are stationary. Hints: the plasma frequency is given by: $\omega_p^2 = (Ne^2/m\varepsilon_0)$.

P4.13 If due to the solar storm, an electric field is generated with $E = 10^{12}e^{j2\pi \times 10^9}$ (V/m) that causes all electrons to be displaced by a distance $d_y = 10$ m in the y-direction, calculate the charge density and the induced electric field.

P4.14 In the outer space, on an average, there is about 2 electron and 2 proton per cubic meter. Determine the lowest frequency of the electromagnetic wave that can propagate through the medium without attenuation. Discuss the result.

P4.15 The lowest frequency detectable, known as the critical frequency, is related to the density of electrons by the equation: $f_c = 9 \times 10^{-3} \times \sqrt{N}$ MHz. In this equation f_c is the critical frequency and N is the electron density per cm^3, $\sqrt{}$ means to take the square root of the electron density. The electron density ranges (bands) from 33300 (electrons/cm^3) (band 1) to 249750 (electrons/cm^3) (band 2) to 552780 (electrons/cm^3) (band 3). Calculate *the plasma frequency* for the three cases.

Mobile Satellite Communications

P4.16 Explain why mobile satellite communication systems use circularly polarized signals. Give a few examples of right hand circularly polarized mobile satellite communications systems that we use in modern time. Also, search for LHCP satellite systems in the Internet.

P4.17 Explain what has happened when a linearly polarized signal incidents on the ionosphere of the earth.

P4.18 Investigate the operating frequencies of GPS, Inmarsat, Singapore-Taiwan (ST) VSAT systems. What handedness of these systems have?

P4.19 Explain the difference between the magnetic field associated with the plane wave and that for the earth.

P4.20 The transverse current density vector has the motion equation defined by $m\frac{d\boldsymbol{J}}{dt} = e^2 N\boldsymbol{E} - e\boldsymbol{J} \times B$. Using the basic definition of the convection current density $J = -eN\frac{dR}{dt}$ show that the vector motion equation for displacement vector $\boldsymbol{R}$ is: $\frac{d^2\boldsymbol{R}}{dt^2} - j\omega_B\frac{d\boldsymbol{R}}{dt} = -\frac{e}{m}\boldsymbol{E}$. Here symbols have their usual meanings.

P4.21 What is gyration frequency ω_B? Explain its physical meaning.

P4.22 Using the three sets of equations $\frac{e}{m}E_0 + [-\omega^2 \pm \omega_B\omega]\, R_0 = 0$; $\beta\, E_0 + j\omega\mu_0\, H_0 = 0$ and $\omega\varepsilon_0\, E_0 + j\beta\, H_0 - \omega e N\, R_0 = 0$, show that the phase constant of the ionosphere is defined as:

$$\beta^2 = \omega^2\mu_0\varepsilon_0 \left[1 - \frac{\omega_p^2}{\omega^2 \mp \frac{\omega_B}{\omega}}\right]$$

P4.23 Show that the relative permittivity (dielectric constant) of the ionosphere is defined as: $\varepsilon_r = 1 - \left(\omega_p^2/\omega^2 \mp (\omega_B/\omega)\right)$.

P4.24 Explain the physical significance of the two solutions of the phase constant β.

P4.25 What is an *extraordinary wave*? In which direction the wave rotates? What is the polarization of the wave?

P4.26 Calculate the Poynting vector of an electromagnetic field in the ionosphere.

P4.27 Explain why the ion concentration changes in ionosphere and what impact it creates in the radio communications.

P4.28 Calculate the polarization loss factor of a signal with the subtended angle of $\theta = 45°$.

P4.29 What is *Faraday rotation*? Explain how it affects the linearly polarized signal as it passes through the ionosphere.

P4.30 Show that when a linearly polarized wave is incident on the ionosphere the transmitted wave is rotated by an angle θ. Hints: The amount of rotation depends on the two different phase constants in the ionosphere and the distance travel by the wave through the ionosphere.

P4.31 Explain the advantages of using circularly polarized wave over the linearly polarized wave in satellite communications in the light of *Faraday rotation*.

P4.32 An electromagnetic wave at 1.6 GHz is propagating through the ionosphere with the electron concentration of 3.3×10^{12} electrons/m^3. The earth's magnetic field $B_0 = 50 \times 10^{-6}$ T. Calculate the gyration frequency ω_B, the plasma frequency ω_p, the refractive indices, phase velocities, and propagation constants for both RHCP and LHCP. The mass of electron $m = 9.109 \times 10^{-31}$ kg.

Electron Theory of Dielectric

P4.33 Discuss the very nature of the dielectric materials in terms of their classical electron theory.

P4.34 Explain the difference between the dipole moment and the polarization vector. Is there any free charge involved in creating the polarization vector?

P4.35 What is the susceptibility of a dielectric material? Show that $\boldsymbol{P} = \chi_e \boldsymbol{E}$.

P4.36 Show that $\nabla . \boldsymbol{P} = -\rho_p$.

P4.37 Show that $\varepsilon_r = 1 + \chi_e$

P4.38 The static relative permittivity of Teflon in 2.1. The applied electric field intensity is 10 (V/m). Calculate the (i) the electric flux density D, (ii) the susceptibility χ_e, and (iii) the polarizability.

P4.39 Explain the effect of insertion of a dielectric material on the applied field and displacement vector.

5

Reflection and Transmission of Uniform Plane Wave

5.1 Introduction

The theory of reflection and transmission of uniform plane waves has a profound impact on electromagnetics. Many applications can be perceived using the theory. This chapter presents the theory of transmission and reflection of a uniform plane wave in a plane boundary of two media. First, the transmission and reflection as the boundary value problem is examined, followed by the wave encounters at the conducting and dielectric boundaries. Next, reflection from a multilayered medium is examined, with many special cases as interesting applications.

The theory of uniform plane wave (UPW) and the advanced topic of propagation through dispersive media were examined in the preceding chapters. There we have assumed that the wave is traveling freely in an unbounded medium. Figure 5.1 illustrates the roadmap of the theory of transmission and reflection of UPW at the interface of two and multiple media as outlined in the chapter. In this chapter, we shall examine many interesting phenomena of electromagnetic field waves when they encounter obstacles. As we are surrounded by buildings and many objects around us, when a wireless transmitter sends signals, then a portion of the incoming signal often gets reflected by these obstacles and rest of the incoming signal propagates through the second medium. This means if a UPW encounters an obstacle, then some energy transmit through the object and some are reflected back by the object toward the transmitter. Similar phenomena are also observed in guided structures such as transmission lines and waveguides. Due to the mismatch of impedances in different sections of the guiding structures and/or terminated with a load that is different than the characteristic impedance of the guiding structure, signal encounters reflection, and transmission. Therefore, irrespective of unbounded media and/or guiding structures, the reflection occurs due to the impedance mismatch between the media and different sections of the guiding structure. Therefore, we can find a nice analogy between the two. For unguided media, the impedance is the ratio of the electric and magnetic fields and is called *the intrinsic impedance*. For the guiding structure, the impedance is the ratio of the voltage (or equivalent electric field in a waveguide) and current (or equivalent magnetic field in a waveguide) and is defined as *characteristic impedance* of the guiding structures.[1] The incoming signal is called *the incident wave*, the reflected

1 The transmission and reflection of the uniform plane waves are many ways analogous to that for the voltage and current waves on the transmission line. As an assembly of a circuit is made of junctions of transmission lines of various impedance. When the voltage and current waves propagates through various sections of the circuit they encounter reflection and transmission in a very similar ways of that for the uniform plane wave's electric and magnetic fields. Therefore, the analogy is a valid encounter to understand the theory of plane wave reflection and transmission. We shall compare these two different entities during the coverage of the topic to enhance our understanding of the abstract nature of the uniform plane wave theory.

Fields and Waves in Electromagnetic Communications, First Edition. Nemai Chandra Karmakar.

Companion website: www.wiley.com/go/fieldsandwavesinelectromagneticcommunications

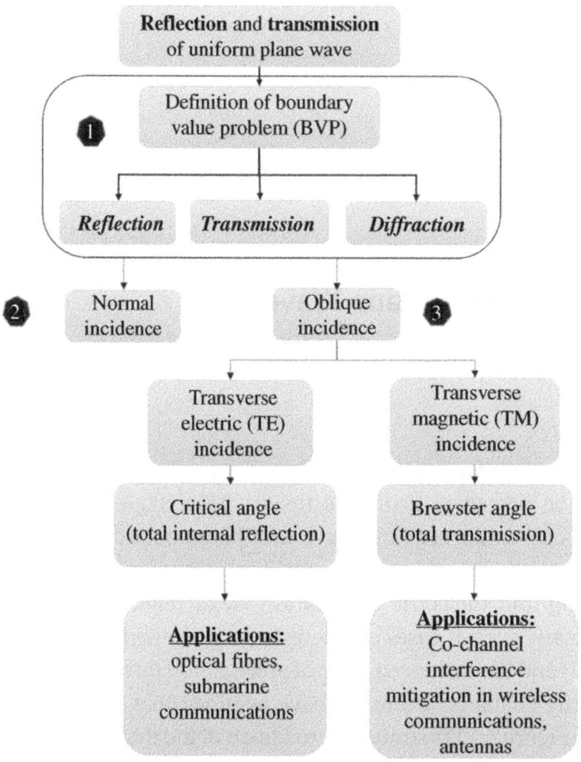

Figure 5.1 Roadmap of theory of transmission and reflection of UPW at the interface of two and multiple media as outlined in the chapter.

signal is called *reflected wave,* and the transmitted signal through the second medium/section of guiding structure is called *the transmitted wave.*[2] In this chapter, we shall also examine how the phase of the field wave is affected through the reflection and transmission processes and the amount of power reflected back in the first medium compared to the amount of power transmitted through the second medium. As we know in the UPWs, the electric field and magnetic field are transverse to the direction of propagation and they are also orthogonal to each other. During the reflection and transmission process of the signal in the interface of two media, the orthogonality between different vector entities still holds. However, the magnetic field reverses its phase with respect to the electric field in reflection. It is analogous to the reversing of current when the polarity of DC source changes. Certainly, in the reflected signal, the electromagnetic energy propagates in the opposite signal with respect to the incident energy in the medium. Therefore, a mixing of two signals – incident and reflected signals – occurs in the process of reflection. Such mixing of signals creates *standing wave patterns.* This means the composite signal's crests and troughs stand still in specific positions in space and do not move. The difference between the crests and troughs of the standing electric field waves defines the amount of reflected signal and is called the voltage standing wave ratio (VSWR). The VSWR of a particular wave depends on the electromagnetic properties of the two media. Here comes the intrinsic wave impedance $\eta = \sqrt{\mu/\varepsilon}$ of the medium. In this chapter,

2 In the plane wave reflection and transmission, we usually consider the transmission medium is semi-infinitely large compared to the wavelength of the electromagnetic signal. However, if the obstacle is small and comparable to the wavelength of the electromagnetic signal, then *diffraction* happens. The obstacle acts like a *secondary source* and it starts to scatter signal in all direction depending on the shape of the object. The incident signal is called *the primary source.*

we deal with the nonmagnetic materials, this means $\mu = \mu_0$. Therefore, the constitutive parameters that define the properties of the two media are the relative permittivity, ε_r, and conductivity σ of the media if the media are lossy. We characterize the electromagnetic fields at the interface of the two media using those constitutive parameters and the boundary conditions. We apply boundary conditions of the electromagnetic fields at the interface of the two media and develop the relationship between the transmitted and reflected signals. The ratio of the reflected and the incident signals is called *the reflection coefficient* and the ratio between the transmitted and incident signals is called *the transmission coefficient*. The coefficients are in terms of the electric fields and that is why they are called *the voltage transmission coefficient* and *the voltage reflection coefficient*, respectively. However, when the power transmission and reflection in concern, we always use the conservation of energy theory to calculate *the power reflection coefficient* and *the power transmission coefficient*.

For the simplistic case, we first assume that the signal is incident normally at the interface of the two media. This case of wave incidence is called *normal incidence*. However, in reality, the signal can be incident at the interface of the two media with an angle. This signal can come from an arbitrary angle and reflects back in the first medium and then transmits through the second medium. These two things happen instantaneously. This case of wave incidence with an angle at the interface of two media is called *oblique incidence*. When the electric field vector is transverse to the direction of wave propagation, the oblique incidence is called *transverse electric* (TE) oblique incidence. When the magnetic field is transverse to the direction of propagation, it is called *transverse magnetic* (TM) oblique incidence. When the signal propagates from a denser medium to the lighter medium after a certain incident angle, the signal totally reflects back into the first medium, and ideally, no signal transmits through the second medium. This angle is called *the critical angle* and the phenomenon of the electromagnetic wave reflection is called *total internal reflection*. The electromagnetic wave propagation in the optical region via the optical fiber exploits the characteristic of *the total internal reflection* of the light wave. In the TM case, at a certain incident angle, total transmission occurs. This angle is called *Brewster's angle* and the phenomenon is called *the total transmission*. We exploit *Brewster's angle* and *total transmission* in many wireless communications systems such as radar and mobile tower antennas to exploit co-channel interferences in wireless mobile communications. We are also surrounded by highly conductive materials such as metal bridges, walls, and infrastructures made of concrete.[3] When a uniform plane wave encounters highly conductive materials, total reflection occurs. This is equivalent to the short and open circuits of a terminated transmission line. We shall also examine this characteristic of wave propagations. Figure 5.1 illustrates the outline of the chapter. The topic that we cover in the chapter is: definition of the boundary value problems. As stated above any electromagnetic field analysis, which exploits the known boundary conditions, is called *boundary value problem*.

Boundary value problems deal with the analysis of the electric and magnetic fields at the interface of two media using boundary conditions. Applications of boundary conditions of the electromagnetic waves touch many practical design problems such as transmission lines, waveguides, antennas, optical fibers, and electromagnetic interference and compatibility. Therefore, the design problems are also called *boundary value problems*. The main goal of applying boundary conditions is to calculate *the reflection* and *transmission coefficients* using Maxwell's equations. The reflection and transmission coefficients based on the electric field solutions are called *voltage reflection coefficient* and *voltage transmission coefficient*, respectively. As stated above, with *the law of conservation of energy*, we calculate the power reflection and transmission coefficients. We also define the

3 With the reinforced steel inside the concrete and high loss sometimes the concrete can be perceived as the conductor for electromagnetic simulations. We shall consider uniform plane wave propagation through multiple reflection at the interface of media in a similar treatment of metallic waveguide in a case study.

standing wave pattern, which is the plot of the resultant non-propagating electric field with distance in the medium of wave incidence. We shall solve interesting problems using the theory learned. Finally, we shall summarize key definitions and answer questions.

During the course of analysis, we shall draw many examples of modern technologies that exploit the theory of reflection and transmission of UPW at the interface of two media. For example, microwave medical imaging is based on the theory of reflection and transmission. Figure 5.2 illustrates an example of image captured with a microwave biomedical imaging device for diagnostics. Our body is made of different dielectric media (different ε_r) and the contrast of the dielectric media reflects signals with different strengths. These returned echoes are superimposed in a 3D colored profile for biomedical imaging. Other applications are ground penetration radars (GPRs), sonars, microwave level gauges, chipless RFIDs, and many more. Magnetic resonance imaging (MRI) is another good example. An array of large coils that carries a very high energy magnetic field in the order of Tesla is passed through the parts of the body via a sequential switch and the transmitted and reflected signals are collected sequentially with other sets of coils. After synthesis of the collected signal, a sharp image of scanned body parts is constructed for diagnostic purposes. Computer tomography (CT) scanning using X-ray signals is another example of electromagnetic wave reflection and transmission for biomedical diagnostics. New emerging microwave imaging technology will replace or coexist with many traditional imaging technologies such as MRI and CT scans. It is cheaper and less harmful to the human body compared to CT scans and X-rays. Figure 5.2a illustrates a microwave medical imaging vest on a patient, (Figure 5.2b) microwave transceiver electronics with antennas, and (Figure 5.2c) a scanned medical image from such a device using reflection and transmission through the torso.

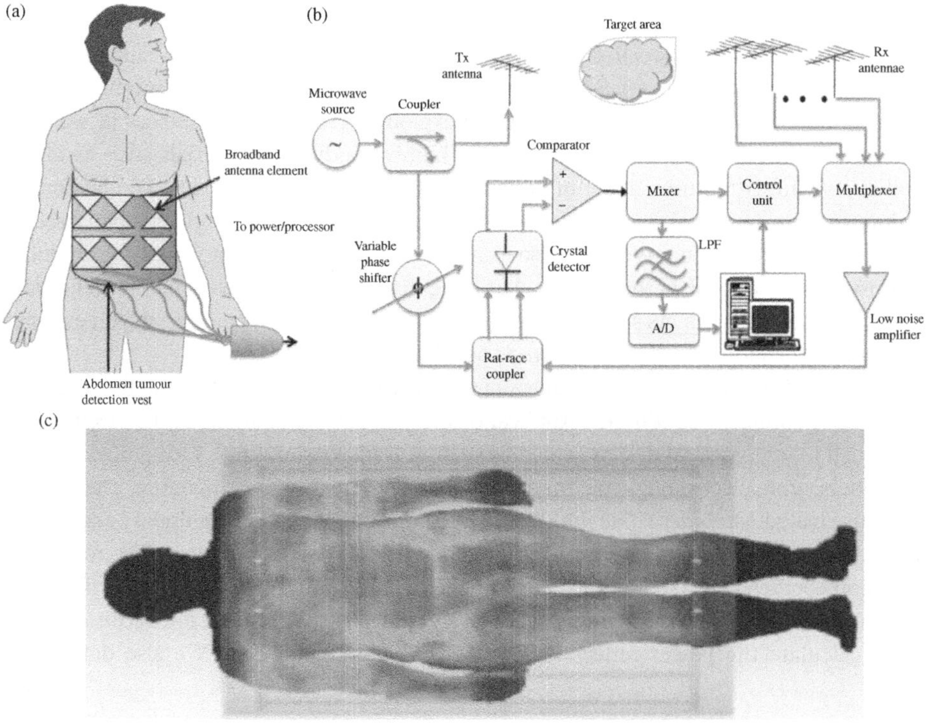

Figure 5.2 Illustration of a microwave medical imaging, (a) vest on a patient, (b) microwave transceiver electronics with antennas, and (c) scanned medical imaging of such a device using reflection and transmission through the torso.

Case Study 1: GPR for coal seam detection

Figure 5.3 illustrates the operating principle of GPR signals reflecting back from multiple underground layers of different electromagnetic properties. A GPR detects coal seams at a depth of 1 m from the soil surface. The radar has an antenna, a cable connected to the transceiver and a graphical user interface (GUI) loaded with a signal processing algorithm. Recent heavy rainfall shows the coal seam disappears for the GUI screen. What is the possible cause of the detection failure?

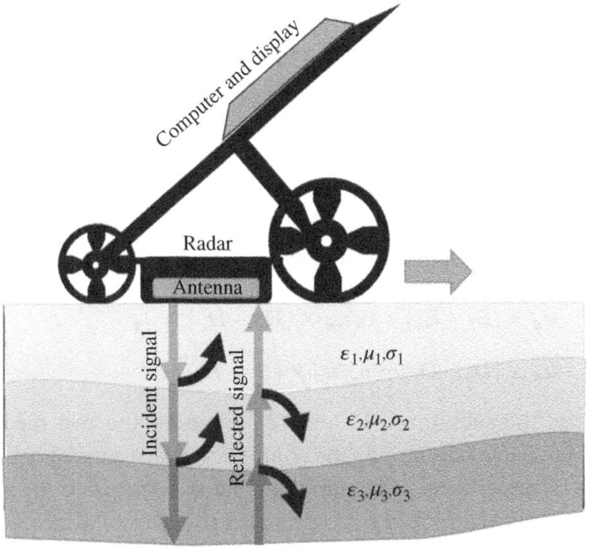

Figure 5.3 Ground penetration radar signals reflect from underground layers.

Answer: See the picture and comprehend the operation of the radar. As can be seen, the transmitted signal goes downward and at each layer, with different wave impedances, it reflects the signal upward. The radar signal processor has been programmed to remove noises from various layers. Sudden rainfall changes the wave impedances of the layer, even the positions and thickness of the layers. Therefore, the detection failed. The remedy is to use a new calibration, including the variations in wave impedances of underground soil layers. Another possible cause may be the water leakage inside the cable assembly that caused a high mismatch. This is most unlikely due to the fact that most ground penetration radars are robust and are made water tight so that no contamination can cause damage to the electronics.

Make your own case studies by reading new technology articles on microwave imaging, submarine communications, etc.

5.2 Electromagnetic Waves Analysis in the Context of Boundary Value Problems

The electromagnetic problems involve the solutions of uniform plane wave equations in free space where we assume the medium is source free, this means $\rho_v = 0$ and $J = 0$. The wave equation in a source free region is called *the homogeneous wave equation.*

$$\nabla^2 \boldsymbol{E} - \gamma^2 \boldsymbol{E} = 0$$
$$\nabla^2 \boldsymbol{H} - \gamma^2 \boldsymbol{H} = 0 \tag{5.1}$$

These conditions are imposed to solve wave equations of both electric and magnetic fields. Otherwise, it is almost impossible to solve such equations. These problems that we encounter are in a homogeneous medium. In practical situations, the problem becomes complex due to multiple reflections and transmissions of incident waves in nonhomogeneous media. Waves frequently undergo successive transmissions and reflections as they progress through complex media. The only way that we can solve the EM field problems is by using the boundary conditions at the interfaces of media that the electric and magnetic fields must obey. These conditions at the boundaries play a very important role in the solution of EM field problems. These problems are often referred to as *the boundary value problem*. Examples of boundary value problems are reflection and transmission of EM fields in nonhomogeneous media, terminated T-lines, waveguides, optical fibers, and antennas.

As shown in the case study of GPR, the EM field problem at the interface of two media is a boundary value problem. Because we know the conditions of the fields at the interface of two media. The boundary conditions of electric and magnetic fields are:

$$\boldsymbol{u_n} \times (\boldsymbol{E}_2 - \boldsymbol{E}_1) = 0, \boldsymbol{u_n} \times (\boldsymbol{H}_2 - \boldsymbol{H}_1) = \boldsymbol{J_s}$$
$$\boldsymbol{u_n} \times (\boldsymbol{D}_2 - \boldsymbol{D}_1) = \rho_s, \boldsymbol{u_n} \times (\boldsymbol{B}_2 - \boldsymbol{B}_1) = 0 \tag{5.2}$$

Here, the subscripts "1" and "2" represent the two media when the signal propagates from medium 2 to medium 1. $\boldsymbol{J_s}$ and ρ_s are the surface current density (A/m) and surface charge density (C/m^2), respectively. The conditions must be satisfied for static as well as time varying electromagnetic fields.

In the analysis, we only consider the rectangular coordinate system for simplicity of analyses. If the boundary value problems involve cylindrically and spherically symmetric cases, then we use cylindrical and spherical coordinate systems, respectively. For example, the electromagnetic wave radiation from an antenna is spherically symmetric.[4] Therefore, the coordinate system is conveniently chosen depending on the symmetry of the matter on which the boundaries can be specified in a simplistic manner. The question is to address how a wave does interact when it encounters the interface of two media. When an EM wave incident from one medium to the boundary of the second medium, a portion of the incident energy reflects back into the first medium and the rest is transmitted through the second medium. The amount of power reflected and transmitted depends on the constitutive parameters of the two media. In most cases, no source such as current density J or charge density ρ_v is present at the interface of the media when we consider the free space propagation far away for the source (transmitter). This makes the boundary conditions much simple and manageable to calculate the fields. Remember, the wave behavior must follow the boundary conditions irrespective of whether the field is static or time varying.

First, we look at the essential boundary conditions of the electric and magnetic fields at the interface of the two media. See the electric field vectors $\boldsymbol{E}_1$ propagate at an arbitrary angle and hit the interface of the two media obliquely, as shown in Figure 5.4a. The field vector is decomposed into two orthogonal components: (i) E_{T1}, a tangential component that is parallel to the plane of the interface of the two media, and (ii) E_{N1}, a normal component that is normal to the plane of the interface of the two media. The normal vector to the plane of the interface of the two media is $\boldsymbol{a_n}$, as shown in

4 We shall analyze the radiation fields of the antennas in spherical coordinate system. Likewise, we shall analyze the optical fibers in cylindrical coordinate system.

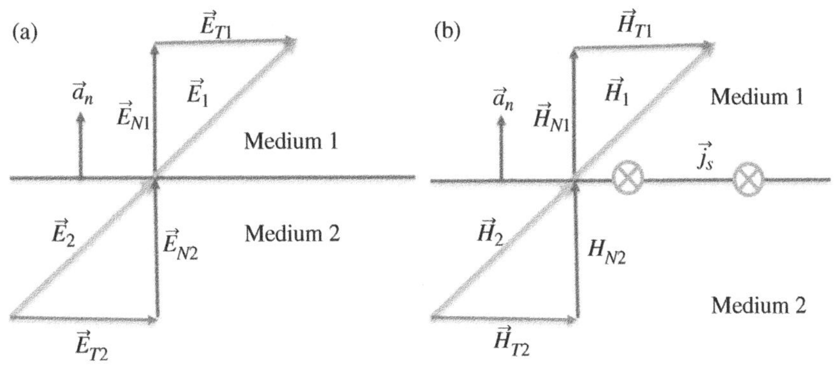

Figure 5.4 Boundary conditions for (a) electric field, and (b) magnetic field at the interface of two media.

the figure. As we can recall, we have taken a closed rectangular contour and made a derivation of the electric field along the closed contour. Based on Kirchhoff's voltage law in a closed contour, the integral becomes zero. See the detailed derivation in any EM textbook. This leads to $\boldsymbol{E}_{T1} = \boldsymbol{E}_{T2}$. Likewise, see the magnetic field vectors $\boldsymbol{H}_1$ and $\boldsymbol{H}_2$ at the interface of the two media in Figure 5.4b with a current density $\boldsymbol{J}_s$ which is present at the plane of the interface of the two media. In this case, we use Ampere's circuital law. For our present problem, there is no current flowing at the interface of the two media as we assume the plane of the interface is far away from the transmitter; therefore, $\boldsymbol{J}_s = 0$. These conditions lead to the tangential magnetic fields being equal at the interface $\boldsymbol{H}_{t1} = \boldsymbol{H}_{t1}$. In this analysis, we do need to consider the flux density vectors for both electrical ($\boldsymbol{D}$) and magnetic ($\boldsymbol{B}$) cases. Because these two vector quantities are linked with the electric and magnetic fields via the constitutive parameters ε_r and μ_r respectively. Applying the boundary conditions in the charge free interface lead to $\boldsymbol{D}_{N1} = \boldsymbol{D}_{N2}$ and $\boldsymbol{B}_{N1} = \boldsymbol{B}_{N2}$. For our current problem, we need only the tangential fields components to solve for the fields in the two media. Therefore, the essential boundary conditions are:

$$\begin{aligned} &\boldsymbol{E}_{t1} - \boldsymbol{E}_{t2} = 0; \varepsilon_1 \boldsymbol{E}_{n1} - \varepsilon_2 \boldsymbol{E}_{n2} = 0 \\ &\boldsymbol{H}_{t1} - \boldsymbol{H}_{t2} = 0; \mu_1 \boldsymbol{H}_{t1} - \mu_2 \boldsymbol{H}_{t2} = 0 \end{aligned} \tag{5.3}$$

In general, the direction of the wave changes as it enters the second medium instantly as the wave incidences obliquely with an angle at the interface of the two media. The angle between the normal line on the plane of interface and incident ray is called *the incident angle* or *the angle of incidence* and the angle between the normal line on the plane of interface and transmitted ray is called *the angle of transmission*. The signal also reflects back with an angle called *the angle of reflection*. All the angles are linked with the above boundary conditions and the law governing the relationships is called *the Snell's law of reflection* and *the Snell's law of refraction*. In the oblique transmission of the signal from one medium to the second medium, the signal bends more toward the normal line of the interface if the signal enters from an electromagnetically lighter medium to a denser medium.[5] This phenomenon is called *refraction*.[6] For the case of the normal incidence, the direction of propagation remains the same in both media. Usually, we assume the media is semi-infinite in extent and the interface of the media is a plane surface. Then the law governing the reflection and transmission is relatively simple. Figure 5.5 shows the example of the refraction of light waves in an aquarium of a gold fish. As can be seen, the light is entering from a denser medium (water, $\varepsilon_r \cong 80$) to a lighter medium

5 We defined earlier the lighter medium has refractive index n_1 less than the denser medium n_2.
6 In optics it is similar to parallax.

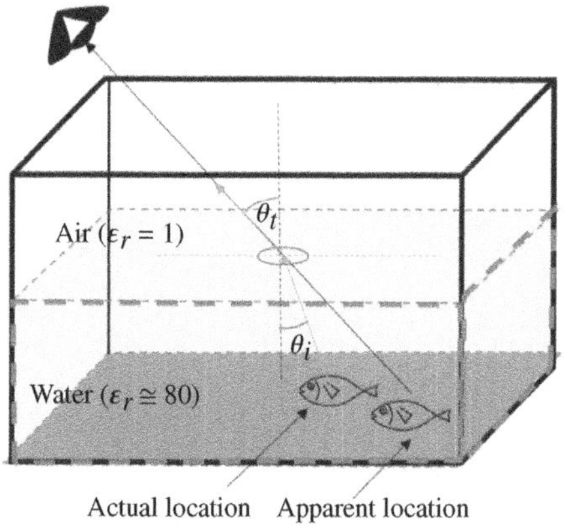

Figure 5.5 Refraction of light wave ray from a denser medium to a lighter medium.

(air $\varepsilon_r \cong 1$). Due to the refraction phenomenon of the transmitted signal, the fish appears at a different location than the actual location due to the fact that $\theta_i < \theta_t$. The opposite happens if the media swap their positions.

However, if the second medium extends only in the order of the wavelength of the signal, then the law governing the reflection and transmission is quite involving. In this case, we call the incident field as *the primary field* and the perturbation of the primary field in the presence of the obstacle as a secondary source. The phenomenon of the scattering of the electromagnetic signal is called *diffraction*. In our daily life, we always encounter such diffraction of light wave, for example, extended light rays in a dark room when light enters from the gap of the door of the room. The gap acts as the secondary source and diffracts light in the room in a larger area than the gap itself. In indoor wireless communications, the diffraction phenomenon of the electromagnetic signal helps to communicate non-line of sight (NLoS) between rooms of concrete walls. Figure 5.6 shows the telecommunications tower behind a hill. The hill crest acts as the secondary source and diffracts radio wave signals at the neighborhood downhill due to the diffraction effect of the radio wave. Therefore, diffraction has many useful applications for NLoS wireless communications. The topic is out of the scope of the book and we concentrate on the reflection and transmission of the uniform plane wave signals for normal and oblique incidence cases.

As long as the boundaries involved do not include a perfect conductor, neither the field nor its derivative is zero on the boundary. So far, we have discussed the wave propagation between two dielectric media. The electromagnetic field analysis that involves dielectric (not a perfect conductor) boundaries and their known field conditions at these boundaries are referred to as *inhomogeneous boundary value problems*. In *inhomogeneous boundary value problems*, the fields in both media are nonzero. However, the electromagnetic fields that involve conducting boundaries we can define the tangential electric field components at the interface of the dielectric and conducting media is zero. The problem becomes much simpler than the inhomogeneous boundary value problems.

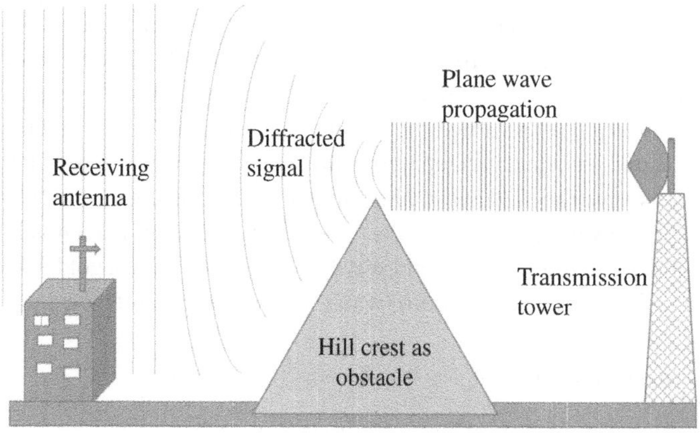

Figure 5.6 Diffraction of signal at the hill's crest helps receive signal at the neighborhood downhill which are off-line of sight of the transmission tower.

5.3 Reflection and Refraction at Plane Surface

We have mentioned that when a plane wave is incident on an interface of two dissimilar homogenous media, a part of the energy is transmitted into the second medium and the remaining portion of the energy is reflected back in the first medium. We also assume that the transmitted and reflected waves are also plane waves. Figure 5.7 illustrates the generic 3D boundary value problem of the uniform plane wave that is obliquely incident at the plane of the interface of two media. We also assume that the xy-plane at $z = 0$ (plane of interface) is extended semi-infinitely in the ±ve xy-directions. A plane wave is incident at the origin (0, 0, 0) with an angle θ_i. As defined above, θ_i is the *angle of incidence*, the angle created with the normal line which is along −ve z-axis. The normal line is the line that is perpendicular to the xy-plane, the plane of the interface of media 1 and 2. Now assume that a linearly polarized xz-directed electric field is propagating along the unit vector $\boldsymbol{a}_i$ and hits the interface at the origin (0, 0, 0) with the angle θ_i. In a UPW, the magnetic field **H** should be orthogonal to the electric field **E** and the direction of propagation. For the forward propagating uniform plane wave, the relation between the electric field, magnetic field, and the direction of propagation is governed by the right-hand rule.

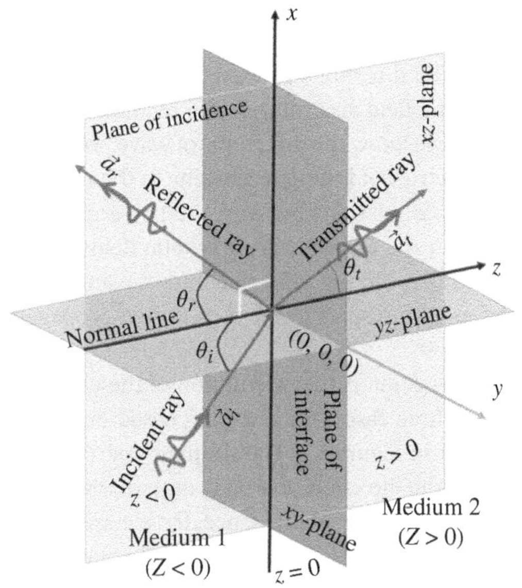

Figure 5.7 Generic plane wave incidence at the interface of two media. Here, we only show the direction of wave propagation without the electromagnetic field interaction.

After the incident wave falls on the interface, it will partly reflect at the direction $\boldsymbol{a}_r$ with an angle θ_r with the normal line. As defined above, this angle θ_r is the *angle of reflection*. Note that the reflected ray resides on the same plane of incidence as well. The direction of the reflected electric field vector changes its direction based on the direction of propagation and likewise, the magnetic field vector does. Some energy with be transmitted in the second medium with the transmitted ray creates an angle θ_t with the normal line along the +ve z-axis. As defined above, this angle is called the *angle of transmission*. Note that the transmitted ray resides on the plane of incidence as well. The direction of the transmitted ray is $\boldsymbol{a}_t$. Note that the incidence and reflection happen in medium 1 (in the region of $z < 0$) and the transmission happens in medium 2 (in the region of $z > 0$).

Note that all three sets of waves (both electric and magnetic fields) follow the right-hand rule for UPW propagation. The directions of the field vectors are governed by the following equations:

$$\boldsymbol{P} = \boldsymbol{E} \times \boldsymbol{H} \tag{5.4}$$

$$\boldsymbol{H} = \frac{1}{\eta}\boldsymbol{a_p} \times \boldsymbol{E} \tag{5.5}$$

$$\boldsymbol{E} = -\eta\boldsymbol{a_p} \times \boldsymbol{H} \tag{5.6}$$

Here, $\boldsymbol{P}$ is the Poynting vector or power follow density vector with a unit (w/m^2), $\boldsymbol{a_p}$ is the unit vector of the direction of propagation and coincides with $\boldsymbol{a}_i$, the unit vector of the incident ray as illustrated in Figure 5.7. The first equation $\boldsymbol{P} = \boldsymbol{E} \times \boldsymbol{H}$ reveals orthogonality between the power density vector $\boldsymbol{P}$, and electric and magnetic field vectors. All are orthogonal to each other. If we know the directions of wave propagation $\boldsymbol{a_p}$ and the electric field vector, then we can calculate magnetic field according to the relation governed by the second equation. Look at the third equation, if we know the directions of wave propagation $\boldsymbol{a_p}$ and the magnetic field, then we can calculate the electric field vector according to the relation. Therefore, it is very important to know this right-hand rule relations between the power flow density vector and its direction $\boldsymbol{a_p}$, and the relations between the electric and magnetic fields. We shall use these relations and compute all the field components in the reflection and transmission problems based on the known polarizations (orientation with respect to the direction of propagation) of the wave. What is the meaning of polarization here?

Polarization is the orientation of the electric field vector in space with respect to time. We shall study three cases of incidences based on the polarization and as outlined in the roadmap of the chapter in Figure 5.1. The simplest case of wave incidence is the normal incidence. In normal incidence, the direction of propagation of the wave is normal to the place of the interface, in this case, xy-plane. In normal incidence, the electric field vector is orthogonal to the normal line and parallel to the plane of the interface. Since the magnetic field is orthogonal to both directions of propagation and the electric field vector, the magnetic field is also normal to the normal line and directed along y-axis if the electric field is directed along x-axis.

Figure 5.7 illustrates the generic and oblique incidence case where the incident wave makes an angle θ_i with the normal line. There are two cases of oblique incidence. When the electric field is transverse to the direction of propagation, it is called *TE polarization* or *perpendicular polarization*. If the magnetic field is transverse to the direction of propagation, it is called *TM polarization* or *parallel polarization*. Obviously, all waves are linearly polarized waves because the tip of the electric field vector traces a line at the plane of wave propagation.

Analyses of the TE and TM polarizations have many significant applications, as stated above. They impose many interesting boundary value problems. Using the known polarizations (orientations) of the electric and/or magnetic field vectors and the direction of propagation, we can calculate the unknown field components and the power densities that are reflected back in the first medium and are being transmitted in the second medium. The same technique can be extended to the boundary value problems of multiple obstacles that have many practical applications. In this aspect, our final goal of the analysis is to find the ratio of the reflected to the incident field components, known as *the reflection coefficient* and the ratio of the transmitted to the incident field components, known as *the transmission coefficient*. From these relations, we finally compute the power reflection and transmission coefficients using the law of conservation of energy.

5.4 Normal Incidence on a Perfect Conductor

As stated before, the simplest analysis case for the reflection and transmission of the uniform plane wave is the case for the signal incidents *normally* on the plane of the interface. Figure 5.8 illustrates the topics covered in the normal incidence of the uniform plane wave. The field can incident on any boundary, both conducting and dielectric. In this case, the incident angle is zero, as shown in Figure 5.9a.

Figure 5.9a illustrates the case for a plane wave propagation on the perfect conductor, which is extended infinitely on the xy-plane ($z = 0$).[7] The incident wave is propagating along the positive

7 The extension of the conducting wave is along the positive and negative x and y directions.

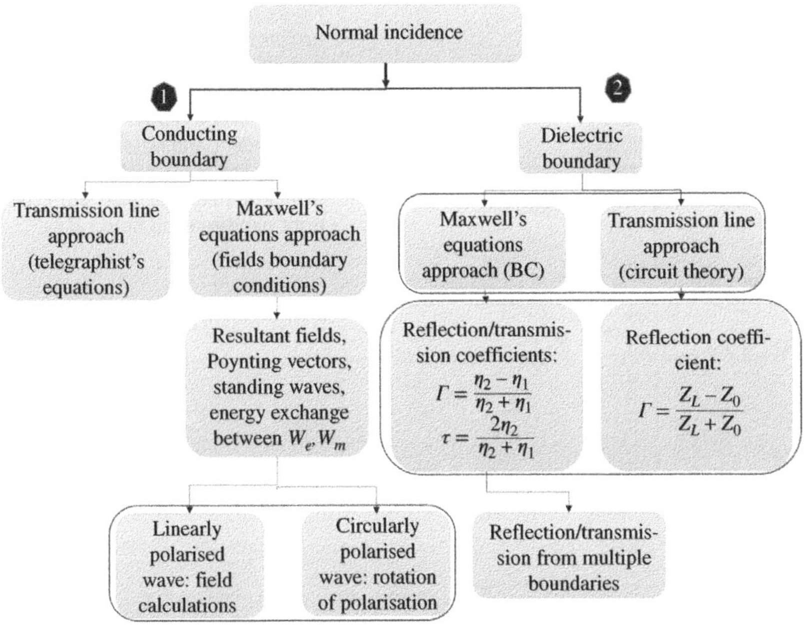

Figure 5.8 Analysis of normal incidence in different dimensions.

z-direction and the reflected wave along the –ve z-direction as shown in Figure 5.9b. The wave is assumed to be coming from $z = -\infty$ hence in the medium 1, z is always negative. It is also assumed that the wave happens to be linearly polarized with the electric field vector vibrates along x-direction, as shown in Figure 5.9a.[8] Hence by the right-hand rule, the magnetic field is oriented along the y-direction and both waves are propagating along the z-direction. Since the second medium is a perfect conductor, the field cannot penetrate into the second medium (no absorption of electromagnetic energy does happen) and the energy carried by the incident wave (ray) is totally reflected back in the opposite (–ve z) direction as shown in the figure with the dark gray arrow heads.[9] Figure 5.10b illustrated the analogy of normal incidence on a perfect conducting boundary with a bouncing shuttlecock from a racket. The motion and trajectories of an incoming and returned shuttlecock are self-explanatory and it has similarities of the motion and trajectories of the uniform plane was that bounces back from a perfect conducting wall (Figure 5.10).

Assume that the first medium is lossless, then the instantaneous electric field can be defined as the combination of the incident (forward traveling) wave and the reflected (reverse traveling) wave as follows:

$$\boldsymbol{E}_x(z, t) = \boldsymbol{E}_s^i + \boldsymbol{E}_s^r = E_0^i \cos(\omega t - \beta_1 z)\boldsymbol{a}_x + E_0^r \cos(\omega t + \beta_1 z)\boldsymbol{a}_x \tag{5.7}$$

8 We arbitrary select the x-polarized electric field. We shall solve problems for fields oriented in other polarization as well.

9 When an electric field is incident on a perfect conducting boundary, it induces the electric current, in this case the induced current is defined as $\boldsymbol{J}_s = J_0\boldsymbol{a}_x$. This current is the secondary source and backscatters electromagnetic field, which is the reflected field wave that carries the full incident energy in the reverse direction. The skin depth of the perfect conduct is zero and hence total reflection happens.

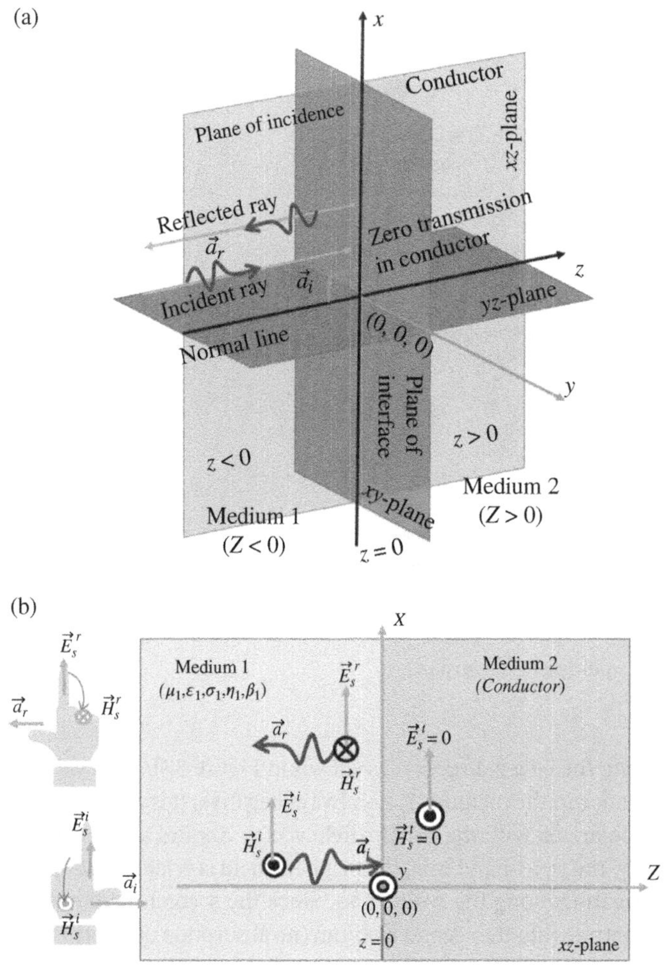

Figure 5.9 Normal incident of a uniform plane wave on a conducting plane: (a) 3D view with all field components and unit vectors, and (b) 2D view that shows the incident and reflected waves have phase reverse of the magnetic field and must obey the right hand-rules in both cases.

where $\beta_1 = \omega\sqrt{\mu_1\varepsilon_1}$, E_{i0} is the peak amplitude of the incident electric field that is propagating along positive z-direction and E_{r0} is the peak amplitude of the reflected electric field that is propagating along –ve z-direction. Therefore, the phasor vector expression of the field wave is defined as:

$$\boldsymbol{E}_{sx}(z) = E_0^i\, e^{-j\beta_1 z}\boldsymbol{a}_x + E_0^r\, e^{j\beta_1 z}\boldsymbol{a}_x \tag{5.8}$$

Here, the forward traveling wave is the incident wave with an amplitude of E_0^i with a propagation phase of $e^{-j\beta_1 z}$ and the reverse traveling wave is the reflected wave with the amplitude of E_0^r with a propagation phase of $e^{j\beta_1 z}$, respectively. Likewise, the phasor vector expression of the magnetic field is defined as:

$$\boldsymbol{H}_{sy}(z) = \boldsymbol{H}_s^i + \boldsymbol{H}_s^r = H_0^i\, e^{-j\beta_1 z}\boldsymbol{a}_y + H_0^r\, e^{j\beta_1 z}\boldsymbol{a}_y \tag{5.9}$$

Or, in terms of the electric field and the wave impedance as:

$$\boldsymbol{H}_{sy}(z) = \frac{E_0^i}{\eta_1} e^{-j\beta_1 z}\boldsymbol{a}_y - \frac{E_0^r}{\eta_1}\, e^{j\beta_1 z}\boldsymbol{a}_y \tag{5.10}$$

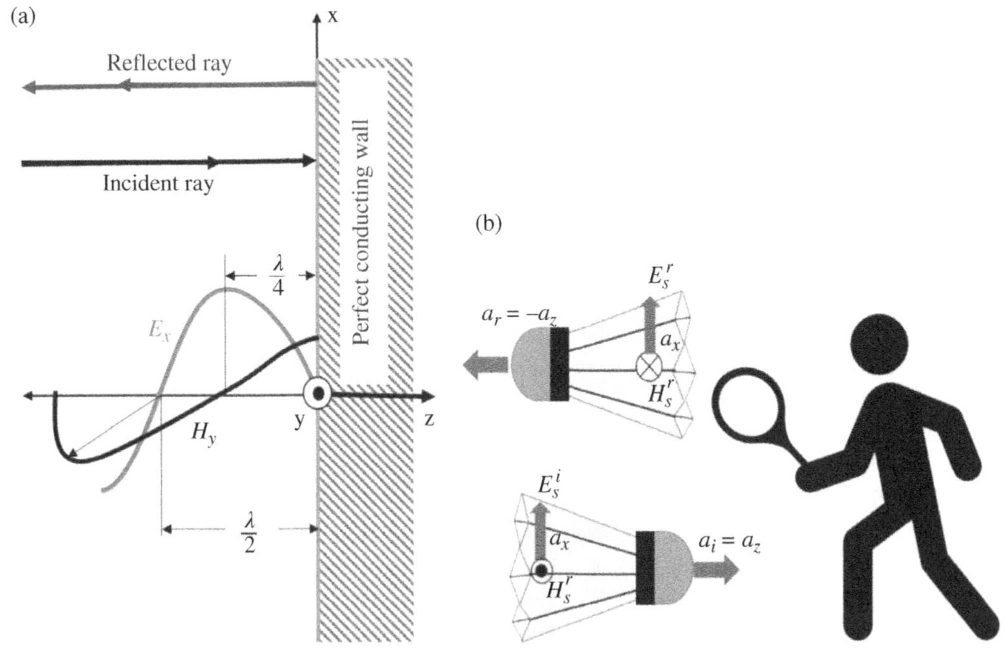

Figure 5.10 (a) A plane wave incident on a perfect conductor and fully reflected back by the conductor is analogous case to, and (b) the motion and trajectories of an incoming and returned shuttlecock.

where, $\eta_1 = \sqrt{\mu_1/\varepsilon_1}$ the intrinsic wave impedance in medium 1, $H_0^i = (E_o^i)/(\eta_1)$ and $H_0^r = -(E_0^r)/(\eta_1)$, the equivalent forward and reverse traveling waves in terms of the respective electric field components and the intrinsic wave impedance, respectively. As you may notice that there is a phase reversal of the reflected magnetic field, and we define the amplitude of the reflected magnetic field as $H_0^r = -(E_0^r)/(\eta_1)$; where the intrinsic impedance in medium 1 is defined as the ratio of the respective electric and magnetic field components with proper signs as

$$\eta_1 = \frac{E_o^i}{H_o^i} = -\frac{E_o^r}{H_o^r} \tag{5.11}$$

Here, we assume that the incident and reflected waves are plane waves. We can develop a set of equations for the incident and reflected field waves in the phasor vector form (with $e^{j\omega t}$ understood) assuming that the medium 1 is lossless ($\alpha = 0$) as follows:

The incident fields in medium 1 ($z < 0$):

$$\boldsymbol{E_s^i}(z) = E_0^i \, e^{-j\beta_1 z} \, \boldsymbol{a_x} \tag{5.12}$$

$$\boldsymbol{H_s^i}(z) = H_0^i \, e^{-j\beta_1 z} \, \boldsymbol{a_y} \tag{5.13}$$

The reflected fields in medium 1 ($z < 0$):

$$\boldsymbol{E_s^r}(z) = E_0^r e^{j\beta_1 z} \boldsymbol{a_x} \tag{5.14}$$

$$\boldsymbol{H_s^r}(z) = -\frac{E_0^r}{\eta_1} \, e^{j\beta_1 z} \, \boldsymbol{a_y} \tag{5.15}$$

And the transmitted electric and magnetic fields are zero in the conducting medium. The best analogy of the normal incidence of a plane wave at the interface of a dielectric (medium 1) and a conductor (medium 2) media is the movement of a shuttlecock when players play badminton, as shown in Figure 5.10b. When a player hits the shuttle with a racket, the direction of the incoming (incident) shuttle is analogous to the forward traveling wave. Note the phasor field vector tips and their direction in Figure 5.10b. When an opponent hits the shuttle, instantly, the shuttle changes its direction of propagation but keeps the position of the arrowhead of electric field intact. This returned shuttle is analogous to the reflected traveling wave. Another way we can say is when the forward traveling shuttle hits the opponent's racket, it instantly changes its direction of propagation (tip of the shuttle), keeping the orientation of the shuttle in the same direction.[10] You may assume this orientation is the polarization or the direction of variation of the electric field vector, not the direction of the wave propagation. Here, the polarization of the uniform plane wave is always x-polarization at all the time and the traveling directions are $\pm z$-directions, in the forward and reverse directions, respectively as shown in Figure 5.10a. However, the magnetic field changes its direction that is similar to the direction of a returned current in a circuit. This also fulfills the condition of the power flow density vector or Poynting vector ($\boldsymbol{P} = \boldsymbol{E} \times \boldsymbol{H}$) in the opposite direction via the reverse traveling wave and must follow the right-hand rule. Here $\boldsymbol{P} = P(-\boldsymbol{a_z})$. Since in this case, the shuttlecock is not divided into two pieces meaning partly traveling in the original direction of propagation as transmission in medium 2 (perfect conductor), and partly traveling in the opposite direction in medium 1 (dielectric), it is analogous to the normal incidence on a conductor. Since through a conductor an electric field cannot pass, a conductor to an incident plane wave is equivalent to a short-circuited transmission line.[11] A perfect conductor does not absorb energy, all energy brought by the incident wave must be reflected back. In the next section, we shall look into the vector analysis of the normal incidence of the electromagnetic fields at the interface of two dielectric media where part of the incident signal propagates through the second medium as the transmitted signal.

Now let us look at what is happening in medium 1 due to the total reflection by the perfect conducting plane. As shown in Figure 5.10a, the electric field $\boldsymbol{E_x}$ is tangential to the interface of the dielectric and conducting media; hence the total field is zero at $z = 0$. This gives rise to the following condition at the interface of the two media:

$$\boldsymbol{E_{sx}}(z)|_{z=-0} = \boldsymbol{a_x}\left[E_0^i\, e^{-j\beta_1 z} + E_0^r\, e^{j\beta_1 z}\right]_{z=-0} = \left[\boldsymbol{E_s^t}\right]_{z=+0} = 0 \tag{5.16}$$

Therefore, we obtain $E_o^i = -E_0^r$. Substitution for $E_0^i = -E_0^r$ in the expression of the resultant field for any arbitrary z in medium 1, we obtain the total x-directed electric field in medium 1 as:

$$\boldsymbol{E_{sx}}(z) = \boldsymbol{a_x} E_0^i\left(e^{-j\beta_1 z} - e^{j\beta_1 z}\right) = -j2E_0^i \sin(\beta_1 z)\boldsymbol{a_x} \tag{5.17}$$

Likewise, we can define the total y-directed magnetic field in medium 1 as:

$$\boldsymbol{H_{sy}}(z) = \frac{\boldsymbol{a_y} E_0^i}{\eta_1}\left(e^{-j\beta_1 z} + e^{j\beta_1 z}\right) = \frac{2E_0^i}{\eta_1} \cos(\beta_1 z)\boldsymbol{a_y} \tag{5.18}$$

Now we can define the instantaneous (space time function) electric field in medium 1 and examine the characteristics of the total wave in the medium as follows:

---

10 You may put a black dot on the shuttlecock and observe if the dot remains at the fixed position for the retuned shuttlecock

11 We shall discuss this case in the transmission line theory and the theory of shielding in EMI/EMC.

$$\mathbf{E}_x(z,t) = Re\left[E_{sx}(z)e^{j\omega t}\right] = Re\left[-j2E_0^i \sin(\beta_1 z)\mathbf{a}_x e^{j\omega t}\right] = Re\left[2E_0^i \sin(\beta_1 z)\mathbf{a}_x e^{j\omega t}e^{-\frac{j\pi}{2}}\right]$$
$$= 2E_0^i \sin(\beta_1 z)\cos\left(\omega t - \frac{\pi}{2}\right)\mathbf{a}_x = 2E_0^i \sin(\beta_1 z)\sin(\omega t)\mathbf{a}_x \tag{5.19}$$

Or, $\mathbf{E}_x(z,t) = 2E_0^i \sin(\beta_1 z)\sin(\omega t)\mathbf{a}_x$ (5.20)

Likewise, we can also define the instantaneous magnetic field in medium one:

$$\mathbf{H}_y(z,t) = \frac{2E_0^i}{\eta_1}\cos(\beta_1 z)\cos(\omega t)\mathbf{a}_y \tag{5.21}$$

We can observe a very interesting phenomenon in the two instantaneous electromagnetic field expressions. Usually, the phase factor for a propagating wave (a space–time function) is defined as $(\omega t \mp \beta_1 z)$. In this case, the wave is frozen in space at $\beta_1 z = 0$. This reveals that although the incident and the reflected waves individually are traveling waves and a function of $(\omega t \mp \beta_1 z)$, respectively, the resultant electric and magnetic waves in medium 1 is a *standing wave* with its crests and troughs are fixed at a defined location as we move along z. The electric field is zero whenever $\sin(\beta_1 z) = 0$. This means:

$$\sin(\beta_1 z) = 0 = \sin(n\pi); \quad \text{where } n = 1,2,3,... \tag{5.22}$$

From the definition of $\beta_1 = 2\pi/\lambda$, we easily obtain the locations where the electric field is zero. Since the wave is coming from $z = -\infty$ and the medium 1 is $z < 0$ region, therefore, we can define the locations of the waveform where the electric field is zero in space as follows[12]:

$$\beta_1 z = \frac{2\pi}{\lambda}z = -n\pi; \quad \text{or,} \quad z = -n\lambda/2 \quad \text{for } n = 1,2,3,... \tag{5.23}$$

Likewise, we can define the locations where the magnetic field is zero as follows

$$\cos(\beta_1 z) = 0 = \cos\left\{(2n+1)\frac{\pi}{2}\right\}; \quad \text{where } n = 1,2,3,... \tag{5.24}$$

$$\beta_1 z = \frac{2\pi}{\lambda}z = -(2n+1)\frac{\pi}{2}; \quad \text{or,} \quad z = -(2n+1)\frac{\lambda}{4} \quad \text{for } n = 1,2,3,... \tag{5.25}$$

Figure 5.11 illustrates the waveforms of the instantaneous electric and magnetic fields in medium 1 at different instances of time-phase (ωt). Such a wave is termed as *the standing wave* and the waveform is called *the standing wave pattern. The standing wave pattern* is formed when two traveling waves from opposite directions are superposed. As can be seen in the figure that the maxima (crests) and minima (troughs) are at the same locations as the time increases. Sometimes it is hard to perceive the physical significance of the standing wave pattern as we consider two independent variables space and time at the same time for the propagating sinusoids. If you submerge your feet in a wavy pool of water where the standing wave is already formed due to the edges of the pool, you may notice that the wave is hitting your feet at different locations as time progresses. If you carefully

12 This technique of finding locations of the maxima and minima of the standing wave pattern is frequently used in calibration of instrument to measure the impedance of any arbitrary load. Since the perfect conductor offers a short circuit condition, the condition of known reflection coefficient of −1, we can use the information to calculate the values of unknown loads by measuring the shift of the locations of the maxima and minima of the reflected voltage waveform with respect to the calibrated ones (short circuit). In microwave measurements, we use three known load conditions of open (reflection coefficient 1), short (reflection coefficient −1) and fully match, also called load, (reflection coefficient 0) for calibration before we start measurements of unknown loads.

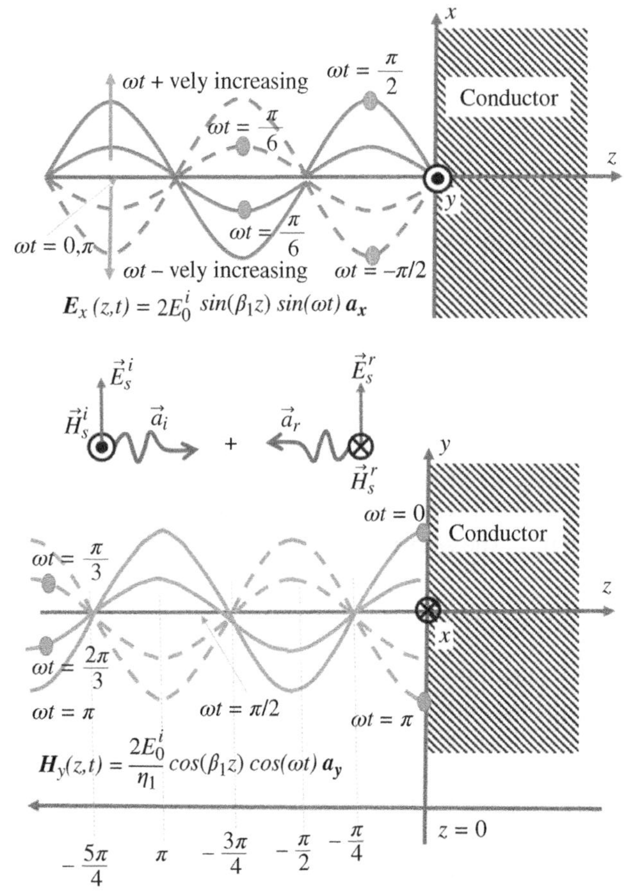

Figure 5.11 Instantaneous electric and magnetic fields in medium at different instances of time-phase (ωt).

observe the water level at the edge of the reservoir, you will also notice that the water level changes with time at the edge line. Figure 5.12 illustrates the analogy of the time variant level of water to illustrate how the standing wave pattern changes its intensity with time at a fixed location as time progresses.

The standing wave does not carry any real electromagnetic energy. The incident energy impinging on the perfect conducting wall is totally reflected back. In the standing wave pattern of the electromagnetic field in medium 1, the electromagnetic energy just exchanges between the electric energy W_e and the magnetic energy W_m.[13] This net zero energy propagation in medium 1 can be verified with the calculation of time average Poynting vector as follows:

$$\begin{aligned}\boldsymbol{P_{av}} &= \frac{1}{2} Re\left[E_{sx} \times H_{sy}^*\right] = \frac{1}{2} Re\left[-j2E_0^i \sin(\beta_1 z)\boldsymbol{a_x} \times \frac{2E_0^i}{\eta_1}\cos(\beta_1 z)\boldsymbol{a_y}\right] \\ &= \frac{1}{2} Re\left[\boldsymbol{a_z}\left(-\frac{j2|E_0^i|^2}{\eta_1}\sin(2\beta_1 z)\right)\right] = 0^{14}\end{aligned} \tag{5.26}$$

13 The detailed derivation of the instantaneous Poynting vector and the electric energy W_e and the magnetic energy W_m can be found in U.S. Inan and A.S. Inan, *Electromagnetic Waves*, Prentice Hall, Upper Saddle River, NJ, 2000.
14 $sin2A = 2sinAcosA$ is used.

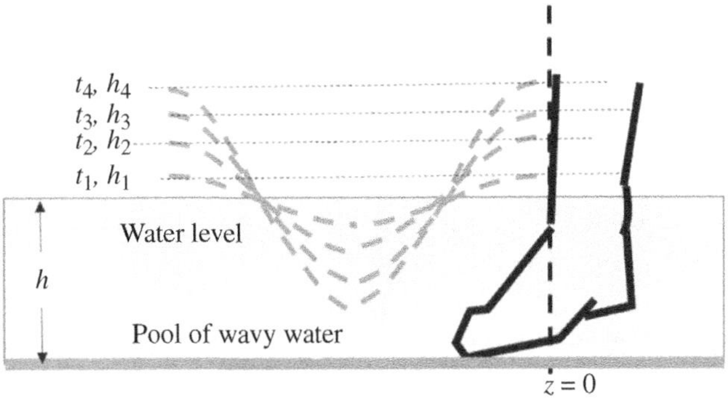

Figure 5.12 Analogy of standing wave pattern and varying level of amplitude at $z = 0$ can be experienced by dipping your leg in a wavy pool of water. You experience different water levels thrusting on your submerged leg at different times.

The physical interpretation of the zero Poynting vector is that the average electromagnetic energy that crosses a cross sectional area in space from one direction is fully bounced back through the same cross sectional area in the opposite direction. Therefore, the net energy flow is zero. This phenomenon can be explained via the induced electric current sheet at the interface of the dielectric and conducting media. The induced current on the conducting wall is defined as:

$$\boldsymbol{J_s} = n \times \boldsymbol{H_{sy}}|_{z=0} = -\boldsymbol{a_z} \times \boldsymbol{a_y}\frac{2E_0^i}{\eta_1} = \boldsymbol{a_x}\frac{2E_0^i}{\eta_1} \tag{5.27}$$

The induced current sheet resides on a very thin layer ($\ll\lambda$) and acts as an antenna of infinite extend along xy plane ($z = 0$). As the current coexists with the field according to Maxwell's equations, it instantly starts reradiating in both $\pm z$-direction. The reradiated electromagnetic energy nullifies the incident wave in both directions. The phenomenon is called *the radiation pressure.*[15] Finally, the reflection of the electromagnetic wave from a perfect conductor can be explained through the transmission line theory. This case is exactly similar to the case for a short circuited transmission line. The analogy of the two entities as illustrated in Figure 5.13 and the equivalent time harmonic parameters are given as follows:

Plane Wave	Transmission line[16]
$E_{sx}(z) = E_o^i e^{-j\beta_1 z} + E_0^r e^{j\beta_1 z}$	$V(z) = V_i e^{-j\beta z} + V_r e^{j\beta z}$
$H_{sy}(z) = \frac{E_0^i}{\eta_1}\left(e^{-j\beta_1 z} - e^{j\beta_1 z}\right)$	$I(z) = \frac{1}{Z_o}\left(V_i e^{-j\beta z} - V_r e^{j\beta z}\right)$
$\beta_1 = \omega\sqrt{\mu_1\varepsilon_1}$	$\beta = \omega\sqrt{LC}$
$\eta_1 = \sqrt{\frac{\mu_1}{\varepsilon_1}}$	$Z_0 = \sqrt{\frac{L}{C}}$ (5.28)

15 For more details about the radiation pressure and alternative view of reflection process on the conductor can be found in the above reference of U.S. Inan and A.S. Inan, *Electromagnetic Waves*, Prentice Hall, Upper Saddle River, NJ, 2000.

16 This part is covered by the Transmission Lined Theory. We have covered the theory of transmission line in limited scope in this chapter and defined the characteristic impedance of the line Z_0. Here L is the inductance per unit length and C is the capacitance per unit length on the transmission line.

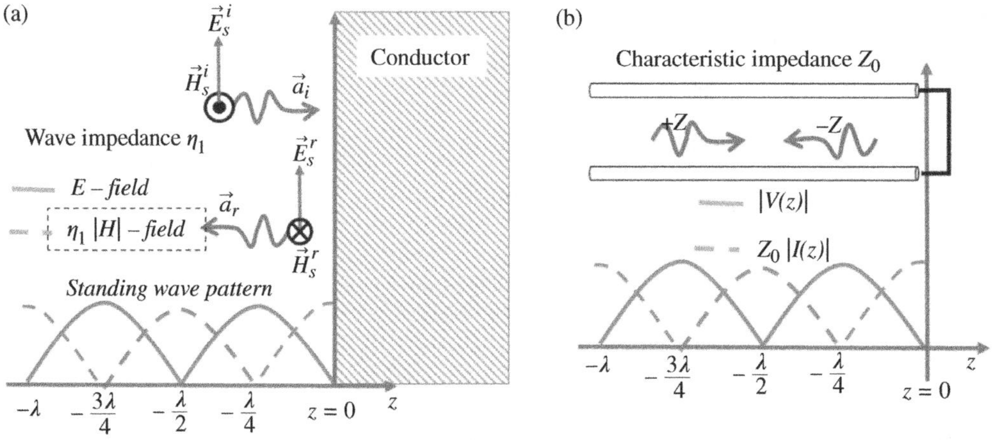

Figure 5.13 Analogy between the standing wave patterns of (a) the electric and magnetic fields of the uniform plane wave impinging normally on a perfect conductor, and (b) the voltage and current waves on a short circuited transmission line of impedance Z_0.

Review Questions 5.1: Normal Incidence on a Perfect Conductor

Q1. What is radiation pressure?
Q2. Explain why at the perfect conducting wall, the resultant time average power is zero in medium 1.
Q3. What is a standing wave pattern? Does it carry any real power?
Q4. Explain how the conducting boundary helps calibrate equipment.

Next we shall solve some examples of the normal incidence on the perfect conductors.

Example 5.1: Normal Incidence on Conductor
An x-polarized uniform plane wave from a GSM[17] mobile phone tower propagating along z-direction and impinges normally on a concrete wall of a nearby tall building, as shown in Figure 5.14. The GSM dual band mobile phone system has an operating frequency of 900 and 1800 MHz and transmits[18] the electric field intensity of 450 V/m for both frequencies. The concrete is located at $y = 0$, as shown in the figure. Write down the phasor vector expression of the resultant electric and magnetic fields in air. Determine the locations of the first minimum and maximum of the standing wave pattern (for both electric and magnetic fields). Make comments on the exchange of electric and magnetic energy via a sketch. Assume that the concrete wall is a perfect conductor.

---

17 GSM stands for Global System for Mobile communications (originally Groupe Spécial Mobile) is a European standard from mobile communications that covers about 90% market share and provides services in 193 countries and territories. Source https://en.wikipedia.org/wiki/GSM (accessed 08 March 2018)
18 A GSM base station transmits around 36–40 dBm of power in both frequency bands that equates around 4 KW of power. If we consider the antenna has a driving point impedance of 50 Ω and is one meter tall, and it distributes radiating energy in air evenly, then we may assume the electric field intensity impinging on the wall is around 450 V/m. In this case we do not consider the gain of the antenna for simplicity and is left for the relevant chapter of antenna theory.

Figure 5.14 A practical problem of normal incidence of uniform.

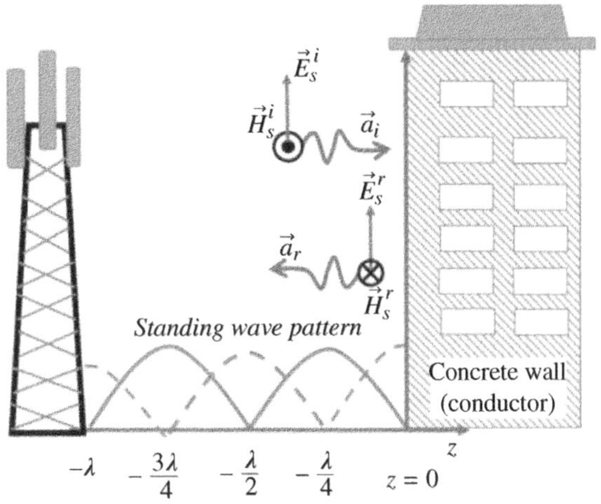

Solution:

First, we shall calculate the wave impedance and the propagation constant of the air medium (medium 1).

$$\eta_1 = \sqrt{\frac{\mu_0}{\varepsilon_0}} = 377\,\Omega; \beta_1 = 2\pi f\sqrt{\mu_0\varepsilon_0} = \frac{2\pi f}{c}\ (\text{rad/m})$$

Since we have dual bands at 800 and 1800 MHz, we designate "1" for 800 MHz and "2" for 1800 MHz and calculate the phase constants for the two waves:

$$\beta_{11} = \frac{2\pi \times 8 \times 10^8\ (\text{Hz})}{3 \times 10^8\ (\text{m/s})} = 16.76\ (\text{rad/m})$$

The wavelength is: $\lambda_{11} = \dfrac{2\pi}{\beta_{11}} = 374.9$ mm

$$\beta_{12} = \frac{2\pi \times 18 \times 10^8\ (\text{Hz})}{3 \times 10^8\ (\text{m/s})} = 37.7\ (\text{rad/m})$$

The wavelength is: $\lambda_{12} = \dfrac{2\pi}{\beta_{12}} = 166.7$ mm

a) The phasor vector expression of the x-polarized incident electric and magnetic fields are:

$$\boldsymbol{E}_{s1}^{i}(z) = E_0^i\, e^{-j\beta_{11}z}\boldsymbol{a_x} = 450e^{-j16.76z}\boldsymbol{a_x}\ (\text{V/m})$$

$$\boldsymbol{E}_{s2}^{i}(z) = E_0^i\, e^{-j\beta_{12}z}\boldsymbol{a_x} = 450e^{-j37.7z}\boldsymbol{a_x}\ (\text{V/m})$$

$$\boldsymbol{H}_{s1}^{i}(z) = \frac{E_0^i}{\eta_1}e^{-j\beta_1 z}\boldsymbol{a_y} = \frac{450}{377}\,e^{-j16.76z}\boldsymbol{a_y} = 1.19e^{-j16.76z}\boldsymbol{a_y}\ (\text{A/m})$$

$$\boldsymbol{H}_{s2}^{i}(z) = 1.19e^{-j37.7z}\boldsymbol{a_y}\ (\text{A/m})$$

The instantaneous (time domain) expressions of the incident field quantities are:

$$\boldsymbol{E}_1^i(z,t) = 450\cos\left(16\pi\times10^8t - 16.76z\right)\boldsymbol{a_x}\ (\text{V/m})$$

$$\boldsymbol{E}_2^i(z,t) = 450\cos\left(36\pi\times10^8t - 37.7z\right)\boldsymbol{a_x}\ (\text{V/m})$$

$$\boldsymbol{H}_1^i(z,t) = 1.19\cos\left(16\pi\times10^8t - 16.76z\right)\boldsymbol{a_y}\ (\text{A/m})$$

$$\boldsymbol{H}_2^i(z,t) = 1.19\cos\left(36\pi\times10^8t - 37.7z\right)\boldsymbol{a_y}\ (\text{A/m})$$

The reflected field quantities are exactly the same except the signal changes as $E_0^i = -E_0^r$:

$$\boldsymbol{E}_{s1}^r(z) = -450e^{j16.76z}\boldsymbol{a_x}\ (\text{V/m})$$

$$\boldsymbol{E}_{s2}^r(z) = -450e^{j37.7z}\boldsymbol{a_x}\ (\text{V/m})$$

$$\boldsymbol{H}_{s1}^r(z) = 1.19e^{-j16.76z}\boldsymbol{a_y}\ (\text{A/m})$$

$$\boldsymbol{H}_{s2}^r(z) = 1.19e^{-j37.7z}\boldsymbol{a_y}\ (\text{A/m})$$[19]

The phasor vector expressions of the resultant field electric and magnetic fields in air (medium 1) are:

$$\boldsymbol{E}_{s1}^1(z) = E_{s1}^i(z) + E_{s1}^r(z) = j900\sin(33.5z)\boldsymbol{a_x}\ (\text{V/m})$$

$$\boldsymbol{E}_{s2}^1(z) = E_{s2}^i(z) + E_{s2}^r(z) = j900\sin(75.4z)\boldsymbol{a_x}\ (\text{V/m})$$

$$\boldsymbol{H}_{s1}^1(z) = H_{s1}^i(z) + H_{s1}^r(z) = 900\cos(33.5z)\boldsymbol{a_y}\ (\text{V/m})$$

$$\boldsymbol{E}_{s2}^1(z) = E_{s2}^i(z) + E_{s2}^r(z) = 900\cos(75.4z)\boldsymbol{a_y}\ (\text{V/m})$$

The corresponding time domain resultant field expressions in air (medium 1) are:

$$\boldsymbol{E}_1^1(z,t) = E_1^i(z,t) + E_1^r(z,t) = 900\sin(33.5z)\sin\left(16\pi\times10^8t\right)\boldsymbol{a_x}\ (\text{V/m})$$

$$\boldsymbol{E}_2^1(z,t) = E_2^i(z,t) + E_2^r(z,t) = 900\sin(75.4z)\sin\left(36\pi\times10^8t\right)\boldsymbol{a_x}\ (\text{V/m})$$

$$\boldsymbol{H}_1^1(z,t) = H_1^i(z,t) + H_1^r(z,t) = 2.38\sin(33.5z)\sin\left(16\pi\times10^8t\right)\boldsymbol{a_x}\ (\text{V/m})$$

$$\boldsymbol{H}_2^1(z,t) = H_2^i(z,t) + H_2^r(z,t) = 2.38\sin(75.4z)\sin\left(36\pi\times10^8t\right)\boldsymbol{a_x}\ (\text{V/m})$$

b) For GSM 1 at 800 MHz, the nearest minimum from $z = 0$ boundary occurs at:
   $33.5z = \pi$ this leads to $z = \dfrac{\pi\ (\text{rad})}{33.5\ (\text{rad/m})} = 93.8$ mm; since the distance between the minimum and the maximum is $\lambda_{11}/4 = 93.7$ mm. Therefore, the first maximum occurs at $l_{max1} = 187$ mm. You can achieve the same results by repeating the calculation in terms of $\lambda_{11}/4\times2 = 187$ mm (avoid the round off error).
   For GSM 2 at 1800 MHz, the nearest minimum from $z = 0$ boundary occurs at:
   $75.4z = \pi$, this leads to $z = (\pi\ (\text{rad}))/(75.4\ (\text{rad/m})) = 41.7$ mm. Therefore, the first maximum occurs at $l_{max2} = 83.4$ mm.
c) Since the time average Poynting vector is zero in medium 2, the electromagnetic wave does not convey any electromagnetic energy, only the electric and magnetic energy alternately change their role as for the case for their respective fields.

19 Calculate the instantaneous (time–space) expressions of the reflected signals.

The perfect conductor reflectors are used in various forms in antenna technology. Examples include parabolic reflectors of dish antennas, corner reflectors for last mile communications and direct to home broadcast satellite antennas, and a circular back reflector dipole antenna for gain improvement as shown in Figure 5.15. While the parabolic dish antennas exploit the geometrical optics theory to focus the beam at the feed horn, the corner reflector and circular flat reflector focus the beam toward the forward direction and enhance the gain of the omnidirectional dipole antenna by a factor of two. The reflectors are placed at quarter wavelength ($d = \lambda/4$) distance so that the single antenna works like an array of two antennas due to the image theory of the source above the ground plane. The phase addition of the reflected and incident waves adds 3 dB gain to the omnidirectional antenna and makes the antenna directional. The advantage is that the radiated power is not wasted in the unwanted direction and efficient communications can be established with reduced cost. The following two examples provide the mechanism of the improvement of gain in both linearly polarized and circularly polarized antennas. For the case of circularly polarized (CP) antennas, when the signal impinges on a conducting wave, the polarization changes to opposite hand polarization. The example also amplifies the mechanism of changing of polarization due to the phase reversal of the electric field in the backscattered signal from the wall.

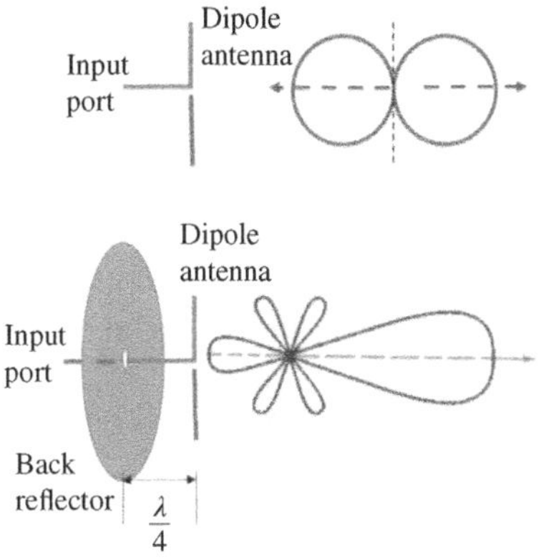

Figure 5.15 A method of beam shaping and gain improvement of a dipole antenna with the introduction of the back reflector at λ/4 distance from the dipole. The patterns change from an omnidirectional (figure-of-eight) pattern to a direction pattern of dumbbell shape main beam pattern with small sidelobes.

Example 5.2: Back Reflector Gain Improvement of Dipole Antenna

A half-wavelength dipole antenna,[20] as shown in Figure 5.15 radiates an x-polarized uniform plane wave at GSM 900 MHz as explained in the above example, assume that the electric field intensity is 450 V/m. A back reflector[21] is added at $d = \lambda/4$ away from the dipole element. This arrangement makes the incident field intensity of the back reflector dipole antenna directive. Calculate (i) the length of the dipole element ℓ in mm, (ii) the distance d in mm, (iii) the incident electric field of the x-polarized dipole element, (iv) the resultant electric field after introducing the reflector, and (v) make comments on gain improvement and the pros and cons of the back reflector.

Solution:

As before, first we calculate the wavelength, impedance and the propagation constant of the air medium at 900 MHz:

$$\lambda_{11} = 374.9\ \text{mm}; \eta_1 = 377\ \Omega; \beta_{11} = 16.76\ (\text{rad/m})$$

a) The length of the dipole element:

$$d = \frac{\lambda}{2} = \frac{374.9\ \text{mm}}{2} = 187.45\ \text{mm}$$

20 Dipole antennas become resonant at $\lambda/2$ length of the operating frequency and produce the omni-directional radiation pattern.

21 Assume large compared to half wavelength to avoid diffraction from the edge of the circular reflector.

b) The distance of the reflector in mm is: $d = \lambda/4 = 374.9$ mm$/4 = 93.7$ mm
c) Since the reflector is at $d = \lambda/4$, the spatial phase factor $\beta_{11}z = (2\pi/\lambda)\times(\lambda/4) = (\pi/2)$
The radiated (incident) electric field is:

$$\boldsymbol{E}^{\boldsymbol{i}}_{\boldsymbol{s1}}(z) = E^i_0 e^{-j\beta_{11}z}\boldsymbol{a_x} = 450e^{-\frac{j\pi}{2}}\boldsymbol{a_x} = -j450\boldsymbol{a_x}\ (\text{V/m})$$

The reflected electric field is: $\boldsymbol{E}^{\boldsymbol{r}}_{\boldsymbol{s1}}(z) = -450e^{(j\pi)/2}\boldsymbol{a_x} = -j450\boldsymbol{a_x}$ (V/m)
The phasor vector expression for the resultant field is:

$$\boldsymbol{E}^{\boldsymbol{1}}_{\boldsymbol{s1}}(z) = E^i_{s1}(z) + E^r_{s1}(z) = -j900\boldsymbol{a_x}\ (\text{V/m})$$

The instantaneous expression for the resultant field is:

$$\boldsymbol{E}^{\boldsymbol{1}}_{\boldsymbol{1s}}(z, t) = E^i_1(z, t) + E^r_1(z, t) = 900\cos\left(16\pi \times 10^8 t\right)\boldsymbol{a_x}\ (\text{V/m})$$

d) As can be seen, the electric field intensity has increased by two fold. This can be considered as the image of equi-phase dipole element at $-\lambda/4$ distance from the reflector and acts as the array of two dipole elements of equal phase. The gain is improved by 3 dB (two times). The advantage is that we can save energy from the radiation of electromagnetic wave in an unwanted direction. The disadvantage is that the antenna coverage is restricted and cannot be used in mobile environments such as Wi-Fi at home.

5.5 Circularly Polarized Wave Incidence on a Conducting Surface

The propagation of the uniform plane wave in the mode of circularly polarization has great significance in satellite communications and mobile wireless communications.[22] In both cases, the polarization of the signal can change arbitrarily. This phenomenon has been discussed in the preceding chapters. In this section, we examine the circularly polarized wave incidence on a conducting surface. If we consider the conducting interface is analogous to the mirror of an incidence light wave, we see that the reflected signal changes its sense of polarization (handedness). We examine this phenomenon for any electromagnetic signal. Assume the incident circularly polarized wave is reflected from a planar conducting boundary at $z = 0$. Then the phasor vector expression of the incident electric field can be written as:

$$\boldsymbol{E}^{\boldsymbol{i}}_{\boldsymbol{s}}(z) = E^i_0\left(\boldsymbol{a_x} \pm j\boldsymbol{a_y}\right)e^{-j\beta_1 z} \tag{5.29}$$

Here, for the right hand circularly polarized (RHCP) wave, $\boldsymbol{a_x}$ leads $\boldsymbol{a_y}$ hence the unit vector is: $(\boldsymbol{a_x} - j\,\boldsymbol{a_y})$. For the left hand circularly polarized (LHCP) wave, $\boldsymbol{a_x}$ lags $\boldsymbol{a_y}$ hence the unit vector is: $(\boldsymbol{a_x} + j\,\boldsymbol{a_y})$. Figure 9.15 illustrates the transmitted and reflected signals for both polarizations. The time domain expression of the incident electric field is:

$$\boldsymbol{E}^{\boldsymbol{i}}(z, t) = E^i_0\left\{\boldsymbol{a_x}\cos(\omega t - \beta_1 z) \pm \boldsymbol{a_y}\cos\left(\omega t - \beta_1 z + \frac{\pi}{2}\right)\right\}$$

$$\text{Or, } \boldsymbol{E}^{\boldsymbol{i}}(z, t) = E^i_0\left\{\boldsymbol{a_x}\cos(\omega t - \beta_1 z) \mp \boldsymbol{a_y}\sin(\omega t - \beta_1 z)\right\} \tag{5.30}$$

22 Revisit the significance of the circularly polarized signal in satellite communications in Section 4.15.

As the incident signal impinges on the conducting boundary, the reflected signal changes its sense of polarization from right to left and vice versa. The reflected electric field, using the condition $E_0^i = -E_0^r$, is defined as:

$$\boldsymbol{E}_s^r(z) = -E_0^i\left(\boldsymbol{a_x} \pm j\boldsymbol{a_y}\right) e^{j\beta_1 z} \tag{5.31}$$

The time domain expression of the reflected electric field is:

$$\boldsymbol{E}^r(z, t) = -E_0^i\left\{\boldsymbol{a_x} \cos(\omega t + \beta_1 z) \pm \boldsymbol{a_y} \cos\left(\omega t + \beta_1 z + \frac{\pi}{2}\right)\right\}$$

$$\text{Or, } \boldsymbol{E}^r(z, t) = E_0^i\{-\boldsymbol{a_x} \cos(\omega t + \beta_1 z) \pm \boldsymbol{a_y} \sin(\omega t + \beta_1 z)\} \tag{5.32}$$

The time domain expressions of the incident and reflected electric fields are placed together for easy comparison as follows:

$$\boldsymbol{E}^i(z, t) = E_0^i\{\boldsymbol{a_x} \cos(\omega t - \beta_1 z) \mp \boldsymbol{a_y} \sin(\omega t - \beta_1 z)\}$$

$$\boldsymbol{E}^r(z, t) = E_0^i\{-\boldsymbol{a_x} \cos(\omega t + \beta_1 z) \pm \boldsymbol{a_y} \sin(\omega t + \beta_1 z)\} \tag{5.33}$$

It can be seen that the orientation of the two fields is opposite to each other. The resultant electric field is the summation of the above two field components and equates to zero at $z = 0$.

$$\boldsymbol{E}_{s1}(z) = E_s^i(z) + E_s^r(z) = E_0^i\left(\boldsymbol{a_x} \mp j\boldsymbol{a_y}\right) e^{-j\beta_1 z} - E_0^i\left(\boldsymbol{a_x} \pm j\boldsymbol{a_y}\right) e^{j\beta_1 z} \tag{5.34}$$

After mathematical manipulations, we obtain:

$$\boldsymbol{E}_{s1}(z) = E_0^i(-j2)\{\boldsymbol{a_x} \sin(\beta_1 z) \mp \boldsymbol{a_y} \cos(\beta_1 z)\} \tag{5.35}$$

The time domain expression of the resultant field in medium 1 is:

$$\boldsymbol{E}_1(z, t) = E_0^i(2)\, Re\left\{\boldsymbol{a_x} \sin(\beta_1 z) e^{j\omega t} e^{-j\pi/2}\right\} \mp Re\left\{\boldsymbol{a_y} \cos(\beta_1 z) e^{j\omega t} e^{-j\pi/2}\right\}$$

$$= 2E_0^i\left\{\boldsymbol{a_x} \sin(\beta_1 z) \cos\left(\omega t - \frac{\pi}{2}\right) \mp \boldsymbol{a_y} \cos(\beta_1 z) \cos\left(\omega t - \frac{\pi}{2}\right)\right\}$$

$$\therefore\ \boldsymbol{E}_1(z, t) = 2E_0^i\{\boldsymbol{a_x} \sin(\beta_1 z) \mp \boldsymbol{a_y} \cos(\beta_1 z)\} \sin(\omega t) \tag{5.36}$$

Remarks

1) From expressions for the incident field wave and for the reflected field wave in (5.33) we can see then at $z = 0$, the RHCP incident wave converts to the LHCP wave and vice versa. Figure 5.16 illustrates the rotation of the electric field at a different time at $z = 0$. Considering the direction of the wave propagation in both the forward and reverse traveling waves, we can see that the incident and the reflected waves change their sense of circular polarization once reflected from the conducting boundary.
2) The resultant field wave in (5.36) is a standing wave in medium 1 with the time phase factor $\sin(\omega t)$[23] only. From the expression, we can see that the standing wave is also formed with the superposition of the two orthogonal waves.
3) The back reflector at $\lambda/4$ away from a circularly polarized dipole antenna also will get a 3 dB gain improvement. We shall analyze such a dipole antenna in the next example. Finally, the resultant field.

23 No spatial phase factor $\beta_1 z$ involves, hence it is not a travelling wave.

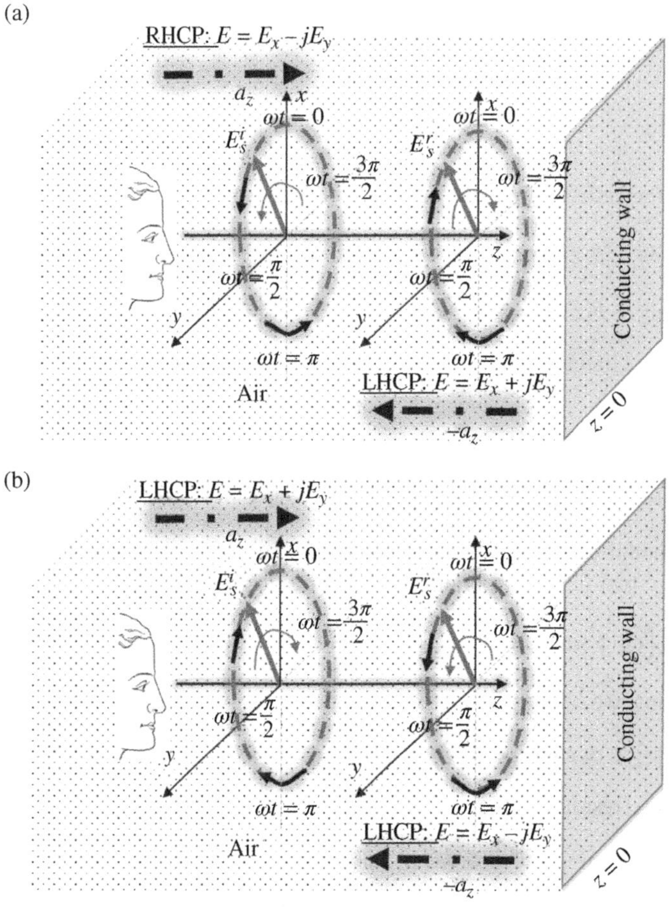

Figure 5.16 Change in the handedness of circularly polarized wave when it encounters the conducting boundary: (a) RHCP-LHCP, and (b) LHCP-RHCP.

$\boldsymbol{E_1}(z, t) = 2E_0^i\{\boldsymbol{a_x} \, sin\,(\beta_1 z) \mp \boldsymbol{a_y} \, cos\,(\beta_1 z) \, cos\} \, sin\,(\omega t)$ is a standing wave with a time function of $sin(\omega t)$ and has twice the amplitude of the original peak amplitude of the incident field.

Example 5.3: Back Reflector Gain Improvement of CP Helical Antenna

A helical antenna produces a circularly polarized signal as shown in Figure 5.17 and radiates a circularly polarized signal. Based on the directions of turns, the handedness of the polarization is determined. A back reflector[24] is added at $d = \lambda/4$ away from the helix element. This arrangement makes the incident field intensity of the back reflector helical antenna directive. Calculate (i) the resultant electric field.

Solution:

The RHCP radiated (incident) electric field is:

$$\boldsymbol{E^i_{s1}}(z) = E_0^i\left(\boldsymbol{a_x} - j\boldsymbol{a_y}\right)e^{-j\beta_1 z}\big|_{z=(\lambda/4)} = -jE_0^i\left(\boldsymbol{a_x} - j\boldsymbol{a_y}\right)$$

24 Assume large compared to half wavelength to avoid diffraction from the edge of the circular reflector.

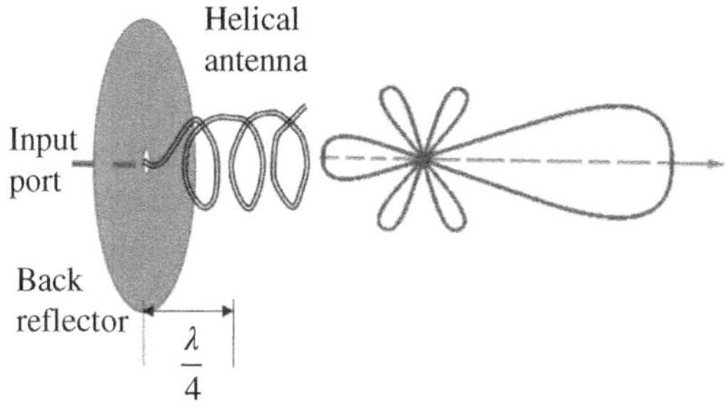

Figure 5.17 The helical antenna with back reflector and its radiation pattern.

The reflected electric field is:

$$\boldsymbol{E}_{s1}^{\boldsymbol{r}}(z) = E_0^r(\boldsymbol{a}_x + j\boldsymbol{a}_y)e^{j\beta_1 z}|_{z=0} = -E_0^i\,(\boldsymbol{a}_x + j\boldsymbol{a}_y)$$

The phasor vector expression for the resultant field is:

$$\boldsymbol{E}_{s1}^{1}(z) = \boldsymbol{E}_{s1}^{\boldsymbol{i}}(z) + \boldsymbol{E}_{s1}^{\boldsymbol{r}}(z)$$

The instantaneous expression for the resultant field is:

$$\boldsymbol{E}_{1s}^{1}(z,t) = \boldsymbol{E}_1^{\boldsymbol{i}}(z,t) + \boldsymbol{E}_1^{\boldsymbol{r}}(z,t) = 900\cos\left(16\pi \times 10^8 t\right)\boldsymbol{a}_x\ (\text{V/m})$$

a) As can be seen, the electric field intensity has increased by two fold. This can be considered as the image of equi-phase helix element at $-\lambda/4$ distance from the reflector and acts as the array of two helix elements of equal phase. This gain is improved by 3 dB (two times). Same as for the reflector backed dipole element, for this case, the advantage is that we can save energy from the radiation of electromagnetic wave in an unwanted direction. The disadvantage is that the antenna coverage is restricted and cannot be used in mobile environments such as Wi-Fi at home.

5.6 Normal Incidence at Dielectric Boundary

In the preceding section, we consider the simplest case for wave incidence at the interface of air and conductor media. The most prominent and known boundary conditions that makes the field analysis simple are: (i) the electric field cannot penetrate the conducting medium; this means the transmitted electric field in medium 2 is: $E_s^t = 0$, and (ii) this also leads to the boundary condition that the tangential electric field components at the interface of the dielectric-conducting media is zero: $E_s^i + E_s^r|_{z=-0} = 0$. This leads to $E_s^i = -E_s^r$. Since $E_s^t = 0$, that forces the reflected field E_s^r in the opposite direction of the incident field E_s^i. Based on the two boundary conditions, we calculated the voltage reflection coefficient $\Gamma = -1$ and voltage transmission coefficient $\tau = 0$. We also considered a few very interesting applications of the theory; they are: (i) gain improvement of the omni-directional dipole antenna with a back reflector by the beam shaping with a planar grounded reflector at $\lambda/4$ away from the active antenna element, and (ii) the change of sense of polarizations of the circularly polarized electromagnetic wave that can be used in microwave radars to detect metallic targets (aircraft) from its clutter such as cloud.[25] We also draw the analogy of a short-circuited transmission line to exemplify the similar phenomena of the circuit theory with the wave theory.[26] However, the incidence of a plane wave on a planar air–conductor interface is a specialized case. In this section, we shall consider a generic case of reflection and transmission of the plane wave on a planar interface of two dissimilar dielectric media. The theory is equally valid for lossless and lossy dielectric media due to the fact that the boundary conditions of the electromagnetic fields that we use for the field analyses are the same at the interface of the two-dielectric media[27] irrespective of lossless and lossy media. Therefore, for generalization, we start with the lossy media for the field analyses and can consider the lossless media are the special cases where we assume the loss factor $e^{\mp\alpha z}$. The main motto of the field analysis of the uniform plane wave at the interface of two dissimilar dielectric media is to calculate the reflection and transmission coefficients. The analysis helps us to determine what part of the incident power is reflected back by the second medium and what part of the power is transmitted through the second medium. Understanding the mismatched condition of the free space waves at the interface of two dielectric media helps us to design many interesting devices. Examples include optical fibers, radomes[28] of antennas either small or big, multilayer microwave monolithic integrated circuits (MMICs), and various multilayer coated optical devices even the reading glasses to stop the glare from your computer monitor when you use it. We shall examine some of these applications after covering the theory.

Figure 5.18 shows the normal incidence of a plane wave at the interface of two dissimilar media. The media are assumed to be linear, homogeneous but lossy.[29] Figure 5.18a illustrates the 3D perspective view with all geometrical planes and (Figure 5.18b) 2D plan view of the electromagnetic waves, the field vectors and directions of wave propagation on the xz-plane for the normal incidence case. The respective constitutive parameters for medium 1 (μ_1, ε_1, σ_1, η_1, β_1) and for Medium 2 (μ_2, ε_2, σ_2, η_2, β_2) are also shown here. The wave impedance is defined as $\eta_{1,2} = \sqrt{\mu_{1,2}/\varepsilon_{1,2}}$ and $\beta_{1,2} = \omega\sqrt{\mu_{1,2}\varepsilon_{1,2}}$. We assume the simplest time harmonic fields $(\vec{E}_s^i, \vec{H}_s^i)$, $(\vec{E}_s^r, \vec{H}_s^r)$, and $(\vec{E}_s^t, \vec{H}_s^t)$ for the incidence, reflected and transmitted cases, respectively. As shown in Figure 5.18b, we assume an x-polarized incident electric field impinges at the interface of the two media without losing the generality of the wave propagation and the coordinate system chosen. The direction of propagation is from medium 1 to medium 2. Since the second medium is a dissimilar dielectric medium, some part of the incident energy propagates through the second medium and the rest energy will be reflected back at the interface of the two media. However, the polarization of the

25 The author also uses the sense of polarization change of vertical linearly polarized (LP) signal into a horizontally polarized signal with a 45° reflecting strip. This reflective strip is called a *polarizer* that changes the polarization of the linearly polarized incoming signal to its orthogonal polarization (also called cross polar signal) and reflects back the signal to its receiver. Usually the clutter and noises are responsive to the co-polar components not in the cross polar components. Thus the noise and clutter can be eliminated in the cross-polar receiver and crispy returned signal is detected by the receiver.

26 In some textbooks the transmission line theory is explained first to draw the analogy and the approach is called *early transmission line approach to electromagnetic theory*. See textbook, S. Wentworth, *Applied electromagnetic: early transmission line approach*, Wiley, 2007.

27 The lossy medium yields complex wave impedance due to the finite conductivity of the medium. As long as we consider the complex impedance in the calculations of the reflection and transmission coefficients, we can use the same theoretical analysis of the field components for the lossless case that we are deriving in this section.

28 A radome is a cover of an antenna that protect the antenna from harsh weather conditions such as snow, foliage and wind drag for cars, supersonic fighter jets and passengers planes.

29 The loss factor is included with the finite conductivity terms for both media σ_1 and σ_2, respectively. This finite conductivity is associated with the attenuation constants α_1 and α_2 and the complex wave impedances η_1 and η_2, for media 1 and 2 respectively.

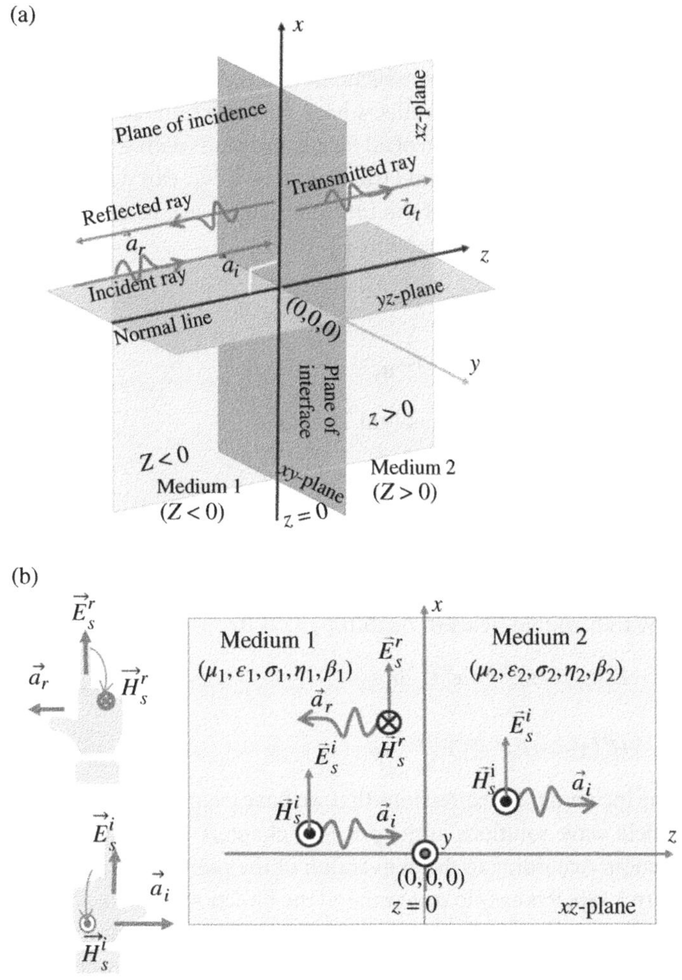

Figure 5.18 Normal Incidence of a uniform plane wave at the interface of two dissimilar dielectric media: (a) 3D, and (b) 2D.

electromagnetic wave in medium 2 will remain the same as for the incident wave. The instantaneous or time domain expression for the incident electric field (with i superscript) is:

$$\mathbf{E^i}(z, t) = E_0^i e^{-\alpha_1 z} \cos(\omega t - \beta_1 z)\mathbf{a_x} \text{ (V/m)} \tag{5.37}$$

where E_o^i is the peak amplitude of the incident electric field at $z = 0$ at the time origin, α_1 is the attenuation constant (Np/m) and β_1 is the phase constant (rad/m) in medium 1, respectively. As stated and also shown in the three forms of vector expressions of the fields,[30] it is far easier to carry out the field calculations in phasor form with $e^{j\omega t}$ understood for this boundary value problem. It

30 The three forms are time domain (space time function) as produced above, the exponential form with $e^{j\omega t \mp \gamma z}$ and the phasor vector form of $e^{\mp \gamma z}$. Among them the easiest way to analyze the complex vector field by using the phasor vector form with $j\omega t$ understood and replacing $\frac{\partial}{\partial t} = j\omega$. Once all fields are calculated we can easily covert the phasor vector fields into the time domain fields by simply multiplying the phasor vector expression in space function with $e^{j\omega t}$ and taking the real part of the product. This is the main advantage of using the phasor vector form of the fields for electromagnetic analyses.

means we only write the exponents of spatial phase components: the exponential loss factor $e^{\pm\alpha z}$ and the oscillating factor $e^{\pm j\beta z}$ along with field amplitudes and unit vectors. We assume that both the media are linear and homogenous, and the polarization of the waves does not change as it propagates in medium 2 and reflects back with the same polarization in medium 1.[31] Now using the time harmonic phasor form of all field expressions with subscript "*s*" and respective superscripts for incident "*i*," reflection "*r*" and transmission "*t*," we can write a new set of field solutions for the case of the normal incidence of the uniform plane wave as follows:

The incident fields in medium 1 ($z < 0$):

$$\boldsymbol{E}_s^i(z) = E_0^i e^{-j\alpha_1 z} e^{-j\beta_1 z} \boldsymbol{a_x} \tag{5.38}$$

$$\boldsymbol{H}_s^i(z) = H_0^i e^{-j\alpha_1 z} e^{-j\beta_1 z} \boldsymbol{a_y} \tag{5.39}$$

The reflected fields in medium 1 ($z < 0$):

$$\boldsymbol{E}_s^r(z) = E_0^r e^{j\alpha_1 z} e^{j\beta_1 z} \boldsymbol{a_x} \tag{5.40}$$

$$\boldsymbol{H}_s^r(z) = -\frac{E_0^r}{\eta_1} e^{j\alpha_1 z} e^{j\beta_1 z} \boldsymbol{a_y} \tag{5.41}$$

The transmitted fields in medium 1 ($z > 0$):

$$\boldsymbol{E}_s^t(z) = E_0^t e^{-j\alpha_2 z} e^{-j\beta_2 z} \boldsymbol{a_x} \tag{5.42}$$

$$\boldsymbol{H}_s^t(z) = H_0^t e^{-j\alpha_2 z} e^{-j\beta_2 z} \boldsymbol{a_y} \tag{5.43}$$

The incident field expressions that we have examined already in the development of the theory of the field wave solutions in the previous chapters. It is traveling in medium 1 along the positive z-direction. According to the convention of the right hand rule, also shown in the hand diagrams in Figure 5.18b, it is easy to comprehend the directionalities of all field components with respect to the directions of the wave propagation of the three sets of fields in both media. They obey the governing equations of right hand rule between the Poynting or power flow density vector, the electric field and the magnetic field vectors irrespective of the direction of propagation. Both incident and reflected waves follow the right hand rule. Observing closely the directions of the incident and reflected magnetic fields, we can easily comprehend that the incident and reflected magnetic fields swap their directions to maintain the relation of the right hand rule. We know that the magnetic field is analogous to the current. As current changes its direction during the returned power flow, the magnetic field also changes its direction when the electromagnetic signal reverses its direction of propagation ($\boldsymbol{H} = -\boldsymbol{a_p} \times \boldsymbol{E}$), here for reflected signal $\boldsymbol{a_p} = \boldsymbol{a_r}$. This also follows the definition of the wave impedance in terms of the reflected (reverse traveling wave) electric and magnetic fields: $\eta = -\frac{E_0^r}{H_0^r}$. Also, as in medium 1 the distance traveled by the returned signal is in $-z$-direction, the exponents with +ve sign for both the attenuation and phase constants in the reverse traveling/reflected wave are used. Therefore, all the above arguments support the reflected electric field is x-directed, the magnetic field is $-y$-directed and the wave propagates along $-z$-direction.

31 Obviously with the direction change from $+\boldsymbol{a_x}$ to $-\boldsymbol{a_x}$ to satisfy the boundary conditions and the change of directions of propagation of the two fields.

The transmitted field continues propagating infinitely with time in medium 2 as long as it does not hit another discontinuity.[32] The polarization and direction of propagation of the transmitted wave in medium 2 are the same as that for the incident wave with the exception of the constitutive parameters ($\mu_2, \varepsilon_2, \sigma_2, \eta_2, \beta_2$). This indicates that the wave in the second medium propagates at different velocities as the phase constant is different. Therefore, when we shall write the expressions for the transmitted electric and magnetic fields in medium 2, we keep the directions of field vectors and propagation the same as those for the incident wave, but writing the subscript 2 instead of 1 in α_n and β_n for $n = 1, 2$ due to the medium changes. Therefore the transmitted field expressions are derived in (5.42–43) as they should be. So far we have obtained the phasor expressions for the three sets of field components. Our objective is to know how the fields interact at the interface of two media at $z = 0$ using the known boundary conditions and develop two important relations between the reflected and incident electric fields called *the voltage reflection coefficient* and the transmitted and incident electric fields called *the voltage transmission coefficient*. We shall also develop the concept of respective power coefficients using the law of conservation of energy.

Next, as you observe that in medium 1 a complex situation happens due to the presence of the incident and reflected signals simultaneously. We also observe that in medium 1 instantly after reflection at the interface of the 2 media, the incident field mixes with the reflected field and they form a standing wave pattern as shown in Figure 5.18b. We shall further the concept of the standing wave pattern in medium 1 in more detail in the next section.

5.6.1 Calculation of Reflection and Transmission Coefficients

So far, we have developed the coordinate system of the vector fields and got some understanding of the standing wave pattern as the resultant field of the incident and reflected waves. Now we proceed to calculate the reflection and transmission coefficients at the interface of the two dielectric media so that the fields must satisfy the fields' boundary conditions at the interface. Unlike the case for the air–conductor interface, the dielectric-dielectric interface has nonzero transmitted electric and magnetic field components. Applying the boundary conditions at the interface ($z = 0$) assuming zero electric current density ($K = 0$), we have:

i) Tangential electric field component (x-directed fields are tangential to the plane of the interface that lies on xy-plane at $z = 0$) is continuous at the interface of the two dielectric media. This leads

$$\boldsymbol{E}_{sx}(z)|_{z=-0} = \boldsymbol{a}_x\left[E_0^i e^{-j\beta_1 z} + E_0^r e^{j\beta_1 z}\right]_{z=-0} = \boldsymbol{a}_x\left[E_s^t e^{-j\beta_2 z}\right]_{z=+0} \tag{5.44}$$

$$\boldsymbol{H}_{sx}(z)|_{z=-0} = \boldsymbol{a}_y\left[H_0^i e^{-j\beta_1 z} + H_0^r e^{j\beta_1 z}\right]_{z=-0} = \boldsymbol{a}_y\left[H_s^t e^{-j\beta_2 z}\right]_{z=+0} \tag{5.45}$$

Substitution for $z = 0$ and considering the same directed scalar field quantities lead to the following expressions

$$E_0^i + E_0^r = E_0^t \tag{5.46}$$

$$H_0^i + H_0^r = H_0^t \text{ or}, \frac{E_0^i}{\eta_1} - \frac{E_0^r}{\eta_1} = \frac{E_0^t}{\eta_2} \tag{5.47}$$

$$\text{or}, \ E_0^i - E_0^r = \frac{\eta_1}{\eta_2} E_0^t \tag{5.48}$$

---

32 At infinity distance, the field will diminish to zero due to the loss factor $e^{-\alpha z}$. We also considered the field is zero at infinity in static case in the development of some fundamental theories. This is due to the fact that the field is proportional to $1/r^2$ in static/dynamic case. We shall also examine the radiated far field is proportional to $1/r$ during the development of antenna theory. Therefore, at infinity distance the signal will die down any way.

Here, the wave impedance definitions for media 1 and 2, $\eta_1 = E_0^i/H_0^i = -E_0^r/H_0^r$ and $\eta_2 = E_0^t/H_0^t$, respectively are used. Addition and subtraction of (5.46) and (5.48) and then dividing both sides by 2 yields the transmission coefficient τ and the reflection coefficient Γ, respectively. The solutions for the reflected and transmitted fields in terms of the incident field are given in the following table:

Addition of (5.46) and (5.48):	Subtraction of (5.46) and (5.48):
$E_0^i + E_0^r = E_0^t$	$E_0^i + E_0^r = E_0^t$
$E_0^i - E_0^r = \frac{\eta_1}{\eta_2} E_0^t$	$E_0^i - E_0^r = \frac{\eta_1}{\eta_2} E_0^t$
$E_o^i = \left(1 + \frac{\eta_1}{\eta_2}\right) \frac{E_o^t}{2} = \frac{\eta_1 + \eta_2}{2\eta_2} E_o^t$	$E_o^r = \left(1 - \frac{\eta_1}{\eta_2}\right) \frac{E_o^t}{2} = \left(\frac{\eta_2 - \eta_1}{2\eta_2}\right) \times \left(\frac{2\eta_2}{\eta_2 + \eta_1}\right) E_o^i$
The transmission coefficient:	The reflection coefficient:
$\tau = \frac{E_o^t}{E_o^i} = \frac{2\eta_2}{\eta_1 + \eta_2}$ (5.49)	$\Gamma = \frac{E_o^r}{E_o^i} = \frac{\eta_2 - \eta_1}{\eta_2 + \eta_1}$ (5.50)

Comparing equation (5.49) and (5.50) yields:

$$\tau = 1 + \Gamma \tag{5.51}$$

From the above field analysis, we can make the following remarks:

1) For lossy dielectric media the wave impedances are complex; hence the transmission coefficient τ and the reflection coefficient Γ are also complex. For lossless media, the values of the wave impedance are real and the transmission and reflection coefficients will also be real.
2) As stated above, although we have examined the field analysis in the lossy media, the theory is equally applicable to the lossless case. The only difference is that we assume $e^{\alpha_1 z}$ and $e^{\alpha_2 z}$ will be unity. The loss factor does not influence the polarization and direction of wave propagation; hence the field analysis is equally applied to any medium.
3) We can also draw a similar analogy of the transmission line for the dielectric–dielectric interface as we did for the dielectric-conductor interface in Figure 5.13b. In this case, the two wave impedances (η_1 and η_2) of the two media represent the two transmission lines of wave impedances (Z_{01} and Z_{02}) connected at the interface of the two mediums. The second impedance can be considered as the load impedance $Z_L = Z_{02}$ of the first transmission line with the characteristic impedance Z_0. The later section of the chapter depicts the analogy of the dielectric–dielectric interface with a terminated transmission line with load impedance Z_{02} which is equivalent to the wave impedance of medium 2. Substitutions for the analogous wave impedances in the above reflection and transmission coefficients equations we can obtain *the voltage reflection coefficient* and *the voltage transmission coefficients*, respectively, as follows:

$$\Gamma = \frac{Z_L - Z_0}{Z_L + Z_0} \tag{5.52}$$

$$\tau = \frac{2Z_L}{Z_L + Z_0} \tag{5.53}$$

Here, we assume both analogous transmission lines are extended infinitely in both directions at the junction at $z = 0$. We shall be developing the theory of terminated transmission line in which the load is assumed to be absorbing all power is matched perfectly so that $Z_L = Z_0$. Otherwise, the mismatched load absorbs partial incident power and reflects back the rest toward the load. This phenomenon follows the conservation of energy theory as we assume the transmission line is lossless. Otherwise, we also need to consider the power dissipation in the lossy transmission line, which is also analogous to the lossy medium that we have considered in our analysis.

4) For dielectric media, the permeability does not change appreciably from its free space value of $\mu_0 = 4\pi \times 10^{-7}$(H/m). Therefore, we can express the reflection and transmission coefficients in terms of the relative permittivity of the medium as follows:

$$\eta_1 = \sqrt{\left(\frac{\mu_0}{\varepsilon_0 \varepsilon_{r1}}\right)} = \eta_0/\sqrt{\varepsilon_{r1}} \text{ and } \eta_2 = \sqrt{\left(\frac{\mu_0}{\varepsilon_0 \varepsilon_{r2}}\right)} = \eta_0/\sqrt{\varepsilon_{r2}} \tag{5.54}$$

$$\Gamma = \frac{\sqrt{\varepsilon_{r1}} - \sqrt{\varepsilon_{r2}}}{\sqrt{\varepsilon_{r1}} + \sqrt{\varepsilon_{r2}}} \text{ and } \tau = \frac{2\sqrt{\varepsilon_{r1}}}{\sqrt{\varepsilon_{r1}} + \sqrt{\varepsilon_{r2}}} \tag{5.55}$$

where the symbols have their usual meanings.

5) One of the prominent topics of the modern electromagnetics is optical waveguides and fibers. In optical application, the electromagnetic properties of the optical materials, which are also dielectric materials, are expressed in terms of their refractive indices. For the optical materials, the refractive index is defined as $n = \sqrt{\varepsilon_r}$. Substitutions for the definition $n = \sqrt{\varepsilon_r}$ in the above two equations lead to the definitions of the transmission and reflection coefficients as:

$$\Gamma = \frac{n_1 - n_2}{n_1 + n_2}; \tag{5.56}$$

$$\tau = \frac{2n_1}{n_1 + n_2} \tag{5.57}$$

In the following, we shall do a few examples based on the theory on the normal incidence.

5.6.2 Calculation of Electromagnetic Power Density

So far, we have dealt with the electromagnetic fields calculations and derived the reflection and transmission coefficients. The main goal is to know the portion of the real power (time average) is reflected and transmitted through the interface of the two mediums.[33] In an ideal situation, we prefer no power is reflected and the theory of matching comes into play. Some active devices as for example low noise amplifiers need to have controlled mismatch for stability. The most convenient way to calculate the time average power density for three field cases and to prove *the conservation of energy theory* at the interface of the two media, we use a complex Poynting vector and the phasor vector expressions of the three sets of the electromagnetic fields. In the following, we calculate the time average power density vectors for the incident, reflected and transmitted electromagnetic fields.

The time average transmitted power density in medium 2 is:

$$\boldsymbol{P}^2_{\boldsymbol{ave}} = \frac{1}{2} Re\left(\boldsymbol{E}^t_s \times \boldsymbol{H}^{t*}_s\right) = \frac{1}{2} Re\left(\tau E^i_0 e^{-j\beta_1 z} \boldsymbol{a}_x \times \frac{\tau E^i_0}{\eta_2} e^{j\beta_1 z} \boldsymbol{a}_y\right) = \frac{{E^i_0}^2}{2\eta_2} \tau^2 \boldsymbol{a}_z \ (\text{W/m}^2) \tag{5.58}$$

The time average power density $\boldsymbol{P}^1_{\boldsymbol{ave}}$ in medium 1 is the resultant power density carried out by the incident and reflected electromagnetic fields. Therefore, the total electric field in medium 1 is defined as:

$$\boldsymbol{E}^1_s(z) = \boldsymbol{a}_x\left[E^i_0 e^{-j\beta_1 z} + E^r_0 e^{j\beta_1 z}\right] = \boldsymbol{a}_x E^i_0 e^{-j\beta_1 z}\left[1 + \Gamma e^{j2\beta_1 z}\right] \tag{5.59}$$

33 The interface of the two media can be assumed as a thin film membrane that reduces the flow of fluid.

Likewise, the total magnetic field in medium 1 is:

$$\boldsymbol{H}_s^1(z) = \boldsymbol{a}_y\left[\frac{E_0^i}{\eta_1}e^{-j\beta_1 z} - \frac{E_0^i}{\eta_1}e^{j\beta_1 z}\right] = \boldsymbol{a}_y\frac{E_0^i}{\eta_1}e^{-j\beta_1 z}\left[1-\Gamma e^{j2\beta_1 z}\right] \tag{5.60}$$

The time average power density in medium 1 is:

$$\begin{aligned}\boldsymbol{P}_{ave}^1 &= \frac{1}{2}Re\left(\boldsymbol{E}_s^1 \times \boldsymbol{H}_s^{1*}\right) = \frac{1}{2}Re\left(\boldsymbol{a}_x E_0^i e^{-j\beta_1 z}\left[1+\Gamma e^{j2\beta_1 z}\right] \times \boldsymbol{a}_y\frac{E_0^i}{\eta_1}e^{j\beta_1 z}\left[1-\Gamma e^{-j2\beta_1 z}\right]\right)\\ &= \boldsymbol{a}_z\frac{E_0^i}{2\eta_1}\,Re\left(\left[1+\Gamma e^{j2\beta_1 z}\right]\times\left[1-\Gamma e^{-j2\beta_1 z}\right]\right) = \boldsymbol{a}_z\frac{E_0^i}{2\eta_1}\,Re\left(\left(1-\Gamma^2\right)+\Gamma\left[e^{j2\beta_1 z}-e^{-j2\beta_1 z}\right]\right)\\ &= \boldsymbol{a}_z\frac{E_0^i}{2\eta_1}\,Re\left[\left(1-\Gamma^2\right)+j2\Gamma sin(j2\beta_1 z)\right]\end{aligned} \tag{5.61}$$

Therefore, we obtain the time average power density vector in medium 1 is:

$$\boldsymbol{P}_{ave}^1 = \boldsymbol{a}_z\frac{E_0^i}{2\eta_1}\left(1-\Gamma^2\right)\ (\mathrm{W/m^2}) \tag{5.62}$$

The electromagnetic system must follow the conservation of energy theory at the interface of the two media. Therefore, we can write without the loss of generality:

$$\boldsymbol{P}_{ave}^1 = \boldsymbol{P}_{ave}^2$$

This leads to the following relationship:

$$1-\Gamma^2 = \frac{\eta_1}{\eta_2}\tau^2 \tag{5.63}$$

Since Γ is defined as *the voltage reflection coefficient*, Γ^2 is called *the power reflection coefficient* and is usually denoted by $\Gamma_p = \Gamma^2$ in many texts. Likewise, we define: $\tau_p = \tau^2$, *the power transmission coefficient*. Therefore we can relate the power reflection and transmission coefficients via the following expression:

$$1-\Gamma_p = \frac{\eta_1}{\eta_2}\tau_p \tag{5.64}$$

We can make the following remarks for the above analysis:

1) The above power coefficient equations can also be directly and independently obtained from the definitions of the voltage reflection and transmission coefficients, $\Gamma = (\eta_2-\eta_1)/(\eta_2+\eta_1)$ and $\tau = 2\eta_2/(\eta_1+\eta_2)$, respectively. To differentiate between the voltage reflection and transmission coefficients from their power reflection and transmission coefficients, usually a subscript "v" is used with Γ as Γ_v and τ as τ_v in many texts. However, we shall drop the subscript in our calculations.
2) In the above power calculation, we always assume E_0^i etc., real quantities. This is a legitimate guess as any complex number can be incorporated with its phase angel with an exponent with $e^{j\theta}$ factor.
3) From the above time average power density calculations, we can see that the incident power is divided into the transmitted power in medium 2 and the reflected power in medium 1. The signs of the unit vectors of the power densities prove that the reflected power is opposite to the incident power, however, the resultant power in medium 1 is z-directed and transmitted power is

propagating in the same direction of the incident power in medium 2. The power densities $P^i_{ave} - P^r_{ave} = P^t_{ave}$ must follow the conservation of energy theory, and therefore, *the power must be conserved across the interface of the two media.*

Review Questions 5.2: Reflection and Transmission of Uniform Plane Wave

Q1. What are the different forms of field expressions?

Q2. Explain, how waves interact when they encounter the interface of two different media.

Q3. Can you explain what does happen when such oppositely directed two field components vectorially sum up? How do we manipulate vectors in this type of problem?

Q4. When we communicate via mobile phones, incoming and outgoing signals always interact with such discontinuities many times. We receive an NLOS signal when we receive or make a mobilephone call inside a building or inside a car. Can you imagine how many such changes of propagation media happen in these calls? Remember that all are time harmonic EM fields with a designated operating frequency. For GSM mobile phone it is 900 and 1800 MHz. Do you think all signals are normally incident in these multiple reflective cases? What are the conditions for oblique incidence cases?

Q5. See Figure 5.18(a) first and then Figure 5.18(b) very closely and note all parameters, directions of propagation and field vectors with respect to (x, y, z) coordinates.

Q6. Can you write down three sets of field expressions with its appropriate signs and notations? Please practice a few times writing these sets of equations and explain each component.

Q7. Can you write the total electric field using the complete solution of the incident and reflected electric fields as we have learned in wave solution theory?
(Hints: It is the summation of the forward and reverse traveling waves as we derived in the complete electric field solution. The complete instantaneous electric field solution is given in the expression.)

Q8. Can you identify what are the components for forwarding traveling and reverse traveling waves, respectively? Wire down the phasor vector and time domain expressions of electric fields in both lossy and lossless media.

Q9. If the electric field vector is in the direction of x-axis and traveling along z-axis, can you tell what is the polarization and the direction of propagation of the forward traveling wave (shuttle)? Draw a reference upper arrowhead on the feather of the shuttle and mark as the direction of electric field.

Q10. What are the attenuation constant and propagation constant? What are the directions of the forward and returned traveling waves? What are the amplitude of the two waves at $z = 0$. In this case what is $z = 0$ position, the position of the racket of the first player or the opponent player?

Q11. For a perfect conductor, what is the value of a tangential electric field? Can you explain why it is zero?

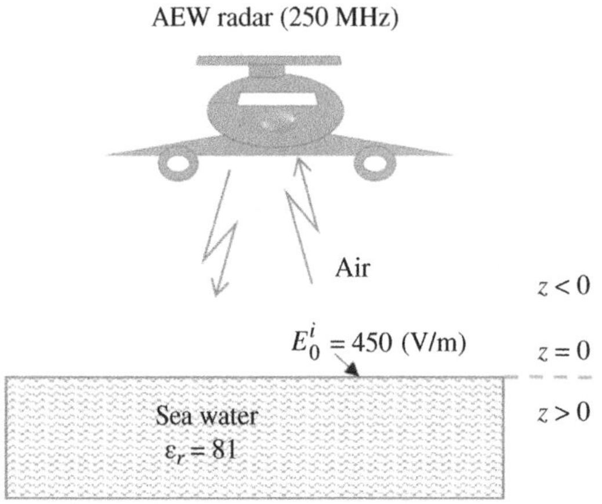

Figure 5.19 AEW radar signal reflected from sea water.

Example 5.4: Normal Incidence and Reflection

An airborne early warning (AEW) radar sends a 250 MHz x-polarized uniform plane wave signal with an amplitude of 450 V/m as shown in Figure 5.19. The wave is normally incident onto sea water with dielectric constant $\varepsilon_r = 81$ ($z > 0$) Calculate: (i) the reflection and transmission coefficients, (ii) the incident electric and magnetic fields, (iii) the reflected electric and magnetic fields, (iv) transmitted electric and magnetic fields, and (v) show that power must be conserved across the interface of the two media. Assume that both media are lossless dielectric media so that $\mu_1 = \mu_2 = \mu_0$.

Answer:

Given: for medium 1 air $\varepsilon_{r1}= 1$; for medium 2 sea water $\varepsilon_{r2} = 81$

i) The reflection coefficient: $\Gamma = \left(\sqrt{\varepsilon_{r1}} - \sqrt{\varepsilon_{r2}}\right)/\left(\sqrt{\varepsilon_{r1}} + \sqrt{\varepsilon_{r2}}\right) = (1-9)/(1+9) = -0.8 = 0.8\angle 180^{\circ}$[34]
   The transmission coefficient: $\tau = 1 + \Gamma = 1 - 0.8 = 0.2$
ii) To perform the calculation of the fields, we need to determine the wave impedances and phase constants of the two media.

$$\eta_1 = \sqrt{\frac{\mu_1}{\varepsilon_1}} = 120\pi\,\Omega; \eta_2 = \sqrt{\frac{\mu_2}{\varepsilon_2}} = \frac{120\pi}{9} = 41.9\,\Omega$$

$$\beta_1 = \omega\sqrt{\mu_1\varepsilon_1} = \frac{\omega}{c_{01}} = \frac{2\pi \times 25 \times 10^7}{3 \times 10^8} = 5.24\,(\text{rad/m})$$

$$\beta_2 = \omega\sqrt{\mu_2\varepsilon_2} = \frac{\omega}{c_{02}} = \frac{2\pi \times 25 \times 10^7 \times 9}{3 \times 10^8} = 47.12\,(\text{rad/m})$$

Using the above two sets of results, we can calculate the incident electric and magnetic fields The incident electric field is:

$$\boldsymbol{E}_s^i(z) = E_0^i e^{-j\beta_1 z}\boldsymbol{a}_x = 450 \times e^{-j5.24z}\boldsymbol{a}_x\,(\text{V/m})$$

The time domain field expression is:

$$\boldsymbol{E}^i(z,t) = Re\left\{E_0^i e^{-j\beta_1 z}e^{j\omega t}\boldsymbol{a}_x\right\} = Re\left\{450 \times e^{-j5.24z} \times e^{j50\pi \times 10^7 t}\boldsymbol{a}_x\right\}$$
$$= 450\cos\left(50\pi \times 10^7 t - 5.24z\right)\boldsymbol{a}_x\,(\text{V/m})$$

The incident magnetic field is:

$$\boldsymbol{H}_s^i(z) = \frac{E_0^i}{\eta_1}e^{-j\beta_1 z}\boldsymbol{a}_y = \frac{450}{377} \times e^{-j5.24z}\boldsymbol{a}_y = 1.19 \times e^{-j5.24z}\boldsymbol{a}_y\,(\text{A/m})$$

34 The calculation can be done using the original formula of $\tau = \dfrac{2\eta_2}{\eta_1 + \eta_2}$ and $\Gamma = \dfrac{\eta_2 - \eta_1}{\eta_2 + \eta_1}$.

The time domain field expression is:

$$H^i(z,t) = Re\left\{1.19 \times e^{-j5.24z} \times e^{j50\pi \times 10^7 t} \boldsymbol{a_y}\right\} = 1.19\cos\left(50\pi \times 10^7 t - 5.24z\right)\boldsymbol{a_y}\ (\mathrm{A/m})$$

iii) The reflected electric field is:

$$\boldsymbol{E_s^r}(z) = \Gamma E_0^i e^{j\beta_1 z}\boldsymbol{a_x} = 450 \times (-0.8) \times e^{j5.24z}(\boldsymbol{a_x}) = -360 \times e^{j5.24z}\boldsymbol{a_x}\ (\mathrm{V/m})$$

The time domain field expression is:

$$\boldsymbol{E}^r(z,t) = Re\left\{E_0^r e^{j\beta_1 z} e^{j\omega t}\boldsymbol{a_x}\right\} = Re\left\{360 \times e^{j5.24z} e^{j50\pi \times 10^7 t} e^{j\pi}\boldsymbol{a_x}\right\}$$
$$= 360\cos\left(50\pi \times 10^7 t + 5.24z + \pi\right)\boldsymbol{a_x}\ \ (\mathrm{V/m})$$

The reflected magnetic field is:

$$\boldsymbol{H_s^r}(z) = \frac{E_0^r}{\eta_1} e^{j\beta_1 z}\left(-\boldsymbol{a_y}\right) = \frac{360}{377} \times e^{j5.24z}\boldsymbol{a_y} = 0.95 \times e^{j5.24z}\boldsymbol{a_y}\ (\mathrm{A/m})$$

The time domain field expression is:

$$\boldsymbol{H}^r(z,t) = Re\left\{0.95 \times e^{j5.24z} e^{j50\pi \times 10^7 t}\boldsymbol{a_y}\right\} = 0.95\cos\left(50\pi \times 10^7 t + 5.24z\right)\boldsymbol{a_y}\ (\mathrm{A/m})$$

iv) The transmitted electric field is:

$$\boldsymbol{E_s^t}(z) = \tau E_0^i\, \mathrm{e}^{-j\beta_2 z}\boldsymbol{a_x} = 0.2 \times 450 \times e^{-j47.12z}\boldsymbol{a_x} = 90e^{-j47.12z}\boldsymbol{a_x}$$

The time domain field expression is:

$$\boldsymbol{E}^t(z,t) = Re\left\{E_0^t e^{-j\beta_2 z} e^{j\omega t}\boldsymbol{a_x}\right\} = Re\left\{90 \times e^{-j47.12z} \times e^{j50\pi \times 10^7 t}\boldsymbol{a_x}\right\}$$
$$= 90\cos\left(50\pi \times 10^7 t - 47.12z\right)\boldsymbol{a_x}\ (\mathrm{V/m})$$

The transmitted magnetic field is:

$$\boldsymbol{H_s^t}(z) = \frac{E_0^t}{\eta_2} e^{-j\beta_2 z}\boldsymbol{a_y} = \frac{90}{41.9} \times e^{-j47.12z}\boldsymbol{a_y} = 2.15 \times e^{-j47.12z}\boldsymbol{a_y}\ (\mathrm{A/m})$$

The time domain field expression is:

$$\boldsymbol{H}^t(z,t) = Re\left\{2.15 \times e^{-j47.12z} \times e^{j50\pi \times 10^7 t}\boldsymbol{a_y}\right\} = 2.15\cos\left(50\pi \times 10^7 t - 47.12z\right)\boldsymbol{a_y}\ (\mathrm{A/m})$$

v) To prove the conservation of energy theory across the interface of the two media, we need to calculate the time average power density for the incident, reflected and transmitted electromagnetic fields.

The time average incident power density is:

$$\boldsymbol{P_{ave}^i} = \frac{1}{2} Re\left(\boldsymbol{E_s^i} \times \boldsymbol{H_s^{i*}}\right) = \frac{1}{2} Re\left(450 \times e^{-j5.24z}\boldsymbol{a_x} \times 1.19 \times e^{j5.24z}\boldsymbol{a_y}\right) = 267.75\boldsymbol{a_z}\ \left(\mathrm{W/m^2}\right)$$

The time average reflected power density is:

$$\boldsymbol{P_{ave}^r} = \frac{1}{2} Re\left(\boldsymbol{E_s^r} \times \boldsymbol{H_s^{r*}}\right) = \frac{1}{2} Re\left(-360 \times e^{j5.24z}\boldsymbol{a_x} \times 0.95 \times e^{-j5.24z}\boldsymbol{a_y}\right) = -171\,\boldsymbol{a_z}\ \left(\mathrm{W/m^2}\right)$$

The time average transmitted power density is:

$$\boldsymbol{P}_{ave}^{t} = \frac{1}{2} Re\left(\boldsymbol{E}_{s}^{t} \times \boldsymbol{H}_{s}^{t*}\right) = \frac{1}{2} Re\left(90e^{-j47.12z}\boldsymbol{a}_{x} \times 2.15 \times e^{j47.12z}\boldsymbol{a}_{y}\right) = 96.75\boldsymbol{a}_{z}\ (\mathrm{W/m^2})$$

From the above three-time average power density calculations, we can see that the incident field is divided into the transmitted power in the water and the reflected power in the air. The signs of the unit vectors of the power densities proved that the reflected power is opposite to the incident power and the transmitted power is propagating in the same direction of the incident power in sea water. Equating the power densities $\boldsymbol{P}_{ave}^{i} = \boldsymbol{P}_{ave}^{r} + \boldsymbol{P}_{ave}^{t}$ we can see that the power must be conserved across the interface of the two media. This proves the conservation of energy theory across the interface of the two media.

In the following example, we generalize the case for normal incidence assuming the sea water is lossy. The loss factor can be included into the example considering the finite conductivity of sea water of 5 (S/m).[35] AT 250 MHz $\frac{\sigma}{\omega\varepsilon} = 7.04$; hence sea water is not a good conductor. Then we need to recalculate the above example using the loss factor $e^{\mp\alpha z}$ in the above example. Let us solve the following example.

Example 5.5: Reflection from Lossy Medium

An airborne early warning (AEW) radar sends a 250 MHz x-polarized uniform plane wave signal with an amplitude of 450 V/m, as shown in Figure 5.20. The wave is normally incident onto sea water with dielectric constant $\boldsymbol{\varepsilon_r} = 81$ and conductivity $\boldsymbol{\sigma} = 5$ (S/m) ($z > 0$) Calculate: (i) the reflection and transmission coefficients, (ii) the incident electric and magnetic fields, (iii) the reflected electric and magnetic fields, (iv) the transmitted electric and magnetic fields, and (v) the time average power transmitted through sea water. Assume sea water is a lossy dielectric media so that $\mu_2 = \mu_0$.

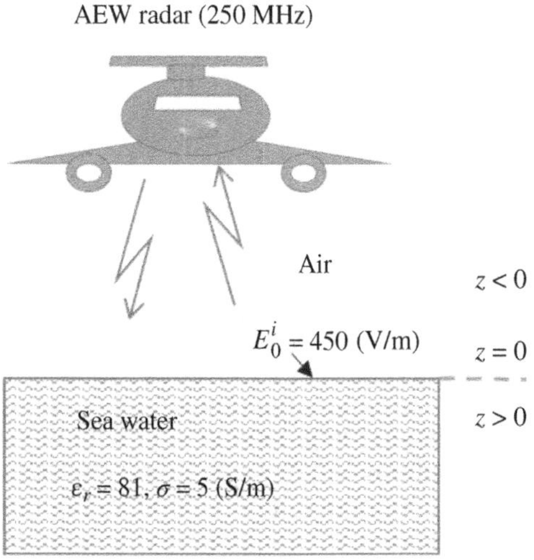

Figure 5.20 AEW radar signal reflected from sea water as a lossy medium.

Answer:

Since sea water is considered to be lossy, we shall first calculate the attenuation constant α in (Np/m), the propagation constant β in (radian/m), and the complex wave impedance of sea water as follows:

$$\alpha_2 = \omega\sqrt{\frac{\mu_2\varepsilon_2}{2}\left(\sqrt{1 + \left(\frac{\sigma_2}{\omega\epsilon_2}\right)^2} - 1\right)}$$

35 See *Conductivity (electrolytic)* in Wikipedia for materials about the conductivity of seawater and other electrolytes source: https://en.wikipedia.org/wiki/Conductivity_(electrolytic), (accessed 15 March 2018)

$$\beta_2 = \omega\sqrt{\frac{\mu_2\varepsilon_2}{2}\left(\sqrt{1+\left(\frac{\sigma_2}{\omega\epsilon_2}\right)^2}+1\right)}$$

We have just calculated at 250 MHz $\sigma_2/\omega\varepsilon_2 = 5/(2\pi\times 25\times 10^7\times 81\times 5.584\times 10^{-12}) = 7.04$.
Let us calculate $\mu_2\varepsilon_2/2 = (4\pi\times 10^{-7}\times 81\times 8.854\times 10^{-12})/2 = 45\times 10^{-16}$
This makes our calculation simple.

$$\alpha_2 = 2\pi\times 25\times 10^7\sqrt{4.5\times 10^{-16}\left(\sqrt{1+7.04^2}-1\right)} = 82.37\ (\mathrm{Np/m})$$

$$\beta_2 = 2\pi\times 25\times 10^7\sqrt{4.5\times 10^{-16}\left(\sqrt{1+7.04^2}+1\right)} = 94.9\ (\mathrm{rad/m})$$

The wave impedance $\eta_2 = \sqrt{(j\omega\mu)/(\sigma + j\omega\varepsilon)} = 54.99 + j386.97 = 391\angle 41°\Omega$
Given: for medium 1 air $\eta_1 = 377\ \Omega$; for medium 2 sea water $\eta_2 = 54.99 + j386.97\ \Omega$

i) The reflection coefficient $\Gamma = (\eta_2 - \eta_1)/(\eta_2 + \eta_1) = (377 - 54.99 - j386.97)/(377 + 54.99 + j386.97) = -0.032 - j0.867 = 0.868\angle -92.1°$
   The transmission coefficient: $\tau = (2\eta_2)/(\eta_1 + \eta_2) = (2(54.99 + j386.97))/(377 + 54.99 + j386.97) = 0.516 + j0434 = 0.673\angle 40° = 0.673e^{j0.7}$
ii) To perform the calculation of the fields we can use the wave impedances and phase constants of the two media that we have already calculated.
   $\eta_1 = 120\pi\ \Omega; \eta_2 = 19.77\angle 41°\Omega, \beta_1 = 5.24\ (\mathrm{rad/m}), \beta_2 = 60.88\ (\mathrm{rad/m})$
   Using the above two sets of results, we can calculate the incident electric and magnetic fields.
   The incident electric and magnetic fields remain the same as the medium 1 (air) remains the same as before. The incident electric field is:

$$\boldsymbol{E}_s^i(z) = 450\times e^{-j5.24z}\boldsymbol{a_x}\ (\mathrm{V/m})$$

   The time domain field expression is:

$$\boldsymbol{E}^i(z,t) = 450\cos\left(50\pi\times 10^7 t - 5.24z\right)\boldsymbol{a_x}\ (\mathrm{V/m})$$

   The incident magnetic field is:

$$\boldsymbol{H}_s^i(z) = 1.19\times e^{-j5.24z}\boldsymbol{a_y}\ (\mathrm{A/m})$$

   The time domain field expression is:

$$\boldsymbol{H}^i(z,t) = 1.19\cos\left(50\pi\times 10^7 t - 5.24z\right)\boldsymbol{a_y}\ (\mathrm{A/m})$$

iii) The reflected fields have major changes due to the complex reflection coefficient. The reflected electric field is[36]:

$$\boldsymbol{E}_s^r(z) = \Gamma E_0^i e^{j\beta_1 z}\boldsymbol{a_x} = 450\times\left(0.868\,e^{-j1.58}\right)\times e^{j5.24z}(\boldsymbol{a_x}) = 390.6\times e^{j(5.24z-1.59)}\boldsymbol{a_x}$$

36 To be consistent we can convert 41° into radian 0.72 rad.

The time domain field expression is:

$$\boldsymbol{E}^r(z, t) = Re\left\{E_0^r e^{j\beta_1 z} e^{j\omega t}\boldsymbol{a_x}\right\} = Re\left\{390.6 \times e^{j(5.24z - 1.59)}\boldsymbol{a_x}\right\}$$
$$= 390.6 \cos\left(50\pi \times 10^7 t + 5.24z - 1.59\right)\boldsymbol{a_x} \text{ (V/m)}$$

The reflected magnetic field is:

$$\boldsymbol{H}_s^r(z) = \frac{E_0^r}{\eta_1} e^{j\beta_1 z}\left(-\boldsymbol{a_y}\right) = \frac{390.6}{377} \times e^{j(5.24z - 1.59)}\boldsymbol{a_y} = 1.036 \times e^{j(5.24z - 1.59)}\boldsymbol{a_y} \text{ (A/m)}$$

The time domain field expression is:

$$\boldsymbol{H}^r(z, t) = Re\left\{1.036 \times e^{j(5.24z - 1.59)}\boldsymbol{a_y}\right\} = 1.036 \cos\left(50\pi \times 10^7 t + 5.24z - 1.59\right)\boldsymbol{a_y} \text{ (A/m)}$$

iv) The transmitted electric field is:

$$\boldsymbol{E}_s^t(z) = \tau E_0^i\, e^{-j\beta_2 z}\boldsymbol{a_x} = 0.673 e^{j0.7} \times 450 \times e^{-j60.88z}\boldsymbol{a_x} = 302.85 e^{-j(60.88z - 0.7)}\boldsymbol{a_x} \text{ (V/m)}$$

The time domain field expression is:

$$\boldsymbol{E}^t(z, t) = Re\left\{E_0^t e^{-j\beta_2 z} e^{j\omega t}\boldsymbol{a_x}\right\} = Re\left\{302.85 \times e^{-j(60.88z - 0.7)} \times e^{j50\pi \times 10^7 t}\boldsymbol{a_x}\right\}$$
$$= 302.85 \cos\left(50\pi \times 10^7 t - 60.88z + 0.7\right)\boldsymbol{a_x} \text{ (V/m)}$$

The transmitted magnetic field is:

$$\boldsymbol{H}_s^t(z) = \frac{E_0^t}{\eta_2}\, e^{-j\beta_2 z}\boldsymbol{a_y} = \frac{302.85}{19.77\angle 41^\circ} \times e^{-j60.88z}\boldsymbol{a_y} = 2.15 \times e^{-j47.12z}\boldsymbol{a_y} \text{ (A/m)}$$

The time domain field expression is:

$$\boldsymbol{H}^t(z, t) = Re\left\{2.15 \times e^{-j47.12z} \times e^{j50\pi \times 10^7 t}\boldsymbol{a_y}\right\} = 2.15 \cos\left(50\pi \times 10^7 t - 47.12z\right)\boldsymbol{a_y} \text{ (A/m)}$$

v) The time average transmitted power density is:

$$\boldsymbol{P}_{ave}^t = \frac{1}{2} Re\left(\boldsymbol{E}_s^t \times \boldsymbol{H}_s^{t*}\right) = \frac{1}{2} Re\left(90 e^{-j47.12z}\boldsymbol{a_x} \times 2.15 \times e^{j47.12z}\boldsymbol{a_y}\right) = 96.75\boldsymbol{a_z} \ \left(\text{W/m}^2\right)$$

5.7 Concept of Standing Waves

Standing wave theory is one of the most important topics in the electromagnetic wave propagation. It has huge industrial applications. Therefore, understanding standing wave theory using fundamental field analysis develops solid foundations and touches many areas of applied electromagnetism. Figure 5.21 depicts different analytical methods of the standing waves. They are: (i) trigonometric analysis[37] of the resultant field that provides the basic understating definitions of the voltage wave ratio (VSWR), the electric field maximum $E_{\max}$ and the electric field minimum $E_{\min}$, and the standing wave patterns, (ii) time domain analysis reinforces our understanding of the conservation of energy

37 M. A. Islam, *Electromagnetic Theorey*, EPUET, Dacca, East Pakistan, 1969.

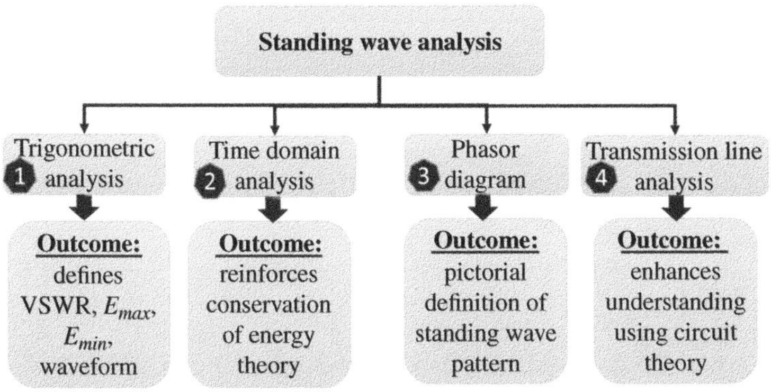

Figure 5.21 Various analytical methods of standing wave theory.

theory and illustrates the incident field energy as the summation of the transmitted and reflected field wave energy,[38] (iii) phasor vector analysis with diagram provides the pictorial definition of the standing wave pattern, and (iv) finally, the transmission line analogy provides us with the understanding in terms of the circuit theory. In the following, we shall examine each method and develop the fundamental concept of the standing wave.

Remember, electric and magnetic fields are orthogonal to each other and propagate along the direction of the Poynting vector in a uniform plane wave. In the analysis of electromagnetic fields at the air-conductor interface, we observed total reflection of the incident wave and the creation of *pure standing wave*[39] in medium 1. A similar phenomenon of the standing wave pattern in medium 1 is also observed at a dielectric-dielectric interface, although a portion of the incident power is transmitted through the second medium. This means we can say that when two opposite field components vectorially mix together in any medium, they create a tiebreaker situation and the wave does not propagate instantaneously after they mix up. This mixing of two waves creates a *standing wave* and the waveform of the standing wave pattern depends on the reflection coefficient.[40] The creation of the standing wave pattern in an electromagnetic wave is analogous to the disturbance created on a water reservoir, as shown in Figure 5.22. The difference between the electromagnetic wave and the water wave is that the electromagnetic wave travels with a speed of light and the reflected and the incident field waves coexist at the interface of the two media to satisfy the boundary conditions derived from Maxwell's equations. On the other hand, the velocity of water wave is too slow and the creation of a standing wave pattern on a calm reservoir takes time until they mix together after the incident wave hits the edge of the reservoir. Thus, we can easily see the progression of the incident wave and then the creation of the standing wave.

If you throw a stone at the center of a calm reservoir, that creates a disturbance. The water wave starts propagating energy toward the edge of the reservoir. The disturbance is analogous to an oscillator or a source. The forward traveling wave is equivalent to the incident wave $\boldsymbol{E}_s^i$. The waveform of

38 U. S. Inan and A.S. Inan, *Electromagnetic Waves*, Prentice Hall, Upper Saddle River, NJ, 2000.

39 The waveform of a standing wave is called standing wave pattern.

40 Surely for matched condition the reflection coefficient is zero and no reflected signal is created and the wave continues to propagate until it hits any discontinuity. For stealth bombers, the composite materials and the shape of the plane are made such that almost zero reflection for the direction of propagation. Thus the plane becomes almost invisible to military radars.

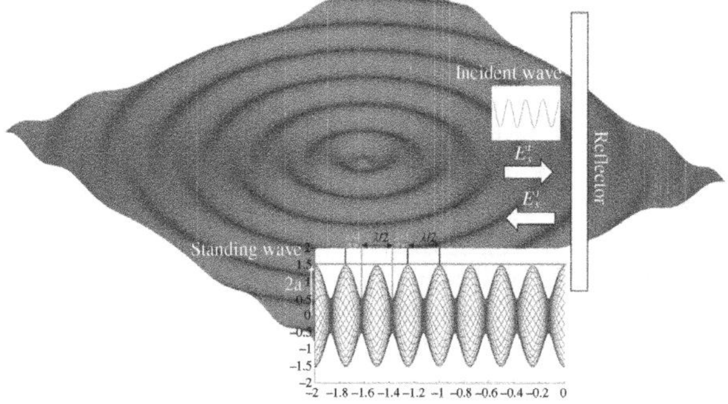

Figure 5.22 Formation of standing wave in a calm reservoir with a disturbance created by a stone or a drop of water. The propagating wave's successive crests/troughs are full wavelength away. Once the incident wave hits the boundary and the reflected wave mixes with the incident wave they create stationary rings as shown in the figure. This is the formation of standing wave pattern. In a standing wave, the successive crests/troughs are half wavelength away. Depending on the amount of reflection (nature of reflector) the location of the first maximum and minimum from the reflector changes. Measuring the distance, we can calculate the wavelength, and from the amplitudes of crests and troughs we can measure reflection coefficient and unknown load for a terminated transmission line.

the incident wave is shown in the top inset figure with its amplitude, crests, troughs, and wavelength. Once the incident wave reaches the edge and is bounced back by the reflector/wall as shown, instantly the forward traveling wave and the reflected or reverse traveling wave start superimposing vectorially and *the standing wave pattern* starts taking shape as shown in the figure. Ultimately when the superposition of the forward and reverse traveling waves is complete, they create stationary concentric rings of water with successive maxima and minima. This waveform of the standing wave is shown in the bottom inset figure. If you closely observe the maxima and minima of the forwarding incident wave in the top inset figure, and that for the standing wave in the bottom inset figure, you can easily observe that the distance of the successive crest/trough of the concentric wrinkles for the standing wave pattern is half that for the forward traveling/incident wave. The "unmixed" forward traveling wave is equivalent to a pure sine wave (similar to a propagating wave that you can see traveling forward from the center toward the shore). The successive maxima or minima of the wave is called the "*wavelength*" of the original wave. Likewise, the reflected wave is also a traveling wave in the opposite direction of the incident wave. However, the standing wave is also a sinusoidal wave, but the distance between the successive maxima or minima of the wave is half the original wavelength. You may assume that the maxima is analogous to *the wave front* of the electromagnetic wave that we define in the theory of the uniform plane wave. Now we are going to analyze the electromagnetic fields at the interface of the two media and develop the theory of the standing wave.

The ratio between the amplitudes of the maxima and minima of the voltage (electric field in V/m) waveform is called VSWR. VSWR is a very significant definition for industrial applications of RF/microwave devices. VSWR varies from 1 to ∞.[41] The industry standard is maximum 2. If you go

---

41 VSWR becomes 1 when the signal is perfectly matched and the two media have the same constitutive parameters. VSWR becomes infinity when total reflection happens such as dielectric–conductor interface. The reflection coefficient is 0 for the perfect matched condition meaning the two media are the same and ± 1 when they encounter total reflection from a conducting wall. In transmission line analogy, the reflection coefficient is defined as $\Gamma = \dfrac{Z_{02} - Z_{01}}{Z_{02} + Z_{01}}$. In the matched condition, at the junction of two transmission line must satisfy the condition $Z_{01} = Z_{02}$, where Z_{01} and Z_{02} are the characteristic impedance of the two lines, and the reflection coefficient is zero. For short circuit the reflection coefficient is -1 and for open circuit condition it is 1.

beyond 2, more than 10% incident energy is reflected back and devices do not meet the standard and need tuning. This tuning aspect we shall learn in the matching theory of the transmission line.

As we have discussed the superposition of the complex incident and reflected waves in medium 1 occurs when the incident wave encounters a discontinuity. We have discussed this phenomenon qualitatively using the analogy of the disturbance of a water reservoir above. We also have observed the formation of a standing wave pattern in medium 1 due to the superposition of incident and reflected traveling waves. Now we shall examine the standing wave phenomenon in more detail with mathematical analyses as depicted in Figure 5.22 in electromagnetics perspectives.

5.7.1 Trigonometric Analysis of Standing Wave

For simplicity, let us assume an ideal medium without any loss, so that $e^{\mp \alpha z} = 1$. The phasor vector expression of the electric field, which is the summation of the incident and transmitted fields in medium 1 is:

$$\boldsymbol{E}_{sx}^{1}(z) = \left(E_{o}^{i} e^{-j\beta_1 z} + E_{0}^{r} e^{j\beta_1 z}\right)\boldsymbol{a_x} = E_{o}^{i} e^{-j\beta_1 z}\left(1 + \Gamma e^{j2\beta_1 z}\right)\boldsymbol{a_x} \tag{5.65}$$

where $\Gamma = E_0^r / E_0^i$.[42] Depending on the value of Γ, the resultant field in medium 1 takes different forms. For example, if there is no reflection ($\Gamma = 0$) of the incident field at the interface of the two media, the field conditions are: $E_0^r = 0; E_0^i = E_0^t$, total transmission occurs in the second medium. The outcome is a pure traveling sinusoidal wave with the time-space phase factor ($\omega t - \beta_1 z$). On the other hand, if ($\Gamma = \mp 1$),[43] in total reflection the field conditions are: $E_0^i = \mp E_0^r$; $E_0^t = 0$, then a pure standing wave is formed in the medium 1. In general cases for the dielectric-dielectric interface of dissimilar materials, ($E_0^i \neq E_0^r$; $E_0^t \neq 0$), we always expect partial transmission and reflection of signals occurs in the normal incidence case.[44] In this expression, we consider E_0^i, E_0^r and E_0^t to be real. There is of course no restriction to consider that they are to be real because E_0^i and E_0^r can be made real by properly choosing the origins of the spatial coordinate z and time coordinate t. We can write the resultant electromagnetic fields components in medium 1 in scalar form without loss of generality as follows:

$$E_s^1 = E_o^i e^{-j\beta_1 z} + E_o^r e^{j\beta_1 z} \tag{5.66}$$

$$H_s^1 = \frac{E_o^i}{\eta_1} e^{-j\beta_1 z} - \frac{E_o^r}{\eta_1} e^{j\beta_1 z} \tag{5.67}$$

We shall work only on the electric field expression in (5.66) as per the convention all calculations are made based on the electric field components.[45] Using Euler's theorem

$$e^{\pm j\theta} = \cos\theta \pm j\sin\theta \tag{5.68}$$

42 In this instance, we can assume both E_0^i and E_0^r are real quantities hence Γ is also a real quantity.

43 For total reflection the sign changes based on the nature of the reflectors, it is better depicted with transmission line analogy where the equivalent transmission line encounters an open circuit condition so that $\Gamma = +1$, contrary to the short circuit condition when $\Gamma = -1$ as if the plane wave hits a perfect conducting boundary.

44 Total internal reflection like this can also occur in oblique incidence cases that will be discussed in the TE and TM oblique incidence theory.

45 This calculation creates confusion when the reflection coefficient is positive ($\Gamma > 0$) and that makes the transmission coefficient $\tau = 1 + \Gamma > 1$. This does not mean power amplification at the interface of the passive media. The magnetic field $H = E/\eta$ will reduces proportionally to maintain the theory of energy conservation.

A complex string of expression comprising the incident and reflected fields in *cos* and *sin* functions

$$E_s^1 = E_o^i\{\cos(\beta_1 z) - j\sin(\beta_1 z)\} + E_o^r\{\cos(\beta_1 z) + j\sin(\beta_1 z)\} \tag{5.69}$$

Rearranging the real and imaginary parts yields:

$$E_s^1 = (E_o^i + E_o^r)\cos(\beta_1 z) - j(E_o^i - E_o^r)\sin(\beta_1 z) = |E_1|\, e^{j\theta} \tag{5.70}$$

Also, we can observe that the polar form expression with amplitude and phase of the resultant field is $|E_1|\, e^{j\theta}$. Where, the amplitude is

$$\begin{aligned} |E_1| &= \sqrt{(E_o^i + E_o^r)^2 \cos^2(\beta_1 z) + (E_o^i - E_o^r)^2 \sin^2(\beta_1 z)} \\ &= \sqrt{(E_o^i)^2 + (E_o^r)^2 + 2E_o^i E_o^r \cos(2\beta_1 z)} \end{aligned} \tag{5.71}$$

And the phase angle is:

$$\theta = \tan^{-1}\left\{\frac{E_o^i - E_o^r}{E_o^i + E_o^r}\tan^2(\beta_1 z)\right\} = \tan^{1}\left\{\frac{E_{min}}{E_{max}}\tan^2(\beta_1 z)\right\}$$

where we define $E_{min} = E_o^i - E_o^r$ and $E_{max} = E_o^i + E_o^r$.

The total electric field in lossless medium 1 is then also defined as:

$$|E_1| = \sqrt{E_{max}^2 \cos^2(\beta_1 z) + E_{min}^2 \sin^2(\beta_1 z)} \tag{5.72}$$

In expressions (5.72), the spatial phase factor or electrical length $2\beta_1 z$, which is double in the standing wave pattern. $tan^2(\beta_1 z)$ tells the same phase doubling effect in the standing wave pattern. Now we plot the resultant field $|E_1|$ versus $\beta_1 z$ to understand the characteristics of the standing wave pattern as shown in Figure 5.23.

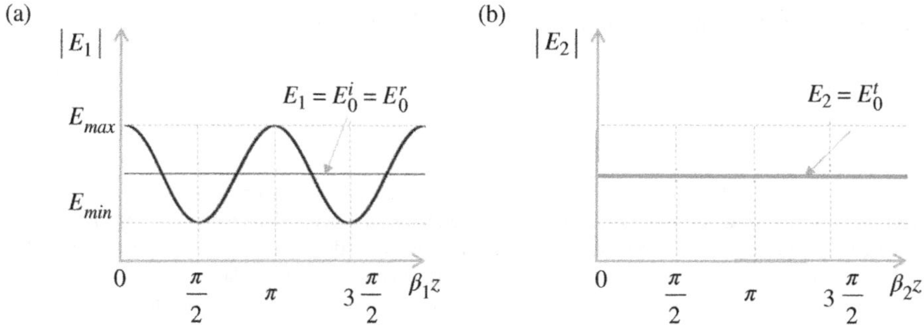

Figure 5.23 Standing wave pattern in frequency in spatial domain ($\omega t = 0$) (a) resultant field and standing wave pattern in medium 1, and (b) standing wave pattern in medium 2.

The plot of the resultant field $|E_1|$ as a function of $\beta_1 z$ is called the "*standing wave pattern*" or *the envelope* of the resultant field $|E_1|$ in medium 1. A typical plot is shown in the figure. The most significant part of the plot is that from the knowledge (both amplitudes and their locations) of the maximum and minimum values of $|E_1|$ we can determine the value of the reflected field amplitude or reflection coefficient $|\Gamma|$ and the location of the maximum and minimum from the reflecting boundary at $z = 0$. This theory is most useful in a terminated transmission line that is covered in transmission line theory.

From the plot, it is apparent that for a pure traveling wave, $|E_1|$ must be equal to either $|E_i|$ or $|E_r|$. This means the standing wave pattern is a straight line as shown in the figure. On the other hand, if $|E_i| = |E_r|$, a *pure standing wave pattern,* it is of the form of $|cos(\beta z)|$, which is similar to the output of a full-wave rectifier. Generally we are interested to know the standing wave pattern. The most important parameter we are interested in is the ratio of the E_{max} to E_{min}. This ratio is called VSWR or in short s as in (5.37)

$$VSWR = S = \frac{E_{max}}{E_{min}} = \frac{E_o^i + E_o^r}{E_o^i - E_o^r} \tag{5.73}$$

If E_i and E_r are not real we just take the modulus of them so that $S = (1 + |\Gamma|)/(1 - |\Gamma|)$. We can also exchange the relationship so that we can determine the reflection coefficient once the VSWR is known as follows:

$$|\Gamma| = \frac{S-1}{S+1} \tag{5.74}$$

Figure 5.24 shows the computer-generated plot of the standing wave patterns at different time for $\Gamma = 0.5$ over two wavelengths. They are plotted at 20° increments for ωt ranges from 0 to 360°. For no reflection $\Gamma = 0$, then VSWR is 1. For total reflection, $\Gamma = \pm 1$, and therefore, the value of VSWR is $\propto$. In general, for a good conductor $\Gamma = -1$ analogous to a short circuit at the interface of two media and $VSWR = \propto$.

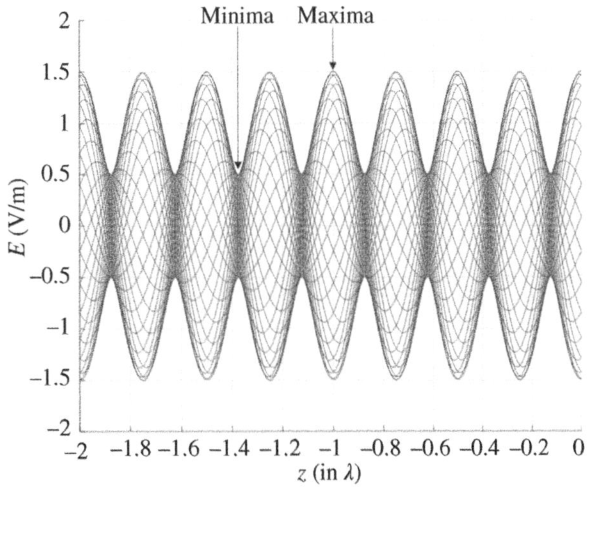

Figure 5.24 Standing wave pattern in time domain with finite reflection coefficient.

Review Questions 5.3: Standing Waves

Q1. Can you explain what does happen when such oppositely directed two field components vectorially sum up? How do we manipulate vectors in this type of problem?
Q2. Calculate the power transmission coefficient τ_P? Do you think simply taking the squared of τ_v gives you τ_p?
Q3. Would you like any power returned back to your transmitter if you want to transmit full incident power from a section to its adjacent section of a system? Justify with reasoning.
Q4. What is the power reflection coefficient of your system?

Example 5.6: Standing Wave Theory: Trigonometric Analysis
The 30 mV/m rms electric field component of a plane wave propagating in free-space in region $z < 0$ is incident on a loss-less dielectric space in region $z > 0$ with a relative dielectric constant is $\varepsilon_r = 4$. Calculate the followings:

i) The time domain expression of the electric field in exponential form
ii) The impedances of the free space and the dielectric field region
iii) The transmission and reflection coefficients and VSWR
iv) The power reflection and transmission coefficients and the percentage of transmitted power into the dielectric region
v) The rms values of the transmitted electric and magnetic fields in the dielectric region
vi) The average power density in the dielectric region
vii) If the frequency is 1 GHz frequency, calculate the wavelength.

Answer:

i) $\boldsymbol{E}^i(z, t) = 30\sqrt{2} \times 10^{-3} e^{j(\omega t - \beta_1 z)} \boldsymbol{a_x}$ (V/m)

ii) $\eta_1 = \eta_0 = \sqrt{\dfrac{\mu_0}{\varepsilon_0}} = 120\pi\ \Omega = 377\ \Omega$

$$\eta_2 = \sqrt{\frac{\mu_0\mu_r}{\varepsilon_0\varepsilon_r}} = \sqrt{\frac{\mu_0}{4\varepsilon_0}} = \frac{1}{2}\sqrt{\frac{\mu_0}{\varepsilon_0}} = \frac{\eta_0}{2} = 188.5\ \Omega$$

iii) $\Gamma = \dfrac{\eta - \eta_0}{\eta + \eta_0} = \dfrac{\eta_0/2 - \eta_0}{\eta_0/2 + \eta_0} = \dfrac{-1/2}{3/2} = -\dfrac{1}{3}$; $\tau = 1 + \Gamma = 1 - \dfrac{1}{3} = \dfrac{2}{3}$; $S = \dfrac{1 + 1/3}{1 - 1/3} = 2.$

iv) $\Gamma_p = \Gamma^2 = \dfrac{1}{9}$; $\tau_p = (1 - \Gamma_p)\dfrac{\eta_2}{\eta_1} = \dfrac{8}{9} \times \dfrac{188.5}{377} = 0.44$;

$$1 - |\Gamma|^2 = 1 - \left(\frac{1}{3}\right)^2 = \frac{8}{9} = 88.9\%$$

This represents with VSWR 2, an electromagnetic system transfers 90% or the incident energy to the second medium.

v) $E_0^i = 30\sqrt{2} \times 10^{-3}\,\text{V/m}; \tau = 1 + \Gamma = 1 - \frac{1}{3} = \frac{2}{3}$

$$E_0^t = \tau E_0^i = \frac{2 \times 30\sqrt{2} \times 10^{-3}}{3 \times \sqrt{2}} = 20 \times 10^{-3}\ (\text{V/m})\ (\text{RMS})$$

$$H_0^t = \frac{E_0^t}{\eta} = \frac{20 \times 10^{-3}}{188.5} = 0.1061 \times 10^{-3}\ (\text{A/m})\ (\text{RMS})$$

vi) $P_{ave}^t = \frac{1}{2}|E_0^t||H_0^t| = \frac{1}{2}\frac{|E_0^t|^2}{\eta}$ (if we use normal magnitude)

$$P_{ave}^t = |E_0^t||H_0^t| = \frac{|E_0^t|^2}{\eta} = 20 \times 10^{-3} \times 0.1061 \times 10^{-3}\ (\text{RMS magnitude})$$

$$= 2.122 \times 10^{-6}\ (\text{W/m}^2)$$

vii) Since the above calculations do not involve the frequency (ω) hence the values remain the same.

5.7.2 Time Domain Analysis of Standing Wave

In the preceding section, we have defined the formation of the standing wave pattern from the derivation of the resultant electric field $|E_1|$ in the medium 1. In this section, we shall examine the time domain analysis of standing wave pattern and understand its behavior. When a uniform plane wave is incident normally at the interface of two dielectric media, a portion of the signal propagates through the second medium as the transmitted wave and the rest is reflected back toward the first medium as the reflected wave. In the development of the theory of the normal incidence at the interface of two media, we also have observed that the reflected wave is superimposed on the incident wave and forms a standing wave pattern in medium 1. The conservation of energy law must be obeyed in the process of the incident wave to be converted to the transmitted wave and reflected wave and formation of the standing wave. To determine the above conservation of energy theory, we take the phasor expression of the resultant wave in medium 1 as before and then covert it to the time domain expression. We shall find that the wave has two components: (i) a traveling wave that sustains the transmitted wave in the second medium, and (ii) the standing wave. The resultant scalar electric field in medium 1 is expressed as:

$$E_s^1(z) = \left(E_o^i e^{-j\beta_1 z} + E_0^r e^{j\beta_1 z}\right) = E_o^i\left(e^{-j\beta_1 z} + \Gamma e^{j\beta_1 z}\right) \tag{5.75}$$

We can expand the above expression as follows[46]:

$$E_s^1(z) = E_o^i\{(1 + \Gamma)e^{-j\beta_1 z} - \Gamma e^{-j\beta_1 z} + \Gamma e^{j\beta_1 z}\} \tag{5.76}$$

Substitutions for $\tau = 1 + \Gamma$ and $-\,e^{-j\beta_1 z} + e^{j\beta_1 z} = j2\sin(\beta_1 z)$, and simple mathematical manipulations yield:

$$E_s^1(z) = E_o^i\{\tau e^{-j\beta_1 z} + j2\Gamma sin(\beta_1 z)\} \tag{5.77}$$

46 Similar derivation is also available in U.S. Inan and A.S. Inan, *Electromagnetic Waves*, Prentice Hall, Upper Saddle River, NJ, 2000.

Substitution for $E_o^t = \tau E_o^i$, the corresponding time domain expression of the field is:

$$E_1(z,t) = E_o^t \cos(\omega t - \beta_1 z) + (-2\Gamma)E_o^i \sin(\beta_1 z)\sin(\omega t) \quad (5.78)$$

The first term in the right hand side of the above equation $\{E_o^t \cos(\omega t - \beta_1 z)\}$ represents a propagating wave in medium 1 with spatial phase factor $(\beta_1 z)$ with an amplitude $E_o^t = \tau E_o^i$ that is traveling toward the forward direction (in the positive z-direction) to support the transmitted wave in the second medium. The second term $\{-2\Gamma E_o^i \sin(\beta_1 z)\sin(\omega t)\}$ represents a standing wave pattern that is formed by the superposition of the incident and reflected waves with a spatial factor $\{2\Gamma sin(\beta_1 z)\}$. For completeness we also derive the resultant magnetic field in medium 1 using the wave impedance definition $\eta_1 = E_0^i/H_0^i = -E_0^r/H_0^r$ as follows:

$$H_s^1(z) = \frac{E_0^i}{\eta_1}\left(e^{-j\beta_1 z} - e^{j\beta_1 z}\right) = \frac{E_0^i}{\eta_1} e^{-j\beta_1 z}\left(1 - \Gamma e^{j2\beta_1 z}\right) \quad (5.79)$$

Similar substitution of $\tau = 1 + \Gamma$ and simple mathematical manipulation yields the following time domain expression for the resultant magnetic field:

$$H_1(z,t) = \frac{E_o^i}{\eta_1}\{\tau\cos(\omega t - \beta_1 z) + (-2\Gamma)\cos(\beta_1 z)\cos(\omega t)\} \quad (5.80)$$

Similar to the electric field case, we can also conclude that the resultant magnetic field in medium 1 is composed of the forward traveling wave and the standing wave. In medium 2, only one traveling wave propagating as the transmitted wave and expressed as follows:

$$E_s^t(z) = E_0^t e^{-j\beta_2 z} = \tau E_0^i e^{-j\beta_2 z} \quad (5.81)$$

$$H_s^t(z) = H_0^t e^{-j\beta_2 z} = \frac{\tau E_0^i}{\eta_2} e^{-j\beta_2 z} \quad (5.82)$$

The time domain expressions for the transmitted field components are:

$$E_t(z) = \tau E_0^i \cos(\omega t - \beta_2 z) \quad (5.83)$$

$$H_{t(z)} = \frac{\tau E_0^i}{\eta_2}\cos(\omega t - \beta_2 z) \quad (5.84)$$

Figure 5.25 depicts the waveforms of the time domain electromagnetic fields for the electromagnetic system of the two media. In reference to the example of the air–water interface in the preceding section, we observe that the transmission coefficient is larger than 1; hence the transmitted electric field is assumed to be amplified with the factor of $\tau = 1.2$ for $\varepsilon_1 = 1$ (air) and $\varepsilon_2 = 81$ (water). This does not violate the conservation of energy law.[47] In this regard, we must consider the Poynting vector in the perspective medium. The time average incident power must be the summation of the transmitted and reflected power. As shown in Figure 5.25 we take snapshots for cases (a–d) of the time domain electric field components, say at $t = 0$. In this case, we assume both media are lossless ($\sigma_1 = \sigma_2 = 0$) dielectric medium ($\mu_1 = \mu_2 = \mu_0$). As can be seen in Figure 5.25(a) when the plane wave travels from the lighter medium to the denser medium (for $\varepsilon_1 < \varepsilon_2$), the reflection coefficient, $\Gamma = (\eta_2 - \eta_1)/\eta_1 + \eta_2 < 0$, is negative as $\eta_1 > \eta_2$ and the transmission coefficient $\tau = 1 + \Gamma = (2\eta_2)/(\eta_1 + \eta_2) < 1$.

---

47 As in the circuit theory, considering a fixed power is sent for a source via the circuit element, if the voltage increases, then the current must decrease to maintain the conservation of energy law. Likewise, if there is an increase in the amplitude of the electric field due to the boundary conditions of dissimilar dielectric materials then there must be a diminishing magnetic field to conserve the energy.

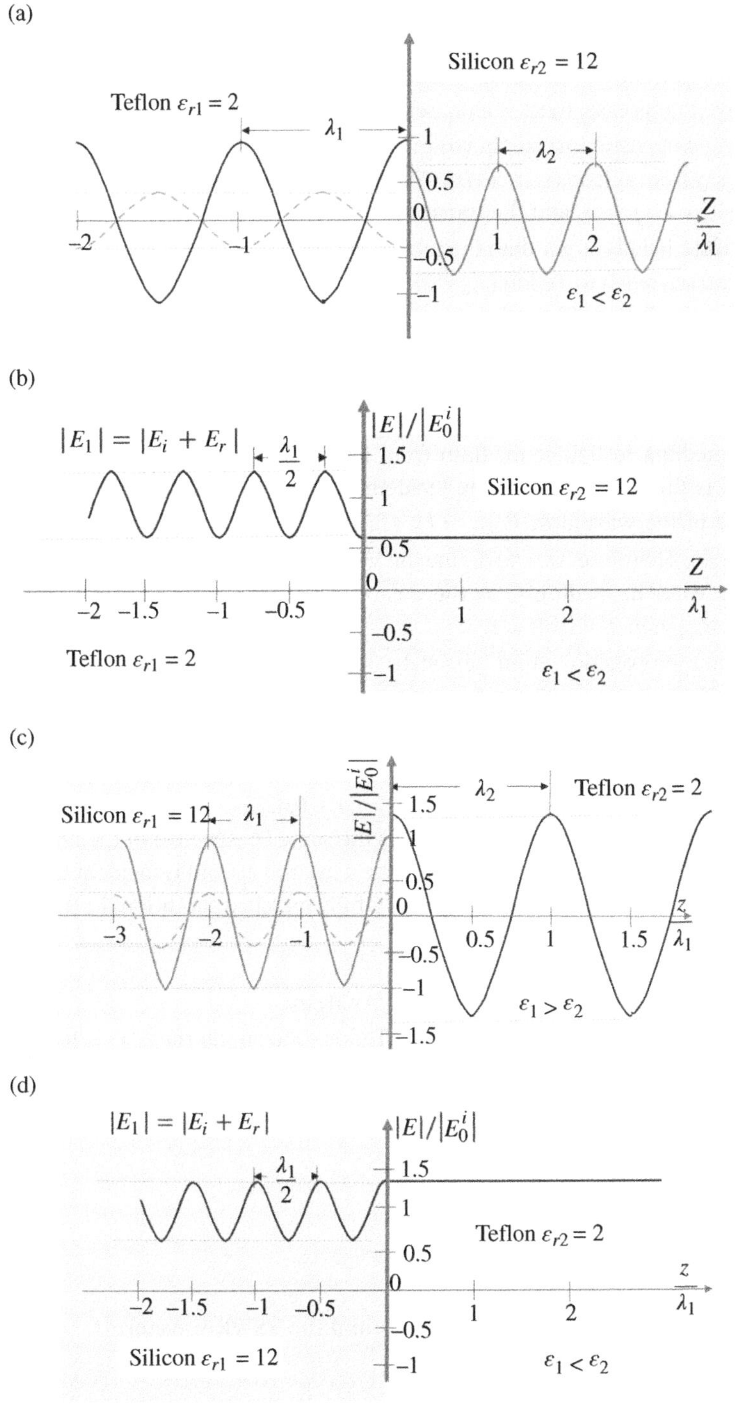

Figure 5.25 The incident electric field mixes with the reflected field and forms a standing wave pattern with fixed locations for crests and troughs with half the wavelength distance $\lambda_1/2$ between the two crests or troughs and the transmitted signal in the second medium is a propagating wave with fill wavelength distance λ_2.

The transmitted electric field is smaller than the incident electric field. This is depicted, as for example, for media with a Teflon ($\varepsilon_{r1} \approx 2$) and silicon ($\varepsilon_{r2} \approx 12$) interface in Figure 5.25(a). The wavelengths of the two media also differ significantly with denser medium has shorter wavelength. Since the reflection coefficient is negative in this case, there is a phase reversal of the reflected field, as shown in the figure ($E_0^r = \Gamma E_0^i$). For the opposite case as shown in Figure 5.25(c), when the wave is traveling from a denser medium, say from silicon ($\varepsilon_{r1} \approx 12$) to a lighter medium, say Teflon ($\varepsilon_{r2} \approx 2$), the reflection coefficient Γ is positive as $\eta_1 < \eta_2$ and the transmission coefficient $\tau = 1 + \Gamma = (2\eta_2)/(\eta_1 + \eta_2) > 1$. The transmitted electric field is larger than the incident electric field ($E_0^t = \tau E_0^i$).[48] This is depicted, for media with silicon ($\varepsilon_{r1} \approx 12$) to Teflon ($\varepsilon_{r2} \approx 2$) interface again. The wavelength of the silicon is smaller than that for the Teflon. Since the reflection coefficient is positive in this case, there is no phase reversal of the reflected field, as shown in Figure 5.25(c). The Poynting vector for both cases is defined as $P_{ave}^t = 1/(2\eta_2)|E_0^t|^2$. Figure 5.25(b) depicts the standing wave patterns for Figure 5.25(a) for wave traveling from lighter medium to denser medium and Figure 5.25(d) for wave traveling from denser medium to lighter medium from Figure 5.25(c). As can be seen, in medium 1 the resultant field is the summation of the incident and reflected fields $|E_1| = |E_i + E_r|$ with the maximum value of the waveform $E_{max} = (1 + |\Gamma|)E_0^i$ and the minimum value of the waveform $E_{min} = (1 - |\Gamma|)E_0^i$. Therefore, we can define the VSWR of the standing wave pattern. Whereas there is no standing wave in medium 2 as there is only one transmitted wave in that medium. In this case $E_{max} = E_{min}$ hence $VSWR = 1$.

For the above example, wave propagation from silicon to Teflon: $\Gamma = \left(\sqrt{\varepsilon_{r1}} - \sqrt{\varepsilon_{r2}}\right)/\left(\sqrt{\varepsilon_{r1}} + \sqrt{\varepsilon_{r2}}\right) = \left(\sqrt{2} - \sqrt{12}\right)/\left(\sqrt{2} + \sqrt{12}\right) = -0.42$. The transmission coefficient is: $\tau = 1 - 0.42 = 0.58$. However, for wave propagation from Teflon to silicon, $\Gamma = 0.42$, $\tau = 1 + 0.42 = 1.42$. Irrespective of the direction of wave propagation from denser and lighter media, the voltage standing ratio is defined as: $VSWR = (|E_{max}|)/(|E_{min}|) = (1 + |\Gamma|)/(1 - |\Gamma|) = (1 + 0.42)/(1 - 0.41) \cong 2.45$. This means, by definition the VSWR is always positive and remains the same irrespective of the reflection coefficient changes its sign depending on the direction of the wave travels and is incident upon the medium. Finally, the Value of VSWR ranges from 1 to ∞. For fully matched condition, $VSWR = 1$ and for total reflection $VSWR = \infty$.

Example 5.7: Voltage Standing Wave Theory: Time Domain Analysis

A y-polarized 100 mV/m uniform plane wave is incident on a Teflon–Silicon interface. The dielectric constant of Teflon is 2.1 and that for Silicon is 11.7. The direction of propagation is along z-direction and the operating frequency is 24 GHz. Calculate the followings:

i) The time domain expression of the incident electric field

ii) The phasor vector expression of the incident electric field

iii) The transmission and reflection coefficients, and VSWR

iv) Draw the time domain waveform of the individual signals and the VSWR waveform

v) The percentage of power reflected back to medium 1

Answer:

The phase constant is: $\beta_1 = \dfrac{2\pi \times 24 \times 10^9 \times \sqrt{2.1}}{3 \times 10^8} = 232\pi\ (\text{rad/m})$

---

48 This voltage gain of the transmitted wave appears to be an alarming violation of the conservation of energy law, basically, it is not the case when we consider the energy balance theory from the previous example; $P_{ave}^i = P_{ave}^r + P_{ave}^t$ must be maintained all the time.

i) The time domain expression of the incident electric field is:

$$\boldsymbol{E}^i(z,t) = 100 \times 10^{-3} \cos\left(48\pi \times 10^9 t - 232\pi z\right)\boldsymbol{a_y} \text{ (V/m)}$$

ii) The phasor vector expression of the incident electric field is:

$$\boldsymbol{E}_s^i(z) = 100 \times 10^{-3} \times e^{232\pi z}\boldsymbol{a_y} \text{ (V/m)}$$

iii) The wave impedance of Teflon is:

$$\eta_1 = \sqrt{\frac{\mu_0}{\varepsilon_0\varepsilon_r}} = 120\pi/\sqrt{2.1} = 260.15\,\Omega$$

$$\eta_2 = 120\pi/\sqrt{11.7} = 110.21\,\Omega$$

iv) The reflection and transmission coefficients:

$$\Gamma = \frac{\eta_2 - \eta_1}{\eta_2 + \eta_1} = \frac{110.21 - 260.15}{110.21 + 260.15} = -0.4; \tau = 1 + \Gamma = 0.6$$

The VSWR: $S = (1 + 0.4)/(1 - 0.4) = 2.33$

v) See Figure 5.25 for the plot. You need to do minor adjustment for the values of the field quantities.

$$\%P_r = \Gamma^2 \times 100\% = 16\%$$

5.7.3 Phasor Vector Analysis of Standing Wave

In the previous analysis, we have used the time domain expressions of the electromagnetic fields and have depicted the sinusoidal waveforms of the incident, transmitted, and reflected waves to understand their interactions due to the presence of a discontinuity in the propagation dielectric medium. This analysis is a bit complex and may raise difficulty in the computation of the voltage (electric field) maximum and minimum points from the interface of the two media. As an aid to understand the characteristic of the standing wave pattern and calculate the values of the maximum and minimum electric fields, a simpler mathematical representation of the standing wave pattern is presented using the phasor vector analysis in this section. The analysis starts with the phasor vector representation of the resultant field wave that is x-polarized and propagating along the positive z-direction, all are happening in medium 1, as follows:

$$\boldsymbol{E}_{sx}^1(z) = E_0^i e^{-j\beta_1 z}\left(1 + \Gamma e^{j2\beta_1 z}\right)\boldsymbol{a_x} \tag{5.85}$$

The expression has two components, the incident electric field $E_0^i e^{-j\beta_1 z}\boldsymbol{a_x}$ and a complex factor $\left(1 + \Gamma e^{j2\beta_1 z}\right)$ that represent the normalized resultant standing wave electric field in medium 1 with respect to the incident field. In this section, we pictorially analyze the normalized resultant field and understand the characteristics of the field vector in more elegant way with the phasor vector diagram in the complex phasor vector plane. The normalized resultant field in a scalar phasor form is expressed as:

$$E^1_{sn}(z) = 1 + \Gamma e^{j2\beta_1 z} \quad (5.86)$$

In this expression, the reflection coefficient can be a complex quantity with amplitude and phase and can be written as:

$$\Gamma = |\Gamma| e^{j\theta} \quad (5.87)$$

where, θ is the phase angle of the reflection coefficient Γ. Using the complex reflection coefficient, the normalized field expression can be written as:

$$E^1_{sn}(z) = 1 + |\Gamma| e^{j(2\beta_1 z + \theta)} \quad \text{for } -\infty < z \leq 0 \quad (5.88)$$

as the medium 1 is assumed to be of semi-infinite extend and located in $z < 0$ region, therefore, we can use the spatial condition for the electromagnetic system as $(-\infty < z \leq 0)$. Figure 5.26a plots the normalized phasor field in the complex plane with real and imaginary axes. The two components are: 1 is extend from the origin (0, 0) on the horizontal axis and the polar form of the phasor component $|\Gamma| e^{j(2\beta_1 z + \theta)}$ with an arrow of length $|\Gamma|$ and variable phase term of $(2\beta_1 z + \theta)$.[49] As can be seen the normalized resultant field is the resultant vector starting at origin (0, 0) in the complex plane and changes its position tracing a circle with the radius of $|\Gamma|$ at the center (1, 0). The instantaneous point of the tip of the normalized electric field vector depends on the phase angle $\psi = 2\beta_1 z + \theta$. For fixed media 1 and 2, the reflection coefficient is constant with a fixed phase angle θ. Therefore, as we move along the negative z-direction in medium 1, the phasor vector of the standing wave field traces the circle in the clockwise direction (to meet the convention of the angle as z is increasing negatively). The standing wave pattern is also shown in Figure 5.26b. As can be seen in the figure, the maxima of the electric field occur when the normalized resultant field vector has amplitude $1 + |\Gamma|$ and lies on the horizontal axis (real axis). At this point, the phase angle $\psi = 2\beta_1 z + \theta = -2m\pi$ for $m = 0, 1, 2$, etc.[50] and the amplitude of the normalized resultant field is $E_{max} = 1 + |\Gamma|$. Since $\beta_1 = 2\pi/\lambda_1$, the maxima repeats every $z = -m(\lambda_1/2)$ distance. For nonzero phase of the reflection coefficient, the first maximum (for $m = 0$) is located at $z' = -(\theta/2\beta_1)$. Therefore, the locations of the maxima of the resultant electric field can be expressed as:

$$|E_1|_{max} = |E^i_o| (1 + |\Gamma|) \quad \text{for } z = -m\frac{\lambda_1}{2} - \frac{\theta}{2\beta_1}; \ m = 0, 1, 2, \text{etc.} \quad (5.89)$$

The normalized electric field maximum points are shown in Figure 5.26b. Sometimes it is hard to access the first maximum point due to the load conditions in a transmission line and waveguide.

---

49 For the real value of reflection coefficient Γ is real and $\theta = 0$. For special conditions, the reflection coefficient can be real such as lossless media, so that the intrinsic impedance η is real, the fields are set so that the phase angle is zero. The transmission line equivalent of the real reflection coefficient occurs when the transmission line is lossless so that the characteristic impedance of the line Z_0 is real and/or the load impedance is real. For complex load impedance the reflection coefficient is complex and the phase angle must be included in the complex reflection coefficient.

50 Since z is increasing negatively, we assume the phase angle ψ is also increasing negatively as we move away from the interface of the two media.

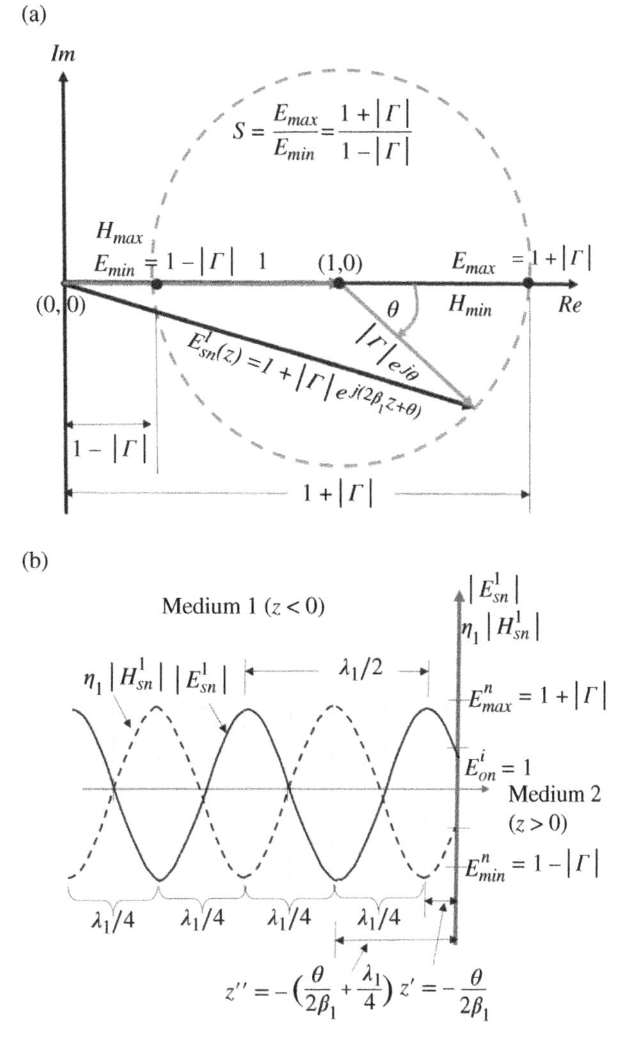

Figure 5.26 Normalized resultant electric field in medium 1 (a) phase representation, and (b) waveforms of electromagnetic fields in medium 1.

Knowing the frequency of operation viz the wavelength and the accessible nearest maximum point from the interface (discontinuity) and the field maximum E_{max}, and the field minimum E_{min} we can determine the unknown VSWR and reflection coefficient.

Now let us examine the minimum point of the normalized resultant field. As can be seen from the phasor diagram, the minimum occurs when the amplitude of the field is real and touches the circle on the left hand side of the horizontal axis. The normalized resultant field vector has amplitude $1-|\Gamma|$ and lies on the horizontal axis (real axis). At this point the phase angle $\psi = 2\beta_1 z + \theta = -(2+1)m\pi$ for $m = 0, 1, 2$ etc.[51] and the amplitude of the normalized resultant field is $E_{min} = 1-|\Gamma|$. Since $\beta_1 = (2\pi/\lambda_1)$, the maxima repeats every $z = -(2m+1)(\lambda_1/2)$ distance. For the nonzero phase of the reflection coefficient, the first minimum (for $m = 0$) is located at $z'' = -((\theta/2\beta_1)+(\lambda_1/4))$.

51 Since z is increasing negatively, we assume the phase angle ψ is also increasing negatively as we move away from the interface of the two media.

This is shown in Figure 5.26b. Therefore, the minima of the resultant electric field repeat every $z = -(2m+1)(\lambda_1/2)$ with the first minimum located at $z' = -(\theta/2\beta_1)$.

Now we can examine what is happening to the magnetic field. The magnetic field in medium 1 is defined as:

$$H_s^1(z) = \frac{E_0^i}{\eta_1} e^{-j\beta_1 z}\left(1 - \Gamma e^{j2\beta_1 z}\right) \tag{5.90}$$

And the normalized resultant magnetic field is defined as:

$$H_{sn}^1(z) = 1 - \Gamma e^{j2\beta_1 z} \tag{5.91}$$

The minus sign in the reflected field component tells that when the electric field has a maximum, then the magnetic field has a minimum and vice versa. This phenomenon is shown in Figures 5.26a, b.

A few remarks can be made from the above analysis:

1) For the real value of reflection coefficient Γ, $\theta = 0$. In this condition, the maximum locates at the interface of two media ($z = 0$).
2) For the dielectric and conductor interface, the first minimum of the electric field is at $z = 0$.
3) For $\theta < 0$ negative, the first maximum locates at $z'' = -(\theta/(2\beta_1) + (\lambda_1)/4)$ from the interface of two media ($z = 0$).
4) Knowing the location of the maximum and/or minimum points in a physical electromagnetic system, we can determine the unknown reflection coefficient and the characteristics of the discontinuities. The discontinuity could be a medium of unknown constitutive parameters or a termination of a transmission line and a waveguide. Therefore, as stated before also, the knowledge of standing wave patterns and the locations of the maximum and minimum with the ratio of the field maximum and minimum help us solve many practical design problems.

Figure 5.27 shows the practical measurement set up of VSWR in an open slit of a coaxial transmission line. The difference is the guided nature of the transmission line. However, the coaxial cable supports a transverse electromagnetic (TEM) wave, which is a plane wave. Therefore, the theory of normal incidence at the interface of two dielectric media is equally applicable to this situation. The transmission line segment can be assumed to be the medium 1 (usually coaxial cable is filled with a Teflon dielectric with relative permittivity $\varepsilon_r = 2.1$) and the second medium is assumed to be the load. The matched load can be another section of the transmission line of semi-infinite length or a 50 Ω load. This is called *the matched condition* and the transmitted signal is exactly the same as the incident signal with zero reflection. In this situation, the VSWR is 1, as shown in the figure. When the short termination is used, then load impedance $Z_L = 0\,\Omega$ and the reflection coefficient is $\Gamma = -1$. The electric field (V/m) is the minimum ($V_{S/C} = 0\,\text{V}$). The magnetic field (A/m) is the maximum. This situation is shown in Figure 5.27b. For short circuit condition, the VSWR if ∞. For the open circuit, the load impedance $Z_L = \infty\,\Omega$. In this case, the reflection coefficient is 1 and *VSWR* ∞. For unknown load impedance, the reflection coefficient has the complex form with its amplitude and phase $\Gamma = |\Gamma|e^{j\theta}$. Therefore, based on the positive and negative phase angle of the reflection coefficient, the standing wave patterns start at $z = 0$. In the current example, the phase angle is negative. Reading the value of E_{max} and E_{min} (equivalent to V_{max} and V_{min}, respectively)

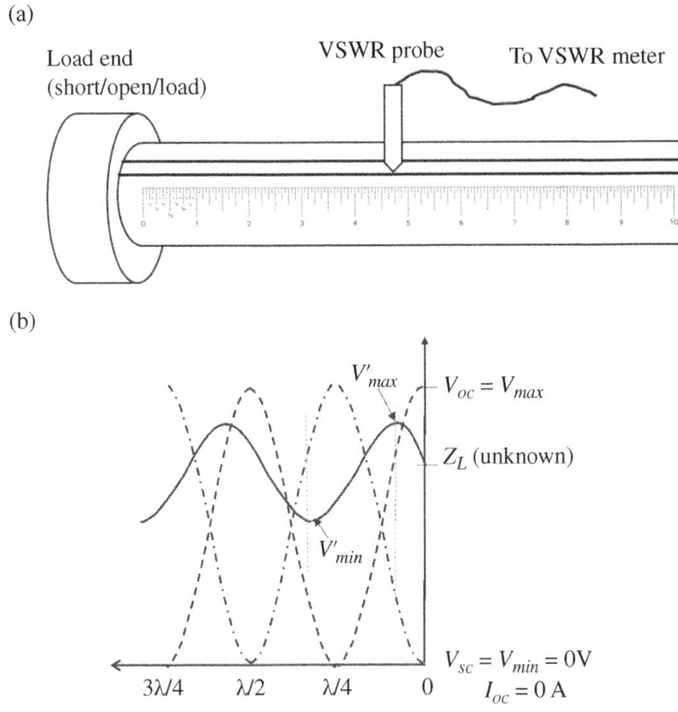

Figure 5.27 (a) A coaxial cable based VSWR measurement system, and (b) waveforms at different load conditions.

from the VSWR meters and their location from the scale on the coaxial cable, we can calculate the VSWR, the locations of E_{max} and E_{min} hence the wavelength of the operating frequency in the dielectric medium. An electric field probe with a pointed head is used to measure the waveform, as shown in Figure 5.27b at different load conditions.

Finally, the analogy of the phasor vector representation is shown with the fly wheel of a steam engine as shown in Figure 5.28. Although we cannot find exact similarities, we can create some analogy of the phasor vector representation of the standing wave pattern with the fly wheel as follows:

i) The end point of the piston cylinder on the left hand side is the origin (0, 0) and the length of the piston cylinder is 1 that is the equivalent amplitude of the normalized incident electric field.
ii) The length and angle of the piston rod as assumed to be the complex reflection coefficient $\Gamma = |\Gamma|e^{j\theta}$. This is the moving part and exerts force to rotate the wheel around its axle. The rotation distance on the ground is the traveling distance z.
iii) The resultant distance from the origin to end of the piston rod is the resultant distance $1 + |\Gamma|e^{j\theta}$ which is the normalized resultant electric field in medium 1.
iv) The voltage (electric field) maximum and minimum points are shown in the figure on the right most and left most points on the horizontal plane of the wheel.

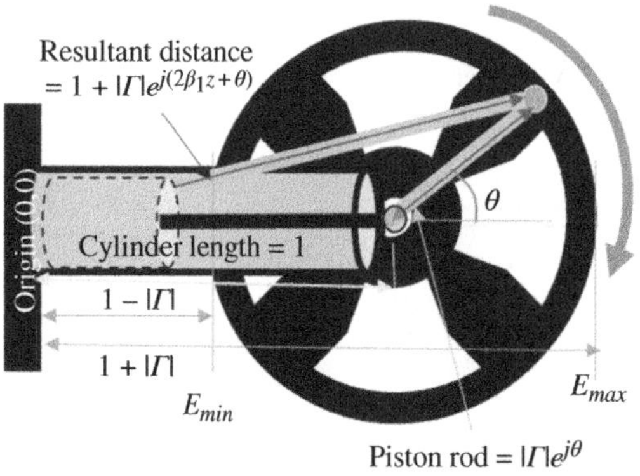

Figure 5.28 Analogy of the phasor vector representation of the standing wave is shown with the locomotive engine's fly wheel assembly.

Review Questions 5.4: Standing Wave Theory: Phasor Vector Representation

Q1. Define reflection and transmission coefficients at the interface of a dielectric and conducting media

Q2. What is standing wave pattern? Draw the standing wave pattern for an incident wave impinges upon a perfect conductor.

Q3. Show that the normalized maximum and minimum electric field intensities are defined as $E^n_{min} = 1 - |\Gamma|$ and $E^n_{max} = 1 + |\Gamma|$.

Q4. Proof that the resultant electric field in medium 1 is defined as:

$$|E_1| = \sqrt{E^2_{max}\cos^2(\beta_1 z) + E^2_{min}\sin^2(\beta_1 z)}$$

Q5. Explain why VSWR is unity in medium 2.

Q6. Explain how standing wave can be measured in a laboratory setting.

Q7. Explain how the unknown impedance can be calculated by the measurement of the standing wave pattern and locations of voltage maximum and minimum.

Q8. When the electric field is the maximum the magnetic field is the minimum. Explain why?

Example 5.8: Standing Wave Theory: Phasor Vector Representation

A base station antenna transmits a y-polarized 450 V/m uniform plane wave via an antenna which is covered with a Teflon–silicon interface at 900 MHz. The dielectric constants are $\varepsilon_r = 2.1$ for Teflon and $\varepsilon_r = 11.7$ for silicon. Assume both the media are semi-infinitely extended and lossless. Calculate the followings:

i) The phasor vector expression of the resultant electric field for the wave propagation along z-direction.

ii) The maximum and minimum electric fields of the standing wave.

iii) Now, consider the fly wheel as shown in the analogy with respect to the phasor vector diagram of the standing wave pattern. Consider the length is equivalent to the electric field intensity (V/m). This means the piston length is 1, equivalent to the normalized amplitude of the incident electric field. If we consider that 1 unit is equivalent to 1 m, then calculate the length of the piston rod, the wheel diameter (assume the piston rod is connected at the end of the wheel's rim) with a rotary joint, the length of the resultant distance from the origin to the end of the piston rod when z is 1 m.

Answer:

The phase constant of medium 1 (Teflon) is: $\beta_1 = (2\pi\times 9\times 10^8)/\left(3\times\left(10^8/\sqrt{2.1}\right)\right) = 27.32(\text{rad/m})$;

i) The phasor vector expression of the incident electric field is:

$$\boldsymbol{E}^{i}_{sy}(z) = 450\times e^{-j27.32z}\boldsymbol{a}_y\ (\text{V/m})$$

From the previous calculation, the reflection coefficient is: $\Gamma = -0.4 = 0.4e^{-j\pi}$; $\therefore \theta = -\pi$ (rad), The phasor vector expression of the resultant electric field is:

$$\boldsymbol{E}^{1}_{sy}(z) = E^{i}_{o}e^{-j\beta_1 z}\left(1+\Gamma e^{j2\beta_1 z}\right)\boldsymbol{a}_y = 450\times e^{-j27.32z}\left(1-0.4e^{j54.64z}\right)\boldsymbol{a}_y\ (\text{V/m})$$

ii) The amplitude of the maximum electric field is: $E_{max} = 450\times(1+0.4) = 630$ (V/m)
The amplitude of the minimum electric field is: $E_{min} = 450\times(1-0.4) = 270$ (V/m)

iii) The length of the piston rod is $|\Gamma| = 0.4$. The diameter is: $E_{max} - E_{min} = 2|\Gamma| = 0.8$ (m)
At $z = 1$ m, the electrical length is: $\theta' = 2\beta_1 z = 54.64$ (rad); $\psi = 2\beta_1 z + \theta = 57.78$ (rad)
The resultant distance is:

$$|R_1| = \sqrt{(1+|\Gamma|)^2\cos^2(\psi) + (1-|\Gamma|)^2\sin^2(\psi)}$$
$$= \sqrt{(1+0.4)^2\cos^2(57.78) + (1-0.4)^2\sin^2(57.78)} = 0.832\ (m)$$

5.7.4 Transmission Line Analogy of Normal Incidence

Similar to the electromagnetic system with the dielectric-conductor interface in the preceding section, we can develop the transmission line analogy for the dielectric–dielectric interface. The transmission line analogy is a simplistic method of systematic analysis of the multiple-layer dielectric interfaces. This method we shall be using in the shielding analysis for electromagnetic compatibility (EMC) procedure.[52] In defining the transmission line analogy of the dielectric–dielectric boundary conditions, we can define the equivalent transmission line impedance for the standing wave in the medium 1 by the ratio of the electric and the magnetic fields at any point in space at an arbitrary distance z as follows:

$$Z_1(z) = \frac{|E_1(z)|}{|H_1(z)|} = \frac{E^{i}_{o}e^{-j\beta_1 z} + E^{r}_{0}e^{j\beta_1 z}}{\left((E^{i}_{0})/(\eta_1)\right)\left(e^{-j\beta_1 z} - e^{j\beta_1 z}\right)} = \eta_1\frac{1+\Gamma e^{j2\beta_1 z}}{1-\Gamma e^{j2\beta_1 z}} \tag{5.92}$$

52 In the EMI/EMC theory the transmission line model is used for *the shielding effectiveness* measurement. Shielding effectiveness is basically a measure of the percentage of electromagnetic energy that can penetrate through the multilayer media shielding. This forward power transmission coefficient in the ratio of incident to transmitted power is measured in isolation in dB, usually a large positive value.

For the case of a lossy medium, η_1 becomes complex. We can generalize the definition of the equivalent transmission line impedance by replacing $j\beta z$ with γz for any lossy medium[53] as:

$$Z_1(z) = \eta_1 \frac{1 + \Gamma e^{2\gamma_1 z}}{1 - \Gamma e^{2\gamma_1 z}} \tag{5.93}$$

The impedance at any point $z = -l$ is called *the input impedance* of the transmission line of length l which is analogous of the first medium of thickness l and the impedance is due to the loading effect of the second medium. Therefore, in this aspect of the transmission line analogy, the impedance of the second medium η_2 is also the characteristic impedance of the second transmission line of infinite extend that is joined with the first line at $z = 0$. This is *the load impedance* of the equivalent transmission line that represents the medium 1. Therefore, the equivalent *input impedance* of a lossless transmission line is defined as:

$$Z_1(-l) = \eta_1 \frac{e^{j\beta_1 l} + ((\eta_2 - \eta_1)/(\eta_2 + \eta_1))e^{-j\beta_1 l}}{e^{j\beta_1 l} - ((\eta_2 - \eta_1)/(\eta_2 + \eta_1))\, e^{-j\beta_1 l}} \tag{5.94}$$

Using Euler's theorem $e^{j\beta_1 l} + e^{-j\beta_1 l} = 2\cos(\beta_1 l)$ and $e^{j\beta_1 l} - e^{-j\beta_1 l} = 2\sin(\beta_1 l)$ and simple mathematical manipulation with trigonometric identities, we obtain the input impedance at any point in medium 1 as:

$$Z_{in} = \eta_1 \frac{\eta_2 + j\eta_1 \tan(\beta_1 l)}{\eta_1 + j\eta_2 \tan(\beta_1 l)} \tag{5.95}$$

Likewise, we can define the wave impedance of medium 2 where only the transmitted wave is propagating infinitely along the positive z-direction as follows:

$$Z_{02}(z) = \frac{|E_2(z)|}{|H_2(z)|} = \frac{E_0^t e^{-j\beta_2 z}}{(E_0^t/\eta_2)e^{-j\beta_2 z}} = \eta_2 \tag{5.96}$$

Likewise, we can also generalized the above definition of $Z_{02}(z)$ for the lossy medium 2 as follows:

$$Z_{02}(z) = \frac{|E_2(z)|}{|H_2(z)|} = \frac{E_0^t e^{-\gamma_2 z}}{(E_0^t/\eta_2)e^{-\gamma_2 z}} = \eta_2 \tag{5.97}$$

As stated above, η_2 becomes complex for the lossy second medium. From the above two definitions, we can define the reflection coefficient and transmission coefficient, respectively, in terms of the equivalent characteristic impedances of the two transmission lines as follows:

$$\Gamma = \frac{Z_{02} - Z_{01}}{Z_{02} + Z_{01}}; \quad \text{and } \tau = \frac{2Z_{02}}{Z_{01} + Z_{02}} \tag{5.98}$$

Substitution for $\eta_1 = Z_{01} = Z_0$ and $\eta_2 = Z_{02} = Z_L$ we obtain the generalized input impedance definition as follow:

$$Z_{in} = Z_0 \frac{Z_L + jZ_0 \tan(\beta l)}{Z_0 + jZ_L \tan(\beta l)} \tag{5.99}$$

Figure 5.29 depicts the electromagnetic system and the equivalent transmission line model of the dielectric-dielectric interface. Figure 5.29a shows the pure electromagnetic system with

53 For lossy medium not only the propagation constant $\gamma = \alpha + j\beta$ is complex but also the intrinsic impedance of both media are complex and as a consequence the reflection coefficient Γ and the transmission coefficient τ are also complex.

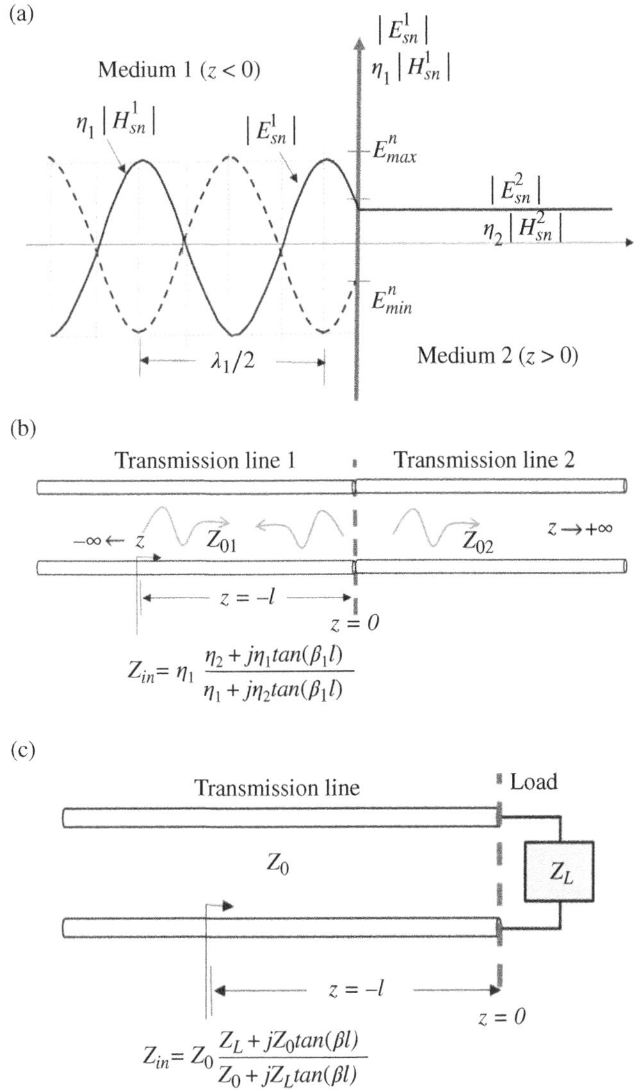

Figure 5.29 (a) The standing wave pattern of at the interface of two media, (b) transmission line analogy with two transmission lines with impedances Z_{01} and Z_{02} extended semi-infinitely in opposite directions along z-axis from the junction at $z = 0$, and (c) the equivalent transmission line circuit model of the two mediums.

the propagation medium in unbounded condition with the two intrinsic impedances η_1 and η_2 meeting at the interface at $z = 0$ with its standing wave patterns in both mediums. Figure 5.29b depicts the semi-infinitely extended equivalent transmission lines with characteristic impedances Z_{01} and Z_{02}, for the media 1 and 2, respectively, joined at $z = 0$. As can be seen in the transmission line analogy in Figure 5.29b, both lines can be assumed to be semi-infinitely extended at both directions at the junction point $z = 0$. Both lines have their respective characteristic impedance defined by the above expressions. Figure 5.29c shows the pure transmission line equivalent circuit model of the two media showing the loading effect of the medium 2 as the equivalent load impedance Z_L.

If the medium is lossy then $j\beta$ will be replaced with γl and the definition of the input impedance of the equivalent lossy transmission line will be generalized as:

$$Z_{in} = Z_0 \frac{Z_L + Z_0\, tanh\,(\gamma l)}{Z_0 + Z_L\, tanh\,(\gamma l)} \tag{5.100}$$

Remarks

We can add the following remarks from the above derivation of the transmission line analogy of the dielectric–dielectric electromagnetic system:

1) The transmission line analogy of the electromagnetic system provides a straight forward method to calculate the reflection and transmission coefficients. The reflection and transmission coefficients are defined as $\Gamma = (Z_{02} - Z_{01})/(Z_{02} + Z_{01})$; and $\tau = (2z_{02})/(Z_{01} + Z_{02})$, respectively.
2) We can have a much simpler closed form expression for the input impedance of the medium 1. Close observation of the expression of the input impedance reveals that the impedance changes both amplitude and phase as it moves along the z-direction. The input impedance repeats the same complex value every $\lambda/2$ distance as we move along the z-direction. This phenomenon is similar to the standing wave patterns of the electric and magnetic fields in medium 1.
3) In the analysis, we must assume both media are semi-infinitely extended so that we can avoid any other loading effects besides the media themselves.

Review Questions 5.5: Standing Wave Patten Theory: Transmission Line Model

Q1. Explain the advantages of transmission line analogy of the wave propagation in two media.

Q2. Explain why we do need to consider the medium semi-infinitely extended to exploit the transmission line analogy.

Q3. Explain why the input impedance repeats the same value for each $\lambda_1/2$ distance along $-z$-direction.

Q4. What are the characteristic impedances of the two transmission lines?

Example 5.9: Standing Wave Patten Theory: Transmission Line Model

For an Teflon-silica interface draw the transmission line analogy. Calculate the input impedance at 5 m from the interface in Teflon. The dielectric constants for Teflon and silica are 2.1 and 11.7 respectively.

$$Z_0 = \eta_1 = \frac{120\pi}{\sqrt{2.1}} = 260.1\,\Omega; Z_L = \eta_2 = 120\pi/\sqrt{11.7} = 110.2\,\Omega$$

The phase constant is: $\beta_1 = (2\pi \times 9 \times 10^8)/\left(3 \times \left(10^8/\sqrt{2.1}\right)\right) = 27.32\,(\text{rad/m})$; $tan(\beta_1 l) = tan$ $(27.32 \times 5) = 16.85$

The input impedance seen at 5 m inside Teflon is: $Z_{in} = 260.1((110.2 + j260.1 \times 16.85)/(260.1 + j110.2 \times 16.85) = 604.2 + j69.2\,\Omega = 608.1\angle 6.5°\,\Omega$.

5.8 Reflection from Multiple Layers

In the preceding section, we have considered the reflection and transmission of electromagnetic wave propagation in only two semi-infinite layers of dissimilar dielectric media that create one interface of impedance discontinuity. There we have analyzed the electromagnetic system using four different methods. In this section, we shall examine an electromagnetic system with more than two dielectric media. Such multilayered dielectric media with multiple dielectric interfaces offer

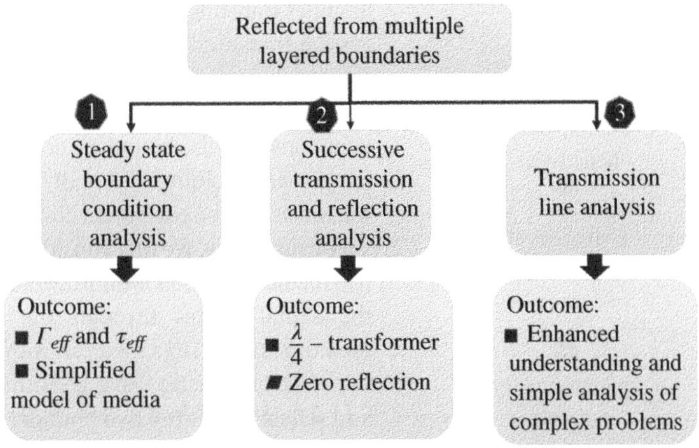

Figure 5.30 Analytical approaches of reflection and transmission from multiple layered boundaries.

many practical applications in industries. The applications include shielding of electronic devices with thin conducting films for EMC measures, radar domes, also called radomes in short, for aircraft radar antennas and mobile ground station satellite antennas,[54] antireflective and anti-glare coating for reading and sunglasses, and efficient transmission of light in optical devices such as lenses, in-chip wireless communication via in-built antennas in multilayered MMICs with multiple dielectric/semiconductor materials,[55] and stealth materials for fighter planes to avoid radar detection. We shall discuss such applications in comprehensive detail in the very last section of the chapter.

We shall examine such a multilayered electromagnetic system in various approaches as shown in Figure 5.30. Such a multilayered system creates very complex field analysis due to boundary conditions applied to many boundaries of dissimilar dielectric media. Therefore, we shall start with the existing knowledge that we have gained in the analysis of a single interface of two media. We extend the complex electromagnetic system with a progression of many such interfaces in tandem. We shall apply the known knowledge of the boundary conditions of the fields and the individual transmission and reflection coefficients at each interface of the complex electromagnetic system. Thus, we simplify the whole system with successive transmission and reflection coefficients in each layer in both directions to represent the forward and reverse waves due to these successive reflections and transmissions in all interfaces. This approach helps field analysis for any number of dissimilar dielectric layers using the fundamental knowledge that we have applied for the two media discontinuity as presented in the preceding section. The transmission line analogy of the complex electromagnetic system helps ease the field problems in the light of equivalent characteristic line impedances for each section. Using the expended knowledge, we shall define the effective reflection and transmission coefficients of the multilayered electromagnetic system. In summary, we shall examine the concept with the conventional steady-state field conditions for the multilayered medium with equivalent reflection and transmission coefficients considering the influences of all dielectric layers. There is a fine difference between the successive transmission and reflection and the steady state approaches. The approach to the successive transmissions and reflections

54 The author extensively worked on such antennas and provides many example related to the mobile ground station satellite antenna system in the textbook.

55 This is a hot topic in recent years due to the wireless inter-chip and intra-chip communications with in-built antennas with the application specific integrated circuit (ASIC) chipset.

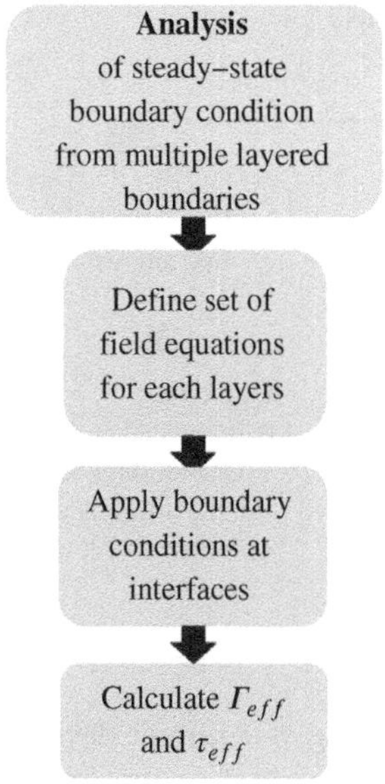

Figure 5.31 Methodology of steady-state field boundary conditions to calculate effective reflection and transmission from multiple layered boundaries.

assumes transmission phase delays in individual rays due to the successive transmissions and reflections at the interfaces, whereas in the steady-state field conditions (time harmonic) we consider all field quantities to be simultaneously present to fulfill Maxwell's equations and the boundary conditions. However, both cases lead to the same solution.[56] Finally, we expand the concept with the transmission line equivalent circuit model for any number of media.

For simplicity, we first consider a three-layered dissimilar dielectric media. For such a multilayered electromagnetic system, once the signal passes through the first semi-infinitely extended medium (say air) to a medium with two boundaries of finite depth, the electromagnetic wave goes through successive transmissions and reflections at the two boundaries in the sandwiched medium. The main aim is to find *the effective reflection coefficient* between medium 1 and medium 3 via the sandwiched medium 2 so that efficient coupling between the end media can be calculated. First we examine the electromagnetic system of three dielectric media in the light of steady-state field boundary conditions and develop the expressions for the effective reflection and transmission coefficients. In doing so, we shall draw the knowledge of the field conditions for the interfaces separately and then, assuming they form a linear system, we shall add the field contributions with their boundary conditions to get the effective values of the reflection and transmission coefficients. In the second approach, we shall derive the effective or total reflection coefficient in the first medium due to the presence of all dielectric media in the multilayered electromagnetic system using the successive transmissions and reflections.

Therefore, for simplicity of analysis and without losing the generality of power transmission and reflection from medium 1 and medium 3 via medium 2, we can assume an effective transmission coefficient τ_{eff} and an effective reflection coefficient Γ_{eff} for the electromagnetic system. In this case, we consider the propagation to be via the steady-state signals that must satisfy Maxwell's boundary conditions, as shown in Figure 5.31. The calculations of the effective reflection and transmission coefficients augment the field analysis and pave the way for efficient design considering the propagation effect of all media in concern. In all cases, we shall use the transmission line analogy to enhance the understanding of the complex fields that go through successful transmissions and reflections. The transmission line analogy helps understand any number of such multiple layered media. Finally, we shall conclude the section with some exciting applications of the theory in industries.

5.8.1 Effective Transmission and Reflection Analysis of Multilayered Dielectric Media Using Steady-State Boundary Conditions

In this section, we start with a simple electromagnetic system of three media, as shown in Figure 5.32a. Medium 1 has constitutive parameters $(\varepsilon_1, \mu_1, \eta_1)$ located in the region $-\infty < z < -d$,

56 Although in U.S. Inan and A.S. Inan, *Electromagnetic Waves*, Prentice Hall, Upper Saddle River, NJ, 2000 advocate the results may differ in some cases.

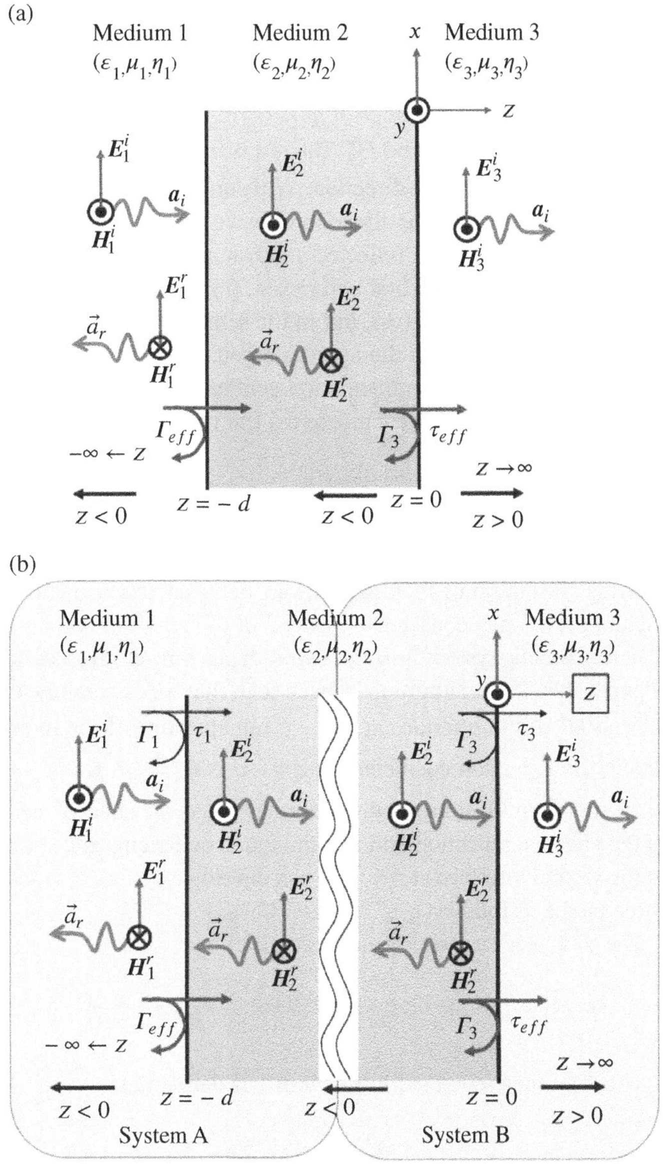

Figure 5.32 Three-layered dielectric medium, (a) electromagnetic system configuration, and (b) analysis of (a) into system configurations: Systems A and B.

medium 2 has constitutive parameters (ε_2, μ_2, η_2) located in the region $-d < z < 0$, and medium 3 has constitutive parameters (ε_3, μ_3, η_3) located in the region $0 < z < \infty$ .The three dielectric media have intrinsic impedances η_1, η_2, and η_3, respectively, that offer impedance discontinuities at the two interfaces located at $z = -d$ and 0, respectively. This impedance discontinuities are defined as the planar interfaces with two thick lines at the two locations at $z = -d$ and 0, respectively.

We also assume that the three media is lossless for simplicity and a uniform plane wave is traveling in the positive z-direction from $z = -\infty$ in medium 1 and hits the first interface of media 1 and 2 at $z = -d$. As shown in the figure, an x-polarized incident electromagnetic wave with field intensities $\boldsymbol{E}_1^i$ and $\boldsymbol{H}_1^i$ are propagating along positive z-direction with unit vector $\boldsymbol{a_i} = \boldsymbol{a_z}$ from $z = -\infty$

and impinges on medium 2 at $z = -d$. A portion of the incident wave is reflected back in medium 1 with electromagnetic fields $\boldsymbol{E}_1^r$ and $\boldsymbol{H}_1^r$ along the unit vector $\boldsymbol{a}_r = -\boldsymbol{a}_z$ and the rest is transmitted through medium 2 with the unit vector $\boldsymbol{a}_i = \boldsymbol{a}_z$ along positive z-direction with fields $\boldsymbol{E}_2^i$ and $\boldsymbol{H}_2^i$. A portion of the incident fields $\boldsymbol{E}_2^i$ and $\boldsymbol{H}_2^i$ in medium 2 are reflected back in medium 2 at the interface $z = 0$ with fields $\boldsymbol{E}_2^r$ and $\boldsymbol{H}_2^r$. The rest is transmitted through medium 3 as the incident fields $\boldsymbol{E}_3^i$ and $\boldsymbol{H}_3^i$ along positive z-direction. With our general conviction, therefore, we may legitimately assume that based on the direction of wave propagation along the z-direction, both media 1 and 2 have two field waves – reflected and transmitted, but medium 3 has only one transmitted field wave. This means in the first two media, there are standing waves due to the existence of the two opposite directed field waves, but in the semi-infinite third media, there is only one forward traveling wave. Assume that the aggregated effect of the electromagnetic system is defined with the effective reflection and transmission coefficients as we did for a single interface with the buffer medium 2 understood, we can express the two coefficients as:

$$\Gamma_{eff} = E_{01}^r / E_{01}^i ; \tau_{eff} = E_{03}^t / E_{01}^i \tag{5.101}$$

where Γ_{eff} is the effective reflection coefficient from medium 2 to medium 1 considering the loading effect of medium 3. τ_{eff} is the effective transmission coefficient from medium 1 to medium 3 considering the presence of medium 2 in between the two-end media. For the field quantities in medium 2 we may consider the extend of the medium is such that we can split the three-layered electromagnetic system into the two independent electromagnetic systems A and B, as shown in Figure 5.32b. The assumption is made such that we can assume the amplitudes of the incident fields $\boldsymbol{E}_2^i$ and $\boldsymbol{H}_2^i$ at the interface at $z = -d$ remain same at the interface at $z = 0$.[57] Therefore, we can define the reflection coefficient at $z = -d$ is $\Gamma_3 = \frac{E_{02}^r}{E_{02}^i} = (\eta_3 - \eta_2)/(\eta_3 + \eta_2)$ that accounts for the presence of semi-infinite medium 3 and neglects the effect of medium 1. Using the above definitions of the effective reflection and transmission coefficients and the unaffected reflection coefficient Γ_3 at the second interface at $z = 0$ we can develop a system of equations for the field quantities in the three media as follows:

The total electric field in medium 1 is:

$$E_1(z) = E_1^i + E_1^r = E_{01}^i \left(e^{-j\beta_1(z+d)} + \Gamma_{eff} e^{j\beta_1(z+d)} \right) \quad \text{for } -\infty \le z \le -d \tag{5.102}$$

The total magnetic field in medium 1 is defined as:

$$H_1(z) = H_1^i + H_1^r = \frac{E_{01}^i}{\eta_1} \left(e^{-j\beta_1(z+d)} - \Gamma_{eff} e^{j\beta_1(z+d)} \right) \quad \text{for } -\infty \le z \le -d \tag{5.103}$$

The total electric field in medium 2[58] is:

$$E_2(z) = E_2^i + E_2^r = E_{02}^i \left(e^{-j\beta_2 z} + \Gamma_3 e^{j\beta_2 z} \right) \quad \text{for } -d \le z \le 0 \tag{5.104}$$

---

57 We may assume this condition for simplicity of analysis. This is possible because we assume the system is lossless and as if there is a phase matched condition occur so that the fields' amplitudes remain the same for both interfaces. We shall examine the situation in the next section for the case of $d = \lambda_2/4$.

58 For the definition of $E_2(z)$ and $H_2(z)$ with Γ_3 it may appear that we are not considering the presence of medium 1. Actually, the is considered in the definition of the field quantities E_{02}^i, which is associated with the incident field E_{01}^i via the transmission coefficient $\tau_1 = 2\eta_2/(\eta_1 + \eta_2)$.

The total magnetic field in medium 2 is defined as:

$$H_2(z) = H_2^i + H_2^r = \frac{E_{02}^i}{\eta_2}\left(e^{-j\beta_2 z} - \Gamma_3 e^{j\beta_1 z}\right) \quad \text{for } -d \leq z \leq 0 \tag{5.105}$$

Assuming that medium 2 is acting as a buffer stage for media 1 and 3 like an impedance transformer of which the field conditions and propagation phenomena are still unknown. In this case, the wave impedances of media 1 and 3 are assumed to be the input impedance η_1 and the load impedance η_3, respectively, for the transformer represented by medium 2. The role of medium 2 is to take the impedance η_3 and present it to medium 1[59] as the buffer unity gain operational amplifier circuit in linear electronics does. The electromagnetic fields in medium 3 is the forward propagating wave along z-direction and can be defined as:

$$E_3(z) = E_3^i(z) = E_{03}^i e^{-j\beta_3 z} = \tau_{eff} E_{01}^i e^{-j\beta_3 z} \quad \text{for } 0 \leq z \leq +\infty \tag{5.106}$$

$$H_3(z) = H_3^i(z) = \frac{\tau_{eff} E_{01}^i}{\eta_1} e^{-j\beta_3 z} \quad \text{for } 0 \leq z \leq +\infty \tag{5.107}$$

Note τ_{eff} is the effective transmission coefficient between media 1 and 2 via medium 2 and links the field quantities in media 1 and 3, bypassing the buffer medium with the impedance η_2. Therefore, $\tau_{eff} \neq \tau_{23}$ which is defined as the transmission coefficient between media 2 and 3, neglecting the effect of medium 1. If we neglect the effect of medium 1 on the forward transmission of the signal from medium 2 to medium 3 as shown in Figure 5.32b, then we can define the forward transmission coefficient at the interface $z = 0$ as the transmission coefficient from medium 2 to medium 3: $\tau_{23} = 2\eta_3/(\eta_2 + \eta_3)$. We can then consider two independent electromagnetic systems: System A in between medium 1 and medium 2 with the interface at $z = -d$ and System B in between medium 2 and medium 3 with the interface at $z = 0$. Then we can find the aggregate effective transmission and reflection coefficients between the media 1 and 3 via the buffer medium 2 by adding the effect of the two linear system parameters. The advantage is that we can use the prior knowledge of the electromagnetic system with the discontinuity created by only two media. From the above assumptions and analysis of the field quantities, we can make the following remarks:

1) $\Gamma_{eff} \neq \Gamma_1 = (\eta_2 - \eta_1)/(\eta_2 + \eta_1)$. At best we can define $\Gamma_{eff} = \left(\eta_2' - \eta_1'\right)/\left(\eta_2' + \eta_{1'}\right) = |\Gamma_{eff}|\, e^{j\theta}$ a complex quantity due to the translation of the wave from medium 3 to medium 1 via medium 2. η_2' and η_1' still need to be determined. Γ_{eff} is a function of all three impedances (η_1, η_2 and η_3).
2) In a similar manner we can also define; $\tau_{eff} = |\tau_{eff}| e^{j\theta'}$, again a complex quantity. Likewise, τ_{eff} is a function of all three impedances (η_1, η_2, and η_3).
3) In both the earlier definitions of the effective reflection and transmission coefficients, we considered the aggregated effect of the three media and their propagation phases. Therefore, the complex effective reflection and transmission coefficients are the natural consequence of the translation of impedances from medium 3 to medium 1 via the buffer medium 2.[60]

Now we are going to calculate the effective transmission and reflection coefficients using the boundary conditions at the interfaces of the three media as follows:

Assuming $E_1(z) = E_{01}^i$ at $z = -d$, we can define the continuity of the x-directed field quantities in media 1 and 2, respectively, as follows:

---

59 In U.S. Inan and A.S. Inan, *Electromagnetic Waves*, Prentice Hall, Upper Saddle River, NJ, 2000.

60 The phase also depends on the choice of the reference plane. We choose the interface between the media 2 and 3 at the origin $z = 0$. Therefore, the phase term $e^{-j\beta z} = 1$. In the process of calculation of τ_{eff} we shall see $e^{j\theta'} = 1$.

$$E_1(z) = E_2(z) \Rightarrow E^i_{01}\left(1 + \Gamma_{eff}\right) = E^i_{02}\left(e^{-j\beta_2 d} + \Gamma_3 e^{j\beta_2 d}\right) \text{ at } z = -d^{61} \tag{5.108}$$

$$H_1(z) = H_2(z) \Rightarrow \frac{E^i_{01}}{\eta_1}\left(1 - \Gamma_{eff}\right) = \frac{E^i_{02}}{\eta_2}\left(e^{-j\beta_2 d} - \Gamma_3 e^{j\beta_2 d}\right) \text{ at } z = -d \tag{5.109}$$

After simple mathematical manipulation, we obtain:

$$\Gamma_{eff} = \frac{\eta_2}{\eta_1}\Gamma_3 e^{-j2\beta_2 d} \text{ at } z = -d \tag{5.110}$$

Substitution for $\Gamma_3 = (\eta_3 - \eta_2)/(\eta_3 + \eta_2)$ we obtain:

$$\Gamma_{eff} = \frac{\eta_2}{\eta_1} \times \frac{\eta_3 - \eta_2}{\eta_3 + \eta_2} e^{-j2\beta_2 d} \text{ at } z = -d \tag{5.111}$$

From the above expression of Γ_{eff}, we observe that it has considered impedances of all three media and is a complex quantity due to the propagation phase factor $e^{-j2\beta_2 d}$ from $-\infty$ to $z = -d$. Therefore, the complex $\Gamma_{eff} = \dfrac{\eta'_2 - \eta'_1}{\eta'_2 + \eta_{1'}} = |\Gamma_{eff}|\, e^{j\theta}$ is a valid assumption.

Similarly assuming $E_2(z) = E^i_{02}$ at $z = -0$,[62] and $E_3(z) = E^i_{03}$, at $z = 0$, we can apply the boundary conditions of continuity of the x-directed fields at the interface of media 2 an 3 at $z = 0$ as follows:

$$E_2(z) = E_3(z) \Rightarrow E^i_{02}(1 + \Gamma_3) = \tau_{eff}\, E^i_{03} \text{ at } z = 0 \tag{5.112}$$

$$H_2(z) = H_3(z) \Rightarrow \frac{E^i_{02}}{\eta_2}(1 - \Gamma_3) = \frac{\tau_{eff}\, E^i_{03}}{\eta_3} \text{ at } z = 0 \tag{5.113}$$

$$\text{Or, } \tau_{eff} = \frac{E^i_{02}}{E^i_{03}} \times \frac{\eta_3}{\eta_2} \times (1 - \Gamma_3) = \frac{E^i_{02}}{E^i_{01}} \times \frac{E^i_{01}}{E^i_{03}} \times \frac{\eta_3}{\eta_2} \times (1 - \Gamma_3)$$

Substitutions for $\tau_1 = (E^i_{02})/(E^i_{01}) = (2\eta_2)/(\eta_2 + \eta_1)$, $\tau_{eff} = (E^i_{03})/(E^i_{01})$ and $\Gamma_3 = (\eta_3 - \eta_2)/(\eta_3 + \eta_2)$ we obtain

$$\tau_{eff} = \frac{2\eta_2}{\eta_2 + \eta_1} \times \frac{\eta_3}{\eta_2} \times \left(1 - \frac{\eta_3 - \eta_2}{\eta_3 + \eta_2}\right) = \frac{4\eta_2\eta_3}{(\eta_2 + \eta_1)(\eta_3 + \eta_2)}$$

Since we calculate the effective transmission coefficient τ_{eff} at $z = 0$ boundary; therefore, the phase term is zero in this instance. However, we observe that τ_{eff} becomes a function of all three impedances as expected. In the following section, we consider the phenomenon of multiples transmissions and reflections in the second medium and develop the effective reflection coefficient in medium 1.

Example 5.10: Reflection and Transmission for Multilayer Media: Steady State State Analysis
A dielectric (Teflon) slab of thickness $d = 30$ cm is inserted into a sand pit of infinite extend. If at 150 MHz a UPW signal is transmitted through a probe, (i) calculate: the reflection coefficient seen by the probe at the sand and dielectric slab interface ($z = -d$), (ii) calculate the transmission coefficient at the other interface ($z = 0$), and (iii) if the incident power intensity is 20 (W/m^2), calculate the % of power reflected and transmitted. Assume the sand is pure and has a dielectric constant of 5.6, and for the slab it is 3.2.

---

61 We can legitimately consider that there is no more effect from a forth medium 4 after the semi-infinitely extended medium 3. Hence we can assume Γ_3 is the casual effect on medium 2 due to the presence of medium 3.
62 Based on the assumption of the two linear electromagnetic systems A and B.

Answer:

The impedance of sand $\eta_1 = \eta_3 = 377/\sqrt{5.6} = 159.31\,\Omega$; the impedance of Teflon slab is: $\eta_2 = 377/\sqrt{3.2} = 210.75\,\Omega$;

The distance phase of dielectric slab of thickness $d = 30$ cm is: $\beta_2 d = \left(2\pi \times 150 \times 10^6 \times \sqrt{3.2}\right)/(3 \times 10^8) \times 0.3 = 1.69$ (rad); $2\beta_2 d = 3.37$ (rad) $= 193.2°$

The effective reflection coefficient seen at $z = -d$ is:

$$\Gamma_{eff} = \frac{\eta_2}{\eta_1} \times \frac{\eta_3 - \eta_2}{\eta_3 + \eta_2} e^{-j2\beta_2 d} = \frac{210.75}{159.31} \times \frac{159.31 - 210.75}{210.75 + 159.31} e^{-j193.2°} = -0.18\angle 193°$$

The effective transmission coefficient seen at $z = 0$ is:

$$\tau_{eff} = \frac{4\eta_2\eta_3}{(\eta_2 + \eta_1)(\eta_3 + \eta_2)} = \frac{4 \times 210.75 \times 159.31}{2(210.75 + 159.31)} = 181.45$$

Percent power reflected: $\%P_r = (1 - 0.18^2) \times 100\,\% = 97\%$
Reflected power density: $P_r = (1 - |\Gamma|^2) \times P_i = 0.97 \times 20 = 19.35$ (W/m^2)

5.8.2 Successive Transmission and Reflection Analysis of Multilayered Dielectric Media

In the above steady-state analysis of the effective reflection and transmission coefficients of three-layered media, we consider our conventional knowledge of transmission and reflection of multiple interfaces and their aggregated effects. There we have considered only one set of incident waves in each medium and one set of reflected waves in media 1 and 2 based on the direction of propagation. Again based on the direction of propagation, medium 3 has only one set of transmitted field waves. In this section, we consider multiple transmissions and reflections of electromagnetic waves at the two interfaces of the three media and develop the expression of the resultant reflection coefficient in medium 1.

The generalized phenomenon of transmission and reflection through multiple media can be analyzed with a ray trace model of successive transmissions and reflections of electromagnetic waves in medium 2 as shown in Figure 5.33. In this analysis, we shall define the field quantities in their normalized scalar amplitudes so that we can define the individual fields in terms of the multiplication of their successive reflection and transmission coefficients. The respective reflection and transmission coefficients in terms of the wave impedances of the three media are defined as:

$$\Gamma_1 = \frac{(\eta_2 - \eta_1)}{(\eta_2 + \eta_1)}; \tau_1 = \frac{2\eta_2}{(\eta_2 + \eta_1)} \tag{5.114}$$

$$\Gamma_2 = \frac{(\eta_1 - \eta_2)}{(\eta_1 + \eta_2)} = -\Gamma_1; \tau_2 = \frac{2\eta_1}{(\eta_1 + \eta_2)} \tag{5.115}$$

$$\Gamma_3 = \frac{(\eta_3 - \eta_2)}{(\eta_3 + \eta_2)}; \tau_3 = \frac{2\eta_3}{(\eta_2 + \eta_3)} \tag{5.116}$$

Here, Γ_1 is the reflection coefficient looking from medium 1 to medium 2; τ_1 are the transmission from medium 1 to medium 2; Γ_2 is the reflection coefficient looking from medium 2 to medium 1; τ_2 is the transmission from medium 2 to medium 1. Note that the definition suggests that $\Gamma_2 = -\Gamma_1$, however, $\tau_2 \neq \tau_1$. Finally, Γ_3 is the reflection coefficient looking from medium 2 to medium 3; and τ_3 is the transmission coefficient from medium 2 to medium 3. With these definitions, we start

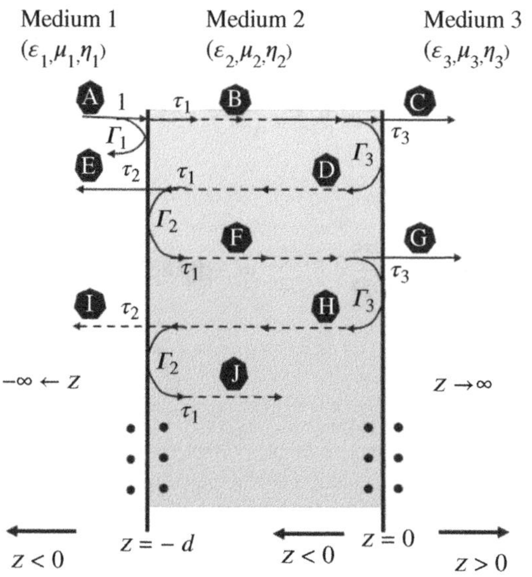

Figure 5.33 Successive transmission and reflection of wave in a three-layered composite medium.

analysis of the field quantities in terms of the successive transmission and reflection of the electromagnetic waves in the ray trace model, as shown in Figure 5.33.

As shown in the figure, the normalized incident wave with amplitude $E_0^i = 1$ is defined as ray (A). Due to the discontinuity in the wave impedances ($\eta_1 \neq \eta_2$) at $z = -d$ a part of the electromagnetic energy penetrates in medium 2 and rest reflects back in medium 1 according to the transmission coefficient (τ_1) and reflection coefficient (Γ_1), respectively.[63] The transmitted signal in medium 2 starts propagating in the medium until it hits the second interface of media 2 and 3 at $z = 0$. Due to the discontinuity in the wave impedances ($\eta_2 \neq \eta_3$), a part of the transmitted wave reflects back in medium 2 with a reflection coefficient Γ_3 and the rest is transmitted in medium 3 with a transmission coefficient τ_3. The first transmitted wave in medium 3 is shown with ray (C) and the second reflected wave is shown with ray (D). Ray (D) continues to propagate toward $-z$-direction and hits the first interface at $z = -d$. Therefore, a part of the signal is transmitted in medium 1 as ray (E) with the transmission coefficient τ_2 along $-z$-direction, rest reflects back in the second medium as ray (F) with the reflection coefficient Γ_2 and continues propagating along positive z-direction toward interface 2 in medium 2. The second transmitted signal in medium 3 is shown as ray (G) and the second reflected signal at interface 2 is shown as ray (H). Ray (H) propagates toward $z = -d$ interface, partly transmits through medium 1 as ray (I) with transmission coefficient τ_2, and partly reflects back in medium 2 as ray (J). Thus, a process of infinite number of reflections and transmissions occurs in medium 2 in between the two interfaces until the signal diminished to zero.[64]

63 Therefore, as we know, a standing wave is formed in medium 1 that we have extensively studied in the previous sections. In this case there are many similar transmitted and reflected rays will join with the first incident and reflected rays. Our goal of the analysis is to find the effective standing wave pattern via the effective reflection and transmission coefficients due to the introduction of the sandwiched second medium in-between the first and third media.

64 This is most unlikely due to the fact that we assume the media are lossless. However, every reflection and transmission at the two interfaces at $z = 0$ and $-d$ diminishes the signal inside medium 2. We define an effective transmission coefficient and an reflection coefficient with some approximations to simplify the analysis of the multiple reflection in medium 2 without loss of generality.

This electromagnetic phenomenon is similar to the multiple reflections that occur when a person stands between two mirrors. Infinite number of images are created on the mirror due to multiple reflections. The difference is that the mirrors are assumed to be fully reflective surfaces for light rays; therefore, the transmission of signal in the two end media is zero. On the other hand, the interfaces of the dielectric material associated with medium 2 is penetrable; therefore, the transmission of signal in the two end media is nonzero. The above analysis of the successive reflection and transmission leads us to the definition of the effective reflection and transmission coefficients between the two end media via the buffer medium 2. In the following section, we shall develop a closed form expression for the resultant/effect reflection coefficient first for a quarter wavelength buffer medium 2 and then it is generalized to any arbitrary thickness of the buffer medium 2.

5.8.3 Successive Transmission and Reflection Analysis Via $\frac{\lambda}{4}$-Thick Dielectric Medium

A $\lambda/4$ thick sandwiched medium in between two media has practical significance. Figure 5.34a shows such an electromagnetic multilayered medium. The generalized definition of multiple transmission and reflection due to the introduction of a finite medium 2 with thickness d is analyzed with a special case for $d = \lambda_2/4$ so that every reflected ray from the second interface at $z = 0$ undergoes a total transmission phase of 180° when it arrives at the interface $z = -d$. We develop an effective reflection coefficient in medium 2 using the electromagnetic fields' boundary conditions at the two interfaces located at $z = -d$ and $z = 0$, respectively. Based on the effective reflection coefficient, we shall develop a transmission line model for simplistic and systematic analysis of the complex electromagnetic system of multiple dielectric layers. Then, we shall derive the expression for the effective reflection coefficient for any arbitrary thickness of the second medium d for the multilayered electromagnetic system in the next section.

As before, consider an x-polarized uniform plane wave propagating along z-direction in medium 1 is normally incident at the interface of media 1 and 2 at $z = -d$. As stated above, the transmitted signal in medium 2, undergoes successive reflections and transmissions in the two interfaces located at $z = 0$ and $-d$, respectively. Figure 5.34 illustrates the successive reflected rays in medium 2 of thickness $d = \lambda_2/4$. Assume the normalized incident electric field intensity (ray A) is $E_0^1 = 1$ as shown in Figure 5.34b. Then the normalized transmitted signal intensity (ray B) in medium 2 is $E_0^t = \tau_1$. This transmitted signal (ray B) propagates in the second medium with the transmission phase of $e^{-j\beta_2 z}$ until it hits the second interface at $z = 0$ and is reflected back in medium 2 with the reflection coefficient Γ_3 and the normalized amplitude of the reverse signal (ray C) in medium 2 after the second reflection is $-\tau_1\Gamma_3$. The transmission phase of the reverse signal is $e^{+j\beta_2 z}$. Therefore, the total transmission phase of the round trip (ray C) is $2\beta_2 d = 2 \times (2\pi/\lambda_2) \times (\lambda_2/4) = \pi$. Some portion of the signal (ray D) is transmitted into medium 1 with the transmission coefficient τ_2 at the interface 1 at $z = -d$. Therefore, the first reversed signal (ray D) transmitted in medium 1 is: $-\tau_1\tau_2\Gamma_3$. Some portion of the reverse transmitted signal is reflected back by interface 1 as (ray E) with reflection coefficient Γ_2 and becomes the second forward transmitted signal (ray E) with the normalized intensity $\tau_1\Gamma_2\Gamma_3$. The portion of the signal is reflected back as (ray F) at interface 2 with reflection coefficient Γ_3 and the rest transmitted in medium 3 as (ray G) with transmission coefficient τ_3. The transmitted normalized signal into medium 3 is defined as: $\tau_1\tau_2\Gamma_3^2\Gamma_2$.The rest of (ray F) is reflected back at interface 2 at $z = 0$ with reflection coefficient Γ_3 becomes (ray H). This signal hits interface 1 at $z = -d$ and transmitted with a transmission coefficient τ_2 in medium 1 as (ray I). The normalized amplitude of ray I is $\tau_1\tau_2\Gamma_2\Gamma_3^2$. Portion of (ray H) returns back with reflection coefficient Γ_2 in medium 2 as (ray J). And this process will continue to go on and on. Examining the multiple reflected signals at interface 1 the total effective reflection coefficient

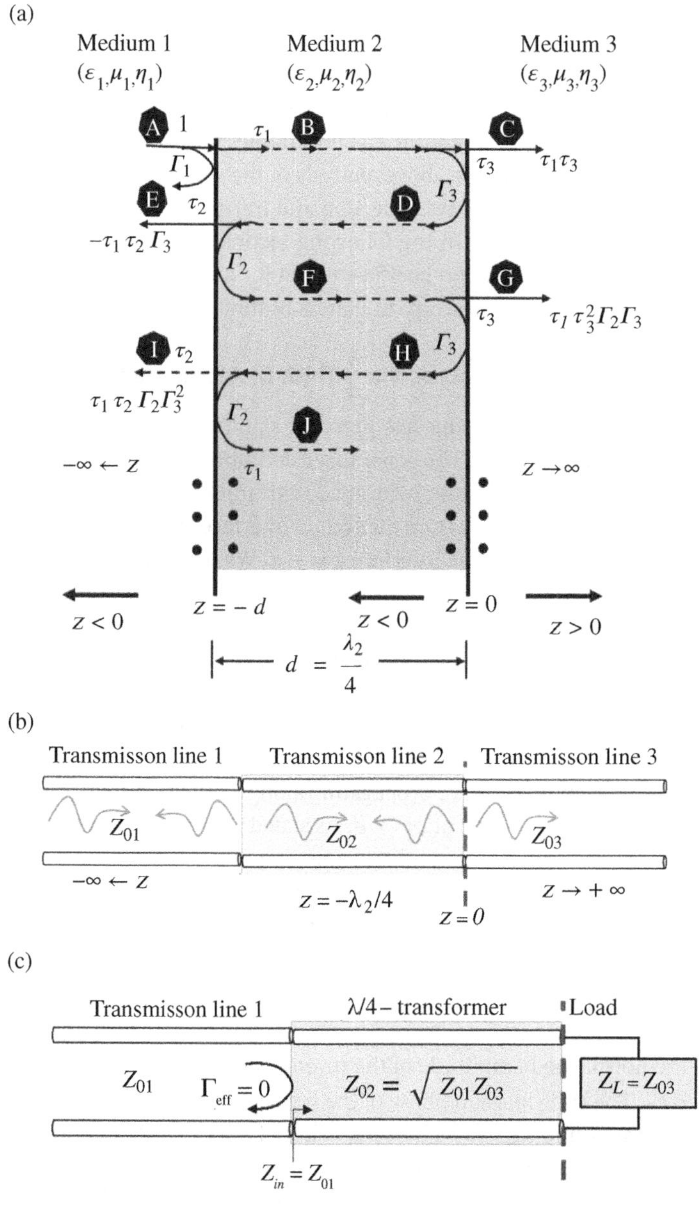

Figure 5.34 Successive transmission and reflection of wave in a three-layered composite medium with $d = \lambda_2/4$ thickness; (b) exact transmission line model of infinitely extended media 1 and 3, and (c) simplified transmission line model of (a).

can be defined by summing all normalized reverse transmitted signals (reflected signals at interface 1 at $z = -d$) in medium 1 as shown with the $-z$-directed arrowheads as:

$$\begin{aligned}\Gamma_{eff} &= \Gamma_1 - \tau_1\tau_2\Gamma_3 + \tau_1\tau_2\Gamma_2\Gamma_3^2 - \tau_1\tau_2\Gamma_2^2\Gamma_3^3 + \dots \\ &= \Gamma_1 - \tau_1\tau_2\Gamma_3\left(1 + \Gamma_2\Gamma_3 - \Gamma_2^2\Gamma_3^2 + \dots\right) \\ &= \Gamma_1 - \tau_1\tau_2\Gamma_3 \sum_{n=0}^{\infty} (-\Gamma_2\Gamma_3)^n \text{ for } n = 0,1,2,3,\dots \end{aligned} \tag{5.117}$$

Since, for the passive electromagnetic system $|\Gamma_2| < 1$ and $|\Gamma_3| < 1$, the infinite series can be added using the geometric series as follows:

$$\sum_{n=0}^{\infty} (x)^n = \frac{1}{1-x} \quad \text{for } |x| < 1 \tag{5.118}$$

Substitution for $x = -\Gamma_2\Gamma_3$, we obtain the total effective reflection coefficient as follows:

$$\Gamma_{eff} = \Gamma_1 - \tau_1\tau_2\Gamma_3 \times \left(\frac{1}{1+\Gamma_2\Gamma_3}\right) \tag{5.119}$$

Therefore, the resultant effective reflection coefficient in a closed form expression is defined as:

$$\Gamma_{eff} = \Gamma_1 - \frac{\tau_1\tau_2\Gamma_3}{1+\Gamma_2\Gamma_3} = \frac{\Gamma_1 + \Gamma_1\Gamma_2\Gamma_3 - \tau_1\tau_2\Gamma_3}{1+\Gamma_2\Gamma_3} \tag{5.120}$$

In many practical applications, we want zero effective reflection ($\Gamma_{eff} = 0$) from the electromagnetic system with a quarter wavelength thick medium 2, which is sandwiched between media 1 and 3. One excellent example is a stealth composite material to avoid radar detection. To do so, we define the numerator of (5.120) equals zero. To fulfill the condition, we must have the intrinsic impedances (hence the constitutive parameters) must follow certain boundary conditions. Using the relation $\Gamma_1 = -\Gamma_2$, the numerator of the effective reflection coefficient can be simplified as follows:

$$A = \Gamma_1 + \Gamma_1\Gamma_2\Gamma_3 - \tau_1\tau_2\Gamma_3 = \Gamma_1 - \Gamma_3\left(\Gamma_1^2 + \tau_1\tau_2\right) \tag{5.121}$$

Now using the definitions of the respective reflection and transmission coefficients in terms of their intrinsic wave impedances of the media from expressions (5.114–5.116) we can deduce the followings:

$$A = \Gamma_1 - \Gamma_3\left\{\left(\frac{\eta_2-\eta_1}{\eta_2+\eta_1}\right)^2 + \frac{4\eta_1\eta_2}{(\eta_2+\eta_1)^2}\right\} = \Gamma_1 - \Gamma_3 \tag{5.122}$$

Now substitution for $\Gamma_1 = (\eta_2 - \eta_1)/(\eta_2 + \eta_1)$ and $\Gamma_3 = (\eta_3 - \eta_2)/(\eta_3 + \eta_3)$, the numerator is defined as:

$$A = \frac{\eta_2-\eta_1}{\eta_2+\eta_1} - \frac{\eta_3-\eta_2}{\eta_3+\eta_2} = \frac{2\left(\eta_2^2 - \eta_1\eta_3\right)}{(\eta_2+\eta_1)\times(\eta_3+\eta_2)} \tag{5.123}$$

Substitution for the numerator in (5.123) we obtain the effective reflection coefficient as:

$$\Gamma_{eff} = 0 \quad \text{if } \eta_2 = \sqrt{\eta_1\eta_3} \tag{5.124}$$

From the above deduction, we can define the intrinsic impedance, which will be the square root of the wave impedance of media 1 and 3. Figure 5.34c shows the generalized transmission line model of the three-layered electromagnetic system. The respective media are represented with the characteristics impedances $Z_{01} = \eta_1$, $Z_{02} = \eta_2$, and $Z_{03} = \eta_3$, respectively. The transmission line 2 is quarter wavelength long. Drawing the analogy of the semi-infinite medium 3 as a load impedance $Z_L = Z_{03}$ we can develop a transmission line system with the input impedance $Z_{in} = Z_{01}$ a perfect matched condition if we can design the $\lambda/4$-transmission line section with an impedance $Z_{02} = \sqrt{Z_{01}Z_{03}}$. Since the transmission line segment of impedance Z_{02} brings a perfect matched condition at the output end of the transmission line 1, the $\lambda/4$-transmission line section is called *quarter wave transformer*.

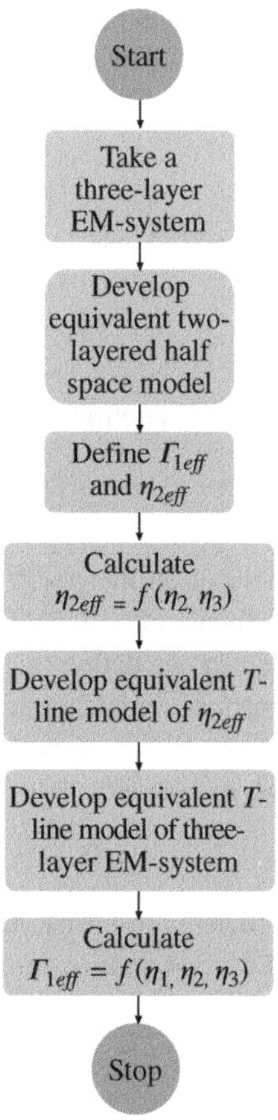

Figure 5.35 Flowchart for analytical steps to calculate Γ_{1eff} of multilayered medium.

Example 5.11: Quarter Wave Transformer

A quarter wavelength think dielectric material is sandwiched between medium 1 ($\varepsilon_r = 10.2$) and medium 3 ($\varepsilon_r = 3.1$) so that the reflection coefficient can be made zero. The operating frequency is 2.45 GHz. Calculate: (i) the effective impedance of the slab so that there is no reflection from the insertion, (ii) the new dielectric constant, and (iii) the thickness of the slab.

Answer:

The intrinsic impedances of media 1 and 3, respectively, are: $\eta_1 = \dfrac{377}{\sqrt{10.2}} = 118\,\Omega$ and $\eta_3 = \dfrac{377}{\sqrt{3.1}} = 214.1\,\Omega$

The effective impedance of quarter wavelength sandwiched medium 2 is:

$$\eta_{eff} = \sqrt{\eta_1 \eta_3} = \sqrt{118 \times 214.1} = 159\,\Omega$$

The dielectric constant of the dielectric slab is: $\varepsilon_r = 377^2/159^2 = 5.62$

The guided dielectric wavelength is:

$\lambda_g = \dfrac{c}{f} = \left(\left(3 \times 10^8\right)/\left(2.45 \times 10^9 \times \sqrt{5.62}\right)\right) = 51.65\text{mm}$. Therefore.

The slab thickness is $\ell = (51.65/4) = 12.91$ mm.

5.8.4 Effective Transmission and Reflection Coefficients of Multilayered Dielectric Media

So far, we have examined the cases for multiple reflections in three different approaches: (i) steady-state field boundary conditions, (ii) successive reflections and transmissions in the sandwiched medium 2, and (iii) a quarter-wavelength thick medium 2 with the concept of a $\lambda/4$ – impedance transformer. In this section, we shall complete the analysis with the calculation of the effective reflection and transmission coefficients with an equivalent wave impedance concept. Figure 5.35 shows the flowchart/outline of the approach developed in the calculation of the effective transmission and reflection coefficients of the multilayered dielectric media with any arbitrary thickness of the second medium. In the approach, the core outcome is the development of a generalized transmission line model for any arbitrary thickness d of the sandwiched medium 2.

Consider the x-polarized wave propagation via the multilayered three media with two interface at $z = -d$ and $z = 0$ and its equivalent two media with a single interface at $z = -d$. Our mission is to find the effective reflection coefficient in medium 1 of the equivalent two semi-infinite extent medium case. In this equivalence, we consider a composite medium of constitutive parameters (ε_{2eff}, μ_{2eff}, η_{2eff}) has semi-infinite extent. The composite medium is the combination of the original second medium (ε_2, μ_2, η_2) and medium 3 (ε_3, μ_3, η_3). In the analysis, we also assume that the reflection coefficient at the second interface at $z = 0$ is: $\Gamma_3 = (\eta_3 - \eta_2)/(\eta_3 + \eta_2)$ remains intact; as if the two media 2 and 3 are semi-infinitely extended so that no effect of medium 1 is considered as shown in Figure 5.36. We shall be using Γ_3 to develop the expression for the effective wave impedance η_{2eff} for the composite medium 2. Now let us define the effective reflection coefficient in medium 1 at the interface $z = -d$, for the equivalent electromagnetic system in Figure 5.36b as:

$$\Gamma_{1eff} = \left(\eta_{2eff} - \eta_1\right)/\left(\eta_{2eff} + \eta_1\right); \tau_{1eff} = 2\eta_{2eff}/\left(\eta_{2eff} + \eta_1\right) \tag{5.125}$$

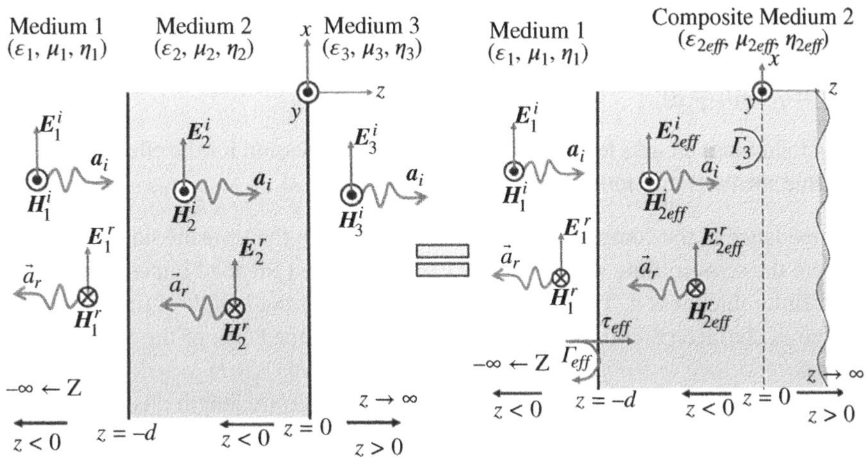

Figure 5.36 The effective composite materials for multilayer materials: (left) electromagnetic system and (right) semi-infinite model.

It is natural to think that Γ_{1eff} must consider the loading effect of media 2 and 3, and η_{2eff} must be a function of the individual intrinsic impedances η_2 and η_3 of medium 2 and medium 3, respectively. Therefore, if we can derive the closed form solution in terms of η_2 and η_3 then we can accurately determine the effective reflection and transmission coefficients in medium 1 due to the presence of media 2 and 3. Note that the electromagnetic system is nonreciprocal as we must consider the direction of wave propagation.

The intrinsic impedance η_{2eff} is the ratio of the electric and magnetic fields at any point in the composite medium 2 and can be defined as the ratio of the total electric and magnetic fields in the medium. The resultant fields are standing wave fields and expressed a:

$$E_{2eff}(z) = E^i_{2eff} + E^r_{2eff} = E^i_{02eff}\left(e^{-j\beta_2 z} + \Gamma_3 e^{j\beta_2 z}\right) \quad \text{for } -d \leq z \leq 0 \tag{5.126}$$

$$H_{2eff}(z) = H^i_{2eff} + H^r_{2eff} = \frac{E^i_{02eff}}{\eta_2}\left(e^{-j\beta_2 z} - \Gamma_3 e^{j\beta_2 z}\right) \quad \text{for } -d \leq z \leq 0 \tag{5.127}$$

Here, we assume $E^i_{2eff} = E^i_{02eff}$ at $z = -d$. Therefore, the wave impedance can be calculated at $z = -d$ as follows:

$$\eta_{2eff} = \frac{E_{2eff}(z = -d)}{H_{2eff}(z = -d)} = \eta_2 \frac{\left(e^{j\beta_2 d} + \Gamma_3 e^{-j\beta_2 d}\right)}{\left(e^{j\beta_2 d} - \Gamma_3 e^{-j\beta_2 d}\right)} \tag{5.128}$$

Now our task is to derive η_{2eff} via the substitution for $\Gamma_3 = (\eta_3 - \eta_2)/(\eta_3 + \eta_2)$, so that we can define η_{2eff} as a function of (η_2, η_3). Simple manipulation and substitution for $sinx = (e^{jx} - e^{-jx})/j2$ and $cosx = (e^{jx} + e^{-jx})/2$, where $x = \beta_2 d$ we obtain the closed form expression for η_{2eff} as follows:

$$\eta_{2eff} = \eta_2 \frac{\eta_3 + j\eta_2 \tan(\beta_2 d)}{\eta_2 + j\eta_3 \tan(\beta_2 d)} \quad {}^{65} \tag{5.129}$$

65 We leave the derivation as an exercise example in the end of the section.

If we consider lossy case then the effective impedance is defined as:

$$\eta_{2eff} = \eta_2 \frac{\eta_3 + \eta_2 \tanh(\gamma_2 d)}{\eta_2 + \eta_3 \tanh(\gamma_2 d)} \tag{5.130}$$

We can make the following remarks for the above closed form expression for the effective impedance of the composite medium 2 as follows:

1) The effective impedance of the composite medium 2 can be put in the transmission line model perspective where the semi-infinite medium three can be assumed the load impedance for the medium 2 with finite thickness *d*. The thickness is equivalent to the length of the equivalent transmission line of characteristic impedance η_2, the intrinsic impedance of the medium 2 of finite thickness *d*.
2) The input impedance of the finite length transmission line of arbitrary length *d* is the effective impedance η_{2eff} of the composite medium 2. Figure 5.37 illustrates the equivalent transmission line model of the composite medium as shown in Figure 5.36, assuming the absence of medium 1 as yet.
3) As can be seen, the equivalent circuit model of the two media has no link with medium 1. The constitutive parameters define the equivalent impedance. While the physical length is *d*, the electrical length of the line is $\theta = \beta_2 d$ in radians.

Now we are warm enough to bring the loading effect of both media 2 and 3 on to medium 1 to calculate Γ_{1eff}. The process is equivalent to assemble a complex load circuit and then fit it into an input circuit to complete the system. The advantage is to generalize the process for any number of the multilayered electromagnetic system in tandem. We shall do some examples in this retrospective. The equivalent circuit model of the complete electromagnetic system with the three media is shown in Figure 5.38. As can be seen now, the equivalent input circuit (transmission line 1) is the

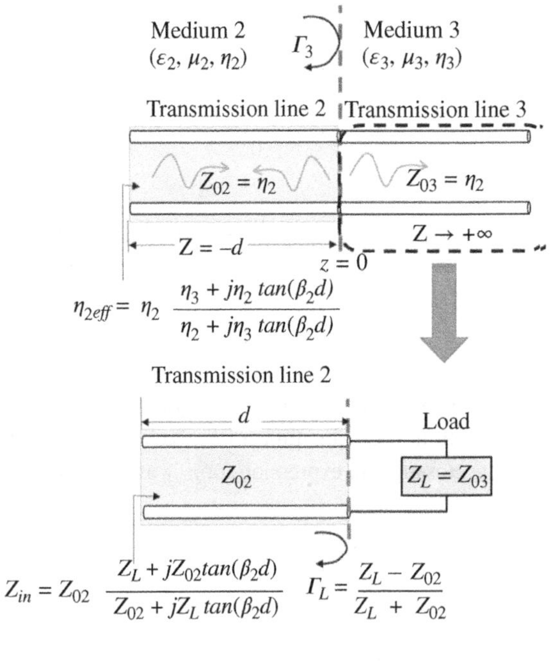

Figure 5.37 Equivalent transmission line model of Figure 5.36.

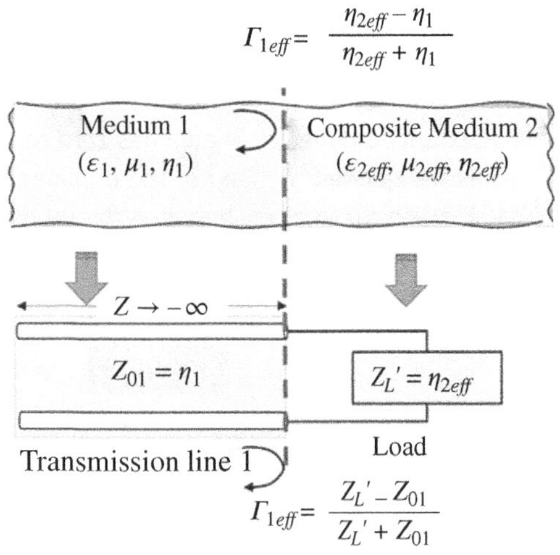

Figure 5.38 Simplified electromagnetic system of three-layered medium with media 2 and 3 as the single composite medium derived from Figure 5.37 (top); and simplified transmission line mode (bottom).

medium 1 with infinite extend along $z = -\infty$ and characteristic impedance is defined as $Z_{01} = \eta_1$. The effective reflection coefficient is also shown in the figure. Now we are going to develop the closed from expression for the equivalent reflection coefficient.

Figure 5.38 (top) shows the equivalent single interface with two half space electromagnetic systems and the bottom one shows the equivalent circuit model. Our original motto is to derive the closed form expression for Γ_{1eff}. That we can find easily the equivalent circuit model (bottom). Let us perform some calculations to derive the closed form expression for Γ_{1eff}. Substitution for, $\eta_{2eff} = \eta_2\left(\left(e^{j\beta_2 d} + \Gamma_3 e^{-j\beta_2 d}\right)/\left(e^{j\beta_2 d} - \Gamma_3 e^{-j\beta_2 d}\right)\right)$ in $\Gamma_{1eff} = (\eta_{2\text{eff}} - \eta_1)/(\eta_{2\text{eff}} + \eta_1)$ and after some lengthy manipulations we obtain:

$$\Gamma_{1eff} = \frac{\{(\eta_2 - \eta_1) \times (\eta_3 + \eta_2) + (\eta_2 + \eta_1) \times (\eta_3 - \eta_2)e^{-j2\beta_2 d}\}}{\{(\eta_2 + \eta_1) \times (\eta_3 + \eta_2) + (\eta_2 - \eta_1) \times (\eta_3 - \eta_2)e^{-j2\beta_2 d}\}} \tag{5.131}$$

Likewise we can calculate the effective transmission coefficient from medium 1 to medium 2

$$\tau_{eff} = \frac{(4\eta_2\eta_3)e^{-j2\beta_2 d}}{(\eta_2 + \eta_1) \times (\eta_3 + \eta_2) + (\eta_2 - \eta_1) \times (\eta_3 - \eta_2)e^{-j2\beta_2 d}}$$

Review Questions 5.6: Reflection and Transmission of Multilayered Medium

Q1. Name different industrials applications of wave propagation and reflection and transmission through multilayered medium

Q2. Explain different analytical approaches of reflection and transmission from multiple layered boundaries. What are the inherent differences between different approaches?

Q3. Explain the steady-state approach to analysis of reflection and transmission from multi-layered medium, Explain what conditions must be fulfilled.

Q4. Explain the basis of assumption made to consider the two interfaces created with a consecutive three-layered media into two independent electromagnetic systems of media with two separate interfaces at the same locations.

Q5. Explain the mechanism for successive transmissions and reflections at the interfaces of 3-ply medium. Where this theory lead us to?

Q6. Show that for a $\lambda/4$ thick dielectric window sandwiched between two semi-infinitely extended dielectric media has zero reflection if the intrinsic impedance of the window is the geometric mean of the impedances of the two end media.

Q7. Explain the analogy between the infinitely extent 3-ply medium with the loaded transmission line. What is the load impedance equivalent in this case? What effect is considered here?

Q8. Show that the effective impedance of a 3-ply lossless medium is defined as: $\eta_{2eff} = \eta_2(\eta_3 + j\eta_2 \tan(\beta_2 d))/(\eta_2 + j\eta_3 \tan(\beta_2 d))$. Convert this expression for a lossy medium case.

Q9. Show the input impedance seen at the end of the equivalent loaded lossless transmission line is defined as: $Z_{in} = Z_{02}(Z_L + jZ_{02} \tan(\beta_2 d))/(Z_{02} + jZ_L \tan(\beta_2 d))$.

Q10. Find the intrinsic impedance against the impedance of the above equation.

Q11. Show that the load reflection coefficient of the equivalent loaded transmission line is defined as: $\Gamma_L = (Z_L - Z_{02})/(Z_L + Z_{02})$.

Example 5.12: Reflection and Transmission from Multilayered Medium: Transmission Line Analysis

Calculate the effective impedance η_{2eff} for the third semi-infinite dielectric medium has the dielectric constant $\varepsilon_{r3} = 10$ and the dielectric window of thickness $d = 10$ cm with the dielectric constant $\varepsilon_{r2} = 5$. The operating frequency is 900 MHz. Assume the wave is incident from air as the medium 1.

Answer:

The intrinsic wave impedance of medium 2: $\eta_2 = \sqrt{\mu_0/(\varepsilon_0\varepsilon_{r2})} = \dfrac{377}{\sqrt{5}} = 168.6\,\Omega$

The intrinsic wave impedance of medium 3: $\eta_3 = \sqrt{\mu_0/(\varepsilon_0\varepsilon_{r3})} = 377/\sqrt{10} = 119.22\,\Omega$

The phase constant of dielectric medium 2 is: $\beta_2 = (2\pi \times 9 \times 10^8)/(3 \times 10^8/\sqrt{5}) = 8.43$ (rad/m)

The electrical length of the dielectric window of thickness d is: $\beta_2 d = 8.43 \times 0.1 = 0.843$ (rad)

The effective input impedance of medium 2 is:

$$\eta_{2eff} = \eta_2 \frac{\eta_3 + j\eta_2 \tan(\beta_2 d)}{\eta_2 + j\eta_3 \tan(\beta_2 d)} = 168.6 \times \frac{119.22 + j168.6 \tan(0.843)}{168.6 + j119.22 \tan(0.843)} = 168.23 + j58.05\,\Omega$$

The effective reflection coefficient is seen from air medium 1:

$$\Gamma_{eff} = \frac{\eta_{2eff} - \eta_0}{\eta_{2eff} + \eta_0} = \frac{168.23 + j58.05 - 377}{168.23 + j58.05 + 377} = -0.37 + j0.15 = 0.4\angle 157.9^\circ$$

5.8.5 Reflection for a Large Number of Multiple Dielectric Media

The above theory of the effective multiple reflection and transmission coefficients can be expanded to any number of wave propagation layers by conversion of the very last two media as a single composite medium, and then imposing the effective impedance of the composite medium to the next left hand medium. The approach is analogous to rolling up a carpet from one end until one reaches to the other end, as shown in Figure 5.39a. At any point of the rolling process, we can define the wave impedance as the ratio of the resultant electric and magnetic fields at that point. Therefore, we can say that the above expressions for the effective reflection and transmission coefficients are completely applicable to any number of dissimilar dielectric layers if we follow the sequential calculation of the loading effect of the previous two layers and impose the equivalent wave impedance to the next layer as we did in the preceding section for the case of three media wave propagation.

Figure 5.39b shows the N-number of multilayer propagation media with dissimilar dielectric constants and thicknesses so that the impedance of each layer is different from those for the adjacent layers on both sides. As shown in the figure, there are N-number of layers with dissimilar

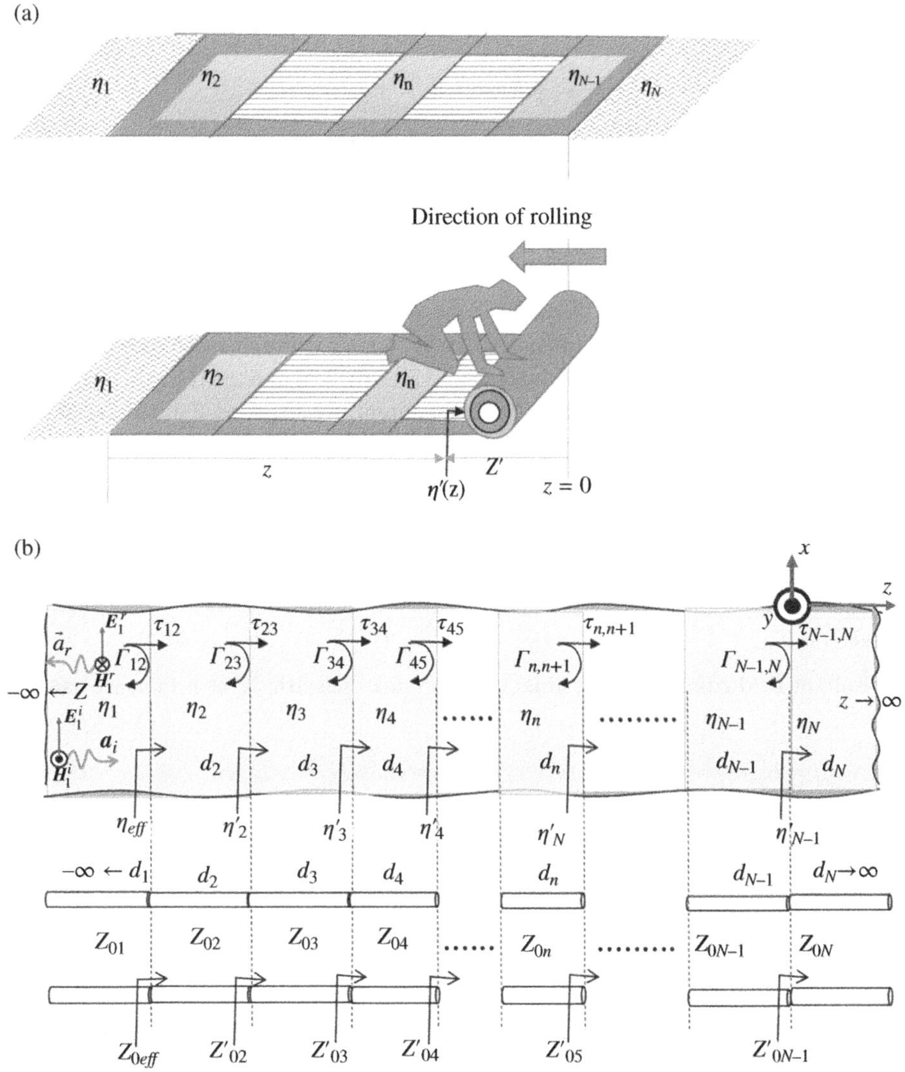

Figure 5.39 (a) N-layered dielectric medium with intrinsic impedance η_n for N = 1, 2, 3..., with carpet rolling method of $n'(z)$ and Γ_{eff}, and (b) equivalent transmission line model with characteristic impedance Z_{On} for N = 1, 2, 3, ...

thicknesses d_n and impedances η_n for $n = 1, 2, 3, \ldots N$. In the layered medium an x-polarized field wave is propagating along the z-direction, and in the propagation process, it partly reflects back from and transmits through each interface until it reaches to the final semi-infinitely extended medium N with the wave impedance η_N. As shown in the figure, the impedance for any intermediate medium n is $\eta_n(z)$. Including the loading effect of the next right dielectric layer, the effective wave impedances are defined as η'_n, where $n = 1, 2, 3, \ldots N$. The respective reflection coefficient between the two adjacent media are defined as: $\Gamma_{n-1,\,n}$, where $n = 2, 3, \ldots N$. Likewise, the respective transmission coefficient between the two adjacent mediums are defined as: $\tau_{n-1,n}$, where $n = 2, 3, \ldots N$. Adding contribution from all media from $n = 2$, to N, the effective wave impedance seen from medium 1 in the multilayered electromagnetic system is defined as η_{eff}. Therefore, using our previous knowledge of the three media propagation, we can generalize the effective reflection and transmission coefficients in medium 1 as:

$$\Gamma_{eff} = \left(\eta_{eff} - \eta_1\right) / \left(\eta_{eff} + \eta_1\right); \tau_{eff} = 2\eta_{eff} / \left(\eta_{eff} + \eta_1\right) \tag{5.132}$$

Figure 5.39b also provides a transmission line analogy of the multiple transmissions and reflections for any number N-media, here we extend the theory using our knowledge for the three layered propagation medium. For each medium, the equivalent transmission line has the characteristic impedance $Z'_{0n} = \eta_n$ where, $n = 1, 2, 3, \ldots N-1$. Such successive impedance transformation provides a systematic and methodical approach to the solution for the effective multiple reflection and transmission of any number of dielectric media.[66]

The core hypothesis is to define the wave impedance at any point z in the electromagnetic system. The wave impedance is defined as the ratio of the resultant electric field and the resultant magnetic field at that point. Therefore, the dielectric layer that encounters any interface of dissimilar media must have a standing wave pattern as the resultant field waves. Hence their wave impedances are not simply defined as $\eta_n = \sqrt{\mu_n/\varepsilon_n}$, rather the impedances are defined as η'_n that considers the loading effects of the successive layers toward the propagation direction. The impedance at any point z in space is defined as:

$$\eta_n = \frac{E_{nx}(z)|_{resultant}}{H_{nx}(z)|_{resultant}} \tag{5.133}$$

The scalar resultant field components at any point z in any dielectric layer n in space can be defined as:

$$E_{nx}(z)|_{resultant} = E^i_{0n} e^{-j\beta_n z}\left(1 + \Gamma_{n-1,n} e^{j2\beta_n z}\right) \quad \text{for } n = 2,3,\ldots N \tag{5.134}$$

$$H_{nx}(z)|_{resultant} = \frac{E^i_{0n}}{\eta_n} e^{-j\beta_n z}\left(1 - \Gamma_{n-1,n} e^{j2\beta_n z}\right) \quad \text{for } n = 2,3,\ldots N \tag{5.135}$$

where, $E^i_{0n} = E_{nx}(z)$, the amplitude of the x-polarized electric field at the interface at $z = nd_n$, for $n = 1, 2, 3, \ldots N-1$. Here, the reflection coefficient at any location z is defined as the ratio of the reflected electric field to the incident electric field as follows:

$$\Gamma_{neff}(z) = \frac{E^r_{0n}}{E^i_{0n}} e^{j2\beta_n z} \tag{5.136}$$

66 We developed similar theory for the reflection and transmission coefficients for multiple junction transmission line with a load impedance at the end.

For simplicity of the calculation, we can define a complex reflection coefficient for each layer with their propagation phase factor $e^{j2\beta_n z}$ in consideration, as follows:

$$\Gamma_{neff}(z) = \Gamma_{n-1,n} e^{j2\beta_n z} \quad \text{for } n = 2,3,...N \tag{5.137}$$

Substitutions for $E_{nx}(z)|_{resultant}$ and $H_{nx}(z)|_{resultant}$, and thereafter, the complex effective reflection coefficient with the propagation phase $\Gamma_{neff}(z) = \Gamma_{n-1,n} e^{j2\beta_n z}$, we obtain the closed form expression for the wave impedance at any point in space z as follows:

$$\eta_n(z) = \eta'_n \frac{1 + \Gamma_{neff}(z)}{1 - \Gamma_{neff}(z)} \tag{5.138}$$

A simple exchange of parameters yields the definition for the effective reflection coefficient at any intermediate layer n as follows:

$$\Gamma_{neff}(z) = \frac{\eta_n(z) - \eta'_n}{\eta_n(z) + \eta'_n} \tag{5.139}$$

The transmission line analogy as shown in the bottom of Figure 5.39 can be obtained by simply a substitution for $Z'_{0n}(z) = \eta_n(z) = \eta'_n\left((1 + \Gamma_{neff}(z))/(1 - \Gamma_{neff}(z))\right)$. It is customary to find the impedance and the reflection coefficient at any arbitrary location z' in a region in terms of the impedance and the reflection coefficient with respect to z of the original coordinate system.[67] The reflection coefficient at the new location z' can be expressed as:

$$\Gamma_{neff}(z') = \frac{E^r_{0n}}{E^i_{0n}} e^{j2\beta_n z'} \tag{5.140}$$

Substitution for the $(E^r_{0n})/(E^i_{0n}) = \Gamma_{neff}(z)e^{-j2\beta_n z}$ from the previous equation yields:

$$\Gamma_{neff}(z') = \Gamma_{neff}(z) e^{j2\beta_n(z'-z)} \tag{5.141}$$

Based on the above definition of the reflection coefficient, we can define the input impedance at any location z' as follows:

$$Z'_{0n}(z') = \eta'_n \frac{1 + \Gamma_{neff}(z')}{1 - \Gamma_{neff}(z')} \tag{5.142}$$

We can make the following remarks:

1) Total impedance is continuous across the interface at location $z = -d_n$; and is defined as:

$$Z'_{0n}(d_n^-) = Z'_{0n}(d_n^+); \text{this condition leads to } \eta_{neff}(d_n^-) = \eta_{neff}(d_n^+) \tag{5.143}$$

67 Such translation of coordinate system from the original position z to a new position z' is very useful in the analysis of terminated transmission line. In a terminated transmission line the load is located as at fixed and known location and the input impedance at any arbitrary point on the transmission line is calculated based on the new coordinate system. We shall be covering the translation of the coordinate system from z to z' in theory of terminated transmission line and input impedance calculation. The new location z' indicates the distance away from the termination. In this theory the last semi-infinitely extended medium with wave impedance η_N represents the load impedance. Hence in this case z' can be assumed to be the length in reference to the last medium and we are looking at the effective impedance and reflection coefficient at any point in space.

This is evident from the continuity of the tangential electric and magnetic fields.

2) However, the effective reflection coefficient is discontinuous across the interface because from the definition: $\Gamma_{neff}(z) = \left(\eta_{neff}(z) - \eta'_n\right)/\left(\eta_n(z) + \eta'_n\right)$, if $\eta_{neff}(z)$ is to be continuous across the interface, η'_n must be different from $\eta_{neff}(z)$ for finite: $\Gamma_{neff}(z)$.
3) With the rolling up processing of calculations of the effective wave impedance, once we reach to the interface between the medium 1 and the equivalent medium 2 as if the whole N-layered electromagnetic system has a single interface at $z = -\sum_{n=2}^{N} d_n$, for $n = 2, 3, \ldots N$. Now we can define the effective reflection coefficient Γ_{eff} for the complete N-layered electromagnetic system as a two-layered medium with the medium 1 and a composite layer that includes effects from all subsequent layers toward the positive z-direction.

In the following, we shall do a few examples to enhance our understanding of wave propagation in the multilayered propagation media.

Review Questions 5.7: Reflection for Large Number of Multilayered Medium

Q1. Explain the carpet rolling analogy to calculate the effective reflection coefficient for a large number of multiple layered dielectric media.

Q2. Draw an equivalent circuit model for a 5-ply dielectric media with intrinsic impedances: η_n where $n = 1, 2, \ldots, 5$ and thicknesses d_n where, $n = 1, 2, \ldots, 5$.

Q3. What is the impedance at n-th layer considering the loading effect of the preceding layers?

Exercise 5.1: Effective Reflection of 4-ply Dielectric Medium

Q1. Calculate the effective reflection coefficient of 4-ply dielectric media for $\varepsilon_{r1} = 1$, $d_1 = -\infty$, $\varepsilon_{r2} = 4.2$, $d_2 = 10$ cm, $\varepsilon_{r3} = 10$, $d_3 = 20$ cm, $\varepsilon_{r4} = 12.5$, $d_4 = \infty$. Assume the operating frequency is 1.6 GHz.

5.9 Special Cases of Reflection from Multiple Layers

In the above section, we have examined the theory of multiple reflections for a large number of dissimilar dielectric media. We have observed that the effective reflection of such a complex medium can be made simple with the folding process from the far right side, and then, find the effective wave impedance and its equivalent input impedance at any point z along the distance of the multilayered media. Using the effective input impedance, we can calculate the effective reflection coefficient at any point along the propagation direction of the multilayered medium, and finally, the effective reflection coefficient of the whole electromagnetic system. As stated before, the theory of the special cases for reflections from such multilayered media has significant industrial applications. Before going to the specific applications of the multilayered electromagnetic system in the very last section of the chapter, let us examine the theories of a few special cases for the reflections from multiple layered medium. Figure 5.40 illustrates the topics that we going to

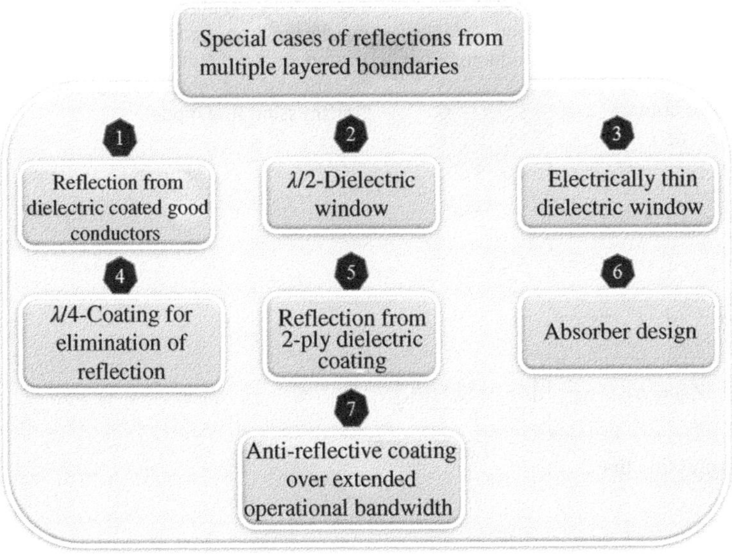

Figure 5.40 Special cases of reflection from multiple layers of practical significances.

cover in the section.[68] As can be seen, we have different special cases of reflections from multiple layers: (i) reflection from dielectric coated good conductors, (ii) $\lambda/2$-dielectric window, (iii) electrically thin dielectric window, (iv) $\lambda/4$-coating for elimination of reflection, (v) reflection from 2-ply dielectric coating, (vi) absorber design, and (vii) antireflection coating over extended operational bandwidth. All the above cases have significant industrial applications.

5.9.1 Reflection from a Dielectric Coated Good Conductor

Consider the case of a dielectric-coated good conductor as shown in Figure 5.41. Microwave superconductive materials design, shielding of circuits for EMC measures, dielectric loaded antennas for gain and bandwidth enhancement, surface wave mode guiding structures and microwave biomedical devices are some applications where we see frequent use of dielectric coated conductors. In the analysis, we find the characteristic impedance of a good conductor, which is defined as $\eta_{gc} = (1 + j)R_s$ where, R_s is the surface resistivity of the conductor.

If we assume the intrinsic impedance of the dielectric coating of finite thickness d is defined as η_2 then we can define the effective impedance at the air–dielectric interface as follows:

$$\eta_{2eff} = \eta_2 \frac{(1 + j)R_s + j\eta_2 \tan(\beta_2 d)}{\eta_2 + (1 + j)R_s \tan(\beta_2 d)} \tag{5.144}$$

For a good conductor, $(R_s/\eta_2) \ll 1$, the effective impedance at the air-dielectric interface at $z = -d$ is reduced to:

$$\eta_{2eff} \cong R_s + j\eta_2 \tan(\beta_2 d) \tag{5.145}$$

68 This section can also be considered as the worked-out examples of the theories of multiple reflections from multilayered boundaries as explained in the above sections.

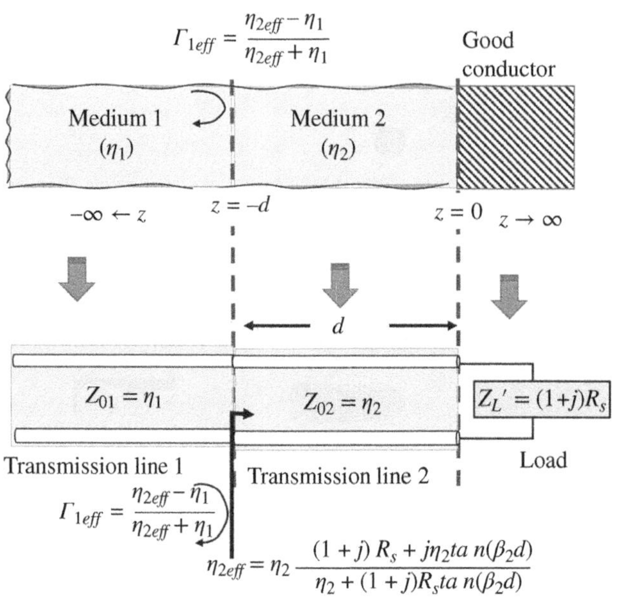

Figure 5.41 Dielectric coated good conductor: (top) good conducting layer coated with a dielectric medium of thickness d and (bottom) equivalent transmission line model.

Then the reflection coefficient is defined as:

$$\Gamma_{eff} \cong \frac{\eta_{2eff} - \eta_1}{\eta_{2eff} + \eta_1} = \frac{R_s + j\eta_2 \tan(\beta_2 d) - \eta_1}{R_s + j\eta_2 \tan(\beta_2 d) - \eta_1} \tag{5.146}$$

Example 5.13: Dielectric Coated Good Conductor

Assume a copper of surface resistivity of $R_s = 2.61 \times 10^{-3}\ \Omega$ at 100 MHz is coated with Teflon with dielectric constant $\varepsilon_r = 2.1$ and thickness $d = 1$ cm. Calculate the reflection coefficient without the Teflon and with Teflon.

First let us consider without Teflon, the reflection coefficient is defined as:

$$\Gamma_{eff} = \frac{\eta_{gc} - \eta_1}{\eta_{gc} + \eta_1} = \frac{(1+j)R_s - \eta_1}{(1+j)R_s + \eta_1} = -\frac{1-(1+j)R_s/\eta_1}{1+(1+j)R_s/\eta_1} \tag{5.147}$$

Here, $\eta_1 = 377\ \Omega$. Given $R_s = 2.61 \times 10^{-3}\ \Omega,\ \eta_1/R_s \ll 1$. We have ground to approximate Γ_{eff} using a binomial expansion as follows:

$$\Gamma_{eff} \cong -\{1-(1+j)R_s/\eta_1\} \tag{5.148}$$

Using binomial expansion with the first order we obtain:

$$\Gamma_{eff} \cong \frac{-1+2(1+j)R_s}{\eta_1} = 1 - \frac{2R_s}{\eta_1} + \frac{j2R_s}{\eta_1} \tag{5.149}$$

The power reflection coefficient:

$$|\Gamma_{eff}|^2 \cong \left(1-\frac{2R_s}{\eta_1}\right)^2 + \left(\frac{2R_s}{\eta_1}\right)^2 \cong 1 - \frac{4R_s}{\eta_1} \tag{5.150}$$

For the above example: The effective reflection coefficient is:

$$|\Gamma_{eff}|^2 \cong 1 - \frac{4 \times 2.61 \times 10^{-3}}{377} \cong 1 \tag{5.151}$$

The % power transmitted is:

$$|\tau_{eff}|^2 = \left(1 - |\Gamma_{eff}|^2\right) \times 100\% \cong \frac{4R_s}{\eta_1} \times 100\% = \frac{4 \times 2.61 \times 10^{-3}}{377} \times 100\% \cong 2.8 \times 10^{-3}\% \tag{5.152}$$

Without the Teflon, we can see that when an electromagnetic energy impinges on a conductor, it totally reflects back almost all signal.

Now consider a Teflon coating on the good conductor and recalculate the effective transmission and reflection coefficients in medium 1.

The wave impedance and propagation constant at 100 MHz in medium 2 are:

$$\eta_2 = \frac{\eta_0}{\sqrt{\varepsilon_r}} = \frac{377}{\sqrt{2.1}} = 260\,\Omega$$

$$\beta_2 = 2\pi \times 10^8 \times \frac{\sqrt{2.1}}{3 \times 10^8} = 3.04\,(\text{rad/m}) \tag{5.153}$$

The effective wave impedance at $z = -d$ is:

$$\begin{aligned}\eta_{2eff} &\cong R_s + j\eta_2 \, tan\,(\beta_2 d) = 2.61 \times 10^{-3} + j260 \times \, tan\left(3.04 \times 10^{-2}\right) \\ &= 2.61 \times 10^{-3} + j7.91\,\Omega\end{aligned} \tag{5.154}$$

Still the value of the effective impedance at the end of the dielectric medium 2 is still quite low. The effective reflection coefficient is:

$$\begin{aligned}\Gamma_{eff} &= \frac{\eta_{2eff} - \eta_1}{\eta_{2eff} + \eta_1} = \frac{2.61 \times 10^{-3} + j7.91 - 377}{2.61 \times 10^{-3} + j7.91 + 377} \cong \frac{-377 + j7.91}{377 + j7.91} = -0.9991 + j0.0419 \\ &\cong 0.999978\angle 177.6^\circ\end{aligned} \tag{5.155}$$

As before, without Teflon coating almost all normally incident power is reflected back in the first dielectric medium as no power is dissipated in Teflon (assuming no loss) and copper is also a bulk perfect conductor.

5.9.2 $\lambda/2$-Dielectric Window for Zero Reflection

The $\lambda/2$-dielectric window is also called "half-wave matching plate" due to its inherent characteristics of zero reflection of the normal incident electromagnetic wave at the particular operating frequency. It is inherently a narrowband design due to the single frequency dependency.

Figure 5.42 illustrates the half-wave dielectric window of wave impedance η_2, which is sandwiched between to semi-infinite dielectric media 1 and 3 with their respective impedances η_1 and η_3, respectively. Let us assume, $d = m\lambda_2/2$, form $m = 1, 2, 3, \ldots$ so that the electrical length of the dielectric window can be defined as $\theta_d = \beta_2 d = (2\pi/\lambda_2) \times m(\lambda_2/2) = m\pi$. We obtain the smallest thickness of the dielectric window considering $m = 1$, so that $\theta_d = \pi$. In this case, medium1 and medium 3 are made of the same material, hence we can define: $\eta_1 = \eta_3 \neq \eta_2$. We shall use the

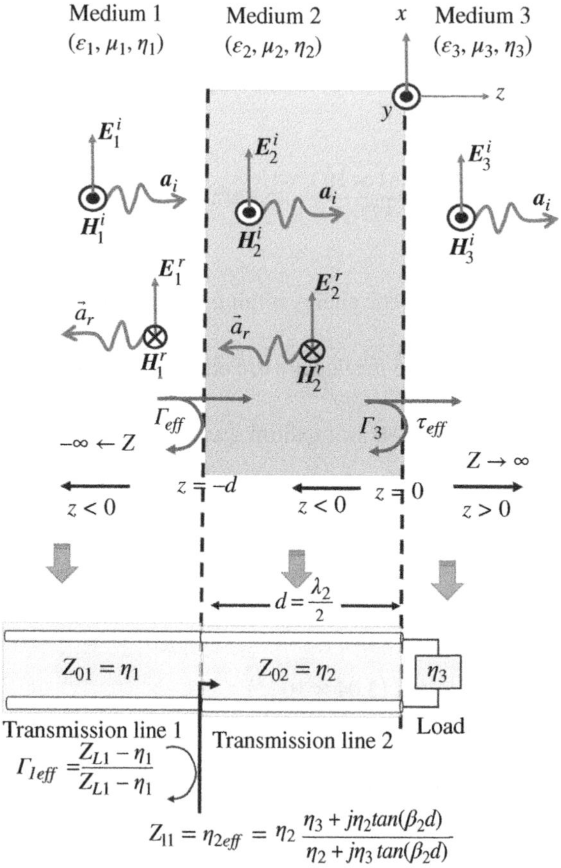

Figure 5.42 A $\lambda/2$-thick dielectric medium sandwiched between two semi-infinite dielectric media.

transmission line analogy as shown in the bottom of Figure 5.42. As can be seen, in this case, then $\eta_1 = \eta_3 = Z_L$. The effective impedance seen at the interface at $z = -d$ can be defined as:

$$Z_{L1} = \eta_2 \times (\eta_3 \cos \beta_2 d + j\eta_2 \sin \beta_2 d)/(\eta_2 \cos \beta_2 d + j\eta_3 \sin \beta_2 d) \tag{5.156}$$

Substitution for $\beta_2 d = \pi$ in the above equation we obtain:

$$Z_{L1} = \eta_3 = \eta_1 \tag{5.157}$$

This condition provides the effective reflection coefficient seen from medium 1:

$$\Gamma_{1eff} = \frac{Z_{L1} - \eta_1}{Z_{L1} + \eta_1} = 0 \tag{5.158}$$

In the modern era of wireless mobile communications, Wi-Fi has become readily available in every household. Communications from one room to another room and even one household to neighboring other household is possible if we can provide such electromagnetically transparent dielectric windows. Consider the example below.

a) Microwave Radome design for mobile satellite communications earth station antenna

You may encounter many hidden or covered antennas at the nose of an aircraft and a fighter jet, on the roof of a modern car, and many in mobile electronic gadgets. This cover is called a radome or a

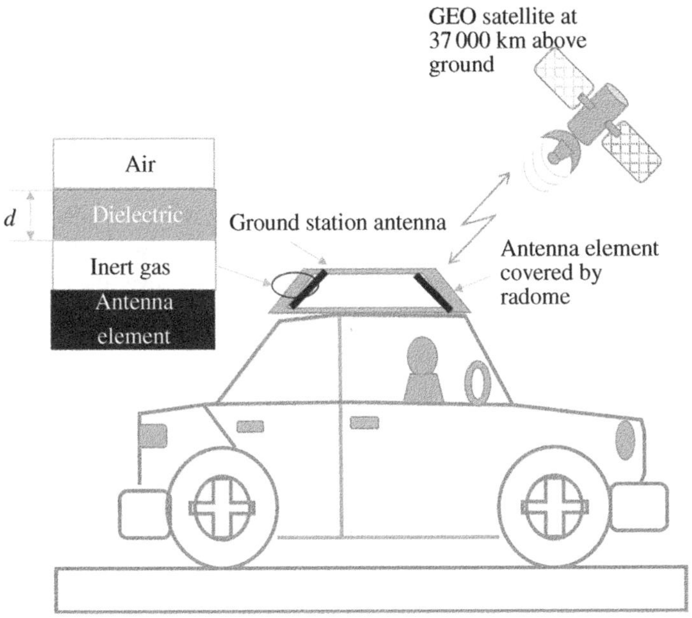

Figure 5.43 Radome used on mobile satellite ground station antenna. Inset figure illustrates radome as multilayered propagation medium.

radar dome. A radome is a conformal, light weight, robust, and microwave transparent (lossless or with negligible loss) protective cover for microwave antennas of various sorts. Figure 5.43 illustrates a mobile satellite communications earth station antenna array with inclined (usually 45°) radiating microstrip patch antenna elements. The antennas are circularly symmetric or cylindrically symmetric so that the antenna can receive normally incident waves from the geostationary earth orbit (GEO) satellites which are located around 37,000 km above the earth. A radome made of a polycarbonate material is made conformal to the antenna structure and sealed airtight with some inert gas so that the electronics avoid corrosion. The expanded cross-sectional view of the electromagnetic system of the radome is shown in the figure.[69]

Assume an x-polarized electromagnetic uniform plane wave is incident on the radome at 1.6 GHz operating frequency.[70] Our goal is to design the radome for minimum reflection loss, light weight, and robustness. Assume the polycarbonate has a dielectric constant $\varepsilon_r = 2.3$ at 1.6 GHz and lossless (microwave transparent). The inert gas has an almost similar dielectric constant of air outside hence $\varepsilon_1 \cong \varepsilon_3 \cong 1.0$. Calculate the thickness of the polycarbonate radome for the minimum reflection and light weight without losing robustness at 1.6 GHz.

Consider the smallest thickness for zero reflection at 1.6 GHz so that

$$d = \frac{\lambda_2}{2} = \frac{\lambda_0}{2\sqrt{\varepsilon_r}} = \frac{3 \times 10^8\ (\text{m/s})}{2 \times 1.6 \times 10^9\ (\text{Hz}) \times \sqrt{2.3}} = 61.8\ (\text{mm})$$

69 Current driverless car is using such satellite signals for GPS position data.

70 Note that many Global Navigation Satellite System (GLONASS) such as Australian Optus Mobilesat™ and international maritime satellite consortium INMARSAT operating frequency fall in the frequency band.

This thickness should protect the antenna for robustness and also provides zero reflection. However, 61.8 mm thick radome would be too bulky; therefore, we chose a thickness of 10 mm, Also note that the condition of zero reflection with $\lambda/2$-window satisfies the operation for a single frequency of 1.6 GHz. Usually, we desire radome design for the following purposes: (i) microwave transparency of the protecting medium for low loss, (ii) lightweight and low cost, and (iii) the mechanical robustness. Since 61.8 cm thick radome is quite bulky, hard to make conformal to the antenna as shown, and adds weight to the mobile ground terminal antenna, we can recalculate the signal reflected from a radome with a thickness 10 mm and examine the power reflected from the radome at 1.6 GHz and its two extreme frequencies at 1.2 and 1.8 GHz to cover the entire band for the GLONASS application.

The wave impedances of media 1, 2 and 3 are:

$$\eta_1 = \eta_3 = 377\,\Omega \text{ and } \eta_2 = \frac{377}{\sqrt{2.3}} = 248.6\,\Omega$$

The electrical lengths and associate parameters at 1.2, 1.6 and 1.8 GHz for a 10 mm thick polycarbonate material, respectively are:

$$\theta_{d-1.2\,\text{GHz}} = \beta_2 d = \frac{2\pi}{\lambda_2} \times d = \frac{2\pi \times 1.2 \times 10^9\,(\text{Hz}) \times \sqrt{2.3}}{3 \times 10^8\,(\text{m/s})} \times 10 \times 10^{-3}(\text{m}) = 0.381\,\text{rad};\ tan\,(\beta_2 d) = 0.4$$

$$\theta_{d-1.6\,\text{GHz}} = \beta_2 d = \frac{2\pi}{\lambda_2} \times d = \frac{2\pi \times 1.6 \times 10^9\,(\text{Hz}) \times \sqrt{2.3}}{3 \times 10^8\,(\text{m/s})} \times 10 \times 10^{-3}\,(\text{m}) = 0.508\,\text{rad};\ tan\,(\beta_2 d) = 0.557$$

$$\theta_{d-1.8\,\text{GHz}} = \beta_2 d = \frac{2\pi}{\lambda_2} \times d = \frac{2\pi \times 1.8 \times 10^9(\text{Hz}) \times \sqrt{2.3}}{3 \times 10^8\,(\text{m/s})} \times 10 \times 10^{-3}\,(\text{m})$$
$$= 0.572\,\text{rad};\ tan\,(\beta_2 d) = 0.662$$

The effective impedances at 1.2, 1.6, and 1.8 GHz seen at the interface $z = -d$, are:

$$Z_{L1-1.2\,\text{GHz}} = 248.6 \times \frac{377 + j248.6 \times 0.4}{248.6 + j377 \times 0.4} = 319.7 - j94.5\,\Omega$$

$$Z_{L1-1.6\,\text{GHz}} = 248.6 \times \frac{377 + j248.6 \times 0.557}{248.6 + j377 \times 0.557} = 288.3 - j105\,\Omega$$

$$Z_{L1-1.8\,\text{GHz}} = 248.6 \times \frac{377 + j248.6 \times 0.662}{248.6 + j377 \times 0.662} = 270 - j106.5\,\Omega$$

The effective reflection coefficient seen from medium 1 at $z = -d$ interface at 1.2, 1.6, and 1.8 GHz are:

$$\Gamma_{1eff-1.2\,\text{GHz}} = \frac{319.7 - j94.5 - 377}{319.7 - j94.5 + 377} = -0.06269 - j0.14414 = 0.157\angle -113.5^\circ$$

$$\Gamma_{1eff-1.6\,\text{GHz}} = \frac{288.3 - j105 - 377}{288.3 - j105 + 377} = -0.10578 - j0.17452 = 0.204\angle -121^\circ$$

$$\Gamma_{1eff-1.8\,\text{GHz}} = \frac{270 - j106.5 - 377}{270 - j106.5 + 377} = -0.13464 - j0.18677 = 0.23\angle 125.8^\circ$$

The percentage of the incident power reflects back at 1.2, 1.6, and 1.8 GHz seen at the interface $z = -d$, are:

$$|\Gamma_{1eff-1.2\,\text{GHz}}|^2 = 0.157^2 \times 100\% = 2.46\%$$

$$|\Gamma_{1eff-1.6\ \mathrm{GHz}}|^2 = 0.204^2 \times 100\% = 4.16\%$$

$$|\Gamma_{1eff-1.8\ \mathrm{GHz}}|^2 = 0.23^2 \times 100\% = 5.29\%$$

From the above results we can conclude that the percentage of reflected power from the radome over the GLONASS frequency band is significantly low. The results also augment the theory of a thin dielectric window that is going to present in the next section.

b) Microwave glass window design for wireless Wi-Fi routers

A 5.8 GHz Wi-Fi router communicates between two rooms of a house via a glass window made of Corning Pyrex® 7070 Borosillicate glass with dielectric constant $\varepsilon_r = 4.1$. Such glass is also used from wireless communications devices using microsystem technology (MST) and radio frequency micro-electromechnacial systems (MEMS) technology. Assume the glass is lossless and has a dielectric constant $\varepsilon_r = 4.1$ at 5.8 GHz. Calculate the thickness of the glass for zero reflection.

Answer:
We consider the smallest thickness so that $m = 1$.

$$d = \frac{\lambda_2}{2} = \frac{\lambda_0}{2\sqrt{\varepsilon_r}} = \frac{3 \times 10^8\ (\mathrm{m/s})}{2 \times 5.8 \times 10^9 (\mathrm{Hz}) \times \sqrt{4.1}} = 12.77\ (\mathrm{mm})$$

Of course the calculated thickness provides zero reflection for a single frequency of 5.8 GHz and provides reflection at other frequency. Also, we obtain zero reflection at 5.8 GHz for a thicker dielectric window which is an integer multiple of $d = m \times \lambda_2/2$ for $m = 1, 2, 3, \ldots$ etc. For structural robustness, we may select a thicker dielectric window without affecting the performance if we obey the above condition.

Now let us consider the reflection coefficient at the lower wi-fi frequency of 2.45 GHz. The electrical length of $d = 12.77$ (mm) at 2.45 is:

$$\theta_d = \beta_2 d = \frac{2\pi}{\lambda_2} \times d = \frac{2\pi \times 2.45 \times 10^9\ (\mathrm{Hz}) \times \sqrt{4.1}}{3 \times 10^8\ (\mathrm{m/s})} \times 12.77 \times 10^{-3}\ (\mathrm{m}) = 1.32\ \mathrm{rad}$$

The wave impedance of medium 2 is: $\eta_2 = \frac{\eta_0}{\sqrt{\varepsilon_r}} = \frac{377}{\sqrt{4.1}} = 186.2\ \Omega$

The effect impedance seen at the interface $z = -d$ is:

$$\begin{aligned} Z_{L1} &= \eta_2 \times (\eta_3 + j\eta_2 \tan \beta_2 d)/(\eta_2 + j\eta_3 \tan \beta_2 d) \\ &= 186.2 \times (377 + j186.2 \tan 1.32)/(186.2 + j377 \tan 1.32) = 96.8 - j35.4\ \Omega \end{aligned}$$

The reflection coefficient seen from medium 1 is:

$$\Gamma_{1eff} = \frac{Z_{L1} - \eta_1}{Z_{L1} + \eta_1} = \frac{96.8 - j35.4 - 377}{96.8 - j35.4 + 377} = 0.59\angle - 168.5^\circ$$

The % power reflected from the dielectric window is defined as: $|\Gamma_1|^2 = 0.59^2 \times 100\,\% = 34.8\%$ This results indicate more than one third power reflects back in medium 1 during the 2.45 GHz signal propagation through the dielectric window.

5.9.3 Electrically Thin Dielectric Window

Electrically thin dielectric windows are used to protect expensive optical and microwave devices. However, we would not like to have a high reflection from the thin dielectric coating.

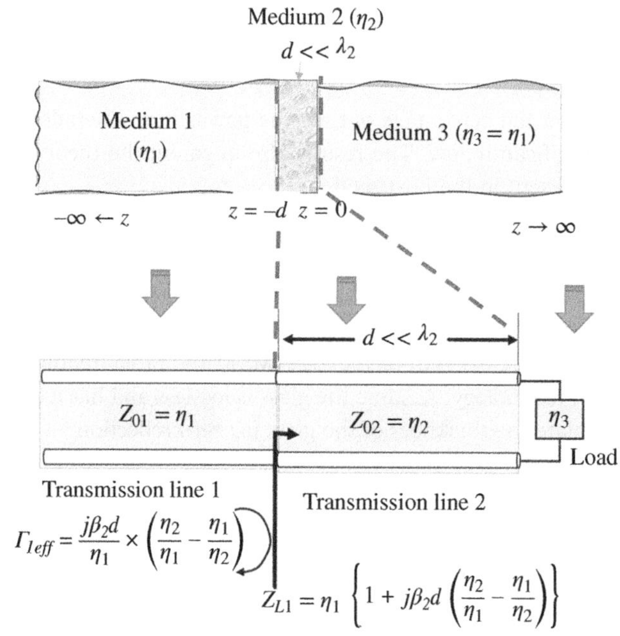

Figure 5.44 Electrically thin dielectric window of thickness $d \ll \lambda$.

Figure 5.44 illustrates the electromagnetic system with thin dielectric window $d \ll \lambda$. For the conditions $\eta_1 = \eta_3$, and $d \ll \lambda$, we can define $\tan(\beta_2 d) \cong \beta_2 d$. We can define the effective impedance at the interface $z = -d$ as follows:

$$Z_{L1} = \eta_2 \times (\eta_1 + j\eta_2\beta_2 d)/(\eta_2 + j\eta_1\beta_2 d)$$

$$= \eta_1 \times \frac{\eta_2 + j\left(\frac{\eta_2^2}{\eta_1}\right)\beta_2 d}{\eta_2 + j\eta_1\beta_2 d}$$

$$\text{Or, } Z_{L1} = \eta_1\left\{1 + j\beta_2 d\left(\frac{\eta_2}{\eta_1} - \frac{\eta_1}{\eta_2}\right)\right\} \tag{5.159}$$

The effective reflection coefficient seen from medium 1 is defined as:

$$\Gamma_1 = \frac{Z_{L1} - \eta_1}{Z_{L1} + \eta_1} = \frac{j\beta_2 d}{\eta_1} \times \left(\frac{\eta_2}{\eta_1} - \frac{\eta_1}{\eta_2}\right) \tag{5.160}$$

A few remarks can be made for the signal conditions via the thin dielectric window:

1) The effective reflection coefficient seen from medium 1 is proportional to d for $\tan(\beta_2 d) \cong \beta_2 d$. Therefore, the power reflection coefficient $|\Gamma_1|^2$ is proportional to d^2. Let us do an example for the same glass with a thin layer of $d = 1$ μm and the operating frequency is 5.8 GHz. Substation for $\eta_1 = 377\ \Omega$ and $\eta_2 = 186.3\ \Omega$ we obtain the reflection coefficient:

$$\Gamma_1 = \frac{j1.32}{377} \times \left(\frac{168.2}{377} - \frac{377}{186.2}\right) = -j188.3 \times 10^{-6} \tag{5.161}$$

The % power reflected from the dielectric window is defined as: $|\Gamma_1|^2 = 3.5 \times 10^{-6}\,\%$.

5.9.4 $\lambda/4$-Dielectric Transformer Window

In the preceding examples, we consider a sandwiched dielectric medium of specific thickness in between two similar dielectric media for zero reflection looking from medium 1. These are specific cases of electromagnetic systems where propagation happens in special circumstances, as delineated in the above examples. Robustness of structures and indoor wireless communications are two examples that use the theory. In generalized cases with dissimilar dielectric media for multilayered electromagnetic system, we can encounter many challenging propagation problems. In those cases, we would like to design the intermediate propagation medium such that we can achieve zero reflection coefficient. Such an intermediate dielectric medium that offers a perfect match and zero reflection is called *impedance transformer*. Usually, the impedance transformer widows are quarter-wavelength long with respect to the operating frequency. Figure 5.45 illustrates a quarter wavelength intermediate medium of wave impedance η_2 in between two end semi-infinite media 1 and 2 with wave impedance η_1 and η_3, respectively. As usual, an x-polarized uniform plane wave propagates from medium 1 to medium 3 via the $\lambda_2/4$-dielectric window. Our goal is to find the value of the wave impedance of medium 2 such that $\eta_2 = \sqrt{\eta_1 \eta_3}$.

In Section 5.8.3, successive transmission and reflection analysis via $\lambda/4$-thick dielectric medium, we have developed a relationship between the wave impedance as: $\eta_2 = \sqrt{\eta_1 \eta_3}$ for zero reflection coefficient using the successive reflection and transmission theory. The condition is called *perfect*

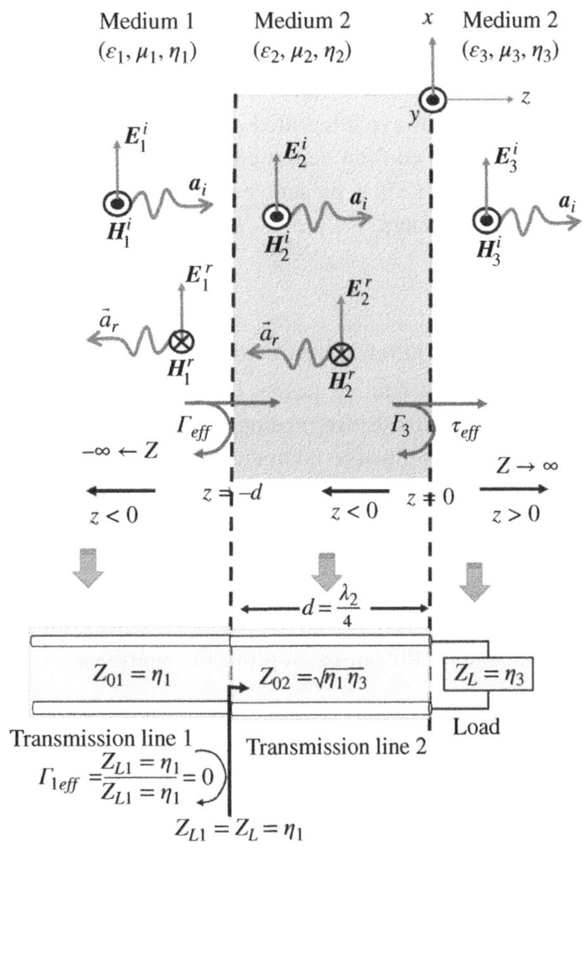

Figure 5.45 $\lambda/4$ – dielectric window and equivalent transmission line model.

matched condition to dielectric 1 that yield $\Gamma_{1eff} = 0$. In this section, we shall use the instantaneous wave theory and field boundary conditions for the development of the reflection coefficient and wave impedance as we have done in the above few examples.

For $d = n\lambda_2/4$, for $n = (2m + 1)$ for $m = 0, 1, 2, 3, \ldots$, the odd multiple of a quarter wavelength, the electrical length of the dielectric window can be defined as $\theta_d = \beta_2 d = (2\pi/\lambda_2) \times (2m + 1)(\lambda_2/4) = (2m + 1)\pi/2$. We obtain the smallest thickness of the dielectric window considering $m = 0$, so that $\theta_d = \pi/2$. In this case, medium 1 and medium 3 are made of dissimilar dielectric material; hence we can define: $\eta_1 \neq \eta_3 \neq \eta_2$. We shall use the transmission line analogy as shown at the bottom of Figure 5.45 for the calculation of the effective impedance at the interface $z = -d$ and the reflection coefficient. As can be seen, in this case, our goal is to make the intermediate medium's wave impedance η_2 is such that $\eta_2 = \sqrt{\eta_1\eta_3}$ so that η_1 becomes Z_{L1}. The effective impedance see at the interface at $z = -d$ can be defined as:

$$Z_{L1} = \eta_2 \times \frac{\eta_3 \cos\beta_2 d + j\eta_2 \sin\beta_2 d}{\eta_2 \cos\beta_2 d + j\eta_3 \sin\beta_2 d} = \frac{\eta_2^2}{\eta_3} \quad \text{for } \Gamma_{1eff} = 0 \tag{5.162}$$

Substitution for $Z_{L1} = \eta_1$ for $\Gamma_{1eff} = 0$, in the above equation and simple mathematical manipulation we obtain:

$$\eta_2 = \sqrt{\eta_1\eta_3} \tag{5.163}$$

This condition provides the effective reflection coefficient seen from medium 1:

$$\Gamma_{1eff} = \frac{Z_{L1} - \eta_1}{Z_{L1} + \eta_1} = 0$$

In the modern era of wireless mobile communications, we use various types of antenna and monolithic microwave integrated circuits with multiple layer dielectric. One good example is low temperature co-fired ceramic (LTCC) circuits with many interleaved dielectric layers. In the following, we shall present an interesting example of the quarter-wavelength dielectric window transformers.

$$Z_{L1} = \eta_3 = \eta_1 \tag{5.164}$$

Example 5.14: Antireflective Coating of Optical Devices

Once you buy new reading glasses, the optometrist always recommends you antireflective reading glasses. These antireflective reading glasses are specifically beneficial in modern day usage when people are always glued to computer monitors and smartphones. Antireflective coating[71] makes more light available to your eyes and reduces the stress on your eyes, and improves vision, especially at low light conditions. Since there is no reflection of incoming light from the glasses, antireflective coated glasses improve eye contact with people as they can see your eyes without reflection. Antireflective coated eyeglasses also look more aesthetically pleasing as the lenses look almost invisible.[72] Antireflective coating virtually eliminates the reflection of light from the front and back sides of the lenses, improve the sharpness of vision, and improves visibility at night.

71 Anti-reflective coating is also called anti-glare coating specifically for camera lenses and sunglasses. For sunglasses the anti-reflective coating are applied at the back side of the lens so that the sun light does not reflect from the back side of the lenses to your eyes.

72 Gary Heiting, O.D. (2018). Anti-reflective coating: see better and look better. http://www.allaboutvision.com/lenses/anti-reflective.htm (accessed 14 April 2018).

High refractive index glass and plastic lenses reflect almost 50% of the incoming light; therefore, less light is available for vision. Therefore, antireflective coated vision glasses are very useful for a low light condition such as at night. Most expensive modern cars have an antireflective coating on their back mirrors that can adaptively adjust the refractive index of the back mirror to stop high glare from the headlights of the cars behind a driver.

Modern antireflective coatings virtually eliminate any reflection of light and allow about 99.5% of visible light passing through the lenses, which is available for vision.[73] The process of applying antireflective coating on reading glasses and optical lenses needs very high precision vacuum deposition technology in a clean room environment so that any contamination cannot damage the efficacy of the coating. Antireflective coating usually comprises multiple layers of metallic oxides with alternating high and low refractive indices so that antireflective coating (as a multiple layered propagation medium) can neutralize light of wide spectrum. The materials used as the antireflective coating are Al_2O_3, MgF_2, silicon nitride, silica (SiO_2), titanium, zinc oxide, and soluble Teflon.

Some high quality antireflective coating can go up to seven layers. However, the total thickness of the multilayered coating is around 0.2–0.3 μm and comprises about 0.02% of the total thickness of a standard eyeglass lens. For heavy computer and smartphone users, and people who are exposed to energy saving bright lights, antireflective coating that filters out blue light is suggested. In this section, we shall work out a couple of examples of the antireflecting coating of optical lenses. First, we examine the theory of a single-layer coating followed by a multiple-layered antireflective coating.

Example 5.15: Single-layer Coating for Reading Glasses for the Elimination of Reflection of Light

Consider the white light wave with a mean wavelength of 545 nm. Calculate the length of the antireflective coating of a reading glass. Assume the refractive index of the glass is 1.52 and lossless. Design an antireflective coating using a quarter-wavelength dielectric window.

Answer:

The conventional reading glass has a dielectric constant 2.31 ($n = 1.52$).

The mean operating frequency at 545 nm wavelength is: $f = (3 \times 10^8\ (\text{m/s}))/(545 \times 10^{-9}\ (\text{m})) = 550.45 \times 10^{12}$ Hz

The wave impedance of air: $\eta_1 = 377\ \Omega$; the wave impedance of glass: $\eta_3 = 377/\sqrt{2.31} = 248\ \Omega$

The wave impedance of the coating is: $\eta_2 = \sqrt{377 \times 248} = 305.8\ \Omega$;

The dielectric constant of the coating is: $\varepsilon_{r2} = \eta_0/\eta_2 = 377/305.8 = 1.23$. This equates the refractive index of the single-layer coating is: $n = \sqrt{1.23} = 1.11$.

The thickness of the coating is: $d = \lambda_2/4 = (3 \times 10^8)/\left(4 \times 550.45 \times 10^{12} \times \sqrt{1.23}\right) = 123\,\text{nm} = 0.123\,\mu\text{m}$.

The calculated thickness provides zero reflection for a single frequency of 550 THz and provides reflection at other frequencies. Also, we obtain zero reflection at 550 THz for a thicker dielectric window which is an integer multiple of $d_m = m \times d$ for $m = 1, 2, 3, \ldots$ etc. For structural robustness, we may select a thicker dielectric window without affecting the performance if we obey the above condition.

---

73 The objective of the section is to theoretically prove that only ~1% of the incoming light is reflected back from a multilayered anti-reflective coating.

Example 5.16: Multilayered Antireflective Coating of Optical Devices for Extended Operational Bandwidth

In the above example, we consider the narrowband design that operates only for the signal frequency at which we can meet the condition of zero reflection with a $\lambda/4$ – thick dielectric window. We can extend the bandwidth of the zero-reflection condition if we can use multiple layered coating of different refractive indices and different thicknesses, as shown in Figure 5.46. In this case, we consider the rounded mean/middle wavelength of visible light $\lambda_0 = 550$ nm. Our goal is to design a multilayered antireflective coating. The procedure of calculation of the effective reflection coefficient is to first find the physical thickness of each layer in *nm*; calculation of the phase constants β_n, the wave impedances η_n and the transmission phase $\beta_n d_n$ and $tan\,(\beta_n d_n)$ for $n = 2, 3, 4, 5$; substitutions for these values in the calculation of effective wave impedance Z'_{04},Z'_{03}, Z'_{02} and Z_{0eff} and then finally calculation of effective reflection coefficient Γ_{1eff} to find the percentage of power reflected back from the multilayered coating from the entire band. The calculation is done from right to left, as shown in the carpet rolling process.

$$d_4 = \frac{\lambda_0}{4n_4} = \frac{550\,(\text{nm})}{4 \times 1.45} = 95\,(\text{nm})$$

$$d_3 = \frac{\lambda_0}{2n_3} = \frac{550\,(\text{nm})}{2 \times 1.68} = 164\,(\text{nm})$$

$$d_2 = \frac{\lambda_0}{4n_2} = \frac{550\,(\text{nm})}{4 \times 1.32} = 104\,(\text{nm})$$

Not that for a smooth transition from air to the dielectric coating outermost layer must have a low reflective index. In this case, we use MgF_2 with refractive index $n_2 = 1.32$ is used. Next, we calculate

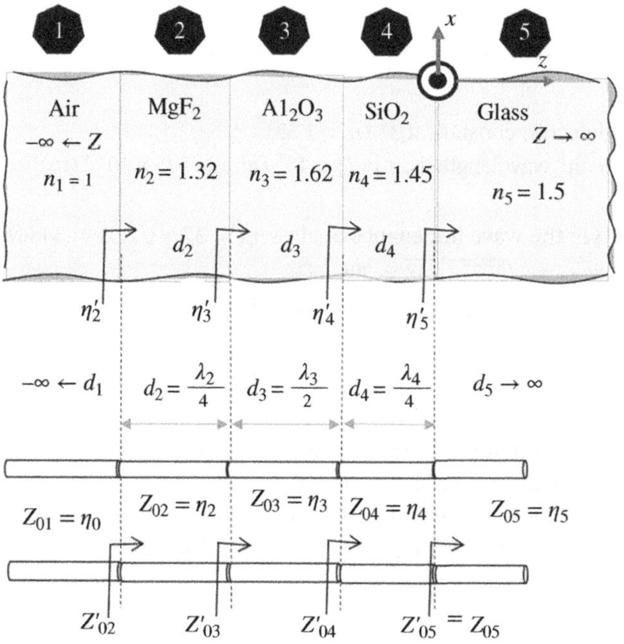

Figure 5.46 Multilayered coating of reading glasses.

the phase constants β_n, the wave impedances η_n and the transmission phase $\beta_n d_n$ and $tan\ (\beta_n d_n)$ for $n = 2, 3, 4, 5$. Let us work on these calculations before calculating the effective impedances Z'_{04}, Z'_{03}, Z'_{02}, and Z_{0eff} and the effective reflection coefficient Γ_{1eff}.

$$\beta_4 = \frac{2\pi n_4}{\lambda_0} = \frac{2\pi \times 1.45}{550 \times 10^{-9}} = 16.56 \times 10^6 \text{ (rad/m)}$$

$$\eta_4 = \frac{\eta_0}{n_4} = \frac{377}{1.45} = 260\ \Omega$$

$$\beta_3 = \frac{2\pi n_3}{\lambda_0} = \frac{2\pi \times 1.68}{550 \times 10^{-9}} = 19.19 \times 10^6 \text{ (rad/m)}$$

$$\eta_3 = \frac{\eta_0}{n_3} = \frac{377}{1.68} = 224\ \Omega$$

$$\beta_4 = \frac{2\pi n_2}{\lambda_0} = \frac{2\pi \times 1.32}{550 \times 10^{-9}} = 15.08 \times 10^6 \text{ (rad/m)}$$

$$\eta_4 = \frac{\eta_0}{n_4} = \frac{377}{1.32} = 286\ \Omega$$

The transmission phases $\beta_n d_n$ are calculated in terms of radians. To simplify the calculation, we use wavelengths λ_n for $n = 2, 3, 4$ in our calculations:

$$d_4 = \frac{\lambda_4}{4}; \beta_4 = \frac{2\pi}{\lambda_4}; d_4\beta_4 = \frac{2\pi}{\lambda_4} \times \frac{\lambda_4}{4} = \frac{\pi}{2}$$

$$d_3 = \frac{\lambda_3}{2}; \beta_3 = \frac{2\pi}{\lambda_3}; d_3\beta_3 = \frac{2\pi}{\lambda_3} \times \frac{\lambda_3}{2} = \pi$$

$$d_2 = \frac{\lambda_2}{4}; \beta_2 = \frac{2\pi}{\lambda_2}; d_2\beta_2 = \frac{2\pi}{\lambda_2} \times \frac{\lambda_2}{4} = \frac{\pi}{2}$$

The effective impedances from right to left are as follows:

$$Z'_{04} = Z_{04} \times \frac{Z_{05}\cos(\beta_4 d_4) + jZ_{04}\sin(\beta_4 d_4)}{Z_{04}\cos(\beta_4 d_4) + jZ_{05}\sin(\beta_4 d_4)} = \frac{\eta_4^2}{\eta_5} \quad \text{for } d_4\beta_4 = \frac{\pi}{2} \text{ and } Z_{04} = \eta_4 \text{ and } Z_{05} = \eta_5$$

$$Z'_{03} = Z_{03} \times \frac{Z'_{04}\cos(\beta_3 d_3) + jZ_{03}\sin(\beta_3 d_3)}{Z_{03}\cos(\beta_3 d_3) + jZ'_{04}\sin(\beta_3 d_3)} = \frac{\eta_4^2}{\eta_5} \quad \text{for } d_3\beta_3 = \pi \text{ and } Z'_{04} = \frac{\eta_4^2}{\eta_5}$$

Note that for a transmission line section of the electrical length $d_3\beta_3 = \pi$ brings back the same load impedance as the input impedance. This means here $Z'_{03} = Z'_{04} = \eta_4^2/\eta_5$. This is due to the fact that the standing wave created in the reflective medium the phase rotates twice as first as the pure incident wave. We shall examine similar behavior of a terminated transmission line of the electrical length $\theta = \beta l = \pi$, where the physical length of the line is defined in terms of wavelength $l = \lambda/2$.

Finally, we calculate the effective impedance Z'_{02} seen at the interface of media 1 and 2 (air-MgF_2) as follows:

$$Z'_{02} = Z_{02} \times \frac{Z'_{03}\cos(\beta_2 d_2) + jZ_{02}\sin(\beta_2 d_2)}{Z_{02}\cos(\beta_2 d_2) + jZ'_{03}\sin(\beta_2 d_2)} = \frac{\eta_2^2\eta_5}{\eta_4^2} \quad \text{for } d_2\beta_2 = \frac{\pi}{2} \text{ and } Z'_{03} = \frac{\eta_4^2}{\eta_5}$$

Substitutions for the respective wave impedances we obtain the numerical value of Z'_{02} as follows:

$$Z'_{02} = \frac{\eta_2^2\eta_5}{\eta_4^2} = \frac{286^2 \times 251}{260^2} = 303\ \Omega$$

Finally, the effective reflection coefficient seen at medium 1 is defined as:

$$\Gamma_{1eff} = \frac{Z'_{02} - Z_{01}}{Z'_{02} + Z_{01}} = \frac{303 - 377}{303 + 377} = -0.11$$

The percentage of power reflected back in medium one is:

$$|\Gamma_{1eff}|^2 = (0.11)^2 \times 100\% \cong 1\% \tag{5.165}$$

This indicates the effectiveness of the multilayered coated optical devices. If plot the above definition of $\Gamma_{1\mathrm{eff}} = \left(Z'_{02} - Z_{01}\right)/\left(Z'_{02} + Z_{01}\right)$ using the $d_n\beta_n$ over the visible optical region that for the entire band of 400–750, extremely low reflection coefficient ~1% of power reflects back in the medium. If no coating is used, the reflection from air to glass is calculated as approximately 4%.

For completeness of the special cases, we shall do two more examples at microwave frequency bands. One is at a 1.6 GHz frequency band for a 2-ply dielectric and the second one is a microwave absorber design with a thin layer of dielectric place at a distance of $(3\lambda_0)/4$ for a perfect electric conducting reflector.

5.9.5 Reflection for 2-Ply Dielectric Window

Consider a 2-ply dielectric medium in between two semi-infinite media 1 and 4, as shown in Figure 5.47. Calculate the effective reflection coefficient seen from medium 1 to the right for a 2-ply radome for the ground mobile satellite antenna. Given an x-polarized wave is incident from medium 1 at the center operating frequency of 1.6 GHz with band end frequencies of 1.2 and 1.8 GHz. The dielectric constant for medium 1 to medium 4 are: $\varepsilon_{r1} = 1.0$ (air); $\varepsilon_{r2} = 2.54$ (polysterence); $\varepsilon_{r3} = 4.2$ (Corning glass); *and* $\varepsilon_{r4} = 1.0$ (air), respectively. The thicknesses of the 2-ply dielectric

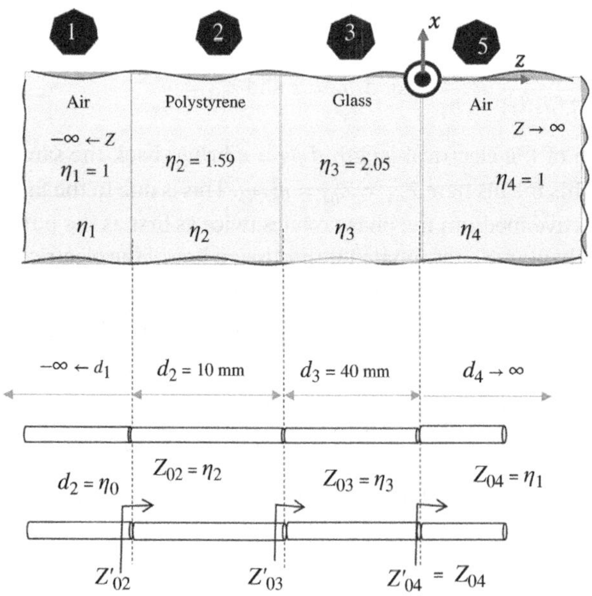

Figure 5.47 Reflection from two-ply dielectric window with finite thickness.

radome are: $d_2 = 1$ cm and $d_3 = 4$ cm. We would like to apply the transmission line model to calculate the effective reflection coefficient of medium 1.

The wave impedances of media 1, 2, 3, and 4 are:

$$\eta_1 = \eta_4 = 377\,\Omega,\ \eta_2 = \frac{377}{1.59} = 237\,\Omega \text{ and } \eta_3 = \frac{377}{2.05} = 184\,\Omega$$

The electrical lengths and associate parameters at 1.2, 1.6 and 1.8 GHz for a 40 mm thick glass material, respectively are:

$$\theta_{d-1.2\,\text{GHz}} = \beta_2 d = \frac{2\pi}{\lambda_2} \times d = \frac{2\pi \times 1.2 \times 10^9\,(\text{Hz}) \times 2.05}{3 \times 10^8\,(\text{m/s})} \times 40 \times 10^{-3}\,(\text{m})$$
$$= 2.06\,\text{rad};\ tan\,(\beta_2 d) = -1.88$$

$$\theta_{d-1.6\,\text{GHz}} = \beta_2 d = \frac{2\pi}{\lambda_2} \times d = \frac{2\pi \times 1.6 \times 10^9\,(\text{Hz}) \times 2.05}{3 \times 10^8\,(\text{m/s})} \times 40 \times 10^{-3}\,(\text{m})$$
$$= 2.75\,\text{rad};\ tan\,(\beta_2 d) = -0.41$$

$$\theta_{d-1.8\,\text{GHz}} = \beta_2 d = \frac{2\pi}{\lambda_2} \times d = \frac{2\pi \times 1.8 \times 10^9\,(\text{Hz}) \times 2.05}{3 \times 10^8\,(\text{m/s})} \times 40 \times 10^{-3}\,(\text{m})$$
$$= 3.09\,\text{rad};\ tan\,(\beta_2 d) = -0.05$$

The effective impedances at 1.2, 1.6, and 1.8 GHz seen at the interface $z = -d_3$, are:

$$Z'_{03-1.2\,\text{GHz}} = 184 \times \frac{377 + j184 \times (-1.88)}{184 + j377 \times (-1.88)} = 107.94 + j69.85\,\Omega$$

$$Z'_{03-1.6\,\text{GHz}} = 184 \times \frac{377 + j184 \times (-0.41)}{184 + j377 \times (-0.41)} = 262.51 + j139.7\,\Omega$$

$$Z'_{03-1.8\,\text{GHz}} = 184 \times \frac{377 + j184 \times (-0.05)}{184 + j377 \times (-0.05)} = 420.03 - j10.25\,\Omega$$

The electrical lengths and associate parameters at 1.2, 1.6, and 1.8 GHz for a 10 mm thick polycarbonate material, respectively, are:

$$\theta_{d-1.2\,\text{GHz}} = \beta_2 d = \frac{2\pi}{\lambda_2} \times d = \frac{2\pi \times 1.2 \times 10^9\,(\text{Hz}) \times 1.59}{3 \times 10^8\,(\text{m/s})} \times 10 \times 10^{-3}\,(\text{m})$$
$$= 0.4\,\text{rad};\ tan\,(\beta_2 d) = 0.38$$

$$\theta_{d-1.6\,\text{GHz}} = \beta_2 d = \frac{2\pi}{\lambda_2} \times d = \frac{2\pi \times 1.6 \times 10^9\,(\text{Hz}) \times 1.59}{3 \times 10^8\,(\text{m/s})} \times 10 \times 10^{-3}\,(\text{m})$$
$$= 0.53\,\text{rad};\ tan\,(\beta_2 d) = 0.49$$

$$\theta_{d-1.8\,\text{GHz}} = \beta_2 d = \frac{2\pi}{\lambda_2} \times d = \frac{2\pi \times 1.8 \times 10^9\,(\text{Hz}) \times 1.59}{3 \times 10^8\,(\text{m/s})} \times 10 \times 10^{-3}\,(\text{m})$$
$$= 0.6\,\text{rad};\ tan\,(\beta_2 d) = 0.54$$

The effective impedances at 1.2, 1.6, 1.8 GHz seen at the interface $z = -d_2$, are:

$$Z'_{02-1.2\,\text{GHz}} = 237 \times \frac{107.94 + j69.85 + j237 \times 0.38}{237 + j(107.94 + j69.85) \times 0.38} = 150.87 + j150.67\,\Omega$$

$$Z'_{02-1.6\ \mathrm{GHz}} = 237 \times \frac{262.52 + j139.7 + j237 \times 0.49}{237 + j(262.52 + j139.7) \times 0.49} = 406.76 + j49.29\ \Omega$$

$$Z'_{02-1.8\ \mathrm{GHz}} = 237 \times \frac{420.03 - j10.25 + j237 \times 0.54}{237 + j(420.03 - j10.25) \times 0.54} = 231.88 + j62.71\ \Omega$$

The effective reflection coefficient seen from medium 1 at $z = -d$ interface at 1.2, 1.6 and 1.8 GHz are:

$$\Gamma_{1eff-1.2\ \mathrm{GHz}} = \frac{150.87 + j150.67 - 377}{150.87 + j150.67 + 377} = -0.32 + j0.38 = 0.45\angle -130^\circ$$

$$\Gamma_{1eff-1.6\ \mathrm{GHz}} = \frac{406.76 + j49.29 - 377}{406.76 + j49.29 + 377} = -0.04 + j0.06 = 0.072\angle 124^\circ$$

$$\Gamma_{1eff-1.8\ \mathrm{GHz}} = \frac{231.88 + j62.71 - 377}{231.88 + j62.71 + 377} = -0.26 + j0.13 = 0.29\angle 153^\circ$$

The percentage of the incident power reflects back at 1.2, 1.6, 1.8 GHz seen at the interface $z = -d$, are:

$$|\Gamma_{1eff-1.2\ \mathrm{GHz}}|^2 = 0.45^2 \times 100\% = 20.25\%$$

$$|\Gamma_{1eff-1.6\ \mathrm{GHz}}|^2 = 0.072^2 \times 100\% = 0.58\%$$

$$|\Gamma_{1eff-1.8\ \mathrm{GHz}}|^2 = 0.23^2 \times 100\% = 5.29\%$$

From the above results, we can conclude that the percentage of reflected power from the radome over the GLONASS frequency band is significantly low except 1.2 GHz. The results also augment the theory of a thin dielectric window that is going to present in the next section.

5.9.6 Electromagnetic Absorber Design with a Thin Dielectric Window Placed $(3\lambda_0)/4$ Distance from a Perfect Electric Conductor

Consider a thin dielectric window (usually lossy) or a good conductor of thickness d with constitutive parameters $(\varepsilon_2, \mu_2, \sigma_2)$ is placed $(3\lambda_0)/4$ distance away from a semi-infinitely extended perfect electric conductor (PEC) as shown in Figure 5.48. Hence the air gap thickness between the PEC and the lossy medium is $(3\lambda_0)/4$. Assume both air on the left most side and the PEC on the right most side are extended semi-infinitely as media 1 and 4. The thin lossy medium 2 has thickness d and the $(3\lambda_0)/4$ thick air gap is medium 3. Design the absorbing medium so that a zero effective reflection coefficient can be seen from air medium 1 to the right for a 3-ply absorber for an anechoic chamber measurement system at the same frequency of the satellite communications antenna of 1.6 GHz.

Answer:
We need to calculate the thickness d_2 and constitutive parameters $(\varepsilon_2, \mu_2, \sigma_2)$ of the lossy dielectric or conducting media 2. For a lossy dielectric, we can assume $\mu_{r2} \cong 1$ and a good conductor we can assume $\varepsilon_2 = 1$, $\mu_{r2} \cong 1$.[74]

Given an x-polarized wave is incident from medium 1 at the center operating frequency of 1.6 GHz. The dielectric constant for medium 1 to medium 4 are: $\varepsilon_{r1} = 1.0$ (air); $\varepsilon_{r2} =$ (lossy medium);

74 Refresh the theory of wave propagation in lossy dielectric and good conducting media in Chapter 2 Wave Propagation in Homogeneous, Nondispersive Lossy Media Sections 2.3 and 2.7 respectively.

Figure 5.48 Absorber design with a thin good conductor placed $(3\lambda_0)/4$ away from a PEC wall.

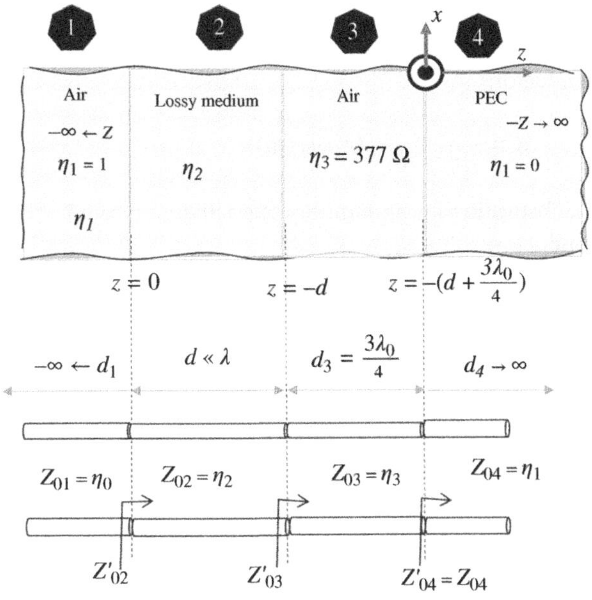

$\varepsilon_{r3} = 1$ (air); and $\varepsilon_{r4} = 1.0$ (*PEC*), respectively. The thicknesses of the 3-ply dielectric-conductor medium are: $d_2 = 1$ cm and $d_3 = (3\lambda_0)/4$. We would like to apply the transmission line model to calculate the effective reflection coefficient of medium 1.

We consider the thin conducting material first.

The wave impedances of media 1, 2, 3 and 4 are:

$$\eta_1 = \eta_3 = 377\,\Omega\,(\text{air}),\ \text{and}\ \eta_4 = 0\,\Omega\,(\text{PEC})$$

For the conducting media of thickness d, the wave impedance at 1.6 GHz is:

$$\eta_2 = \sqrt{\frac{\mu}{\varepsilon}\left(1 + \frac{\sigma}{j\omega}\right)} = \sqrt{\frac{4\pi \times 10^{-7}}{8.854 \times 10^{-12}}\left(1 + \frac{4.7 \times 10^5}{j2\pi \times 1.6 \times 10^9}\right)} = 376.7 - 0.0088j\,\Omega$$

The electrical length at the right side of the interface 2–3 ($z = -d$) is:

$$\theta_3 = \beta_3 d_3 = \frac{2\pi}{\lambda_0} \times \frac{3\lambda_0}{4} = \frac{3\pi}{2}\ (\text{rad})$$

The effective impedance is seen at the interface $z = -(3\lambda_0)/4$ is:

$$Z'_{03} = \lim_{tan\,(3\pi/2) \to \infty} 377 \times \frac{0 + j377 \times tan\,(3\pi/2)}{377 + j0 \times tan\,(3\pi/2)} = \infty\,\Omega$$

The zero impedance of the PEC is basically transferred to as the infinite impedance at $z = -(3\lambda_0)/4$. This transformation of impedance from zero to infinity and then infinity to zero repeats alternately every $((2n + 1)/4)\lambda_0$ distance where, $n = 0, 1, 2, 3, \ldots$ as every odd integer of $(\pi/2)$ makes $\tan(\beta_3 d_3) = \infty$. Next looking at the impedance at the interface 1–2 at $z = 0$ at the left side of the thin good conductor of thickness d, we obtain the input impedance Z'_{02} as:

$$Z'_{02} = \lim_{Z'_{03} \to \infty} \eta_2 \times \frac{\eta_2 + jZ'_{03} \times tanh\,(\gamma_2 d)}{Z'_{03} + j\eta_2 \times tanh\,(\gamma_2 d)} = \frac{\eta_2}{\gamma_2 d}\,\Omega$$

where, $\gamma_2 = \alpha_2 + j\beta_2$ and for the lossy thin layer $\tanh(\gamma_2 d) \cong \gamma_2 d$ for $d \ll \lambda$.

Now for the condition of zero reflection, $Z'_{02} = Z_{01} = \eta_0$. Using the condition for a good conductor,

$$d = \frac{\eta_2}{\gamma_2 \eta_0} \cong \frac{1}{\sigma_2}\sqrt{\varepsilon_0/\mu_0}\ (\text{m})$$

Substitution for aluminum conductivity, $\sigma_2 = 4.7 \times 10^5$ (S/m), we obtain the thickness of good conductor:

$$d = \frac{1}{4.7 \times 10^5} \times \frac{1}{377} \cong 5.6 \times 10^{-9}\ (\text{m}) = 5.6\ (\text{nm})$$

It is really thin compared to the household aluminum foil of a typical thickness of 16×10^{-6} (m).

Exercise 5.2: Electromagnetic Absorber Design with a Thin Dielectric Window
Q1. Repeat the example for a lossy dielectric of $\varepsilon_r = 3 - j2$. (Hints, calculate γ_2 at 1.6 GHz).

5.9.7 Absorbers in Anechoic Chamber: Antenna Measurement

Usually, pyramidal cone absorbers are used in anechoic chambers for antenna measurement to create an interference-free environment. Figure 5.49a shows the photograph of a typical installation of commercially available absorbers in an anechoic chamber.[75] These absorbers are used in EMI/EMC measurement for electronic gadgets and vehicles to examine if they emit spurious electromagnetic interferences beyond the limits assigned by the regulatory body. Usually, some specific standards are followed based on the country where the anechoic chamber is used for EMI/EMC measurements.

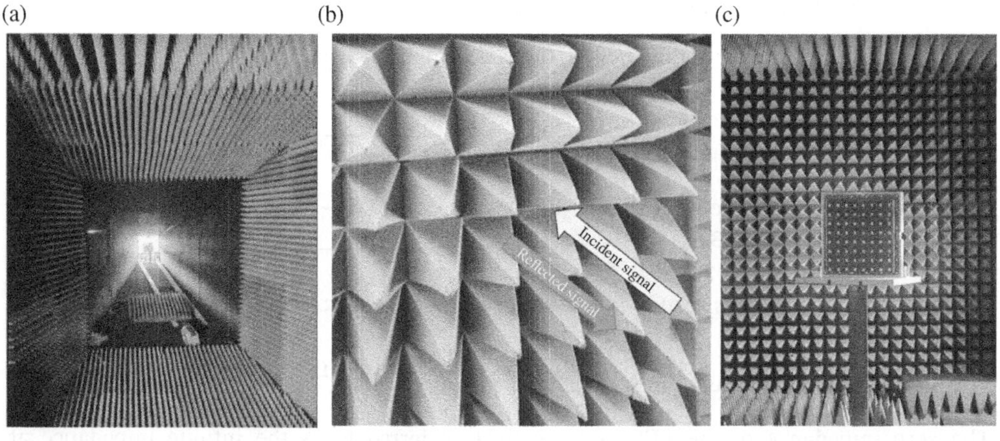

Figure 5.49 (a) Monash University conical anechoic chamber with pyramidal absorber, (b) incident and reflected wave from a piece of absorber, and (c) photograph of the inside of an anechoic chamber used for antenna array radiation pattern measurement.

75 Top 10 Anechoic Absorber Considerations For RF And Microwave Applications, source, http://www.ets-lindgren.com/sites/etsauthor/General_Brochures/Top%2010%20Anechoic%20Absorber%20Considerations.pdf (accessed 07 February 2020)

The pyramidal shape of the absorber provides a gradual change of free space impedance of 377 Ω to the termination impedance, which is the back-supporting material of the absorber, which is usually assumed to be a perfect electric conductor of zero impedance. Therefore, any deformation of the pyramidal absorber, e.g. broken tips of the absorbers, affect the transition of the impedance and matching. The conical shape is to make the incident angle close to 90°. Usually, absorbers are designed for wideband operation using full wave electromagnetic solvers such as HFSS, CST Microwave Studio and similar software tools. Also, Multiphysics software tools are used to profile the thermal absorption by the absorber. Appropriately designed absorber reduces reflection from emitting sources significantly by absorbing incident power as heat inside the absorber. Therefore, thermal mapping is a significant consideration along with electromagnetic considerations.

The effectiveness of the absorber is defined with the reflectivity as a function of frequency at the normal incident angle. The reflectivity is defined, in this case, as the ratio of the incident signal reflected by the absorber rest on a metal surface compared to that by a metallic surface without the absorber. Figure 5.49b illustrates the incident and reflected signals from a piece of the absorber. Hence signal received at a larger incident angle than the normal incident case the performance of the absorber is overly compromised.

At low frequency ferrite sheet absorber is used. Ferrite absorbers are not efficient at microwave and millimeter wave frequencies. Pyramidal absorbers are designed based on their performances at the lowest operational frequency in mind; how good the reflectivity of the absorber at that frequency. The length is proportional to the wavelength of the lowest frequency of operational bandwidth. For the normal incidence assumption, the length of the pyramidal absorber is designed based on the specifications of the reflectivity. For example, for 20 dB reflectivity, the absorber length is 0.4λ, for 30 dB 0.8λ and for 40 dB 1.2λ. Here λ is the free space wavelength which is defined as $\lambda = 300/f$ (m) where, f is the operating frequency in MHz. As stated above, advanced numerical modeling using full wave electromagnetic solvers as well as precise controls of material properties of the absorbers and computerised manufacturing processes are used to design and manufacture the absorbers. These advanced design and manufacturing processing make anechoic chamber absorbers an expensive item. Usually, the length and the sides of the base of the pyramidal absorber are made similar in size. Commercial absorbers typically are available in lengths of 3″, 5″, 8″, 12″, 18″, 24″, 36″, 48″ and 72″. Absorbers are made of foams filled with electromagnetic lossy materials such as carbon graticules and coated with rubberized paint for protection. Figure 5.49c shows the photograph of an array antenna radiation pattern measurement in an anechoic chamber. The turntable is rotated every 1° interval and received power is recorded with the angle. The data is plotted manually for radiation patterns. For an automated system, a backend signal processing is done with a highly powerful computer with an advanced GUI capable of 2D and 3D illustrations of the radiation patterns, and electromagnetic fields and power intensity profiles. Such an installation may cost a few hundred to a few million dollars.

5.10 Final Remarks

The composite structure made of multiple propagation medium layers is often aimed at reducing the reflection of the incident signal for a broadband operation.[76] If the intrinsic impedances or the refractive indices of these layers progressively increase or decrease from layer to layer, then the composite propagation medium becomes less sensitive to the design frequency. For example, in

76 W. H. Hayt and J. A. Buck, *Engineering Electromagnetics*, 7, McGraw Hill, Boston, 2006, p 483.

the antireflection multilayered coating of a camera lens or a pair of reading glasses, as we examined in the above examples, the layer on the lens would be of the impedance, which is very close to the glass. Successively the outer layers toward air are given successively high impedances. Thus with a large number of layers fabricated in this way for antireflective lenses and reading glasses, and anti-glare and sunglasses, the condition reaches almost the ideal matched conditions in which the impedance of the leftmost hand (top) layer reaches to that for the air and the impedance of the rightmost (bottom) layer reaches to that for the glass. With this gradual change in impedance as the electromagnetic wave propagates in the multilayered medium from left to right, there is no interface from which the electromagnetic signal reflects and, therefore, the electromagnetic signal (in this case light) is totally transmitted. Multiple layered coating designed delicately with the aid of the transmission and reflection theory thus produces excellent broadband transmission characteristics. This theory is equally applicable to the broadband impedance matching in transmission lines. These are the feedlines that are used as the wideband beamforming feed networks for broadband antennas and matching sections for the microwave active and passive devices. For prescribed bandwidth with minimum ripples in the broadband design, various polynomial approximations for impedance tapering techniques such as Chebyshev and normal distribution etc. are used. Thus, tapering of cased transmission lines can be either continuous or step impedances based on the polynomial approximations.

5.11 Problems

Normal Incidence on Conductor and Applications

P5.1 A y-polarized uniform plane wave is propagating in air (medium 1) along z-direction and impinges normally on a conducting boundary at $z = 0$ cm. (i) Write the general expression of the electric field in both phasor and time domains. If the successive minima (zero) of the resultant electric field in air is 10 cm and the maximum of the electric field is measured 120 (V/m) at $z = -15$ cm, calculate (ii) the wavelength, propagation constant and frequency of operation, (iii) the time domain expression of the resultant electric field, (iv) the time domain expression of the resultant magnetic field, and (v) the resultant time average power density in air.

P5.2 A z-polarized uniform plane wave is propagating in air (medium 1) along the x-direction and impinges normally on a conducting boundary at $y = 0$ cm. (i) Write the general expression of the electric field in both phasor and time domains. If the successive minima (zero) of the resultant electric field in air is 15 mm and the maximum of the electric field is measured 250 (mV/m) at $z = -22.5$ mm, calculate (ii) the wavelength, propagation constant and frequency of operation, (iii) the time domain expression of the resultant electric, (iv) the time domain expression of the resultant magnetic fields, and (v) the induced current density on the conducting boundary.

P5.3 A conducting circular reflector is added to a dipole antenna that operates at a 5.8 GHz ISM frequency band for biomedical imaging applications. Calculate the wave number and dipole length. Calculate the distance between the dipole element and the back reflector so that the gain can be enhanced by 3 dB.

P5.4 In P5.3 the diploe antenna incident an electric field with amplitude 240 (V/m) along x direction and the antenna is directed along z-direction, calculate the (i) incident, (ii) reflected, (iii) resultant electric fields of the antenna, and (iv) observing the resultant field's amplitude and phase make comments on the gain improvement of the antenna.

P5.5 In P5.3 assume the antenna is radiating a power density of 3 (mW/cm^2) to meet the biomedical specific absorption rate (SAR) limits. The resultant phasor magnetic field expression is given by $H_i(z) = \boldsymbol{a_y} H_0 \cos(\beta_1 z)$ (A/m). Calculate the (i) magnitude of the magnetic field H_0 in (A/m), and (ii) at the antenna surface the resultant electric field.

P5.6 A left-hand circularly polarized wave impinges from air region $z < 0$ on a perfect conducting boundary at $z = 0$. If the amplitude of the electric field is 120 (V/m) and the operating frequency is 2.45 GHz, calculate the followings: (i) the phasor and time domain expressions of the electric field in air, (ii) the associated incident magnetic field in air, (iii) the reflected electric and the magnetic fields in air, and (iv) resultant electric and magnetic field in air.

P5.7 A right-hand circularly polarized wave impinges from a dielectric medium of $\varepsilon_r = 10.2$ in region $z < 0$ on a perfect conducting boundary at $z = 0$. If the amplitude of the electric field is 50 (mV/m) and the operating frequency is 18 GHz, calculate the followings in dielectric region: (i) the phasor and time domain expressions of the electric field, (ii) the associated incident magnetic field, (iii) the reflected electric and the magnetic fields, and (iv) resultant electric and magnetic fields.

P5.8 The theory of reflection for a perfect conductor is used in the calibration of microwave measuring equipment. Because the reflection coefficient is −1, which is a known standard for calibration. This condition is called short circuit standard. Explain how the −1 reflection condition at the perfect conducting boundary helps to measure unknown load impedances.

P5.9 An arbitrary polarized signal is represented with the following expression: $\boldsymbol{E^i}(x, t) = \boldsymbol{a_y} E_{0y} \cos(\omega t - \beta x) + \boldsymbol{a_z} E_{0z} \sin(\omega t - \beta x)$ (V/m). The signal is normally incident on a perfect conducting boundary wall located at $x = 0$. (a) Determine the time domain expression for the reflected field. If the operating frequency is 1.8 GHz, $E_{x0} = 24$ (V/m) and $E_{oy} = 48$ (V/m), calculate the expression of the resultant field in air. (b) As for the circularly polarized field case do you think there is any change of sense of polarization?

Normal Incidence on Arbitrary Medium (Measuring of Dielectric Constant and Relative Permeability)

P5.10 From expressions (5.49) and (5.50) show that $\Gamma = (n_1 - n_2)/(n_1 + n_2)$; $\tau = (2n_1)/(n_1 + n_2)$. for the normal incidence the relationship between the transmission and reflection coefficients is $\tau = 1 + \Gamma$.

P5.11 The normal incidence theory on an arbitrary medium is be used in instrumentation to determine the properties of the medium. Assume a uniform plane wave incidents from air on an unknown lossless medium. The reflection coefficient is measured $\Gamma = -0.1$ and the phase velocity $u_p = 2 \times 10^8$ (m/s), calculate the relative permittivity and permeability of the material media. Assume the medium is lossless for simplicity in the calculation.

P5.12 In P5.11 assume the medium is a lossless dielectric, calculate the dielectric constant of the unknown medium.

P5.13 If both media are lossless dielectric media having a reflection coefficient of 0.25, find the ratio of the dielectric constants of the two media. If the second medium is air, calculate the dielectric constant of medium 1.

P5.14 A uniform plane wave is normally incident from a lossless medium with ($\mu_r = 2.5$, $\varepsilon_r = 4.2$) to media 2 with ($\mu_r = 12.0$, $\varepsilon_r = 2.45$). Calculate the reflection and transmission coefficients at the interface of the media. Comment on the dielectric interface system if the positive reflection coefficient create an amplification of the signal.

Normal Incidence from Dielectric to Dielectric

P5.15 A uniform plane wave from medium 1 ($z < 0$) is PTFE $\varepsilon_r = 2.45$ to medium 2 ($z > 0$) has $\varepsilon_r = 10.2$. The transmitted magnetic field intensity is defined as $\boldsymbol{H}^t(z, t) = \boldsymbol{a}_y\, 250 \cos(2\pi \times 2.45 \times 10^9 t - \beta_2 z)$ (mA/m). Calculate the followings: (i) the time domain expression of the incident and transmitted electric fields, (ii) the incident electric field, (iii) the transmitted and the reflected average power density in (W/m^2), and (iv) show that the conservation of power law is intact.

Normal Incidence from Air to Lossy Dielectric

P5.16 A uniform plane wave $\boldsymbol{E}^i(z, t) = 235 \cos(2\pi \times 10^{10} t - \beta_1 z)\boldsymbol{a}_x$ (V/m) is incident from the air (at $z < 0$) to a nonmagnetic lossy medium at $z > 0$ with conductivity $\sigma = 0.006$ (S/m) and dielectric constant $\varepsilon_r = 4.2$. Calculate (i) the VSWR, (ii) the time domain expressions for the reflected and transmitted electric and magnetic fields, and (iii) the average power densities for the incident, reflected and transmitted waves.

P5.17 A uniform plane wave $\boldsymbol{E}^i(z, t) = 235 \cos(2\pi \times 10^{10} t - \beta_1 z)\boldsymbol{a}_x$ (V/m) is incident from the air (at $z < 0$) to a nonmagnetic lossy medium at $z > 0$ with conductivity $\sigma = 0.006$ (S/m) and dielectric constant $\varepsilon_r = 4.2$. Calculate (i) the VSWR, (ii) the time domain expressions for the reflected and transmitted electric and magnetic fields, and (iii) the average power densities for the incident, reflected and transmitted waves.

P5.18 Repeat P5.17 when the frequency of operation is 50 MHz. Make comments on the nature of impedance, attenuation and propagation constants and the power transfer.

P5.19 An airborne early warning (AEW) radar sends a 125 MHz x-polarized uniform plane wave signal with an amplitude of 250 (V/m). The wave is normally incident onto sea water with dielectric constant $\varepsilon_r = 81$ and conductivity $\sigma = 4$ (S/m) ($z > 0$). Calculate: (i) the reflection and transmission coefficients, (ii) the time domain expressions for the incident electric and magnetic fields, (iii) the reflected electric and magnetic fields, (iv) transmitted electric and magnetic fields, and (v) the time average power transmitted through sea water. Assume sea water is a lossy dielectric media so that $\mu_2 = \mu_0$.

P5.20 In a semiconductor die of a Ku-band active phase shifter an x-polarized 234 mV/m uniform plane wave is incident on a Gallium arsenide (GaAs)–Silicon–Germanium (SiGe) interface. The dielectric constant of GaAs is 12.88 and that for SiGe is 13.95. The direction of propagation is along z-direction and the operating frequency is 17.8 GHz (Ku-band). Calculate the followings:

i) The time domain expression of the incident electric field
ii) The phasor vector expression of the incident electric field
iii) The transmission and reflection coefficients, and VSWR
iv) Draw the time domain waveform of the individual signals and the VSWR waveform
v) The percentage of power reflected back to medium 1

Applications of Normal Incidence: Anechoic Chamber Design, Radome Design, Shielding and Concrete Health Monitoring

P5.21 An 1800 MHz base station panel antenna transmits an x-polarized 180 V/m uniform plane wave via an antenna that is covered with a Rogers AD1000™ laminate and polycarbonate radome. The dielectric constants are: $\varepsilon_{r1} = 10.2$ for AD1000 and $\varepsilon_r = 2.56$ for polycarbonate. Assume both the media are semi-infinitely extended and lossless. Calculate the followings:

i) The phasor vector expression of the resultant electric field for the wave propagation along z-direction.
ii) The maximum and minimum electric fields of the standing wave.

P5.22 Now, consider the fly wheel analogy of the phasor vector diagram of the standing wave pattern. Consider the length is equivalent to the electric field intensity (V/m). This means the piston length is 1 equivalent to the normalized amplitude of the incident electric field. If we consider that 1 unit is equivalent to 1 m, then calculate the length of the piston rod, the wheel diameter (assume the piston rod is connected at the end of the wheel's rim), the length of the resultant distance from the origin to the end of the piston rod when z is 1 m.

P5.23 For P5.21 above, (i) draw the transmission line analogy of the two media: Rogers AD1000™ laminate and polycarbonate, (ii) calculate the load impedance if the reflection occurs from AD1000 laminate at the interface of AD1000 and polycarbonate, and (iii) calculate the impedance seen from the polycarbonate at 11 m from the interface.

P5.24 A concrete slab made of cement and reinforced rods is assumed to be a perfect conductor. Due to movements, cracks and weathering the dielectric constant changed to 5.3. A drone based radar is used to monitor the heath of the concrete. The distance of the radar antenna is about 5.2 m above the ground and the operating frequency of the radar is

10.2 GHz. Calculate the impedances seen by the radar when the concrete was laid the recently. Make a comment on the measurement system.

Transmission Through Multiple Layer and Applications

P5.25 A Teflon slab of thickness d = 20 cm is inserted in a sand pit of infinite extend. If at 560 MHz a UPW signal is transmitted through a probe calculate the reflection coefficient seen by the probe at the sand and dielectric slab interface ($z = -d$), calculate the transmission coefficient at the other interface ($z = 0$). If the incident power intensity is 10 (W/m^2), calculate the % of power reflected and transmitted. Assume the sand is pure silica with dielectric constant of 11.68. and Teflon has 2.1.

P5.26 A quarter wavelength thick dielectric window is sandwiched between silica ($\varepsilon_r = 12$) and Teflon ($\varepsilon_r = 2.1$) so that the reflection coefficient can be made zero. Calculate (i) the effective impedance of the slab so that there is no reflection from the insertion, (ii) the dielectric constant, and (iii) the thickness of the slab.

P5.27 Design a stealth material shield so that a 12.5 GHz military radar cannot detect the stealth bomber's fuselage of the dielectric constant of 24 (the fuselage assume to be infinite extend).

P5.28 Calculate the effective impedance for the third semi-infinite dielectric medium has the dielectric constant ε_{r3} = 4.2 and the dielectric window of thickness d = 3 mm with the dielectric constant ε_{r3} = 2.45. The operating frequency is 3.2 GHz. Draw the equivalent transmission line model.

P5.29 Consider the mean wavelength of blue light emanating out from a computer monitor is 472.5 nm. Calculate the length of the antireflective coating of blue light for designing a reading glass. Assume the refractive index of the glass is 1.42 and lossless. Design an antireflective coating using a quarter-wavelength dielectric window. For robustness and manufacturing ease, consider multiple quarter wavelength thickness so that the design thickness is larger than 1 μm.

P5.30 A wideband antireflective 4-ply coating for a sunglass made of a transparent dielectric laminate is to be designed. Assume the refractive index of the transparent dielectric laminate is 1.79 and lossless. The other three dielectric layers from right to left are as follows: $n_4 = 1.4$, $d_4 = \lambda_4/4$; $n_3 = 1.7$, $d_3 = \lambda_3/2$; $n_2 = 1.4$ and $d_2 = \lambda_2/4$. Design an antireflective coating for the wavelength 540 nm with following information: (i) the thickness of each layer, (ii) the wavenumber and impedance of each layer, (iii) the effective impedance seen by the plane wave at the interface of air and first dielectric, and (iv) the effective reflection coefficient and the percentage of power reflected back.

P5.31 Design a 2-ply dielectric radome for a military fighter aircraft's attack and reconnaissance radars operating in the 8.5–11 GHz frequency band. (i) Calculate the effective reflection coefficient seen from medium 1 (air) to the right for the 2-ply radome. Given an x-polarized wave is incident from medium 1 air at the center operating frequency. The dielectric constant for medium 1 to medium 4 are: $\varepsilon_{r1} = 1.0$ (air); $\varepsilon_{r2} = 2.54$ (polysterence); $\varepsilon_{r3} = 4.2$ (Corning glass); and $\varepsilon_{r4} = 1.0$ (air), respectively. The thicknesses of the 2-ply dielectric radome are: $d_2 = 1.64$ mm and $d_3 = 6.6$ mm. Draw the transmission line model of the radome. Calculate the effective reflection coefficient of medium 1.

Normal Incidence on Conductor and Applications

P5.32 A y-polarized uniform plane wave is propagating in air (medium 1) along z-direction and impinges normally on a conducting boundary at $z = 0$ cm. (i) Write the general expression of the electric field in both phasor and time domain. If the successive minima (zero) of the resultant electric field in air is 10 cm and the maximum of the electric field is measured 120 (V/m) at $z = -15$ cm, calculate (ii) the wavelength, propagation constant and frequency of operation, (iii) the time domain expression of the resultant electric field, (iv) the time domain expression of the resultant magnetic field, and (v) the resultant time average power density in air.

P5.33 A z-polarized uniform plane wave is propagating in air (medium 1) along x-direction and impinges normally on a conducting boundary at $y = 0$ cm. (i) Write the general expression of the electric field in both phasor and time domain. If the successive minima (zero) of the resultant electric field in air is 15 mm and the maximum of the electric field is measured 250 (mV/m) at $z = -22.5$ mm, calculate (ii) the wavelength, propagation constant and frequency of operation, (iii) the time domain expression of the resultant electric, (iv) the time domain expression of the resultant magnetic fields, and (v) the induced current density on the conducting boundary.

P5.34 A conducting circular reflector is added to a dipole antenna that operates at 5.8 GHz ISM frequency band for biomedical imaging applications. Calculate the wave number and dipole length. Calculate the distance between the dipole element and the back reflector so that the gain can be enhanced by 3 dB.

P5.35 In P5.3 the diploe antenna incident an electric field with an amplitude 240 (V/m) along x direction and the antenna is directed along the z-direction, calculate the (i) incident, (ii) reflected, (iii) resultant electric fields of the antenna, and (iv) observing the resultant field's amplitude and phase. Make comments on the gain improvement of the antenna.

P5.36 In P5.3 assume the antenna is radiating a power density of 3 (mW/cm^2) to meet the biomedical specific absorption rate (SAR) limits. The resultant phasor magnetic field expression is given by $H_i(z) = \mathbf{a}_y H_0 \cos(\beta_1 z)$ (A/m). Calculate the (i) magnitude of the magnetic field H_0 in (A/m), and (ii) at the antenna surface the resultant electric field.

P5.37 A left-hand circularly polarized wave impinges from air region $z < 0$ on a perfect conducting boundary at $z = 0$. If the amplitude of the electric field is 120 (V/m) and the operating frequency is 2.45 GHz, calculate the followings: (i) the phasor and time domain expressions of the electric field in air, (ii) the associated incident magnetic field in air, (iii) the reflected electric and the magnetic fields in air, and (iv) resultant electric, and magnetic field in air.

P5.38 A right-hand circularly polarized wave impinges from a dielectric medium of $\varepsilon_r = 10.2$ in region $z < 0$ on a perfect conducting boundary at $z = 0$. If the amplitude of the electric field is 50 (mV/m) and the operating frequency is 18 GHz, calculate the followings in dielectric region: (i) the phasor and time domain expressions of the electric field, (ii) the associated incident magnetic field, (iii) the reflected electric and the magnetic fields, and (iv) resultant electric and magnetic field.

P5.39 The theory of reflection for a perfect conductor is used in the calibration of microwave measuring equipment. Because the reflection coefficient is −1, which is a known standard for calibration. This condition is called short circuit standard. Explain how the −1 reflection condition at the perfect conducting boundary helps to measure unknown load impedances.

P5.40 An arbitrary polarized signal is represented with the following expression: $\boldsymbol{E}^i(x, t) = \boldsymbol{a}_y E_{0y} \cos(\omega t - \beta x) + \boldsymbol{a}_z E_{0z} \sin(\omega t - \beta x)$ (V/m). The signal is normally incident on a perfect conducting boundary wall located at $x = 0$. (a) Determine the time domain expression for the reflected field. If the operating frequency is 1.8 GHz, $E_{x0} = 24$ (V/m) and $E_{oy} = 48$ (V/m), calculate the expression of the resultant field in air. (b) As for the circularly polarized field case, do you think there is any change of sense of polarization?

Normal Incidence from Dielectric to Dielectric

P5.41 A uniform plane wave from medium 1 ($z < 0$) is PTFE ($\varepsilon_r = 2.45$) to media 2 ($z > 0$) has $\varepsilon_r = 10.2$. The transmitted magnetic field intensity is defined as $\boldsymbol{H}^t(z, t) = \boldsymbol{a}_y\, 250 \cos(2\pi \times 2.45 \times 10^9 t - \beta_2 z)$ (mA/m). Calculate the followings: (i) the time domain expressions of the incident and transmitted electric fields, (ii) the reflected electric field, (iii) the transmitted and the reflected average power density in (W/m^2), and (iv) show that the conservation of power law is intact.

Normal Incidence from Air to Lossy Dielectric

P5.42 A uniform plane wave $\boldsymbol{E}^i = 235 \cos(2\pi \times 10^{10} t - \beta_1 z)\boldsymbol{a}_x$ (V/m) is incident from air (at $z < 0$) to a nonmagnetic lossy media at $z > 0$ with conductivity $\sigma = 0.006$ (S/m) and dielectric constant $\varepsilon_r = 4.2$. Calculate (i) the (VSWR), (ii) the reflected and transmitted electric and magnetic fields, and (iii) the average power densities for the incident, reflected and transmitted waves.

P5.43 A uniform plane wave $\boldsymbol{E}^i(x, t) = 250 \cos(2\pi \times 10^{10} t - \beta_1 x)\boldsymbol{a}_y$ (V/m) is incident from air (at $x < 0$) to a nonmagnetic lossy media at $x > 0$ with conductivity $\sigma = 0.03$ (S/m), relative permeability $\mu_r = 6.5$ and dielectric constant $\varepsilon_r = 10.2$. Calculate: (i) the incident magnetic field and time average power density, (ii) the reflection, and transmission coefficients and VSWR,

(iii) the reflected and transmitted electric and magnetic fields, and (iv) the average power densities for the incident, reflected and transmitted waves.

P5.44 Repeat P5.17 when the frequency of operation is 50 MHz. Make comments on the nature of impedance, attenuation and propagation constants and the power transfer.

P5.45 An airborne early warning (AEW) radar sends a 125 MHz x-polarized uniform plane wave signal with an amplitude of 250 (V/m). The wave is normally incident onto sea water with dielectric constant $\varepsilon_r = 81$ and conductivity $\sigma = 4$ (S/m) ($z > 0$) Calculate: (i) the reflection and transmission coefficients, (ii) the incident electric and magnetic fields, (iii) the reflected electric and magnetic fields, (iv) the transmitted electric and magnetic fields, and (v) the time average power transmitted through sea water. Assume sea water is a lossy dielectric media so that $\mu_2 = \mu_0$.

P5.46 In a semiconductor die of a Ku-band active phase shifter an x-polarized 234 mV/m uniform plane wave is incident on a Gallium arsenide (GaAs)–Silicon–Germanium (SiGe) interface. The dielectric constant of GaAs is 12.88 and that for SiGe is 13.95. The direction of propagation is along z-direction and the operating frequency is 17.8 GHz (Ku-band). Calculate the followings:

i) The time domain expression of the incident electric field
ii) The phasor vector expression of the incident electric field
iii) The transmission and reflection coefficients, and VSWR
iv) Draw the time domain waveform of the individual signals and the VSWR waveform
v) The percentage of power reflected back to medium 1

Applications of Normal Incidence: Anechoic Chamber Design, Radome Design, Shielding, Concrete Health Monitoring

P5.47 A 1800 MHz base station panel antenna transmits a x-polarized 180 V/m uniform plane wave via an antenna which is covered with a Rogers AD1000™ laminate and polycarbonate radome. The dielectric constants are: $\varepsilon_{r1} = 10.2$ for AD1000 and $\varepsilon_r = 2.56$ for polycarbonate. Assume both the media are semi-infinitely extended and lossless. Calculate the followings:

i) The phasor vector expression of the resultant electric field for the wave propagation along z-direction.
ii) The maximum and minimum electric fields of the standing wave.
iii) Now, consider the fly wheel analogy of the phasor vector diagram of the standing wave pattern. Consider the length is equivalent to the electric field intensity (V/m). This means the piston length is 1 equivalent to the normalized amplitude of the incident electric field. If we consider that 1 unit is equivalent to 1 m, then calculate the length of the piston rod, the wheel diameter (assume the piston rod is connected at the end of the wheel's rim), the length of the resultant distance from the origin to the end of the piston rod when z is 1 m.

P5.48 For P5.21 above, (i) draw the transmission line analogy of the two media Rogers AD1000™ laminate and polycarbonate, (ii) calculate the load impedance if the reflection occurs from AD1000 laminate at the interface of AD1000 and polycarbonate, and (iii) calculate the impedance seen from the polycarbonate at 11 m from the interface.

P5.49 A concrete slab is made of cement and reinforced rods is assumed to be a perfect conductor. Due to movements, cracks and weathering the dielectric constant changed to 5.3. A drone-based radar is used to monitor the heath of the concrete. The distance of the radar antenna is about 5.2 m above the ground and the operating frequency of the radar is 10.2 GHz. Calculate the impedances seen by the radar. Make a comment on the measurement system.

Transmission Through Multiple Layer and Applications

P5.50 A Teflon slab of thickness $d = 20$ cm is inserted in a sand pit of infinite extend. If at 560 MHz a UPW signal is transmitted through a probe calculate the reflection coefficient seen by the probe at the sand and dielectric slab interface ($z = -d$), calculate the transmission coefficient at the other interface ($z = 0$). If the incident power intensity is 10 (W/m^2), calculate the % of power reflected and transmitted. Assume the sand is pure silica with a dielectric constant of 11.68. and Teflon has 2.1.

P5.51 A quarter wavelength thick dielectric window is sandwiched between silica ($\varepsilon_r = 12$) and Teflon ($\varepsilon_r = 2.1$) so that the reflection coefficient can be made zero. Calculate: (i) the effective impedance of the slab so that there is no reflection from the insertion, (ii) the new dielectric constant, and (iii) the thickness of the slab.

P5.52 Design a stealth material shield so that a 12.5 GHz military radar cannot detect the stealth bomber's fuselage of the dielectric constant of 24 (the fuselage assume to be infinite extend).

P5.53 Calculate the effective impedance seen from air as medium 1 for the third semi-infinite dielectric medium has the dielectric constant $\varepsilon_{r3} = 4.2$ and the dielectric window of thickness $d = 3$ mm with the dielectric constant $\varepsilon_{r2} = 2.45$. The operating frequency is 3.2 GHz. Draw the equivalent transmission line model.

P5.54 Consider the mean wavelength of blue light emanating out from a computer monitor is 472.5 nm. Calculate the length of the antireflective coating of blue light for designing a reading glass. Assume the refractive index of the glass is 1.42 and lossless. Design an antireflective coating using a quarter-wavelength dielectric window. For robustness and manufacturing ease consider multiple quarter wavelength thickness so that the design thickness is larger than 1 μm.

P5.55 A wideband antireflective 4-ply coating for a sunglass made of a transparent dielectric laminate is to be designed. Assume the refractive index of the transparent dielectric laminate is 1.79 and lossless. The other three dielectric layers from right to left are as follows: $n_4 = 1.4$,

$d_4 = \lambda_4/4$; $n_3 = 1.7$, $d_3 = \lambda_3/2$; $n_2 = 1.4$ and $d_2 = \lambda_2/4$. Design an antireflective coating for the wavelength 540 nm with the following information: (i) the thickness of each layer, (ii) the wavenumber and impedance of each layer, (iii) the effective impedance seen by the plane wave at the interface of air and first dielectric, and (iv) the effective reflection coefficient and the percentage of power reflected back.

P5.56 Design a 2-ply dielectric radome for a military fighter aircraft's attack and reconnaissance radars operating in the 8.5.11 GHz frequency band. (i) Calculate the effective reflection coefficient seen from medium 1 (air) to the right for the 2-ply radome. Given an x-polarized wave is incident from medium 1 air at the center operating frequency. The dielectric constant for medium 1 to medium 4 are: $\varepsilon_{r1} = 1.0$ (air); $\varepsilon_{r2} = 2.54$ (polysterence); $\varepsilon_{r3} = 4.2$ (Corning glass); and $\varepsilon_{r4} = 1.0$ (air), respectively. The thicknesses of the 2-ply dielectric radome are: $d_2 = 1.64$ mm and $d_3 = 6.6$ mm. Draw the transmission line model of the radome. Calculate the effective reflection coefficient of medium 1.

6

Oblique Incidence of Uniform Plane Wave

6.1 Introduction

This chapter presents the theory of oblique incidence of uniform plane wave, firstly, at the interface of two media, and then, extended to multiple media. Oblique incidence can be classified into two types: (i) Transverse electric (TE), and (ii) Transverse magnetic (TM). First, the coordinate systems and vector analyses of the oblique wave are covered. This is the backbone of further analyses of the oblique incidence phenomena. In the development of the theory, firstly, the oblique incident wave on a conducting boundary is analyzed for both polarizations followed by the penetrable dielectric boundaries. In the analysis of conducting boundaries, we develop *the waveguide concept* via multiple oblique reflections of signal. In the dielectric boundaries, we shall develop the concept of total internal reflection for the TE and TM cases. This is the backbone of *the optical fiber concept* in which wave propagates inside the core due to the total internal reflection. This occurs due to the wave incidences at an incident angle greater than a certain angle, which is called *the critical angle Also the* wave is launched from an electromagnetically denser medium to a lighter medium for total internal reflection to occur. Finally, we shall develop the theory of total transmission for TM polarization. The angle at which total transmission of TE polarization occurs is called *the Brewster angle.* Brewster angle helps us to understand the wave propagation in a particular polarization, in this case full transmission of TM polarized wave and partial reflection of TE polarized wave. Applications of Brewster angle are antiglare sunglasses, mobile communications, and lasers. We examine the oblique incidence phenomena using Maxwell's field equations in the last part of the chapter. The oblique incidences in multiple dielectric boundaries and their applications are also explained in the final part of the chapter.

In the preceding chapter, we comprehensively analyzed the normal wave incidence of the uniform plane wave. In the chapter, we defined boundary value problems, reflection, refraction, and diffraction of uniform plane waves when they encounter boundaries of dissimilar dielectric media. We utilized the boundary conditions and field analyses for normal incidence cases and their applications in calculations of transmission and reflection coefficients. We also developed the theory of reflection and transmission via multiple boundaries using the transmission line analogy. We went through many interesting problems during the course of the development of the theory of normal incidence. The normal incidence case is a special case of wave incidence. The general case is the oblique incidence in arbitrary polarization where we do not know in which angle and field orientation the wave hits an interface. In this chapter, we shall develop the theory of oblique incidence utilizing the knowledge and foundation gained from the normal incidence cases. In contrast to the normal incidence case when the field has definite orientation, in the oblique incidence cases, the wave fields impinge on the boundaries from an arbitrary direction (angle) with an arbitrary

Fields and Waves in Electromagnetic Communications, First Edition. Nemai Chandra Karmakar.

Companion website: www.wiley.com/go/fieldsandwavesinelectromagneticcommunications

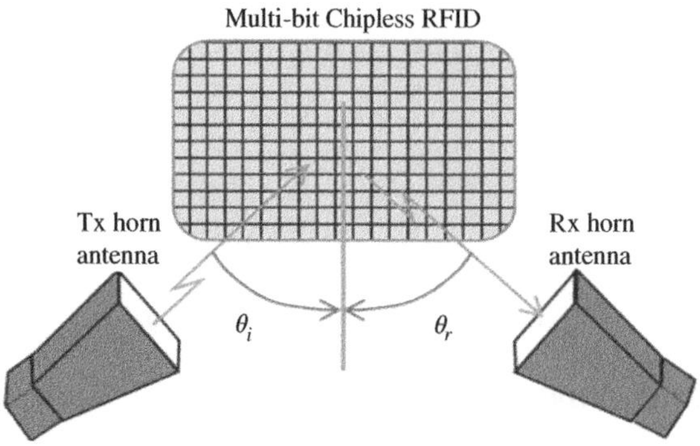

Figure 6.1 A chipless RFID system. The multi-bit passive tag is illuminated with a UWB horn antenna and the returned echo is received by the horn antenna. According to Snell's law of reflection, the incident angle $\theta_i = \theta_r$, the reflected signal.

polarization (electric field orientation). Thus, the theory of oblique incidence becomes more complex compared to the normal incidence case. To ease the complexity in calculation of the reflection and transmission coefficients at the oblique incidence cases, we categorize the field polarizations into two types: TE and TM cases.[1] Any deviation from these two polarizations can also be analyzed with the two types with addition to some phase angle for the arbitrary nature of the incident wave. Before we classify our analysis and derive the reflection and transmission coefficients, we shall go through a few practical case studies to highlight the geneses and applications of the theory.

Figure 6.1 illustrates a multi-bit chipless RFID tag, which is basically a passive reflector with a few passive modulating elements, is illuminated with a transmitting horn antenna that launches an ultra-wide band (UWB)[2] linearly polarized UPW at an incident angle θ_i. The tag processes signal with its multi-bit identification data[3] and then the reflected modulated signal is received by the receiving horn antenna, which is oriented at the angle of reflection θ_r. The received data is processed in the reader electronics for extraction of ID data. This is a very good example of the application of oblique incidence for emerging identification and RF sensing market. The signal can come from an arbitrary angle and the receiver receives the returned modulated echoes from the tag at another angle. This concept is also the foundation for the emerging intelligent reflective surface that has the capability to stop eavesdropping of the enemy signal.[4]

There are a few significant, known BCs that we shall utilize in field analysis and synthesis for the oblique incident fields. As you may recall that we apply the two prominent boundary conditions: (i) the tangential electric field at the boundary of the two dissimilar medium is continuous ($E_{T1} = E_{T2}$), and (ii) the tangential magnetic field at the boundary of the two dissimilar medium is continuous ($H_{T1} = H_{T2}$) provided that there is surface current source J_s is present at

1 TE polarized field can be compared with the movement of humpback of a caterpillar (the electric field vibrates vertically at the plane of incidence) and the TM polarization with a snake (the electric field vibrates parallel to the plane of incidence). For the TM polarization, the electric field is decomposed into two components along the two principal axes on which the plane of incidence lies. In many textbooks, TE is called the perpendicular polarization and TM is called the parallel polarization due to their orientation of the electric field vectors.

2 We shall analyze the theory of oblique incidence case for a single continuous wave (CW) frequency. Therefore, the UWB case will be a linear combination of the field analysis if one wants to get the aggregate outcome of the UWB system.

3 In a frequency domain chipless RFID, the UWB frequency modulation happens with a set of passive resonant structures. If the resonator produces resonance at a particular frequency, the frequency modulation is recorded as a binary bit "1" and absence of resonance is recorded as binary "0" bit. Thus we obtain a 1 : 1 correspondence of binary data via the UWB frequency modulation. Since the bandwidth of microwave UWB spectrum is extremely large spanning from 3 to 10.5 GHz, the tag can accommodate up to 124 data bits and even more.

4 https://encyclopedia.pub/entry/21383 (accessed 31 December 2022).

the boundary.[5] We also assume that the plane of incidence is on the xz-plane (plane of paper). Therefore, for the TE field case, the electric field is perpendicular to the direction of wave propagation and has only one field component along y-direction. To satisfy the orthogonality relationship between the direction of wave propagation and the electric and magnetic fields, we must consider the magnetic field vibrating along the xz-plane on which the plane of incidence lies, which can be decomposed into two components along the two-principal x- and z-axes. We shall decompose this arbitrarily polarized magnetic field into two orthogonal components along x- and z-axes to facilitate the application of the above boundary condition (ii). This boundary conditions along with the continuous y-directed electric field component (perpendicular to the plane of paper for TE-case) provides the relations of the transmitted and reflected fields with respect to the incident electric field. Thus, we obtain the transmission and reflection coefficients of TE case.

For the case of the TM field incidence, the magnetic field is perpendicular to the xz-plane hence has only y-directed components. However, the electric field has an arbitrary direction on the xz-plane. In this case, the electric field is decomposed into x- and z-directed field components. Again, the analyses of the field components are performed with the boundary conditions: (i) the tangential field components at the boundary are continuous along with the continuous y-directed magnetic field components. Finally, we calculate the transmission and reflection coefficients of TM case. We shall start our analysis with the TE case first and then develop many interesting concepts as delineated above. Then we shall move to the TM case and develop many interesting concepts related to the total transmission. Let us discuss another case study that supports the above concepts.

Figure 6.2 illustrates another case study of a ground penetration radar. Instead of considering the normal incidence case in the previous chapter we assume that the radar scans underground layers

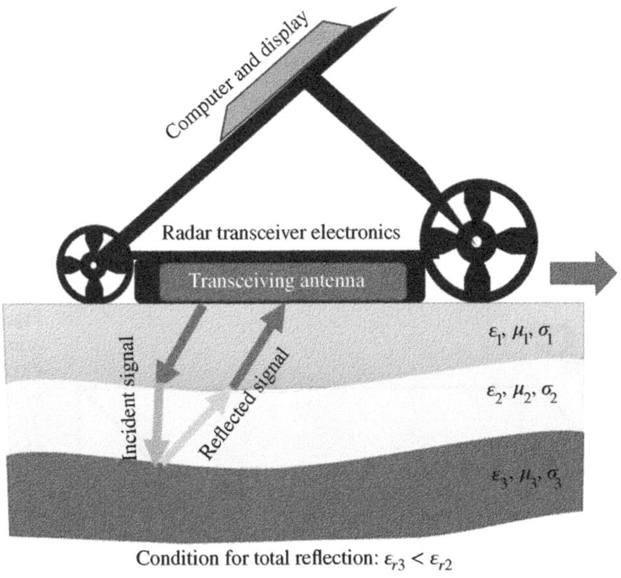

Figure 6.2 Oblique incidence with total internal reflection. A ground penetration radar sends signal towards underground soil layers with different dielectric constants. Due to the oblique incidence angle in layer 2 at an angle larger that the critical angle, the signal is totally reflected internally inside layer 2 and no signal penetrates in layers 1 and 3. This happens when an electromagnetic wave enters from a denser medium to a lighter medium. For the first two layers 1 and 2, we do not have the incidence angle more than the critical angle however, from medium 2 to medium 3 the signal enters from the denser medium to the lighter medium with an angle greater that the critical angle so that total internal reflection happens. Thus a conduit of signal in the second medium occurs and bounces back and forth at the boundaries of the first and third media.

5 Since there is an induced current on the conducting boundary, we apply another known boundary condition at this case the tangential electric field is zero at the conducting boundary and the transmitted signal is zero as there is no field penetration in conductor.

with an oblique incidence ray of EM signal. However, it cannot see more than two layers. Explain the possible cases of limitation of radar detection. Examine the transmitted and reflected ray carefully at the interface of various layers with their respective constitutive parameters. Can you tell which layer will be denser than the other by seeing the parameters? See the conditions of various layers. Can you explain why *total internal reflection* is happening at the interface of layers 2 and 3? If we assume the soil layers are good dielectrics, what are the values of σ's and μ_rs?

As you can see in the figure, the signal is internally reflected back at the interface of the second and third layers. When a signal incidences from an electromagnetically denser to a lighter layer with an angle greater than a certain incidence angle also called *the critical angle*, the signal is totally reflected back to the incident layer. That is why the radar cannot see the reflected signals after second layer. Such internal reflection can occur in both TE and TM cases. We shall introduce the theory of the internal reflection in the analysis for TE case. The theory of internal reflection is the basis of optical waveguide and optical fiber communications. Can you explain how we can correct the limitation? The answer is simple, by orienting the radar transmitter antenna so that it does not launch the signal that can cause the incident angle in layer 2 larger than the critical angle.

Figure 6.3 shows a good example of total internal reflection which light rays incident from the electromagnetically denser medium to the lighter medium. In this case, a gold fish in an aquarium can be hidden completely if an opaque disc is placed at the top of the cover. After certain incidence angle $\theta_i \geq \theta_c$, total internal reflection occurs and the incident ray from the fish can be seen by an observer. We shall solve many interesting problems like the above three case studies.

Figure 6.4 illustrates the topics covered in the chapter. As can be seen we shall first define the TE and TM fields incident on a conducting boundary and investigate the characteristics of the reflected field at any arbitrary incident angle. As we know that there is no transmission in the conducting medium, we shall use the known boundary condition of the transmitted field is zero in addition to the tangential electric field at the conducting boundary ($\boldsymbol{E}_{T1} = 0$) is also zero. By this analysis, we shall develop the concept how the wave propagates inside a metallic waveguide due to the multiple reflections of the obliquely incident waves on the conducting walls. After we shall analyze the electromagnetic field at the conducting boundary, we shall move next to the field analyses of oblique incidence on the dielectric boundary and derive the reflection and transmission coefficients. During the course of the theoretical development we shall analyze the field theory and find a few important

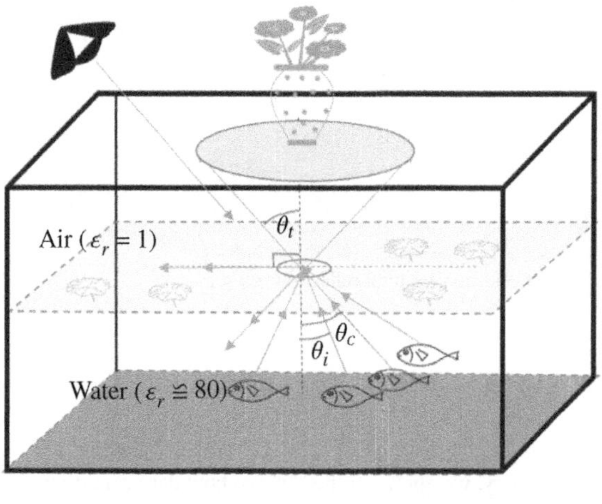

Figure 6.3 Total internal reflection from an aquarium. The fish cannot be detected if a circular opaque disc is placed on the cover with a radius equal to the angular spread of $2\theta_t$ corresponds to the critical angle θ_c as shown in the figure. Light rays coming from the gold fishes at positions greater than the critical angle $\theta_i \geq \theta_c$ undergo total internal reflection and the observer cannot see any gold fish.

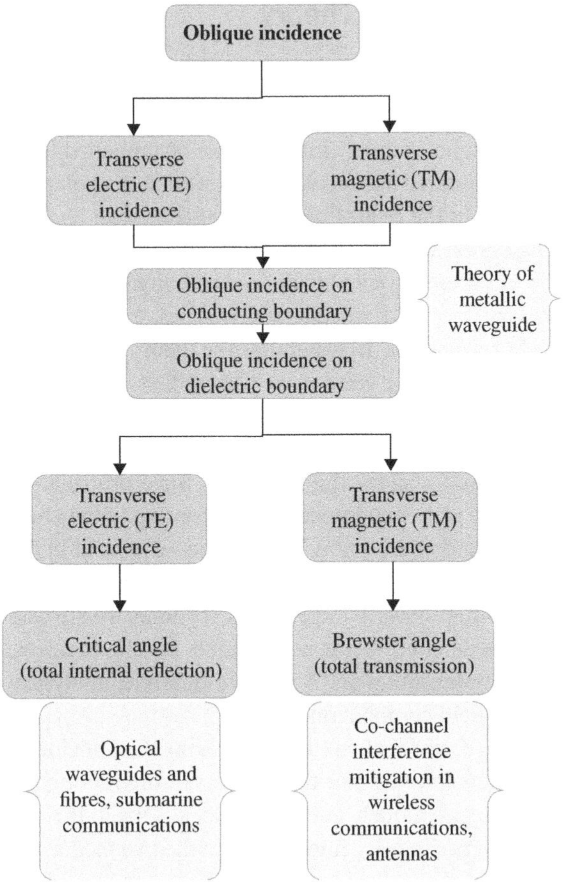

Figure 6.4 Roadmap of topics covered in oblique incidence cases. Oblique incidence cases are broadly classified as the TE and TM based on the direction of the electromagnetic fields' orientations with respect to the plane of incidence. Based on the less to intense complexity, first we develop the theory of oblique incidence on a dielectric boundary. We develop the concept of metallic waveguide. Then we move on to more generic cases of oblique incidence on dielectric boundary and develop the concept of total internal reflection and total transmission. Total internal reflection provides the basis of optical waveguide and fibers and submarine communications, and total transmission for co-channel interference mitigation in mobile wireless communication, radars, and antennas.

relationships such as *Snell's law of reflection* and *Snell's law of refraction*. Snell's law of reflection tells that the angle of incidence is equal to the angle of reflection. Snell's law of transmission relates the incident and transmission angles with the wave impedances of the two media. Snell's law of refraction also leads to special case of total internal reflection and total transmission as delineated above.

During the course of analysis, we shall draw many practical examples of modern technologies that exploit the theory of oblique incidence and resulting reflection and transmission of UPW at the interface of multiple media. During the development of the theoretical concepts will shall use multiple approaches to reinforcing our understanding of the complex theory of oblique incidence and the resulting transmission and reflection.

6.2 Methodologies Used in Oblique Incidence Theory

Figure 6.5 illustrates methods and techniques used to exemplify the oblique incidence theories. We shall use four distinctive techniques to analyze the fields in oblique incidence cases. Complexity occurs as the obliquely incident wave propagation on an arbitrary interface can be launched from any angle and with any orientation (polarization). As stated above, we have classified the polarizations into vertical (TE) and parallel (TM) directions. In the case of oblique incidence with its arbitrary nature of wave propagation and launching on the boundary, the situation becomes complex once reflected and transmitted rays into media of different wave incidences and angles of propagation. Therefore, we need to accurately analyze the field vectors with respect to their own polarizations (orientations) and the directions of propagation. To generalize the theory of arbitrary angle of incidence and arbitrary polarizations of the field wave vectors, we need to first examine the wave field vector and its propagation direction in accurate details. In this occasion we shall take shelter of *complex coordinate system* based on which we shall develop our vector analysis. This is the methodology we shall utilize to understand and analyze the fields and wave propagation in any arbitrary angle and polarization. The outcome is the fundamental understating of the characteristics of the field wave in terms of its amplitude and phase at any point in the wave front in the propagation medium. Once the analysis of the wave field vector at any arbitrary polarization, angle of incidence and direction of propagation, and two different wave impedances are done, we are well dressed to solve for the field wave equations in its detailed vector components for the incident, reflected, and transmitted field waves for any complex situation. There are different analytical methods that we shall pursue to understand those field components.

The first and most simple method is *the ray trace model* of the field waves in their incident, reflected, and transmitted modes. In the ray trace model, we assume that the wave front is bounded by two equi-phase rays of the field waves and impinges obliquely at the boundary.[6] Due to the discontinuity of the media at the interface and the difference in wave impedances of the two media, the

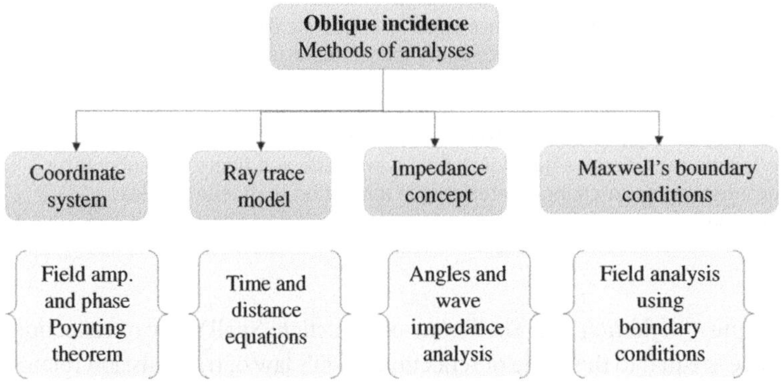

Figure 6.5 Methods of analyses of oblique incidence. The analysis of coordinate system for wave propagation from arbitrary direction involves Poynting theorem. Field amplitudes and phases in TE and TM polarizations are determined with the direction of wave propagation via the Poynting theorem and right-hand rule. The ray trace model of impinging wave provides useful information and Snell's laws of reflection and refraction without using complex Maxwell's equations and boundary conditions. Impedance variation concept due to dissimilar media is another method to find the reflection and transmission coefficients for oblique incidence cases by simply deriving the equivalent impedances for both TE and TM polarizations and using the generic definition of reflection and transmission coefficients for normal incidence.

6 In the ray trace model, the plane wave can be assumed as if the total wave propagates through an invisible and fictitious cylinder. The two rays are the two parallel extreme lines defining the cylindrical surface in 2D planar setting that conveys the total electromagnetic energy and impinges on the interface of two dissimilar media at an arbitrary angle and polarization.

rays reflect and transmit at two different paths at the interface. However, the time of flight of the two rays in their reflection and refraction modes remains the same. We equate the distances of the incident, reflected, and transmitted rays, and find the relations between the three sets of electromagnetic fields. Thus, we can obtain the basic laws — Snell's laws of reflection and refraction with the help of trigonometry only. Thus we can develop the fundamental concepts of the oblique incidence cases bypassing complex field analyses using Maxwell's equations and field boundary conditions.

The third method is *the impedance concept*. In the impedance concept we use the basic definition of the wave impedance as the ratio of the electric and magnetic fields at any point in space. Since the field is modulated with its angles of incidence, reflection, and transmission, the wave impedance of the respective medium, and finally, the polarization of the incident field waves with respect to the plane of incidence, we can modify the definition of the wave impedances based on the above fundamental wave defining parameters.[7] Once the impedances are calculated based on the definite polarization and angle of incidence, we can use the basic transmission line analogy and definition of reflection and transmission coefficients as we used in the case of normal incidence. *The impedance concept* eases the effective reflection and transmission coefficients calculations for any number of media.

Finally, we shall take the approach to *Maxwell's field wave equations* and resulting boundary conditions. This analysis, although too complex, gives the comprehensive details of the field conditions and the resulting transmission and reflection coefficients. In the following sections, we shall follow the same sequence of theoretical analysis as delineated above. First, we define the coordinate system for the general oblique incident case. Then we shall analyze the fields at the conducting boundary and develop the concept of wave propagation in waveguides. Then we shall analyze the fields at the dielectric boundaries using the three methods: (i) Ray trace model, (ii) impedance concept, and (iii) Maxwell's boundary conditions.

Review Questions 6.1: Introduction to Oblique Incidence

Q1. Explain why oblique incidence theory is more complex than the normal incidence theory.
Q2. Define TE and TM polarizations for oblique incidence.
Q3. Explain how signal from any arbitrary polarization oblique incident field can be analyzed.
Q4. Explain different methods and approaches to analyze the oblique incidence cases.
Q5. Explain different applications of the oblique incidence theory in modern telecommunications and optics.

Hints: In analyzing linearly polarized fields incident at any arbitrary polarization and angle, we decomposed the arbitrary field components into two orthogonal polarized fields. Here again, we shall apply the known boundary conditions at the interface of two media.

7 Basically, the wave impedance of the oblique wave is defined in terms of the actual wave impedance in a medium times the sin and cosine of the angles of incidence and transmission. For TE case the wave impedances are $\eta_{1p} = \eta_1 sec\theta_i;\ \eta_{2p} = \eta_2 sec\theta_t$ where subscripts 1 and 2 represent media 1 and 2, respectively. For TM case the wave impedances are $\eta_{1p} = \eta_1 cos\theta_i;\ \eta_{2p} = \eta_2 cos\theta_t$ where subscripts 1 and 2 represent media 1 and 2, respectively. Using the two impedances we can apply the same expressions for transmission and reflection coefficients for both TE and TM case as for normal incidence case: $\Gamma_{TE/TM} = \frac{\eta_{2p} - \eta_{1p}}{\eta_{2p} + \eta_{1p}}$ and, $\eta_{TE/TM} = \frac{2\eta_{2p}}{\eta_{2p} + \eta_{1p}}$. The *wave impedance concept* thus facilitate to compute the transmission and reflection coefficients without remembering the complex expressions of these coefficient as we shall develop in Maxwell's field analysis using the field boundary conditions.

6.3 Coordinate System for Oblique Incidence Cases

In this section, our motto is to provide mathematical interpretation of a uniform plane wave that propagates at any direction and hit a boundary at an arbitrary angle and polarization. This mathematical analysis helps us analyze any obliquely incident fields at the boundaries. In this context, we can say that the normal incidence of the previous chapter is a special case of the oblique incidence of a plane wave. To analyze the field problem with an arbitrary incident angle and polarization, we need to establish a coordinate system so that we can obtain the complete field solutions that involve all three coordinate axes (x, y, z) for completeness.[8] To exemplify the coordinate system, we begin with a reference antenna which is a natural source of a free space plane wave propagation and a detector which is a nature observation point as shown in Figure 6.6a (below).[9] In this analysis, we consider the medium is lossless, homogeneous, isotropic, and linear medium so that the constitutive parameters μ and ε of the medium do not depend of the field orientation (polarization) and spatial coordinates. The magnitude of the propagation constant k of the lossless medium presents the phase shift per unit distance along the direction of propagation with unit (rad/m). Assume a z-directed wave propagation as a choice of convenience. In this case, we usually characterize a wave with its phase at any location in space with a factor $e^{\mp jkz}$, where it represents a $\pm z$-directed propagating wave with a phase constant or wave number k in (rad/m).[10] In the phasor form of the field, we analyze the wave with its phase variation along the direction of propagation with a phase factor, also called *the electrical length*, $\theta = kz$ with respect to a reference point, say origin. Now consider that the propagation is along any arbitrary direction, then the phase factor of the wave is defined as $e^{\mp j\beta\zeta}$, where ζ represents any direction that can be decomposed with the three coordinate axes x, y, and z. Since the direction associated with the propagation phase term $\theta = k\zeta$, we can use the vector nature of k to represent its directionality with respect to space.[11] Therefore, as shown in Figure 6.6 if we define a generic position vector $\boldsymbol{r} = \boldsymbol{a}_x x + \boldsymbol{a}_y y + \boldsymbol{a}_z z$, in space that represents the observation point, $P(x, y, z)$ then we can generalize the phase factor of the field wave at any arbitrary location, with reference to the origin from which the position vector $\boldsymbol{r}$ is originated, with the dot product of the vector propagation constant $\boldsymbol{k}$ and $\boldsymbol{r}$ as $\boldsymbol{k}.\ \boldsymbol{r}$. This is done solely by assuming the vector nature of

---

8 The most beautiful aspects of electromagnetic field analysis that involves the three principal coordinates are extractions of the detailed characteristics of the fields solutions at any point in space in minute details. Electromagnetic software companies are exploiting these features taking the advantages of the modern extra-ordinary computing power with powerful animations and the advanced numerical techniques of the advanced electromagnetics theories. The electromagnetic software tools such as CST Microwave Studio, HFSS, FEKO, and Kesight Advanced Design Systems are sold in billions of dollars. With the advent of new techniques and resources with cloud computing the software market becomes as prominent as measuring equipment market. Vendors are aggressively delving their ways to merge the two domains into one to exploit more market opportunities using augmented reality (AR) in their software tools.

9 At this stage we do not consider the polarization of field wave and leave for the analyzes of the special cases of TE and TM polarizations. Thus we only consider the propagation phase at an arbitrary direction with respect to the observation point of the wavefront of the plane wave.

10 We interchangeably use the two terms: phase constant and wavenumber for k conveniently. There is a subtle difference between the phase constant with the wave number k in free space (unbounded) wave propagation. Many textbooks use the phase constant of an electromagnetic wave for both free space and guided cases and k for the free space case, this textbook prefers to use k from the beginning for the unguided waves. Note both phase constant β and k have the similar concept and same unit in (rad/m).

11 Again, the directionality of k is a totally new concept for the oblique incidence case and the genesis of the guides wave theory.

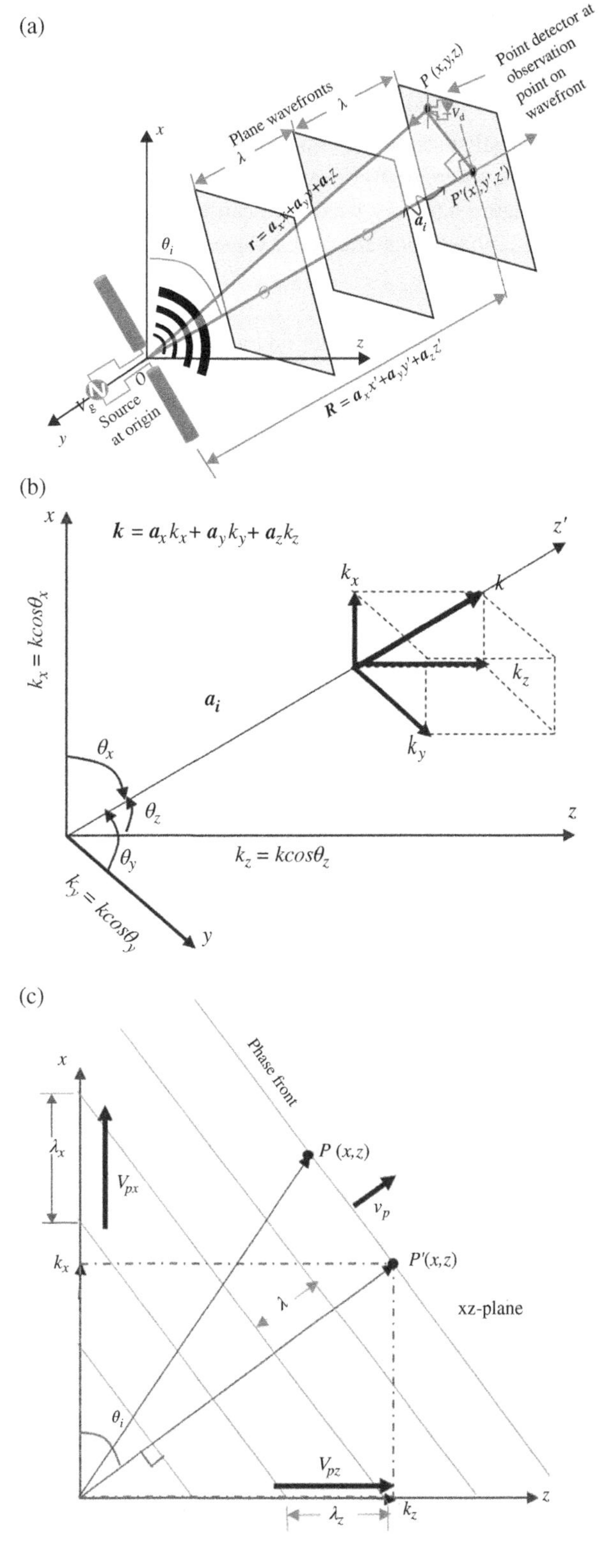

Figure 6.6 Coordinate system for oblique incidence case: (a) 3D perspective view, (b) 2D view, and (c) components of k. The incident field for an antenna at an arbitrary angle θ_i with x-axis creates plane wave fronts separated by wavelength λ. A point of the plane wave at a distance with position vector $\boldsymbol{R} = \boldsymbol{a_x}x' + \boldsymbol{a_y}y' + \boldsymbol{a_z}z'$ is defined as $P'(x', y', z')$. We use $'$ for source related coordinated system. A point detector is placed at an observation point $P(x, y, z)$ at a distance with position vector $\boldsymbol{r} = \boldsymbol{a_x}x + \boldsymbol{a_y}y + \boldsymbol{a_z}z$. Then $OP'P$ creates as right-angle rectangle. The position vector $\boldsymbol{R}$ is normal to the wavefront at P'. The general direction of wave propagation is defined by the unit vector $\boldsymbol{a_i}$. All the unit vectors for electromagnetic fields and the direction of wave propagation are on the same plane of $OP'P$.

the propagation constant k. Using the vector nature of the propagation constant we can define the field wave in any general direction as follows:

$$\boldsymbol{E}(z) = \boldsymbol{E}_0 e^{-j\boldsymbol{k}.\boldsymbol{r}} \tag{6.1}$$

We are mostly used to the two-dimensional field analysis. For the two-dimensional case of wave field analysis, where we freeze a coordinate system or discretely vary one coordinate system as illustrated in Figure 6.6b, say y-axis, we can use the vector nature of the propagation constant k and consider the phase constant of the wave at a general location as:

$$\boldsymbol{k} = \boldsymbol{a}_x k_x + \boldsymbol{a}_z k_z \tag{6.2}$$

And the position vector is defined as:

$$\boldsymbol{r} = \boldsymbol{a}_x x + \boldsymbol{a}_z z \tag{6.3}$$

Therefore, we can define, the field wave in its phasor form at any arbitrary direction on the xz-plane ($y = 0$) as:

$$\boldsymbol{E}(x, z) = \boldsymbol{E}_0 e^{-j\boldsymbol{k}.\boldsymbol{r}} = \boldsymbol{E}_0 e^{-j\left(k_x x + k_y z\right)} \tag{6.4}$$

In the following, we are going to explain the vector nature of the propagation phase for any arbitrary directed field wave with the help of Figure 6.6. In this explanation, we are going in details to show how the arbitrary incident field wave at any arbitrary incident angle and polarization can be generated and propagated.

Consider a dipole antenna oriented in an arbitrary angle with respect to the rectangular coordinate system at the origin as shown in Figure 6.6a. The antenna after certain distance (far field) launches periodic plane wavefronts shown with a wavelength distance λ in free space. In the context of oblique incidence such an arbitrarily directed plane wave impinges on a planar boundary.[12] Since we illustrate the uniform plane wave with a ray along the direction of propagation,[13] we can assume that the ray, which represents the plane wave, makes a pin hole through the plane of incidence and it maintains the orthogonality with the wavefront all the time. It is our choice of convenience that we can place the planar boundary along any of the coordinate planes, as for example in all previous cases, we place the boundary along xz-plane of the coordinate system with origin included. Figure 6.6a (above) shows such a coordinate system to exemplify the oblique incidence problem on any arbitrary plane of the wavefront that contains two points of interests: (i) a point P' (x', y', z') that is linked with the source at the origin O to the wavefront with the position vector $\boldsymbol{R} = \boldsymbol{a}_x x' + \boldsymbol{a}_y y' + \boldsymbol{a}_z z'$; note that the position vector $\boldsymbol{R}$ is normal to the wavefront at point P',[14] and all previous and subsequent wavefronts. The incident ray at an arbitrary angle is represented by the position vector, and (ii) $P(x, y, z)$ is the point of observation and is linked with a position vector $\boldsymbol{r} = \boldsymbol{a}_x x + \boldsymbol{a}_y y + \boldsymbol{a}_z z$ that defines an arbitrary observation direction OP. Using the vector relationship between the unit vector along the direction of propagation $\boldsymbol{a}_i$, which is normal to the wavefront, and the position vector $\boldsymbol{R}$, which is now defined as: $\boldsymbol{R} = \boldsymbol{a}_i R = \boldsymbol{a}_x x' + \boldsymbol{a}_y y' + \boldsymbol{a}_z z'$ here $R = \sqrt{x'^2 + y'^2 + z'^2}$.

---

12 For most practical cases of wireless communications systems, we do not know in which direction and/or angle and at which polarization a plane wave impinges on the boundaries of various dielectric media.

13 The direction of Poynting vector of a uniform plane wave is also a natural choice of the direction of propagation. Knowing the field components and its polarization we can define the direction of propagation via the equation of Poynting vector $\boldsymbol{P} = \boldsymbol{E} \times \boldsymbol{H}$.

14 The orthogonality of $\boldsymbol{R}$ and the wavefront can be nicely depicted with the animation feature of modern electromagnetic software simulation tools such as Computer Simulation Technologies Inc. (CST) Microwave Studio.

Now we are in a position to have the relationship between the two position vectors $\boldsymbol{R}$ and $\boldsymbol{r}$, via the right-angle triangle $OP'P$ as $\boldsymbol{R}.\ \boldsymbol{r} = R^2 = x'^2 + y'^2 + z'^2$; this means the projection of $\boldsymbol{r}$ on $\boldsymbol{R}$, is basically the modulus of $|R|^2$. Therefore, the triangle created by points $\triangle OP'P$ is a always right angle triangle as long as the two points P and P' are on the phase plane as shown in Figure 6.6a. This means the condition of right-angle triangle $\triangle OP'P$ is maintained as long as observation point $P(x, y, z)$ resides on the plane of the incident wavefront. The angle created with the incident ray and/or the position vector $\boldsymbol{R}$ with the x-axis is considered to be the angle of incident θ_i. We consider the direction of the wave propagation in the medium (assumed lossless) with a unit vector $\boldsymbol{a_i}$ along which the propagation constant is defined as $\beta = k = \omega\sqrt{\mu\varepsilon}$. Here we consider that the propagation constant is a three-dimensional vector and is represented with $\boldsymbol{k} = \boldsymbol{a_x}k_x + \boldsymbol{a_y}k_y + \boldsymbol{a_z}k_z = \boldsymbol{a_i}k$, where $\boldsymbol{a_i}$ is the unit vector along the direction of wave propagation. Therefore, the modulus of k can be defined as: $k = \sqrt{k_x^2 + k_y^2 + k_z^2}$, where, k_x, k_y, and k_z, respectively, are the x-, y-, and z-directed components of the vector propagation constant $\boldsymbol{k}$. As stated above, the vector $\boldsymbol{k}$ has the same direction of the power flow density vector or Poynting vector $\boldsymbol{P} = \boldsymbol{E} \times \boldsymbol{H}$.

So far, we have defined the three-dimensional aspect of the wave propagation at any direction. For simplicity and ease of understanding we can consider the two-dimensional aspect of the problem as shown in Figure 6.6b. In this case, we assume that the phase front makes an angle with x-axis and lines along y-axis.[15] Here, we can see more information about defining the oblique incidence of the plane wave. The equi-phase wavefronts, which are at angle θ_i with x-axis, are represented with periodic lines separated by the wavelength λ. The line OP' is normal to the lines representing the planar wavefronts and along the phase velocity v_p of the wave. The x- and z-directed phase constants along x- and z-axes, are k_x and k_z, respectively. Likewise, the phase velocities along x- and z-axes are v_{px} and v_{pz}, respectively. Note that the component phase velocities along the principal axes x, y, and z are not equal to each other and the actual phase velocity v_p of the obliquely incident wave, meaning $v_{px} \neq v_{pz} \neq v_p$. This can be easily visible from the differently stretched wavelengths λ_x and λ_z along x- and z-axes, respectively. If the incident angle $\theta_i = 90°$, then we shall have the special case of the normal incidence and the phase velocity will be $v_{pz} = v_p$ and $k = k_z$. For simplicity, let us start with the two-dimensional field that is oriented along x-axis and is propagating along z-axis is defined as:

$$\boldsymbol{E}(z) = \boldsymbol{a_x}E_0e^{-jkz} \tag{6.5}$$

Now consider that the field is rotated such that Figure 6.6b can be applicable in our analysis. We define a unit vector $\boldsymbol{a_i}$ of the wave that represents the direction of wave propagation and is along a new axis z'. The direction of the phase velocity v_p of the wave is also along z'. Then the electric field that makes an angle θ_i with x-axis can be defined as:

$$\boldsymbol{E}(z') = \boldsymbol{a_e}E_0e^{-jkz'} \tag{6.6}$$

Here, $\boldsymbol{a_e}$ is the unit vector of the obliquely incident electric field at an arbitrary incident angle θ_i. As you may recall $k = \boldsymbol{a_i}k$, which is the $\boldsymbol{a_i}$ directed propagation phase constant in free space at an arbitrary direction that makes an angle θ_i with x-axis. We can define the arbitrary z' axis in the same token as for $\boldsymbol{R}.\ \boldsymbol{r} = R^2 = x'^2 + y'^2 + z'^2$ as if $\boldsymbol{a_i}.\ r = z'$. Therefore, we can define the new field at any arbitrary direction as:

$$\boldsymbol{E}(z') = \boldsymbol{a_e}E_0e^{-j\boldsymbol{k}.\boldsymbol{r}} \tag{6.7}$$

15 We may assume $y = 0$ meaning xz-plane.

Now we are in a situation to generalize the definition of the electric field at any direction that considers the three coordinate axes as follows:

$$\boldsymbol{E}(x, y, z) = \boldsymbol{a_e} E_0 e^{-j\boldsymbol{k}.\boldsymbol{r}} = \boldsymbol{a_e} E_0 e^{-j\left(\boldsymbol{a}_x k_x + \boldsymbol{a}_y k_y + \boldsymbol{a}_z k_z\right).\left(\boldsymbol{a}_x x + \boldsymbol{a}_y y + \boldsymbol{a}_z z\right)} = \boldsymbol{a_e} E_0 e^{-j\left(k_x x + k_y y + k_z z\right)} \tag{6.8}$$

We have not resolved the unit vector of the electric field $\boldsymbol{a_e}$ as yet. We shall solve the issue when we consider the particular polarization of the field wave.

We can also generalize the definition of each components of the phase constant if we define the respective angles of the incident ray with the individual principal coordinate axes. In this case, we define $\theta_x = \theta_i$, and θ_y and θ_z, yet to be determined, then we can have the definitions of the individual components of the wave number as follows:

$$k_x = k\cos\theta_x;\ k_y = k\cos\theta_y;\ k_z = k\cos\theta_z. \tag{6.9}$$

Where k is the actual wave number of the obliquely incident wave. Figure 6.6b illustrates the three components of the phase constant and their respective direction along x-, y-, and z-axes. The actual wavelength that separates the phase fronts of the plane wave along z'-direction as depicted in Figure 6.6a is defined as:

$$\lambda = \frac{2\pi}{k} \tag{6.10}$$

As shown in the two-dimensional illustration in Figure 6.6b, the wavefronts (equi-phase planes) intersect x- and z-axes with distances λ_x and λ_z, which are different than the actual wavelength λ. We can generalize the three wavelengths along x, y, and z axis as follows:

$$\lambda_x = \frac{2\pi}{k_x} = \frac{2\pi}{k\cos\theta_x};\ \lambda_y = \frac{2\pi}{k_y} = \frac{2\pi}{k\cos\theta_y};\quad \text{and } \lambda_z = \frac{2\pi}{k_z} = \frac{2\pi}{k\cos\theta_z}. \tag{6.11}$$

These are the intersecting wavelengths corresponding to the respective components of the phase constants along x-, y-, and z-axes, and all are greater than the actual wavelength λ. These important relations tell us that the guided waves are stretched out along the guiding axes.[16] The phase velocities associated with the respective wavelengths and the phase constants along x-, y-, and z-axes are defined as:

$$\begin{aligned} v_{px} &= \frac{\omega}{k_x} = \frac{\omega}{k\cos\theta_x} = \frac{v_p}{\cos\theta_x};\ v_{py} = \frac{\omega}{k_y} = \frac{\omega}{k\cos\theta_y} = \frac{v_p}{\cos\theta_y};\quad \text{and} \\ v_{pz} &= \frac{\omega}{k_z} = \frac{\omega}{k\cos\theta_z} = \frac{v_p}{\cos\theta_z}. \end{aligned} \tag{6.12}$$

Again the phase velocities along the principal coordinate axes are larger than that for the actual phase velocity of the obliquely incident wave. They are associated with the stretched wavelengths along the principal axes. We can put an analogy of the higher component phase velocities compared to the actual wave velocity with the velocities of sea waves that are obliquely incident on the shoreline. This wave with the apparent higher velocity glides through the principal axis and does not carry any true energy hence does not violate the relativity theory of Einstein.

Thus far we have discussed about the direction of propagation of the obliquely incident wave. The second important aspect is the polarization of the incident field waves.[17] As stated earlier, any arbitrarily polarized, obliquely incident wave fields can be analyzed with two polarization states: (i) parallel polarization, also called TM due to the fact that the magnetic field is transverse to the

16 We shall comprehend the elongated wavelength (also called *guide wavelength*) in the waveguide theory.

17 The direction of field vibration is referred to as the polarization of the field wave of the direction of the field. It is not associated with the direction of wave propagation and usually transverse to the direction of the wave propagation. In waveguide theory, we shall look into the aspect where a component of the field wave can be along the direction of propagation.

direction of propagation, and (ii) perpendicular polarization also called TE due to the fact that the electric field is transverse to the direction of propagation. Since the polarization state is solely defined with the direction of vibration of the electric field vector, we define the polarization states in terms of an arbitrary unit vector $\boldsymbol{a_e}$ of the electric field. Figure 6.7 illustrates the vector analyses of the oblique incident wave in both polarization states. As can be seen for the case of the parallel polarization the electric field $\boldsymbol{E}(x, z)$ is parallel to the phase front hence the field amplitude vector $\boldsymbol{a_e}$ can be

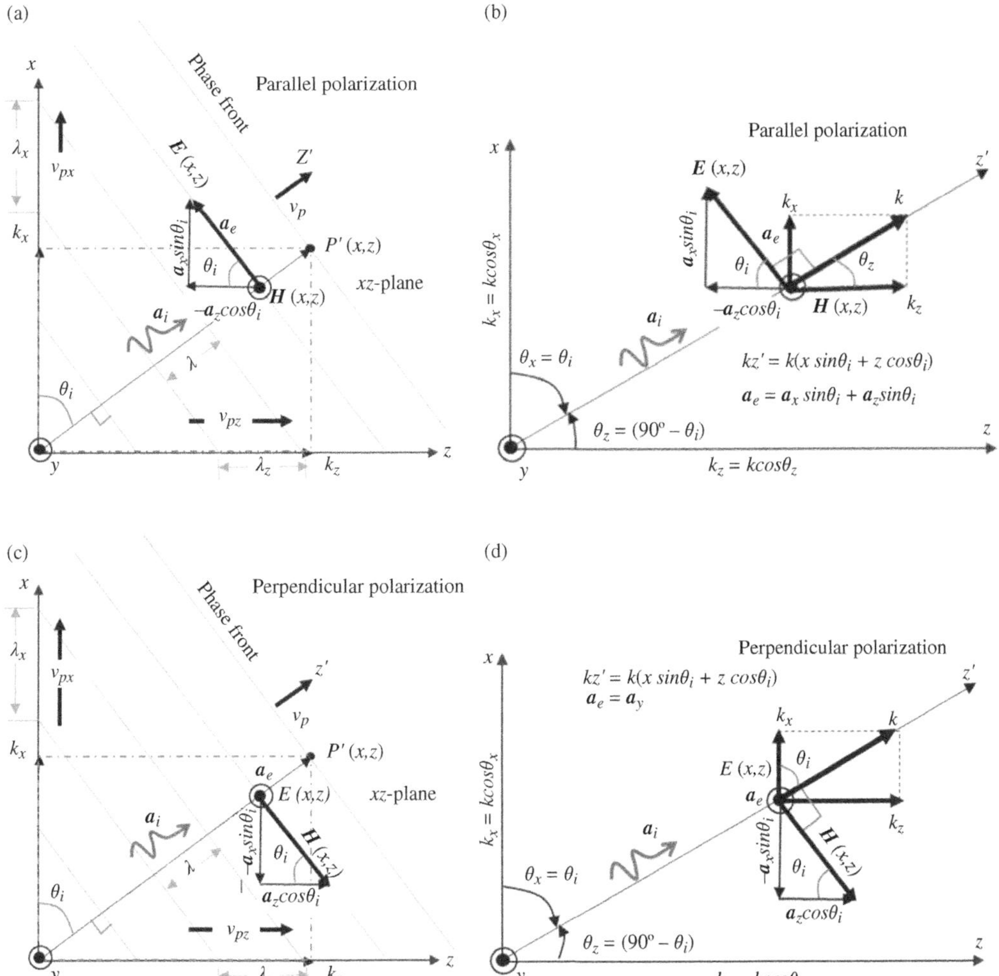

Figure 6.7 Field coordinate systems for: (a, b) parallel polarization, and (c, d) perpendicular polarization. In parallel polarization, the electric field $\boldsymbol{E}(x, z)$ resides on the phase front hence is decomposed into two components along positive x- and negative z-axes as a function of unit vectors and incident angle θ_i. By virtue of the orthogonality of the electric and magnetic fields and the direction of propagation vector $\boldsymbol{a_i}$, the magnetic field $\boldsymbol{H}(x, z)$ has only y-directed components as shown in (a) and (b). Whereas in perpendicular polarization the electric field $\boldsymbol{E}(x, z)$ is perpendicular to the phase front hence it has only y-directed components. But the magnetic field lies on the phase front hence it is decomposed into two components along x- and z-axes as a function of unit vectors and incident angle θ_i. The orthogonality relation between the electric and magnetic fields and the direction of propagation still holds. In both cases, the propagation phase factor remains the same as $kz' = k(x \sin\theta_i + z \cos\theta_i)$.

decomposed into two principal components, one along positive x-axis and one along negative z-axis. Therefore, $\boldsymbol{a_e}$ is the function of the respective unit vectors $\boldsymbol{a_x}$ and $\boldsymbol{a_z}$, and the incident angle θ_i. By virtue of the orthogonality of the electric and magnetic fields and the direction of propagation vector $\boldsymbol{a_i}$, the magnetic field $\boldsymbol{H}(x, z)$ must have only y-directed components as shown in Figure 6.7a, b. As shown in Figure 6.7c, d, the electric field $\boldsymbol{E}(x, z)$ is perpendicular to the phase front hence it has only y-directed components in the perpendicular polarization case. Again with the virtue of orthogonality of the fields and direction of propagation along z', the magnetic field must lie on the phase front as shown in Figure 6.7c, d. Again we can decompose the magnetic field into two components along x- and z-axes with respective unit vectors and as the function of the incident angle θ_i. The orthogonality relation between the electric and magnetic fields and the direction of propagation must hold in all polarization states. In both cases the propagation phase factor remains the same as $kz' = k(x \sin\theta_i + z\cos\theta_i)$. In summary, the fields are expressed as:

TM Polarization field expressions:	TE Polarization field expressions:
$\boldsymbol{E}(x,z) = \boldsymbol{a}_e E_0 e^{-jkz'}$ $= (\boldsymbol{a_x}\sin\theta_i - \boldsymbol{a_z}\cos\theta_i)e^{-jk(x\sin\theta_i + z\cos\theta_i)}$	$\boldsymbol{E}(x,z) = \boldsymbol{a}_y E_0 e^{-jkz'}$ $= \boldsymbol{a}_y e^{-jk(x\sin\theta_i + z\cos\theta_i)}$
$\boldsymbol{H}(x,z) = \boldsymbol{a_i} \times \dfrac{\boldsymbol{E}(z')}{\eta}$ $= \boldsymbol{a_y}\dfrac{E_0}{\eta}e^{-jk(x\sin\theta_i + z\cos\theta_i)}$	$\boldsymbol{H}(x,z) = \boldsymbol{a_i} \times \dfrac{\boldsymbol{E}(z')}{\eta}$ $= (-\boldsymbol{a_x}\sin\theta_i + \boldsymbol{a_z}\cos\theta_i)\dfrac{E_0}{\eta}e^{-jk(x\sin\theta_i + z\cos\theta_i)}$

In the following, we attempt review questions and do an example.

Review Questions 6.2: Vector Decomposition of Oblique Incidence

Q1. Explain why the coordinate system is so important for the analysis of oblique incidence cases.

Q2. Explain the vector nature of the propagation constant $k = \omega\sqrt{\mu\varepsilon}$. Why do we consider propagation constant as the phase constant? What is the difference between β phase constant and wave number k in this case?

Q3. Show that an electric field arbitrarily incident on a planar interface can be defined as $E(x, y, z) = a_e E_0 e^{j(k_x x + k_y y + k_z z)}$. Does this field expression include the vector nature of k?

Q4. What is the direction of an obliquely incident field wave? How many polarization states are there in oblique incident cases? Explain their physical orientations with respect to the direction of wave propagation.

Q5. Explain why the actual wave phase velocity is different from the component phase velocities along the principal axes. Give an analogy of the higher component phase velocity. Does this violate the relativity theory of Einstein?

Q6. Explain why wavelength along the principal axes are stretched out. What happens to the phase velocities along the principal coordinate axes?

Q7. If the angle of incidence is $\theta_i = 60°$, calculate the phase velocities along x- and z-axes. Assume the medium is lossless and has a dielectric constant $\varepsilon_r = 10.2$.

Q8. Calculate the actual wavelength of the same medium at 2.45 GHz ISM band. Calculate the component wavelengths.

Example 6.1: Oblique Incidence Coordinate System

A 1 GHz uniform plane wave with electric field amplitude 10 V/m is propagating in a lossless medium of dielectric constant $\varepsilon_r = 2.45$. The wave propagating in the *xz*-plane ($y = 0$) creates an angle $\theta_i = 30°$ with *x*-axis. Assume the electrical field is polarized parallel to the *xz*-plane. Calculate: (i) the component phase constants, (ii) the component wavelengths, (iii) the component phase velocities, and (iv) the electric and magnetic fields in both phasor form and time domain. Figure 6.8a illustrates the 2D view of TM oblique incidence on *xz*-plane with the unit vector decompositions for $E(x, z)$ and the *x*- and *z*-directed velocity vectors and wavelengths. Figure 6.8b shows the decomposition of propagation phase factors along *x* and *z*-axes.

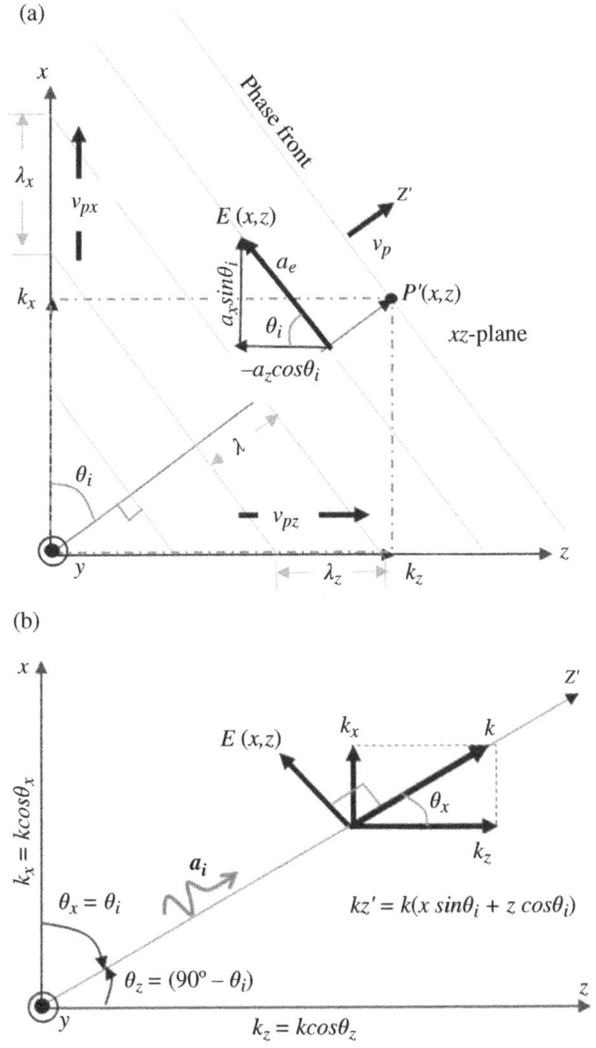

Figure 6.8 2D view of TM oblique incidence on *xz*-plane with *z*-directed vector decomposition of (a) electric field $E(x, z)$, phase velocities and wavelengths, and (b) the *x*- and *z*-directed propagation phase factors.

Answer:

Given parameters: frequency $f = 1$ GHz $= 10^9$ Hz

The dielectric constant of the medium $\varepsilon_r = 2.45$

The incident angle $\theta_i = 30°$

The angular frequency: $\omega = 2\pi f = 2\pi \times 10^9$ (rad/s)
The phase constant of the wave: $k = (\omega/c_0) \times \sqrt{\varepsilon_r} = \left(2\pi \times 10^9 \times \sqrt{2.45}\right)/(3 \times 10^8) = 32.78\,(\text{rad/m})$
The wavelength of the wave: $\lambda = 2\pi/k = 2\pi/32.78 = 0.192$ (m)
The phase velocity of the wave: $v_p = \omega/k = (2\pi \times 10^9)/32.78 = 191.66 \times 10^6$ (m/s)

i) The component phase constants:

$$k_x = k\cos\theta_x = 32.78 \times \cos\cos(30°) = 28.39\,(\text{rad/m})$$

$$k_z = k\cos\theta_z = 32.78 \times \cos\cos(60°) = 16.39\,(\text{rad/m})$$

ii) The component wavelengths:

$$\lambda_x = \frac{\lambda}{\cos\theta_x} = \frac{0.192}{\cos(30°)} = 0.221\,(\text{m})$$

$$\lambda_z = \frac{\lambda}{\cos\theta_z} = \frac{0.192}{\cos(60°)} = 0.383\,(\text{m})$$

iii) The component phase velocities:

$$v_{px} = \frac{v_p}{\cos\theta_x} = \frac{191.66 \times 10^6}{\cos(30°)} = 221.31 \times 10^6\,(\text{m/s})$$

$$v_{pz} = \frac{v_p}{\cos\theta_z} = \frac{191.66 \times 10^6}{\cos(60°)} = 383.33 \times 10^6\,(\text{m/s})$$

iv) The general expressions of the electromagnetic fields in the phasor form are:

$$\boldsymbol{E}(z') = \boldsymbol{a_e} E_0 e^{-jkz'}; \boldsymbol{H}(z') = \boldsymbol{a_i} \times \frac{\boldsymbol{E}(z')}{\eta} = \boldsymbol{a_y}\frac{E_0}{\eta}e^{-jkz'}$$

Now we calculate the unit vector of the electric field that lies on the xz-plane as shown in the figure:

$$\boldsymbol{a_e} = \boldsymbol{a_x}\sin\theta_i - \boldsymbol{a_z}\cos\theta_i = \boldsymbol{a_x}\sin(30°) - \boldsymbol{a_z}\cos(30°) = 0.5\boldsymbol{a_x} - 0.866\boldsymbol{a_z}$$

The spatial phase factor of the wave is:

$$kz' = k_x x + k_z z = 28.39x + 16.39z\,(\text{rad})$$

Now we have all ingredients to calculate the electromagnetic fields in both phasor form and time domain.

The phasor form electric field:

$$\boldsymbol{E_s}(x, z) = 10 \times (0.5\boldsymbol{a_x} - 0.866\boldsymbol{a_z})e^{-j(28.39x + 16.39z)}\,(\text{V/m})$$

The time domain electric field expression:

$$\boldsymbol{E}(x, z, t) = (5\boldsymbol{a_x} - 8.66\boldsymbol{a_z})\cos(2\pi \times 10^9 t - 28.39x - 16.39z)\,(\text{V/m})$$

The phasor form magnetic field:

$$\boldsymbol{H_s}(x, z) = \boldsymbol{a_y}\frac{10 \times \sqrt{2.45}}{377}e^{-j(28.39x + 16.39z)} = \boldsymbol{a_y}41.5\,e^{-j(28.39x + 16.39z)}\,(\text{mA/m})$$

The time domain magnetic field expression:

$$\boldsymbol{H}(x, z, t) = \boldsymbol{a_y}41.5\cos(2\pi \times 10^9 t - 28.39x - 16.39z)\,(\text{mA/m})$$

Remarks

i) As can be seen, the phase constant is quite large at 1 GHz frequency and it becomes even higher with the dielectric constant of the medium. However, the component phase factors along the principal axes shrink due to the elongated component wavelengths.
ii) The component phase velocities along the principal axes are larger than the wave phase velocity along z' direction. This is as expected. However, this phase velocities do not carry any real power. This can be explained with an analogy of obliquely incident waves breaking at a higher speed than the wave at the shoreline.
iii) The polarization of the incident electric field that lies on the phase front of the incident wave is called *the parallel polarization*. In this case, the unit vector $\boldsymbol{a_e}$ is orthogonal to the wave front of incident wave and is decomposed into x and z components. However, due to the orthogonality of the electric and magnetic fields with respect to the direction of propagation $\boldsymbol{a_i}$, the magnetic field has only one component along y-axis.
iv) Finally, the propagation phase constant is decomposed into x- and z-directed components. Since the phase factor is defined in terms of (rad/m), the component phase factors vary with distance, it is a scalar phase quantity, and is multiplied with the distances along the principal axes x and z respectively.

6.4 Oblique Incidence on Conducting Boundary

In the preceding two sections we analyzed the coordinate system for the propagation phase at arbitrary angle of wave incidence. There we observed that this propagation phase factors $e^{-j\zeta}$ and $e^{-j\zeta'}$ remain unchanged for the change of the field's polarization. We also develop the vector analysis of the electromagnetic fields for both TE and TM polarizations of oblique incidence cases. In this section, we shall utilize the coordinate system and develop the theory of the oblique incidence on the conducting boundary. Figure 6.14 illustrates an obliquely incident electromagnetic wave impinges on a perfect conducting boundary. Based on the polarizations of the field wave, we consider both cases: (i) TM, when the electric field is parallel to the plane of incidence and is denoted with $E_{\|}$, and TE, in the case when the electric field is normal to the plane of incidence and is denoted with $E_{\perp}$. As stated before, for the analysis of an arbitrary polarization besides TE and TM cases, the field solutions can be found by summing up the components in the parallel and perpendicular planes, and any difference in phase due to the arbitrary polarization can be taken care of by making the electric field complex. Therefore, $E_{\|}$ and $E_{\perp}$ suffice the analysis of any obliquely incident field wave for completeness. Figure 6.9a (below) shows the TE polarized oblique incidence on a conducting boundary along xy-plane. The TE polarized electric field is oriented vertically with the plane of incidence which is on the xz-plane (plane of paper), and the vector field hits parallelly the plane of interface or boundary plane which is on the xy-plane. This is analogous to as if you throw a horizontally oriented rod with two hands along the direction of incident field that hits the conducting wall making zero angle with the line of intersection of yz-plane and xy-plane at the origin, and as a consequence, the rod reflects back with the same zero angle with the intersection line, but along the direction of the reflected signal with the angle of reflection. The field orientation of the TE polarization suggests the hump back movement of a caterpillar as shown on the top of the figure. As we can see also, the angle of incidence is θ_i that makes the angle between the direction of the incident wave propagation $\boldsymbol{a_i}$ and the normal line to the plane of interface along z-axis. The angle of

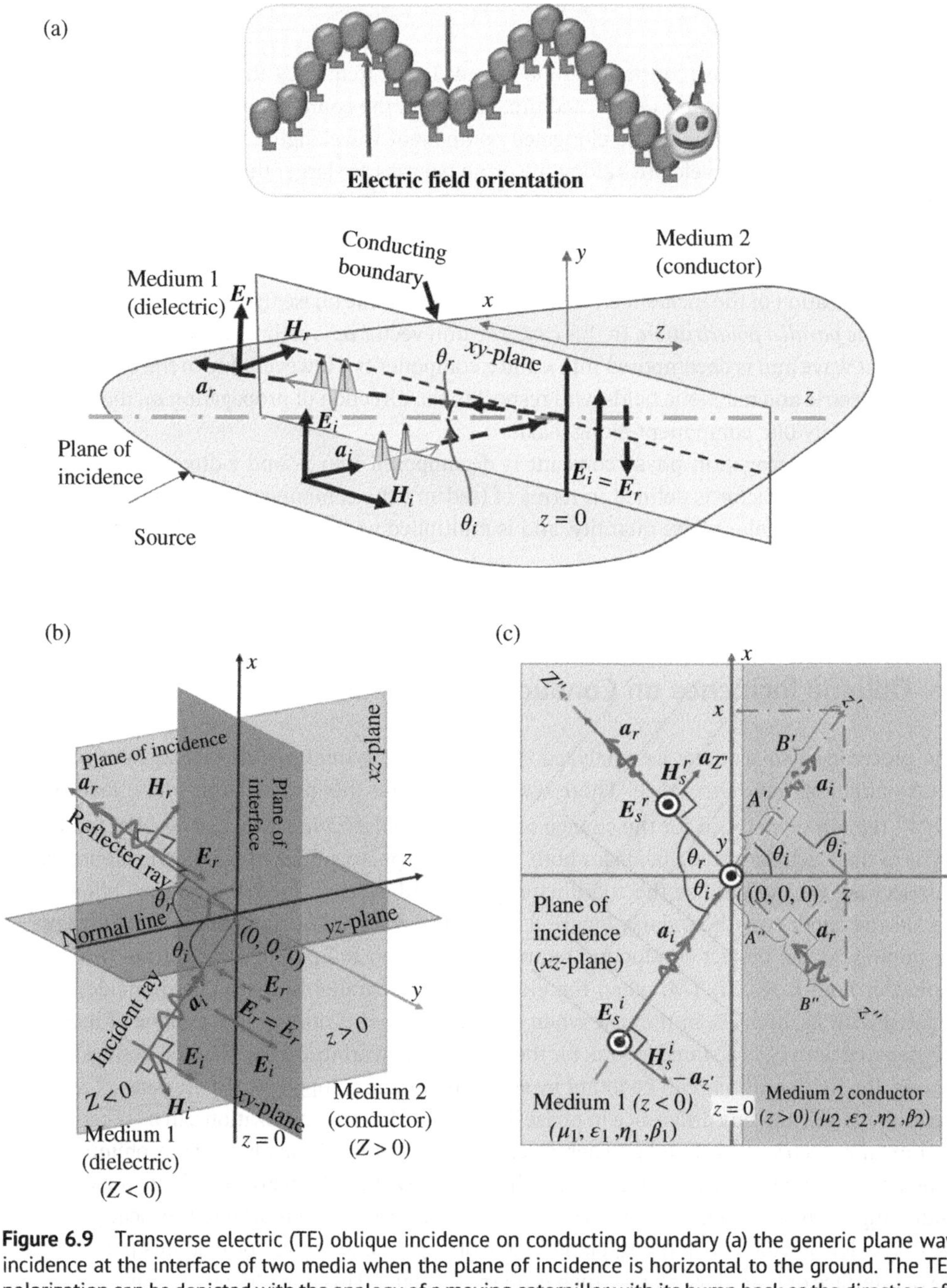

Figure 6.9 Transverse electric (TE) oblique incidence on conducting boundary (a) the generic plane wave incidence at the interface of two media when the plane of incidence is horizontal to the ground. The TE polarization can be depicted with the analogy of a moving caterpillar with its hump back as the direction of the electric field. At the interface of the dielectric and conductor media, incident electric field is equal and opposite to make the tangential electric field zero at the conducting boundary. This is an equipotential surface with zero tangential electric field, (b) 3D plan view of the same illustration as for (a) with vertically oriented plane of incidence, and (c) 2D plan view of the same illustration as for (a) with plane of incident is on the paper.

reflection is θ_r, which is the angle between the reflected wave's unit vector $\boldsymbol{a_r}$ and the z-axis. The incident magnetic field resides on the xz-plane with an arbitrary angle which is perpendicular to the direction of $\boldsymbol{a_i}$. The relationship between the incident electric field, magnetic field and the direction of wave propagation must obey the right-hand rule as shown in the figure. As can be seen, the directions for the incident and reflected electric fields remain unchanged along y-axis. They are vibrating along the y-axis as it propagates in both incident and reflected conditions. However, the magnetic field changes its direction and still remain on the xz-plane (plane of paper) to maintain the right-hand rule with the electric field and the direction of wave propagation. The 2D propagating waves in both incident and reflected conditions are shown with sinusoidally varying red curves in the figure. Our motto of the analysis is to find the field solutions of the incident and reflected waves. Note that the transmitted field is zero in the conducting medium hence it imposes the boundary condition such that the incident and reflected field amplitudes are equal and opposite ($\boldsymbol{E_i} = -\boldsymbol{E_r}$) at the interface on the xy-plane (at $z = 0$) as shown in the figure. The other field conditions that we shall prove leads to *Snell's law of reflection*. The law says that the incident angle is equal to the reflected angle ($\theta_i = \theta_t$). Figure 6.7b shows the oblique incidence case of Figure 6.7a, in this case the plane of incident is placed vertically for clear depiction of the electromagnetic system. Figure 6.7c shows 2D view of the electromagnetic system where the plane of incidence on the plane of the paper (xz-plane) and the electric fields are piercing out of the paper while the magnetic fields and the unit vectors of the incident and reflected waves are residing on the plane of the paper. The vector analyze of the fields and the directions of propagation in both incident and reflected field waves along z' and z'' are also shown on the right side of the plane of interface at $z = 0$. Note that this vector diagram is for field analysis purposes only and no transmitted fields are propagating in medium 2 (conductor). This transmitted field zero condition simplifies analysis and reduces field solutions for the incident and reflected fields only. Note that since the TE incident field vector is perpendicular to the plane of incident (xz-plane), the field is also called the field of *perpendicular polarization*. Contrary to this when the electric field is parallel to the plane of interface (plane of paper) the wave field is called *parallel polarization*. Both polarizations, although the direction of propagation vectors are the same, bring different solutions to the wave fields. We start our analysis with the TE oblique incidence case.

Review Questions 6.3: On Oblique Incidence

Q1. Define perpendicular and parallel polarizations of uniform plane waves incident obliquely on an interface of two media.

Q2. Explain why we need not consider these two polarizations for the case of normal incidence.

Q3. What is the Snell's law of reflection.

Q4. For both polarizations do the orthogonality relationship and right-hand rule of the uniform plane wave hold?

Q5. For TE polarization what analogy do we bring and why?

6.5 TE Polarization on Conducting Boundary

As stated above, we start the field analysis of the oblique incidence case with the TE polarization as the electric field is well defined on only y-direction. The phasor vector form of the incident electric field in medium 1 can be defined as

$$\boldsymbol{E}^i_{s\perp}(x, z) = \boldsymbol{a}_y E_{0i} e^{-jk_1 z'} \tag{6.13}$$

The vector decomposition of the propagation phase factor, which is linked with the distance along the direction of propagation, is shown in Figure 6.9c. This is a summation of two lengths A' and B'. These lengths are basically the projections of z and x axes with their signs, respectively, on the distance along the direction of wave propagation with unit vector $\boldsymbol{a_i}$.[18] Therefore, examining the geometry in Figure 6.9c, these component distances are defined as $A' = z\cos\theta_i$ and $B' = x\sin\theta_i$. Adding A' and B' we obtain the expression of the distance phase factor of the incident wave. The detailed decomposition of the propagation phase factor of the incident field wave is illustrated separately in Figure 6.9a and is expressed as:

$$k_1 z' = k_1(x\sin\theta_i + z\cos\theta_i) \tag{6.14}$$

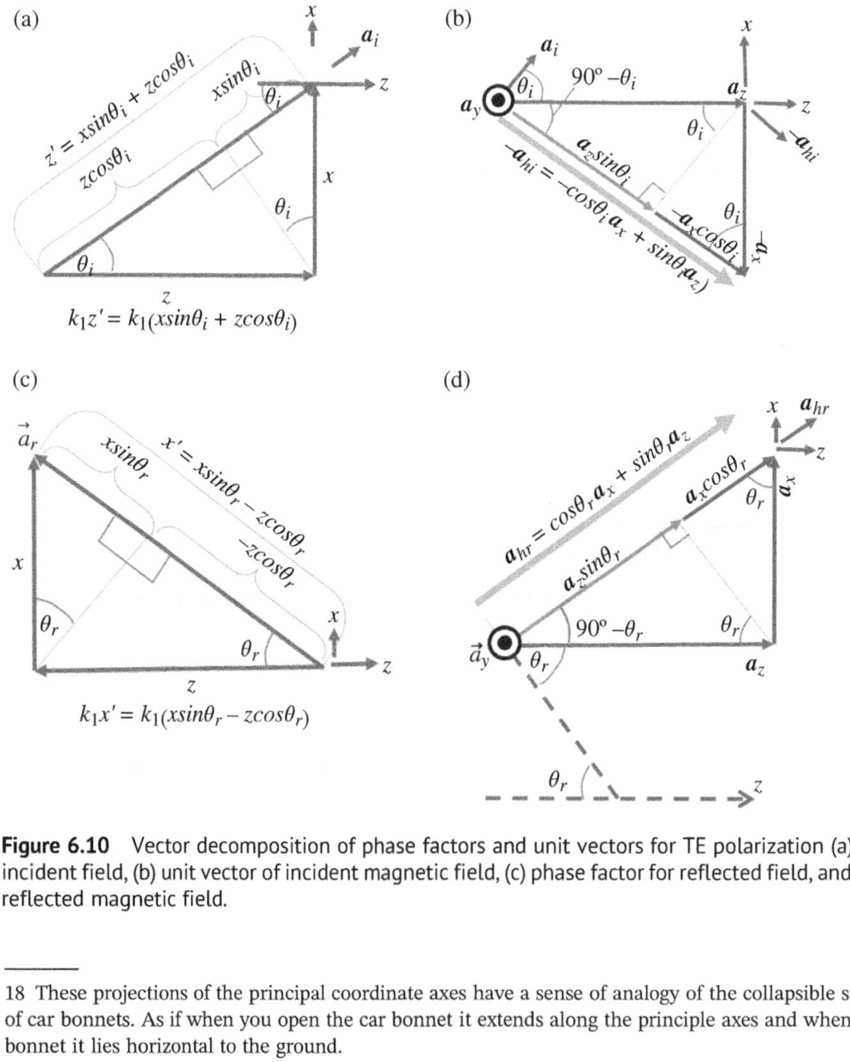

Figure 6.10 Vector decomposition of phase factors and unit vectors for TE polarization (a) phase factor for incident field, (b) unit vector of incident magnetic field, (c) phase factor for reflected field, and (d) unit vector of reflected magnetic field.

18 These projections of the principal coordinate axes have a sense of analogy of the collapsible supporting arms of car bonnets. As if when you open the car bonnet it extends along the principle axes and when you close the car bonnet it lies horizontal to the ground.

Substitution for (6.14) in (6.13) we obtain the incident electric field expression for the TE polarized field as:

$$\boldsymbol{E}^{i}_{s\perp}(x, z) = \boldsymbol{a_y} E_{0i} e^{-jk_1(x \sin\theta_i + z\cos\theta_i)} \tag{6.15}$$

The magnetic field lies on the xz-plane (the plane of paper) and is a function of both x and z. Note that the direction of the magnetic field vector $\boldsymbol{a}_{hi}$ can be easily determined using the right-hand rule of the electric field, magnetic field, and the direction of propagation, which is represented by the unit vector $\boldsymbol{a_i}$. This means the direction of the Poynting vector $\boldsymbol{P} = \boldsymbol{E}\times\boldsymbol{H}$ must be along the direction of propagation. We apply the unit vector decomposition of the magnetic field as shown in Figure 6.10b. The unit vector of the magnetic field is defined as: $\boldsymbol{a}_{hi} = -\sin\theta_i \boldsymbol{a_x} + \cos\theta_i \boldsymbol{a_z}$ as decomposed in the figure. Since the magnetic field vector has both x- and z-directed vector components, it is defined as:

$$\boldsymbol{H}^{i}_{s\perp}(x, z) = (-\cos\theta_i \boldsymbol{a_x} + \sin\theta_i \boldsymbol{a_z})\frac{E_{0i}}{\eta_1} e^{-jk_1(x \sin\theta_i + z\cos\theta_i)} \tag{6.16}$$

Next, the reflected electric and the magnetic fields can be defined as follows:

$$\boldsymbol{E}^{r}_{s\perp}(x, z) = \boldsymbol{a_y} E_{0r} e^{-jk_1(x \sin\theta_r - z\cos\theta_r)} \tag{6.17}$$

The magnetic field vector has both x- and z-directed vector components hence is defined as:

$$\boldsymbol{H}^{r}_{s\perp}(x, z) = (\boldsymbol{a_x} \cos\theta_r + \boldsymbol{a_z} \sin\theta_r)\frac{E_{0r}}{\eta_1} e^{-jk_1(x \sin\theta_r - z\cos\theta_r)} \tag{6.18}$$

As stated above, the second medium is a conducting medium, therefore, no transmitted field exists in the medium and $\boldsymbol{E}^{t}_{s\perp}(x, z) = \boldsymbol{H}^{t}_{s\perp}(x, z) = 0$.

Now we can apply the boundary conditions at the interface of the two media at $z = 0$. Since, the boundary is a perfect conductor, the total electric field which is tangential to the interface at $z = 0$ must vanish. This leads to the following expression:

$$\boldsymbol{E}_{s\perp 1}(x, 0) = \boldsymbol{E}^{i}_{s\perp}(x, 0) + \boldsymbol{E}^{r}_{s\perp}(x, 0) = \boldsymbol{a_y}\left\{E_{0i} e^{-jk_1(x \sin\theta_i)} + E_{0r} e^{-jk_1(x \sin\theta_r)}\right\} = 0$$

$$\text{Or, } E_{0i} e^{-jk_1(x \sin\theta_i)} = -E_{0r} e^{-jk_1(x \sin\theta_r)} \tag{6.19}$$

The above boundary condition must be true for any value of x at $z = 0$ plane. This means the amplitude and phase terms must be equal to satisfy the above condition. This condition leads to the following important relations between the incident and reflected fields:

$$E_{0i} = -E_{0r}; \quad \text{and } x \sin\theta_i = x \sin\theta_r; \quad \text{or, } \theta_i = \theta_r \tag{6.20}$$

The second equation $\theta_i = \theta_r$ of the equality of the incident and transmitted angles is called *Snell's law of reflection*. The electric and the magnetic fields are linked with the wave impedance at any point in space and is defined by:

$$\eta = \frac{E_i}{H_i} = \frac{E_r}{H_r} \tag{6.21}$$

In the above equation, we do not change the sign for the reflected field ratio because the reflected electric field E_r changes its direction compared to the incident field on the conducting boundary.

Using (6.20) and Euler's theorem, we can define the total electric field in medium 1 as

$$\boldsymbol{E}_{s\perp 1}(x, z) = \boldsymbol{a_y} E_{0i}\left\{e^{-jk_1 z\cos\theta_i} - e^{jk_1 z\cos\theta_i}\right\} e^{-jk_1 x\sin\theta_i}$$

$$\boldsymbol{E}_{s\perp 1}(x, z) = -\boldsymbol{a_y} j2E_{0i} \sin(k_1 z\cos\theta_i)\, e^{-jk_1 x\sin\theta_i} \tag{6.22}$$

The total magnetic field in medium 1 is defined as:

$$H_{s\perp 1}(x,z) = \frac{E_{0i}}{\eta_1}\{(-\cos\theta_i \mathbf{a_x} + \sin\theta_i \mathbf{a_z})e^{-jk_1 z\cos\theta_i} - (\cos\theta_r \mathbf{a_x} + \sin\theta_r \mathbf{a_z})e^{jk_1 z\cos\theta_i}\}e^{-jk_1 x\sin\theta_i} \tag{6.23}$$

Again using (6.20), and simple mathematical manipulations we obtain the expression for the total magnetic field in medium 1 as

$$H_{s\perp 1}(x,z) = -\frac{2E_{0i}}{\eta_1}\{\mathbf{a_x}\cos\theta_i \cos(k_1 z\cos\theta_i) + j\mathbf{a_z}\sin\theta_i \sin(k_1 z\cos\theta_i)\}\, e^{-jk_1 x\sin\theta_i} \tag{6.24}$$

We can make a few remarks about the characteristics of the individual field components:

1) For the TE-polarization, the resultant electric field has only y-directed component, whereas the resultant magnetic field has both x- and z-directed components. Both fields have spatial variations of amplitudes as they propagate along x- and z-axes, besides oscillations of their amplitude along their respective unit vectors. As stated before such field waves are called *nonuniform waves*. Thus the wave fields create a standing wave pattern along z-axis and a propagating wave along x-axis as shown in Figure 6.11. This phenomenon of the oblique field wave can be explained with a water jet obliquely incident upon a smooth surface (such as a car body) in a sunny day. You may notice that along the vertical direction of the body there is mist of wave with a rainbow pattern as the standing wave and some part of the wave with a faint trace glides along the body. The rainbow pattern created along the vertical direction is equivalent to the standing wave along z-axis and the thin layer of wave propagating along the body of the smooth surface is equivalent to the propagating wave along x-axis. This gliding wave along x-axis is called *the surface wave*.

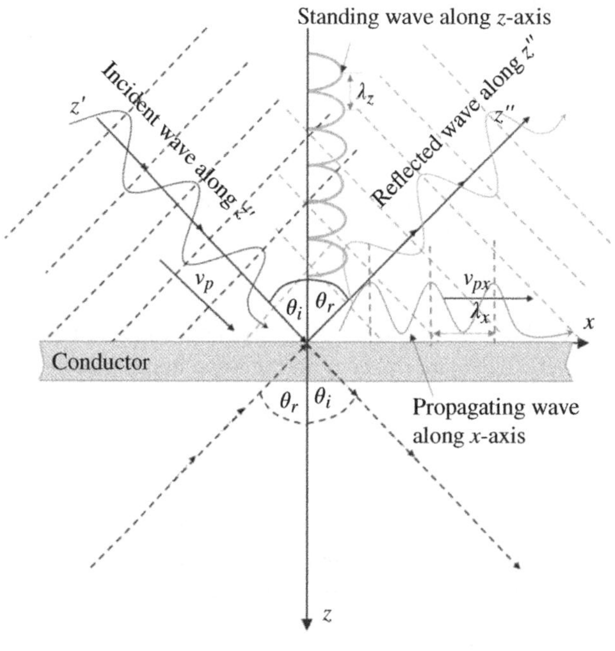

Figure 6.11 Analysis of oblique incident wave on a perfect conducting boundary. The figure illustrates an obliquely oriented electromagnetic field creating a standing wave pattern along z-axis and a propagating wave pattern along x-axis. The original wave amplitude is different than the axial wavelength. The phase velocity of the propagating wave along x-axis is v_{px}, which is larger than the actual wave phase velocity v_p.

2) Looking at propagation phase term $e^{-jk_1 x \sin\theta_i}$ of the electromagnetic fields, we can say that the fields are propagating along x-direction with a phase constant $k_{1x} = k_1 \sin\theta_i$. This axial phase constant is smaller than the phase constant of the actual propagation. Thus, the propagation wavelength is stretched along x-axis for both electric and magnetic field waves. The stretched wave length can be referred to the guide wavelength. Since the guided wave travels longer distance than the actual obliquely incident signal at the same time, the guided phase velocity along the axis is also larger than the actual wave phase velocity ($v_{px} > v_p$).
3) The resultant electric field in medium 1 is zero at the boundary plane ($z = 0$), and creates a standing wave pattern along z-direction with the amplitude variation along with a factor of $2E_{0i} \sin(k_1 z \cos\theta_i)$. This field must satisfy the boundary condition of the tangential electric field meaning it is zero at the conducting interface at $z = 0$ as stated in (6.20). It can also be explained in a contrary statement that since the tangential fields $E_y^{\pm}$, and $H_x^{\pm}$ are zero on the conducting boundary, the characteristics wave impedance on the surface is zero. This means the z-directed transverse impedance $Z_z = 0$ on the conducting boundary leading to a z-directed standing wave pattern with zeros at multiples of $\lambda/2$ distance away. This wavelength can be calculated from the z-directed phase velocity v_{pz}. Therefore, using the ratio of the forward and reverse traveling electric and the magnetic field waves in (6.22) and (6.23) we can define the characteristic wave impedance of TE-polarization as follows:

$$Z_{TE} = -\frac{E_y^i}{H_x^i} = \frac{E_y^r}{H_x^r} = \eta_1 \sec\theta_i \tag{6.25}$$

The –ve sign in the first ratio is to make the wave impedance positive. We have noticed that $Z_{TE} > \eta_1$. This is due to the fact that in TE polarization, the electric field has only one y-directed component, whereas the magnetic field has both x- and z-directed components. This means that the magnetic field has only a x-directed component lying transvers to z-direction (on xy-plane) and contributing to the value of the characteristic wave impedance. Thus, the ratio gives larger value of characteristic wave impedance compared to the free space wave impedance η_1 of medium 1.[19]
4) The resultant magnetic field vibrates along both x- and z-axes with amplitude variations of $(2E_{0i})/(\eta_1) \cos\theta_i \cos(k_1 z \cos\theta_i)$ and $(2E_{0i})/(\eta_1) \sin\theta_i \sin(k_1 z \cos\theta_i)$, respectively. Thus, we can define directional (guided) equivalent impedances in terms of the actual wave impedance and incident angle as follows:

$$\eta_x = \eta_1 \cos\theta_i \tag{6.26}$$

This relationship tells us that the guided wave impedance along the principal x-axis is less than the actual wave impedance η_1 of medium 1. We shall utilize these guided wave impedances to calculate the reflection and transmission coefficients for the oblique incidence cases when we shall develop *the wave impedance concept* for the obliquely incident plane wave analyses as in Section 6.10.4.

6.5.1 Poynting Vector in TE Polarization

So far, we have analyzed the TE-polarized electromagnetic field waves at the conducting boundary. Now we are going to examine how the electromagnetic power flows at the TE oblique incidence

19 For the same reason we shall see that in the waveguide the TE-mode wave impedance is larger than the unguided or free space wave impedance.

case. The time average power flow density vector can be defined with the help of the resultant electromagnetic fields in (6.22) and (6.23).

$$\begin{aligned}\boldsymbol{P}_{ave} &= \frac{1}{2}Re\{\boldsymbol{E}_{s\perp 1}(x,z)\times\boldsymbol{H}^*_{s\perp 1}(x,z)\} = \frac{1}{2}Re\{-\boldsymbol{a_y}j2E_{0i}\sin(k_1z\cos\theta_i)e^{-jk_1x\sin\theta_i}\}\\ &\times\left\{-2\frac{E_{0i}}{\eta_1}\{\boldsymbol{a_x}\cos\theta_i\cos(k_1z\cos\theta_i)+j\boldsymbol{a_z}\sin\theta_i\sin(k_1z\cos\theta_i)\}e^{+jk_1x\sin\theta_i}\right\}\\ &= \frac{1}{2}Re\left[\frac{4E_{0i}^2}{\eta_1}\{j\boldsymbol{a_z}\cos\theta_i\sin(k_1z\cos\theta_i)\cos(k_1z\cos\theta_i)+\boldsymbol{a_x}\sin\theta_i\sin^2(k_1z\cos\theta_i)\}\right]\end{aligned}$$

$$\text{Or, } \boldsymbol{P}_{ave} = \frac{1}{2}Re\left\{-j\boldsymbol{a_z}\frac{2E_{0i}^2}{\eta_1}\cos\theta_i\sin(2k_1z\cos\theta_i)+\boldsymbol{a_x}\frac{4E_{0i}^2}{\eta_1}\sin\theta_i\sin^2(k_1z\cos\theta_i)\right\} \tag{6.27}$$

Assuming E_{0i} and η_1 are positive real quantities, we can have the following closed form expression of the time-average power flow density vector as:

$$\boldsymbol{P}_{ave} = \boldsymbol{a_x}\frac{2E_{0i}^2}{\eta_1}\sin\theta_i\sin^2(k_1z\cos\theta_i) \tag{6.28}$$

A few remarks can be made on the calculations of the time average power flow density vector as follows:

1) Examining (6.28) we observe that as expected a real power is flowing along x-axis with a variation along z-axis as a function of $\sin^2(k_1z\cos\theta_i)$ in medium 1. Such a wave is called *a nonuniform* wave due to its amplitude variation along the direction of propagation. There is an imaginary power flow density vector along z-axis. This represents no real power flow along z-axis. This phenomenon can be explained as follows.
2) There is no real power flow along z-axis and a standing power wave pattern along z-axis with the factor $\sin(2k_1z\cos\theta_i)$ is created. The non-zero power propagation of the standing wave pattern along z-axis is the inherent phenomenon and can be explained with the orthogonality between the y-directed electric field and the x-directed magnetic field. This is expected as no power is absorbed in the conductor along z-axis and the z-directly power component is totally reflected back along negative z-axis in medium 1 and create a pure standing wave pattern. If we compare the pure standing wave created in the normal incidence on the conducting boundary with the oblique incidence case here, we find that there is always a pure standing wave pattern along z-direction for normal incidence case, whereas in latter case, there is an additional propagating wave glides along the surface meaning along x-axis. In the following we shall examine the phase velocity of the propagating wave and the standing wave pattern along z-axis.

6.5.2 Phase Velocity Calculation

Examining the field analysis we have found that there is a standing wave along z-axis and a propagating wave along x-axis. First we shall calculate the phase velocity of the propagating wave. The time harmonic field expression of (6.15) has the phase factor in the exponential form with the time dependence factor $e^{j\omega t}$ and is expressed as:

$$\boldsymbol{E}^i_\perp(x,z,t) = Re\left\{\boldsymbol{a_y}E_{0i}e^{j(\omega t-k_1x\sin\theta_i-k_1z\cos\theta_i)}\right\} = \boldsymbol{a_y}E_{0i}\cos(\omega t-k_1x\sin\theta_i-k_1z\cos\theta_i) \tag{6.29}$$

The instantaneous phase along x-axis ($\omega t - k_1 x \sin\theta_i$) must be kept constant as we move along x-axis if we want to calculate the phase velocity v_{px} as shown in Figure 6.11. Therefore, we can write

$$\omega t - k_1 x \sin\theta_i = c, \text{ where } c \text{ is a constant} \tag{6.30}$$

The phase velocity along x-axis is defined as the derivative of x with respect to time t. This leads to the definition of the x-directed phase velocity as follows:

$$v_{px} = \frac{\partial x}{\partial t} = \frac{\omega}{k_1 \sin\theta_i} = \frac{v_p}{\sin\theta_i} \tag{6.31}$$

Here, v_p is the wave phase velocity along the direction of oblique incidence as is defined as $v_p = \frac{\omega}{k_1} = 1/\sqrt{\mu\varepsilon}$. We find that the phase velocity v_{px} along x-axis is larger than the actual phase velocity of the incident wave.[20] The corresponding wave number and wavelength are defined, respectively, as:

$$k_{1x} = k_1 \sin\theta_i;\ \lambda_x = \frac{\omega}{k_{1x}} = \frac{\lambda}{\sin\theta_i} \tag{6.32}$$

where, the wavelength of the oblique incident wave field has a wavelength which is defined as $\lambda_1\omega = \omega/k_1$. As can be seen from (6.32) that the x-directed wavelength along the conducting boundary is larger than that for the obliquely incident wavelength. Likewise, we can also calculate the phase velocity along z-axis in the similar manner:

$$v_{pz} = \frac{\omega}{k_1 \cos\theta_i} = \frac{v_p}{\cos\theta_i} \tag{6.33}$$

The corresponding wave number and wavelength are defined, respectively, as:

$$k_{1z} = k_1 \cos\theta_i;\ \lambda_z = \frac{\omega}{k_{1z}} = \frac{\lambda}{\cos\theta_i} \tag{6.34}$$

In both cases, we observe a few interesting phenomena. The component phase velocities are larger than the actual phase velocities and this occurs when $\theta_i > 0°$. The rays are touching the conducting boundary at an angle at any point x and glides along the surface as *the surface wave*. Therefore, v_{px} represents the movement of a fictitious point along the surface at the intersection point of x and z' and no real energy is propagating along x-axis. Again the x- and z-directed phase constants, which are the phase changes with distance, reduce by certain factors $\sin\theta_i$ and $\cos\theta_i$, respectively, due to the consequence of the wavelengths λ_x and λ_z are stretched out by the factors as defined in (6.32) and (6.34), respectively. This phenomenon is illustrated in Figure 6.12. As can be seen in the figure, the obliquely incident wavefront $\chi\chi$ moves along the normal of z'-axis a wavelength λ and becomes the wavefront $\chi'\chi'$. The distance it travels along z'-axis (normal to the wavefront) is $\Delta z' = z_1' z_2'$. At the same time, the wavefront glides along x-axis a distance $\Delta x = x_1 x_2$. This distance is defined as the x-directed wavelength λ_x in (6.32). At the same time, the obliquely incident wavefronts $\chi\chi$ and $\chi'\chi'$ touches the z-axis at z_1 and z_2 points respectively. As if we can put a conducting boundary along z-axis, then the incident wavefront glides along z-axis at distance $\Delta z = z_1 z_2$. This distance is the wavelength λ_z along z-axis as defined in (6.34). In conclusion, in the obliquely incident wave fields, the phase constant k at a particular direction reduces by sine and cosine of the incident angle. The x- and z-directed phase velocities and the wavelengths increase by the same factor. The concept of the guided phase velocity and wavelength are the

20 Note that this larger velocity does not violate any fundamental principle of wave energy propagation as such no real power or information carrying signal propagates at that speed along x-axis.

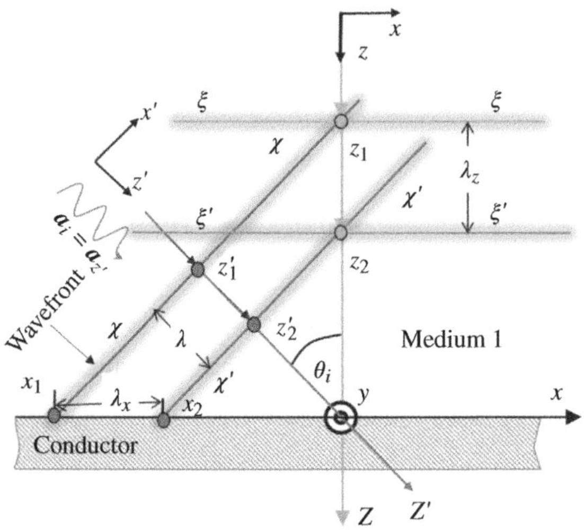

Figure 6.12 Obliquely incident wave impinges on a conductor boundary at an angle θ_i.

foundations for the waveguide concept. In the following section we are examining the concept of waveguides.

6.5.3 Waveguide Concept

Now consider the electromagnetic field expressions in (6.22) and (6.24). As we can see that the electric field $\boldsymbol{E}_{s\perp 1}(x, z)$ is zero as $sin\,(k_1 z \cos\theta_i) = 0$. This condition leads to the following expression

$$sin\,(k_1 z \cos\theta_i) = sin\,(n\pi) \quad \text{for } n = 0, 1, 2, 3... \tag{6.35}$$

This gives a particular periodic distance h_{zn} along z-axis at which the obliquely incident electric field is zero. The distance is defined as:

$$k_1 h_{zn} \cos\theta_i = n\pi;\ h_{zn} = \frac{n\pi}{k_1 \cos\theta_i} \tag{6.36}$$

Substituting $k_1 = 2\pi/\lambda$ leads to the closed form expression for h_{zn} at which the electric field is zero as follows:

$$h_{zn} = \frac{n\lambda}{2\cos\theta_i} \quad \text{for } n = 0, 1, 2, 3... \tag{6.37}$$

Figure 6.13 illustrates the distance at which the electric field is null at periodic planes parallel to xy-plane. Here n = 1, 2, 3, and 4, four different locations are illustrated, for n = 0 represents the conducting boundary itself. Thus, if we place a thin metallic sheet thicker than a few skin depths[21] then we can enforce the boundary condition of having zero electric field at the thin metallic sheet without affecting the electric field distributions in between the conducting boundary and the metallic sheet. This configuration is called *the parallel plate waveguide*. In a parallel plate waveguide, the obliquely incident wave propagates back and forth as the wave propagates along the waveguide along x-axis.

21 Recall why we do need to use a few skin depths to avoid any electric field leakage from the conducting sheet and field distortions inside the parallel plate waveguide.

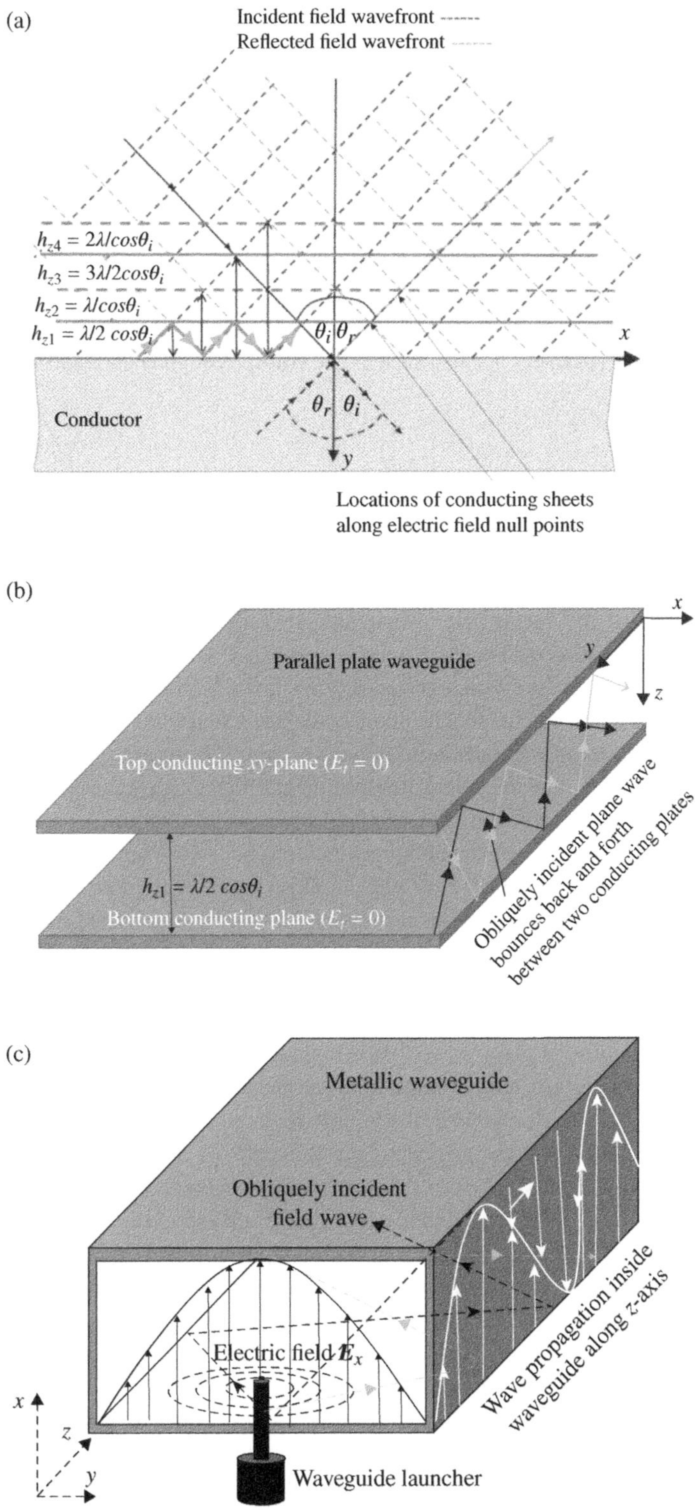

Figure 6.13 Electric field nulls and periodic locations ($h_{zn} = n\lambda/(2\ cos\ \theta_i)$ for $n = 0, 1, 2, 3...$) of conducting sheets above conducting boundary. The genesis of waveguide theory from the obliquely incident field theory. (a) The electric field nulls at distances ($h_{zn} = (n\lambda)/(2\ cos\ \theta_i)$ for $n = 0, 1, 2, 3...$). A conducting sheet can be placed at $h_{z1} = \lambda/(2\ cos\ \theta_i)$ above the conducting boundary at $z = 0$ to get the minimum distance parallel plate waveguide without affecting the field distribution. (b) the parallel plate waveguide with conducting sheets, and (c) the rectangular waveguide is a hollow tube makes the boundary condition of the plane wave that bounces back and forth along the side walls as it propagates along z-axis.

Obliquely incident plane wave bounces back and forth between the two conducting plates as shown in Figure 6.13a. The genesis of waveguide theory starting from the concept of the obliquely incident field waves reflects between the two parallel plates as shown in Figure 6.13b. Note, though the two obliquely incident rays bounce back and forth on the two parallel conducting boundaries, the electromagnetic energy propagates along z-direction. A conducting sheet can be placed at the minimum distance $h_{z1} = \lambda/(2\cos\theta_i)$ above the conducting boundary at $z = 0$ to get the most compact parallel plate waveguide without affecting the field distribution inside the waveguide. Since h_{1z} is a function of the operating frequency f or wavelength λ, the waveguide is a frequency/wavelength dependent device. According to (6.37), as θ_i varies h_{z1} also varies. The cutoff frequency corresponding to h_{z1} is:

$$f_{c1} = \frac{v_p}{2h_{z1}\cos\theta_i} \tag{6.38}$$

where v_p is the actual phase velocity of wave propagation.

The waveguides are high pass filters and to get wave propagation, the operating frequency must be above the certain operating frequency which is defined as *the first cut off frequency*.

Figure 6.13c shows the evolution of the rectangular waveguide by shielding the two semi-infinitely extended parallel plates with two vertical walls at a distance $a = \lambda/2$. The waveguide thus is made of a hollow tube. A waveguide launcher which is basically a coaxial feed probe launches electromagnetic wave as shown with the dotted circles. The plane wave generated from the probe is obliquely incident at the side walls at an angle θ_i and then bounces back and forth at the vertical walls. The x-polarized electric field makes a half sine wave distribution and must satisfy the boundary conditions of field nulls at the side walls. In conclusion, the field wave meets all boundary conditions at the four walls of the metallic hollow tubes, the obliquely incident field wave bounces back and forth along the side walls and propagates along z-axis without affecting the field distribution inside the waveguide. Thus undistorted field wave propagates along the waveguide as the guided wave with minimum loss and distortion.

In the above example, we have analyzed the perpendicularly polarized, obliquely incident wave propagation on a conducting boundary. We have a fair bit of understanding about the characteristics of the perpendicularly polarized uniform plane wave that is incident upon a planar conducting boundary. The electromagnetic fields are analyzed in both propagation phase due to its dependence on the principal x and z coordinates as well as the unit vectors of the magnetic field is decomposed along the two-unit vectors along the principal x- and z-axes.

Thus far the TE polarized field wave is compared with the caterpillar humpback movement as shown in Figure 6.8a. In the analysis, the understanding of the coordinate analysis of the preceding section was utilized and the unit vector and propagation phase decomposition are explained in details with illustration. The boundary conditions are applied and the relationship between the amplitudes and angles of incident and reflected fields are derived. It has been shown that at the interface, the incident field is equal and opposite of the reflected field. There is no transmission through the conducting medium hence the transmitted field is zero. However, as we studied for the case of the normal incidence, there must be an induced current on the conducting boundary. That we shall analyze in the following section. From the electromagnetic field in the light of their vector decompositions we observed that the field waves are non-uniform field waves with spatial variations along both principal coordinates. The amplitude of the electromagnetics wave varies with cosine and sine of the incident angle and the distance z as it propagates along z' oblique direction. Thus, the wave fields can be decomposed into components: (i) a propagating wave along x-axis which is also called *the surface wave*, and (ii) a *standing wave* along z-axis. We also computed the Poynting vector and observed that the real power is propagating as *the surface wave* along x-axis. We further analyzed the phase velocity of the obliquely incident TE wave and found that the phase velocities v_{px} and v_{pz} along the principal x- and z-axes, respectively, due to the stretched out of the wavelengths along x- and z-axes. The consequence is the decrement in the phase constants β_x and β_z along the principal axis or vice versa. The decomposition of the wavelengths along the principal axes with respect to the actual wavelength of the obliquely incident wave provides us the insight of *the guided wave* propagation along the conducting boundary. Therefore, the wavelength λ_x is called the guided wavelength along the conducting planar boundary. Finally, based on the understanding of the guided wavelength and propagation of the guided wave along the conducting boundary we develop the concept of waveguides. We have analyzed the wave propagation in the metallic parallel plate and hollow rectangular waveguides. We also examined that the wave propagation in the guided structure happens due the bouncing back and forth of the obliquely incident fields in the conducting walls. As long as the boundary conditions of the electric field nulls at the conducting boundary is maintained, the plane wave will propagate without distorting the field distribution in the guided structure. Thus a new theory of metallic waveguide has evolved. Now let us examine the induced surface current as this contributes to the loss term for an imperfect metallic boundary. We shall analyze such Ohmic loss in the waveguide theory.

6.5.4 Surface Current Calculation on Metallic Boundary

As derived in (6.24) we have found that the magnetic field has two components along the two principal axes which can be decomposed into two separate fields as follows

$$\begin{aligned} \boldsymbol{H}_{s\perp 1x}(x, z) &= -\frac{2E_{0i}}{\eta_1}\{\boldsymbol{a}_x \cos\theta_i \cos(k_1 z \cos\theta_i)\}\, e^{-jk_1 x \sin\theta_i} \\ \boldsymbol{H}_{s\perp 1z}(x, z) &= -\frac{2E_{0i}}{\eta_1}\{j\boldsymbol{a}_z \sin\theta_i \sin(k_1 z \cos\theta_i)\}\, e^{-jk_1 x \sin\theta_i} \end{aligned} \tag{6.39}$$

Therefore, there is a nonzero tangential component $H_{\perp 1x}(x, z)$ that contributes to the induced surface current on the conducting boundary at $z = 0$. The induced current on the conducting boundary at $z = 0$ is defined as:

$$\boldsymbol{J}(x)|_{z=0} = \boldsymbol{n} \times \boldsymbol{H}_{\perp 1x}(x, z)|_{z=0} \tag{6.40}$$

The normal to the conducting surface is along $-z$-axis in medium 1, hence $\boldsymbol{n} = -\boldsymbol{a}_z$. We can now define the induced surface current due $\boldsymbol{J}_s(x)$ to the incident magnetic field as:

$$\boldsymbol{J}_s(x) = (-\boldsymbol{a}_z \times \boldsymbol{a}_x)\frac{-2E_{0i}}{\eta_1}\cos\theta_i e^{-jk_1 x \sin\theta_i} = \boldsymbol{a}_y \frac{2E_{0i}}{\eta_1}\cos\theta_i e^{-jk_1 x \sin\theta_i} \tag{6.41}$$

A few remarks can be made on the calculation of the induced surface current due to the non-zero tangential magnetic field along the conducting boundary as follows:

1) The induced surface current $\boldsymbol{J}_s(x)$ resides on the interface and flows perpendicular to the plane of incidence (plane of paper).
2) The surface current is essential to support the field wave in medium 1. As analyzed for the case of normal incidence, the reflected field is being reradiated by the surface current on the conducting boundary.[22]

22 Recall the calculation of induced sheet current in normal incidence case and the explanation of radiation pressure.

3) If the conductor is not perfect, there must be Ohmic loss and power will be dissipated inside a guided structure such as a metallic waveguide. In reality, the perfect conductor does not exist and in radio transmitters, the loss factor is of special significance as the waveguides convey power in the order of kilowatts. We shall discuss the power loss in guiding structures such as in transmission lines and waveguides.
4) Finally, the time domain expression of the inducted current can be defined as:

$$\boldsymbol{J}_s(x, t) = \boldsymbol{a}_y \frac{2E_{0i}}{\eta_1} \cos\theta_i cost(\omega t - k_1 x \sin\theta_i) \tag{6.42}$$

The unit of the surface current is defined as (A/m). We shall be using the power loss calculations due to the finite conductivity and the induced surface current of the metallic waveguide.

Finally, we are also interested to know the induced charge due to the electric field incident on the conducting boundary. The induced charge density is defined by $\rho_s = \boldsymbol{n}.(\varepsilon\boldsymbol{E})$. Since for the case of TE polarization the electric field has only tangential components along y-axis, the dot product of $\boldsymbol{n}.(\varepsilon\boldsymbol{E})$ leads to zero induced surface charge. If we apply Maxwell's current continuity equation we have $-(\partial\rho_s)/(\partial t) = \nabla.\boldsymbol{J}_s = 0$ as there is no induced surface charge density in TE polarization. This means the induced current as the surface current only resides on the conducting boundary as divergence-less and does not contribute to the displacement current. Therefore, it has no killer effect of spurious radiation or an additional source of radiation.

Review Questions 6.4: TE-Polarization

Q1. Explain the phenomenon of the standing and surface wave creations when the signal hits a conducting boundary at an oblique incidence.

Q2. Show that the incident electric field in TE polarization reverses its direction at the boundary wall.

Q3. Explain Snell's law of reflection.

Q4. Show that the total electric field in TE polarization in medium 1 is defined as $\boldsymbol{E}_{\perp 1}(x, z) = -\boldsymbol{a}_y j2E_{0i}\sin(k_1 z\cos\theta_i)\, e^{-jk_1 x\sin\theta_i}$. Explain why the field wave of such a nature is called a non-uniform wave.

Q5. Show that the magnetic field in medium 1 in TE polarization is defined as $\boldsymbol{H}_{\perp 1}(x, z) = -\frac{2E_{0i}}{\eta_1}\{\boldsymbol{a}_x \cos\theta_i \cos(k_1 z\cos\theta_i) + j\boldsymbol{a}_z \sin\theta_i \sin(k_1 z\cos\theta_i)\}\, e^{-jk_1 x\sin\theta_i}$. Identify the standing wave and the travelling wave components.

Q6. Explain why Z_{TE} is larger than the free space characteristic impedance.

Q7. Explain why the power flow is only along the surface of the conducting boundary.

Example 6.2: TE Polarized Oblique Incidence on Conducting Boundary

A uniform plane wave at 100 MHz operating frequency incidences upon a conducting boundary from air as medium 1. The direction of propagation is along z-axis. The electric field is TE polarized and amplitude of 1 V/m. The incidence angle is 40°. Calculate the followings: (i) the angle of reflection and the reflection coefficient, the phasor vector and time domain expressions for (ii) incident electric and magnetic fields, (iii) the reflected electric and reflected magnetic fields, (iv) the resultant electric field in medium 1, (v) the resultant magnetic field in medium 1, (vi) the phase velocities, (vii) wavelengths along x- and z-axes, (viii) the impedance Z_{TE}, (ix) the time average power density, (x) the induced current on the conducting surface, and (xi) the conductor loss due to finite conductivity of the second medium (copper, $\sigma = 5.8 \times 10^7$ (S/m)) and surface current.

Hints: use the coordinate decomposition method to calculate the field expressions.

Answer:

First we need to calculate the phase constant in medium 1 at 100 MHz:

$$k_1 = 2\pi f\sqrt{\mu_1\varepsilon_1} = 2\pi \times \frac{10^8}{3 \times 10^8} = 2.09 \text{ (rad/m)}$$

The angular frequency $\omega = 2\pi f = 628.3 \times 10^6$ (rad/s)

i) Based on Snell's law of reflection $\theta_i = \theta_r$; hence the angle of reflection is: $\theta_r = 40°$.
Since $E_{0i} = -E_{0r}$, the reflection coefficient is defined as:

$$\Gamma = \frac{E_{0i}}{E_{0r}} = -1$$

ii) The expression for the incident electric field is defined

$$\boldsymbol{E}^i_{s\perp} = \boldsymbol{E}_0 e^{-jk_1 z'} = \boldsymbol{a}_y 1 e^{j2.09(x\sin 40° + z\cos 40°)} = 1e^{-j(1.34x + 1.6z)} \text{ (V/m)}$$

The time domain expression for the electric field is:

$$\boldsymbol{E}^i_{\perp}(x, z, t) = 1\cos\left(628.3 \times 10^6 t - 1.34x - 1.6z\right) a_y \text{ (V/m)}$$

The incident magnetic field is defined as:

$$\begin{aligned}
\boldsymbol{H}^i_{s\perp}(x, z) &= (-\cos\theta_i \boldsymbol{a}_x + \sin\theta_i \boldsymbol{a}_z)\frac{E_{0i}}{\eta_1} e^{-jk_1(x\sin\theta_i + z\cos\theta_i)} \\
&= (-\cos 40° \boldsymbol{a}_x + \sin 40° \boldsymbol{a}_z)\left(\frac{1}{377}\right) e^{-j(1.34x + 1.6z)} \\
&= \left(-20.3 \times 10^{-3} \boldsymbol{a}_x + 1.7 \times 10^{-3} \boldsymbol{a}_z\right) e^{-j(1.34x + 1.6z)} \text{ (A/m)}
\end{aligned}$$

The time domain expression for the magnetic field is:

$$\boldsymbol{H}^i_{\perp}(x, z, t) = \left(-20.3 \times 10^{-3} \boldsymbol{a}_x + 1.7 \times 10^{-3} \boldsymbol{a}_z\right) \cos\left(628.3 \times 10^6 t - 1.34x - 1.6z\right) \text{ (A/m)}$$

iii) Now we calculate the reflected electromagnetic fields. The expression for the reflected electric field is defined as:

$$\boldsymbol{E}^r_{s\perp}(x, z) = \boldsymbol{a}_y E_{0r} e^{-jk_1(x\sin\theta_r - z\cos\theta_r)} = \boldsymbol{E}^r_{\perp}(x, z) = -\boldsymbol{a}_y 1 e^{-j(1.34x - 1.6z)} \text{ (V/m)}$$

The time domain expression for the reflected electric field is:

$$\boldsymbol{E}^r_{\perp}(x, z, t) = -1\cos\left(628.3 \times 10^6 t - 1.34x + 1.6z\right) \boldsymbol{a}_y \text{ (V/m)}$$

The reflected magnetic field is defined as:

$$\begin{aligned}
\boldsymbol{H}^r_{s\perp}(x, z) &= (\cos\theta_r \boldsymbol{a}_x + \sin\theta_r \boldsymbol{a}_z)\frac{E_{0r}}{\eta_1} e^{-jk_1(x\sin\theta_r - z\cos\theta_r)} \\
&= (\cos 40° \boldsymbol{a}_x + \sin 40° \boldsymbol{a}_z)\left(\frac{-1}{377}\right) e^{-j(1.34x - 1.6z)} \\
&= -\left(20.3 \times 10^{-3} \boldsymbol{a}_x + 1.7 \times 10^{-3} \boldsymbol{a}_z\right) e^{-j(1.34x - 1.6z)} \text{ (A/m)}
\end{aligned}$$

The time domain expression for the reflected magnetic field is:

$$H^i_\perp(x, z, t) = -\left(20.3 \times 10^{-3}\boldsymbol{a_x} + 1.7 \times 10^{-3}\boldsymbol{a_z}\right) cos\left(628.3 \times 10^6 t - 1.34x + 1.6z\right)\ (\text{A/m})$$

iv) The resultant electric field in medium 1 is:

$$\boldsymbol{E}_{s\perp 1}(x, z) = -\boldsymbol{a_y} j2E_{0i}\, sin\,(k_1 z\, cos\,\theta_i)\, e^{-jk_1 x\, sin\,\theta_i} = \boldsymbol{E}_{\perp 1}(x, z) = -\boldsymbol{a_y} j2\, sin\,(1.6z)\, e^{-j1.34x}\ (\text{V/m})$$

The time domain expression for the resultant electric field is:

$$\boldsymbol{E}_{\perp 1}(x, z, t) = -\boldsymbol{a_y} j2\, sin\,(1.6z)\, cos\left(628.3 \times 10^6 t - 1.34x\right)\ (\text{V/m})$$

The resultant magnetic field in medium 1 is:

$$\begin{aligned}\boldsymbol{H}_{s\perp 1}(x, z) &= -\frac{2E_{0i}}{\eta_1}\{\boldsymbol{a_x}\, cos\,\theta_i\, cos\,(k_1 z\, cos\,\theta_i) + j\boldsymbol{a_z}\, sin\,\theta_i\, sin\,(k_1 z\, cos\,\theta_i)\}\, e^{-jk_1 x\, sin\,\theta_i}\\ &= -\left\{\boldsymbol{a_x} 40.6 \times 10^{-3}\, cos\,(1.6z) + j\boldsymbol{a_z} 3.4 \times 10^{-3}\, sin\,(1.6z)\right\} e^{-j1.34x}\ (\text{A/m})\end{aligned}$$

The time domain expression for the resultant magnetic field is:

$$\boldsymbol{H}_{\perp 1}(x, z, t) = -\left\{\boldsymbol{a_x} 40.6 \times 10^{-3} cos(1.6z) + j\boldsymbol{a_z} 3.4 \times 10^{-3} sin(1.6z)\right\} cos\left(628.3 \times 10^6 t - j1.34x\right)\ (\text{A/m})$$

v) The component phase velocities calculations

The phase velocity for the incident wave in medium 1 is:

$$v_p = \frac{1}{\sqrt{\mu_1 \varepsilon_1}} = 3 \times 10^8\ (\text{m/s})$$

The phase velocity along x-axis is:

$$v_{px} = \frac{v_p}{sin\,\theta_i} = \frac{3 \times 10^8}{sin\,40^\circ} = 4.67 \times 10^8\ (\text{m/s})$$

$$v_{pz} = \frac{v_p}{cos\,\theta_i} = \frac{3 \times 10^8}{cos\,40^\circ} = 3.92 \times 10^8\ (\text{m/s})$$

vi) Wavelengths along x and z-axes,

The wavelength in medium 1 is: $\lambda = v_p/f = (3 \times 10^8)/10^8 = 3$ (m) ; $\lambda_x = 3/(sin\ 40^\circ) = 4.67$ (m): $\lambda_z = 3/(cos\ 40^\circ) = 3.91$ (m)

vii) The impedance Z_{TM}.

$$Z_{TM} = \eta_1 Sec\theta_i = \frac{377}{cos\,40^\circ} = 492\ (\Omega)$$

viii) The time average power

$$\boldsymbol{P}_{ave} = \boldsymbol{a_x}\frac{2E_{0i}^2}{\eta_1}\, sin\,\theta_i (k_1 z\, cos\,\theta_i) = \boldsymbol{a_x}\frac{2}{377} \times sin\,40^\circ \times (1.6z) = \boldsymbol{a_x} 3.4 \times 10^{-3}(1.6z)\ \left(\text{W/m}^2\right)$$

ix) The induced current:

$$J_s(x) = a_y \frac{2E_{0i}}{\eta_1} \cos\theta_i e^{-jk_1 x \sin\theta_i} = a_y 4.06 \times 10^{-3} e^{-j1.34x} \text{ (A/m)}$$

The time domain expression of the induced current:

$$J(x, t) = a_y 4.06 \times 10^{-3} \cos\left(628.3 \times 10^6 t - 1.34x\right) \text{ (A/m)}$$

x) The conductor loss due to finite conductivity and surface current is: $R_s = \frac{1}{\delta\sigma_2}$. Where the skin depth at 100 MHz is: $\delta = 1/\left(\sqrt{(\pi f \mu_2 \sigma_2)}\right) = 1/\left(\sqrt{(\pi \times 10^8 \times 4\pi \times 10^{-7} \times 5.8 \times 10^7)}\right)$ $= 6.61 \times 10^{-6}$ (m)

The surface resistance is: $R_s = \frac{1}{6.61 \times 10^{-6} \times 5.8 \times 10^7} = 2.6 \times 10^{-3}$ (Ω)

The power loss: $P_{loss} = \frac{1}{2}|J_s|^2 R_s = \frac{1}{2} \times \left(4.06 \times 10^{-3}\right)^2 \times 2.6 \times 10^{-3} = 2.15 \times 10^{-8} \left(\text{W/m}^2\right)$

The loss due to the induced current is too low.

6.6 Parallel (TM) Polarization on Conducting Boundary

In parallel (TM) polarization the roles of the electric and the magnetic fields are swapped as for the case of perpendicular (TE) polarization. Figure 6.14a shows the TM polarized oblique incidence on a conducting boundary along *xy*-plane. In this case, the electric field orientation is parallel to the plane of incidence (plane of paper) and incident upon the conducting boundary with an incident angle θ_i with the normal line along *z*-axis and propagates along z' with unit vector (a_i) direction. The reflected electric field is propagating along z'' and makes an angle of reflection θ_r with the normal line along *z*-axis. The magnetic field is perpendicular to the plane of incidence and has only one component along *y*-direction.[23] The orthogonal relationship between the electric field, magnetic field, and the direction of propagation suggests a *y*-directed magnetic field in both incidence and reflection. The TM polarized magnetic field is oriented vertically with the plane of incidence which is on the *xz*-plane (plane of paper), and it hits parallel to the plane of interface or the boundary plane which is on the *xy*-plane. However, the electric field propagates on the plane of incidence and hits the conducting boundary with an angle θ_i as shown in the figure. This is analogous to a series of balls placed on a line along E_i field vector that is orthogonal to the unit vector a_i and makes an angle θ_i with the normal line along *z*-axis. The first ball is at the intersection point of vectors E_i and a_i and the last ball at the tip of E_i. If the balls are thrown parallel to a_i with a regular interval of time then the first ball hits the boundary at the origin (0, 0, 0) and bounces back on the conducting boundary and travels along unit vector a_r on its returned path. Then the second ball and subsequent balls continue until the last ball which resides on the tip of the electric field vector. The returned/reflected balls are arranged on the vector line of the reflected electric field E_r; if we take the fixed time of travel for each ball. We shall use the time equality and travel distance to establish the ray trace model of oblique incident in a later stage. The field orientation (direction of the field vibration) of the time domain/instantaneous electric field of the TM polarized field wave suggests

23 That is why it is called TM polarization.

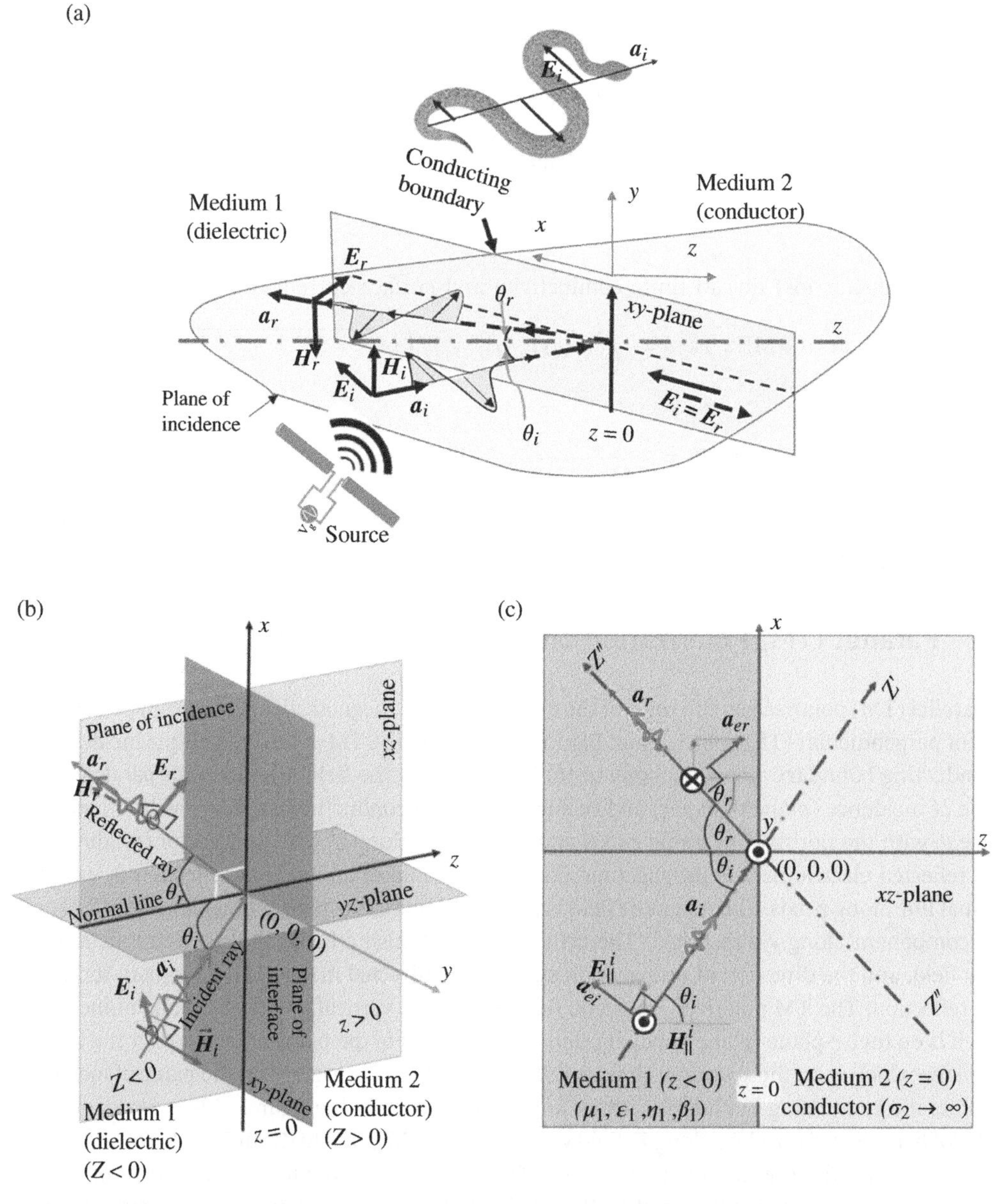

Figure 6.14 Transverse Magnetic (TM) Oblique incidence (a) the generic plane wave incidence at the interface of two media when the plane of incidence is horizontal to the ground. The TM polarization can be depicted with the analogy of a moving snake with its hump back as the direction of the electric field. At the interface of the dielectric and conductor media, incident electric field is equal and opposite to make the tangential electric field zero at the conducting boundary. This is an equipotential surface with zero tangential electric field, (b) 3D plan view of the same illustration as for (a) with vertically oriented plane of incidence, and (c) 2D plan view of the same illustration as for (a) with plane of incident is on the paper.

an analogy of horizontal movement of a snake as shown on the top of the figure. As a snake moves lateral to the ground plane and makes a shape of sine wave, the field wave at time instances vibrates only on *xz*-plane as shown in the figure. The head of the snake is along the direction of wave field propagation along $\boldsymbol{a_i}$. As always, the relationship between the incident electric field, the incident magnetic field, and the direction of wave propagation must obey the right-hand rule as shown in the figure. Likewise, the reflected electric field, the reflected magnetic field and the direction of wave propagation must obey the same rule. Based on the right-hand rule $\boldsymbol{P} = \boldsymbol{E} \times \boldsymbol{H}$, we set the directions of the incident and returned electromagnetic fields. If we compare the TM case with the TE case in Figure 6.7, we can easily see that the roles of electric and magnetic fields are exchanged in terms of their orientations keeping the direction of propagation unchanged. For completeness and ease of depiction we also produce three different perspectives of depictions here. Figure 6.14b shows the 3D view with vertical orientation of the electromagnetic system. Figure 6.14c shows the 2D view where the electric field lies on the plane of incidence. Based on the vector decomposition in Figure 6.15a, b, the electric field is given by

$$\boldsymbol{E}^i_{s\|}(x, z) = (\boldsymbol{a_x} \cos\theta_i - \boldsymbol{a_z} \sin\theta_i)E_{0i}\, e^{-jk_1(x \sin\theta_i + z\cos\theta_i)} \tag{6.43}$$

Where the unit vector of the incident electric field is defined as $\boldsymbol{a}_{ei} = \boldsymbol{a_x} \cos\theta_i - \boldsymbol{a_z} \sin\theta_i$. As can be seen Figure 6.15a shows the vector decomposition of the phase factor for incident field which is propagating along $\boldsymbol{a_i}$ direction; and Figure 6.15b shows the unit vector decomposition of parallel polarization incident electric field. The magnetic field has only y-directed component and is defined as

$$\boldsymbol{H}^i_{s\|}(x, z) = \boldsymbol{a_y} H_{0i} e^{-jk_1(x \sin\theta_i + z\cos\theta_i)} \tag{6.44}$$

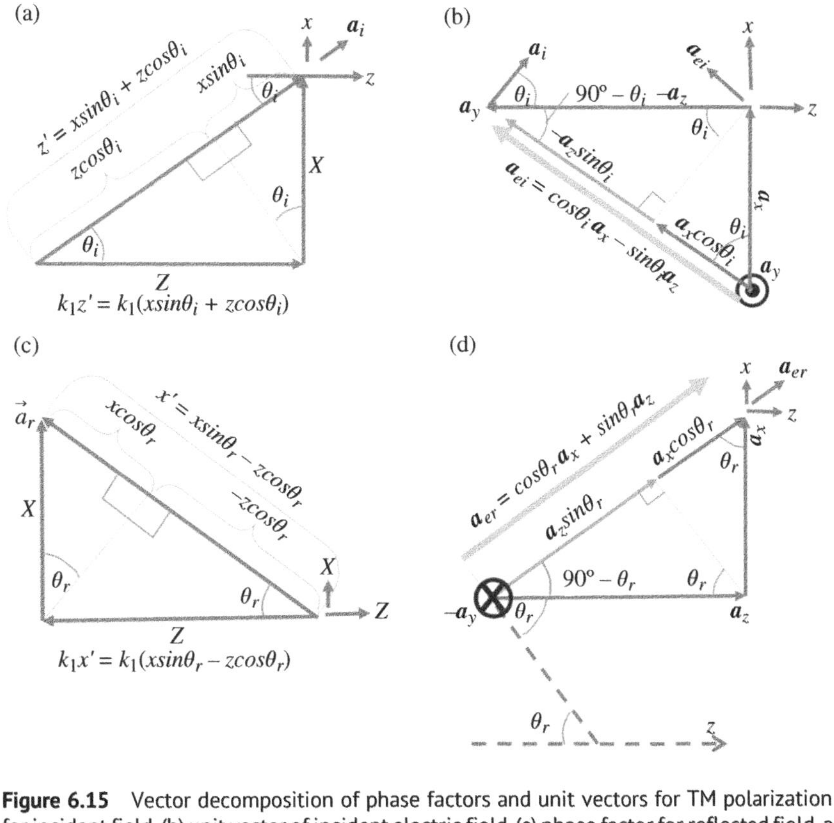

Figure 6.15 Vector decomposition of phase factors and unit vectors for TM polarization (a) phase factor for incident field, (b) unit vector of incident electric field, (c) phase factor for reflected field, and (d) unit vector of reflected electric field.

Figure 6.15c illustrates the vector decomposition of the phase factor for the reflected fields along unit vector $\boldsymbol{a_r}$; and Figure 6.15d shows the vector decomposition of the unit vector of the reflected electric field. We can easily deduce the negative y-direction of the reflected magnetic field to maintain the direction of wave propagation and the direction of the reflected electric field.

The reflected electric and magnetic fields are defined as

$$\boldsymbol{E}^r_{s\|}(x, z) = (\cos\theta_r \boldsymbol{a_x} + \sin\theta_r \boldsymbol{a_z}) E_{0i} e^{-jk_1(x\sin\theta_r - z\cos\theta_r)} \tag{6.45}$$

$$\boldsymbol{H}^r_{s\|}(x, z) = -\boldsymbol{a_y}\frac{E_{0i}}{\eta_1} e^{-jk_1(x\sin\theta_r - z\cos\theta_r)} \tag{6.46}$$

where, the unit vector of the reflected electric field is $\boldsymbol{a}_{er} = \cos\theta_r \boldsymbol{a}_x + \sin\theta_r \boldsymbol{a}_z$. As stated above, the second medium is a conducting medium, therefore, no transmitted field exists in the medium and $\boldsymbol{E}^t_{\|}(x, z) = \boldsymbol{H}^t_{\|}(x, z) = 0$.

Now we can apply the boundary conditions at the interface of the two media at $z = 0$. Since, the boundary is a perfect conductor, the total electric field which is tangential to the interface at $z = 0$ must vanish. This leads to the following expression:

$$\begin{aligned}\boldsymbol{E}_{s\|1}(x, 0) &= \boldsymbol{E}^i_{s\|}(x, 0) + \boldsymbol{E}^r_{s\|}(x, 0)\\ &= \left\{\boldsymbol{a_x}\cos\theta_i E_{0i} e^{-jk_1(x\sin\theta_i)} + \boldsymbol{a_x}\cos\theta_r E_{0r} e^{-jk_1(x\sin\theta_r)}\right\} = 0\end{aligned}$$

$$\text{Or, } E_{0i} e^{-jk_1(x\sin\theta_i)} = -E_{0r} e^{-jk_1(x\sin\theta_r)} \tag{6.47}$$

Again, equating the phase terms $k_1(x\sin\theta_i) = k_1(x\sin\theta_r)$ for any value of x at the boundary wall at $z = 0$ leads to *Snell's law of reflection* $\theta_i = \theta_r$. And the equality of amplitude leads to $E_{0i} = -E_{0r}$. This means that as for the case of the TE polarization, the TM polarization maintains the same relationship in between the incident and reflected electric fields.

As for the case of TM-polarization, we can define the total electric and magnetic fields in medium 1 as follows:

$$\begin{aligned}\boldsymbol{E}_{s\|1}(x, z) &= \boldsymbol{E}^i_{\|}(x, z) + \boldsymbol{E}^r_{\|}(x, z)\\ &= \boldsymbol{a_x}\cos\theta_i E_{0i}\left\{e^{-jk_1 z\cos\theta_i} - e^{jk_1 z\cos\theta_i}\right\}e^{-jk_1 x\sin\theta_i}\\ &\quad - \boldsymbol{a_z}\sin\theta_i E_{0i}\left\{e^{-jk_1 z\cos\theta_i} - e^{jk_1 z\cos\theta_i}\right\}e^{-jk_1 x\sin\theta_i}\\ \therefore\ \boldsymbol{Es}_{\|1}(x, z) &= -2E_{0i}\,\{j\boldsymbol{a_x}\cos\theta_i\sin(k_1 z\cos\theta_i)\\ &\quad + \boldsymbol{a_z}\sin\theta_i\cos(k_1 z\cos\theta_i)\}\,e^{-jk_1 x\sin\theta_i}\end{aligned} \tag{6.48}$$

$$\boldsymbol{H}_{s\|1}(x, z) = \boldsymbol{H}^i_{\|}(x, z) + \boldsymbol{H}^r_{\|}(x, z) = \boldsymbol{a_y}\frac{2E_{0i}}{\eta_1}\cos(k_1 z\cos\theta_i)\, e^{-jk_1 x\sin\theta_i} \tag{6.49}$$

From the above analysis for the TM oblique incidence case the following remarks can be made:

1) We can make similar observation as for the TE case for the electric field components that for the magnetic field in the TM case. We see that with respect to z-direction there is a standing wave with the amplitude factor $2E_{0i}\cos(k_1 z\cos\theta_i)$. With respect to x-direction there is a *nonuniform* travelling wave with the phase factor $e^{-jk_1 x\sin\theta_i}$.
2) The characteristic wave impedance of TM-polarization as follows:

$$Z_{TM} = \frac{E^i_x}{H^i_y} = -\frac{E^r_x}{H^r_y} = \eta_1\cos\theta_i \tag{6.50}$$

The –ve sign in the second ratio is to make the wave impedance positive. We have noticed that $Z_{TM} < \eta_1$. This is due to the fact that in TM polarization, the electric field has been divided into x- and z-directed components and the magnetic field has only one y-directed component. This means the electric field has only a x-directed component lying transvers to z-direction (on xy-plane) and contributes to the value of the characteristic wave impedance. Thus, the ratio gives smaller value of the characteristic wave impedance compared to the free space wave impedance $\eta_1$of medium 1.[24]

3) The phase velocity and wave number components remain the same as that for the TE case.
4) The field nulls $h_{zn} = n\lambda/(2\ cos\ \theta_i)$ and the concept of waveguide also remain similar to that for the TE case.
5) The Poynting vector for the TM case can be calculated in a similar manner as for the TE case derived in (6.27). Using similar calculation steps as for (6.27), we obtain the time average power density vector for TM case as follows:

$$\boldsymbol{P}_{ave} = \boldsymbol{a_x}\frac{2E_{0i}^2}{\eta_1}\ sin\ \theta_i(k_1 z\ cos\ \theta_i) \tag{6.51}$$

As expected, a real power is flowing along x-axis with a variation along z-axis as a function of $(k_1 z\ cos\ \theta_i)$ in medium 1 with the same phase velocity as before.

6.6.1 Surface Current and Induced Electric Charge Calculations on Metallic Boundary

In the TM polarization, the magnetic field has only y-directed component. Therefore, using the similar procedure as for the TE case, we can derive the induced current on the conducting boundary at $z = 0$ as follows:

$$\boldsymbol{J}(x)|_{z=0} = (-\boldsymbol{a_z} \times \boldsymbol{a_y})\frac{2E_{0i}}{\eta_1}e^{-jk_1 x\ sin\ \theta_i} = \boldsymbol{a_x}\frac{2E_{0i}}{\eta_1}e^{-jk_1 x\ sin\ \theta_i} \tag{6.52}$$

We can comprehend that there is an x-directed surface current which is parallel to the plane of incidence along xz-plane. This is a contrasting difference with respect to the surface current in TE case that flows along y-axis and perpendicular to the plane of incidence.

Another contrasting observation is the induced electric charge at the conducing surface. Since the electric field has a component along z-axis which is normal to the conducting boundary, the non-zero electric field induces surface charge. Applying Gauss' law at the interface we obtain the induced charge density as:

$$\rho_s|_{z=0} = \boldsymbol{n}.(\varepsilon_1 \boldsymbol{E}_1(x, 0)) = 2\varepsilon_1 E_{0i}\ sin\ \theta_i e^{-jk_1 x\ sin\theta_i} \tag{6.53}$$

The surface charge must satisfy Maxwell's current continuity equation $-\dfrac{\partial \rho_s}{\partial t} = \nabla \cdot \boldsymbol{J_s} \neq 0$.

We shall repeat the example for the case of TE-polarization for TM polarization at 100 MHz.

---

24 For the same reason we shall see that in the waveguide the TM-mode wave impedance is less than the unguided or free space wave impedance.

Review Questions 6.5: Field Analyses of TM-Polarization

Q1. Show that the total electric field in TM polarization in medium 1 is defined as $\boldsymbol{E}_{\parallel 1}(x,z) = -2E_{0i}\{j\boldsymbol{a_x}\cos\theta_i \sin(k_1 z\cos\theta_i) + \boldsymbol{a_z}\sin\theta_i\cos(k_1 z\cos\theta_i)\}e^{-jk_1 x\sin\theta_i}$. Identify the standing wave and the traveling wave components.

Q2. Show that the magnetic field in medium 1 in TE polarization is defined as $\boldsymbol{H}_{\parallel 1}(x,z) = \boldsymbol{a_y}((2E_{0i})/\eta_1)\cos(k_1 z\cos\theta_i)e^{-jk_1 x\sin\theta_i}$.

Q3. Explain why Z_{TM} is smaller than the free space characteristic impedance.

Example 6.3: TM Polarized Oblique Incidence on Conducting Boundary

A uniform plane wave at 100 MHz operating frequency incidences upon a conducting boundary from air as medium 1. The direction of propagation is along z-axis. The electric field is TM polarized and amplitude of 1 V/m. The Incidence angle is 40°. Calculate the followings: The phasor vector and time domain expressions for: (i) incident electric and magnetic fields, (ii) the reflected electric and magnetic fields, (iii) the resultant electric field in medium 1, and (iv) the resultant magnetic field in medium 1, (v) the impedance Z_{TM}, (vi) the time average power, (vii) the induced current on the conducting surface, and (viii) the conductor loss due to finite conductivity of the second medium (copper, $\sigma = 5.8\times10^7$ (S/m)) and the surface current.

Hints: use the coordinate decomposition to calculate the field expressions.

Answer:

We shall use the results of the first parts which are common to both polarizations for the TM polarization example.

The phase constant in medium 1 at 100 MHz: $k_1 = 2.09$ (rad/m), the angular frequency $\omega = 628.3\times10^6$ (rad/s), the angle of reflection $\theta_r = 40°$ and $E_{0i} = -E_{0r}$.

i) The expression for incident electric field is defined as follows:

$$\boldsymbol{E}^i_{\parallel}(x,z) = (\boldsymbol{a_x}\cos\theta_i - \boldsymbol{a_z}\sin\theta_i)E_{0i}\,e^{-jk_1(x\sin\theta_i + z\cos\theta_i)} = (0.766\boldsymbol{a_x} - 0.642\boldsymbol{a_z})e^{-j(1.34x+1.6z)}\,(\text{V/m})$$

The time domain expression for the electric field is:

$$\boldsymbol{E}^i_{\parallel}(x,z,t) = (0.766\boldsymbol{a_x} - 0.642\boldsymbol{a_z})\cos\left(628.3\times10^6 t - 1.34x - 1.6z\right)\,(\text{V/m})$$

The incident magnetic field is defined as:

$$\boldsymbol{H}^i_{s\parallel}(x,z) = \boldsymbol{a_y}\frac{E_{0i}}{\eta_1}e^{-jk_1(x\sin\theta_i + z\cos\theta_i)} = \boldsymbol{a_y}2.65\times10^{-3}e^{-j(1.34x+1.6z)}\,(\text{A/m})$$

The time domain expression for the magnetic field is:

$$\boldsymbol{H}^i_{\parallel}(x,z,t) = \boldsymbol{a_y}2.65\times10^{-3}\cos\left(628.3\times10^6 t - 1.34x - 1.6z\right)\,(\text{A/m})$$

ii) Now we calculate the reflected electromagnetic fields. The expression for the reflected electric field is defined

$$\boldsymbol{E}^r_{s\parallel}(x,z) = (\cos\theta_r\boldsymbol{a_x} + \sin\theta_r\boldsymbol{a_z})E_{0i}e^{-jk_1(x\sin\theta_r - z\cos\theta_r)} = (0.766\boldsymbol{a_x} + 0.642\boldsymbol{a_z})e^{-jk_1(1.34x-1.6z)}\,(\text{V/m})$$

The time domain expression for the reflected electric field is:

$$\boldsymbol{E}_{\|}^{r}(x, z, t) = (0.766\boldsymbol{a_x} + 0.642\boldsymbol{a_z})\cos\left(628.3 \times 10^6 t - 1.34x + 1.6z\right) \text{ (V/m)}$$

The reflected magnetic field is defined as:

$$\boldsymbol{H}_{s\|}^{r}(x, z) = -\boldsymbol{a_y}\frac{E_{0i}}{\eta_1}e^{-jk_1(x\sin\theta_r - z\cos\theta_r)} = -\boldsymbol{a_y}2.65 \times 10^{-3}e^{-j(1.34x - 1.6z)}\text{(A/m)}$$

The time domain expression for the reflected magnetic field is:

$$\boldsymbol{H}_{\perp}^{i}(x, z, t) = -\boldsymbol{a_y}2.65 \times 10^{-3}\cos\left(628.3 \times 10^6 t - 1.34x + 1.6z\right) \text{ (A/m)}$$

iii) The resultant electric field in medium 1 is:

$$\begin{aligned}\boldsymbol{E}_{s\|1}(x, z) &= -2E_{0i}\left\{j\boldsymbol{a_x}\cos\theta_i\sin(k_1 z\cos\theta_i) + \boldsymbol{a_z}\sin\theta_i\cos(k_1 z\cos\theta_i)\right\}e^{-jk_1 x\sin\theta_i} \\ &= -\left\{\boldsymbol{a_x}1.532\sin(1.6z) + j\boldsymbol{a_z}(1.6z)\right\}e^{-j1.34x}\text{(V/m)}\end{aligned}$$

The time domain expression for the resultant electric field is:

$$\boldsymbol{E}_{\|1}(x, z, t) = -\left\{\boldsymbol{a_x}1.532\sin(1.6z) + j\boldsymbol{a_z}(1.6z)\right\}\cos\left(628.3 \times 10^6 t - 1.34x\right) \text{ (V/m)}$$

iv) The resultant magnetic field in medium 1 is:

$$\boldsymbol{H}_{s\perp 1}(x, z) = \boldsymbol{a_y}\frac{2E_{0i}}{\eta_1}\cos(k_1 z\cos\theta_i)\,e^{-jk_1 x\sin\theta_i} = -\boldsymbol{a_y}5.36 \times 10^{-03}\cos(1.6z)\,e^{-j1.34x}\text{(V/m)}$$

The time domain expression for the resultant magnetic field is:

$$\boldsymbol{H}_{\|1}(x, z, t) = -\boldsymbol{a_y}5.36 \times 10^{-03}\cos(1.6z)\cos\left(628.3 \times 10^6 t - 1.34x\right) \text{ (A/m)}$$

v) The impedance Z_{TM} is:

$$Z_{TE} = \eta_1\cos\theta_i = 377 \times \cos 40^\circ = 288.8\ (\Omega)$$

vi) The time average power is:

$$\boldsymbol{P}_{ave} = \boldsymbol{a_x}\frac{2E_{0i}^2}{\eta_1}\sin\theta_i(k_1 z\cos\theta_i) = \boldsymbol{a_x}\frac{2}{377} \times \sin 40^\circ \times (1.6z) = \boldsymbol{a_x}3.4 \times 10^{-3}(1.6z)\ \left(\text{W/m}^2\right)$$

vii) The time domain expression of the induced current:

$$\boldsymbol{J}_s(x, t) = \boldsymbol{a_x}\frac{2E_{0i}}{\eta_1}\cos t(\omega t - k_1 x\sin\theta_i) = \boldsymbol{a_x}5.31 \times 10^{-3}\cos\left(628.3 \times 10^6 t - 1.34x\right) \text{ (A/m)}$$

viii) The conductor loss due to finite conductivity and surface current is:

$$P_{loss} = \frac{1}{2}|J_s|^2 R_s = \frac{1}{2} \times \left(5.31 \times 10^{-3}\right)^2 \times 2.6 \times 10^{-3} = 7.32 \times 10^{-8}\ \left(\text{W/m}^2\right)$$

The loss due to the induced current is again too low.

ix) The induced charge density on the conducing surface

$$|\rho_s|_{z=0} = 2\varepsilon_1 E_{0i}\sin\theta_i = 2 \times 8.854 \times 10^{-12} \times 0.642 = 11.38\ \left(\text{pC/m}^2\right)$$

6.7 Characteristic Wave Impedances

So far, we have analyzed both TE and TM polarized oblique incidence cases on a conducting boundary. We have observed different wave impedances for the two polarizations. We have defined a z-directed wave impedance in terms of the ratio of the electric and magnetic field components which are parallel to the boundary plane. The continuity conditions of the tangential electric and the magnetic fields at the interface of two media bring about a unique definition of the impedance of the two regions. As per the equivalent impedance circuit analogies of the normal incidence cases, we can define the impedance on the left side of the interface in medium 1 as the input impedance and the impedance of the right-hand side region in medium 2 can be defined as the load impedance. Due to the impedance mismatch of the two regions we also expect a standing wave pattern along z-direction. We also found this phenomenon in the field analysis using the boundary conditions for the oblique incidence cases. We utilize the continuity conditions of the $\boldsymbol{E}_x$, $\boldsymbol{E}_y$, $\boldsymbol{H}_x$, and $\boldsymbol{H}_y$ component electric and magnetic fields at the conducting boundary. Therefore, we can define a z-directed impedance which is *the characteristic impedance of the wave* as the ratio of the transverse field components of the normal line z. The obliquely incident field waves make an incident and reflected angle $\theta_i = \theta_r$ with the normal line. The characteristics wave impedance is the ratio of the transverse electromagnetic field components to the z-direction and is defined as:

$$Z_{TE} = -\frac{E_y^i}{H_x^i} = \frac{E_y^r}{H_x^r} = \eta_1 \sec\theta_i \tag{6.54}$$

As stated above the negative sign in the first ratio is to make the impedance positive. As we can observe that $Z_{TE} > \eta_1$. This is due to the fact that in TE-polarization the electric field has only y-directed component in full strength whereas the magnetic field is divided into x- and z-directed components and only x-directed field component contributes to the impedance calculation.

The *characteristic wave impedance* for TM-polarization can be defined as:

$$Z_{TM} = \frac{E_x^i}{H_y^i} = -\frac{E_x^r}{H_y^r} = \eta_1 \cos\theta_i \tag{6.55}$$

Again the negative sign in the second ratio is to make Z_{TM} positive. We observe that $Z_{TM} < \eta_1$. We can place similar argument that the electric field is divided into two components and only the transverse component to z-axis is contributing to the impedance calculation. Whereas the magnetic field has only y-directed component. Finally, about the standing wave pattern along z-axis we can say that all TE field components E_x and E_y are zero at the conducting boundary which is on the xy-plane at $z = 0$. Therefore, we expect a standing wave pattern with zero load impedance and Z_{TE} and Z_{TM} impedances, respectively, in medium 1 for TE- and TM-polarized obliquely incident wave on the conducting boundary at an angle of incidence θ_i. The standing wave pattern in the z-direction has nulls at the multiple of the half wavelength distances $h_{zn} = (n\lambda)/(2\cos\theta_i)$ as depicted in Figure 6.13a.

6.8 Oblique Incidence on Dielectric Boundary

The oblique incidence on the conducting boundary is the special case of oblique incidence where we assume no transmitted electromagnetic field wave propagates in the conducting region. The examination of the conducting boundary has provided us many useful theoretical understandings that we can bank on and proceed to develop the theory of generalized oblique incidence cases for the dissimilar dielectric boundary. In nature and in-built environment, we regularly observe the phenomena of oblique incidence. Best example of natural oblique incidence is the rainbow in sky and in-built environment the wireless propagation in mobile communications between towers and mobile phones. In the analysis we first approach the ray trace model and understand some basic principle of Snell's laws of reflection and refraction. Then we shall perform Maxwell's field analyses and apply appropriate boundary conditions to understand the detailed phenomena of the oblique incident fields. We shall also utilize the impedance concept that we have derived in the conducting boundary.

6.8.1 Ray Trace Model of Generalized Oblique Incidence Field

Consider a uniform plane wave impinges on the planar boundary of dissimilar dielectric media with constitutive parameters (μ_1, ε_1) and (μ_2, ε_2) at $z = 0$ plane as shown in Figure 6.16a. The plane of incident is on xz-plane (plane of paper). The angles of incidence, reflection, and transmission are θ_i, θ_r, and θ_t, respectively. The dotted lines OA, BC and CD represent the wavefronts of the incident, transmitted and reflected wave with the propagation unit vectors $\boldsymbol{a_i}$, $\boldsymbol{a_t}$ and $\boldsymbol{a_r}$, respectively. The incident and reflected waves are in medium 1 and propagating with the same phase velocity $v_{p1} = 1/\left(\sqrt{(\mu_1\varepsilon_1)}\right)$. On the other hand, the transmitted wave is in medium 2 and propagates with a different phase velocity $v_{p2} = 1/\sqrt{(\mu_2\varepsilon_2)}$. However, the two-phase velocities have equal components along the x-axis at the interface at $z = 0$. This means all three waves travel the same distance OB at the same time due to the equality holds for the x-directed phase velocities for the three waves. Therefore, the distance travelled by the three wave fronts is the same at that instance of time. This means the incidence ray travels from A to B, the transmitted ray travels O to D and the reflected ray travels O to C at the same time. This time equality leads to the following relationship: $t_1 = ((AB)/(v_{p1})) = ((OC)/(v_{p1})) = ((OD)/(v_{p2}))$. However, from the trigonometry, we can define the lengths in term of the common distance OB that holds the equality for the time instance along x-axis. Figures 6.16b–d illustrate the enlarged and decomposed views of individual rays that travel distances at the same time. First, we consider the case for medium 1 and find the relation between the incident and reflected waves. The distances AB and OC travelled by the incident and the reflected waves, respectively, are the same as they are in medium 1 and have the same phase velocity v_{p1}. Therefore, we can equate:

$$AB = OB \sin\theta_i = OC = OB \sin\theta_r$$

This equality leads to *the Snell's law of reflection* $\theta_i = \theta_r$. This we have derived before with the field analysis for the case for the conducting boundary.

Next we are going to develop the relationship between the incident and the transmitted waves. Observing the geometries in Figures 6.16b, and d we can see that from ΔOBD we have $OD = OB \sin\theta_t$, and from ΔOAB we have $AB = OB \sin\theta_t$. The two distances are different as the rays are in two different media and travel at two different phase velocities. However, they travel at the same time instance hence $t_1 = (AB/v_{p1}) = (OD/v_{p2})$. We get three sets of equations and with simple mathematical manipulation we can achieve the relationship between the two field waves as follows:

$$OD = OB \sin\theta_t; OA = OB \sin\theta_t \text{ and } \frac{AB}{v_{p1}} = \frac{OD}{v_{p2}}$$

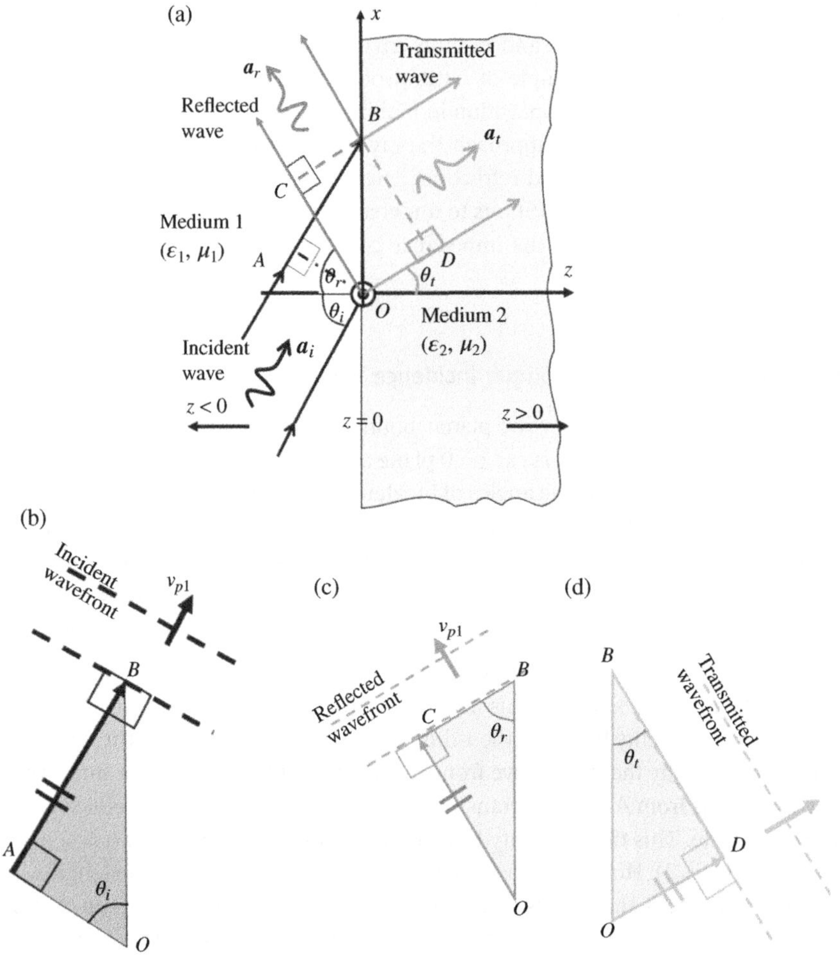

Figure 6.16 Ray trace model of oblique incidence of plane wave at dielectric boundary (a) we can assume a packet of waves are confined in a tube with the two rays as the end lines and is incident upon the dielectric boundary at $z = 0$ plane. The dotted lines perpendicular to the ray are the wavefronts for the respective waves, and (b–d) enlarged views of the incident, reflected, and transmitted rays with their respective angles.

$$\text{Or } \frac{OD}{AB} = \frac{v_{p2}}{v_{p1}} = \frac{\sin\theta_t}{\sin\theta_i}$$

Now substituting $v_{p1} = 1/\left(\sqrt{(\mu_1\varepsilon_1)}\right) = \omega/\beta_1$ and $v_{p2} = 1/\left(\sqrt{(\mu_2\varepsilon_2)}\right) = \omega/\beta_2$ we obtain the Snell's law of refraction as follows: $\sin\theta_t/\sin\theta_i = \beta_1/\beta_2$. Substitutions for $\beta_1 = \omega\sqrt{(\mu_1\varepsilon_1)}$ and $\beta_2 = \omega\sqrt{(\mu_2\varepsilon_2)}$ we obtain

$$\sin\theta_t = (\beta_1/\beta_2)\sin\theta_i = \sqrt{\frac{\mu_1\varepsilon_1}{\mu_2\varepsilon_2}}\sin\theta_i \tag{6.56}$$

If the two media are dielectric media then $\mu_1 = \mu_2 = \mu_0$, hence the law can be written in terms of the relative permittivity as follows:

$$sin\,\theta_t = sin\,\theta_i\sqrt{\frac{\varepsilon_{r1}}{\varepsilon_{r2}}} = sin\,\theta_i\frac{n_1}{n_2} \tag{6.57}$$

where $n_1$and n_2 are the refractive indices of media 1 and 2, respectively.[25]

From the above very simple ray trace model analysis based on the time equality we can make the following remarks:

1) The ray trace model of the oblique incidence of the uniform plane wave at the interface of the two media does not consider any specific polarization of the field waves as we did in *the field wave analysis* using Maxwell's field wave equations. However, the derivation of *Snell's laws of reflection and refraction* leads to the same results without considering the polarization of the plane wave in this case. This tells us that Snell's laws of reflection and refraction are independent of the polarization of the signal.
2) We have derived the two laws assuming that the media are lossless. The theory is equally applicable to the lossy medium considering the loss factors of the media with finite conductivity σ_1 and σ_2.
3) The results derived from the ray trace model are equally applicable for both the TE and TM polarizations of the oblique incidence cases. Therefore, the results obtained thus far in this section such as the total internal reflection and critical angle are equal phenomena for both TE and TM polarizations of the oblique incidence cases.

6.9 Total Internal Reflection

Recall the case study of the gold fish in an aquarium in Figure 6.3. The fish is not visible to the observer who is standing outside in air due to the total internal reflection of the incident light wave from the gold fish inside the water. In this case the incident light wave comes from the gold fish with an oblique incident angle θ_i, which is larger than *the critical angle* θ_c and propagates from the electromagnetically denser water medium and hit the water-air interface. As a consequence, the angle of transmission θ_t becomes larger than 90° and a total internal reflection occurs inside water. Now we shall examine the very phenomenon of the total internal reflection in details with the use of ray trace model. We shall work on *Snell's law of refraction* in (6.57) for the case when $\varepsilon_{r1} > \varepsilon_{r2}$. When an electromagnetic wave propagates from an electromagnetic denser medium to the lighter medium, observing expression (6.57), we can say that $sin\,\theta_t > sin\,\theta_i$ as the ratio of the two refractive indices of the dielectric media $n_1/n_2 = \sqrt{\varepsilon_{r1}/\varepsilon_{r2}} > 1$. An interesting thing happens as we increase the incident angle θ_i, the angle of transmission increases at a larger rate and reaches to $\theta_t = \pi/2$ quicker. In this condition, the transmitted ray in medium 2 has to glaze along the plane of interface at $z = 0$ and thus there is no propagating wave in medium 2 as shown in Figure 6.17a. This is evident looking at the constant phase plane meaning the wavefront of the transmitted wave which is now along x-axis when $\theta_t = \pi/2$. This incident angle is called *the critical angle* θ_c. Mathematically, we define θ_c by setting $\theta_t = \pi/2$ in (6.57) as:

$$sin\,\theta_c = sin\left(\frac{\pi}{2}\right) \times \sqrt{\frac{\varepsilon_{r2}}{\varepsilon_{r1}}} = \sqrt{\frac{\varepsilon_{r2}}{\varepsilon_{r1}}} \tag{6.58}$$

25 As stated, before that the refractive index expressions are used in optical materials.

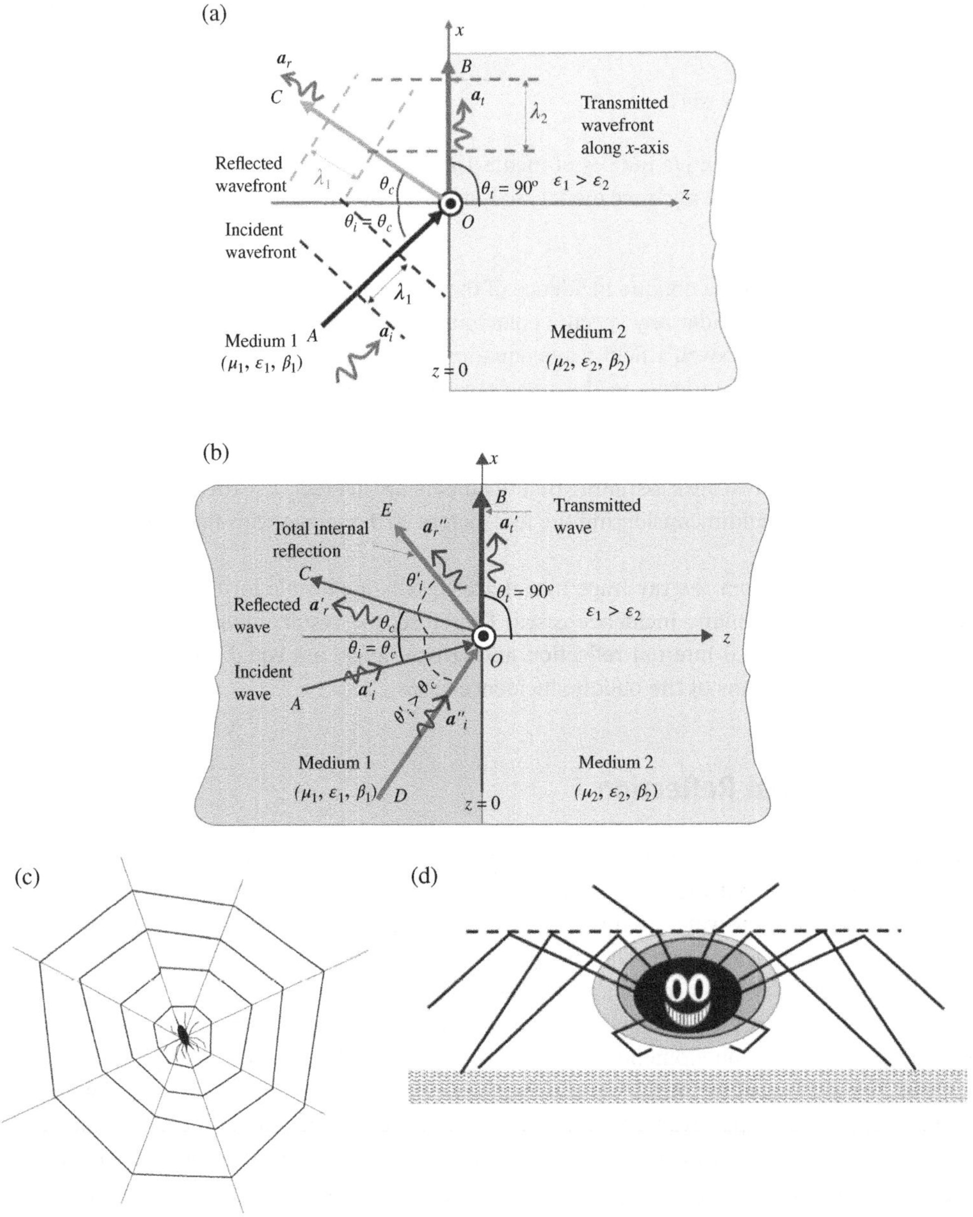

Figure 6.17 Plane wave incident from denser medium to the lighter medium ($\epsilon_1 > \epsilon_2$) (a) total internal reflection happens when the incident angle is $\theta_i = \theta_c = \sin^{1}\left(\sqrt{\varepsilon_{r2}/\varepsilon_{r1}}\right)$. This is shown with the ray *AOB*. At this condition, the angle of transmission $\theta_t = \pi/2$ and the transmitted wave glides the interface along *x*-axis. At this condition, there is also a reflected signal along *OC*, (b) the unit vectors are redefined at this condition are $\boldsymbol{a}_i'$, $\boldsymbol{a}_t'$ and $\boldsymbol{a}_r'$ for the incident, transmitted, and reflected waves, respectively. When $\theta_i > \theta_c$, the incident ray is totally reflected back into medium 1 with ray path of *DOE* with the propagation unit vectors $\boldsymbol{a}_i''$ and $\boldsymbol{a}_r''$, this means we lost the transmitted wave. The transmitted wave along *x*-axis is called *the surface wave*, (c) analogy of the total internal reflection rays make contours like a spider web once seen from the top, and (d) a spider when looking from the side. The tentacles can be assumed as the transmitted wave for the case $\theta_i < \theta_c$ and the legs can be assumed as the total internally reflected wave at different incident angles for the case $\theta_i > \theta_c$.

Therefore, the *critical angle* is defined as:

$$\theta_c = sin^{-1}\sqrt{\frac{\varepsilon_{r2}}{\varepsilon_{r1}}} = sin^{1}\left(\frac{n_2}{n_1}\right) \tag{6.59}$$

Note that in the above derivation we assume that both media are dielectric media hence $\mu_1 = \mu_2$. Total internal reflection starts happening when the incident angle is $\theta_i = \theta_c = \sin^{1}\left(\sqrt{\varepsilon_{r2}/\varepsilon_{r1}}\right)$. This is shown with the ray *AOB* in Figure 6.17a. As stated above, with the ray *AOB*, the angle of transmission is $\theta_t = \pi/2$ and the transmitted wave glides along *x*-axis on the interface. The transmitted wave *OB* that glazes along *x*-axis is called *the surface wave.* At this condition, there is also a reflected signal along *OC* that obeys Snell's law of reflection ($\theta_i = \theta_r = \theta_c$). Now we consider cases when $\theta_i = \theta_c$ as shown in Figure 6.17b. The unit vectors at this condition are redefined as $\boldsymbol{a}_i'$, $\boldsymbol{a}_t'$, and $\boldsymbol{a}_r'$ for the incident, transmitted, and reflected waves. As shown in Figure 6.17b, when $\theta_i > \theta_c$, the incident ray is totally reflected back into medium 1 with the ray path of *DOE* with the propagation unit vectors $\boldsymbol{a}_i''$ and $\boldsymbol{a}_r''$, and apparently there is no transmitted wave in medium 2.

Therefore, we can conclude that a further increase in the incident angle ($\theta_i > \theta_c$) will result in the total internal reflection and the transmitted angle is: $sin\,\theta_t = sin\,\theta_i\sqrt{\varepsilon_{r1}/\varepsilon_{r2}} > 1$. Analogy from the nature can be drawn for the total phenomenon of how the total internal reflection forms in this case is shown with a spider sitting at the center of its web looking from the top and the spider itself looking from the side. As can be seen in Figure 6.17c, the spider is resting at the center of its created spider web. The closed contours can be perceived as the top view of the light rays interacting with the interface of the two media whereas the solid diagonal line can be perceived as the plane of incidence on which the incident rays impinges at the interface and reflects back into medium 1.[26] The side view of the spider itself provides another type of analogy of the total internal reflection. As shown in Figure 6.17d we draw a horizontal line along the head of the spider that represents the plane of interface of the two media. The pair of antennas of the spider extended in medium 2 can be assumed as the transmitted waves in medium 2 for the case of $\theta_i < \theta_c$. The legs of the spider represent the total internal reflection of the incident waves for the case of $\theta_i > \theta_c$. As we observe in Figure 6.17, there is no wave propagation in medium 2 once $\theta_i > \theta_c$. What happened to the transmitted wave? How does it totally vanish in medium 2? Now we are going to examine if really there is no field wave propagation in medium 2 as for $\theta_i > \theta_c$. We get one pertinent clue from (6.57). For large enough difference between n_1 and n_2 and sufficiently large angle of incidence angle θ_i, the angle of transmission θ_t becomes such that $sin\,\theta_t > 1$. As for example for the case study of the gold fish, water has dielectric constant $\varepsilon_{r1} = 81$, and air $\varepsilon_{r1} = 1$. The ratio $\sqrt{\varepsilon_{r1}/\varepsilon_{r2}} = 9$. According to (6.59) the critical angle is $\theta_i = (1/9) = 6.4\,°$. For a small angle $\theta_i \geq 6.4°$ makes $sin\,\theta_t > 1$. In this condition, although $sin\,\theta_t$ is real $cos\,\theta_t$ is complex, hence θ_t does not yield real solution. This leads to the propagation constant being complex and exponential attenuation happening as the transmitted wave propagates in medium 2. This is examined in the next section.

6.9.1 Wave Phenomenon for $\theta_i > \theta_c$

In this section, we shall examine the characteristics of the transmitted field wave for the incident angle θ_i is larger than the critical angle θ_c. When $\theta_i > \theta_c$ a nonuniform plane wave is created in medium 2 as shown in Figure 6.18a. We also examine a nonuniform plane wave created in medium 1 for the conducting boundary, but in that case, we assume no penetration of signal in medium 2

26 The ray also is transmitted in medium 2 provided is $\theta_i < \theta_c$, there must be transmitted wave upward in medium 2 if the condition is met, but all three rays must be on the same plane which is in this case represented with the diagonal lines.

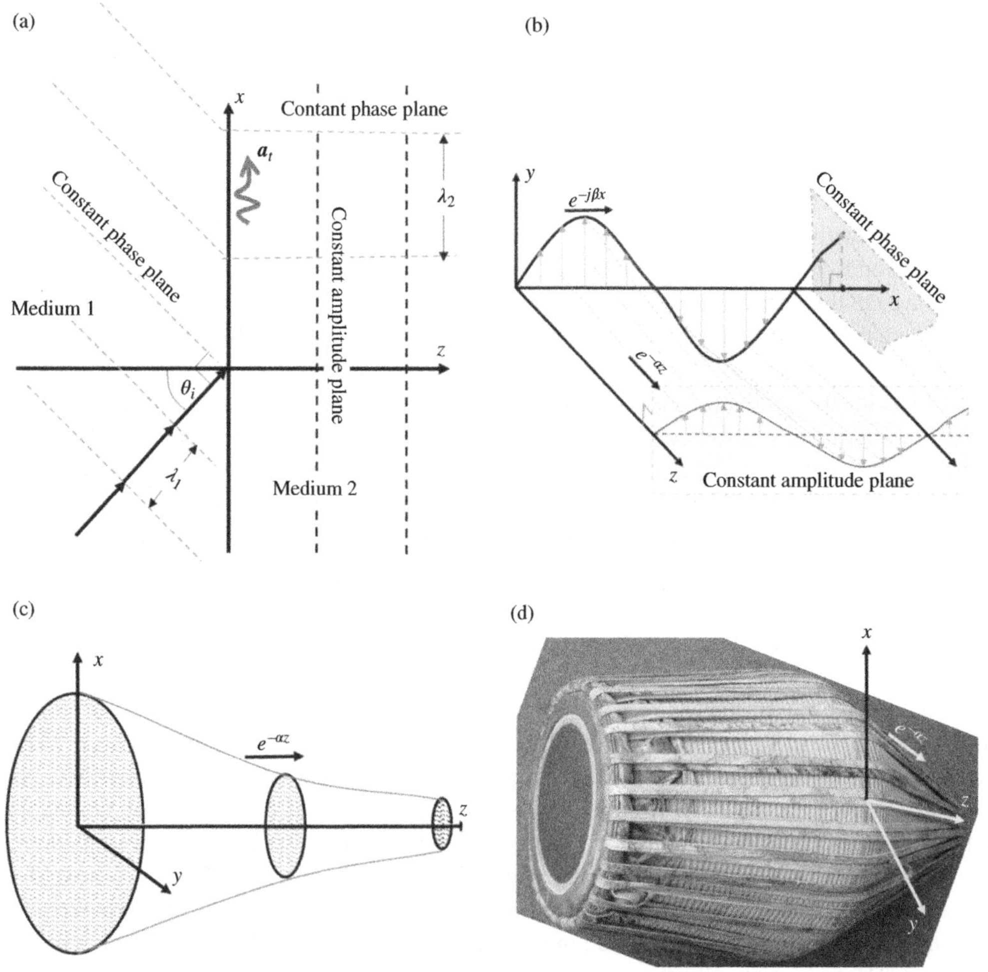

Figure 6.18 Total internal reflection and exponential decay of transmitted signal in medium 2 (a) when the angle of incidence surpasses the critical angle there is no wave front in medium 2. The wave front (constant phase plane) is along x-axis and the constant amplitude plane is along z-axis. As the wave propagates along z-axis exponential decay of the transmitted signal occurs. This phenomenon is illustrated in (b) 3D and (c) in perspective 3D view, respectively. Constant amplitude plane along z-axis and constant phase plane (wave front) are clearly visible in these figures, and (d) an south-east Indian percussion instrument called mridanga ((1970). *Indian Literature*. Sāhitya Akademi, 84. https://en.wikipedia.org/wiki/Khol (accessed 3 May 2018). Retrieved 23 December 2012 (cited in the same Wikipedia site)). *Source:* Photo of mridanga showing the big head by Jan Kraus/Wikimedia Commons/CC BY-SA 3.0.

due to the perfect conducing second medium. In the dielectric boundary case, due to the penetrable second medium there must be a transmitted signal in the second medium. Note that there is no Ohmic loss in the medium as we assume the medium is lossless. Therefore, we must need to know what is happening to the transmitted field wave in the medium 2. The only assumption we can make is that the condition of $\theta_i > \theta_c$ heavily attenuates the transmitted field wave and the only assumption we can make is that there is an exponential decay with factor $e^{-\alpha z}$, where α is the attenuation constant. From (6.57) we observe that θ_t is a real angle value if $sin\theta_t \le 1$. This means θ_t to be a real angle value $\theta_i \le \theta_c$. For $\theta_i \ge \theta_c$, substitution for (6.58) into (6.57) yields

$$sin\,\theta_t = \frac{sin\,\theta_i}{sin\,\theta_c} = \sqrt{\frac{\varepsilon_{r1}}{\varepsilon_{r2}}} > 1 \tag{6.60}$$

Equation (6.60) does not provide a real solution for θ_t rather it becomes complex. Although $sin\,\theta_t$ is real in (6.60), $cos\,\theta_t$ becomes imaginary in this situation as shown below:

$$cos\,\theta_t = \sqrt{1 - sin^2\theta_t} = \pm j\sqrt{sin^2\theta_t - 1} = \pm j\sqrt{\frac{\varepsilon_{r1}}{\varepsilon_{r2}} sin^2\theta_i - 1} \tag{6.61}$$

Figure 6.18a illustrates the typical wave propagation in medium 2 with the unit vector $\boldsymbol{a_t}$ along x-axis. Our main aim of this analysis is to find the propagation phase of the transmitted wave along the unit vector $\boldsymbol{a_t}$ in its generalized expression as if no internal reflection has happened as yet as shown in Figure 6.16a. We have already derived the propagation factor with the angle of transmission in the following form.[27]

$$\boldsymbol{a_t} = \boldsymbol{a_x}\, sin\,\theta_t + \boldsymbol{a_z}\, cos\,\theta_t \tag{6.62}$$

The propagation phase factor for the electromagnetic wave irrespective of their polarization can be defined as:

$$e^{-jk_2'z'} = e^{-jk_2\boldsymbol{a_t}.\boldsymbol{R}} = e^{-jk_2(x\,sin\,\theta_t\, +\, z\,cos\,\theta_t)} \tag{6.63}$$

Since the positive solution of $cos\,\theta_t$ in (6.61) grows exponentially as the wave propagates along $z \to \infty$, and goes beyond control, we naturally select the negative solution of (6.61). Substitutions for (6.60) and (6.61) into (6.63) yield the propagation phase factor of the transmitted wave in medium 2 as follows:

$$e^{-jk_2'z'} = e^{-j(\alpha_2 x + \beta_2 z)} = e^{-jk_2(x\,sin\,\theta_t\, +\, z\,cos\,\theta_t)} \tag{6.64}$$

Here the attenuation constant is obtained by substituting for the negative solution in (6.61) as follows:

$$\alpha_{2z} = k_2\, cos\,\theta_t = k_2\sqrt{\frac{\varepsilon_{r1}}{\varepsilon_{r2}} sin^2\theta_i - 1} \tag{6.65}$$

And the propagation constant is defined after substitution for (6.57) into (6.64) as follows:

$$\beta_{2x} = k_2\sqrt{\frac{\varepsilon_{r1}}{\varepsilon_{r2}}}\, sin\,\theta_i \tag{6.66}$$

A few remarks can be made based on the observation and analysis of the transmitted wave in medium 2 for $\theta_i \geq \theta_c$ as follows:

1) The nonuniform wave propagation in medium 2 can be separated into x-directed propagating wave[28] and z-directed attenuating wave. The expression of z-directed attenuation constant $\alpha_{2z} = k_2\sqrt{\varepsilon_{r1}/\varepsilon_{r2}\, sin^2\theta_i - 1}$ represents a heavily attenuated non-propagating wave with an exponential factor of $e^{-\alpha_{2z}z}$. Such a non-propagating wave is called an *evanescent wave.*
2) There is a surface wave that glazes with a propagation constant $\beta_{2x} = k_2\sqrt{\varepsilon_{r1}/\varepsilon_{r2}}\, sin\,\theta_i$.
3) Figure 6.18b shows the 3D perspective view of the resultant wave in medium 2. As can be shown there are two constant planes[29]: (i) a constant phase plane along x-axis which is parallel to yz-plane, and (ii) a constant amplitude plane along z-axis which is parallel to xy-plane. The surface wave glazes along x axis for $z = 0$ and heavily attenuated with an exponential decaying factor $e^{-\alpha_{2z}z}$ as it moves along z-axis inside the medium 2. Figure 6.18c shows the simplified view as seen through the xy-plane along z-axis into medium 2. The evanescent mode wave is heavily attenuated with factor $e^{-\alpha_{2z}z}$.

27 See Figures 6.9b, d for the coordinate decomposition for a general case of wave propagation for any arbitrary angle at any polarization.
28 Also called the surface wave.
29 C. A. Johnk, *Engineering Electromagnetic Fields and Waves*, Wiley, 1988 P. 378.

4) An analogy of the attenuated wave in medium 2 for the case of $\theta_i > \theta_c$ can be drawn with a south-east Indian percussion instrument called Mridanga. Details of the instrument can be found in caption of Figure 6.18. The profile of the instrument can be perceived as *the evanescent mode wave*. The most interesting characteristics of the instrument is that when the narrow side of the drum is hit it creates a sound with very short pitch of short duration compared to that from the opposite wider end. This instrument is used for devotional songs and music in Bangladesh and south eastern part of India.
5) Finally, the wave inside medium 2 is a nonuniform plane wave and no real power is transported in medium 2 at this condition. Inan and Inan[30] have suggested a method to measure such power in the evanescent mode wave with an antenna placed adjacent to the interface of the two media. The negligible receive power by the antenna proves the non-transmission of power in medium 2 at the total internal reflection condition. We shall solve a few problems to exemplify the characteristics of the total internal reflection.

Review Questions 6.6: Critical angle in Oblique Incidence

Q1. Define *critical angle*. Explain what happens to the transmitted angle when the incident angle surpasses the critical angle.
Q2. What is a *surface wave*? Explain the phenomenon of the surface wave at dielectric boundary.
Q3. Draw the wave front of the surface wave and indicate its direction of propagation.
Q4. Derive the expression for attenuation constant α_{2z} and propagation constant β_{2x} in medium 2 at total internal reflection.
Q5. Explain will there be a reflected wave in medium 1 at the critical angle. What would have happened to the reflected wave when the incident angle surpasses the critical angle?

Example 6.4: Total Internal Reflection – Gold Fish in Aquarium
Consider the case study of a gold fish in an aquarium as shown in Figure 6.3. In visible optical region wavelength 589 nm the refractive index of water $n_w = 1.33$.[31] Assume that the opaque disc is placed at the top of the aquarium is 100 cm in radius. Calculate the followings: (i) the minimum depth of the aquarium so that the fish laying on the bottom cannot be seen, (ii) if the incident wave makes an angle $\theta_i = 30°$, the angle of reflection, the angle of refraction, and the propagation constants in water and air, and (iii) the attenuation constant and propagation constant in air at 75° of incident angle.

Answer:
Given parameters are: $n_1 = 1.33$, $n_2 = 1$, wavelength $\lambda_w = 589$ nm, opaque desic radius $OP = 100$ cm.

Using (6.59) the critical angle is: $\theta_c = sin^{-1}\ (n_2/n_1) = sin^{-1}\ (1/1.33) = 48.75°$ A triangle ΔFOP along the plane of incidence can be drawn, where the depth $d =$ OF.

Using ΔFOP, the depth of aquarium is: $d = OP/\ tan\ \theta_c = 100\ \text{cm}/tan\ (48.75°) = 87.7$ cm

Observation: A reflected ray and a surface wave ray are also visible in Figure 6.3. For $\theta_i < \theta_c$, the incident wave is partly reflected back in water and partly transmitted in medium 2. For $\theta_i > \theta_c$, total internal reflection happens and the incident ray is totally reflected back into water and a surface wave along dielectric boundary exists. The wave is heavily attenuated into the air as it is propagated along the normal direction of the dielectric boundary (water–air interface). See also Figure 6.17.

30 U. S. Inan and A. S. Inan, *Electromagnetic Waves*, Prentice Hall, Upper Saddle River, NJ, 2000, p. 198 Figure 3.37(b).
31 To learn more about the refractive index of water and the refractive index in general refer to https://en.wikipedia.org/wiki/Refractive_index (accessed 2 June 2018). You will get useful information about the refractive indices of other materials.

i) The reflected angle is the same of the incident angle as per *Snell's law of reflection*: $\theta_r = \theta_i = 30\,°$.
The angle of refraction is: $\theta_t = sin^{-1}(1.33\,sin\,30°) = 41.68°$
The phase constant in water: $\beta_1 = 2\pi/\lambda_1 = 2\pi/(589\times10^{-9})\times1.33 = 14.19\times10^6$ (rad/m)
The phase constant in air: $\beta_2 = 2\pi/\lambda_2 = 2\pi/(589\times10^{-9}) = 10.67\times10^6$ (rad/m)
As we can see the phase constant viz the wave number is large in the denser medium.

ii) given $\theta_i = 75°$, using (6.65) the attenuation constant is: $\alpha_{2z} = 10.67\times10^6\times\sqrt{1.33^2 sin^2(75°)-1} = 6.96\times10^6$ (NP/m)
This is huge attenuation that normal scientific calculator cannot calculate the total attenuation even in 1 mm length inside air.
The propagation consonant of the surface wave using (6.66) is:

$$\beta_{2x} = 10.67\times10^6\times1.33\times sin\,(75°) = 13.71\times10^6\ (\text{rad/m})$$

Example 6.5: Total Internal Reflection – Microstrip Patch Antenna
Consider a microstrip patch antenna is designed on Taconic TLX0 substrate with relative permittivity of $\varepsilon_{r1} = 2.45$. The thickness of the substrate is 0.787 mm. The patch antenna is a half wavelength resonator and operates at the upper wi-fi frequency of 4.9 GHz. The patch antenna is energied with a coaxial probe as shown in Figure 6.19 where the probe is radiating omni-direction meaning in all direction. Calculate the followings:

i) The wavelength of the operating frequency and the length of the patch antenna.
ii) The critical angle of the substrate material
iii) The surface wave attenuation constant at 65° of incidence angle
iv) The propagation constant in air.

Assume that the dielectric material is lossless.

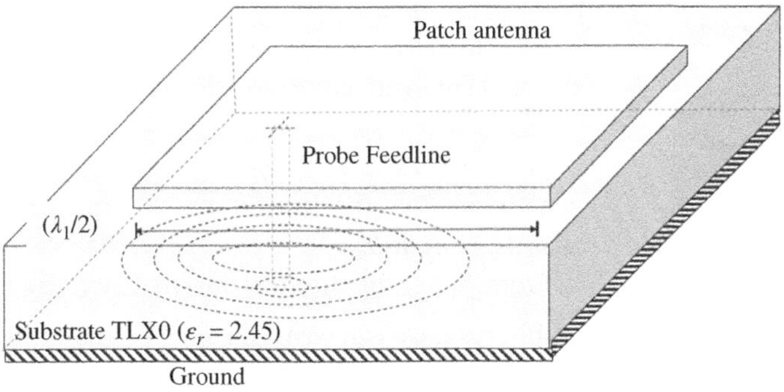

Figure E6.5 Total Internal Reflection – Microstrip Patch Antenna.

Answer:
Given frequency $f = 4.9$ GHz, $\varepsilon_{r1} = 2.45$, $d = 0.787$ mm

i) The wavelength in TLX0 is: $\lambda_1 = \dfrac{3\times10^8/\sqrt{2.45}}{4.9\times10^9} = 39.11\times10^{-3}\,(\text{m})$; wavelength in air $\lambda_2 = \dfrac{3\times10^8}{4.9\times10^9} = 61.22\times10^{-3}\,(m)$

The patch length: $\dfrac{l=\lambda_1}{2} = \dfrac{39.11\times10^{-3}}{2} = 19.55\,(mm)$

ii) The critical angle inside the substrate is: $\theta_c = \sin^{-1}\left(\dfrac{1}{\sqrt{2.45}}\right) = 39.7°$

iii) The surface wave attenuation of the signal in air is:

$$\alpha_{2z} = \frac{2\pi}{39.11 \times 10^{-3}} \times \sqrt{2.45\ sin^2(65°) - 1} = 65.97\left(\frac{Np}{m}\right)$$

iv) The phase constant:

$$\beta_{2x} = \frac{2\pi}{61.22 \times 10^{-3}} \times \sqrt{2.45} \times sin\,(65°) = 145.64\left(\frac{rad}{m}\right)$$

Example 6.6: Total Internal Reflection - Optical Fiber Design

The optical waveguide and optical fiber work solely on the basis of total internal reflection inside the substrate and core, respectively. There must be certain incidence angle also called *the acceptance angle* for the light ray to impinge on the air and core interface to get propagation of the light wave inside the optical waveguide and optical fiber. Figure E6.6 below shows construction of the optical waveguide and the ray trace model of the devices. To understand the operating principle of optical fiber which has enormous applications in high speed data communications using transoceanic submarine optical fiber cables we first analyze the dielectric and/or optical waveguides. Consider a dielectric waveguide as shown in the figure below accepts an incident ray at an angle θ_a from air. The refractive indices of the core in 1.465 and cladding 1.463. Derive and calculate the followings:

i) An expression of $sin\theta_a$ in terms of the refractive indices.
ii) The maximum of $sin\theta_a$ is called *the numerical aperture NA* and a significant design parameter for optical waveguides and optical fibers. Calculate *NA* of the optical waveguide.
iii) An expression of the minimum refractive index n_1 that supports the wave propagation via total internal reflection.

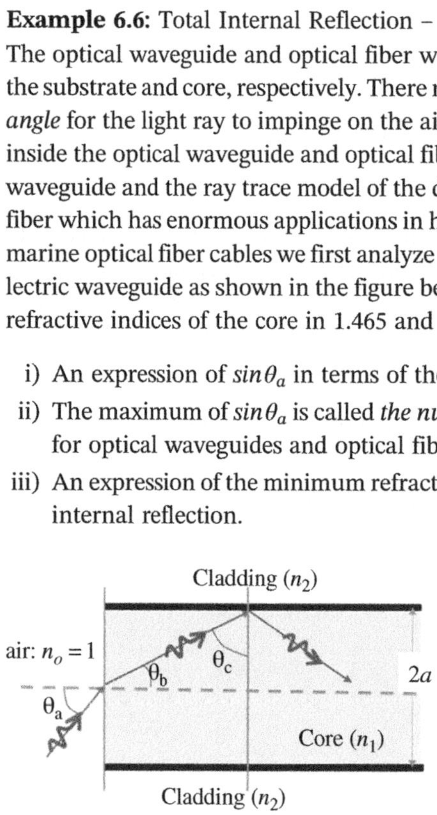

Figure E6.6 Total Internal Reflection - Optical Fiber Design .

Answer:

Given parameters, $n_1 = 1.465$, $n_2 = 1.463$. The refractive index of air: $n_0 = 1$.

i) Total internal reflection happens when the angle of incidence $\theta_i \geq \theta_c$, the critical angle, where, $sin\theta_c = n_2/n_1$.

From *Snell's law of refraction* in (6.57) we can write for the ray from air to core: $n_0 sin\,\theta_a = n_1 sin\,\theta_b$.

From (6.58) for the total internal reflection within core, we can write $sin\,\theta_c = \dfrac{n_2}{n_1}$ and from the trigonometry: $sin\,\theta_b = sin\left(\dfrac{\pi}{2} - \theta_c\right) = cos\,\theta_c$.

Therefore, linking the rays in air and core considering the total internal reflection that supports the wave propagation within the core only, we can write:

$$n_0\, sin\,\theta_a = n_1\, cos\,\theta_c = n_1\sqrt{1 - sin^2\theta_c} = n_1\sqrt{1 - \left(\frac{n_2}{n_1}\right)^2} = \sqrt{n_1^2 - n_2^2}.$$

Therefore, $sin\,\theta_a = \sqrt{n_1^2 - n_2^2}/n_0$ 0

ii) Since, for air $n_0 = 1$, the maximum value of $sin\theta_a$, also called the *numerical aperture:* $NA = \sqrt{n_1^2 - n_2^2}$.

For the current example: $NA = \sqrt{1.465^2 - 1.463^2} = 0.0765$

iii) For total internal reflection, first condition is: $cos\,\theta_b \geq sin\,\theta_c$ (1)

We also obtain from Snell's law of refraction: $n_0\,sin\,\theta_a = n_1\,sin\,\theta_b$; $sin\,\theta_b = (n_0/n_1)\,sin\,\theta_a = (sin\,\theta_a)/n_1$ (2)

From (1) we get: $cos\,\theta_b = \sqrt{1 - sin^2\theta_b} = \sqrt{1 - sin^2(\theta_a)/(n_1^2)} \geq sin\,\theta_c = n_2/n_1$

Squaring: $n_1^2 \geq n_2^2 + sin^2\theta_a$

Since the largest value of n_1 could be achieved for $sin\,\theta_a = 1$; therefore, we require the core medium is at least $n_1^2 \geq n_2^2 + 1$ so that guiding via the core is possible.

For air cladding;

$n_1 \geq \sqrt{2} = 1.14$

Most glass and quartz have refractive indices in that order and light incident at an angle more than $\theta_c = Sin^{-1}\,(1/n_1) = 45°$totally internally reflected inside the media.

So far, we have analyzed the oblique incidence cases using the ray trace model which is independent of the polarization of the incoming signal. We have observed that we can achieve *the Snell's laws of reflection and refraction* using the ray trace model. We also observed that the phenomenon of the total internal reflection is adequately explained using the ray trace model and the phenomenon of the total internal reflection irrespective of the polarization of the incoming signal. However, the reflection and transmission coefficients of the oblique incidence and the wave impedances are polarization dependent and differ with the nature of signal polarizations of the incoming signal. In the following sections, we are going to present the field analysis and impedance concept for the two separate polarizations of oblique incidence cases.

6.10 TE Polarization of Oblique Incidence on Dielectric Boundary

In the preceding section, the ray trace model describes the various hypotheses of the oblique incidence cases which are applicable to any polarization of the incoming signal. However, some electromagnetic phenomena are polarization dependent. As for example though as proved in the previous section, *the total internal reflection* phenomenon hence critical angle θ_c is polarization independent, but *the total transmission* of a signal at a particular incident angle is polarization dependent. The angle at which the total transmission of the incident signal occurs is called *Brewster angle.*[32] Likewise, as stated above the reflection and transmission coefficients and their respective characteristic wave impedances are polarization dependent. To examine the difference of these parameters in different polarizations, we need the Maxwell's field analysis and the boundary conditions. In this section, we shall examine TE polarization of oblique incidence on dielectric boundaries using Maxwell's field analysis and boundary conditions, and then we shall move to the TM polarization in the next Section 6.11. TE polarized oblique incidence has applications in dielectric waveguides and optical fibers utilizing the total internal reflections.

Figure 6.19a shows the TE polarized oblique incidence on a dielectric boundary along *xy*-plane. As shown in the figure the TE polarized electric field is oriented perpendicular to the plane of incidence which is on the *xz*-plane (plane of paper), and the vector field hits parallel to the plane of interface or the boundary plane which is on the *xy*-plane. The difference between the analysis

32 We shall be discussing *the total transmission and Brewster angle* θ_B in TM polarization theory in the next section.

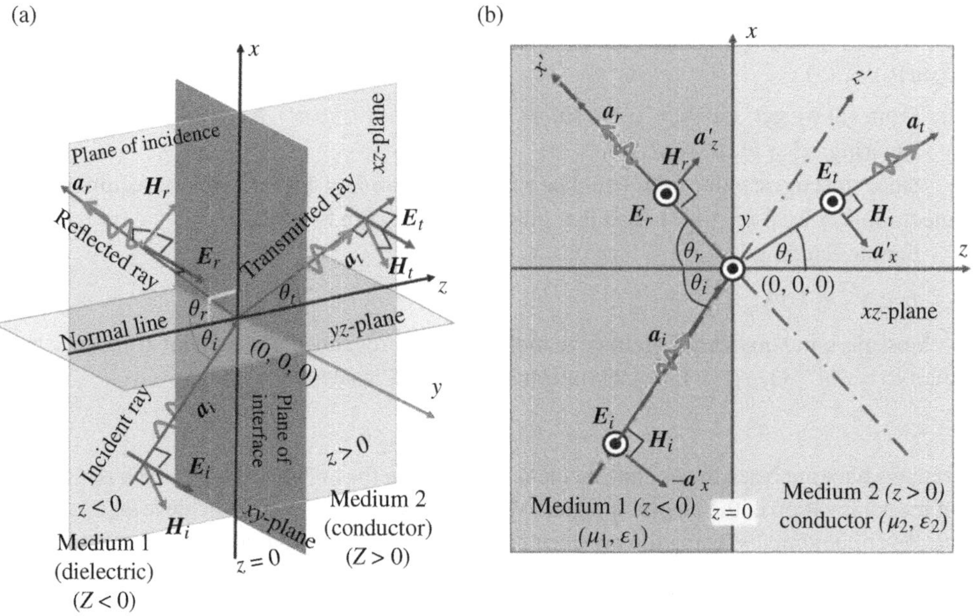

Figure 6.19 TE oblique incidence in dielectric boundary (a) 3D view, and (b) 2D view on *xz*-plane. The electric field is oriented along *y*-axis which is perpendicular to the plane of incident *xz*-plane and the magnetic field is on the plane of incident *xz*-plane.

of TE polarization on the conducting boundary and the dielectric boundary is that in the current situation the transmitted electromagnetic fields are not zero. Therefore, we have to consider the field analysis for the transmitted electromagnetic fields in medium 2 using Maxwell's field analysis as shown in the figure. As usual, the angle of incidence is θ_i that makes the angle between the direction of the incident wave propagation $\boldsymbol{a_i}$ and the normal line to the plane of interface along z-axis. The angle of reflection is θ_r, which is the angle between the reflected wave's unit vector $\boldsymbol{a_r}$ and the z-axis. The angle of transmission is θ_t that makes the angle between the direction of the transmitted wave propagation $\boldsymbol{a_t}$ and the normal line to the plane of interface along z-axis. As usual the incident, reflected and transmitted magnetic fields reside on the xz-plane with an arbitrary angle. Due to the arbitrary polarization of the magnetic field, we need to decompose it into their respective components along the principle axes x and z, as we did before for the case of oblique incidence on the conducting boundary. The electromagnetic fields and their respective directions of wave propagation must obey *the right-hand rule* as shown in the figure. As can be seen the directions for the incident, reflected, and transmitted electric fields remain unchanged along y-axis as if it is vibrating along y-axis as they propagate along their respective directions. However, the magnetic field changes its direction due to its arbitrary polarization on the xz-plane and always remains on the xz-plane (plane of paper) with different directions to maintain the right-hand rule with the electric field and the direction of wave propagations in the three phases on incidence, reflection, and transmission. The 2D propagating waves in incident, transmitted, and reflected conditions are shown with sinusoidally varying red curves in the figure. Our motto of the field analysis is to derive the transmission and reflection coefficients using the electromagnetic field boundary conditions and relate the transmitted and reflected field amplitudes in terms of the incident field amplitude. We have done the comprehensive vector decompositions of the incident, reflected and transmitted electromagnetic fields in Section 6.3. The detailed vector field decompositions are shown in as we shall go along with the field analysis.

In the derivations of the transmission and reflection coefficients, the following rules are obeyed:

i) For incidence angle $\theta_i \neq 0$; the "oblique incidence" is TE/perpendicular polarization when the electric field is y-directed and normal to the plane of incidence.
ii) The magnetic field is arbitrarily polarized on xz-plane. We decompose the magnetic field to its equivalent x- and z-directed components.
iii) In all cases right-hand rule prevails. In the reflected wave, the electric field remains y-polarized, but the magnetic field changes its direction to meet the condition of the propagation direction of the reflected wave, and the unchanged orientation of the reflected electric field.
iv) In the transmitted wave, the electric field remains y-polarized in medium 2, but the magnetic field changes its direction due to the different constitutive parameters in medium 2. The simple approach is to find the transmitted electric and magnetic fields in medium 2 is to just change the signs of the constitutive parameters β_1 to β_2 and the angle of transmission θ_t in medium 2. The vector decomposition follows the same orientations of vectors that for the incident wave.
v) Decompositions of the position vectors of the three field waves are performed in terms of the electrical lengths along the directions of propagation $\boldsymbol{a_i}$, $\boldsymbol{a_r}$, and $\boldsymbol{a_t}$, for the incident, reflected, and transmitted fields, respectively, and the unit vectors of the electromagnetic fields along the principal axes x and z.

Our goals are to find the TE polarized electromagnetic field solutions, reflection and transmission coefficients, critical angle and power balance equation using Maxwell's field analyses, and the boundary conditions as the interface of the two dissimilar dielectric media at $z = 0$. The overall process of the field solutions using the field analysis is quite involving hence need a systematic procedure to compute the fields and derive the relationships between the three sets of electromagnetic fields. In the following, the procedure to analyze the three sets of fields are described with illustration.

The procedure for analysis of the oblique incidence is depicted in Figure 6.20 and the steps are as follows. The procedure is equally applicable to the TM polarization oblique incidence.

1) Draw the 2D configuration of TE-polarized oblique incidence UPW propagation with all vector notations of field vectors on the plane of incidence xz-plane as shown in Figure 6.21.
2) Perform the vector decompositions of the magnetic fields in incidence, reflection, and transmission.
3) Write down phasor vector fields $(\boldsymbol{E}_s^i, \boldsymbol{H}_s^i)$, $(\boldsymbol{E}_s^r, \boldsymbol{H}_s^r)$, and $(\boldsymbol{E}_s^t, \boldsymbol{H}_s^t)$ for incidence, reflected, and

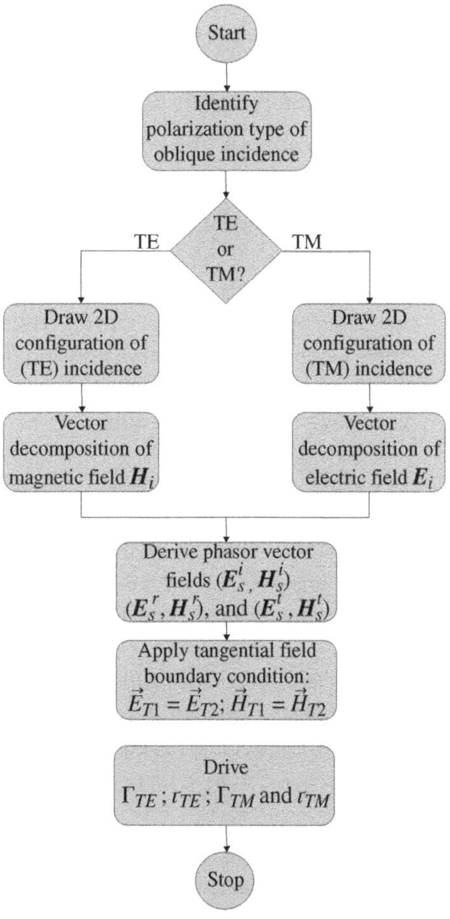

Figure 6.20 Analysis of oblique incidence.

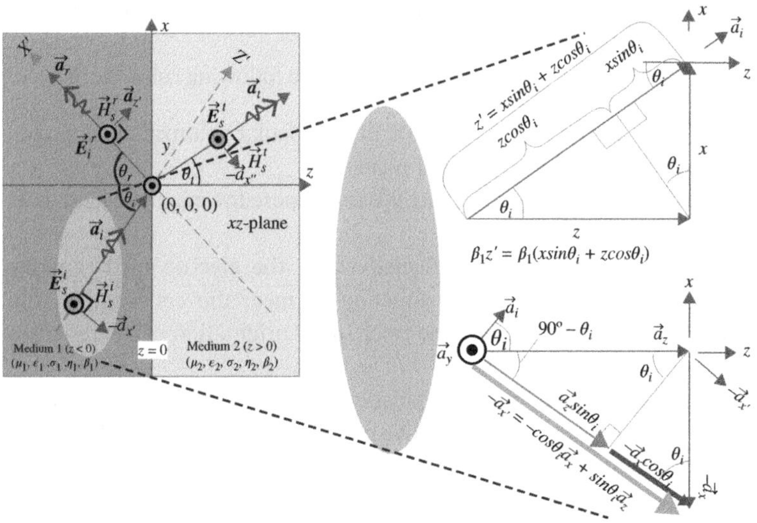

Figure 6.21 Vector decompositions of TE oblique *incidence* fields on a dielectric boundary. Vector decompositions in $\beta z'$ and unit vector $-\boldsymbol{a}_{x'}$.

transmitted cases, respectively, with the propagation electrical lengths: $e^{-j\beta_1 x'}$,$e^{-j\beta_1 z'}$ and $e^{-j\beta_1 x''}$ for all three cases.

4) Once the complete field expressions are obtained, apply boundary conditions at the interface of two media at $z = 0$. The boundary conditions are: $\left\{ \begin{array}{c} E_{T1} = E_{T2} \\ H_{T1} = H_{T2} \end{array} \right\}$. Here, subscript T represents the total tangential field components in media "1" and "2" as the second subscripts.[33] The boundary conditions give relations between the complex electric and magnetic field components in the two media.
5) The phase equality of complex fields leads to *Snell's law of reflection* $\theta_i = \theta_t$ and *Snell's law of refraction* $\theta_i = \sin^{-1}\{(\beta_2/\beta_1)\sin\theta_t\}$. The special condition of incident angle leads to the *critical angle* when the angle of transmission $\theta_t = 90°$.
6) Finally, we shall calculate the voltage reflection coefficient Γ_{TE} and voltage transmission coefficient τ_{TE} using above results.

We can also calculate the power reflection coefficient $|\Gamma_{TE}|^2$ that represents the fraction of power reflected back in medium 1, and the power transmission coefficient $\{1 - |\Gamma_{TE}|^2\}$ using the energy balance theory.[34] Now let us perform the step-by-step procedure to calculate the field solutions and the voltage reflection coefficient Γ_{TE} and voltage transmission coefficient τ_{TE}. The three sets of electromagnetic fields and their phasor vector decompositions are produced one by one as follows.

Incident electromagnetic fields: The phasor vector expressions of the electric and magnetic fields are derived from the vector decompositions according to Figure 6.21. The decompositions are performed in two aspects: (i) the equivalent electrical length along the direction of incident wave propagation $\zeta_{inc} = \beta_1 z' = \beta_1 (x\sin\theta_i + z\cos\theta_i)$, where the equivalent electric length ζ_{inc} has the unit of

33 Do not confuse T with the subscript associated with the lower case subscript t for transmitted field waves.
34 We shall also provide a detailed power balance calculation in Sections 6.10.7 and 6.11.11 for TE and TM oblique incidence cases respectively.

radian, and (ii) the unit vector of the arbitrarily polarized magnetic field on the plane of incidence (xz-plane): $-\boldsymbol{a}_{x'} = -\cos\theta_i \boldsymbol{a}_x + \sin\theta_i \boldsymbol{a}_z$. Figure 6.21 illustrates the enlarged view of the decompositions. Now we have all ingredients of the phasor vector expressions for the TE-polarized incident electric and magnetic fields and are produced as follows.

Incident electromagnetic fields (where β_1 is used in place of k_1)

$$\boldsymbol{E}_s^i = \boldsymbol{a}_y E_o^i e^{-j\beta_1 z\prime} = \boldsymbol{a}_y E_o^i e^{-j\beta_1(x\sin\theta_i + z\cos\theta_i)} \tag{6.67}$$

$$\boldsymbol{H}_s^i = -\boldsymbol{a}_{x'} H_o^i e^{-j\beta_1 z} = (-\cos\theta_i \boldsymbol{a}_x + \sin\theta_i \boldsymbol{a}_z) H_o^i e^{-j\beta_1(x\sin\theta_i + z\cos\theta_i)} \tag{6.68}$$

In this case, the incident electric field has only one component along y-direction and the magnetic field is arbitrarily polarized on the plane of incidence (plane of paper) hence has two components in $-x$ and z directions as shown in Figure 6.21. The boundary conditions are applied only on the tangential components of the incident electric and magnetic fields on the xy-plane at $z = 0$, hence it is worth calculating the two components as follows:

$$\boldsymbol{E}_{T1}^i\big|_{z=0} = \boldsymbol{a}_y E_o^i e^{-j\beta_1(x\sin\theta_i)} \tag{6.69}$$

$$\boldsymbol{H}_{T1}^i\big|_{z=0} = -\boldsymbol{a}_x \cos\theta_i H_o^i e^{-j\beta_1(x\sin\theta_i)} \tag{6.70}$$

Reflected electromagnetic fields: Again, the phasor vector expressions of the reflected electric and magnetic fields in medium 1 are derived from the vector decompositions according to Figure 6.22 and are produced as follows. The decompositions are performed in two aspects: (i) the electrical length along the direction of incident wave propagation $\zeta_{refl} = \beta_1 x' = \beta_1(x\sin\theta_r - z\cos\theta_r)$, where equivalent electric length ζ_{refl} has the unit of radian, and (ii) the unit vector of the arbitrarily polarized magnetic field on the plane of incidence (xz-plane): $-\boldsymbol{a}_{z'} = \cos\theta_r \boldsymbol{a}_x + \sin\theta_r \boldsymbol{a}_z$. Figure 6.22 illustrates the enlarged view of the decompositions as well.

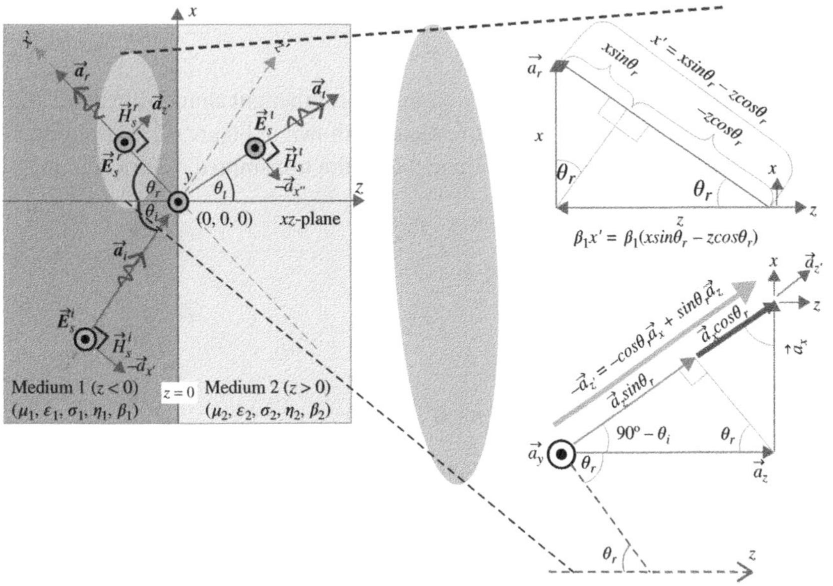

Figure 6.22 Vector decompositions of TE oblique *reflected* fields on a dielectric boundary.

Reflected electromagnetic fields

$$E_s^r = a_y E_o^r e^{-j\beta_1 x\prime} = a_y E_o^r e^{-j\beta_1(x\sin\theta_r - z\cos\theta_r)} \tag{6.71}$$

$$H_s^r = a_{z\prime} H_o^r e^{-j\beta_1 x\prime} = (\cos\theta_r a_x + \sin\theta_r a_z) H_o^r e^{-j\beta_1(x\sin\theta_r - z\cos\theta_r)} \tag{6.72}$$

In this case, the reflected electric field has only one component along y-direction and the magnetic field is arbitrarily polarized on the plane of incidence (plane of paper) hence has two components in x and z directions as shown in Figure 6.22.

The boundary conditions are applied only on the tangential components of the reflected electric and magnetic fields as follows:

$$E_{T1}^r|_{z=0} = a_y E_o^r e^{-j\beta_1(x\sin\theta_r)} \tag{6.73}$$

$$H_{T1}^r|_{z=0} = a_x \cos\theta_r H_o^r e^{-j\beta_1(x\sin\theta_r)} \tag{6.74}$$

Lastly, we perform the similar vector decompositions for the transmitted electromagnetic fields as follows.

Transmitted electromagnetic fields: Finally, the phasor vector expressions of the transmitted electric and magnetic fields are derived from the vector decompositions according to Figure 6.23. Again the decompositions are performed in two aspects: (i) the equivalent electrical length along the direction of transmitted wave propagation $\zeta_{trans} = \beta_2 z'' = \beta_2(x\sin\theta_t + z\cos\theta_t)$, where the equivalent electric length ζ_{tans} has the unit of radian, and (ii) the unit vector of the arbitrarily polarized magnetic field on the plane of incidence (xz-plane): $-a_{x''} = -\cos\theta_t a_x + \sin\theta_t a_z$. Figure 6.23 illustrates the enlarged view of the decompositions as well. Now we have all ingredients of the phasor vector expressions for the TE-polarized transmitted electric and magnetic fields and are produced as follows.[35]

Transmitted electromagnetic fields (where β_2 is used in place of k_2)

$$E_s^t = a_y E_o^t e^{-j\beta_2 z\prime\prime} = a_y E_o^t e^{-j\beta_2(x\sin\theta_t + z\cos\theta_t)} \tag{6.75}$$

$$H_s^t = -a_{x\prime\prime} H_o^t e^{-j\beta_t z} = (-\cos\theta_t a_x + \sin\theta_t a_z) H_o^t e^{-j\beta_2(x\sin\theta_t + z\cos\theta_t)} \tag{6.76}$$

In this case, as usual, the transmitted electric field has only one component along y-direction and the magnetic field has two components in $-x$- and z-directions as shown in Figure 6.23. The boundary conditions are applied only on the tangential components of the transmitted electric and magnetic fields as follows:

$$E_{T2}^t|_{z=0} = a_y E_o^t e^{-j\beta_2(x\sin\theta_t)} \tag{6.77}$$

$$H_{T2}^t|_{z=0} = -a_x \cos\theta_t H_o^t e^{-j\beta_2(x\sin\theta_t)} \tag{6.78}$$

6.10.1 Applications of Boundary Conditions at $z = 0$

So far, we have rigorously performed all vector decompositions of the three sets of electromagnetic fields and also expressed the x-directed tangential field components of the three sets of the fields that will hold the equality. This equality will produce the amplitudes of the reflected and

35 Compare the incident electromagnetic field expressions in (6.67), (6.68) with those for the transmitted one (6.75), (6.76), you can easily find similarities in the vector field decompositions. Since you have first attempted the field solutions for the incident fields, you just need to replace β_1 to β_2 and θ_i to θ_t to derive the transmitted fields. If you would follow the complete procedure, you can at least check the final results with the comparisons.

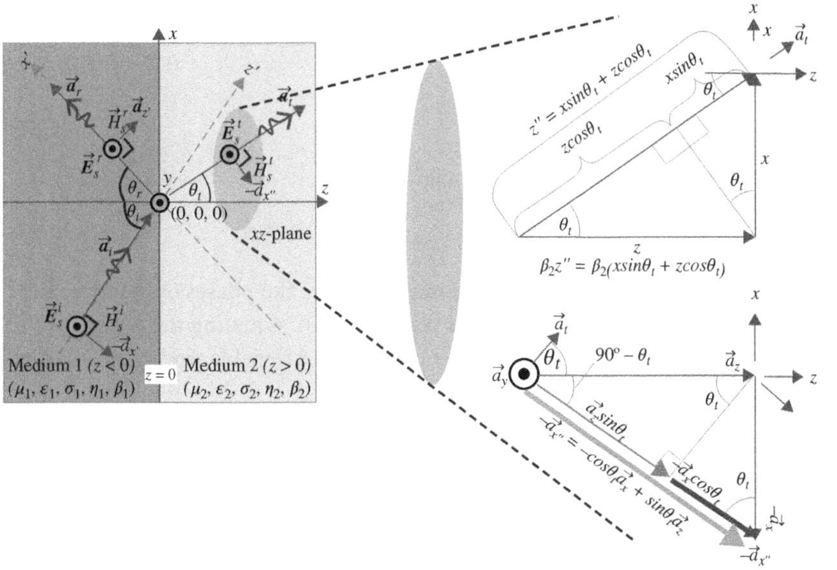

Figure 6.23 Vector decompositions of TE oblique *transmitted* fields on a dielectric boundary.

Figure 6.24 TE polarization oblique incidence field boundary conditions for tangential field components along x-axis.

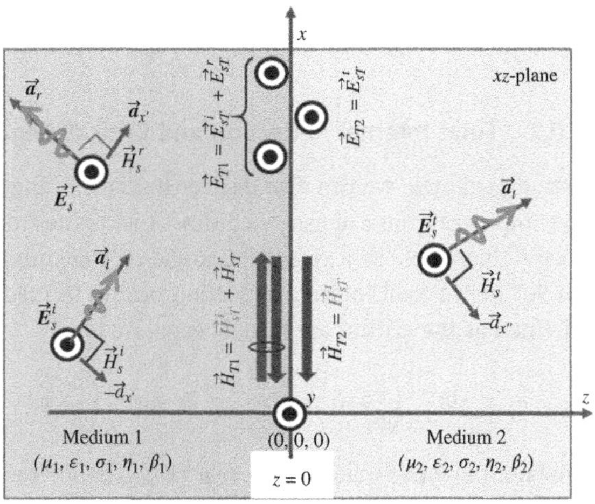

transmitted fields in terms of the incident field amplitudes so that we can calculate the reflection and transmission coefficients for the TE polarized oblique incident fields, respectively. Figure 6.24 illustrates the equality of the three sets of the tangential electromagnetic fields with thick arrows. Applying two boundary conditions for the tangential electric field ($E_{T1} = E_{T2}$) and the tangential magnetic field ($H_{T1} = H_{T2}$) on both sides of the interface at $z = 0$ yields:

$$\boldsymbol{E}^{\rm i}_{T1}|_{z=0} + \boldsymbol{E}^{r}_{T1}|_{z=0} = \boldsymbol{E}^{t}_{T2}|_{z=0} \tag{6.79}$$

$$\boldsymbol{H}^{\rm i}_{T1}|_{z=0} + \boldsymbol{H}^{r}_{T1}|_{z=0} = \boldsymbol{H}^{\rm t}_{T2}|_{z=0} \tag{6.80}$$

Note that the magnetic field boundary condition holds true only if there is no induced electric current at the interface of the two dielectric media at $z = 0$. Now substitutions for (6.69), (6.73) and (6.77) into (6.79), we obtain:

$$E_o^i e^{-j\beta_1(x\sin\theta_i)} + E_o^r e^{-j\beta_1(x\sin\theta_r)} = E_o^t e^{-j\beta_2(x\sin\theta_t)} \tag{6.81}$$

And substitutions for (6.70), (6.74) and (6.78) into (6.80) yield:

$$-\cos\theta_i H_o^i e^{-j\beta_1(x\sin\theta_i)} + \cos\theta_r H_o^r e^{-j\beta_1(x\sin\theta_r)} = -\cos\theta_t H_o^t e^{-j\beta_2(x\sin\theta_t)} \tag{6.82}$$

As derived before also for *Snell's law of reflection*, we must consider the phases of the three field components must be equal for any value of x on the xy-plane at $z = 0$. Equating the phases of the incident and reflected fields in the left hand side of either (6.81) or (6.82) yields *Snell's law of reflection* as in expression (6.83):

$$\sin\theta_i = \sin\theta_r \;\therefore \theta_i = \theta_r. \tag{6.83}$$

Now equating the phases of the incident and transmitted fields in either (6.81) or (6.82) yields *Snell's law of refraction* as in expression in (6.84).

$$\beta_1 \sin\theta_i = \beta_2 \sin\theta_t \;\therefore \theta_i = \sin^{-1}\left(\frac{\beta_2}{\beta_1}\sin\theta_t\right) \quad \text{and}\; \theta_t = \sin^{-1}\left(\frac{\beta_1}{\beta_2}\sin\theta_i\right) \tag{6.84}$$

So far, we have analyzed three sets of the electromagnetic fields, applying the boundary conditions to derive the two laws of Snell. In the following section, we shall examine the total internal reflection in the light of field analysis before deriving the reflection and transmission coefficients.

6.10.2 Total Internal Reflection and Critical Angle θ_c

For completeness, we can also derive the critical angle for the TE polarization assuming that the wave impinges from a denser medium 1 to a lighter medium 2 so that $\beta_1 > \beta_2$. As the angle of incidence θ_i increases, as a result the angle of transmission θ_t increases even faster and reaches to $\theta_t = 90°$ hence total internal reflection occurs in medium 1. The incident angle at this condition is defined as the critical angle as is expressed as:

$$\theta_c = \sin^{-1}\left(\frac{\beta_2}{\beta_1}\sin\theta_t\right)\Big|_{\theta_t = 90°} = \sin^{-1}\left(\frac{\beta_2}{\beta_1}\right). \tag{6.85}$$

For nonmagnetic materials $\mu_1 = \mu_2 = \mu_o$, hence the critical angle reduces to (6.86):

$$\theta_c = \sin^{-1}\sqrt{\frac{\varepsilon_2}{\varepsilon_1}} = \frac{n_2}{n_1}. \tag{6.86}$$

In Section 6.9.1, in the detailed analysis of the ray trace model, we have proven of the exponentially decaying electromagnetic fields along z-axis at the total internal reflection condition that leads to *the evanescent* mode in medium 2 for $\theta_i \geq \theta_c$. Therefore, we are not going to detail this issue here. We also stated the applications of the theory of *critical angle* in optical fibers and dielectric waveguides.

As shown in Figure 6.25a, the photograph illustrates a dollar optical fiber toy with a coin battery and a LED light at the base to launch the light wave via the optical fiber. A bounce of optical fibers is illuminated with the LED that refracts light to the input end of the optical fibers. Since all lights are not at the appropriate acceptance angle naturally some lights are refracted from the base of the

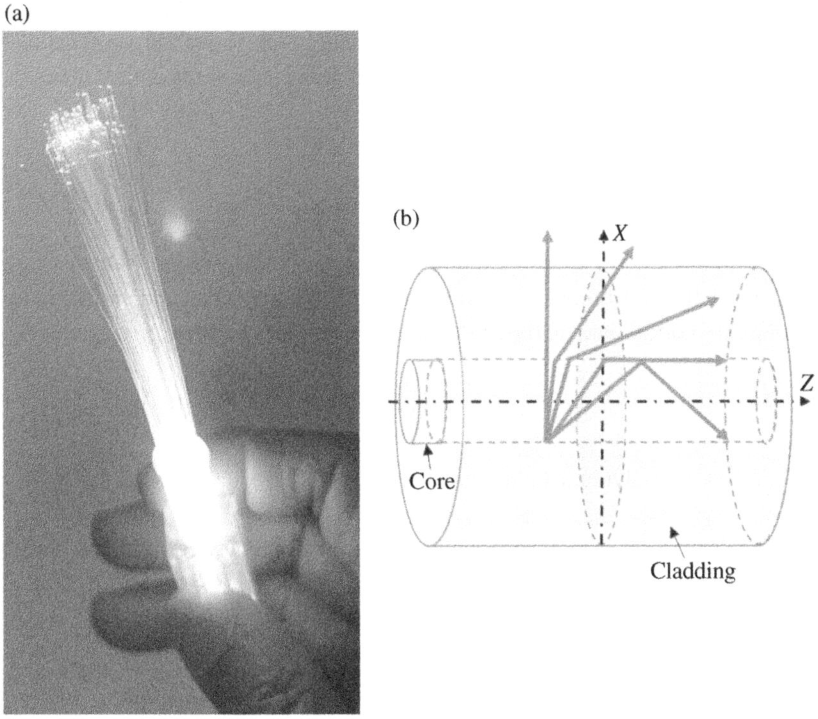

Figure 6.25 Optical fiber in action (a) a dollar optical fiber toy with a coin battery and a LED light at the base illuminates the optical fiber. Although in the bottom there are refraction of lights at the top end it is clearly visible that the light wave is efficiently propagated to the end of the optical fibers. The acceptance (incidence) angle is crucial for the light wave to be internally totally reflected inside the core of the optical waveguide, and (b) the construction of the optical fiber comprises a core made of silica which has radius in the order of a few microns for the single mode and about a decade microns for multimode optical fibers. The cladding has a plastic cover to protect the core and gives stability from bending and breakage. Cladding is many more order larger than the core. As the angle of incidence inside the core increases from the normal incidence (the straight green line parallel to x-axis) to the oblique incidence, once the incidence angle surpasses the critical angle (the green ray glazing along z-axis) the total internal reflection occurs inside the core. The zigzag path of the totally internally reflected rays conveys light energy along the optical fiber length.

optical fiber. Although the bottom parts of the fibers are full of refracted light from the LED, at the top end it is clearly visible that the light wave is guided via the optical fibers. The acceptance (incidence) angle is crucial for the light wave to be internally totally reflected inside the core of the optical fiber. The design of source and coupling with the optical fiber need huge design consideration. Figure 6.25b illustrates the basic configuration and the operating principle of the optical fiber which is based on the theory of the total internal reflection. As shown in the figure, the fiber comprises a core made of silica which has radius in the order of a few microns for the single mode, and about a one to two tens of microns for multimode optical fibers. The cladding has a plastic cover to protect the core and gives stability from bending and breakage. The cladding is optically lighter material, and many orders larger than that of the core. As the angle of incidence inside the core increases from the normal incidence (the straight line parallel to x-axis) to the oblique incidence, the angle of transmission increases in much larger rate than the angle of incidence following *the Snell's law of refraction* $\theta_t = \sin^{-1}((\beta_1/\beta_2)\sin\theta_i)$ for $\beta_1 > \beta_2$. Once the incidence angle surpasses the critical angle (the ray glazing along z-axis) then the total internal reflection occurs inside the core and the evanescent mode fields are created inside the cladding with exponential decay. The zigzag path of the totally internally reflected ray inside the core conveys light energy (also called photons) along the optical fiber length. This is clearly visible as the diffracted light at the ends of the optical fibers in the photograph of Figure 6.25a.

6.10.3 Calculations of Γ_{TE} and τ_{TE}

The derivations of the TE polarization reflection coefficient $\Gamma_{TE} = E_0^r/E_0^i$ and the transmission coefficient $\tau_{TE} = E_0^t/E_0^i$ need the solutions of the amplitudes of the reflected and transmitted electric fields.[36] The boundary conditions yield (6.79) that is true for any value of x. considering $x = 0$ leads to:

$$E_0^i + E_0^r = E_0^t \tag{6.87}$$

All phases associated with the individual field components must be equal to hold the above equality.

By simply dividing both side of (6.87) with E_0^i gives an important relation between Γ_{TE} and τ_{TE} as follows:

$$\frac{E_0^t}{E_0^i} = 1 + \frac{E_0^r}{E_0^i} \qquad \therefore \tau_{TE} = 1 + \Gamma_{TE} \tag{6.88}$$

Similarly substitution for $x = 0$ in (6.82) yields:

$$-\cos\theta_i H_o^i + \cos\theta_r H_o^r = -\cos\theta_t H_o^t \tag{6.89}$$

Now expressing the magnetic fields in terms of the electric fields and associated characteristic impedances in media 1 and 2, we obtain:

$$-\cos\theta_i \frac{E_o^i}{\eta_1} + \cos\theta_r \frac{E_o^r}{\eta_1} = -\cos\theta_t \frac{E_o^t}{\eta_2} \tag{6.90}$$

Substitution for Snell's law of reflection $\theta_i = \theta_r$ from (6.83) into (6.90) and then dividing both sides with $\cos\theta_i/\eta_1$ yield:

$$E_o^i - E_o^r = \frac{\eta_1}{\eta_2} \frac{\cos\theta_t}{\cos\theta_i} E_o^t \tag{6.91}$$

Now it is matter of simple mathematical manipulation to arrive to the solutions for Γ_{TE} and τ_{TE}. Adding (6.87) and (6.91) yields

$$2E_o^i = \left(1 + \frac{\eta_1}{\eta_2} \frac{\cos\theta_t}{\cos\theta_i}\right) E_0^t \tag{6.92}$$

Now simple exchange and division result in the transmission coefficient as follows:

$$\tau_{TE} = \frac{E_o^t}{E_o^i} = \frac{2\eta_2 \cos\theta_i}{\eta_2 \cos\theta_i + \eta_1 \cos\theta_t} \tag{6.93}$$

---

36 The equations relating the amplitudes of the incident and the transmitted field waves are known as Fresnel formulas. A.J. Fresnel derived these formulas for the first time on the basis of his elastic theory of light in 1823.

Now to find the reflection coefficient we can use either (6.88) and (6.93) or we use the similar mathematical manipulation to derive τ_{TE}, but in this case we subtract (6.87) from (6.91). Whichever way we follow we land at the same solution as expressed in (6.94):

$$\Gamma_{TE} = \frac{E_o^r}{E_o^i} = \frac{\eta_2 \cos\theta_i - \eta_1 \cos\theta_t}{\eta_2 \cos\theta_i + \eta_1 \cos\theta_t} \tag{6.94}$$

Substitution for (6.94) into (6.93) yields an important relation between the TE polarized reflection and transmission coefficients as in (6.95):

$$\tau_{TE} = 1 + \Gamma_{TE} \tag{6.95}$$

So far, we have derived many important concepts about the TE polarized oblique incidence fields. Now we shall go through a few review question followed by a few worked out a few examples.

Review Questions 6.7: TE-polairzed Oblique Incidence

Q1. In derivations of transmission and reflection coefficients why do we analyze TE and TM polarization oblique incidence cases separately?
Q2. What is *Brewster angle*? Is it applicable to TE polarization?
Q3. Explain the rules of the oblique incidence cases to follow and derive the transmission and reflection coefficients.
Q4. Explain the steps used in calculating the field components and the transmission and reflection coefficients in oblique incidence cases.
Q5. In the vector decompositions of the incident, reflected and transmitted fields, why do we need to consider both unit vector and propagation length calculations separately?
Q6. Explain the easy way to calculate the transmitted electromagnetic field in medium 2 if the incident electromagnetic field solutions are known in medium 1.
Q7. Explain how the light wave is guided in an optical fiber.
Q8. Derive Γ_{TE} and τ_{TE} from the field expressions and boundary conditions.

Example 6.7: Field Analysis of TE Polarization: Air–Glass Interface
A uniform plane wave at 100 MHz operating frequency incidences upon a glass medium ($n_1 = 1.45$) from air as medium 1. The direction of propagation is along z-axis. The electric field is TE polarized and amplitude of 150 V/m. The incidence angle is 30°. Calculate the followings: (i) the angle of refraction, (ii) the reflection and transmission coefficients, and then the phasor vector and time domain expressions for (iii) the incident electric and magnetic fields, (iv) the reflected electric and magnetic fields, and (v) the transmitted electric and magnetic fields.

Hints: use the coordinate decomposition to calculate the field expressions.

Answer:
At 100 MHz: the angular frequency: $\omega = 628.3 \times 10^6$ (rad/s)

The phase constant in medium 1: $\beta_1 = \omega/c_1 = (2\pi \times 10^8)/(3 \times 10^8) = 2.09$ (rad/m)

The phase constant in medium 2: $\beta_2 = \omega/c_2 = (2\pi \times 10^8 \times 1.45)/(3 \times 10^8) = 3.04$ (rad/m)

The wave impedances in air medium 1: $\eta_1 = 377\ \Omega$; and in glass medium 2: $\eta_2 = 377/1.45 = 260\ \Omega$.

We have all ingredients to solve the problems as follows:

i) Based on Snell's law of reflection $\theta_i = \theta_r$; hence the angle of reflection is: $\theta_r = 30°$.
Based on Snell's law of refraction (6.84): $\theta_t = sin^{-1}((\beta_1/\beta_2)\, sin\,\theta_i) = sin^{-1}(sin 30° / 1.45) = 20.17°$

ii) Substitutions for η_1, η_2, θ_i, and θ_t into (6.94) yield the reflection coefficient:

$$\Gamma_{TE} = \frac{\eta_2 \cos\theta_i - \eta_1 \cos\theta_t}{\eta_2 \cos\theta_i + \eta_1 \cos\theta_t} = \frac{260 \times \cos 30^\circ - 377 \times \cos 20.17^\circ}{260 \times \cos 30^\circ + 377 \times \cos 20.17^\circ} = -0.22$$

Substitution for η_1, η_2, θ_i, and θ_t into (6.94) yield the transmission coefficient:

$$\tau_{TE} = \frac{2\eta_2 \cos\theta_i}{\eta_2 \cos\theta_i + \eta_1 \cos\theta_t} = \frac{2 \times 260 \times \cos 30^\circ}{260 \times \cos 30^\circ + 377 \times \cos 20.17^\circ} = 0.78$$

The negative sign of the reflection coefficient represents the phase reversal of the reflected wave at the interface of the two media.[37]

iii) The expression for incident electric field (6.67):

$$\boldsymbol{E}_s^i(x, z) = \boldsymbol{a_y} E_o^i e^{-j\beta_1(x\sin\theta_i + z\cos\theta_i)} = \boldsymbol{a_y} 150 e^{j2.09(x\sin 30^\circ + z\cos 30^\circ)} = \boldsymbol{a_y} 150 e^{-j(1.05x + 1.81z)}\,(\text{V/m})$$

The time domain expression for the electric field is:

$$\boldsymbol{E^i}(x, z, t) = 150\cos\left(628.3 \times 10^6 t - 1.05x - 1.81z\right)\boldsymbol{a_y}\ (\text{V/m})$$

The incident magnetic field (6.68):

$$\boldsymbol{H}_s^i(x, z) = \left(-\cos\theta_i \boldsymbol{a_x} + \sin\theta_i \boldsymbol{a_z}\right)\frac{E_0^i}{\eta_1} e^{-j\beta_1(x\sin\theta_i + z\cos\theta_i)}$$

$$= \left(-\cos 30^\circ \boldsymbol{a_x} + \sin 30^\circ \boldsymbol{a_z}\right)\left(\frac{150}{377}\right) e^{-j(1.05x + 1.81z)}$$

$$= \left(-0.344\boldsymbol{a_x} + 0.2\boldsymbol{a_z}\right) e^{-j(1.05x + 1.81z)}\,(\text{A/m})$$

The time domain expression for the magnetic field is:

$$\boldsymbol{H^i}(x, z, t) = \left(-0.344\boldsymbol{a_x} + 0.2\boldsymbol{a_z}\right)\cos\left(628.3 \times 10^6 t - 1.05x - 1.81z\right)\ (\text{A/m})$$

iv) Now we calculate the reflected electromagnetic fields. From (6.71) the expression for the reflected electric field is:

$$\boldsymbol{E}_s^r(x, z) = \boldsymbol{a_y}\Gamma_{TE} E_0^i e^{-jk_1(x\sin\theta_r - z\cos\theta_r)} = -0.22 \times 150\,\boldsymbol{a_y} e^{-j(1.05x - 1.81z)} = -33\,\boldsymbol{a_y} e^{-j(1.05x - 1.81z)}\,(\text{V/m})$$

The time domain expression for the reflected electric field is:

$$\boldsymbol{E^r}(x, z, t) = -33\cos\left(628.3 \times 10^6 t - 1.05x - 1.81z\right)\boldsymbol{a_y}$$

$$= 33\cos\left(628.3 \times 10^6 t - 1.05x + 1.81z - 180^\circ\right)\boldsymbol{a_y}(\text{V/m})$$

From (6.72) the reflected magnetic field is defined as:

$$\boldsymbol{H}_s^r(x, z) = \left(\cos\theta_r \boldsymbol{a_x} + \sin\theta_r \boldsymbol{a_z}\right)\frac{E_0^r}{\eta_1} e^{-j\beta_1(x\sin\theta_r - z\cos\theta_r)}$$

$$= \left(\cos 30^\circ \boldsymbol{a_x} + \sin 30^\circ \boldsymbol{a_z}\right)\left(\frac{-33}{377}\right) e^{-j(1.05x + 1.81z)}$$

$$= -\left(75.8 \times 10^{-3}\boldsymbol{a_x} + 43.8 \times 10^{-3}\boldsymbol{a_z}\right) e^{-j(1.05x + 1.81z)}\,(\text{A/m})$$

The time domain expression for the reflected magnetic field is:

$$\boldsymbol{H^r}(x, z, t) = \left(75.8 \times 10^{-3}\boldsymbol{a_x} + 43.8 \times 10^{-3}\boldsymbol{a_z}\right)\cos\left(628.3 \times 10^6 t - 1.05x + 1.81z - 180^\circ\right)\ (\text{A/m})$$

---

37 You can arrive to the same transmission coefficient results using (6.95).

v) The expression for transmitted electric field (6.75):

$$E_s^t = a_y \tau_{TE} E_o^i e^{-j\beta_2(x\sin\theta_t + z\cos\theta_t)} = a_y(0.78 \times 150)e^{j3.04(x\sin 20.17° + z\cos 20.17°)}$$
$$= a_y 117 e^{-j(1.05x + 2.85z)} \text{(V/m)}$$

The time domain expression for the electric field is:

$$E^t(x, z, t) = 117\cos\left(628.3 \times 10^6 t - 1.05x - 2.85z\right) a_y \text{ (V/m)}$$

The transmitted magnetic field is defined as:

$$H_s^t(x, z) = (-\cos\theta_t a_x + \sin\theta_t a_z)\frac{E_0^t}{\eta_2} e^{-j\beta_2(x\sin\theta_t + z\cos\theta_t)}$$
$$= (-\cos 20.17° a_x + \sin 20.17° a_z)\left(\frac{150}{260}\right) e^{-j(1.05x + 2.85z)}$$
$$= (-0.54 a_x + 0.2 a_z) e^{-j(1.05x + 2.85z)} \text{(A/m)}$$

The time domain expression for the magnetic field is:

$$H^i(x, z, t) = (-0.54 a_x + 0.2 a_z)\cos\left(628.3 \times 10^6 t - 1.05x - 2.85z\right) \text{ (A/m)}$$

6.10.4 Effective Impedance Concept of TE Polarized Oblique Incidence

So far, we have used the ray trace model in Section 6.8.1 and examined many interesting phenomena of oblique incidence.[38] Next, we have examined the field analysis in Section 6.10 to derive the reflection coefficient Γ_{TE} and the transmission coefficient τ_{TE} for the specific polarization. We also know that these two coefficients are polarization dependent. In this section, we shall develop *the impedance concept* of TE oblique incidence theory. We shall use (6.90) to develop the impedance concept and give new definitions of effective impedances for the media for a particular polarization[39] as follows:

$$\frac{E_o^i}{\eta_{1TE}} - \frac{E_o^r}{\eta_{1TE}} = \frac{E_o^t}{\eta_{2TE}} \tag{6.96}$$

Here we have defined new impedances in terms of the intrinsic impedances of the two media, for $E_0^i/H_0^i = -\left(E_0^r/H_0^r\right) = \eta_1$ and $E_0^2/H_0^2 = \eta_2$, which are assumed to be real and positive quantities, for the TE oblique incidence case:

$$\eta_{1TE} = \frac{\eta_1}{\cos\theta_i} \quad \text{and} \quad \eta_{2TE} = \frac{\eta_2}{\cos\theta_t} \tag{6.97}$$

Here, the subscripts 1 and 2 represent media 1 and 2 respectively and *TE* represents TE or perpendicular polarization. As expected, the TE effective impedances are larger the intrinsic

---

38 W. H. Hayt and J. A. Buck, *Engineering Electromagnetics*, McGraw-Hill, Boston, 7 2006

39 In the end of the development of the ray trace model we have suggested that the impedances, reflection and transmission coefficients are polarization dependent. In this section we shall develop the theory of the effective impedances of both media for TE polarization. During the theory of TM polarization, we shall develop the theory of the effective impedances of both media for TM polarization.

impedances of the media. Using the definitions of the new effective impedances in (6.97) into (6.93) and (6.94), respectively, we obtain the new expressions for the transmission and reflection coefficients in terms of the new impedances η_{1TE} and η_{2TE} which are also called *the effective polarization impedances* as follows:

$$\tau_{TE} = \frac{E_o^t}{E_o^i} = \frac{2\eta_{2TE}}{\eta_{2TE} + \eta_{1TE}} \tag{6.98}$$

$$\Gamma_{TE} = \frac{E_o^r}{E_o^i} = \frac{\eta_{2TE} - \eta_{1TE}}{\eta_{2TE} + \eta_{1TE}} \tag{6.99}$$

These expressions for τ_{TE} and Γ_{TE} in terms of η_{1TE} and η_{2TE} effective polarization impedances are much easier to recall and have similarities with the expressions of the reflection and transmission coefficients for the normal incidence case.[40] The relationship of (6.95) still holds true.

$$\tau_{TE} = 1 + \Gamma_{TE} \tag{6.95}$$

In the above derivations, we assume the effective impedances are real and positive quantities. In the next section, we shall examine the effective impedance η_{2TE} of medium 2 and find that it becomes a complex quantity in the case for the total internal reflection and the evanescent mode propagation in the medium.

6.10.5 Total Internal Reflection in the Light of Impedance Concept

We have examined the total internal reflection using the ray trace model in Section 6.9 and the field analysis in Section 6.10.2, respectively. In this section, we shall examine the same theory of the total internal reflection using *the effective impedance concept*. We know that the internal reflection occurs in special combination of media when a signal propagates from denser to lighter media ($n_1 > n_2$) and at the incident angle exceeding *the critical angle* ($\theta_i > \theta_c$). The ray trace model also suggests that it occurs in both TE and TM polarizations. Mathematically, the conditions for total internal reflection is such that the total power reflection coefficient $|\Gamma_{TE}|^2 = \Gamma_{TE}\Gamma_{TE}^* = 1$ [4]. Here, the voltage reflection coefficient Γ_{TE} is for the TE polarization oblique incidence.[41] This suggests that the reflection coefficient can be complex and involves the complex impedance for medium 2. As we have found in Section 6.9 that when the total internal reflection occurs, the condition $\theta_i > \theta_c$ brings the solutions for θ_t complex as in (6.61). According to (6.97) the effective impedance η_{2TE} has $cos\theta_t$ component hence the definition of η_{2TE} suggests a complex impedance that makes the reflection coefficient Γ_{TE} complex at the total internal reflection. Hence replacing η_{2TE} with $j\,|\,\eta_{2TE}|$ in (6.99), we obtain a new definition for the complex Γ_{TE} as follows:

$$\Gamma_{TE} = \frac{j\,|\,\eta_{2TE}\,| - \eta_{1TE}}{j\,|\,\eta_{2TE}\,| + \eta_{1TE}} = -\frac{\eta_{1TE} - j\,|\,\eta_{2TE}\,|}{j\,|\,\eta_{2TE}\,| + \eta_{1TE}} = -\frac{\zeta_{TE}}{\zeta_{TE}^*} \tag{6.100}$$

Where, a complex impedance $\zeta_{TE} = \eta_{1TE} - j\,|\,\eta_{2TE}|$ is used. Therefore, the total power reflection coefficient in the total internal reflection condition ($\theta_i \geq \theta_c$) is defined as:

$$|\Gamma_{TE}|^2 = \Gamma_{TE}\Gamma_{TE}^* = -\frac{\zeta_{TE}}{\zeta_{TE}^*} \times \left(-\frac{\zeta_{TE}^*}{\zeta_{TE}}\right) = 1 \tag{6.101}$$

40 The ray trace model advocates that the critical angle and total internal reflection are polarization independent. We shall perform similar calculations for the TM polarization and shall observe that the reflection and transmission coefficients can be expressed in a similar wave with the TM polarization impedances.

41 We shall discuss the internal reflection and critical angle for TM polarization in TM oblique incidence theory.

We have highlighted many examples of total internal reflection specifically applications to beam steering prism, optical waveguides and optical fibers. Besides this total internal reflection is also used for other areas of wireless communications. In the followings, we shall do a few worked out examples.

Example 6.8: Field Analysis - TE Polarization Oblique Incidence
A uniform plane wave from a 2.45 GHz Wi-Fi router on the ceiling of a room impinges on a glass window ($n_2 = 1.45$) from air as medium 1. The direction of propagation is along z-axis. The electric field is TE polarized and amplitude of 25 V/m. The incidence angle is 50°. Calculate the followings: (i) the angle of transmission, (ii) the effective impedances η_{1TE} and η_{2TE}, (iii) the reflection and transmission coefficients, and (iv) the fraction of power reflected in air and transmitted in glass.

Answer:
Given parameters: the operating frequency $f = 2.45 \times 10^9$ (Hz); and the angular frequency $\omega = 2\pi f = 1.54 \times 10^{10}$ (rad/m)

The refractive index of medium 1: $n_1 = 1$; and medium 2 glass is: $n_2 = 1.45$.

The amplitude of the incident electric field: $E_0^i = 25$ (V/m); the angle of incidence: $\theta_i = 50°$.

i) According to (6.57), the angle of transmission: $sin\,\theta_t = sin\,\theta_i \times (n_1/n_2) = sin\,(50°) \times 1/1.45 = 0.528$
   The angle of transmission $\theta_t = sin^{-1}\,(0.528) = 31.89°$
ii) The effective impedances: $\eta_{1TE} = \eta_1/\,cos\,\theta_i = 377/(cos\,(50°)) = 586.5\ (\Omega)$

$$\eta_{2TE} = \frac{\eta_1}{cos\,\theta_t} = \frac{377/1.45}{cos\,(31.89°)} = 306.22\ (\Omega)$$

iii) According to (6.99), the reflection coefficient: $\Gamma_{TE} = \dfrac{\eta_{2TE} - \eta_{1TE}}{\eta_{2TE} + \eta_{1TE}} = \dfrac{306.22 - 586.5}{306.22 + 586.5} = -0.31$
   The negative sign of Γ_{TE} suggests a phase reversal of the reflected field at the air–glass interface.
   According to (6.98), the transmission coefficient: $\tau_{TE} = (2\eta_{2TE})/(\eta_{2TE} + \eta_{1TE}) = (2 \times 306.22)/(306.22 + 586.5) = 0.69$
   We can arrive at the same results for τ_{TE} using (6.95)
iv) The fraction of reflected power: $|\Gamma_{TE}|^2 = |0.31|^2 = 0.0961 = 9.61\%$
   The fraction of transmitted power: $1 - |\Gamma_{TE}|^2 = 1 - 0.0961 = 0.904 = 90.4\%$
   More than 90% incident power is transmitted via the glass window.[42]

6.10.6 Special Cases of Γ_{TE}

6.10.6.1 Reflection Coefficient Γ_{TE} for Perfect Conductor

We have examined the field analysis for TE polarized oblique incidence in Section 6.5. However, we have not derived the reflection coefficients though we developed the relation between the incident and reflected electric fields in (6.20). Now we are going to derive the reflection coefficient for medium 2 as the conductor using (6.94) for conducting medium $\eta_2 = 0$. Hence (6.94) reduces to

42 You may observe so many signals from routers from your neighbors. To provide privacy frequency selective surface can be layered on the widow to protect the privacy. The frequency selective surface is made of planar metamaterials that offers notches to the particular signal frequency and prevents transmission.

$$\Gamma_{TE} = \frac{E_o^r}{E_o^i} = \frac{0 - \eta_1 \cos\theta_t}{0 + \eta_1 \cos\theta_t} = -1 \tag{6.102}$$

This is the result that we expect for a perfect conducting boundary that tells that the reflection coefficient Γ_{TE} for the TE polarized oblique incidence case is unity with a phase reversal of 180°. This is equivalent to a short-circuited termination for a transmission line.

6.10.6.2 Both Medium Lossless and Non-magnetic Media

If both media are lossless and non-magnetic ($\mu_1 = \mu_2 = \mu_0$), applying Snell's law of refraction in (6.84) and the definition of Γ_{TE} in (6.94) we obtain

$$\begin{aligned}\Gamma_{TE} &= \frac{\cos\theta_i - (\eta_1/\eta_2)\cos\theta_t}{\cos\theta_i + (\eta_1/\eta_2)\cos\theta_t} = \frac{\cos\theta_i - \sqrt{(\varepsilon_2/\varepsilon_1)}\sqrt{1 - \sin^2\theta_t}}{\cos\theta_i + \sqrt{(\varepsilon_2/\varepsilon_1)}\sqrt{1 - \sin^2\theta_t}} \\ &= \frac{\cos\theta_i - \sqrt{(\varepsilon_2/\varepsilon_1) - (\varepsilon_2/\varepsilon_1) \times (\varepsilon_1/\varepsilon_2)\sin^2\theta_i}}{\cos\theta_i + \sqrt{(\varepsilon_2/\varepsilon_1) - (\varepsilon_2/\varepsilon_1) \times (\varepsilon_1/\varepsilon_2)\sin^2\theta_i}} \\ &= \frac{\cos\theta_i - \sqrt{(\varepsilon_2/\varepsilon_1) - \sin^2\theta_i}}{\cos\theta_i + \sqrt{(\varepsilon_2/\varepsilon_1) - \sin^2\theta_i}}\end{aligned} \tag{6.103}$$

Equation (6.103) reveals very interesting derivation. If the wave travels from electromagnetically lighter medium (ε_1) to a denser medium (ε_2); $\varepsilon_2 > \varepsilon_1$ then $\sqrt{(\varepsilon_2/\varepsilon_1) - \sin^2\theta_i}$ becomes positive for any value of the angle of incidence θ_i and Γ_{TE} becomes real. This means the rays travel from optically lighter medium to denser medium and does not provide any internal reflection. The above example on the wave propagation from air to glass is a case for $\varepsilon_2 > \varepsilon_1$. If $\varepsilon_1 > \varepsilon_2$ this means the rays travel from optically denser medium to lighter medium and then $\sqrt{(\varepsilon_2/\varepsilon_1) - \sin^2\theta_i}$ becomes imaginary for $\theta_i > \theta_c$, where θ_c is *the critical angle*. Hence Γ_{TE} becomes complex when rays travel from an optically denser medium to a lighter medium. In this case, the modulus of the reflection coefficient $|\Gamma_{TE}| = 1$ and *total internal reflection* occurs.

6.10.6.3 Critical Angle and Submarine Communications

The theory of *total internal reflection* and *the critical angle* is important for design of submarine wireless communications system because the signal travels from electromagnetic denser medium water to the lighter medium air. There we need to precisely determine the incidence angle below the critical angle so that the total internal reflection can be avoided.

Figure 6.26 illustrates a surface missile radar system. As can be seen there a small window of radius R through which the signal only can transmit from water to air. The radius is defined with the critical angle θ_c and the distance of the transmitter from the boundary. Usually, the acceptance angle over which the effective communication may occur is around 70% of the critical angle. Therefore, we can define the acceptance angle as:

$$\theta_a = 0.7\theta_c = 0.7 \times \sin^{-1}\left(\sqrt{\frac{\varepsilon_{r2}}{\varepsilon_{r1}}}\right) \tag{6.104}$$

If the distance of the transmitter from the boundary in medium 1 is h, then the radius of coverage area can be defined with simple trigonometry:

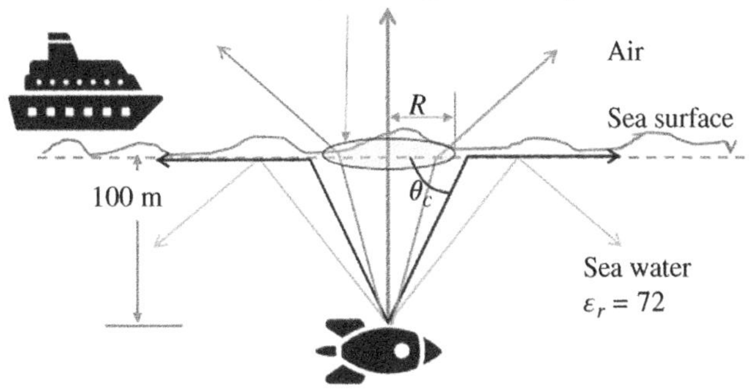

Figure 6.26 Submarine to surface missile radar system at 100 MHz.

$$R = \frac{h}{tan\,(\theta_a)} \tag{6.105}$$

Then the effective coverage area for submarine wireless communications is defined as:

$$A = \pi R^2 = \frac{\pi h^2}{tan^2(\theta_a)} \tag{6.106}$$

Note that since we use only the ray trace model and Snell's law of refraction, the theory is equally applicable to the TM polarized oblique incidence cases. Let us do a numerical example for the submarine wireless communications as follows.

Example 6.9: Total Internal Reflection: Submarine Wireless Communications
A submarine to surface missile radar system operating at 100 MHz is illustrated in Figure (6.25). The depth of the submarine is 100 m from the surface of the seawater. The communication happens from the denser medium to the lighter medium in air. Assume that the effective coverage area of effective communications and guidance is bounded by 70% of the critical angle θ_c. We are going to calculate the coverage area in (m^2) of the missile system for 100 MHz radar. In this case we assume the dielectric constant of the sea water is $\varepsilon_{r2} = 72$.

The critical angle: $\theta_c = sin^{-1}\left(\sqrt{1/72}\right) = 6.77°$

As shown in the figure acceptance angle for efficient communications is: $\theta_a = 0.7 \times 6.77\,° = 4.73°$

The radius of coverage as shown by the circle: $R = 100\text{ m}/tan\,(4.73°) = 1.208\text{ km}$

Coverage area: $A = \pi \times (1.208\text{ km})^2 = 4.59\text{ km}^2$. Therefore, the effective communications and guidance of the radar is confined within 4.59 km^2 above the submarine.

6.10.6.4 TE Oblique Incidence on Multiple Dielectric Layers

So far, we have examined the TE oblique incidence case for only two dielectric media. We can expand the theory for any number of media as we have done for the normal incidence cases in Section 5.8.5. In this section, we shall prove that *the entrance angle is equal to the exit angle* in a multilayered transmission medium. We shall calculate the total displacement of the ray when the ray comes out of the last medium to air. Figure 6.27 shows the transmission of the ray via

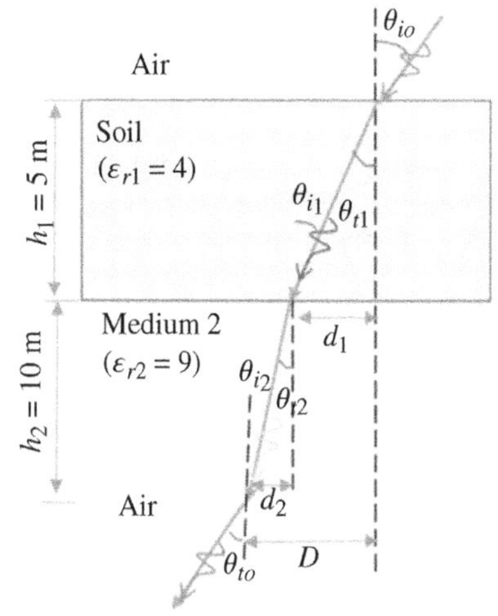

Figure 6.27 TE oblique incidence of obliquely incident fields: Air–soil–dielectric–air transmission.

multilayered media. We shall prove that for the multilayer reflection from air to the two media, the angle of incidence is equal to the angle of transmission $\theta_{io} = \theta_{to}$. This means *the entrance angle is equal to the exit angle*. Since we shall apply the ray trace model with the aid of Snell's law of refraction, the theory is equally applicable to TM polarized incidence cases. Applying Snell's law of refraction at the interfaces:

$$\beta_o \sin\theta_{oi} = \beta_1 \sin\theta_{t1} \tag{6.107}$$

$$\beta_1 \sin\theta_{i1} = \beta_2 \sin\theta_{t2} \tag{6.108}$$

$$\beta_2 \sin\theta_{i2} = \beta_o \sin\theta_{to} \tag{6.109}$$

Since the ray follows a straight line of transmission in individual layers and hits the next right boundary, from angular symmetries on various rays:

$$\theta_{t1} = \theta_{i1};\ \theta_{i2} = \theta_{t2} \tag{6.110}$$

Substitutions for these equalities in (6.110) into (6.107), (6.108), and (6.109) consecutively yield:

$$\beta_o \sin\theta_{io} = \beta_o \sin\theta_{to} \tag{6.111}$$

$$\therefore \theta_{io} = \theta_{to} \tag{6.112}$$

This means the entrance angle is equal to the exit angle. Let us do a numerical example:

Example 6.10: Transmission Through Multilayered Medium
As shown in Figure 6.27, Medium 1 has $\varepsilon_{r1} = 4$ and height $h_1 = 5$ m, Medium 2 has $\varepsilon_{r2} = 9$ and height $h_2 = 10$ m, calculate the total drift D of ray at $\theta_{io} = 45°$. Applying Snell's law of refraction at the interface of air and Medium 1 yields:

$$\therefore\ \theta_{t1} = sin^{-1}\left(\frac{\beta_o}{\beta_1} sin\,\theta_{io}\right) = sin^{-1}\left(\frac{1}{\sqrt{\varepsilon_{r1}}} sin\,\theta_{io}\right) = sin^{-1}\left(\frac{1}{\sqrt{4}} sin\,45^\circ\right)$$
$$= 20.7^\circ \therefore\ d_1 = h_1\, sin\,(20.7^\circ) = 5\,\text{m} \times sin\,(20.7^\circ) = 1.77\,\text{m}$$

Applying Snell's law of refraction at the interface of Media 1 and 2 yields:

$$\theta_{t2} = sin^{-1}\left(\frac{\beta_1}{\beta_2} sin\,\theta_{i1}\right) = sin^{-1}\left(\sqrt{\frac{4}{9}} sin\,(20.7^\circ)\right) = 13.7^\circ$$

The depth $d_2 = h_2\, sin\,(13.7^\circ) = 10\text{ m} \times sin\,(13.7^\circ) = 2.36\text{ m}$

$$\text{Total drift} : D = d_1 + d_2 = 4.13\,\text{m}.$$

Therefore, the ray drifts by 4.13 m after it comes out of the multilayered medium.

So far, we have done the field analyses of the TE polarized oblique incidence in special cases. Now we are going to analyze the power balance calculation in the TE oblique incidence cases.

Remarks

Since we are solely using the ray trace model to calculate the drift after three interfaces this theory is equally valid for TM polarization case.

6.10.7 Power Balance in TE Oblique Incidence

The motto of the section is to first define the power reflection and power transmission coefficients and then find the relationship between the two coefficients with the help of the law of conservation of energy. In an electromagnetic system with boundary value problems, we are more interested in the power incident on a boundary and resulting transmitted and reflected power due to the dissimilar dielectric medium and their resulting boundary conditions. This power balance tells about the efficiency of the electromagnetic system either it could be a wave propagation and bouncing back from a discontinuity, or a terminated transmission line reflecting its matching condition. Or an antenna for its radiation efficiency due to its input impedance mismatch and inherent losses inside the antenna system. The analysis of the oblique incidence is an ideal theory to explain all these generic losses due to reflection. The power definition of the reflection and transmission coefficients links with the power ratios of the reflected and transmitted power with respect to the incident power, respectively. On the other hand there is an energy conservation issue as in reality we cannot create any new energy or sink an existing energy. Hence, the analysis of the power balance in the TE oblique incidence cases, and likewise for TM cases that we shall examine in the next section, is a twofold analysis as shown in Figure 6.28. The first layer comes with the definitions of the power reflection and transmission coefficients, which are the ratios of the reflected power portion versus the incident power, and the transmitted power portion versus the incident power, respectively. Next, we move to relate the two power coefficients with the help of the law of conservation of energy. The law tells us that the summation of the power reflection and transmission must be equal to the incident power.

Now consider the case of the ray trace model in Figure 6.16 in Section 6.8.1, the incident, reflected, and transmitted waves are considered to be confined in a surface area on xy – plane at $z = 0$. This surface area is generated by OB as shown in Figure 6.29. The surface area has a normal vector along z-direction $\boldsymbol{a_z}$. Note that $\boldsymbol{a_z}$ changes its sign depending on the direction of the wave

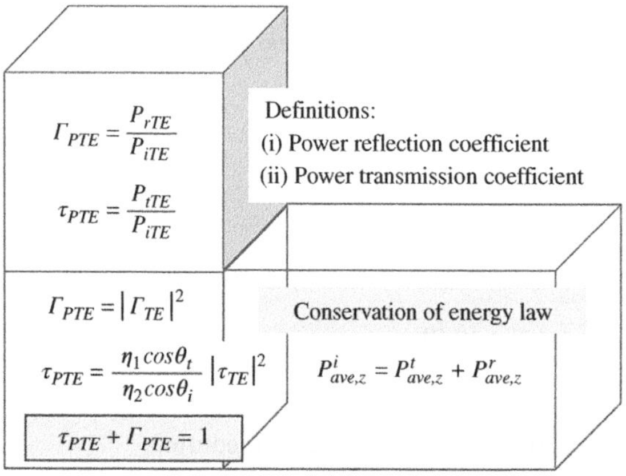

Figure 6.28 Two fold definitions of power balance analysis of TE incidence case. In the first layer of energy balance equation, the definition of power reflection and transmission coefficients are the ratios of the reflected power portion and transmitted power portion with respect to the incident power. In the second layer the law of conservation of energy tells that the power reflection and transmission coefficients are unity which represents the normalized incident power.

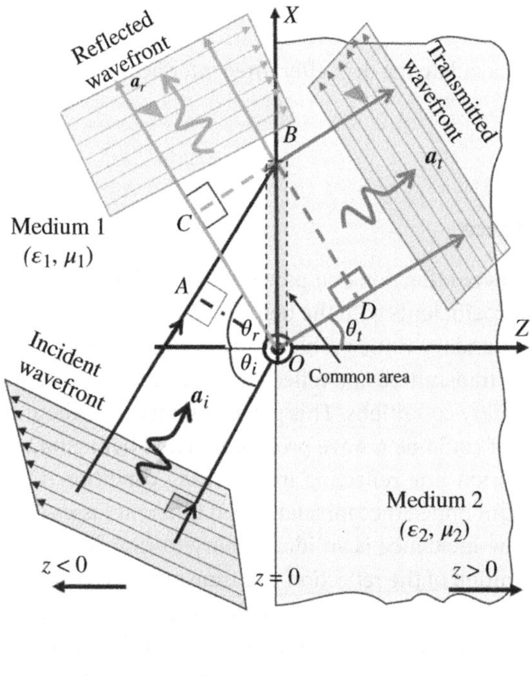

Figure 6.29 Obliquely incident transmitted, reflected, and transmitted wavefronts impinges at a common area confined with line OB on $z = 0$ plane.

propagation of the incident, reflected, and transmitted waves. At OB, the x-directed phase velocities v_{pxi}, v_{pxr} and v_{pxt}, for incident, transmitted and reflected rays, respectively, are the same.

For each of the three waves, the time average power densities propagate towards their respective directions $\boldsymbol{a}_i$, $\boldsymbol{a}_r$, and $\boldsymbol{a}_t$ are defined by $|E_0^i|^2/\eta_1$, $|E_0^r|^2/\eta_1$ and $|E_0^t|^2/\eta_2$, respectively. Therefore, the power densities striking, reflecting, and emanating out of the surface area created by OB as shown with dotted thin rectangle in Figure 6.29 at $z = 0$ can be defined with the three time average power densities as follows:

$$\text{The incident power density on } OB\text{: } \boldsymbol{P}^i_{ave,z} = \left(|E_0^i|^2\right)/\eta_1 \boldsymbol{a_i} \cdot \boldsymbol{a_z} = \left(|E_0^i|^2\right)/\eta_1 \cos\theta_i \ (\text{W/m}^2) \tag{6.113}$$

The reflected power density on OB: $P^r_{ave,z} = \frac{|E^r_0|^2}{\eta_1}\boldsymbol{a}_r.(-\boldsymbol{a}_z) = \frac{|E^r_0|^2}{\eta_1}\cos\theta_r\ (\text{W/m}^2)$ (6.114)

The transmitted power density on OB: $P^t_{ave,z} = \frac{|E^t_0|^2}{\eta_1}\boldsymbol{a}_t.\boldsymbol{a}_z = \frac{|E^t_0|^2}{\eta_1}\cos\theta_t\ (\text{W/m}^2)$ (6.115)

As can be seen from the above three expressions,[43] we only consider z-directed power components which are normal to interface of two media. By the law of conservation of energy, *the z-directed incident power density* must be equal to the summation of *the reflected* and *transmitted power densities* as follows:

$$P^i_{ave,z} = P^t_{ave,z} + P^r_{ave,z} \tag{6.116}$$

Substitutions for (6.113)–(6.115) into (6.116) lead to (6.117) as follows:

$$\frac{|E^i_0|^2}{2\eta_1}\cos\theta_i = \frac{|E^r_0|^2}{2\eta_1}\cos\theta_r + \frac{|E^t_0|^2}{2\eta_2}\cos\theta_t \tag{6.117}$$

Now we can define the power reflection coefficient as $\Gamma_{PTE} = P^r_{ave,z}/P^i_{ave,z}$ in terms of their respective field components as usual:

$$\text{Power reflection coefficient: } \Gamma_{PTE} = \frac{P^r_{ave,z}}{P^i_{ave,z}} = \frac{\left(|E^r_0|^2/2\eta_1\right)\cos\theta_r}{\left(|E^i_0|^2/2\eta_1\right)\cos\theta_i} \tag{6.118}$$

Using Snell's law of reflection $\theta_i = \theta_r$ and the definition of voltage reflection coefficient $\Gamma_{TE} = E^r_o/E^i_o$, we can express the power reflection coefficient in term of the voltage reflection coefficient as follows:

$$\text{Power reflection coefficient: } \Gamma_{PTE} = \frac{|E^r_0|^2}{|E^i_0|^2} = |\Gamma_{TE}|^2 \tag{6.119}$$

(6.119) represents the fraction of power reflected from a dissimilar dielectric interface when the signal is incident in TE-polarization and at an angle of incidence θ_i. Although the definition is power ratio, it become independent of the incident angle θ_i and the impedance η_1 by the virtue of *Snell's law of reflection* and both waves are travelling in the same dielectric medium with the wave impedance η_1. However, this is not the case for the power transmission coefficient and hence the complexity arises.

Usually we are interested more on the percentage of the incident power is reflected back into medium 1 to measure *the reflectivity* or *reflectance* of the media. The percentage power refection coefficient is defined as by taking ×100% of expression (6.119) as follows:

$$\%\text{Power reflected} = \Gamma_{PTE} \times 100\% = |\Gamma_{TE}|^2 \times 100\% \tag{6.120}$$

---

43 We can arrive at the same expression in (6.113), (6.117) using the instantaneous incident power density or Poynting vector along z-axis as follows:

$$\boldsymbol{P}^i_z = \boldsymbol{E}_i \times \boldsymbol{H}_i = E^i_0\boldsymbol{a}_y \times \left(-\frac{E^i_o}{\eta_1}\cos\theta_i\ \boldsymbol{a}_x\right) = \frac{(E^i_o)^2}{\eta_1}\cos\theta_i\boldsymbol{a}_z \therefore P^i_{ave,z} = \frac{|E^i_0|^2}{2\eta_1}\cos\theta_i$$

Similar derivations of (6.113) and (6.114) can be done for reflected and transmitted power, respectively, and substitutions for all components in (6.116) lead to (6.117).

Note that the power reflection coefficient which is the ratio of power reflected versus the power incident is defined as *refractivity*, and is also called *reflectance* in optics. As stated above the reflection of power happens in medium 1 hence with the help of *Snell's law of reflection* we have obtained a very simple expression of the power reflection coefficient which is just the square of the modulus of the voltage reflection coefficient. However, this is not the case for the power transmission coefficient as it involves two dissimilar dielectrics with two different phase velocities, impedances, and angles. Now we are going to derive the power transmission coefficient using the similar procedure as that for the power reflection coefficient as follows. As shown in Figure 6.28 we define the power transmission coefficient as $\tau_{PTE} = P^t_{ave,z}/P^i_{ave,z}$. Substitution for $P^t_{ave,z} = |E^t_o|^2/2\eta_2 \cos\theta_t$ from (6.113) and $P^i_{ave,z} = |E^i_o|^2/2\eta_1 \cos\theta_i$ from (6.115) into $\Gamma_{PTE} = P^t_{ave,z}/P^i_{ave,z}$ we obtain, the power transmission coefficient as follows:

$$\tau_{PTE} = \frac{P^t_{ave,z}}{P^i_{ave,z}} = \frac{\left(|E^t_o|^2/2\eta_2\right)\cos\theta_t}{\left(|E^i_o|^2/2\eta_1\right)\cos\theta_i} = \frac{|E^t_o|^2}{|E^i_o|^2} \times \frac{\eta_1 \cos\theta_t}{\eta_2 \cos\theta_i} \tag{6.121}$$

Substitution for the voltage transmission coefficient $\tau_{TE} = E^t_o/E^i_o$ we obtain the expression for the power transmission coefficient:

$$\tau_{PTE} = \frac{P^t_{ave,z}}{P^i_{ave,z}} = \frac{\eta_1 \cos\theta_t}{\eta_2 \cos\theta_i} \times |\tau_{TE}|^2 \tag{6.122}$$

(6.122) represents the fraction of power transmitted through a dielectric interface when the signal is incident in TE-polarization at an angle θ_i in medium 1 and transmitted at an angle θ_t in medium 2. As usual, we are interested more on the percentage of the incident power that is transmitted through medium 2 to justify *the transmissivity* of the medium. The percentage power transmission coefficient is defined as by taking × 100% of expression (6.122) as follows:

$$\%\text{Power transmitted} = \tau_{PTE} \times 100\% = \left(\frac{\eta_1 \cos\theta_t}{\eta_2 \cos\theta_i} \times |\tau_{TE}|^2\right) \times 100\% \tag{6.123}$$

Note that the power transmission coefficient is defined as *refractivity*, and is also called *transmissivity or transmittance* in optics. So far, we have examined the definition aspect of the power reflection and transmission coefficients. Now we shall apply the conservation of energy theory to develop a relation between the two coefficients.

Now we are going to develop the relationship between the power reflection and transmission coefficients via (6.116). Dividing both sides (6.116) with $P^i_{ave,z}$, and substitutions for (6.119) and (6.122) into (6.116) we obtain this relationship as follows:

$$\tau_{PTE} + \Gamma_{PTE} = 1 \tag{6.124}$$

Figure 6.30 shows representative plot of the power reflection and power transmission coefficients that follows the unity relationship in (6.124) with the incident angle θ_i. We shall also prove that the same relation is true for the TM polarization oblique incidence case. Although the power balance equation of (6.116) holds true for any circumstances as it obeys the law of conservation of energy, but the electromagnetic fields do not have to obey such law of conservation of energy theory. Depending on the conditions of incidence and transmission in dissimilar media, the transmitted field can be larger than the incident field. We shall examine the situation in the total transmission in the TM polarized oblique incident case when the transmitted electric field will be larger the incident field.

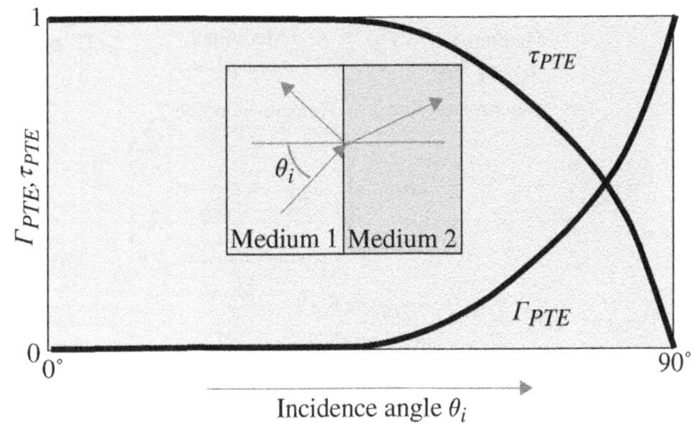

Figure 6.30 Conservation of energy law relates the transmission and reflection coefficients with a unity factor.

Example 6.11: Power Balance Calculation
Calculate the percentage and actual reflected and transmitted power of the GPR. The incident field intensity is 150 V/m.

Answer:

The average incident power: $P_{ave} = \frac{1}{2\eta_1}|E_0^i|^2 = \frac{150^2}{2 \times 377} = 29.84\ (\text{W})$

i) The fraction of transmitted power: $1 - |\Gamma_{TE}|^2 = 1 - 0.22^2 = 0.95 = 95\%$
Actual power transmitted: $0.95 \times 29.84 = 28.35\ (\text{W})$
The fraction of reflected power: 5%
Actual power reflected: $29.84 - 28.35 = 1.49\ (\text{W})$

6.10.8 Equivalent Impedance Concept in Power Balance Equation

The definition of the power transmission coefficient can be implied with the equivalent impedance concept that we have derived in Section 6.10.4. Substitutions for (6.96) into (6.122) we obtain

$$\tau_{PTE} = \frac{\eta_{1TE}}{\eta_{2TE}} \times |\tau_{TE}|^2 \tag{6.125}$$

If we define a *normalized wave impedance* $\eta_{2TE}^n = \eta_{2TE}/\eta_{1TE}$ for medium 2 with respect to the impedance η_{1TE} of medium 1 then we obtain a new definition of the power transmission coefficient as follows:

$$\tau_{PTE} = |\tau_{TE}|^2/\eta_{2TE}^n \tag{6.126}$$

Using the above definition of the power transmission coefficient and the equivalent impedance of medium 2 we can draw an equivalent transmission line model for the TE oblique incidence case as shown in Figure 6.31. Based on the model as shown in the figure we can also calculate the input impedance at a distance d inside medium 1 as follows:

$$Z_{in} = Z_{01}\frac{Z_L + jZ_{02}\,tan\,(\beta_{1z}d)}{Z_{02} + jZ_L\,tan\,(\beta_{1z}d)} \tag{6.127}$$

where, $\beta_{1z} = \beta_1 \cos\theta_i$, the z-directed wave number inside medium 1.

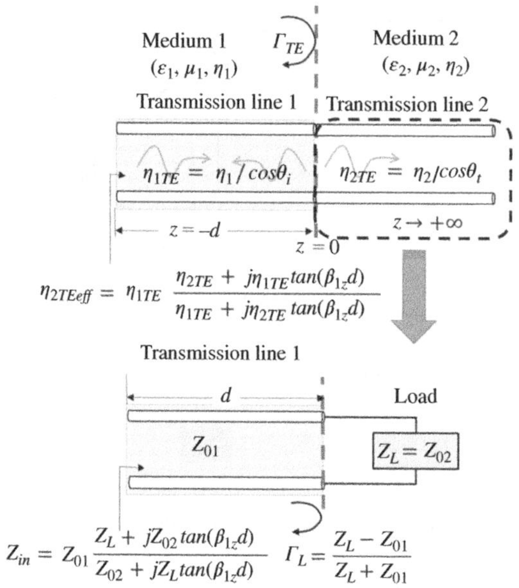

Figure 6.31 Equivalent transmission line model of TE polarized oblique incidence case.

6.10.9 Summary of TE Polarized Oblique Incidence Case

- We have introduced TE oblique incidence and technology perspective of emerging chipless RFID for TE polarization.
- In outlines, learning objectives, and outcome, we have again reinforced the technology perspectives with the hints of many case studies.
- We have given four case studies started with internal reflection and critical angle definition in the context of GPR.
- We have defined the TE oblique incidence case with 3D and 2D representations and highlighted the significance of vector decompositions.
- A novel procedure is devised with step-by-step solution approaches to boundary value problems of the TE oblique incidence case.
- Detailed derivations of incident, reflected and transmitted fields and their tangential components that obey the BCs at the interface are performed.
- We have summarized the complete sets of field solutions in TE polarization.
- We applied the BCs and equality principle of phase and amplitudes to derive Snell's laws of reflection and refraction.
- In the special context of critical angle, we have given two case studies – first one is the case for a perfect conductor and second one is the lossless dielectric.
- A submarine radar has been explained with the context of critical angle. We also examined how the critical angle limits underwater communications.
- A very interesting case study is extended to multilayered medium case for TE polarization.
- We have examined the power balance equation at the interface with the aid of the law of conservation of energy.
- We also did a very advanced level worked-out exercise on TE polarization that highlights broad spectrum of the theory covered here.

In the following section, we are having review questions on the TE polarization theory followed by the detailed analyses of the TM oblique incidence theory.

Review Questions 6.8: TE Oblique Incidence

Q1. What type of oblique incidence case a chipless RFID use? How data is encoded in the reflecting surface of chipless RFID? Hints: Research in the Internet.
Q2. A ground penetration radar work in what principle? In Case Study 1: GPR, why the GPR cannot see more than two layers? Can you devise some method to solve this problem?
Q3. Define TE polarization? What is the difference from the normal incidence?
Q4. In TE-polarization why do the electric and magnetic fields change their directions at the interface of two media? What are the governing parameters for these changes?
Q5. We use two vector decompositions to solve the electromagnetic boundary value problems in TE polarization. What are they? Can you comprehend their differences?
Q6. Can you explain why in TE polarization case the magnetic field is always on the plane of incidence though it has two components in all three field components in incidence, transmission, and reflection?
Q7. What are the components of the magnetic field used in the boundary conditions along with the electric field to derive γ_{TE} and τ_{TE}?
Q8. What is an acceptance angle in the case of submarine communication? How you can increase the area of coverage of the submarine radar?
Q9. Explain when γ_{TE} becomes real and complex at the interface of two dielectric media.
Q10. What are optically denser and lighter media?
Q11. What are the six steps to solve the boundary value problems in TE polarized oblique incidence?
Q12. What is a critical angle in TE polarization?
Q13. Is it possible to have total transmission at the interface of two dielectric media in TE polarization? What angle is called Brewster angle? Hints: use $\gamma_{TE} = 0$.
Q14. Give examples of devices which use the concept of the critical angle.
Q15. How does the power balance equation work in TE-polarization?
Q16. Why do we only consider normal directed power balance equation in TE polarization case?

6.11 TM Polarization Oblique Incidence

In the preceding section, TE polarized oblique incidence has been described in details. Finally, the energy balance equation and the effective impedance concept have been presented with significant practical applications to dielectric waveguides and optical fibers utilizing the total internal reflections. In this section, we shall explain the TM polarized oblique incidence theory using Maxwell's field analyse and boundary conditions. In this case the electric field obliquely lies on the plane of incidence hence can be decomposed into field components along two principal axes. Before doing the field analysis of TM polarization oblique incidence theory we shall highlight a few exciting applications of the TM polarization in modern wireless communications. Figure 6.32 illustrates a wireless mobile communications base station tower on top of a hill that covers a rough terrain and provides services to customers in mobile as well as fixed locations. As shown in the figure the electromagnetic signals bounce back from various terrains of the mountain and still provide robust communications links to customers. The rays at different trajectories provide multipath interferences in both constructive and destructive modes and yield both line-of-sight (LOS) and

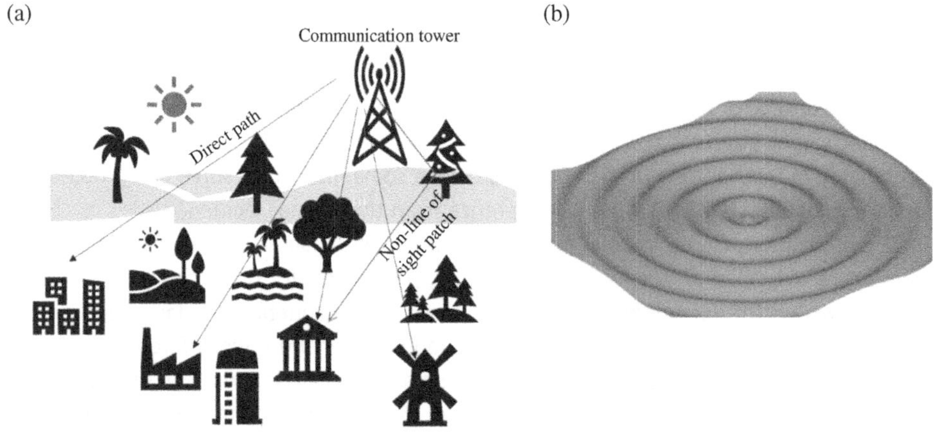

Figure 6.32 Wireless communications from a mobile base station tower in a region of difficult terrain, (a) illustrates the ray trace model of the transmitted signals to various mobile and fixed customers from a base station vis direction line of sight (LOS) and non-line-of-sight (NLOS) paths, and (b) Illustrates the surface plot of constructive and destructive interference to cover a region with both LOS and NLOS signals. *The Brewster angle* of total transmission is in action showing the bright circles.

non-line-of-sight (NLOS) wireless communications. This is possible due to the signals' constructive and destructive interferences when they mix with direct LOS and other NLOS signals. The surface plot of signal coverage in a contour map of the region in Figure 6.32b shows such constructive and destructive interferences with consecutive crests and troughs signal profiles, respectively. *The Brewster angle* of TM polarization for full transmission and zero reflection of signals from the ground and the surrounding obstacles is in play to create such NLOS communications viable. Not only in such a difficult wireless communications environment, but also *the Brewster angle concept* is conveniently used in radars for target detection, futuristic THz communications technology using LED for indoor data communications, minimizing bit-error-rate (BER) in radio wave propagation, polarization diversity for wireless communications and also the 60 GHz chipless RFID tag design using printed polarizers as a passive tag which is illuminated with obliquely incident interrogation signals. In this section, we shall learn about the complete solution of the TM oblique incidence cases of the uniform plane wave. *Brewster angle* and TM oblique incidence theory have enormous applications and touch upon many applications including: (i) 60 GHz chipless RFID, also called EM barcode, (ii) optical barcode technology, (iii) futuristic THz communications technology using both NLOS/LOS communications, (iv) co-channel interference mitigation using *Brewster angle* concept, (v) indoor visible light communications using polarization diversity, (vi) minimizing BER using Brewster angle in radio communications, (vii) small omnidirectional biconical antenna for wireless communications, and (viii) polarization diversity in wireless communications. The readers are encouraged to search for applications of *Brewster angle* in TM polarization in online resources and will be easily overwhelmed with the number and diversity of application areas.

6.11.1 Field Analysis of TM Polarization Oblique Incidence

Figure 6.33a shows the TM polarized oblique incidence on a dielectric boundary along *xy*-plane. As shown in the figure the electromagnetic fields have changed their role (orientations) in TM polarization when compared with the TE polarization oblique incidence case as shown in Figure 6.19.

Here, the TM polarized magnetic field is oriented perpendicular to the plane of incidence which is on the *xz*-plane (plane of paper), and the vector field hits parallel to the plane of interface or the boundary plane which is on the *xy*-plane. As usual the angle of incidence is θ_i that makes the angle between the direction of the incident wave propagation $\boldsymbol{a_i}$ and the normal line to the plane of interface along *z*-axis. The angle of reflection is θ_r, which is the angle between the reflected wave's unit vector $\boldsymbol{a_r}$ and the *z*-axis. The angle of transmission is θ_t that makes the angle between the direction of the transmitted wave propagation $\boldsymbol{a_t}$ and the normal line to the plane of interface along *z*-axis. In TM polarization, the incident, reflected and transmitted electric fields reside on the *xz*-plane with arbitrary angles. Due to the arbitrary polarization of the electric field, we need to decompose it into their respective components along the principle axes *x* and *z*, as we did before for the case of TE oblique incidence on the conducting boundary.[44] The fundamental principle of wave propagation and the orthogonality relation between the electromagnetic fields and the direction of propagation remain unchanged. The electromagnetic fields and their respective directions of wave propagation must obey *the right-hand rule* as shown in the figure. As can be seen the directions of the tips of the incident, and transmitted magnetic fields remain unchanged along *y*-axis, but the reflected field changes its direction to *y*-axis to maintain *the right-hand rule*. As shown in the figure, the electric field changes its direction due to its arbitrary polarization of the oblique incidence on the *xz*-plane and always remains on the *xz*-plane (plane of paper) with different directions to maintain *the right-hand rule* with the magnetic field and the direction of wave propagations in the three phases of incidence, reflection, and transmission. The 2D propagating waves in incident, transmitted, and reflected conditions are shown with sinusoidally varying thick curves in Figure 6.33b. Like the TE polarization case,

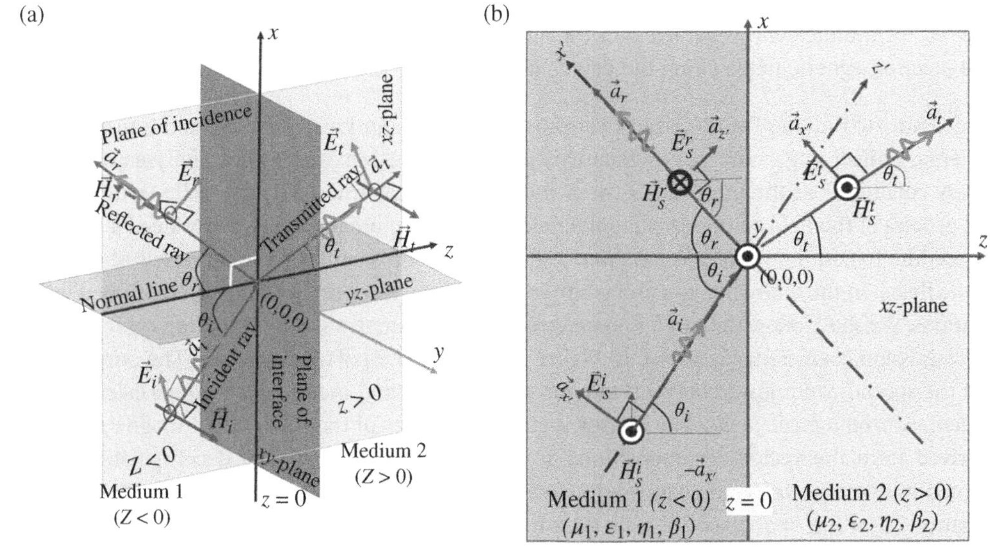

Figure 6.33 TM oblique incidence- (a) 3D plot shows both the electric and magnetic fields change their directions in reflection mode, and (b) the detailed 2D vector field decompositions of the TM polarization.

44 Here the complexity arises due to the fact that the electric field is decomposed into two components hence the definitions of the transmission and reflection coefficients and associated parameters need to deal with the two components to find their exact solutions.

our motto of the field analysis is to derive the transmission and reflection coefficients using the electromagnetic field boundary conditions and relate the transmitted and reflected field amplitudes in terms of the incident field amplitude. Again, to iterate, we have done the comprehensive vector decompositions of the generic obliquely incident, reflected, and transmitted electromagnetic fields in Section 6.3.

In the derivations of the transmission and reflection coefficients, the following rules are obeyed:

i) For incidence angle $\theta_i \neq 0$; the "oblique incidence" is TM/parallel polarization when the magnetic field is y-directed and normal to the plane of incidence.
ii) The electric field is arbitrarily polarized on xz-plane. We decompose the electric field to its equivalent x- and z-directed components.
iii) In all cases, the right-hand rule prevails. In the reflected wave, the magnetic field remains perpendicular to yz-plane but $-y$-directed. The electric field changes its direction to meet the condition of the propagation direction of the reflected wave, and with the changed orientation of the reflected magnetic field to maintain *the right-hand rule.*
iv) In the transmitted wave, the magnetic field remains y-polarized in medium 2, but the electric field changes its direction due to the different constitutive parameters in medium 2. As for the case for TE polarization, we advise a short cut to express the transmitted magnetic field, here in the case of TM polarization, the simple approach is to find the transmitted electric and magnetic fields in medium 2 is to just change the subscripts of the constitutive parameters from β_1 to β_2 and the angle θ_i to θ_t of the incident electric field. The vector decomposition follows the same orientations of vectors as that of the incident wave.
v) As for the case for TE polarization, here decompositions of the position vectors of the three field waves are performed in terms of the electrical length along the directions of propagation $\boldsymbol{a_i}$, $\boldsymbol{a_r}$, and $\boldsymbol{a_t}$, for the incident, reflected, and transmitted fields, respectively, and the unit vectors of the electromagnetic fields along the principal axes x- and z.

Our goals are to find the TM polarized electromagnetic field solutions, reflection and transmission coefficients, critical angle and power balance equation using Maxwell's field analyses, and the boundary conditions as the interface of the two dissimilar dielectric media at $z = 0$. As usual, the overall process of the field solutions using the field analysis is quite involving hence need a systematic procedure to compute the fields and derive the relationships between the three sets of electromagnetic fields. In the following, the procedure to analyze the three sets of fields are described with illustrations. We shall follow the similar procedure for the field analysis and calculations of reflection and transmission coefficients as shown in Figure 6.20 for the TE polarization case. The only difference is the special condition of *the total internal transmission* and *the Brewster's angle* calculation.

Incident electromagnetic fields: The phasor vector expressions of the electric and magnetic fields are derived from the vector decompositions according to Figure 6.34. The decompositions are performed in two aspects: (i) the equivalent electrical length along the direction of incident wave propagation $\zeta_{inc} = \beta_1 z' = \beta_1(x \sin\theta_i + z\cos\theta_i)$, where the equivalent electric length ζ_{inc} has the unit of radian, and (ii) the unit vector of the arbitrarily polarized electric field on the plane of incidence (xz-plane): $-\boldsymbol{a_{x'}} = \cos\theta_i \boldsymbol{a_x} - \sin\theta_i \boldsymbol{a_z}$. Figure 6.34 illustrates the enlarged view of the decompositions of the unit vector $-\boldsymbol{a_{x'}}$. Now we have all ingredients of the phasor vector expressions for the TM-polarized incident electric and magnetic fields and are produced as follows.

Incident electromagnetic fields (where β_1 is used in place of k_1):

$$\boldsymbol{E}_s^i = -\boldsymbol{a_{x'}} E_o^i e^{-j\beta_1 z} = (\cos\theta_i \boldsymbol{a_x} - \sin\theta_i \boldsymbol{a_z}) E_o^i e^{-j\beta_1 (x\sin\theta_i + z\cos\theta_i)} \tag{6.128}$$

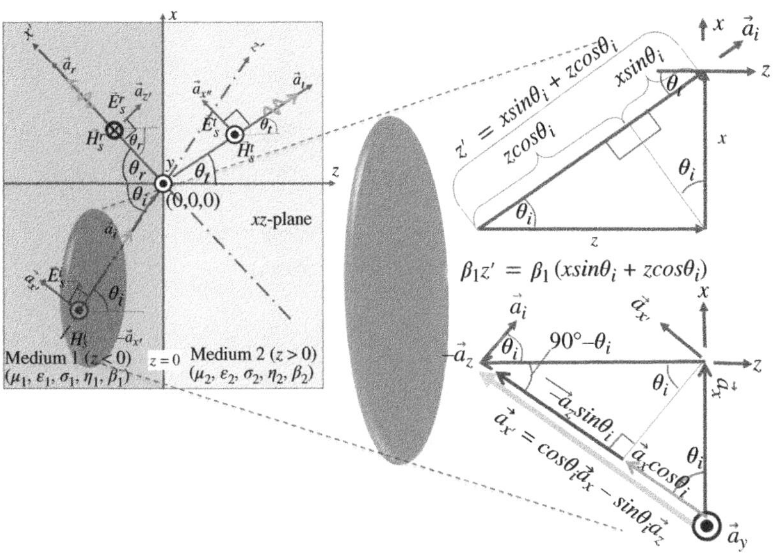

Figure 6.34 Vector decompositions of TM oblique *incidence* fields on a dielectric boundary.

$$H_s^i = a_y H_o^i e^{-j\beta_1 z\prime} = a_y \frac{E_o^i}{\eta_1} e^{-j\beta_1(x\sin\theta_i + z\cos\theta_i)} \quad (6.129)$$

In this case, the arbitrarily polarized incident electric field on the plane of incidence (plane of paper) has two components in x- and -z-directions whereas the magnetic field has only one component along y-direction as shown in Figure 6.34. The boundary conditions are applied only on the tangential components of the incident electric and magnetic fields on the xy-plane at $z = 0$, hence it is worth calculating the two components as follows:

$$E_{T1}^i|_{z=0} = a_x \cos\theta_i E_o^i e^{-j\beta_1(x\sin\theta_i)} \quad (6.130)$$

$$H_{T1}^i|_{z=0} = a_y \frac{E_o^i}{\eta_1} e^{-j\beta_1(x\sin\theta_i)} \quad (6.131)$$

Vector decompositions in $\beta z'$ and unit vector a_x' are performed above.

Reflected electromagnetic fields: The phasor vector expressions of the reflected electric and magnetic fields in medium 1 are derived from the vector decompositions according to Figure 6.35 and are produced as follows. The decompositions are performed in two aspects: (i) the electrical length along the direction of incident wave propagation $\zeta_{refl} = \beta_1 x' = \beta_1(x\sin\theta_r - z\cos\theta_r)$, where the equivalent electric length ζ_{refl} has the unit of radian, and (ii) the unit vector of the arbitrary polarized magnetic field on the plane of incidence (xz-plane): $a_{z'} = \cos\theta_r a_x + \sin\theta_r a_z$. Figure 6.35 illustrates the enlarged view of the decompositions. The TM reflected electromagnetic field expressions as follows:

Reflected electromagnetic fields

$$E_s^r = a_{z'} E_o^r e^{-j\beta_1 x\prime} = (\cos\theta_r a_x + \sin\theta_r a_z) E_o^r e^{-j\beta_1(x\sin\theta_r - z\cos\theta_r)} \quad (6.132)$$

$$H_s^r = -a_y H_o^r e^{-j\beta_1 x'} = -a_y \frac{E_o^r}{\eta_1} e^{-j\beta_1(x\sin\theta_r - z\cos\theta_r)} \quad (6.133)$$

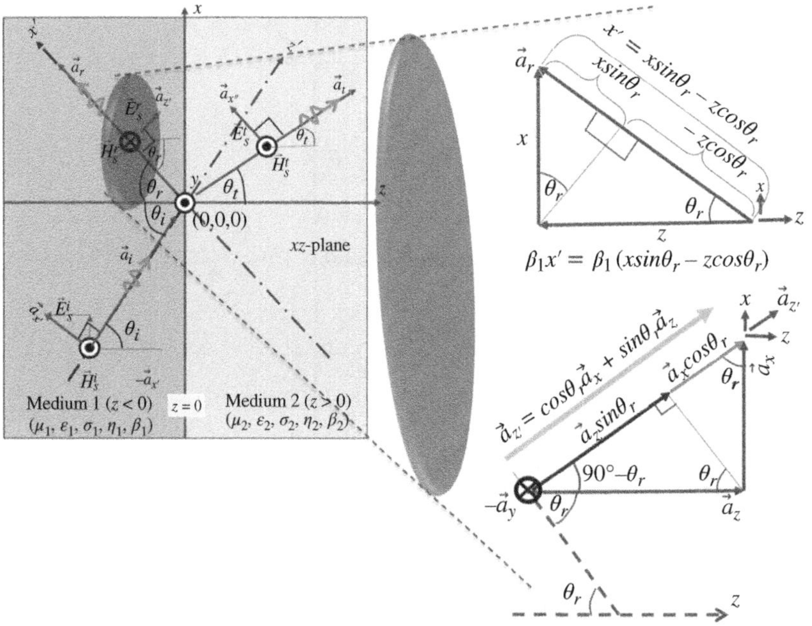

Figure 6.35 Vector decompositions of TM oblique *reflected* fields on a dielectric boundary.

In this case, the reflected magnetic field has only one component along $-y$-direction and the electric field is arbitrarily polarized on the plane of incidence (plane of paper) hence has two components in x and z directions as shown in Figure 6.35.

The boundary conditions is applied only on the tangential components of the reflected electric and magnetic fields and the tangential electromagnetic fields at $z = 0$ are expressed as follows:

$$\boldsymbol{E}^r_{T1}\big|_{z=0} = \boldsymbol{a_x} \cos\theta_r\, E^r_o\, e^{-j\beta_1(x\sin\theta_r)} \tag{6.134}$$

$$\boldsymbol{H}^r_{T1}\big|_{z=0} = -\boldsymbol{a_y}\frac{E^r_o}{\eta_1}\, e^{-j\beta_1(x\sin\theta_r)} \tag{6.135}$$

Lastly, we perform the similar vector decompositions for *the transmitted* electromagnetic fields as follows.

Transmitted electromagnetic fields: Finally, the phasor vector expressions of the transmitted electric and magnetic fields are derived from the vector decompositions according to Figure 6.36. The decompositions are performed in two aspects: (i) the equivalent electrical length along the direction of transmitted wave propagation $\zeta_{trans} = \beta_2 z'' = \beta_2(x\sin\theta_t + z\cos\theta_t)$, where the equivalent electric length ζ_{inc} has the unit of radian, and (ii) the unit vector of the arbitrary polarized magnetic field on the plane of incidence (xz-plane): $\boldsymbol{a_{x''}} = \cos\theta_t\boldsymbol{a_x} - \sin\theta_t\boldsymbol{a_z}$. Figure 6.36 illustrates the enlarged view of the decompositions as well. Now we have all ingredients of the phasor vector expressions for the TM-polarized transmitted electric and magnetic fields and are produced as follows.

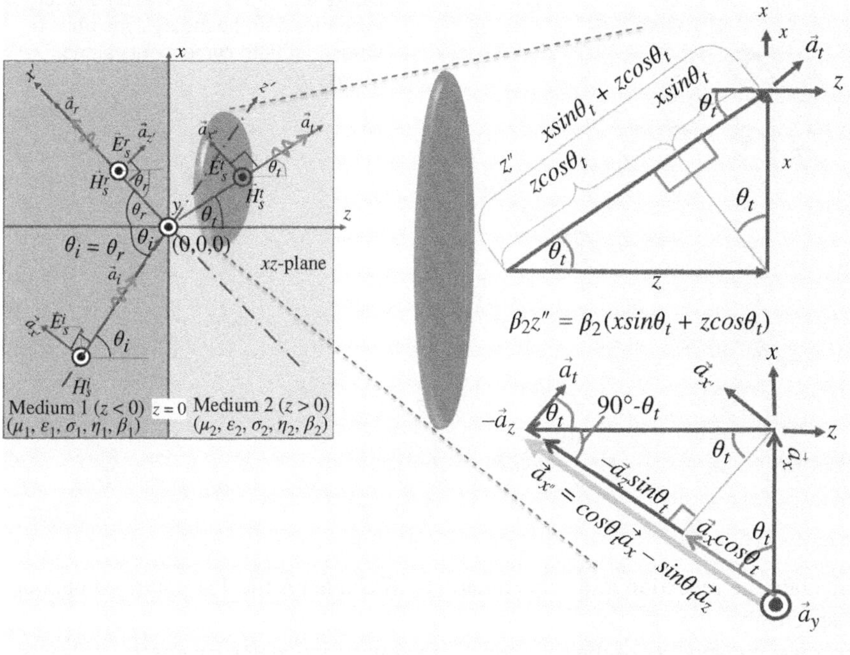

Figure 6.36 Vector decompositions of TM oblique *transmitted* fields on a dielectric boundary.

Transmitted electromagnetic fields (where β_2 is used in place of k_2)

$$\boldsymbol{E}_s^t = \boldsymbol{a}_{\boldsymbol{x}''} E_o^t e^{-j\beta_t z''} = (\cos\theta_t \boldsymbol{a}_{\boldsymbol{x}} - \sin\theta_t \boldsymbol{a}_{\boldsymbol{z}}) E_o^t e^{-j\beta_2(x\sin\theta_t + z\cos\theta_t)} \tag{6.136}$$

$$\boldsymbol{H}_s^t = \boldsymbol{a}_{\boldsymbol{y}} H_o^t e^{-j\beta_2 z''} = \boldsymbol{a}_{\boldsymbol{y}} \frac{E_o^t}{\eta_2} e^{-j\beta_2(x\sin\theta_t + z\cos\theta_t)} \tag{6.137}$$

In this case, as usual, the transmitted electric field has two components in x and $-z$ directions and the magnetic field has only one component along y-direction as shown in Figure 6.36. The boundary conditions is applied only on the tangential components of the transmitted electric and magnetic fields as follows:

$$E_{T2}^t\big|_{z=0} = \boldsymbol{a}_{\boldsymbol{x}} \cos\theta_t \, E_o^t \, e^{-j\beta_2(x\sin\theta_t)} \tag{6.138}$$

$$H_{T2}^t\big|_{z=0} = \boldsymbol{a}_{\boldsymbol{y}} \frac{E_o^t}{\eta_2} e^{-j\beta_2(x\sin\theta_t)} \tag{6.139}$$

6.11.2 Applications of Boundary Conditions at $z = 0$

So far, we have rigorously performed all vector decompositions of the three sets of electromagnetic fields and also expressed the x-directed tangential field components of the three sets of the electromagnetic fields that will hold the equality. As for the case of TE-polarization, this equality will produce the amplitudes of the reflected and transmitted fields in terms of the incident field amplitudes

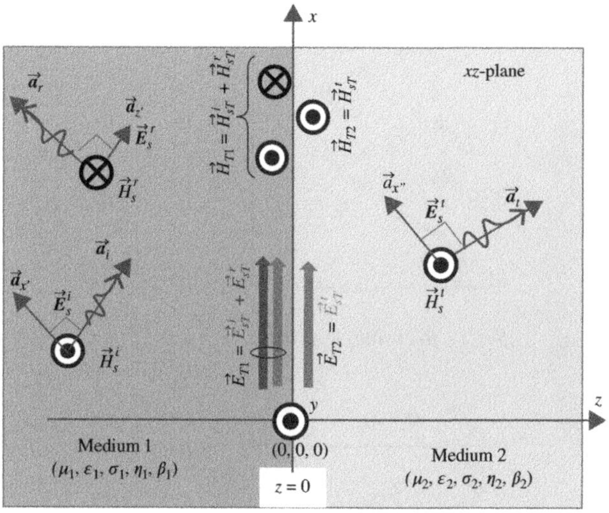

Figure 6.37 TM polarization oblique incidence field boundary conditions for tangential field components along x-axis.

so that we can calculate the reflection and transmission coefficients for the TM polarized oblique incident fields, respectively.

Figure 6.37 illustrates the equality of the three sets of the tangential electromagnetic fields with thick arrows. If you compare the field orientation on the 2D plane (*xy*-plane) for the case of TM polarization with those for TE-polarization in Figure 6.24, we can easily comprehend the differences of the field orientations. These differences in the two polarizations bring significant differences in the calculations of the reflection and transmission coefficients. And this cannot be appreciated with the ray trace model, but only when we perform the complete vector decompositions of the field components using the field analyses of the three sets of electromagnetic fields. Applying two boundary conditions for the tangential electric field ($\boldsymbol{E}_{T1} = \boldsymbol{E}_{T2}$) and the tangential magnetic field ($\boldsymbol{H}_{T1} = \boldsymbol{H}_{T2}$) on both sides of the interface at $z = 0$ yields:

$$\boldsymbol{E}^i_{T1}|_{z=0} + \boldsymbol{E}^r_{T1}|_{z=0} = \boldsymbol{E}^t_{T2}|_{z=0} \tag{6.140}$$

$$\boldsymbol{H}^i_{T1}|_{z=0} + \boldsymbol{H}^r_{T1}|_{z=0} = \boldsymbol{H}^t_{T2}|_{z=0} \tag{6.141}$$

Again note that the magnetic field boundary condition holds true only if there is no induced electric current at the interface of the two dielectric media at $z = 0$. Remember that the equality holds for any value of x and hence all phase components associated with the three sets of field equations must be equal. The phase equality leads to *the Snell's law of reflection* $\theta_i = \theta_r$. Now substitutions for (6.130), (6.134) and (6.138) into (6.140) at $x = 0$, we obtain:

$$E^i_o \cos\theta_i + E^r_o \cos\theta_r = E^t_o \cos\theta_t \tag{6.142}$$

Substitution for $\theta_i = \theta_r$ and dividing both sides of (6.142) with $\cos\theta_i$ yield

$$E^i_o + E^r_o = \frac{\cos\theta_t}{\cos\theta_i} E^t_o \tag{6.143}$$

By simply dividing both side of (6.143) with E_0^i gives an important relation between Γ_{TE} and τ_{TE} as follows

$$\frac{E_0^t}{E_0^i}\frac{\cos\theta_t}{\cos\theta_i} = 1 + \frac{E_0^r}{E_0^i} \quad \therefore \tau_{TM} = (1+\Gamma_{TM})\left(\frac{\cos\theta_i}{\cos\theta_t}\right) \tag{6.144}$$

Comparing the same relationship for TE-polarization in (6.88) with (6.144), we can observe a contrasting difference. This is due to the fact that the definitions of the two coefficients are defined in terms of the incident, reflected, and transmitted electric fields which are on the plane of incidence and changes their direction in the second medium.

And substitutions for (6.131), (6.135), and (6.139) into (6.141) yield:

$$\frac{E_o^i}{\eta_1}e^{-j\beta_1(x\sin\theta_i)} - \frac{E_o^r}{\eta_1}e^{-j\beta_1(x\sin\theta_r)} = \frac{E_o^t}{\eta_2}e^{-j\beta_2(x\sin\theta_t)} \tag{6.145}$$

As before, the phase equality of the three field components in (6.145) of the incident and transmitted fields yields *Snell's law of refraction* $\theta_t = \sin^{-1}((\beta_1/\beta_2)\sin\theta_i)$ as expressed in (6.84). Also (6.145) is true for any value of x. Setting $x = 0$ and multiplying both sides with η_1 we obtain

$$E_o^i - E_o^r = \frac{\eta_1}{\eta_2}E_o^t \tag{6.146}$$

Now we have two expressions of summation and subtraction of amplitudes of the incident and reflected electric fields in terms of the amplitude of the transmitted electric fields in (6.143) and (6.146). Simple mathematical manipulations of the two expressions yield the reflection and transmission coefficients for the TM oblique incidence case. So far, we have analyzed the three sets of the electromagnetic fields applying the boundary conditions to derive the two laws of Snell for the TM case. In the following section, we shall calculate the reflection and transmission coefficients.

6.11.3 Calculations of Γ_{TM} and τ_{TM}

The derivations of the TM polarization reflection coefficient $\Gamma_{TM} = E_0^r/E_0^i$ and the transmission coefficient $\tau_{TM} = E_0^t/E_0^i$ need the solutions for the amplitudes of the reflected and transmitted electric fields in terms of the incident electric field E_0^i. Adding (6.143) and (6.146) and simple mathematical manipulation yield the transmission coefficient:

$$\tau_{TM} = \frac{E_o^t}{E_o^i} = \frac{2\eta_2\cos\theta_i}{\eta_2\cos\theta_t + \eta_1\cos\theta_i} \tag{6.147}$$

Subtraction of (6.146) from (6.143) yields the reflection coefficient

$$\Gamma_{TM} = \frac{E_o^r}{E_o^i} = \frac{\eta_2\cos\theta_t - \eta_1\cos\theta_i}{\eta_2\cos\theta_t + \eta_1\cos\theta_i} \tag{6.148}$$

We have already derived the relationship between τ_{TM} and Γ_{TM} in (6.144). We can also land to the same expression with simple mathematical manipulation using (6.147) and (6.148). So far, we have analyzed many important concepts about the TM polarized oblique incidence fields. Now we shall do a few review questions and an example.

Review Questions 6.9: TM Polarization

Q1. In TM-polarization, why do the electric and magnetic fields change their directions in media 1 and 2? What are the governing parameters for these changes?

Q2. We use similar vector decompositions of TE case to solve EM boundary value problems in TM polarization. What are they? Can you comprehend their differences in TM and TE cases?

Q3. Can you explain why the electric field is always on the plane of incidence though it has two components in all three cases: incidence, transmission, and reflection?

Q4. What are the components of the electric field used in the boundary conditions along with the magnetic field to derive Γ_{TM} and τ_{TM}?

Q5. Derive Γ_{TM} and τ_{TM} from the field expressions from the boundary conditions expressed in (6.143) and (6.146).

Example 6.12: Field Analysis of TM Polarized Oblique Incidence

A uniform plane wave at 100 MHz operating frequency incidences upon a lossless medium ($\varepsilon_1 = 25$) from air as medium 1. The electric field is TM polarized and amplitude of 150 V/m. The incidence angle is 40°. Calculate the followings: (i) the phase constants in media 1 and 2, (ii) the angle of refraction, θ_t (iii) wave impedances in both media η_1 and η_2, (iv) the TM polarized reflection and transmission coefficients, Γ_{TM} and τ_{TM}, respectively, and (v) the phasor vector and time domain expressions for incident, reflected, and transmitted electric and magnetic fields.

Answer:

As before for TE polarization (i) and (ii) remain the same.

At 100 MHz: the angular frequency: $\omega = 628.3 \times 10^6$ (rad/s)

The phase constant in medium 1: $\beta_1 = \omega/c_1 = (2\pi \times 10^8)/(3 \times 10^8) = 2.09$ (rad/s)

The phase constant in medium 2: $\beta_2 = \omega/c_2 = \left(2\pi \times 10^8 \times \sqrt{25}\right)/(3 \times 10^8) = 10.45$ (rad/s)

The wave impedances in air medium 1: $\eta_1 = 377\ \Omega$; and in glass medium 2: $\eta_2 = 377/\sqrt{25} = 75.4\ \Omega$.

We have all ingredients to solve the problems as follows.

i) Based on Snell's law of reflection $\theta_i = \theta_r$; hence the angle of reflection is: $\theta_r = 40°$.

The ratio of phase constants: $\dfrac{\beta_1}{\beta_2} = \sqrt{\dfrac{1}{\varepsilon_{r2}}} = \dfrac{1}{5}$

Based on Snell's law of refraction (6.84): $\theta_t = sin^{-1}\left(\dfrac{\beta_1}{\beta_2} sin\,\theta_i\right) = sin^{-1}(sin\,40°/5) = 7.4°$

ii) Substitutions for η_1, η_2, θ_i, and θ_t into (6.148) yield the reflection coefficient:

$$\Gamma_{TM} = \frac{\eta_2 \cos\theta_t - \eta_1 \cos\theta_i}{\eta_2 \cos\theta_t + \eta_1 \cos\theta_i} = \frac{75.4 \times \cos 40° - 377 \times \cos 7.4°}{75.4 \times \cos 40° + 377 \times \cos 7.4°} = -0.589$$

Substitutions for η_1, η_2, θ_i, and θ_t into (6.148) yield the transmission coefficient:

$$\tau_{TM} = \frac{2\eta_2 \cos\theta_i}{\eta_2 \cos\theta_t + \eta_1 \cos\theta_i} = \frac{2 \times 75.4 \times \cos 40°}{260 \times \cos 40° + 377 \times \cos 74°} = 0.318$$

The negative sign of the reflection coefficient represents the phase reversal of the reflected wave at the interface of the two media. You can verify the result above using $\tau_{TM} = (1 + \Gamma_{TM})(\cos\theta_i/\cos\theta_t)$.

iii) The expression for the incident electric field (6.128):

$$\begin{aligned}\boldsymbol{E}_s^i(x,z) &= (\cos\theta_i\boldsymbol{a_x} - \sin\theta_i\boldsymbol{a_z})E_0^i e^{-j\beta_1(x\sin\theta_i + z\cos\theta_i)}\\ &= (\cos 40^\circ\boldsymbol{a_x} - \sin 40^\circ\boldsymbol{a_z}) \times 150 \times e^{-j(1.34x + 1.60z)}\\ &= (114.9\,\boldsymbol{a_x} - 96\boldsymbol{a_z})e^{-j(1.34x + 1.60z)}\,(\text{V/m})\end{aligned}$$

The time domain expression for the electric field is:

$$\boldsymbol{E^i}(x,z,t) = (114.9\,\boldsymbol{a_x} - 96\boldsymbol{a_z})\cos\left(628.3\times10^6 t - 1.34x - 1.60z\right)\,(\text{V/m})$$

The incident magnetic field (6.129):

$$\boldsymbol{H}_s^i(x,z) = \boldsymbol{a_y}\frac{E_0^i}{\eta_1}e^{-j\beta_1(x\sin\theta_i + z\cos\theta_i)} = \boldsymbol{a_y}\left(\frac{150}{377}\right)e^{-j(1.34x + 1.60z)} = 0.4\,\boldsymbol{a_y}\,e^{-j(1.34x + 1.60z)}\,(\text{A/m})$$

The time domain expression for the magnetic field is:

$$\boldsymbol{H}^i(x,z,t) = 0.4\,\boldsymbol{a_y}\cos\left(628.3\times10^6 t - 1.34x - 1.60z\right)\,(\text{A/m})$$

iv) Now we calculate the reflected electromagnetic fields. From (6.132) the expression for the reflected electric field is:

$$\begin{aligned}\boldsymbol{E}_s^r(x,z) &= \boldsymbol{a_y}\Gamma_{TM}\,E_0^i e^{-jk_1(x\sin\theta_r - z\cos\theta_r)}\\ &= -0.589\times150\,(\cos 40^\circ\boldsymbol{a_x} + \sin 40^\circ\boldsymbol{a_z})\,e^{-j(1.34x - 1.60z)}\\ &= -(67.68\boldsymbol{a_x} + 43.5\boldsymbol{a_z})\,e^{-j(1.34x - 1.60z)}\,(\text{V/m})\end{aligned}$$

The time domain expression for the reflected electric field is:

$$\boldsymbol{E^r}(x,z,t) = (67.68\boldsymbol{a_x} + 43.5\boldsymbol{a_z})\cos\left(628.3\times10^6 t - 1.34x + 1.60z - 180^\circ\right)\,(\text{V/m})$$

From (6.133) the reflected magnetic field is defined as:

$$\boldsymbol{H}_s^r(x,z) = \boldsymbol{a_y}\frac{E_0^r}{\eta_1}e^{-j\beta_1(x\sin\theta_r - z\cos\theta_r)} = \left(\frac{-88.35}{377}\right)\boldsymbol{a_y}\,e^{-j(1.34x - 1.60z)} = -0.234\,\boldsymbol{a_y}\,e^{-j(1.34x - 1.60z)}\,(\text{A/m})$$

The time domain expression for the reflected magnetic field is:

$$\boldsymbol{H^r}(x,z,t) = 0.234\,\boldsymbol{a_y}\cos\left(628.3\times10^6 t - 1.34x + 1.60z - 180^\circ\right)\,(\text{A/m})$$

v) The expression for the transmitted electric field (6.136):

$$\begin{aligned}\boldsymbol{E}_s^t &= (\cos\theta_t\boldsymbol{a_x} - \sin\theta_t\boldsymbol{a_z})\tau_{TM}E_0^i e^{-j\beta_2(x\sin\theta_t + z\cos\theta_t)} = (\cos 7.4^\circ\boldsymbol{a_x} - \sin 7.4^\circ\boldsymbol{a_z})\\ &\quad\times(0.318\times150)e^{j10.45(x\sin 7.4^\circ + z\cos 7.4^\circ)} = (47.33\boldsymbol{a_x} - 6.17\boldsymbol{a_z})\,e^{-j(1.35x + 10.36z)}\,(\text{A/m})\end{aligned}$$

The time domain expression for the electric field is:

$$\boldsymbol{E}^t(x,z,t) = (47.33\boldsymbol{a_x} - 6.17\boldsymbol{a_z})\cos\left(628.3\times10^6 t - 1.35x - 10.36z\right)\boldsymbol{a_y}\,(\text{V/m})$$

The transmitted magnetic field (6.137) is defined as:

$$\boldsymbol{H}_s^t(x,z) = \boldsymbol{a}_y \frac{E_0^t}{\eta_2} e^{-j\beta_2(x\sin\theta_t + z\cos\theta_t)} = \boldsymbol{a}_y \left(\frac{0.318\times 150}{75.4}\right) e^{-j(1.35x+10.36z)} = \boldsymbol{a}_y 0.63\, e^{-j(1.35x+10.36z)}\,(\mathrm{A/m})$$

The time domain expression for the transmitted magnetic field is:

$$\boldsymbol{H}^i(x,z,t) = \boldsymbol{a}_y 0.63\cos\left(628.3\times 10^6 t - 1.35x - 10.36z\right)\ (\mathrm{A/m})$$

So far, we have done comprehensive study of the field analysis of the TM oblique incidence. The big and constructing difference between the characteristics of the TM and TM polarization oblique incidence cases is the total transmission while the total reflection is common to both polarizations and can be solved with the ray trace model. It is not possible to obtain total transmission in TE polarization using the ray trace model, and therefore, is left for the reader to answer the question. In the following section, we are examining the total transmission of TM polarization.

6.11.4 Total Transmission and Brewster Angle θ_B

While total internal reflection and critical angle is a common phenomenon for oblique incidence of an electromagnetic wave from a denser medium to a lighter medium, the total transmission and Brewster angle is a special case for TM polarization. This unique phenomenon of Brewster angle can separate the TM polarized signal from the TE polarized signal for any arbitrary polarized[45] incident signal. For completeness, we can also derive *the Brewster angle* for the TM polarization assuming that the wave impinges from any medium 1 to another medium 2. At a particular incident angle θ_{BA}, the reflection coefficient derived in (6.148) becomes zero ($\Gamma_{TM} = 0$). This means no reflection at the interface of the two media so that total transmission occurs. Substitution for $\Gamma_{TM} = 0$ in (6.148) leads to the following expression

$$\Gamma_{TM} = \frac{\eta_2\cos\theta_t - \eta_1\cos\theta_i}{\eta_2\cos\theta_t + \eta_1\cos\theta_i} = 0; \quad \text{or,}\ \eta_2\cos\theta_t - \eta_1\cos\theta_i = 0 \tag{6.149}$$

Simple mathematical manipulation using Snell's Law of refraction $\sin\theta_t = (\beta_1/\beta_2)\sin\theta_{AB}$ for a particular incidence angle $\theta_i = \theta_{BA}$; *Brewster angle* at which the total transmission and zero reflection occurs, leads to

$$\sin\theta_{BA} = \sqrt{\frac{\beta_2^2(\eta_2^2 - \eta_1^2)}{\eta_2^2\beta_1^2 - \eta_1^2\beta_2^2}} \tag{6.150}$$

For non-magnetic ($\mu_1 = \mu_2 = \mu_0$) and lossless materials Brewster angle simplifies to

$$\sin\theta_{BA} = \frac{1}{\sqrt{1 + \varepsilon_{r1}/\varepsilon_{r2}}};$$

or,

$$\theta_{BA} = \sin^{-1}\left(1 + \frac{\varepsilon_{r1}}{\varepsilon_{r2}}\right)^{-1/2} = \tan^{1}\sqrt{\left(\frac{\varepsilon_{r2}}{\varepsilon_{r1}}\right)} = \tan^{-1}\left(\frac{n_2}{n_1}\right) \tag{6.151}$$

45 The arbitrarily polarized signal can be decomposed into the TE and TM polarized signals. In this sense, it is a *convoluted signal*. Brewster angle filters out TM polarized signal from the TE polarized signal in the total transmission condition.

where, ε_{r1} and ε_{r2} are the dielectric constants of medium 1 and medium 2. Likewise, n_1 and n_2 are the refractive indices of medium 1 and medium 2, respectively. Brewster angle is called *the polarization angle* as it allows to transmit only TM polarized signal and blocks TE-polarized signal.

6.11.5 Total Transmission for Arbitrary Polarized Signal at Plane Interface Between Dissimilar Perfect Dielectric

In the previous section, we examined the defined polarized signals which are incident on the plane interface of two dissimilar, perfect dielectric media. For a perfect dielectric medium $\mu_1 = \mu_2 = \mu_0$. In the absence of losses, both dielectric media are transparent to the electromagnetic signal.[46] The refractive index is defined in terms of the ratio of the velocities of the wave in the two media. Assume the velocity is v_1 in medium 1 with permittivity value ε_1 and the velocity is v_2 in medium 2 with permittivity value ε_2. Therefore, we can define the refractive index $n_{12} = v_1/v_2$ which refers to the refractive index from medium 1 to medium 2. We also define *the absolute refractive index* of medium 1 which is defined as: $n_1 = c_0/v_1 = \sqrt{\mu_1\varepsilon_1/\mu_0\varepsilon_0} = \sqrt{\varepsilon_1/\varepsilon_0}$, where c_0 is the velocity of light in free space or vacuum and if $\mu_1 = \mu_0$. Likewise, we can define the refractive index for medium 2 $n_2 = c_0/v_2 = \sqrt{\mu_2\varepsilon_2/\mu_0\varepsilon_0} = \sqrt{\varepsilon_2/\varepsilon_0}$, if $\mu_2 = \mu_0$. We have shown that *Snell's law of refraction* in terms of n_1 and n_2 in (6.57). We can rewrite the equation in a little bit different form:

$$\frac{\sin\theta_i}{\sin\theta_t} = \frac{\beta_2}{\beta_1} = \frac{v_1}{v_2} = \frac{n_2}{n_1} = n_{12} = \sqrt{\frac{\varepsilon_2}{\varepsilon_1}} \tag{6.152}$$

The reason to write the same equation in the new form is to understand the relative variations of the absolute refractive indices of the two media and their effects in wave propagation and total transmission. For oblique wave propagation from an electromagnetically lighter medium 1 to a denser medium 2 we have $n_2 > n_1$ and $\varepsilon_2 > \varepsilon_1$; and in this case, $n_{12} > 1$. Substitution for $n_{12} = \sqrt{\varepsilon_2/\varepsilon_1}$ and simple manipulations of (6.152) we obtain

$$\sin\theta_t = n_{12}^{-1}\sin\theta_i < \sin\theta_i \tag{6.153}$$

In this section, we shall examine the Fresnel formula only for the case in which θ_i has the real value within the range of 0 to $(\pi/2)$. Consider the case for TM polarization where the electric fields (E_o^r) and (E_o^i) are in the plane of incidence:

$$\Gamma_{TM} = \frac{E_o^r}{E_o^i} = \frac{\eta_2\cos\theta_t - \eta_1\cos\theta_i}{\eta_2\cos\theta_t + \eta_1\cos\theta_i} \tag{6.154}$$

Dividing numerator and denominator with η_1 and expressing the ratio of $\eta_2/\eta_1 = \beta_1/\beta_2$ and *Snell's law of refraction* $\eta_2/\eta_1 = \beta_1/\beta_2 = \sin\theta_t/\sin\theta_i$ in (6.148) we obtain:

---

46 We also mentioned many times earlier that the loss factor does not affect the calculation of the reflection and transmission theory except for the fact that it will make the coefficients complex.

$$\Gamma_{TM} = \frac{\eta_2/\eta_1 \cos\theta_t - \cos\theta_i}{\eta_2/\eta_1 \cos\theta_t + \cos\theta_i} = \frac{\sin\theta_t \cos\theta_t - \sin\theta_i \cos\theta_i}{\sin\theta_t \cos\theta_t + \sin\theta_i \cos\theta_i} = \frac{\sin 2\theta_t - \sin 2\theta_i}{\sin 2\theta_t + \sin 2\theta_i} = \frac{\tan(\theta_i - \theta_t)}{\tan(\theta_i + \theta_t)} \tag{6.155}$$

From (6.155) we can observe that in TM polarization, the electric field is in the plane of incidence, the denominator of the expression for Γ_{TM} is finite except for the case where $\theta_i + \theta_t = \pi/2$ results in $\tan(\theta_i + \theta_t) = \infty$. The consequence is $\Gamma_{TM} = 0$; the reflected field amplitude vanishes. This means for this case, the transmitted and reflected rays' propagation unit vectors $\boldsymbol{a_r} \perp \boldsymbol{a_t}$, respectively, hence the reflected and the transmitted rays are perpendicular to each other. In this case, *Snell's law of refraction* in (6.152) reduces to:

$$\frac{\sin\theta_i}{\sin\theta_t} = \frac{\sin\theta_i}{\sin(\pi/2 - \theta_i)} = \tan\theta_i = \frac{\beta_2}{\beta_1} = \frac{v_1}{v_2} = \frac{n_2}{n_1} = n_{12} = \sqrt{\frac{\varepsilon_2}{\varepsilon_1}} \tag{6.156}$$

Simply (6.156) is defined as follows:

$$\tan\theta_i = n_{12} = \sqrt{\frac{\varepsilon_2}{\varepsilon_1}};\theta_{BA} = \tan^{1}\sqrt{\left(\frac{\varepsilon_{r2}}{\varepsilon_{r1}}\right)} \tag{6.157}$$

The angle θ_i for which (6.157) is satisfied is *the Brewster angle* that was derived in (6.151) using the assumption $\Gamma_{TM} = 0$ for *the total transmission* in medium 2. We have assumed $\varepsilon_2 > \varepsilon_1$ in the above derivation. We can also get a solution for (6.157) for the case of $\varepsilon_2 < \varepsilon_1$. Therefore, there is always some angle θ_i exists for the case of the electric field in the plane of incidence (TM polarization) which provides reflection free transmission. This can further justify that for any arbitrary polarized oblique incident field which can be decomposed into both TE and TM polarized waves; the TM polarized wave has no reflected field wave component in medium 1 at the incident angle equals to Brewster *angle*, and *the total transmission* happens in medium 2. The TE polarized wave has both transmitted and reflected field in both media. This means an unpolarized wave incident on a dielectric boundary becomes linearly (perpendicularly) polarized in its reflection. This principle is used in many optical devices such as anti-glare sunglasses and 3D movie glasses. In wireless communications in microwave frequencies, there are many applications which are discussed in the case studies in the following sections. Inan and Inan [1] provide the physical interpretation of the total transmission phenomenon in terms of *the electron theory* of dynamic electromagnetics.[47] The incident electric field vibrates that produces the constituent atoms in second medium so that a TE field is generated that reflects back in the first medium. It is well known that dipoles do not emit radiation exactly in the direction of their oscillation [2]. Since the reflected field wave is at right angle to the transmitted wave hence the reflected wave does not receive any energy for propagation in the plane of incidence (Figure 6.38).

Example 6.13: Total Transmission – Brewster Angle

Consider a parallel polarized signal entering from an electromagnetically lighter medium to a denser medium with a relation between the relative permittivities of the two media $\varepsilon_{r1} = 8\varepsilon_{r2}$, calculate θ_{BA}. Recalculate θ_{BA} for the signal travels from opposite direction. Assume the medium is non-magnetic and lossless for simplicity.

47 Chapter 5 has presented *the electron theory* of dynamic electromagnetic fields in terms of quantum mechanics of atoms in details.

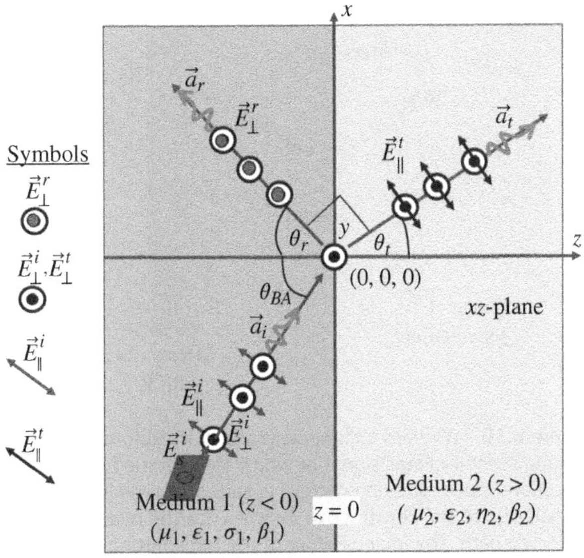

Figure 6.38 Arbitrarily polarized incident wave transmission at *Brewster angle.*

Answer:

Given the ratio of the dielectric constants of the two media as: $\varepsilon_{r2}/\varepsilon_{r1} = 8$. Using (6.150) we obtain

i) $\theta_{BA1} = tan^{1}\sqrt{(8)} = 70.5°$

ii) For the signal traveling from opposite direction we just swap the ratio $\varepsilon_{r1}/\varepsilon_{r2} = 1/8$. This yields

$$\theta_{BA2} = tan^{1}\sqrt{\left(\frac{1}{8}\right)} = 19.5°$$

Remarks

Observe that $\theta_{BA1} + \theta_{BA2} = 90°$; they are complementary to each other as expected.

6.11.6 Brewster Angle and Wireless Communications

As stated above, the total transmission and Brewster angle play significant role in wireless communications. Antennas are used for wireless communications that radiates field in all directions. Such radiation patterns are called omnidirectional radiation pattern or figure-of-eight radiation pattern and are launched by a dipole antenna or a monopole antenna on a ground plane as the reflector.[48] Figure 6.39 shows a vertically polarized dipole antenna which is erected at h_1 above the ground plane. A mobile phone user with the antenna at height h_2 receives the signal when the user is obstructed with buildings and trees. Such wireless communications are called non-line of sight (NLOS) communications. There is also a line of sight (LOS) path if there is no obstacle in the propagation path.

48 We shall discuss these types of radiation patterns in antenna theory.

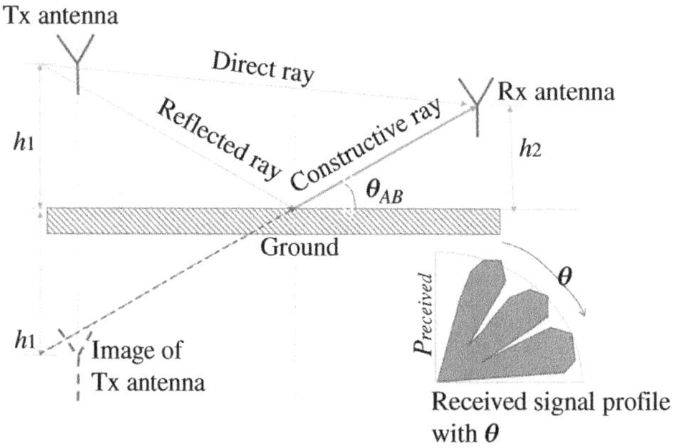

Figure 6.39 Wireless communications that exploits Brewster angle. A vertically polarized antenna above a ground plane radiates signal of which the electric field is also vertically oriented and propagates in all directions. Such radiation pattern is called omnidirectional radiation pattern. A distant receiver receives the signal. Due to the constructive interference the receiver receives more signal from both direct path and the reflected path. The image theory of the transmitting antenna acts like a total transmission of signal from the ground (medium 1) to air (medium 2). *Source:* Adopted from Kraus and Fleisch [3].

Explain how this is possible. Figure inset shows the resultant fields with the azimuth angle. With the phase difference between the LOS and NLOS signal paths the constructive and destructive interface occur. Thus the radiation pattern received by the user is like a palm leaf with nulls and maxima. The maxima can be perceived as the total transmission of the wave as it is due to the image theory the signal is totally transmitted from reflective ground medium to air.

6.11.7 Chipless RFID Polarizer Exploits Brewster Angle

A 60 GHz chipless RFID (also called EM barcode) is shown in Figure 6.40. The tag is comprised of two types of electromagnetic polarizers: (i) a 45°inclined straight stripline of $\lambda/2$ resonant length at 60 GHz, and (ii) a meander line version of the stripline. When an arbitrary polarized signal from the RFID reader antenna illuminates the EM barcode, it in returns sends signal to the reader only one polarized signal at certain inclined angle and separate the signal into two orthogonal polarizations. Thus, only one polarized signal is received by the receiving antenna. As shown in the figure, the arbitrary polarized transmitted signal is incident at the Brewster angle θ_{AB}. Due to total

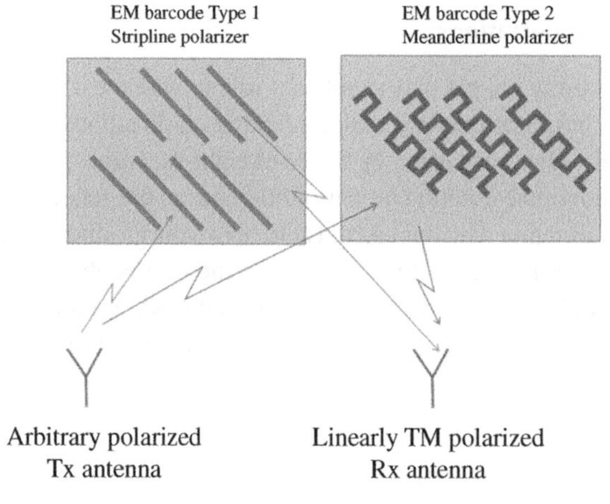

Figure 6.40 Brewster angle in action in EM barcodes.

transmission of TM polarized signal, TM polarized signal is received by the receiving antenna. This technique is used in many applications to mitigate the noise as the noise only amplifies the co-polar component leaving the cross-polar components interference free. This method is very useful for cross-polar interference mitigation in radars and invariably in all wireless communications systems.

6.11.8 Effective Impedance Concept of TM Polarized Oblique Incidence

We can also perform the effective impedance concept for TM polarization as we did for the TE polarized oblique incidence case in Section 6.10.4. We shall use (6.142) and the definitions of the intrinsic impedances of the two media to develop the impedance concept for TM polarization [4] and derive the expressions of effective impedances for the media in the case for the TM polarization. Substations for $E_0^i/H_0^i = -E_0^r/H_0^r = \eta_1$ and $E_0^2/H_0^2 = \eta_2$ in (6.142) we obtain:

$$H_o^i \eta_1 \cos\theta_i - H_o^r \eta_1 \cos\theta_i = H_o^t \eta_2 \cos\theta_t \tag{6.158}$$

As usual, we use minus sign for the reflected magnetic field due to the reversal of its direction as shown in Figure 6.37. Here we have defined the new impedances for the TM oblique incidence case, which are assumed to be real and positive quantities:

$$\eta_{1TM} = \eta_1 \cos\theta_i \text{ and } \eta_{2TM} = \eta_2 \cos\theta_t \tag{6.159}$$

Here the subscripts 1 and 2 represent media 1 and 2, respectively, and TM represents the TM/parallel polarization. In contrast to the TE polarization impedances, the TM polarization impedances η_{1TM} and η_{2TM} are smaller than the intrinsic impedance. As discussed earlier, this is due to the fact the in TM polarization, the electric field is divided into two components and we are considering the definition with one component of the field divided by the y-directed magnetic field component. Using the definitions of the new effective impedances in (6.159) into (6.147) and (6.148), respectively, we obtain the new expressions for the transmission and reflection coefficients for the TM polarization oblique incidence case in terms of the new impedances η_{1TM} and η_{2TM} as follows:

$$\tau_{TM} = \frac{E_o^t}{E_o^i} = \frac{2\eta_{2TM}}{\eta_{2TM} + \eta_{1TM}}\left(\frac{\cos\theta_i}{\cos\theta_t}\right) \tag{6.160}$$

$$\Gamma_{TM} = \frac{E_o^r}{E_o^i} = \frac{\eta_{2TM} - \eta_{1TM}}{\eta_{2TM} + \eta_{1TM}} \tag{6.161}$$

These expressions for τ_{TM} and Γ_{TM} in terms of η_{1TM} and η_{2TM} polarization impedances are much easier to recall and have similarities with the expressions of the reflection and transmission coefficients for the normal incidence case. The relationship of (6.144) still holds true.

$$\tau_{TM} = 1 + \Gamma_{TM}\left(\frac{\cos\theta_i}{\cos\theta_t}\right) \tag{6.162}$$

In the above derivations, we assume the effective impedances are real and positive quantities. We shall do one problem using the impedance concept for TM polarization.

Example 6.14: TM Polarized Field Calculation Using Impedance Concept
A uniform TM-polarized wave at 100 MHz impinges from air to glass interface at an angle of incidence of 35°. Calculate: (i) effective impedances of the two media, (ii) the reflection and transmission coefficients, and (iii) the fraction of transmitted power in medium 2. The refractive index of glass $n_2 = 1.43$.

Answer:

First we calculate the angle of transmission: $\theta_t = sin^{-1}(sin35° /1.45) = 23.3°$

i) Using (6.159) the effective impedances for TM-polarization in media 1 (air) and 2 (glass) respectively the impedances are:

$$\eta_{1TM} = 377 \times cos\,(35°) = 308.82\,\Omega$$

$$\eta_{2TM} = 377 \times cos\,(23.3°) = 346.25\,\Omega$$

ii) Using (6.160) and (6.161), the transmission and reflection coefficients, respectively, are:

$$\tau_{TM} = = \frac{2 \times 346.25}{346.25 + 308.82}\left(\frac{cos\,35°}{cos\,23.3°}\right) = 0.942$$

$$\Gamma_{TM} = \frac{E_o^r}{E_o^i} = \frac{346.25 - 308.82}{346.25 + 308.82} = -0.057$$

iii) The fraction of the incident power reflected in medium 1 (air) is:

$|\Gamma_{TM}| = |0.057|^2 = 3.26 \times 10^{-3}$; % of fractional reflected power is: 0.326%

6.11.9 Total Transmission in the Light of Impedance Concept

We have examined the total internal reflection in the light of impedance concept in Section 6.9.[43] In this section, we shall examine the theory of *total transmission* using *the effective impedance concept* for TE and TM polarizations [4]. As we have examined in Section 6.11.4, the condition for *the total transmission* simply considers the reflection coefficient $\Gamma_{TM} = 0$. First let us examine if the theory of *total transmission* for the case of TE-polarization is possible. In this case, we also consider $\Gamma_{TE} = 0$ assuming *total transmission* is possible in TE polarization; and (6.94) reduces to:

$$\eta_2\, cos\,\theta_i = \eta_1\, cos\,\theta_t \tag{6.163}$$

Simple mathematical manipulations and using the definitions of $\eta_{1TE} = \eta_1/\,cos\,\theta_i$; and $\eta_{2TE} = \eta_2/\,cos\,\theta_t$ from (6.96), we can easily show that $\eta_{1TE} = \eta_{2TE}$. Which is not possible if the medium 1 and medium 2 to be different and *Snell's law of refraction* to be true ($\theta_1 \neq \theta_2$). Again, using Snell's law of refraction from (6.84) into (6.163) and simple manipulation we obtain:

$$\eta_2\sqrt{1 - sin^2\theta_i} = \eta_1\sqrt{1 - sin^2\theta_t} = \eta_1\left\{\sqrt{1 - \left(\frac{\beta_1}{\beta_2}\right)^2 sin^2\theta_i}\right\} \tag{6.164}$$

Further simplification for $\left(\frac{\beta_1}{\beta_2}\right)^2 = \left(\frac{n_1}{n_2}\right)^2$ for optical media we obtain:

$$\eta_2\left\{1 - \left(\frac{n_1}{n_2}\right)^2 sin^2\theta_i\right\}^{-1/2} = \eta_1\left(1 - sin^2\theta_i\right)^{-1/2} \tag{6.165}$$

Examining (6.165), we can conclude that there is no value of the angle of incidence θ_i in TE-polarization that satisfies the above equality. Therefore, *no total transmission* is possible in TE polarization.

Now consider the TM polarization with definitions of η_{1TM} and η_{2TM} in (6.159), Snell's law of refraction in (6.84) and substitute them in the definition of $\Gamma_{TM} = 0$ in (6.161), we obtain [4]:

$$\eta_2\left\{1 - \left(\frac{n_1}{n_2}\right)^2 sin\,\theta_i\right\}^{1/2} = \eta_1\left(1 - sin^2\theta_i\right)^{1/2} \tag{6.166}$$

Also (6.166) for TM polarization has a solution, which helps derive sine of Brewster angle as the solution [4]. This is derived as follows.

To derive the solution for (6.166), we follow exactly the similar procedure for derivation steps of (6.165). Squaring both sides of (6.166) and simple manipulations and substitutions for $\theta_i = \theta_{BA}$ we obtain exactly the same expression for (6.151) as follows:

$$\theta_{BA} = sin^{-1}\left(1 + \frac{\varepsilon_{r1}}{\varepsilon_{r2}}\right)^{-1/2} = tan^{1}\sqrt{\left(\frac{\varepsilon_{r2}}{\varepsilon_{r1}}\right)} = tan^{-1}\left(\frac{n_2}{n_1}\right) \tag{6.167}$$

We have highlighted many examples of *total transmission* specific applications to wireless communications, chipless RFID and polaroid sunglasses. Besides these, there are many other applications of *total transmission* for other areas of wireless communications and optics. Readers are encouraged to find such applications in open sources. In the followings, we shall do a worked-out example and then move on to some special cases for TM polarized oblique incidence cases.

Example 6.15: TM Polarized Field Calculation Using Impedance Concept
A uniform plane wave from a 2.45 GHz Wi-Fi router on the ceiling of a room impinges on a glass window ($n_2 = 1.45$) from air as medium 1. The direction of propagation is along z-axis. The electric field is TM polarized and amplitude of 25 V/m. The incidence angle is 50°. Calculate the followings: (i) the angle of transmission, (ii) the effective impedances η_{1TM} and η_{2TM}, (iii) the reflection and transmission coefficients, (iv) the fraction of power reflected in air and transmitted in glass, and (v) compare the results of power transmission for the case of TE-polarization.

Answer:
We have done this problem in TE-polarization in Example 8.9 for the TE polarized oblique incidence case. We repeat the problem for the case of TM polarization to examine the difference between the TE and TM polarization oblique incidence cases.

Given parameters: operating frequency $f = 2.45 \times 10^9$ (Hz); the angular frequency $\omega = 1.54 \times 10^{10}$ (rad/m)

The refractive index of medium 1: $n_1 = 1$; and medium 2 glass is: $n_2 = 1.45$.

The amplitude of the incident electric field: $E_0^i = 25$ (V/m); the angle of incidence: $\theta_i = 50°$.

i) The Snell's law of refraction is the same for both TE and TM polarizations, hence according to (6.84), the angle of transmission:
   $\theta_t = (sin\,(50°) \times (1/1.45)) = 31.89°$
ii) The effective impedances: $\eta_{1TM} = \eta_1 \cos\theta_i = 377 \times \cos\,(50°) = 242.33\ (\Omega)$

$$\eta_{2TE} = \eta_2 \cos\theta_t = \frac{377}{1.45} \times \cos\,(31.89°) = 220.76\ (\Omega)$$

iii) According to (6.161), the reflection coefficient: $\Gamma_{TM} = (\eta_{2TM} - \eta_{1TM})/(\eta_{2TM} + \eta_{1TM}) = (220.76 - 242.33)/(220.76 + 242.33) = -0.05$

   The negative sign of Γ_{TM} suggests a phase reversal of the reflected field at the air–glass interface.

   According to (6.160), the transmission coefficient: $\tau_{TM} = (2\eta_{2TM})/(\eta_{2TM} + \eta_{1TM})$ $(\cos\theta_i/\cos\theta_t) = (2 \times 220.76)/(220.76 + 242.33) \times \left(\frac{\cos 50°}{\cos 31.89°}\right) = 0.72$

iv) The fraction of reflected power: $|\Gamma_{TE}|^2 = |0.05|^2 = 0.0025 = 0.25\%$
   The fraction of transmitted power: $1 - |\Gamma_{TE}|^2 = 1 - 0.0025 = 0.9975 = 99.75\%$
   More than 99% incident power is transmitted via the glass window.
v) Brewster angle is:

$$\theta_{BA} = tan^{-1}\left(\frac{n_2}{n_1}\right) = tan^{-1}(1.45) = 55.4°$$

The angle of incidence $\theta_i = 50°$ is very close to the Brewster angle $\theta_{BA} = 55.4°$. This suggests almost 100% transmission for the case of the TM polarized signal. While for TE polarized signal the transmitted power is 90.4%.

6.11.10 Special Cases of Γ_{TM}

6.11.10.1 Reflection Coefficient Γ_{TM} for Perfect Conductor

We have examined the Reflection coefficient Γ_{TM} for perfect conductor for TE polarized oblique incidence in Section 6.10. We found that for the perfect conductor $\Gamma_{TE} = -1$ (6.102) and $\tau_{TE} = 0$. Now we are going to derive the reflection coefficient for medium 2 as the conductor using (6.148) for a conducting medium $\eta_2 = 0$. Hence (6.148) reduces to

$$\Gamma_{TM} = \frac{E_o^r}{E_o^i} = \frac{0 - \eta_1 \cos\theta_i}{0 + \eta_1 \cos\theta_i} = -1 \tag{6.168}$$

Likewise using (6.147), we can also show that $\tau_{TM} = 0$. Here, the boundary condition of the tangential component of the total electric field on the conducting interface at $z = 0$ vanishes. This is the result that we expect for a perfect conducting boundary. It tells us again that the reflection coefficient Γ_{TM} for the TM polarized oblique incidence case is unity with a phase reversal of 180°. This is equivalent to a short-circuited termination for a transmission line. Also note that the direction chosen in reference to the directions of the reflected electric field E_r and the reflected magnetic field H_r in Figure 6.33 is arbitrary and were justified with the right-hand rule of wave propagation. The sign of the reflection coefficient also satisfied the direction chosen for the reflected electromagnetic fields.

6.11.10.2 Both Medium Lossless and Non-magnetic Media

If both media are lossless and non-magnetic ($\mu_1 = \mu_2 = \mu_0$) applying *Snell's law of refraction* in (6.84) and the definition of Γ_{TM} in (6.148) we obtain

$$\Gamma_{TM} = -\frac{\cos\theta_i - (\eta_2/\eta_1)\cos\theta_t}{\cos\theta_i + (\eta_2/\eta_1)\cos\theta_t} = -\frac{\cos\theta_i - \sqrt{(\varepsilon_1/\varepsilon_2)}\sqrt{1 - \sin^2\theta_t}}{\cos\theta_i + \sqrt{(\varepsilon_1/\varepsilon_2)}\sqrt{1 - \sin^2\theta_t}}$$

$$= -\frac{\cos\theta_i - \sqrt{(\varepsilon_1/\varepsilon_2)}\sqrt{1 - (\varepsilon_1/\varepsilon_2)\sin^2\theta_i}}{\cos\theta_i + \sqrt{(\varepsilon_1/\varepsilon_2)}\sqrt{1 - (\varepsilon_1/\varepsilon_2)\sin^2\theta_i}} = -\frac{\cos\theta_i - (\varepsilon_1/\varepsilon_2)\sqrt{(\varepsilon_2/\varepsilon_1) - \sin^2\theta_i}}{\cos\theta_i + (\varepsilon_1/\varepsilon_2)\sqrt{(\varepsilon_2/\varepsilon_1) - \sin^2\theta_i}} \tag{6.169}$$

We have obtained very similar form of expression for the Γ_{TE} in (6.103). Comparing (6.103) and (6.169) we can justify that the critical angle, and total internal reflection with the reflection coefficient $|\Gamma_{TM}| = 1$ is still valid for TM-polarization.

6.11.10.3 Brewster Angle and Laser Beam with TM Polarization

Lasers have multitude of applications such as laser printers, computer discs, barcode scanners, and industrial laser cutters and drills in industrial applications. The application of *the total transmission* is illustrated in Figure 6.41 for a gas laser beam [5]. In a laser, the light wave travels between mirrors that are placed at Brewster angle and the TE polarized light is filtered out in the early stage of the laser cavity (not shown here). The end mirror is put at Brewster angle to the optic axis as shown in Figure 6.41. The TM Polarized wave travels with *total transmission* at the other end of the end

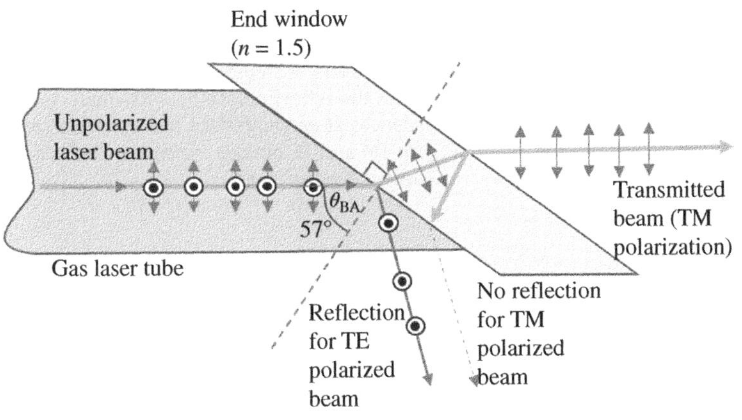

Figure 6.41 Brewster angle as the end window of a laser source. The end window placed at Brewster angle at the end of the leaser tube reflects the TE polarized light and offers full transmission to TM polarized light. *Source:* Adapted from Arieli [5].

mirror. Therefore, there is no reflection loss and only TM polarized signal is oscillating as the output of the laser. The Brewster angle in this example is: $\theta_{BA} \cong 57°$.

6.11.10.4 Calculations of Γ_{eff} for TM and TE Oblique Incidence on Multiple Dielectric Layer

So far, we have examined the TM oblique incidence case for only two dielectric media. Similar to TE polarization case in Section 6.6, we can expand the theory for any number of medias for the TM oblique incidence case. In the case for TE polarized multilayered dielectric media we used Snell's law of refraction to calculate the total displacement. There we have found that the entrance angle is equal to the exit angle. Since we used the ray trace model in the Snell's law of refraction, the theory was equally applicable to the TM polarized oblique incidence case. Here we shall use the effective impedance concept by successively transforming the impedance like the carpet folding technique depicted in Section 5.8.5. Since we are using the effective impedance concept for the effective reflection coefficient calculation we shall treat the TM polarization case first followed by the TE polarization case. Figure 6.42 illustrates the step by step procedure to calculate the reflection coefficients, input impedances at each interface and the effective Γ_{eff} for both the TM and TE oblique incidence cases on a multilayered dielectric medium with N layers. As can be seen in the figure, to start with the calculation of the effective reflection coefficient in medium 1 at the interface 1, we need the input data in the form of the original incidence angle in θ_i, the permittivity ε_n, the permeability μ_n and the thickness d_n for each layer. For perfect dielectric media $\mu_n = \mu_0$, for $n = 1, 2, 3, \ldots N$. The input data can be obtained from Figures 6.43 and 6.44 for TM and TE polarized oblique incidence cases on multiple dielectric layered media, respectively. Since the entrance angle θ_1 in medium 1 is equal to the exit angle θ_N in medium N, we use Snell's law of refraction to calculate all the angles θ_n for $n = 1, 2, 3, \ldots N$ as we have done in Section 6.10. We calculate the intrinsic wave impedance $\eta_n = \sqrt{\mu_n/\varepsilon_n}$ for $n = 1, 2, 3, \ldots N$. The next step is to compute the mode wave impedance of the end layer as follows:

$$\text{Calculate } N\text{-th layer impedance}: \quad Z_{zN}^{TM} = \eta_N \cos\theta_{iN}; Z_{zN}^{TE} = \eta_N \sec\theta_{iN} \tag{6.170}$$

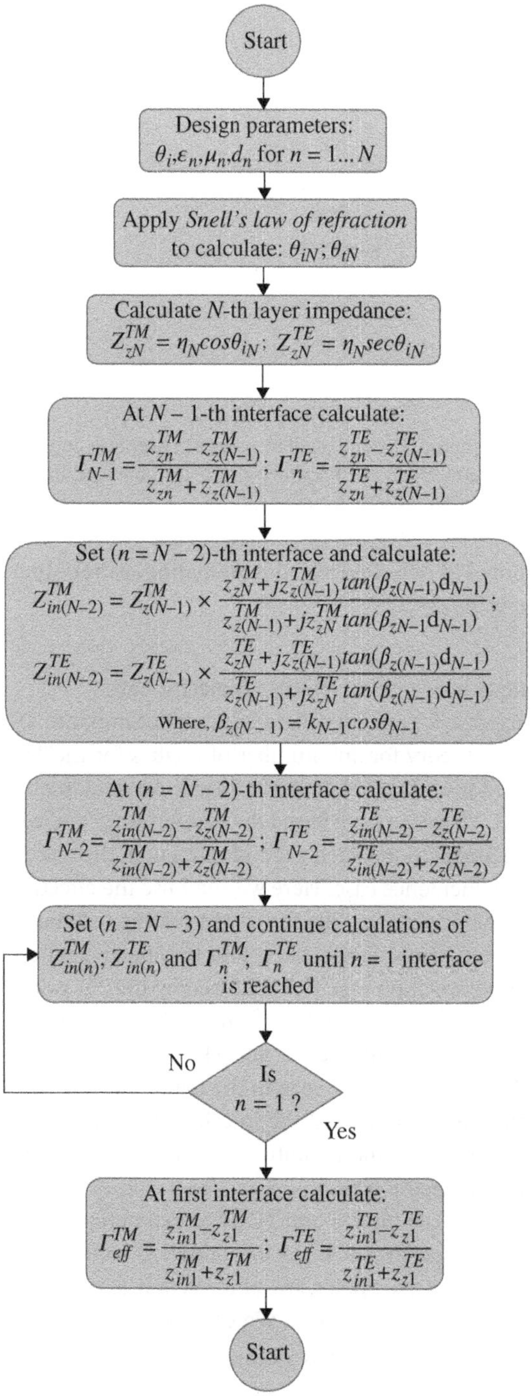

Figure 6.42 Flowchart for calculations of Γ_{eff} for TM and TE Oblique incidences on multiple dielectric layers. A step-by-step procedure to calculate the reflection coefficients, input impedances at each interface and the effective Γ_{eff} for TM and TE oblique incidence cases on multiple dielectric layers.

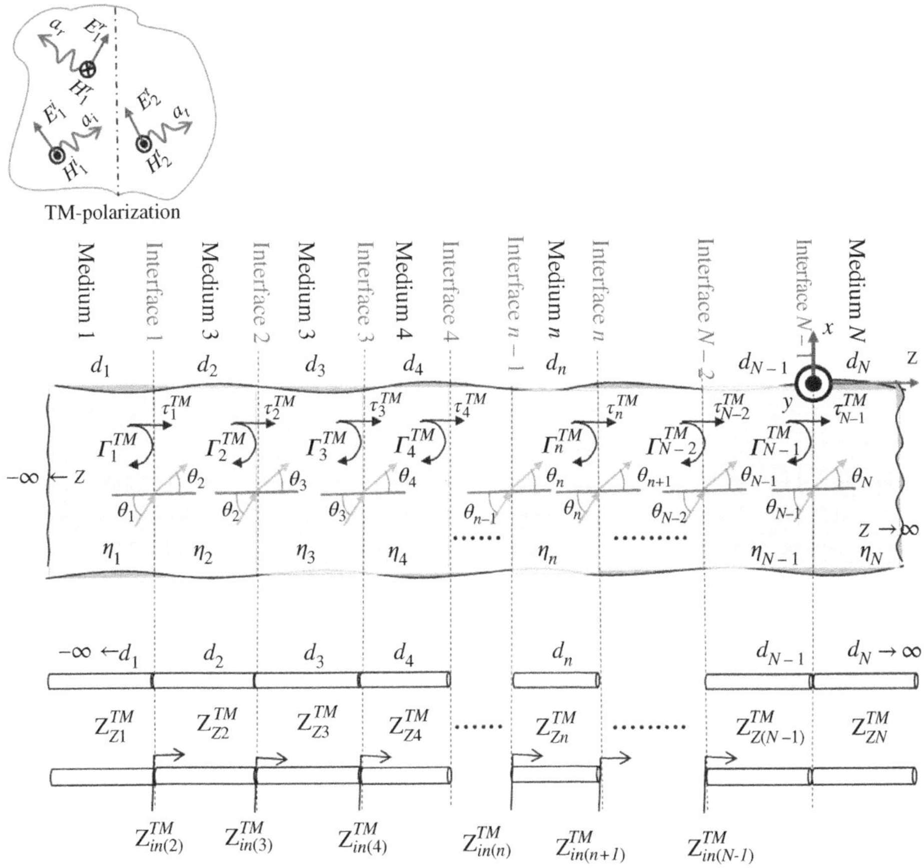

Figure 6.43 TM polarized oblique incidence case on multiple dielectric layered media.

Since the wave impedance is different in TM and TE cases we need to calculate them separately. We use the Z_{zN}^{TM} and Z_{zN}^{TE} in (6.170) to calculate the reflection coefficients for both cases as follows:

At N-1th interface calculate :

$$\Gamma_{N-1}^{TM} = \frac{Z_{zn}^{TM} - Z_{z(N-1)}^{TM}}{Z_{zn}^{TM} + Z_{z(N-1)}^{TM}};\ \Gamma_{n}^{TE} = \frac{Z_{zn}^{TE} - Z_{z(N-1)}^{TE}}{Z_{zn}^{TE} + Z_{z(N-1)}^{TE}} \tag{6.171}$$

Using Z_{zN}^{TM} and Z_{zN}^{TE} and the successive values we calculate the input impedances and the reflection coefficients at n-th interfaces using the following expressions

Set $(n = N-2)$-th interface and calculate :

$$Z_{in(N-2)}^{TM} = Z_{z(N-1)}^{TM} \times \frac{Z_{zN}^{TM} + jZ_{z(N-1)}^{TM}\, tan\left(\beta_{z(N-1)}d_{N-1}\right)}{Z_{z(N-1)}^{TM} + jZ_{zN}^{TM}\, tan\left(\beta_{z(N-1)}d_{N-1}\right)};$$

$$Z_{in(N-2)}^{TE} = Z_{z(N-1)}^{TE} \times \frac{Z_{zN}^{TE} + jZ_{z(N-1)}^{TE}\, tan\left(\beta_{z(N-1)}d_{N-1}\right)}{Z_{z(N-1)}^{TE} + jZ_{zN}^{TE}\, tan\left(\beta_{z(N-1)}d_{N-1}\right)} \tag{6.172}$$

where, $\beta_{z(N-1)} = k_{N-1} \cos\theta_{N-1}$

And

At $(n = N-2)$-th interface calculate:

$$\Gamma^{TM}_{N-2} = \frac{Z^{TM}_{in(N-2)} - Z^{TM}_{z(N-2)}}{Z^{TM}_{in(N-2)} + Z^{TM}_{z(N-2)}};\ \Gamma^{TE}_{N-2} = \frac{Z^{TE}_{in(N-2)} - Z^{TE}_{z(N-2)}}{Z^{TE}_{in(N-2)} + Z^{TE}_{z(N-2)}} \tag{6.173}$$

We continue to do the computation of Z^{TM}_{zn} and Z^{TE}_{zn} and Γ^{TM}_{n} and Γ^{TE}_{n} until we reach the interface layer 1. At the first interface we compute the effective reflection coefficient as follows:

At first interface calculate :

$$\Gamma^{TM}_{eff} = \frac{Z^{TM}_{in1} - Z^{TM}_{z1}}{Z^{TM}_{in1} + Z^{TM}_{z1}};\ \Gamma^{TE}_{eff} = \frac{Z^{TE}_{in1} - Z^{TE}_{z1}}{Z^{TE}_{in1} + Z^{TE}_{z1}} \tag{6.174}$$

Figures 6.43 and 6.44 show the equivalent transmission line models for TM and TE oblique incidence cases on multilayered medium. Using the technique, we can calculate the input wave impedance and the reflection and transmission coefficients at any interfaces from both TM and TE polarizations.

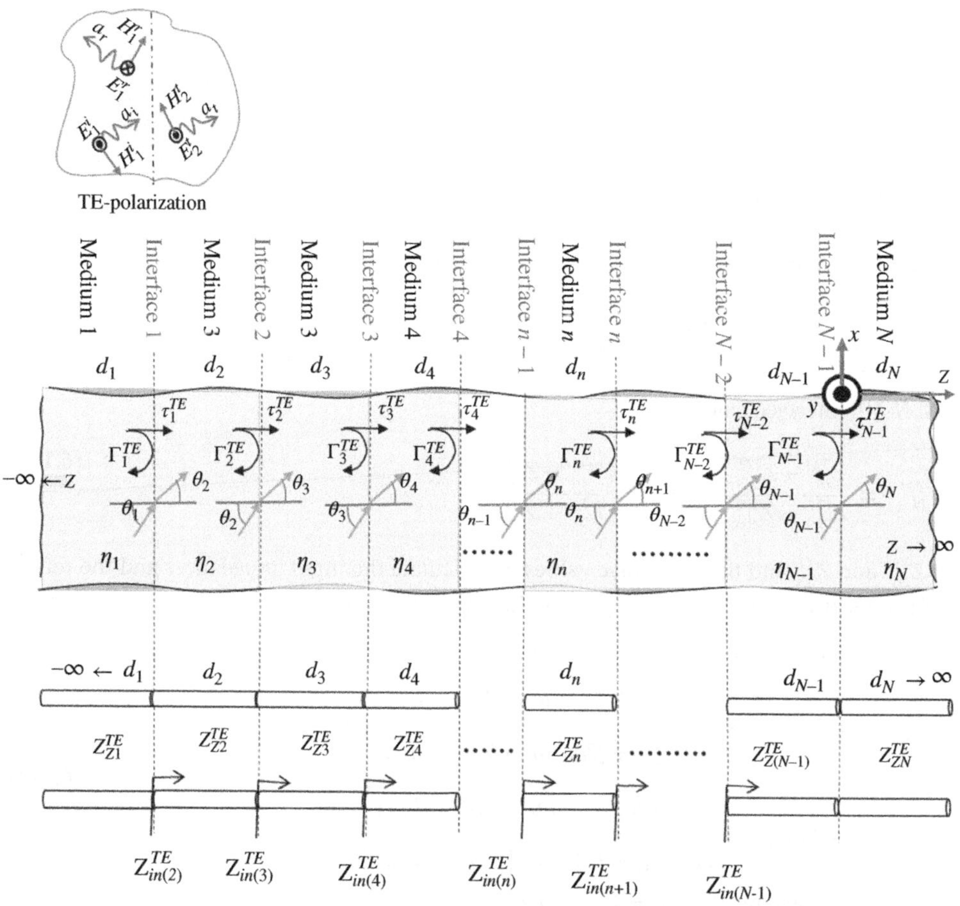

Figure 6.44 TE polarized oblique incidence case on multiple dielectric layered media.

Review Questions 6.10: Analysis of Oblique Incidence in Multilayered Medium

Q1. Explain the difference in the working procedure between the oblique and normal incidence cases on a multilayered medium.

Q2. Explain why we need to consider different phase constants β_z^{TM} and β_z^{TE} instead of β only for the normal incidence case.

Q3. Explain the significance of Z_{zTM} and Z_{zTE}, why they are different than the normal intrinsic wave impedance η in individual layers?

Q4. Why Z_{zTM} is smaller than η the intrinsic wave impedance of the medium?

Q5. Why Z_{zTE} is larger than η the intrinsic wave impedance of the medium?

Q6. Derive the equivalent impedance at the interface 1.

Q7. Derive the effective reflection coefficient of the multilayered medium: Γ_{eff}.

Q8. Calculate the percentage of the reflected power and transmitted power in terms of the incident power for both TE and TM polarized wave incidence of N-layered medium.

Example 6.16: TM and TE Oblique Incidences in Multilayered Medium

A microstrip frequency selective surface is comprised of two dielectric media: Taconic TLX0 with dielectric constant $\varepsilon_r = 2.45$ and thickness 1.6 mm followed by RF4 laminate with dielectric constant $\varepsilon_r = 4.2$ and thickness 4 mm. The operating frequency is 10 GHz. The angle of incidence is 45°. Calculate the followings:

(i) the wave impedance in each layer, (ii) the input impedance in each interface from the right end, (iii) the effective reflection coefficient Γ_{eff}, and (iv) the percentage power reflected.

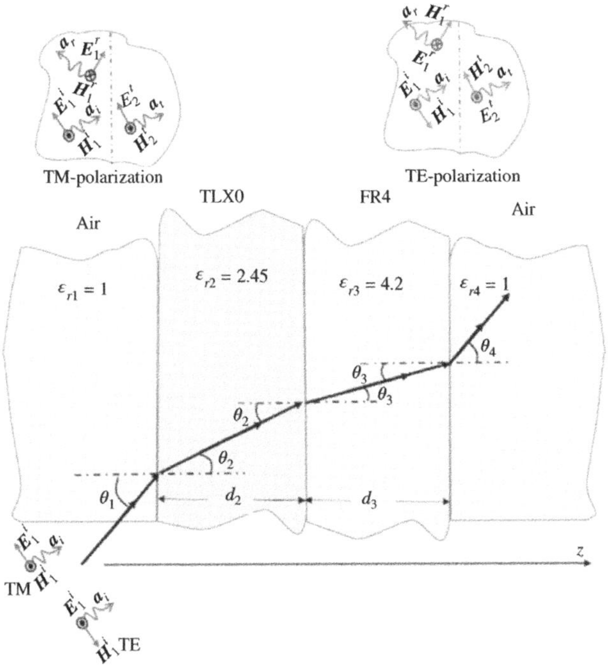

Figure E6.16 A frequency selective surface with two dielectric layers.

Answer:

i) The wave impedance for TM polarization:

$$\theta_1 = 45°; Z_{z1}^{TM} = \eta_1 \cos\theta_1 = 377 \times \cos 45° = 266.6\ (\Omega)$$

$$\theta_2 = \left(\frac{\sin 45°}{\sqrt{2.45}}\right) = 26.9°; Z_{z2}^{TM} = \eta_2 \cos\theta_2 = \frac{377}{\sqrt{2.45}} \times \cos 26.9° = 214.8\ (\Omega)$$

$$\theta_3 = \left(\frac{\sin 26.9°}{\sqrt{4.2}}\right) = 12.8°; Z_{z3}^{TM} = \eta_3 \cos\theta_3 = \frac{377}{\sqrt{4.2}} \times \cos 12.8° = 179.4\ (\Omega)$$

$$\theta_4 = \theta_1 = 45°; Z_{z4}^{TM} = \eta_4 \cos\theta_4 = 377 \times \cos 45° = 266.6\ (\Omega)$$

The wave impedances for TE polarization:

$$Z_{z1}^{TE} = \eta_1 \sec\theta_1 = \frac{377}{\cos 45°} = 533.2\ (\Omega)$$

$$Z_{z2}^{TE} = \eta_2 \sec\theta_2 = \frac{377/\sqrt{2.45}}{\cos 26.9°} = 270.1\ (\Omega)$$

$$Z_{z3}^{TE} = \eta_3 \sec\theta_3 = \frac{377/\sqrt{4.2}}{\cos 12.8°} = 188.6\ (\Omega)$$

$$Z_{z4}^{TE} = \eta_4 \sec\theta_4 = \frac{377}{\cos 45°} = 533.2\ (\Omega)$$

ii) The input impedance calculation

First, we need to calculate the wave number (phase constant) along z-axis β_{zn}^{TM} and β_{zn}^{TE} for $n = 3$ and 2 for both TM and TE polarizations, respectively. Then we calculate the phase in each layer in radians.

The z-directed phase constant of medium 3:

$$\beta_{z3}^{TM} = \frac{2\pi}{\lambda_3} \cos\theta_3 = \frac{2\pi \times 10^{10}\ (\text{Hz}) \times \sqrt{4.2}}{3 \times 10^8\ (\text{m/s})} \times \cos(12.8°) = 418.6\ (\text{rad/m});$$

The electrical length of medium 3:

$$\beta_{z3}^{TM} d_3 = 416.8\ (\text{rad/m}) \times 4 \times 10^{-3}\ (\text{m}) = 1.67\ (\text{rad})$$

The input impedance at interface 3:

$$Z_{in3}^{TM} = Z_{z4}^{TM} = 266.6(\Omega)$$

The z-directed phase constant of medium 2:

$$\beta_{z2}^{TM} = \frac{2\pi}{\lambda_2} \cos\theta_2 = \frac{2\pi \times 10^{10}\ (\text{Hz}) \times \sqrt{2.45}}{3 \times 10^8\ (\text{m/s})} \times \cos(26.9°) = 292.35\ (\text{rad/m});$$

The electrical length of medium 2:

$$\beta_{z2}^{TM} d_2 = 292.35\,(\mathrm{rad/m}) \times 1.6 \times 10^{-3}\,(\mathrm{m}) = 0.468\,(\mathrm{rad})$$

The impedance seen at the interface 2:

$$Z_{in2}^{TM} = 214.8 \times \frac{266.6 + j214.8 \times tan\,(1.67)}{214.8 + j266.6 \times tan\,(1.67)} = 173.7 + j7.5\,(\Omega)$$

The impedance seen at the interface 1:

$$Z_{in1}^{TM} = 533.2 \times \frac{173.7 + j7.5 + j533.2 \times tan\,(0.468)}{533.2 + j(173.7 + j7.5) \times tan\,(0.468)} = 215.3 + j243.3\,(\Omega)$$

The effective reflection coefficient (6.174):

$$\Gamma_{eff}^{TM} = \frac{215.3 + j243.3 - 266.6}{215.3 + j243.3i + 266.6} = 0.12 + j0.4 = 0.4\angle 73$$

Similar calculation for TE polarization is left as an exercise.

6.11.11 Power Balance in TM Oblique Incidence

As examined the power balance equation for TE polarization in Section 6.10.7, the motto of the section is to first define the power reflection and power transmission coefficients for the TM polarization, and then find the relationship between the two coefficients with the help of the law of conservation of energy. The significance of working with the power reflection and transmission coefficients is derived from many practical applications of active and passive design that involves boundary value problems.

The detailed explanation of the generic attributes of the power balance equation and the relevant theory have been explain in Section 6.10.7 hence is not repeated here. Figure 6.45 illustrates the power balance equation and the procedure to calculate the relation between the power transmission and reflection coefficients for the TM polarization case. They are very different than those for the TE case due the orientation of the electromagnetic fields with respect to the plane of incidence and the direction of propagation.

You may easily observe the difference of the TM polarization case from the TE polarization case in Figure 6.28. The first layer of energy balance equation, the definition of power reflection and transmission coefficients are the ratios of the reflected power portion and transmitted power portion with respect to the incident power, respectively. In the second layer, the law of conservation of energy tells that the power reflection and transmission coefficients are unity which represents the normalized incident power.

Using the similar procedure as in Section 6.10.7 we define the three-time average power densities for the TM polarization as follows:

$$\text{The incident power density on } OB : P_{TMave,z}^{i} = \frac{|E_0^i|^2}{\eta_1} \boldsymbol{a_i}.\boldsymbol{a_z} = \frac{|E_0^i|^2}{\eta_1} cos\,\theta_i\,(\mathrm{W/m^2}) \tag{6.175}$$

$$\text{The reflected power density on } OB : P_{TMave,z}^{r} = \frac{|E_0^r|^2}{\eta_1} \boldsymbol{a_r}.(-\boldsymbol{a_z}) = \frac{|E_0^r|^2}{\eta_1} cos\,\theta_r\,(\mathrm{W/m^2}) \tag{6.176}$$

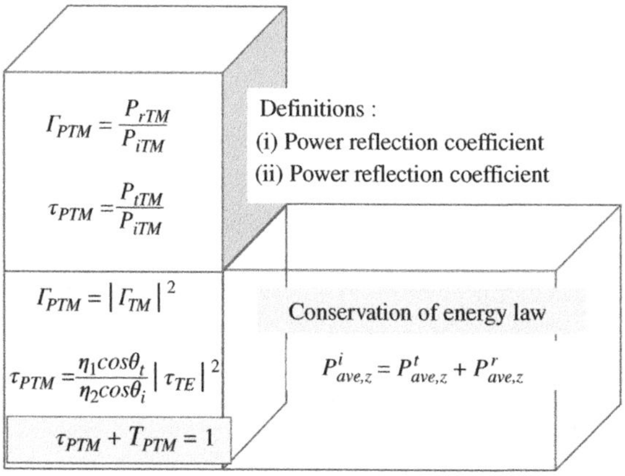

Figure 6.45 Twofold of definitions of power balance analysis of TM incidence case.

$$\text{The transmitted power density on } OB: P^t_{TMave,z} = \frac{|E^t_0|^2}{\eta_1} \boldsymbol{a}_t . \boldsymbol{a}_z = \frac{|E^t_0|^2}{\eta_1} \cos\theta_t \ (\text{W/m}^2) \tag{6.177}$$

Applying the law of conservation of energy, we derive the relation between the three average power densities as follows:

$$P^i_{TMave,z} = P^t_{TMave,z} + P^r_{TMave,z} \tag{6.178}$$

Substitutions for (6.175)–(6.177) into (6.178) lead to (6.179) as follows:

$$\frac{|E^i_o|^2}{2\eta_1} \cos\theta_i = \frac{|E^r_o|^2}{2\eta_1} \cos\theta_i + \frac{|E^t_o|^2}{2\eta_2} \cos\theta_t \tag{6.179}$$

This relation is exactly similar to (6.117). The similarity between the power balance equations in both TE and TM cases can be justified with the ray trace model as depicted in Figure 6.29. Now we can define the power reflection coefficient as $\Gamma_{PTM} = P^r_{TMave,z}/P^i_{TMave,z}$ in terms of their respective field components as usual:

$$\text{Power reflection coefficient}: \Gamma_{PTM} = \frac{P^r_{TMave,z}}{P^i_{TMave,z}} = \frac{(|E^r_o|^2/2\eta_1) \cos\theta_r}{(|E^i_o|^2/2\eta_1) \cos\theta_i} = |\Gamma_{TM}|^2 \tag{6.180}$$

(6.180) represents the fraction of power reflected. The percentage of reflected power is defined as:

$$\%\text{Power reflected} = \Gamma_{PTM} \times 100\% = |\Gamma_{TM}|^2 \times 100\% \tag{6.181}$$

Note that the power reflection coefficient which is the ratio of power reflected versus the power incident is defined as *refractivity*, and is also called *reflectance* in optics. As stated above the reflection of power happens in medium 1 hence with the help of *Snell's law of reflection* we have obtained a very simple expression of the power reflection coefficient which is just the square of the modulus of the voltage reflection coefficient. However, this is not the case for the power transmission coefficient as it involves two dissimilar dielectrics with two different phase velocities, impedances, and angles. Now we are going to derive the power transmission coefficient using the similar procedure as that for the power reflection coefficient as follows. The power transmission coefficient is defined as follows:

$$\tau_{PTM} = \frac{P^t_{TMave,z}}{P^i_{TMave,z}} = \frac{|E^t_o|^2}{|E^i_o|^2} \times \frac{\eta_1 \cos\theta_t}{\eta_2 \cos\theta_i} \tag{6.182}$$

Substitution for the voltage transmission coefficient $\tau_{TM} = E^t_o/E^i_o$ from (6.147) into (6.182) we obtain the expression for the power reflection coefficient:

$$\tau_{PTM} = \frac{P^t_{TMave,z}}{P^i_{TMave,z}} = \frac{\eta_1 \cos\theta_t}{\eta_2 \cos\theta_i} \times |\tau_{TM}|^2 \tag{6.183}$$

(6.183) represents the fraction of power transmitted through a dielectric interface from medium 1 to medium 2. The percentage power transmission coefficient is defined by taking ×100% of expression (6.183) as follows:

$$\%\text{Power transmitted} = \tau_{PTM} \times 100\% = \left(\frac{\eta_1 \cos\theta_t}{\eta_2 \cos\theta_i} \times |\tau_{TM}|^2\right) \times 100\% \tag{6.184}$$

Note that the power transmission coefficient is defined as *refractivity*, and is also called *transmissivity or transmittance* in optics. So far we have examined the definition aspect of the power reflection and transmission coefficients. The relation between the two power coefficients is:

$$\tau_{PTM} + \Gamma_{PTM} = 1 \tag{6.185}$$

Comparing the relationship between the power coefficients for TE case in (6.124) with that for (6.185) we observe that they are the same. This case be justified with the ray trace model as we can define both the TE and TM polarization cases with the same model in Figure 6.29.

Review Questions 6.11: Power Balance Equation for TE and TM Oblique Incidences
Q1. Explain why the power balance equation for the TM and TE cases are the same.

6.11.12 Equivalent Impedance Concept in Power Balance Equation

Similar to the TE oblique incidence case, we can define the power transmission coefficient with the equivalent impedance concept that we have derived in Section 6.10.4. Substitution for (6.159) into (6.183) we obtain

$$\tau_{PTE} = \frac{\eta_{1TM}}{\eta_{2TM}} \left(\frac{\cos\theta_t}{\cos\theta_i}\right)^2 \times |\tau_{TM}|^2 \tag{6.186}$$

If we define a *normalized wave impedance* $\eta^n_{2TM} = \eta_{2TM}/\eta_{1TM}$ for medium 2 with respect to the impedance η_{1TE} of medium 1 then we obtain a new definition of the power transmission coefficient as follows:

$$\tau_{PTE} = \frac{|\tau_{TM}|^2}{\eta^n_{2TM}} \left(\frac{\cos\theta_t}{\cos\theta_i}\right)^2 \tag{6.187}$$

We have defined the equivalent impedance concept for multilayered media in Section 6.10.8 and henceforth are not repeating the concept again. In collusion we can say that the equivalent impedance concept is one of the most effective tool to deal with the oblique incidence cases for any number of layers of dielectric media.

6.11.13 Summary of TM Polarized Oblique Incidence Cases

- We have introduced TM oblique incidence and technology perspectives of emerging chipless RFID and many exciting applications.
- In outlines, learning objectives, and outcomes, we reinforced the technology perspectives with the hints of many case studies.
- We have given examples of many case studies that use total transmission and Brewster angle concept.
- We defined the TM oblique incidence case with 3D and 2D representations and highlighted the significance of vector decompositions.
- Again, a novel procedure is devised with step-by-step solution approaches to boundary values problems of TM oblique incidence case.
- Detailed derivations of incident, reflected, and transmitted fields and their tangential components that obey the BCs at the interface.
- Summarized the complete sets of field solutions in TM polarization.
- We applied the BCs and equality principle of phase and amplitudes to derive TM polarized reflection and transmission coefficients.
- We derived Brewster angle using Snell's law of refraction.
- In the special context of Brewster angle we have given two case studies – first one is multipath interference and second one is EM barcodes.
- Power balance equation as the interface with the aid of the law of conservation of energy.
- We also did a very advanced level worked-out exercise on TM polarization that highlights broad spectrum of the theory covered here.

Review Questions 6.12: TE Oblique Incidence

Q1. What type of oblique incidence case an EM RFID use?

Q2. A quartz window is used at the end of a laser. Explain its functionality in term of Brewster angle.

Q3. Define TM polarization? What is the difference from the normal and TE polarized oblique incidences?

Q4. In TM-polarization why do the electric and magnetic fields change their direction in Media 1 and 2? What are the governing parameters for these changes?

Q5. We use similar vector decompositions of TE case to solve EM boundary value problems in TM polarization. What are they? Can you comprehend their differences in TM and TE cases?

Q6. Can you explain why electric field is always on the plane of incidence though it has two components in all three cases: incidence, transmission, and reflection?

Q7. What are the components of the electric field used in the boundary conditions along with the magnetic field to derive Γ_{TM} and τ_{TM}?

Q8. Why is Brewster angle called "polarizing angle?" Justify your answer in the light of the fundamental definition of θ_{BA}?

Q9. In two types of EM barcodes explain how polarization diversity is exploited.

Q10. Which EM barcode is more efficient? Justify.

Q11. A radar cannot receive signals continuously in elevation angle. Can you explain why?

Q12. What is the "image theory" in Brewster angle?

Q13. Is it possible to have total reflection $|\Gamma_{TM}| = 1$ in dielectric media?

Q14. Give examples of devices which use the concept of Brewster angle.

Q15. Is there any difference in the power balance equation in TE and TM-polarizations? If not explain why?

Q16. Why do we only consider normal directed power balance equation in TM polarization case?

6.12 Problems

Oblique Incidence Coordinate System

P6.1 A 2.54 GHz uniform plane wave with an electric field amplitude of 120 V/m is propagating in a lossless medium of dielectric constant $\varepsilon_r = 3.02$. The wave propagating in the xy-plane ($z = 0$) creating an angle $\theta_i = 35°$ with x-axis. Assume the electrical field is polarized parallel to the xy-plane. Calculate: (i) the component phase constants, (ii) the component wavelengths, (iii) the component phase velocities, and (iv) the electric and magnetic fields in both phasor form and time domain.

P6.2 A 12 GHz uniform plane wave with an electric field amplitude of 550 mV/m is propagating in a lossless medium of dielectric constant $\varepsilon_r = 4.2$. The wave propagating in the yz-plane ($x = 0$) creating an angle $\theta_i = 45°$ with a y-axis. Assume the electrical field is polarized parallel to the yz-plane. Calculate: (i) the component phase constants, (ii) the component wavelengths, (iii) the component phase velocities, and (iv) the electric and magnetic fields in both phasor form and time domain.

Oblique Incidence on Conducting Boundaries

P6.3 A uniform plane wave at 500 MHz operating frequency occurs upon an aluminum boundary from air as medium 1. The plane and the direction of propagation are on the xz-plane and along the z-axis, respectively. The electric field is TE polarized and has an amplitude of 115 V/m. The incidence angle is 30°. Based on Snell's law of reflection, determine: (i) the angle of reflection, and (ii) the amplitude of the reflected electric field.

Calculate the phasor vector and time domain expressions for: (iii) the incident electric and magnetic fields, (iv) the reflected electric and reflected magnetic fields, (v) the resultant electric field in medium 1, (vi) the resultant magnetic field in medium 1, (vii) the impedance Z_{TM}, (viii) the time average power density, and (ix) the induced current on the conducting surface. Find also: (x) the conductor loss due to finite conductivity of the second medium (aluminium, $\sigma = 3.69 \times 10^6$ (S/m)), and (xi) surface current.

P6.4 Recalculate P6.3 for TM polarization.

Total Internal Reflection in Prism, Optical Waveguides and Optical Fibers

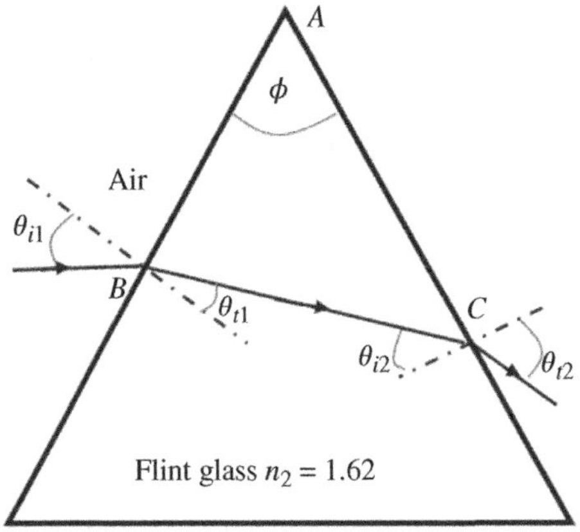

Figure P6.5 Prism made of high refractive index flint glass.

P6.5 A prism is made of high refractive index flint glass. The light ray is incident on the left inclined surface and passes through the other surface. Calculate the incident angle θ_{i1} that makes total internal reflection for $\phi = 60°$.

P6.6 Due to material dispersion the optical medium has different refractive indices for different colored light. For pure silica, the refractive index is approximated by curve fitting as follow:

$$n(\lambda) = c_1 + c_2\lambda^2 + c_3\lambda^{-2},$$

where, λ = the wavelength of the optical signal, $c_1 = 1.45084$, $c_2 = -0.00343\ \mu\text{m}^{-2}$, and $c_3 = 0.00292\ \mu\text{m}^2$. A ray of white light impinges on the left surface of the prism made of pure silica with $\phi = 60°$ and constituent red, orange and yellow light rays disperse from the other surface of the prism. Calculate the angular spread for an incident angle of 45°, also called *angular dispersion*, of the light. The average wavelengths of red, orange and yellow light rays are 685, 605 and 580 nm respectively.

P6.7 In visible optical region wavelength 764 nm, the refractive index of pure glycerol is 1.47. Assume an opaque disc placed at the top of a glycerol transparent tank is 120 cm in radius. Calculate the following: (i) the minimum depth of the tank so that a coin laying on the bottom cannot be seen. Comment on: (ii) if the incident wave makes an angle $\theta_i = 40°$, the angle of reflection, the angle of refraction, and the propagation constants in glycerol and air, and (iii) the attenuation constant and propagation constant in air at 68° incident angle.

P6.8 Consider a microstrip patch antenna is designed on a substrate with relative permittivity of $\varepsilon_{r1} = 2.45$. The thickness of the substrate is 1.5 mm. The patch antenna is a half wavelength resonator and operates at the upper Wi-Fi frequency of 5.8 GHz. The patch antenna is energized with a coaxial probe. Calculate the followings:

i) The operating frequency wavelength and the length of the patch antenna.
ii) The critical angle of the substrate material
iii) The surface wave attenuation constant at 75° of incidence angle
iv) The propagation constant in air.

Assume the dielectric material is lossless.

P6.9 For a dielectric waveguide accepting an incident ray at an angle θ_a from the air. The refractive indices of the core in 1.455 and cladding 1.453. Calculate the numerical aperture.

Field Analysis of TE Polarization: Air–Dielectric Interface

P6.10 A TE polarized uniform plane wave obliquely incidents from a dielectric medium 1 of relative permittivity $\varepsilon_{r1} = 10.2$ to a second dielectric medium 2 with relative permittivity $\varepsilon_{r2} = 2.45$ via a plane interface. Determine the incident angle for which a total internal reflection occurs in medium 1.

P6.11 A uniform plane wave at 950 MHz operating frequency incidences upon a polystyrene medium with refractive index 1.59 from air as medium 1. The direction of propagation is along the z-axis. The electric field is TE polarized and has an amplitude of 250 V/m. The incidence angle is 45°. Calculate the followings: (i) the angle of refraction, (ii) the reflection and transmission coefficients, the phasor vector and time domain expressions for: (iii) incident electric and magnetic fields, (iv) the reflected electric and magnetic fields, and (v) the transmitted electric and magnetic fields.

P6.12 A TE-polarized obliquely incident electric field is defined as $\boldsymbol{E}^i(x, z, t) = \boldsymbol{a}_y 135 \cos(\omega t - 5\pi x - 3\pi z)$ (V/m) from air to a second medium with dielectric constant 4.2. Calculate the following: (i) the resultant phase constants in air and dielectric media (ii) the angle of incident, (iii) the instantaneous resultant electric fields in medium 1, and (iv) in medium 2.

P6.13 A TE polarized obliquely incident uniform plane wave from a 400 MHz behind the wall detection radar is incident on a plasterboard wall ($n_2 = 4.47$) from air as medium 1. The radar emits 12.6 dBm power. The direction of propagation is along the z-axis. The incidence angle is 45°. Calculate the following: (i) the amplitude of the incident electric field intensity, (ii) the angle of transmission, (iii) the effective impedances η_{1TE} and η_{2TE}, (iv) the reflection and transmission coefficients, and (v) the fraction of power reflected in air and transmitted in plasterboard.

P6.14 Draw an equivalent transmission line model of the electromagnetic system in P6.13 and calculate the impedance seen in medium 1 (air) at a distance of 10 cm from the interface. Calculate the reflection coefficient at that point and show if it is different from that in P6.13. If different, explain why?

P6.15 A submarine to surface missile radar system is operating at 50 MHz. The depth of the submarine is 130 m from the surface of the seawater. Assume that the effective coverage area of effective communications and guidance is bounded by 70% of the critical angle. Calculate the coverage area in (m^2) of the missile system for the radar. In this case we assume the dielectric constant of the sea water is $\varepsilon_{r1} = 81$.

Propagation Through Multiple Boundaries

P6.16 A two-ply dielectric medium is made of FR4 ($\varepsilon_r = 4.3$) thickness 4 cm and the second medium TLX0 ($\varepsilon_r = 2.45$) thickness 8 cm. A TE polarized uniform plane wave impinges on the composite medium at an angle of 35°. Calculate the total drift of the signal when it comes out from the medium.

Field Analysis of TM Polarization: Air–Dielectric Interface

P6.17 A 1.0 GHz 125 V/m obliquely incident TM polarized uniform plane wave incidences upon a lossless medium ($\varepsilon_1 = 10.2$) from air as medium 1. The incidence angle is 50°. Calculate the followings: (i) the phase constants in media 1 and 2, (ii) the angle of refraction, θ_t (iii) wave impedances in both media η_1 and η_2, (iv) the TM polarized reflection and transmission coefficients, Γ_{TM} and τ_{TM}, respectively, (v) the phasor vector and time domain expressions for incident, reflected and transmitted electric and magnetic fields, and (vi) power transmitted and reflected.

P6.18 A uniform plane wave at 950 MHz operating frequency incidences upon a polystyrene medium with refractive index 1.59 from air as medium 1. The direction of propagation is along the z-axis. The electric field is TM polarized and has an amplitude of 250 V/m. The incidence angle is 45°. Calculate the followings: (i) the angle of refraction, (ii) the reflection and transmission coefficients, the phasor vector and time domain expressions for: (iii) incident electric and magnetic fields, (iv) the reflected electric and magnetic fields, and (v) the transmitted electric and magnetic fields.

P6.19 A TE-polarized obliquely incident electric field is defined as $\boldsymbol{E}^i(x, z, t) = \boldsymbol{a}_y 135 \cos(\omega t - 5\pi x - 3\pi z)$ (V/m) from air to a second medium with dielectric constant 4.2. Calculate the followings: (i) the resultant phase constants in air and dielectric media, (ii) the angle of incident, (iii) the instantaneous resultant electric fields in medium 1, and (iv) in medium 2.

P6.20 A TM-polarized obliquely incident electric field is defined as $\boldsymbol{H}^i(x, z, t) = \boldsymbol{a}_y 1.35 \cos(\omega t - 5\pi x - 3\pi z)$ (A/m) from air to a second medium with dielectric constant 4.2. Calculate the followings: (i) the resultant phase constants in air and dielectric media, (ii) the angle of incident, (iii) the instantaneous resultant magnetic fields in medium 1, and (iv) in medium 2.

P6.21 An 850 MHz TM polarized wave with amplitude 12.5 V/m is obliquely incident from air ($z < 0$) onto a slab of lossless, nonmagnetic material with $\varepsilon_r = 10.2$ ($z > 0$). The angle of incidence is 50°. Calculate the followings: (i) the phase constants in medium 1 and 2: β_1 and β_2, (ii) the angle of transmission θ_t, (iii) the wave impedances in media 1 and 2, η_1 and η_2, (iv) γ_{TM} and τ_{TM}, (v) expressions for $\boldsymbol{E_i}$, $\boldsymbol{E_r}$ and $\boldsymbol{E_t}$ fields, (vi) the resultant electric field in medium 1, and (vii) the time average power transmitted in medium 2.

Total Transmission – Brewster Angle Calculation

P6.22 A TM polarized signal is entering from an electromagnetically lighter medium to a denser medium with a relation between the relative permittivities of the two media $\varepsilon_{r1} = 12\varepsilon_{r2}$, calculate θ_{BA}. Recalculate θ_{BA} for the signal travels from the opposite direction. Assume the medium is non-magnetic and lossless for simplicity.

P6.23 For a TM polarized obliquely incident signal impinges from medium 1 with $\varepsilon_{r1} = 2.45$ on to a second medium with $\varepsilon_{r1} = 22.05$ at the Brewster angle. Calculate the angle of transmission.

P6.24 For P6.20, calculate the reflection coefficients from both TE and TM polarized uniform plane waves passing from canola oil to air at the Brewster angle. Assume the dielectric constant of canola oil is 4.5.

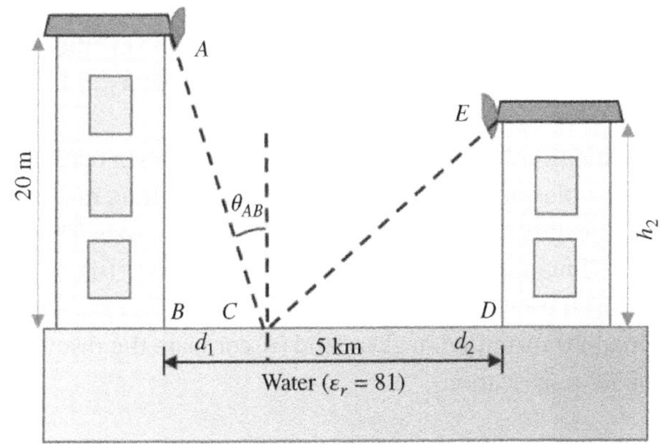

Figure P6.25 A microwave link between two buildings.

P6.25 A microwave link communicated via two dish antennas installed at the rooftop of two buildings located at the two sides of the lake as shown in Figure P6.25. Assume the water surface is lossless and flat with $\varepsilon_r = 81$. Calculate the height h_2 of the second building so that no TM polarized signal reflects back to antenna 2.

P6.26 A randomly polarized UPW can be decomposed into equal TE and TM parts each with an incident electric field amplitude of 125 V/m. The operating frequency is 1800 MHz from a mobile tower incident at the Brewster's angle from air onto a thick dielectric material with dielectric constant 50. Calculate the time domain expressions for the composite fields: (i) the incident, (ii) reflected, and (iii) transmitted electric fields.

P6.27 A randomly polarized UPW can be decomposed into equal TE and TM parts each with an incident electric field amplitude of 125 V/m. The operating frequency is 1800 MHz from a mobile tower. The signal is incident at an angle of 50° from air onto a thick dielectric material with dielectric constant 25. Calculate the time domain expressions for the composite fields: (i) the incident, (ii) reflected, and (iii) transmitted electric fields.

P6.28 A randomly polarized signal incidents at an angle 56° from air to soil surface of dielectric constant 16. It splits the power equally in both TE and TM polarizations. The transmitter launches a signal with power density 24 (dBm/m^2). Calculate the fraction of the incident power reflected back by the soil surface. And the resultant power density reflected back in (mW/m^2). Assume the soil surface is lossless and nonmagnetic.

P6.29 Repeat P6.28 for the incident angle is the Brewster angle. And calculate power transmitted through air–soil interface.

P6.30 A randomly polarized signal incidents at an angle 46° from air to soil surface of dielectric constant 16. It splits the power equally in both TE and TM polarization. The transmitter launches a signal with power density 10 (dBm/m^2). Calculate the fraction of the incident power reflected back by the soil surface. And the resultant power density reflected back in (mW/m^2). Assume the soil surface is lossless nonmagnetic.

P6.31 A uniform TM-polarized wave at 400 MHz impinges from air to plasterboard interface at an angle of incidence 55° from a see-through wall radar. Calculate: (i) effective impedances of the two media, (ii) the reflection and transmission coefficients, and (iii) the fraction of transmitted power in medium 2. The refractive index of plasterboard: $n_2 = 4.47$.

P6.32 A TM polarized obliquely incident uniform plane wave from a 400 MHz behind the wall detection radar is incident on a plasterboard wall ($n_2 = 4.47$) from air as medium 1. The radar emits 12.6 dBm power. The direction of propagation is along the z-axis. The incidence angle is 45°. Calculate the followings: (i) the angle of transmission, (ii) the effective impedances η_{1TM} and η_{2TM}, (iii) the reflection and transmission coefficients, (iv) the fraction of power reflected in air and transmitted in glass, and (v) compare the results of power transmission for the case of TE-polarization.

P6.33 A see-through wall radar operating at 400 MHz would detect wooden studs behind the plasterboard wall. The plasterboard has dielectric constant $\varepsilon_{r1} = 20$ of thickness 2 cm followed by the dry wooden stud of dielectric constant $\varepsilon_r = 2.9$ of thickness 9 cm. The TM polarized signal has an angle of incidence of 35°. Calculate the followings:

i) the wave impedance in each layer;
ii) the input impedance in each interface from the right end,
iii) the effective reflection coefficient Γ_{eff}, and
iv) the percentage power reflected.

References

1 Inan, U.S. and Inan, A.S. (2000). *Electromagnetic Waves*. Upper Saddle River, NJ: Prentice Hall.

2 Paschotta, R. (2018). *RP Photonics Encyclopedia*. RP Photonics Consulting GmbH [Online]. https://www.rp-photonics.com/brewsters_angle.html (accessed 22 June 2018).

3 Kraus, J.D. and Fleisch, D.A. (1999). *Electromagnetics with Applications*, 5e. Boston: WCB McGraw-Hill.

4 Hayt, W.H. and Buck, J.A. (2006). *Engineering Electromagnetics*, 7e. Boston: McGraw-Hill.

5 Arieli, R. (2018). End windows of gas laser. [Online]. https://perg.phys.ksu.edu/vqm/laserweb/Ch-7/F7s5t1p6.htm (accessed 22 June 2018).

7

Incidence of Uniform Plane Wave in Lossy Media

7.1 Introduction

This chapter presents the electromagnetic wave phenomena at the interface of multilayered lossy media. This chapter, firstly, introduces some of those emerging technologies where the theory of reflection and transmission of plane wave propagation in lossy media are used. Then the chapter covers the theory of normal incidence of the uniform plane wave in an imperfect conducting medium followed by normal incidence in multilayered lossy media. Then oblique incidence on the interface of the lossy media for both transverse electric (TE) and transverse magnetic (TM) polarized cases is covered using Maxwell's field analysis and effective impedance concept. The theory is extended to multilayered lossy medium. The pseudo-Brewster angle of lossy medium has significant implication in wireless communications and microwave biomedical applications. Other applications of oblique incidence in lossy media are indoor and outdoor wireless communications at an arbitrarily polarized signal propagation and reflection through glass windows, buildings, roads, and public installations. The chapter completes with many interesting applications.

In the preceding two chapters we comprehensively analyzed the normal and oblique wave incidences of the uniform plane wave. For simplicity, we considered the propagation media were lossless. However, in reality there is no lossless medium. Even small loss in a medium can cause serious concerns in device design at microwave and millimeter wave frequencies. Besides these, there are highly lossy medium such as biological tissues and soil. With the emergence of new wireless technologies for microwave biomedical devices for diagnostics and treatments, wireless mobile communications for urban in-built environments and in-door communications and positioning, mobile satellite communications, RF/microwave nano-fabrications and nano-technologies, microwave printed electronics with nanoparticle conducting inks, RF/microwave shielding for electromagnetic compatibility, wireless identifications and tracking of goods in shipments and logistics, retails, warehouse, personnel accesses and movements in office buildings and hospitals, vehicle to vehicle communications, and microwave precision agricultures using wireless sensors and remote sensing active and passive radiometers, the significance of modeling of plane wave propagation in lossy multilayered media has been increasing tremendously. Electrical and electronics engineers are always on demand to solve such problems.

Electromagnetic wave propagation and reflection from such lossy medium have opened up new fields of microwave biomedical engineering for imaging, diagnostics, and treatment. The local real power loss and generation of heat in a controlled and localized area has opened up the new field of treatment such as diffusion of tumors, repair of malignant tissues and killing of cancerous cells [1].

Fields and Waves in Electromagnetic Communications, First Edition. Nemai Chandra Karmakar.

Companion website: www.wiley.com/go/fieldsandwavesinelectromagneticcommunications

In this chapter, we examine the wave propagation in both normal and oblique incidence cases in imperfect conductor and lossy dielectric media. Although we stated that the same theory of normal and oblique incidences on lossless media can be applicable to the lossy medium by simply considering the complex wave impedance in field analyses. The difference was that we did not consider the attenuation constant α (Np/m) in the analysis of the uniform plane wave in the lossless medium. Hence, we obtain the real values of the reflection and transmission coefficients.[1] However, in the lossy medium the wave attenuates with the attenuation constant $e^{-\alpha d}$ as the wave propagates a distance d in the medium. Hence the wave impedance η and the resulting reflection Γ and transmission τ coefficients becomes complex. Also, in reality there is no perfect conductor even considering the bulk metal such as copper and gold have finite conductivities in the order of 10^7 (S/m). The imperfect conductors have finite conductivity value. Hence, we expect wave penetration in the imperfect conducting medium and associated real power loss as heat. We need to treat the conductor as a lossy penetrable medium to examine the field propagation and associated losses. In microwave and mm-wave frequencies, these losses are significant and needs calculations of those losses. Also stated before, this theory is useful to understand the shielding effect for electromagnetic compatibility and interference. With the emergence of new wireless technologies for microwave biomedical devices for diagnostics and treatments, wireless mobile communications in urban in-built environment and indoor wireless communications, mobile satellite communications, RF/microwave nano-fabrications and nano-technologies, RF/microwave shielding for electromagnetic compatibility, wireless identifications and tracking of goods in shipments and logistics, retails, warehouse, personnel accesses, and movements in office buildings and hospitals, vehicle to vehicle communications, the importance of plane wave propagation in lossy multilayered media is increasing tremendously. This chapter, first, introduces some of those emerging technologies where the theories of reflection and transmission of plane wave propagation in lossy media is used. Then the chapter covers the theory of normal incidence of the uniform plane wave in an imperfect conducting medium followed by normal incidence in multilayered lossy media. This theory of reflection from an imperfect conducting medium is useful to understand the shielding effect that is covered in the electromagnetic compatibility and interference of the electronics devices. Then oblique incidence on the interface of the lossy media for both TE and TM polarized cases is covered using Maxwell's field analysis and the effective impedance concept. The theory is extended to multilayered lossy medium. As delineated above that pseudo-Brewster angle of lossy medium has significant implications in wireless communications and microwave biomedical applications. Other applications of oblique incidence in lossy medium are indoor and outdoor wireless communications at an arbitrarily polarized signal propagation and reflection through glass windows, buildings, roads, and public installations. The chapter completes with many interesting applications as stated above. Figure 7.1 illustrates the generic flow chart of plane wave propagation in lossy media. Two cases are considered here: the normal and oblique incidence cases, and both cases are treated with detailed theories in sequential manner. The theories learned in the previous chapters are applied to examine the field and wave phenomena in both cases.

1 We always get real values of the reflection and transmission coefficients for lossless media unless we consider any phase shift of the electric and magnetic fields in the uniform plane wave due to either the arbitrary polarization or considering the transmission over a distance. The wave propagation adds phase with the distance even if the medium is lossless as we witnessed in the equivalent impedance calculations in multilayered medium.

7.2 Applications

The emerging applications of the theory of plane wave reflection and transmission in lossy media have created high impact in modern society. Scientists and researchers of large institutions have invested enormous resources in the field. The core essence is to understand the electromagnetic waves phenomena in lossy media starting from a few KHz to a few hundred GHz frequency bands. Figure 7.2 illustrates some emerging applications those the author has conducted in his research laboratory. The applications can be broadly categorized into eight fields.[2] Note that this is not an exhaustive study of applications. The readers are encouraged to find more applications in the emerging fields. As can be seen, the biomedical field has been heavily invested in the fields of microwave biomedical imaging of brain, breast cancer patients and abdominal tumors diagnostics, hyperthermia for tumors treatment and hypothermia treatments for animals, bone healing and treatments using wireless energy harvesting technology, and imaging of various sorts such as X-ray and MRI all use electromagnetic spectra at different frequency bands. Figure 7.3a

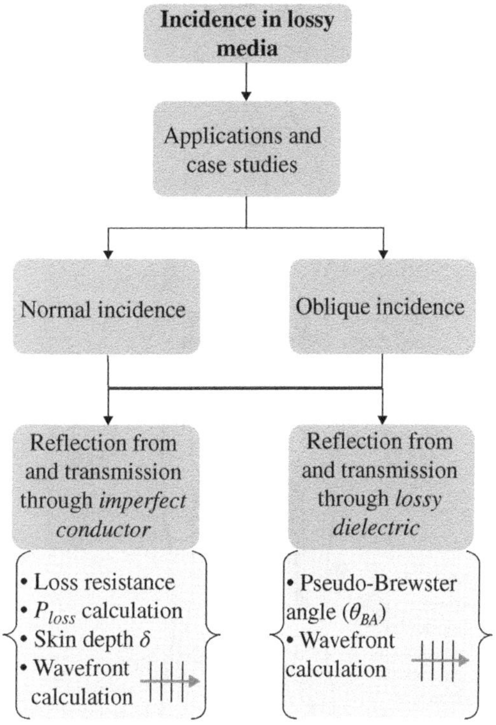

Figure 7.1 Road map of wave propagation in lossy medium.

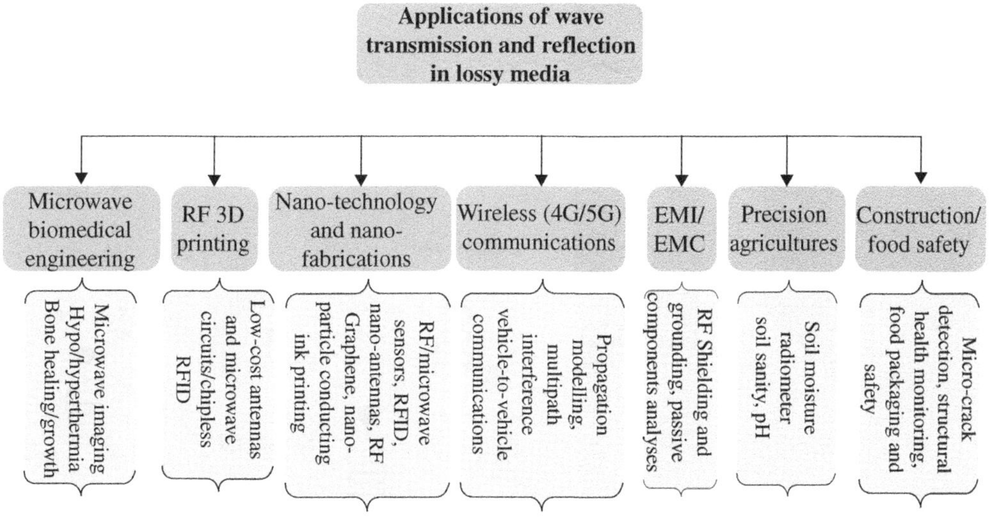

Figure 7.2 Emerging application areas of wave propagation in lossy media.

2 The classification is a rough estimate. Readers are encouraged to find more applications in the field.

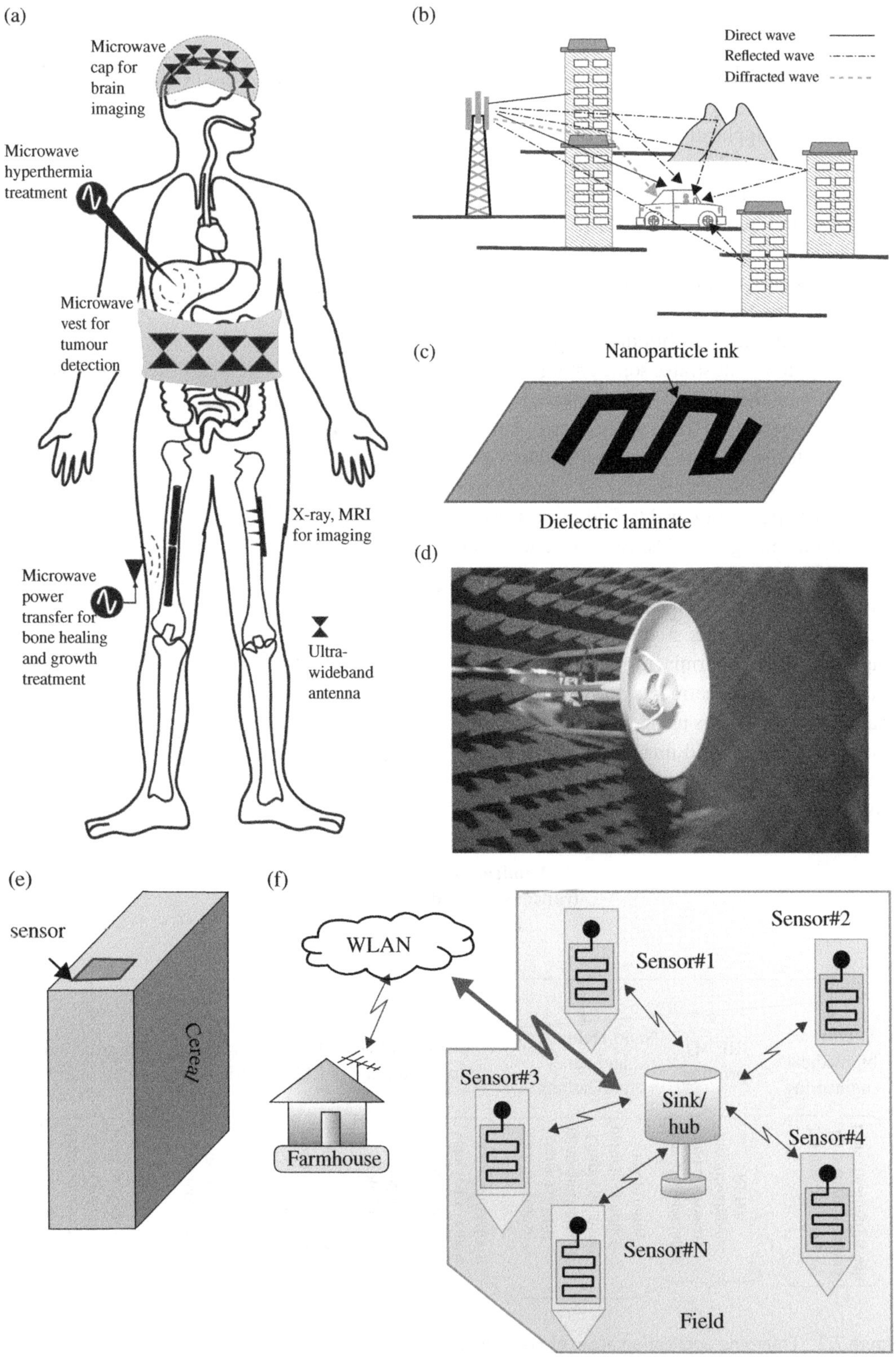

Figure 7.3 Applications of wave transmission and reflection in lossy media in (a) biomedical engineering, (b) 4/5G wireless communications, (c) nano-printing/nano-fabrication, (d) 3D printed antenna (*Source:* ESA–G. Porter/European Space Agency), (e) food safety and security, and (f) precision agriculture, and construction/structural health monitoring.

illustrates some of the applications that are directly linked to the human body. Figure 7.3b illustrates existing 4G and emerging 5G wireless mobile communications. The mobile users constantly encounter obstacles in the field of wave propagation from buildings, mountains, and sounding objects. The refraction and diffraction phenomena from these lossy media are a huge field of wireless wave propagation study. This field needs proper understanding of wave reflection and transmission through lossy medium. Also, different climatic conditions such as rain frost, moisture contents and precipitation in atmosphere play a big part in wireless propagation. With the advent of nontechnology and nanofabrication, many new fields such as nanoparticle ink printing, graphene, gold nanoparticle for cancer treatment have emerged as the new field of research. The wave propagation and interaction with those nanostructures is a big field and the investigation needs the knowledge of wave propagation and resulting reflection and refraction in lossy medium. Figure 7.3c illustrated a nanoparticle ink printed chipless RFID sensor. The nanoparticle ink could be a silver ink with glue contents that constitute a lossy medium of propagation. *The skin effect* will play a significant role in the study of such devices. Nowadays low cost 3D printed antennas are emerging as the competitive technology for the traditional bulky metallic antennas. The 3D antenna structure can be made of light weight, robust and highly conducting alloy and nanoparticle conducting ink coated plastic molds. In both cases, we also need to investigate *the skin effect* due to the finite conductivity of the materials used in constructing the antennas. Figure 7.3d illustrates a 3D printed parabolic reflector antenna that will be used in space applications [2]. For space applications, weight is a critical factor. Therefore, such a light weight 3D printed antenna will impact the space industry.

Smart Packaging with conducting foils, cardboards, plastics, and metallic containers can be studied as the lossy media when wireless sensors are used for condition monitoring of foods. Trillions of dollars' worth of foods are wasted due to expiry dates. Using smart sensors and monitoring the pH, moisture content and temperature of the food we can elongate the expiry date of food and improve its safety. Therefore, there is a great prospect to understand the wireless propagation and behavior of those sensors on food packaging for food safety and security. Figure 7.3e illustrates a cereal box label with such sensors for smart food packaging.

Figure 7.3f illustrate a precision agriculture system using chipless RFID sensor network. Chipped/chipless RFID moisture and nutrient sensors are pegged in the field. A sink/hub collects data wirelessly and sends the data to a farmhouse for real-time monitoring of the condition of the soil. This network helps for planning for irrigation and saves water. The soil is considered as the lossy medium and the wave reflection and transmission through the soil provides physical data of the soil. There are many other enormous applications of the theory of wave propagation, reflection, and transmission in lossy media. Readers are encouraged to delve through resources to find more applications to comprehend the significance of the theory of wave propagation in lossy media of the chapter. In the following section, we are going to examine the normal incidence of the electromagnetic wave on imperfect conductors.

7.3 Normal Incidence on Imperfect Media

In this section, we shall examine the normal incidence on imperfect boundaries. The normal incidence case on lossy media can be broadly classified with single boundary with two lossy media and multiple boundaries in multilayered lossy media. We shall examine the single boundary loss media for three cases: (i) imperfect conducting boundary when the electromagnetic wave impinges from a lossless dielectric, say air, to an imperfect conducting boundary, (ii) imperfect dielectric boundary, and (iii) imperfect metal-metal boundary. Last, we shall investigate the multilayer lossy dielectric media and the phenomenon of wave propagation in such a multilayered medium. The applications of reflection from an imperfect conducting medium are microwave shielding, which are further studied in the theory of electromagnetic interference and compatibility. The theory will also revisit the application of the knowledge of the surface resistance, *skin effect*, the induced current in the imperfect conductor that we also studied in Chapter 2 on wave propagation in imperfect conductors. The vast applications of normal incidences on imperfect boundaries can be found in microwave biomedical imaging, diagnostics, and treatments. We shall examine the electromagnetic systems of air-muscle interface, and skin-fat-muscle interface as examples [3]. For hyperthermia tumor treatments we need to know how much heat is absorbed as loss in the malignant tissue to reduce the growth of tumors. Figure 7.4 shows the section outlines. Some more applications are shown in the bottom of the picture.

As stated in Section 7.2, the emergence of 3D printing, nano-technology and nano-fabrication used in microwave and millimeter wave bands have brought new opportunities for low cost and highly efficient active and passive designs and antennas for emerging wireless communications systems.[3] With the availability of such new resources in modern days researchers and scientists have been investigating new structures such as metamaterials for frequency selective surfaces that help in efficacy improvement of antennas and microwave and millimeter wave active and passive devices and systems. As for example very low cost and high-performance 3D printed antennas and microwave and mm-wave passive circuits can be built with plastic modules and thin coatings of conducting paints and inks, respectively. The nanoparticle based conducting ink layer can be considered as the imperfect conducting medium for plane wave propagation, and becomes a subject of study for the normal incidence on an imperfect conducting medium. Therefore, normal incidence of uniform plane wave on such an imperfect medium can be referred to as a penetrable medium. We need to analyze the field characteristics of propagating waves in those media. The

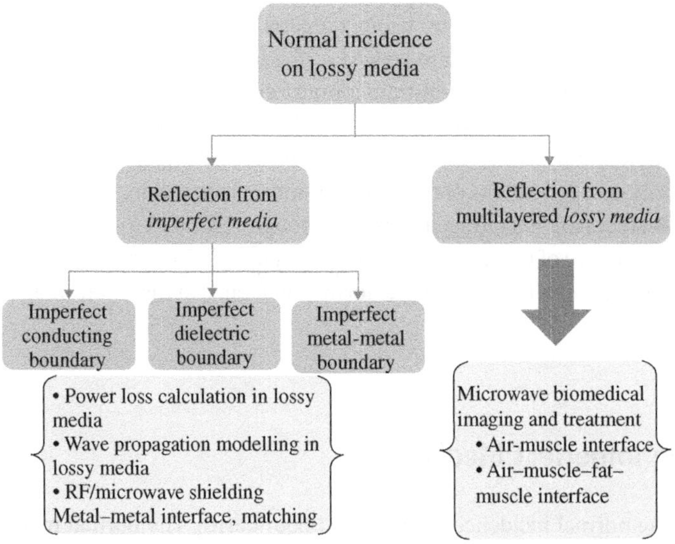

Figure 7.4 Normal incidence in lossy medium with classification for single and multiple boundaries.

3 Emerging 5G wireless communications will use microwave and mm-wave frequency bands from a few hundred MHz to more than 100 GHz.

theory of normal incidence and resulting reflection and transmission can provide information about *the skin effect* and induced loss resistance of that medium. As we have examined in Chapter 2 that if the thickness of the conducting medium is less than the skin depth, conducting current carrying structure cannot guide electromagnetic signal properly and huge transmission and reflection losses occur through those devices due to the leakage. Associated high Ohmic power loss occurs due to the finite conductivity. Even thicker conductors also have more associated power loss as heat to the finite conductivity. For efficient design we need to accurately analyze the field behavior inside such media and associated losses. In this section, we shall study the surface resistance, induced current and the associated losses in the light of the penetrable wavefronts in the imperfect conducting medium.

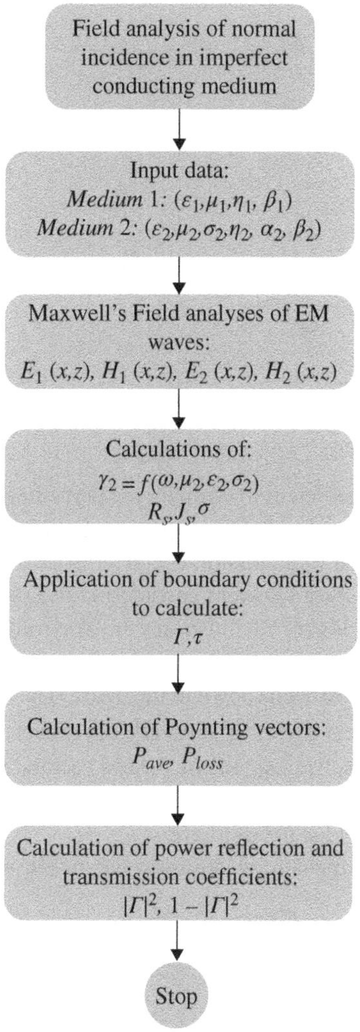

Figure 7.5 Flow chart of field analysis of normal incidence on imperfect conducting boundary.

Figure 7.5 illustrates the flow chart of field analysis of normal incidence on imperfect conducting boundary. First, we need the input data in the forms of the constitutive parameters for both media of the electromagnetic system. Assuming medium one is lossless having the constitutive parameters $(\varepsilon_1, \mu_1, \eta_1, \beta_1)$. The lossy medium 2 has the finite conductivity σ_2 and associated attenuation constant α_2. Therefore, medium 2 has extra constitutive parameters $(\varepsilon_2, \mu_2, \sigma_2, \eta_2, \alpha_2, \beta_2)$. In the field analysis, we need to consider those parameters associated with finite conductivity as the loss factor. Next, we draw the electromagnetic system of the imperfect conducting boundary and perform Maxwell's field analysis for the uniform plane wave. We apply either the boundary conditions or use the equivalent impedance concept to calculate the reflection and transmission coefficients. In this aspect, an equivalent transmission line model of the electromagnetic system can be constructed. We expressed all field quantities and obtained the expression for the propagation constants $\gamma_1 = j\beta_1$ in medium 1 and $\gamma_2 = \alpha_2 + j\beta_2$ in medium 2. While γ_1 is a simple expression for the lossless medium 1, γ_2 is a complex quantity as we need to consider the loss factor α_2 in the analysis. Next, we solve for the reflected and transmitted electric fields in terms of the incident field to calculate the reflection and transmission coefficients at the imperfect conducting boundary. To do so as usual we take help of the boundary conditions of the imperfect conducting boundary. The boundary conditions provide the reflection coefficient Γ and the transmission coefficient τ. The most concerning part is the losses associated with imperfect conducting boundary as the signal penetration occurs in the imperfect conducting medium 2. But, do not fear! We have already examined the wave propagation in the good/imperfect conducting medium in Chapter 2. In this section, we shall study comprehensively the field wave fronts and associated losses using the Poynting theorem. We shall calculate the time average power propagation inside the medium and associated losses using the surface resistance R_s and the induced current $\boldsymbol{J}_s$. Lastly, we calculate the power reflection

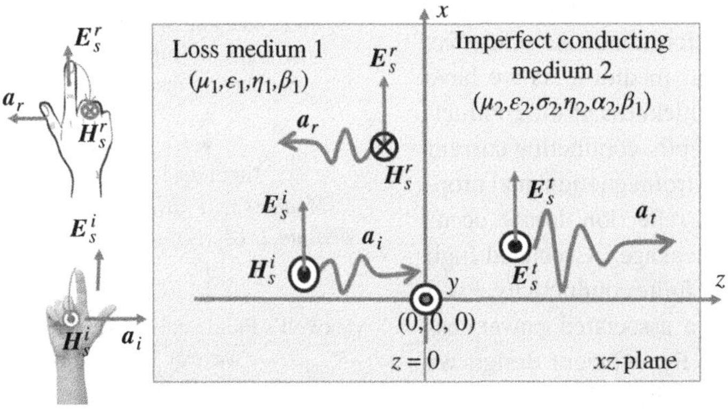

Figure 7.6 Normal incidence of uniform plane wave at imperfect conducting boundary.

coefficient to estimate the percentage of incident power reflected back from the imperfect conducting boundary. Our goal of the field analysis is to find the field solutions to compute the reflection and transmission coefficients followed by the power loss due to surface resistance and associated induced current considering the second medium as the penetrable and lossy medium.

Figure 7.6 illustrates the electromagnetic system with an imperfect conducting boundary at $z = 0$ on which a uniform plane wave impinges normally. The constitutive parameters of both media have been shown in the figure. The figure also illustrates the phasor vector incident electromagnetic field waves: $\boldsymbol{E}_s^i$, and $\boldsymbol{H}_s^i$, the reflected electromagnetic field waves: $\boldsymbol{E}_s^r$ and $\boldsymbol{H}_s^r$ and the transmitted electromagnetic field wave vectors: $\boldsymbol{E}_s^t$ and $\boldsymbol{H}_s^t$. As can be seen the transmitted wave is heavily attenuated as it propagates in the imperfect conducting medium 2. The phasor vector electromagnetic field expression are:

Incident electromagnetic fields in lossless medium 1

$$\boldsymbol{E}_s^i(z) = E_0^i e^{-j\beta_1 z} \boldsymbol{a_x} \tag{7.1}$$

$$\boldsymbol{H}_s^i(z) = H_0^i e^{-j\beta_1 z} \boldsymbol{a_y} = \frac{E_0^i}{\eta_1} e^{j\beta_1 z} \boldsymbol{a_y} \tag{7.2}$$

Reflected electromagnetic fields in lossless medium 1

$$\boldsymbol{E}_s^r(z) = E_0^r e^{j\beta_1 z} \boldsymbol{a_x} \tag{7.3}$$

$$\boldsymbol{H}_s^r(z) = H_0^r e^{j\beta_1 z} \boldsymbol{a_y} = -\frac{E_0^r}{\eta_1} e^{j\beta_1 z} \boldsymbol{a_y} \tag{7.4}$$

Transmitted electromagnetic fields in lossy medium 2

$$\boldsymbol{E}_s^t(z) = E_0^t e^{-\gamma_2} \boldsymbol{a_x} = E_0^t e^{-\alpha_2 z} e^{-j\beta_2 z} \boldsymbol{a_x} \tag{7.5}$$

$$\boldsymbol{H}_s^t(z) = H_0^t e^{-\gamma_2} \boldsymbol{a_y} = \frac{E_0^t}{\eta_2} e^{-\alpha_2 z} e^{-j\beta_2 z} \boldsymbol{a_y} \tag{7.6}$$

where, E_0^i, H_0^i, E_0^r, H_0^r, E_0^t and H_0^t are the peak amplitudes of the respective field components, the propagation constant in the lossless medium 1 is defined as $\gamma_1 = j\beta_1 = j\omega\sqrt{\mu_1\varepsilon_1}$. As for example if we consider medium 1 is air, then the intrinsic wave impedance for medium 1 is defined as: $\eta_1 = \sqrt{\mu_1/\varepsilon_1} = 120\pi\,\Omega$. For lossy (imperfect conducting) medium 2, the propagation constant

$\gamma_2 = \alpha_2 + j\beta_2 = \sqrt{j\omega\mu_2(\sigma_2 + j\omega\varepsilon_2)}$; the wave impedance in medium 2 is: $\eta_2 = \sqrt{j\omega\mu_2/(\sigma_2 + j\omega\varepsilon_2)}$. We have done comprehensive analysis of uniform plane wave propagation in lossy medium in Chapter 2 too. Therefore, readers are referred back to revise Chapter 2 for the expressions of the wave parameters.

We also have examined in Chapter 2 for the propagation constant and the wave impedance of a good conductor:

$$\gamma_2 = \alpha_2 + j\beta_2 = \frac{1+j}{\delta} \tag{7.7}$$

$$\eta_2 = R_2 + jx_2 = \frac{1+j}{\sigma_2\delta} \tag{7.8}$$

where, $\delta = 1/\sqrt{\pi f \mu_2 \sigma_2}$, the skin depth of medium 2.

We can draw an equivalent transmission line model with the equivalent impedance concept as shown in Figure 7.7a.[4] The main ingredient is the surface resistance R_s which is shown in Figure 7.7b and is defined as:

$$R_s = \frac{l}{\sigma_2 A} = \frac{1}{\sigma_2\delta} = \sqrt{\frac{\pi f \mu_2}{\sigma_2}} \tag{7.9}$$

where $A = l\delta$, the vertical cross-sectional area through which the surface current density $\boldsymbol{J_2}$ passes toward x-axis as shown with the thick arrows. Here, $l = 1$ for per unit length resistance R_s, in the equivalent load impedance $Z_2 = \eta_2 = R_2 + jX_2$, the loss component R_2 is represented by the surface resistance R_s as in expression (7.9). We also provided detailed theory of the current wave

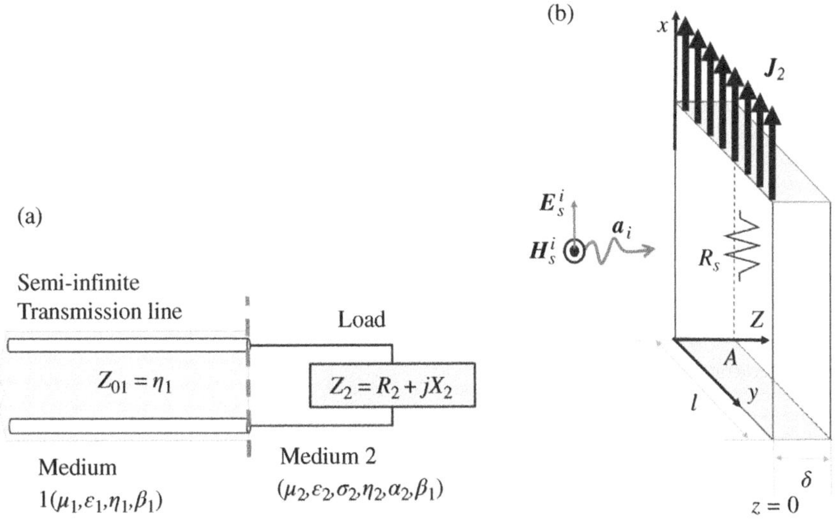

Figure 7.7 (a) Equivalent transmission line model of the electromagnetic system of a conducting boundary, and (b) concept of surface resistance.

4 Using straight away the equivalent transmission line model with the derived intrinsic impedance of both media provides the definition of the reflection and transmission coefficients. Since the equivalent impedance have been derived from the field analysis and boundary conditions at the conducting boundary, we can avoid the field analysis for the same calculations. However, for completeness we shall examine the fields in both media and their relevant derivations to compute the reflection and transmission coefficients and power loss in lossy medium 2.

propagation in good conductors in Chapter 2. For a good conductor, we have also deduced that $R_2 = X_2 = \frac{1}{\sigma_2 \delta}$ Since we have the definitions of the equivalent impedances of the transmission line model for the electromagnetic system as shown in Figure 7.7a, we can easily deduce the reflection and the transmission coefficients as follows:

$$\Gamma = \frac{E_0^r}{E_0^i} = \frac{(Z_2 - \eta_1)}{Z_2 + \eta_1} = |\Gamma| \angle \theta_\Gamma \tag{7.10}$$

$$\tau = \frac{E_0^r}{E_0^i} = \frac{2Z_2}{Z_2 + \eta_1} = |\tau| \angle \theta_\tau \tag{7.11}$$

We can also derive the same definitions of Γ and τ in (7.10) and (7.11) using the field boundary condition as done for the dielectric boundary field analysis. This exercise is left for readers to examine. Note that due to the losses in imperfect conductor generally the reflection and transmission coefficients are complex as the wave impedance is complex in good conducting medium 2.

Example 7.1: Conducting Boundary and Surface Resistance
A UWB chipless RFID is printed with silver ink of finite conductivity of $\sigma_2 = 4 \times 10^5$ (S/m). The operating frequency is centered at 6 GHz with frequency span of 4–8 GHz. Calculate the followings: (i) the skin depth at 6 GHz, (ii) the surface resistance, (iii) the surface impedance, (iv) the reflection coefficient, and (v) percent of power reflected back in air:

Answer:

i) The skin depth of the silver ink at 6 GHz is:

$$\delta = \frac{1}{\sqrt{\pi f \mu_2 \sigma_2}} = \frac{1}{\sqrt{\pi \times 6 \times 10^9 \times 4\pi \times 10^{-7} \times 4 \times 10^5}} = 10.3\ \mu\text{m}$$

ii) The surface resistance is:

$$R_2 = \frac{1}{\sigma_2 \delta} = \frac{1}{4 \times 10^5 \times 10.3 \times 10^{-6}} = 0.24\ \Omega$$

iii) The resistance is quite large and causes signal losses.

$$Z_2 = R_2 + jX_2 = 0.24 + j0.24\ \Omega$$

The intrinsic impedance of medium 1 is:

$$\eta_1 = 377\ \Omega$$

iv) The reflection coefficient is:

$$\Gamma = \frac{(Z_2 - \eta_1)}{Z_2 + \eta_1} = \frac{\{(0.24 + j0.24) - 377\}}{\{(0.24 + j0.24) + 377\}} = 0.9987 \angle 179^\circ.$$

As we can see, the reflection coefficient is close to unity. Next, we calculate the % power reflected back.

v) % of power reflected back in air is:

$$|\Gamma|^2 \times 100\% = 0.9987^2 \times 100\% = 99.7\%$$

About 100% of the incident power reflects back from the conducting silver ink interface and less than 1% power is either absorbed or transmitted through the silver ink.

Example 7.2: Conducting Boundary and Surface Resistance in UWB Frequency
Repeat Example 7.1 for the lower and upper bounds of the UWB frequency band.

Let us repeat calculation for the lower bound frequency at 4 GHz:

i) $$\delta = \frac{1}{\sqrt{\pi f \mu_2 \sigma_2}} = \frac{1}{\sqrt{\pi \times 4 \times 10^9 \times 4\pi \times 10^{-7} \times 4 \times 10^5}} = 12.6\ \mu\text{m}$$

ii) $$R_2 = \frac{1}{\sigma_2 \delta} = \frac{1}{4 \times 10^5 \times 12.6 \times 10^{-6}} = 0.2\ \Omega$$

iii) The resistance is quite large and causes signal losses.

$$Z_2 = R_2 + jX_2 = 0.2 + j0.2\ \Omega$$

The intrinsic impedance of medium 1 is:

$$\eta_1 = 377\ \Omega$$

iv) The reflection coefficient is:

$$\Gamma = \frac{(Z_2 - \eta_1)}{Z_2 + \eta_1} = \frac{\{(0.2 + j0.2) - 377\}}{\{(0.2 + j0.2) + 377\}} = 0.999\angle 179.9^\circ.$$

v) % of power reflected back in air is:

$$|\Gamma|^2 \times 100\% = 0.999^2 \times 100\% = 99.9\%$$

About 100% of the incident power reflects back from the conducting silver ink interface and almost no power is either absorbed or transmitted through the silver ink.

Now let us repeat the calculation for the upper bound frequency at 8 GHz:

i) $$\delta = \frac{1}{\sqrt{\pi f \mu_2 \sigma_2}} = \frac{1}{\sqrt{\pi \times 8 \times 10^9 \times 4\pi \times 10^{-7} \times 4 \times 10^5}} = 8.9\ \mu\text{m}$$

ii) $$R_2 = \frac{1}{\sigma_2 \delta} = \frac{1}{4 \times 10^5 \times 8.9 \times 10^{-6}} = 0.28\ \Omega$$

iii) The resistance is quite large and causes signal losses.

$$Z_2 = R_2 + jX_2 = 0.28 + j0.28\ \Omega$$

The intrinsic impedance of medium 1 is:

$$\eta_1 = 377\ \Omega$$

iv) The reflection coefficient is:

$$\Gamma = \frac{(Z_2 - \eta_1)}{Z_2 + \eta_1} = \frac{\{(0.28 + j0.28) - 377\}}{\{(0.28 + j0.28) + 377\}} = 0.9985\angle 179.9^\circ.$$

v) % of power reflected back in air is:

$$|\Gamma|^2 \times 100\% = 0.9985^2 \times 100\% = 99.7\%$$

About 100% of the incident power reflects back from the conducting silver ink interface and 0% power is either absorbed or transmitted through the silver ink.

Usually the reflected power loss is more in higher frequency. This low conductivity causes less power reflection meaning less backscattered power from the tag.

Exercise 7.1: Surface Resistance of Chipless RFID

Q1. Repeat the above example for a chipless RFID with a 3 μm thick conducting ink with 70 μm dielectric laminate to calculate the Γ_{eff}.

The induced current density in a good conductor has already been comprehensively examined in Section 2.7. In medium 2 due to the wave propagation, the transmitted electric field is defined as:

$$\mathbf{E}_t(z) = \mathbf{a}_x \tau E_0^i e^{-\gamma_2 z} \tag{7.12}$$

The induced current is defined as the conductive current due to the finite conductivity σ_2 in the medium using Ohm's law as follows:

$$\mathbf{J}_2(z) = \sigma_2 \mathbf{E}_t(z) = \mathbf{a}_x \sigma_2 \tau E_0^i e^{-\gamma_2 z} \tag{7.13}$$

The total current per unit cross-sectional width is defined as:

$$\mathbf{I}_2(z) = \int \mathbf{J}_2(z) dz = \mathbf{a}_x \sigma_2 \tau E_0^i \int_0^\infty e^{-\gamma_2 z} dz = \mathbf{a}_x \frac{\sigma_2 \tau E_0^i}{\gamma_2} \text{ (A/m)} \tag{7.14}$$

As before in the theory of surface current, we can relate the magnetic field $\mathbf{H}_t(z)$ with the total current $\mathbf{I}_2(z)$. They are basically equal but work in orthogonal directions. While $\mathbf{H}_t(z)$ is along y-axis, the current is along x-axis. Therefore, we can write the amplitudes are equal at $z = 0$ interface: $I_2(z = 0) = H_t(z = 0)$.

The magnetic field in medium 2 is defined in terms of the transmitted electric field $E_0^t = \eta E_0^i$, and the wave impedance η_2 as:

$$\mathbf{H}_t(z) = \mathbf{a}_y \frac{\tau}{\eta_2} E_0^i e^{-\gamma_2 z} \text{ (A/m)} \tag{7.15}$$

We have performed similar derivation of the magnetic field inside the good conductor in Section 2.7. The difference is that we need to consider the good conductor as medium 2 and a subscript "2" is used in all variables related to that medium.

We have already shown in Section 5.4 that the surface current is defined as $\mathbf{J}_s = \mathbf{n} \times \mathbf{H}_1 (z = 0)$, where, $\mathbf{H}_1 (z = 0)$ is the total tangential magnetic field at the conducting boundary at $z = 0$.

The continuity of the tangential component suggests: $H_1 (z = 0) = H_t (z = 0)$. We also examined in the wave equation of the current in good conductor for $\sigma \to \infty$, the skin depth $\delta \to 0$ and observed that the induced current is a sheet current residing on the surface of the conducting boundary at $z = 0$. We also derived in Section 2.18 the time average real power loss in a good conductor. Using the proper notation, we can define the time average real power loss in a good conductor as follows:

$$\mathbf{P}_{lave} = \frac{1}{2} Re\left(\mathbf{E}_t \times \mathbf{H}_t^*\right) = \frac{1}{2} Re\left(E_x^t H_y^{t*}\right) \mathbf{a}_z = \frac{1}{2} Re\left(Z_2 I_2 I_2^*\right) \mathbf{a}_z = \frac{1}{2} R_2 I_2^2 \mathbf{a}_z \tag{7.16}$$

This is the well-known equation of the time average power loss in terms of the resistance and current. Now we shall do an example to exemplify the theory.

Example 7.3: Conducting Boundary, Surface Resistance and Power Loss

A 6 GHz TE polarized interrogation wave signal from an RFID reader antenna impinges normally on a chipless RFID good conductor from air. The chipless RFID is made of silver ink with conductivity 5×10^5 (S/m). Assume the ink is a few sink depths thick and can be assumed semi-infinitely extended. The surface current is 10 mA. Calculate (i) the skin depth, (ii) the surface resistance, and the equivalent surface impedance, (iii) the real power loss on the conductor, (iv) the % power reflected from the conducting surface, and finally, and (v) draw a conclusion on the performance of the microwave signal which are backscattered from the chipless RFID tag to the reader's receiving antenna. (Assume $\mu_2 = 4\pi \times 10^{-7}$ (H/m))

Answer:

Given parameters: $f = 6 \times 10^9$ Hz; conductivity $\sigma_2 = 5 \times 10^5$ (S/m);

i) First, we calculate the skin depth in silver ink:

$$\delta = \frac{1}{\sqrt{(\pi\mu_2 f\sigma_2)}} = \frac{1}{\sqrt{(\pi \times 4\pi \times 10^{-7} \times 6 \times 10^9 \times 5 \times 10^5)}} = 9.2 \times 10^{-6} \text{ (m)}$$

ii) The surface resistance in medium 2 is: $R_{s2} = 1/\sigma_2\delta = 1/(5 \times 10^5 \times 9.2 \times 10^{-6}) = 0.22$ (Ω)
   The surface impedance is $Z_{s2} = R_{s2} + jX_{S2} = 0.22 + j0.22$ (Ω)
iii) The time average real power loss in medium 2 is:

$$P_{lave} = \frac{1}{2} R_{s2} I_2^2 = \frac{1}{2} \times 0.22 \times \left(10^{-2}\right)^2 = 11 \times 10^{-6} \text{ (W/m}^2\text{)}$$

   Although the time average real power loss looks small, it is substantial due to the fact that very low power in the range of (mW/m^2) of power density impinges on the tag.
iv) Now we can examine the % of power reflected from the tag.

$$\Gamma_{eff} = \frac{Z_2 - \eta_1}{Z_2 + \eta_2} = \frac{0.22 + j0.22 - 377}{0.22 + j0.22 + 377} = 0.999 - j0.0011 = 0.999\angle 0.0063^\circ \cong 1$$

   % of power backscattered if we assume the tag surface is fully printed with the conducting ink

   $|\Gamma|^2 = 0.999^2 = 0.998 = 99.8\%.$
v) We can see apparently all power is reflected back to the reader.

7.3.1 Normal Incidence on Imperfect Dielectric Boundary

In the preceding section, we have examined the normal incidence case on an imperfect conducting boundary. In the analysis, we have used the boundary conditions and the theory of wave propagation in good conducting medium that are comprehensively covered in Sections 2.7 and 5.4. We have found that *the sheet resistance* and the induced *surface current* play the vital role in determining the real power loss in the medium. Apparently, the current wave is confined within a few micrometers thin layer of the conducting medium. This phenomenon of the induced current due to the surface resistivity is defined as *the skin effect* in electromagnetic nomenclature. Since in the preceding section we started the theory with a generic lossy medium 2 with finite conductivity σ_2, and derived the propagation constant $\gamma_2 = \alpha_2 + j\beta_2$ in terms of the constitutive parameters of medium 2, the theory is equally applicable to any lossy dielectric medium. The difference is therefore, just to repeat with the same field solutions as in (7.1) to (7.6) and apply the appropriate boundary conditions for the lossy dielectric medium. Here we do not consider the induced current and surface resistance, instead we calculate the attenuating transmitted field in the second medium. Therefore, the electromagnetic field solutions are equally valid for the dielectric boundary. We also stated that we can use the effective impedances to calculate the reflection and transmission coefficients bypassing the field boundary conditions, and solutions of the reflected and transmitted fields in terms of the incident field. Therefore, the easiest way to calculate the reflection and the transmission coefficients is to use the effective impedances of the two media. We have already derived the impedances of the two media as follows:

i) The intrinsic impedance in lossless medium 1 is: $Z_1 = \eta_1 = \sqrt{\mu_1/\varepsilon_1}$, again for air $\eta_1 = 377\ \Omega$
ii) The intrinsic impedance of medium 2 is: $Z_2 = \eta_2 = \sqrt{(j\omega\mu_2)/(\sigma_2 + j\omega\varepsilon_2)}$.

We shall be using them to calculate the reflection and transmission coefficients, and the time average power loss in the second medium.

The reflection coefficient is defined as:

$$\Gamma = \frac{Z_2 - Z_1}{Z_2 + Z_1} = |\Gamma|e^{j\theta_\Gamma} \tag{7.17}$$

The transmission coefficient is defined as:

$$\tau = \frac{2Z_2}{Z_2 + Z_1} = |\tau|e^{j\theta_\tau} \tag{7.18}$$

As we can see due to the complex impedance in medium 2, both the reflection and transmission coefficients are complex with amplitudes and phases.

7.3.1.1 Time Average Power Loss in Lossy Dielectric Medium

The associate time average power loss can be calculated as follows:

$$\boldsymbol{P}_{lave}(z) = \frac{1}{2} Re\left(\boldsymbol{E}_t \times \boldsymbol{H}_t^*\right) = \frac{1}{2} Re\left(E_x^t H_y^{t*}\right)\boldsymbol{a}_z = \frac{1}{2}\left|\frac{\tau E_0^i e_2^{-\alpha_2 z}}{\eta_2}\right|^2 \boldsymbol{a}_z = \frac{1}{2}\left|\frac{\tau E_0^i}{\eta_2}\right|^2 e_2^{-2\alpha_2 z}\boldsymbol{a}_z \tag{7.19}$$

Since η_2 is a complex quantity we need to consider the modulus of it hence once the complete solution of $(\tau E_0^i e_2^{-\alpha_2 z})/\eta_2$ is obtained, we take the absolute value of it to find the time average power loss at a particular distance in the lossy medium. The contrasting difference from the good conductor is that since the wave transmission is happening in the lossy dielectric the loss factor is a function of distance in medium 2. As the medium thickness is larger the loss is accumulating too and the signal gets weaker. Let us do an example of lossy dielectric media.

Example 7.4: Power Loss in Lossy Dielectric Boundary

A uniform plane wave with amplitude $E_0^i = 250$ (mV/m) is incident on an air-lossy dielectric interface at 1 GHz. The loss tangent is 0.02 and dielectric constant is 3.2. Calculate the power loss at 10 mm from the interface inside the lossy dielectric.

Answer:

i) $tan\delta = \frac{\sigma_2}{\omega\varepsilon_2} = 0.02; \sigma_2 = 0.02 \times 2\pi \times 10^9 \times 3.1 \times 8.854 \times 10^{-12} = 3.45 \times 10^{-3}$ (S/m)

The attenuation constant: $\alpha_2 = \omega\sqrt{(\mu_2\varepsilon_2)/2\left[\sqrt{1 + (\sigma_2/\omega\varepsilon_2)^2} - 1\right]} = 2\pi \times 10^9 \times$

$$\sqrt{(4\pi \times 10^{-7} \times 8.854 \times 10^{-12} \times 3.2)/2\left[\sqrt{1 + (0.02)^2} - 1\right]} = 0.347\ (\text{Np/m})$$

$$\eta_2 = \sqrt{\frac{j\omega\mu_2}{\sigma_2 + j\omega\varepsilon_2}} = \sqrt{\frac{j2\pi \times 10^9 \times 4\pi \times 10^{-7}}{3.45 \times 10^{-3} + j2\pi \times 10^9 \times 3.1 \times 8.854 \times 10^{-12}}} = \sqrt{44\,886\angle 90^\circ} = 212\angle 45^\circ$$

$= 149.9 + j149.9\,\Omega$

The transmission coefficient: $\tau = (2\eta_2)/(\eta_2 + \eta_2) = ((149.9 + j149.9) - 377)/((149.9 + j149.9) + 377) \cong 0.77 \angle 29^\circ$

The power loss intensity at $=10^{-2}$ m: $\boldsymbol{P_{lave}}(z) = 1/2|\left(\tau E_0^i\right)/\eta_2|^2 \times e^{-2\alpha_2 z}\boldsymbol{a_z} = \frac{1}{2}|(0.77 \times 0.25)/212|^2 \times e^{-2\times 0.347\times 10^{-2}} = 408 \times 10^{-9}\boldsymbol{a_z}\,(\mathrm{W/m^2})$

7.4 Applications of Normal Incidences on Lossy Dielectric Boundary

7.4.1 Microwave Biomedical Engineering

The theory of normal incidence of a uniform plane wave in lossy media finds potentials in many fields including the microwave biomedical engineering, backscattering RFID, and many microwave passive devices. As for example the author has done a few interesting projects in microwave biomedical imaging and treatment. In both cases we need to consider the uniform plane wave incidence from air to lossy media such as air to skin. Skin is a highly reflective medium at microwave frequency, and therefore, we need to know the reflection coefficient exactly at the interface. Using microwave medical imaging needs ultra-wide band (UWB) frequency spectra for high resolution images. Knowing the reflection coefficient at different frequencies, a compensation technique is applied in the signal processing algorithm. The entry point of the electromagnetic signal from air to skin can be thought as the single layer interface of normal incidence. The next step is to find the reflections from multiple layers of fat, muscles and bone, and finally, bone marrow. Understanding the effective reflection coefficient of the biomedical electromagnetic system we can accurately know how much power is reflected back from the first layer due to the loading effects of the subsequent layers. Thus, we can have a nice link budget calculation for the microwave biomedical system and transmit appropriate amount of power so that we will not cook tissues unnecessarily. By calculations of the transmission and reflection coefficients in each layer we can predict how much power is reflected in the layer under interest. As for example the author has designed an orthopedic needle which is implanted in the bone marrow of a femur for bone treatment and growth as shown in Figure 7.8. For such a complex electromagnetic system we can use the carpet rolling method in Section 5.8.5 to calculate the individual reflection and transmission coefficients. Figure 7.8 illustrates the wireless orthopedic system and the cross section of a human thigh. The advanced and modern microwave design tools such as CST Microwave Studio™ has many routines for accurate modeling of human anatomy with appropriate constitutive parameters at different microwave and mm-wave frequencies. Therefore, understanding of the fundamental theory of wave propagation in such complex and lossy multilayered media has significant impact in efficient design of microwave biomedical devices and systems for imaging and treatments. In the following we shall do a few interesting problems to examines the wave propagation and associated losses in air–skin and air–skin–fat–muscle–bone interfaces for a wireless orthopedic needle design.

Example 7.5: Reflection from Human Skin

Figure 7.9 illustrates the cross section of human skin with different layers. The epidermis (E), Dermis (D), hypodermis (HYP), and muscle (M). We first determine the reflection coefficient considering air-skin interface. The effective dielectric constant and conductivity are $\varepsilon_{reff} \cong 60$

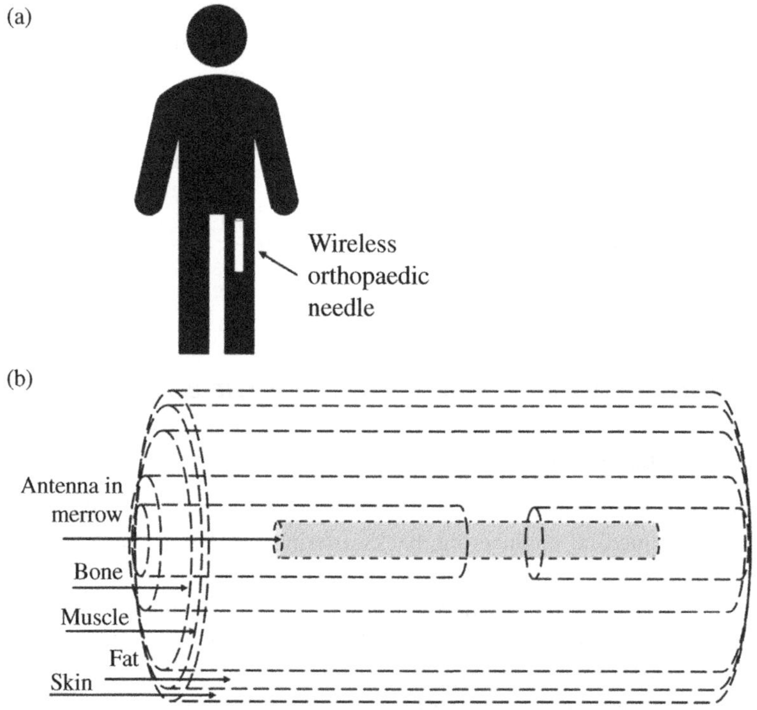

Figure 7.8 A wireless orthopedic needle using wireless power transfer. A wireless orthopedic needle needs microwave power from air to the deep in bone marrow of a femur. (a) Shows the conceptual drawing, and (b) the model of propagation medium to reach the microwave signal from air to the needle in bone marrow [4].

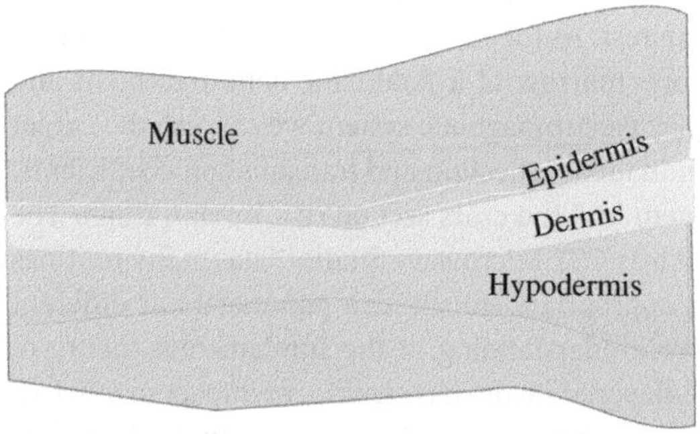

Figure 7.9 Skin layers of a dorsal upper left arm of a human.

and $\sigma_{eff} = 0.5$ (S/m), respectively, at 100 MHz. Calculate the reflection coefficient and % of power reflected back from a microwave medical imaging transmitter.

Assume skin is a dielectric medium so that $\mu_r = 1$.

Answer:

Given parameters:

the operating frequency $f = 100$ MHz; the effective relative permittivity of skin $\varepsilon_{reff} = 60$; the effective conductivity $\sigma_{eff} = 0.5$; and the intrinsic impedance of air: $\eta_1 = 377\ \Omega$.

First, we calculate the intrinsic impedance of skin:

$$Z_2 = \eta_2 = \sqrt{\frac{j\omega\mu_2}{\sigma_2 + j\omega\varepsilon_2}} = \sqrt{\frac{j2\pi \times 10^8 \times 4\pi \times 10^{-7}}{0.5 + j2\pi \times 10^8 \times 60 \times 8.854 \times 10^{-12}}} = 32 + j17.18\,\Omega$$

The reflection coefficient is:

$$\Gamma = \frac{(Z_2 - \eta_1)}{Z_2 + \eta_1} = \frac{\{(32 + j17.18) - 377\}}{\{(32 + j17.18) + 377\}} = -0.84 + j0.08 = 0.84\angle 175^\circ.$$

As we can see the reflection coefficient is close to unity hence skin is highly reflective as stated in (7.4). Next, we calculate the % power reflected back.

$$|\Gamma|^2 \times 100\% = 0.84^2 \times 100\% = 70.56\%$$

About 71% of the incident power reflects back from the skin and only 29% power is either absorbed or transmitted through the skin.

Exercise 7.2: Reflection for Human Skin at High Frequency

Q1. Repeat the above example for 1 and 10 MHz and draw the conclusions.

Exercise 7.3: Wireless Orthopedic Pin

Q1. Figure 7.10 illustrates the cross section of human thigh with a simple model of skin, fat, muscle, compact bone, and bone marrow in which the antenna of the wireless orthopedic pin resides and receives power from a wireless transmitter. The allowable power transmission is 3 mW/cm^2. Calculate the following: (i) draw an equivalent transmission line model of the electromagnetic system of thigh, and calculate impedance in each layer, (ii) the effective reflection coefficient, (iii) the effective transmission coefficient, (iv) the incident electric field at the air–skin interface 1, and (v) the received power density by the antenna. The impedance data for each layer is given at 100 MHz.[5] Assume all biological layers are lossy dielectric media so that $\mu_{rn} = 1$, where $n = 1, 2, \ldots 6$.

Note that the antenna is an impedance transformer that transforms free space impedance of 377 to 50 Ω input impedance of the antenna. Hence, we assume that the antenna offers 377 Ω to the electromagnetic signal. Also assume the electromagnetic wave is an x-polarized signal.

7.4.2 RF/Microwave Shielding for EMC Measures

In the preceding section, we have examined the theory of normal incidence in microwave biomedical applications. In this section, we shall examine the theory of RF/Microwave Shielding for EMC measures. EMC is defined as the technique to protect the devices from surrounding interference so that the device does not malfunction. The second and the most important aspect of the EMC

5 https://www.arpansa.gov.au/sites/default/files/legacy/pubs/rps/tissues5.xls, accessed 30 November 2022.

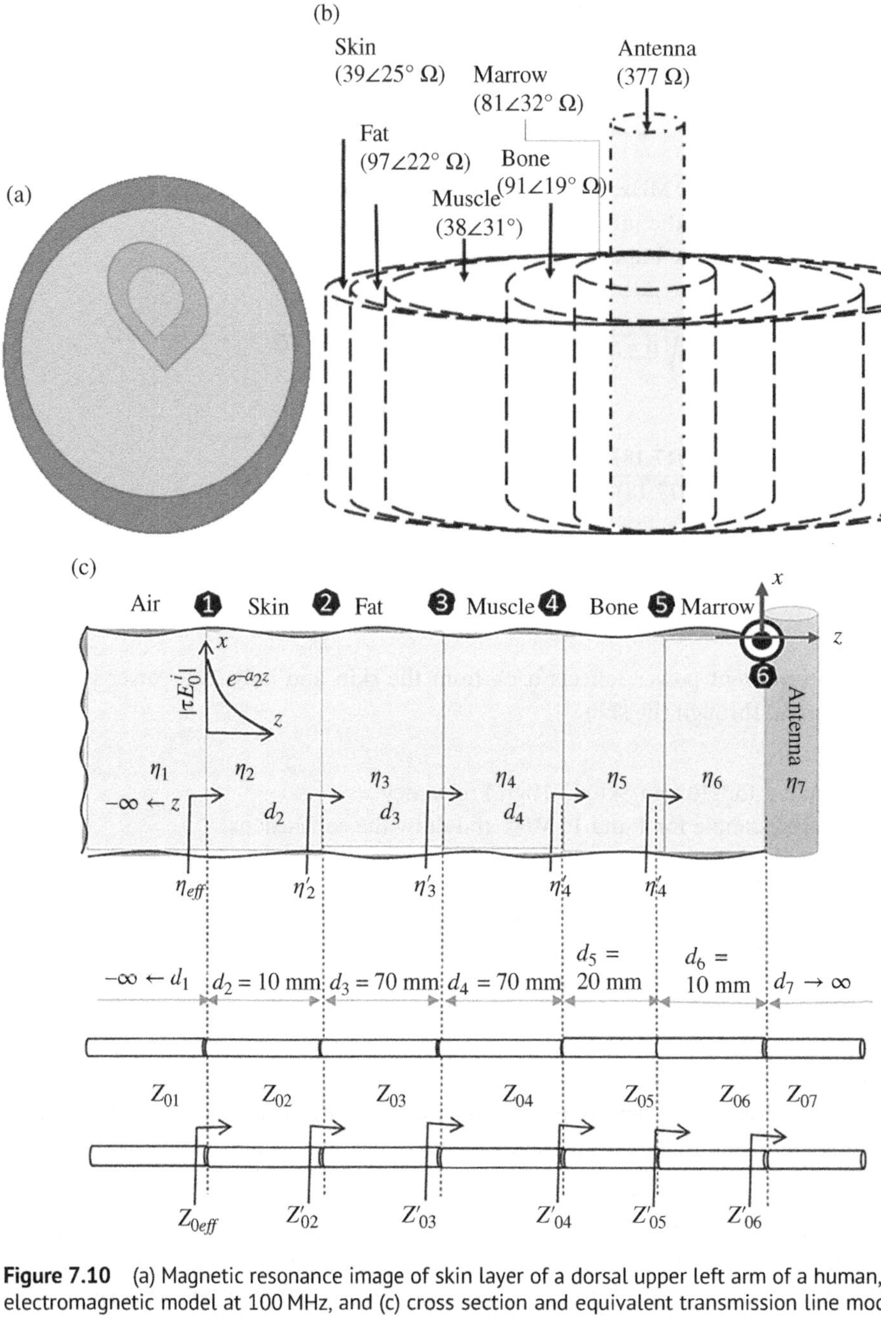

Figure 7.10 (a) Magnetic resonance image of skin layer of a dorsal upper left arm of a human, (b) simplified electromagnetic model at 100 MHz, and (c) cross section and equivalent transmission line model.

measure is that the device should also perform as ideal as possible and would not cause any interference to neighboring devices. In brief, EMC is such a discipline that makes the device a good citizen of the electronics community so that it should behave as an ideal device. The mission statement is "Do not disturb other and also do not be disturbed by other, a harmonious society of the electronics world." This is becoming more and more important due to the emergence of the modern wireless communications as delineated in Chapter 1. Our modern society has been transformed to chaotic electromagnetic traffic congestion due to personalized wireless communications systems via smartphones. Electromagnetic pollution is a buzzword in modern society. You may notice that all your favorite electronics are wireless mobile gadgets. They are communicating via the wireless channel which is nothing but the uniform plane wave propagation. Each electronic gadget is marked with so many international and regional EMC standards. The common standards are FCC (federal communications commission) for USA and CE, the European Union Commission standard. Other standards are also available such as military standard MIL-STD-416D. Any new device appearing in market must meet the standards as mandated by the regional and international regulatory bodies. There are three main important measures for the EMC measures. They are grounding, shielding, and filtering. While grounding and filtering are more on guided natures of electromagnetic signals, shielding is more on the wireless electromagnetics. In simple term, we would not allow any leakage of electromagnetic signal as interference from a device and/or to a device that causes interference to neighboring devices. The consequence of any interference means malfunctions. You may observe that the wireless router in our homes may cause interference to your TV or other mobile gadgets that work in the same or nearby frequencies or their harmonics. You may not appreciate the recommended data speed due to the interferences caused by different channels that are assigned to the Wi-Fi routers. In this section, we shall examine the shielding effectiveness of an imperfect conducting casing, as the imperfect conducting medium, in the light of normal incidence theory. Shielding effectiveness is too important for EMC measure in the modern society where approximately 7 billion people are using mobile phones. Here we are giving some theoretical prospective and applications of the normal incidence wave and its interaction with the imperfect medium that we have covered in Section 7.3. With the advent of additive manufacturing and 3D printing, industries are investing in making very low-cost, light weight and highly robust antennas, passive devices and shielding for the RF/microwave devices. Such additive manufacturing is also used by European Space Agency to make antennas for their space missions as depicted in Figure 7.3d. With the advent of the additive manufacturing and use of nonideal materials such as plastic modules with conducting ink coating or lightweight and composite conducting materials made of new alloys, it becomes significantly important to understand the theory of normal incidence, reflection, and propagation of electromagnetic waves in such non-ideal and exotic material media. The study of the shield effectiveness becomes so significant. Because tremendous possibilities of these new manufacturing precincts of additive manufacturing and nano-fabrication and nanotechnologies are pushing the limits of design and fabrication in microwave, mm-wave, and THz technologies. If successfully developed and standardized the developments will impact upon multifunctional and high data capacity and data speed devices. These new technological challenges make the scientists and engineers reinvent the applications of the theory of normal incidence in a new paradigm. Understanding the core mechanism of those new devices in the emerging applications in the perspective of electromagnetic theory is the only pathway to solve such technical challenges. Certainly, the opportunities are limitless if we really understand the fundamental concepts of electromagnetics for the specific applications. In this section we shall examine the normal incidence theory for two simple cases: (i) air–copper, and (ii) air–aluminum–plastic–aluminum–air for shield casing. We shall examine the shielding effectiveness which is the measure of the incident to transmitted power in the medium.

Example 7.6: Shielding Effectiveness

Consider a plastic casing made of polyimide with dielectric constant $\varepsilon_r = 3.4$, loss tangent $tan\delta = 0.0018$, and the surface resistivity, $R_s = 10^{16}\ \Omega$/square. The plastic casing is 5 mm thick. Both sides of the casing are covered with household aluminum adhesive of thickness 10 μm. The conductivity of Al foil with adhesive is $\sigma_2 = 4 \times 10^6$ (S/m). Assume the adhesive aluminum foil has

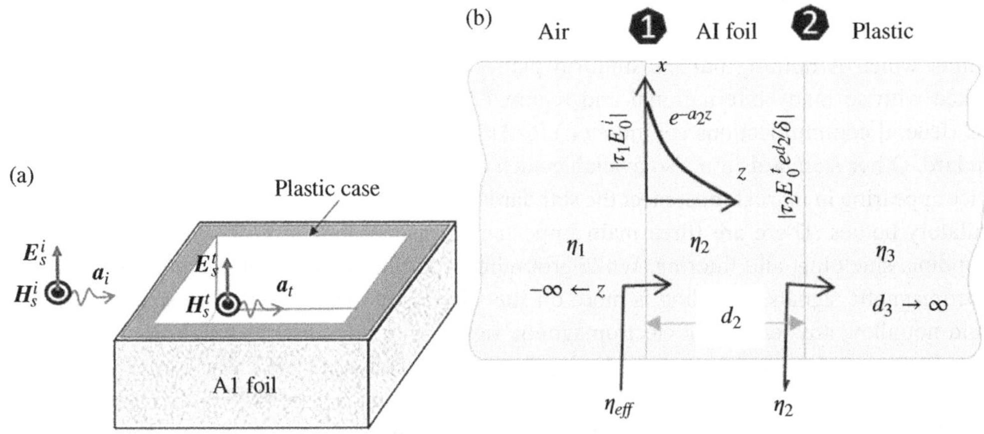

Figure 7.11 Plastic casing coated with aluminum foil: (a) perspective view, and (b) cross-sectional view.

permittivity $\varepsilon_{r2} = 1$ and $\mu_{r2} = 1$. (i) Derive the transmitted field in plastic as medium 3 in terms of the incident field and the constitutive parameters of the media. We shall work on 6 GHz for microwave applications, and (ii) calculate the shielding effectiveness of the casing as shown in Figure 7.11a.

Answer:

i) The boundary condition at interface (1) ($z = 0$) relates the incident field with the transmitted field as follow:

$$|E_{02}^t|_{z=0} = |\tau_1 E_{01}^i|_{z=0} \tag{a}$$

The transmission coefficient at interface (1) with media 1 and 2 is defined as:

$$\tau_1 = \frac{2\eta_2}{\eta_1 + \eta_2}$$

where, $\eta_1 = 377\ \Omega$ and $\eta_2 = \dfrac{1+j}{\sigma_2 \delta_2}$

As shown in Figure 7.11b, the transmitted wave attenuates exponentially with factor e^{-z/δ_2} as it propagates in Al foil. The field in medium 3 at interface (2) $z = d_2$ is defined as:

$$|E_{03}^t|_{z=d_2} = |E_2|_{z=d_2} = |\tau_2 E_{02}^t|_{z=0}\, e^{-d_2/\delta_2} \tag{b}$$

The transmission coefficient at interface (2) with media 2 and 3 is defined as:

$$\tau_2 = \frac{2\eta_3}{\eta_2 + \eta_3}$$

where for a good dielectric medium, plastic, $\eta_3 = \sqrt{\mu/\varepsilon'} \times (1 - j(\varepsilon'')/(2\varepsilon'))$ and $\eta_2 = (1+j)/\sigma_2\,\delta_2$ for aluminum. As we have known the relative permittivity and the loss tangent of medium 3, plastic, we can easily calculate ε' and ε'' and hence:

$$\eta_3 = \sqrt{\frac{4\pi \times 10^{-7}}{8.854 \times 10^{-12} \times 3.4}} \times \left(1 - j\frac{0.00181}{2}\right) = 204.3 - j0.2 = 204.3\angle 0°\Omega$$

Since $\eta_3 = \eta_0/\sqrt{\varepsilon_{r3}} = 377/\sqrt{3.4} = 204.5\,\Omega$ is real. Therefore, we can assume that the plastic is a lossless medium, and the wave propagation involves negligible attenuation $\alpha_3 = 0$.

Therefore, the reflection coefficient at interface (2) is given by:

$$\Gamma_2 = \frac{\eta_3 - \eta_2}{\eta_2 + \eta_3} = \frac{204.5 - 0.77 - j0.77}{0.77 + j0.77 + 204.5} = -1\angle 0^\circ$$

The transmission coefficient is: $\tau_2 = 1 + \Gamma_2 = 0$.

Virtually no signal will propagate inside the plastic casing. Therefore, at 6 GHz we do not need any second shield. However, for wave propagation from inside the plastic casing it is wise to also coat an aluminum foil so that the signal cannot penetrate inside the plastic.

Substitution of (a) into (b) yields the expression of the transmitted electric fields in medium 3 in terms of the incident field as follows:

$$|E_{30}^t|_{z=d_2} = |\tau_2 E_{02}^t|_{z=0}\, e^{-d_2/\delta_2} = |\tau_1 \tau_2 E_{01}^i|_{z=0}\, e^{-d_2/\delta_2} \tag{c}$$

For Equation (b) to calculate the shielding effectiveness, we need to put numerical values in the above derivations of the pertinent variables.

The skin depth of the aluminum foil at 6 GHz is:

$$\delta_2 = \frac{1}{\sqrt{\pi f \mu_2 \sigma_2}} = \frac{1}{\sqrt{\pi \times 6 \times 10^9 \times 4\pi \times 10^{-7} \times 4 \times 10^6}} = 3.25\,\mu\text{m}$$

ii) The surface resistance is:

$$R_2 = \frac{1}{\sigma_2 \delta} = \frac{1}{4 \times 10^6 \times 3.25 \times 10^{-6}} = 0.77\,\Omega$$

You can see that the difference in surface resistance of the aluminum foil with high conductivity in the order of 10^6 (S/m) provides much less resistance value compared to the previous case of Examples 7.1 and 7.2 where silver ink has conductivity in the order of 10^4 (S/m). It has huge boost in microwave performance providing very high quality factor in frequency response for a microwave passive resonator designed using the high conductivity materials that has conductivity close to bulk conductors. The resistance is reasonably small although not ideal.

The intrinsic impedance of aluminum foil at 6 GHz is:

$$\eta_2 = R_2 + jX_2 = 0.77 + j0.77\,\Omega$$

The intrinsic impedance η_1 of air medium 1 is:

$$\eta_1 = 377\,\Omega$$

iii) The transmission coefficient at air-Al interface (1) is:

$$\tau_1 = \frac{2\eta_2}{\eta_2 + \eta_1} = \frac{2 \times (0.77 + j0.77)}{0.77 + j0.77} = 0.004 + j0.0041 = 5.7 \times 10^{-3}\angle 45^\circ.$$

As we can see the transmission coefficient is in the order of 10^{-3} and with a phase angle of $\angle 45^\circ$.

The intrinsic impedance of medium 1 is:

$$\eta_1 = 377\,\Omega$$

iv) The transmission coefficient at interface (1) is:

$$\tau_1 = \frac{2\eta_2}{\eta_2 + \eta_1} = \frac{2 \times (0.77 + j0.77)}{0.77 + j0.77 + 377} = 0.004 + j0.0041 = 5.7 \times 10^{-3} \angle 45^\circ.$$

As we can see, the transmission coefficient is in the order of 10^{-3} and with a phase angle of $\angle 45^\circ$. The percentage of power reflected is:

$$|\Gamma|^2 \times 100\% = 0.98^2 \times 100\% = 96.04\%$$

About 96% of the incident power reflects back from the conducting silver ink interface and 4% power is either absorbed or transmitted through the silver ink.

Exercise 7.4: Improved Shielding Effectiveness
Q1. Repeat the above example for aluminum foil coatings on both sides of the plastic casing.

7.5 Oblique Incidence in Lossy Medium

So far, we have discussed the theories and applications of normal incidence in lossy media. In this section, we shall examine the theory and applications of oblique incidence in lossy media. With the emergence of modern wireless mobile communications, automotive radars and biomedical applications, the significance of the theories of oblique incidence in lossy media has increased tremendously. The knowledge helps accurate propagation modeling of challenging wireless channel in mobile communications. Examples include: (i) heat produced in the hyperthermia applications, (ii) EM signal reflection for imperfect conducting ground interferes with the direct signals between a transmitter and receiver, (iii) indirect signal propagation in an urban non-line-of-sight propagation environment, and (iv) the wireless router at home and office buildings that create similar challenging propagation environment that can be examined with the oblique incidence in lossy medium.

The analysis of oblique incidence in lossy medium is rather complex and needs high level of mathematical treatments. Due to the loss factors of the second medium, the refractive index becomes complex, and the resulting transmitted wave is highly complex with nonuniform field distribution. By Snell's law of refraction, the complex refractive index is a function of incident angle, and can be approximated with an effective refractive index for simplicity of calculations. Such complex propagation can be treated numerically. However, some special cases can be understood with the generalized theory of transmission in lossy medium. In the following sections, we first treat the propagation in lossy medium with the general theory of oblique incidence followed by a case study of such theory in remote sensing radiometer.

7.5.1 General Theory of Oblique Incidence from Air to Lossy Medium

Assume that a uniform plane wave is obliquely incident at the boundary of a lossy media at an incident angle θ_i and is transmitted into the lossy medium with constitutive parameters ε_2, μ_2, and σ_2 as shown in Figure 7.12. The phasor vector expression of the incident field, and the reflected electric field in medium 1, and the transmitted electric field in medium 2, respectively, can be expressed:

$$\boldsymbol{E_i}(z) = \boldsymbol{a_y} E_0^i e^{-j\beta_1 (x sin\theta_i + z cos\theta_i)} \tag{7.20}$$

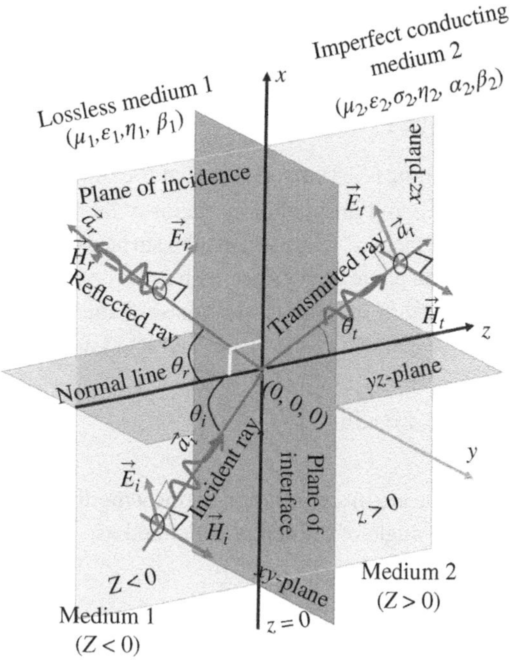

Figure 7.12 Oblique incidence from a lossless medium 1 to a lossy medium 2 with an incident angle θ_i.

$$\boldsymbol{E_r}(z) = \boldsymbol{a_y} E_0^r e^{-j\beta_1(x sin\theta_r - z cos\theta_r)} \tag{7.21}$$

$$\boldsymbol{E_t}(z) = \boldsymbol{a_y} E_0^t e^{-\gamma_2(x sin\theta_t + z cos\theta_t)} \tag{7.22}$$

where the propagation constant in medium 2: $\gamma_2 = \alpha_2 + j\beta_2$, τ is the transmission coefficient and θ_t is the angle of transmission. Apply the boundary conditions at $z = 0$, the field equality holds for the tangential fields in media 1 and 2:

$$\boldsymbol{a_y} E_0^i e^{-j\beta_1(x sin\theta_i)} + \boldsymbol{a_y} E_0^r e^{-j\beta_1(x sin\theta_r)} = \boldsymbol{a_y} E_0^t e^{-\gamma_2(x sin\theta_t)} \tag{7.23}$$

The equality of the phases of (7.23) yields Snell's laws of reflection and refraction, respectively, as follows:

$$\theta_i = \theta_r \tag{7.24}$$

$$j\beta_1 \sin\theta_i = \gamma_2 \sin\theta_t \tag{7.25}$$

Equation (7.25) reveals a very interesting fact: although $\gamma_2 = \alpha_2 + j\beta_2$ is a complex quantity $\gamma_2 \sin \theta_t$ is fully imaginary. Therefore, by virtue of Snell's law of refraction in (7.25) substitution for $\gamma_2 \sin\theta_t = j\beta_1 \sin\theta_i$ and, trigonometric identity $\cos\theta_t = \sqrt{1 - \sin^2\theta_t}$, we can express the product of $\gamma_2 \cos\theta_t$ as a complex quantity as follows:

$$\gamma_2 \cos\theta_t = \gamma_2\sqrt{1 - \sin^2\theta_t} = \sqrt{\gamma_2^2 + \beta_1^2 \sin^2\theta_i} = a + jb \tag{7.26}$$

where we can separate the real and imaginary coefficients a and b as follows:

$$a = Re\left(\sqrt{\gamma_2^2 + \beta_1^2 \sin^2\theta_i}\right) \text{ and } b = Im\left(\sqrt{\gamma_2^2 + \beta_1^2 \sin^2\theta_i}\right) \tag{7.27}$$

Here, the propagation constant $\gamma_2 = \alpha_2 + j\beta_2$ itself is a complex quantity. Otherwise, β_1 and incident angle θ_i are in general real quantity in a lossless medium 1. We can rewrite the transmitted field in (7.22) with the new coefficients a and b, (7.25) and (7.26) as follows:

$$\mathbf{E}_t(x, z) = \mathbf{a}_y E_0^t e^{-az} e^{-j(\beta_1 \sin\theta_i x + bz)} \tag{7.28}$$

Examining (7.28) we can see that the refracted wave has a decaying envelop with a factor e^{-az} in medium 2 and an oscillatory factor $e^{-j(\beta_1 \sin\theta_i x + bz)}$ which is a function of incident angle θ_i. The decaying factor e^{-az} represents the surface of constant amplitude with $az = constant$, which is parallel to the plane of interface with $z = z_1, z_2, z_3$ ... for a fixed incident angle θ_i. Figure 7.13 illustrates both constant amplitude and phase planes of the refracted wave. Observing the constant phase term, we can define the "true" refracted angle [3][6] as:

$$\theta_{true} = tan^{-1}\left(\frac{\beta_1 \sin\theta_i}{b}\right) \tag{7.29}$$

Now we are in a position to define a modified Snell's law of refraction using simple trigonometry. The true angle of refraction is defined as:

$$\sin\theta_{true} = \frac{\beta_1 \sin\theta_i}{\sqrt{b^2 + \beta_1^2 \sin^2\theta_i}} \tag{7.30}$$

The modified Snell's law of refraction is then defined as:

$$\beta_1 \sin\theta_i = \beta_2' \sin\theta_{true} \tag{7.31}$$

Where β_2' is the modified propagation constant of lossy medium 2 and is defined as:

$$\beta_2' = \sqrt{b^2 + \beta_1^2 \sin^2\theta_i} \tag{7.32}$$

We can also define the truc refractive index of the lossy medium 2 as the ratio of the sine of angles θ_i and θ_{true} as follows as shown in Figure 7.13:

$$n_2(\theta_i) = \frac{\sin\theta_i}{\sin\theta_{true}} = \frac{\sqrt{b^2 + \beta_1^2 \sin^2\theta_i}}{\beta_1} \tag{7.33}$$

Note that the modified refractive index $n_2(\theta_i)$ is a function of the incident angle θ_i as the only independent variable assuming the constitutive parameters of both media and the operating angular frequency ω are fixed. Next, we can also define the phase velocity of the lossy medium 2 as the ratio of the angular frequency and the modified propagation constant β_2 in medium 2 as defined in (7.32) as follows:

$$v_{p2} = \frac{\omega}{\beta_2'} = \frac{\omega}{\sqrt{b^2 + \beta_1^2 \sin^2\theta_i}} \tag{7.34}$$

6 In reality, the angle of transmission is a complex quantity and can be expressed as $\theta_t = \theta_{tr} + j\theta_{ti}$. The detailed derivation can be found in [3].

Figure 7.13 The constant amplitude and phase planes of a wave incident on and refracted in a lossy medium of constitutive parameters ($\mu_2, \varepsilon_2, \sigma_2, \eta_2, \beta_2$).

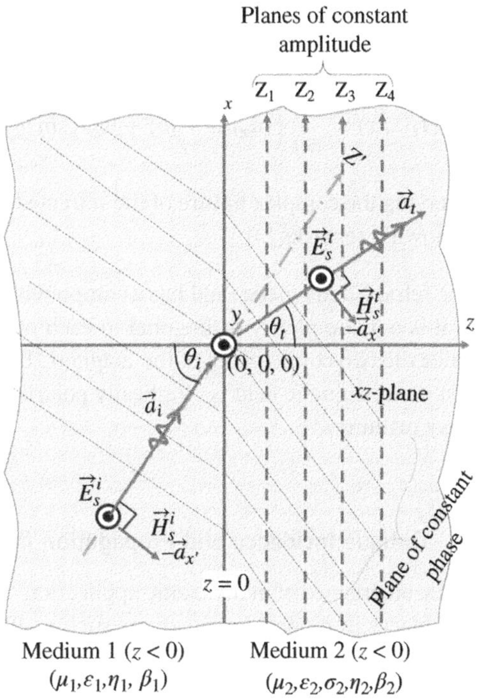

Substitutions for (7.31) and (7.33) into (7.34) yield the definition in terms of the phase velocity $v_{p1} = \omega/\beta_1$, of medium 1 and $n_2(\theta_i)$, the refractive index of medium 2:

$$v_{p2}(\theta_i) = \frac{\omega}{\beta_1 n_2(\theta_i)} = \frac{v_{p1}}{n_2(\theta_i)} \tag{7.35}$$

Examining (7.35) reveals that the phase velocity in lossy medium 2 is a function of the incident angle θ_i and can exceed the velocity of light in medium 2. The wavelength in medium 2 is defined as:

$$\lambda_2 = \frac{2\pi}{\beta_2} = \frac{2\pi}{\sqrt{b^2 + \beta_1^2 \sin^2\theta_i}} \tag{7.36}$$

Again, as can be seen in (7.36), the wavelength in the lossy medium 2 is also a function of the incident angle θ_i. Some useful relations are defined as follows [3]:

$$\beta_1^2 n_2(\theta_i) = \beta_2^2 - \alpha_2^2; ab = \alpha_2\beta_2; \beta_1 a(\theta_i) n_2(\theta_i) = \frac{\alpha_2\beta_2}{\cos\theta_{true}} \tag{7.37}$$

From Equation (7.37) we can determine the attenuation constant α_2 of the lossy medium 2. Finally, we can calculate the magnetic field in medium 2 using Maxwell's equations:

$$\nabla \times \boldsymbol{E}_t(x, z) = j\omega\mu \boldsymbol{H}(x, z) \tag{7.38}$$

Since $E_t(x, z)$ is a y-directed vector field, the refracted magnetic field has two components: $H_{tx}(x, z) = -(1/(j\omega\mu))(\partial E_{ty}/\partial z)$ and $H_{tz}(x, z) = \partial E_{ty}/\partial x$.

Therefore, substation for (7.28) in the above partial derivative expressions of the two magnetic field components, the resultant vector refracted magnetic field is defined as:

$$\boldsymbol{H}_t(x, z) = \frac{E_0^t}{\omega\mu}\left\{(\boldsymbol{a}_x j(a + jb) + \boldsymbol{a}_z \beta_1 \sin\theta_i\right\} e^{-az} e^{-j(\beta_1 \sin\theta_i x + bz)} \tag{7.39}$$

Observing the complex nature of the refracted magnetic field in the lossy medium, we can make the following remarks [3]:

1) The refracted magnetic field has a component along the direction of refracted wave propagation as it is not completely orthogonal to each other.
2) Since the two components of the magnetic fields along x- and z-directions are not in phase, the resultant magnetic field is elliptically polarized and traces an ellipse as it propagates inside the lossy medium.

7.5.2 Oblique Incidence and Propagation in Good Conductor

With the emergence of many exotic applications as stated in Section 7.2, the good conductors have special places in modern applications such as RF/microwave/mm-wave frequency shielding and printed electronics. Also, the theory is equally applicable to earth science, remote sensing, geology, and mining industries as each surface and underground mineral are highly conductive. Therefore, it is worthwhile to examine the wave propagation in good conductors as the lossy medium 2 when a signal incidents obliquely on the interface. Figure 7.14 illustrates an electromagnetic system in which the wave propagates in the lossy medium for the case of an oblique incidence. For a good conductor the conductivity is very high: $\sigma_2 \gg \omega\epsilon_2$, and the propagation and attenuation constants are defined as:

$$\alpha_2 = \beta_2 = \sqrt{\frac{\omega\mu_2\sigma_2}{2}} \tag{7.40}$$

Using the definition of the complex propagation constant in medium 2: $\gamma_2 = \alpha_2 + j\beta_2$, substitution for (7.40) into (7.25), and assuming both media are nonmagnetic ($\mu_1 = \mu_2 = \mu_0$), we can define Snell's law of refraction for a good conducting boundary as follows:

$$j\beta_1 \sin\theta_i = \gamma_2 \sin\theta_t = \sqrt{\frac{\omega\mu_2\sigma_2}{2}}(1 + j)\sin\theta_t \tag{7.41}$$

Substitution for $\beta_1 = \omega\sqrt{\mu_1\epsilon_1}$ in (7.41), we obtain the following relationship:

$$\sin\theta_t = \sqrt{\frac{\omega\epsilon_1}{\sigma_2}} e^{\frac{j\pi}{4}} \sin\theta_i \tag{7.42}$$

Examining (7.42), we can easily comprehend that for a good conductor $\sigma_2 \gg \omega\epsilon_1$ also, therefore, we can approximate that for a good conductor, $sin\theta_t \to 0$, hence $cos\theta_t \to 1$. This condition gives us new expressions for a and b in (7.26) as follows:

$$a = b = \sqrt{\frac{\omega\mu_2\sigma_2}{2}} \tag{7.43}$$

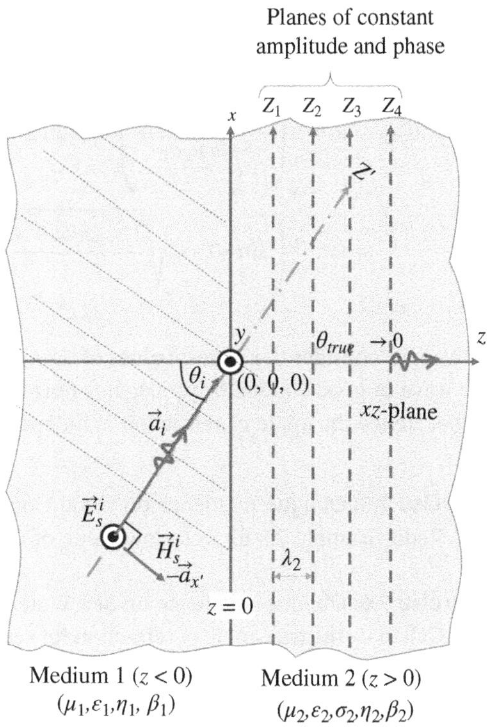

Figure 7.14 Oblique incidence on a good conductor $\sigma_2 \gg \omega\epsilon_2$.

Now we can modify the true angle of refraction θ_{true} in (7.29) with the substitution for (7.43) as follows:

$$\theta_{true} = tan^{-1}\left(\frac{\omega\sqrt{\mu_1\epsilon_1}\, sin\,\theta_i}{\sqrt{\frac{\omega\mu_2\sigma_2}{2}}}\right) \rightarrow 0 \text{ for a good conductor where } \sigma_2 \gg \omega\epsilon_{1,2} \tag{7.44}$$

Equation (7.44) is valid for large σ_2 as for good conductors. This deduction reveals that the true angle of refraction in a good conductor is extremely small and is independent of the incident angle.[7] In such circumstance, both the planes of the constant amplitude and phase angle are parallel to the plane of interface of the two media. Figure 7.14 illustrates the constant planes of the phase and amplitude of the refracted signal in medium 2 with $\theta_{true} \rightarrow 0$. In the following example we shall show that for any practical good conductor the true angle of refraction is very small.

Example 7.7: Oblique Incidence on Good Conductor
Calculate the true angle of refraction in a shield made of silver ink paint of sufficient thickness at 9 GHz. Assume $\mu_1 = \mu_2 = \mu_0$. The angle of incidence is 80°.

7 This situation is very similar to the total internal reflection of TM polarized oblique incidence where both the constant phase and amplitude fronts are parallel to the interface.

Answer:

$$\theta_{true} = tan^{-1}\left(\frac{\omega\sqrt{\mu_1\epsilon_1}\,sin\,\theta_i}{\sqrt{\frac{\omega\mu_2\sigma_2}{2}}}\right) \cong tan^{-1}\left(sin\,\theta_i\sqrt{\frac{2\omega\epsilon_1}{\sigma_2}}\right)$$

$$= tan^{-1}\left(sin\,80^\circ \times \sqrt{\frac{2\times 2\pi\times 9\times 10^9\times 8.854\times 10^{-12}}{6.2\times 10^6}}\right) = 0.007^\circ$$

Even we consider the largest values of $sin\theta_i = 1$, we get θ_{true} very close to zero. This signifies that the wave in good conductor propagates normal to the interface plane irrespective of the incident angle. Hence the angle of refraction is independent of the angle of incidence.

Exercise 7.5: Oblique Incidence on Good Conductor

Q1. Redo Example 7.8 for very low value of θ_i.

Exercise 7.6: Oblique Incidence on Sea Water as Good Conductor

Q1. Calculate the true angle of refraction for seawater at 100 MHz. Consider see water conductivity of 4 (S/m) and relative permittivity of 81 as usual.

We can make the following remarks for the above theoretical development and the practical examples:

1) The true angle of refraction is close to zero for even with much lower conductivity materials such as seawater and like. Hence, we can rewrite the generalized refracted wave in (7.28) with the substitution of (7.43) as follows:

$$\boldsymbol{E_t}(x,z) = \boldsymbol{a_y}E_0^t e^{-\alpha_2 z}e^{-j(\beta_1\,sin\,\theta_i x + \beta_2 z)} \tag{7.45}$$

Equation (7.45) reveals that the refracted wave inside the good conductor indeed heavily attenuates with the attenuation constant $\alpha_2 = \sqrt{(\omega\mu_2\sigma_2)/2} = 1/\delta$; where δ is the skin depth of the good conductor. Therefore, it is a valid measure that *the skin effect* concept is extremely valid even for an oblique incidence case as the parameter is independent of the incident angle θ_i.
2) In this token we can also iterate that the surface resistance concept and power in a good conductor is equally valid. The theory of surface resistance is then discussed in previous sections.
3) Since the incidence angle does not influence the true angle of refraction and wave transmission in good conductor as the lossy medium 2, we can use the theory of normal incidence to calculate the power loss in medium 2 as presented in Section 7.3.1.
4) In the similar token we can also define the refractive index of medium 2 in (7.33) with the substitutions for (7.43) and $\beta_1 = \omega\sqrt{\mu_1\epsilon_1}$ as follows [3]:

$$n_2(\theta_i) = \frac{\sqrt{b^2 + \beta_1^2\,sin^2\theta_i}}{\beta_1} = \sqrt{\frac{\mu_2\sigma_2}{2\omega\mu_1\epsilon_1}} \tag{7.46}$$

Equation (7.46) reveals that the refractive index is independent of the incident angle θ_i. We shall conclude this theory with an example.

Example 7.8: Oblique Incidence on Silver Ink Layer
Calculate refractive index $n_2(\theta_i)$ of sliver ink in Example 7.7. Then calculate the wavelength in medium 2 using (7.36). Make inferences on the incidence angle and the refractive index.

Answer:
Assuming silver ink is non-magnetic hence $\mu_1 = \mu_2 = \mu_0$, the expression for the refractive index of medium 2 is: $n_2 = \sqrt{\sigma_2/(2\pi f \times \epsilon_1)} = \sqrt{(6.2 \times 10^6)/(2\pi \times 9 \times 10^9 \times 8.854 \times 10^{-12})} \cong 3{,}520$. This is a large number, as a consequence the wave number β_2 inside the good conductor is extremely large (7.40). According to (7.36) we can easily deduce the wavelength inside the silver ink at 9 GHz: $\lambda_2 = 2\pi/\beta_2 = 2\pi/(\sqrt{\pi f \mu_2 \sigma_3}) = 2\pi/(469\,350) = 13.4 \times 10^{-6}$ (m), which is extremely small.

7.5.3 Oblique Incidence and Reflection from Lossy Medium

So far, we have examined the theory of refraction through lossy medium including good conductors as the special case. In this section, we shall examine the reflection from lossy medium. As illustrated in the applications in Figure 7.3, the theory of reflection from lossy medium such as buildings, RFID tags, human tissues, 3D printed parabolic reflectors and printed electronics have special practical significance. Modern wireless communications, microwave biomedical engineering, and sensing technologies for IoT have been penetrating in accelerated pace in our day-to-day life. Hence the significance of the theory of oblique incidence and reflection from lossy medium is enormous. In this section the general theory of reflection from lossy medium is examined first followed by the special case of good conductor and a case study of radiometer remote sensing for soil moisture measurement by satellites. We have done comprehensive treatment of reflection in Section 7.3.1. We can easily use the results of the complex impedance of the lossy medium as derived in (7.8): $\eta_2 = R_2 + jx_2 = (1+j)/(\sigma_2\delta)$ in the derivation of complex reflection coefficient in the case for oblique incidences. The complex value of $cos\theta_t$ in (7.26) and η_2 are used to derive the complex reflection coefficient Γ_{TE} in (7.47) for the TE polarized oblique incidence and Γ_{TM} in (7.48) for the TM polarized oblique incidence as follows:

$$\Gamma_{TE} = (\eta_2 \cos\theta_i - \eta_1 \cos\theta_t)/(\eta_2 \cos\theta_i + \eta_1 \cos\theta_t) = |\Gamma_{TE}| e^{j\theta_{\Gamma TE}} \tag{7.47}$$

$$\Gamma_{TM} = \frac{\eta_2 \cos\theta_t - \eta_1 \cos\theta_i}{\eta_2 \cos\theta_t + \eta_1 \cos\theta_i} = |\Gamma_{TM}| e^{j\theta_{\Gamma TM}} \tag{7.48}$$

To derive (7.47) and (7.48), we need the expression for $cos\theta_t$ from (7.26) as follows:

$$\cos\theta_t = \sqrt{1 + \frac{\beta_1^2 \sin^2\theta_i}{\gamma_2^2}} \tag{7.49}$$

Substitution for $\beta_1^2 = \omega^2\mu_1\epsilon_1$ and $\gamma_2^2 = \omega^2\mu_2\epsilon_{2eff}$ where, $\epsilon_{2eff} = \epsilon_2 + (\sigma_2/j\omega)$ in (7.49) we obtain the following expression for the complex $cos\theta_t$ function as follows

$$\cos\theta_t = \sqrt{1 + \frac{j\omega\mu_1\epsilon_1 \sin^2\theta_i}{\mu_2(\sigma_2 + j\omega\epsilon_2)}} \tag{7.50}$$

We can further simplify (7.50) for the case of nonmagnetic material ($\mu_1 = \mu_2 = \mu_0$) as follows:

$$\cos\theta_t = \sqrt{1 + \frac{j\omega\epsilon_1 \sin^2\theta_i}{\sigma_2 + j\omega\epsilon_2}} \tag{7.51}$$

Now substitutions for (7.51) and (7.8) into (7.47) and (7.48) yields the solutions for $\Gamma_{TE} = |\Gamma_{TE}|e^{j\theta_{\Gamma TE}}$ and $\Gamma_{TM} = |\Gamma_{TM}|e^{j\theta_{\Gamma TM}}$. As can be seen both reflection coefficients are complex, and for the same incident angle they are not equal, especially the inequality in phase angles ($\theta_{\Gamma TE} \neq \theta_{\Gamma TM}$) has special significance. Therefore, when a linearly polarized signal impinges on the interface of a lossy medium 2, the reflected wave become such that it is neither parallelly nor perpendicularly polarized; the wave becomes rather elliptical polarized. For the special case of $\theta_i = 45°$, the phase difference between the two polarized signals ($\Delta\theta = \theta_{\Gamma TE} - \theta_{\Gamma TM} = 90°$), the resulting wave becomes a circularly polarized wave with $|\Gamma_{TE}| = |\Gamma_{TM}|$. A dual polarized conducting ink I-shaped resonator [5] and a polarizer work in that principle and is well utilized in the chipless RFID at 60 GHz [6].

Now we have all ingredients to evaluate the characteristics of Γ_{TE} and Γ_{TM}. [3] has done some numerical evaluations for a few lossy materials ranging from $2 \leq \epsilon_r' < 80$ and $0.2 < \epsilon_r''$. Amplitudes $|\Gamma_{TE}|$ and $|\Gamma_{TM}|$ and phases $\theta_{\Gamma TE}$ and $\theta_{\Gamma TM}$ of the two reflection coefficients are produced in a graphical form. A few remarks can be made on the characteristics of Γ_{TE} and Γ_{TM} as follows:

1) The variation of $|\Gamma_{TE}|$ and $|\Gamma_{TM}|$ are substantial with the incident angle θ_i. Especially, $|\Gamma_{TM}|$ has profound impact.
2) There is a pseudo-Brewster angle θ_B near $\theta_i = 60°$ for low loss materials with $\epsilon_r' - \epsilon_r'' = 2.0 - j0.2$. As the loss increases both amplitude and psudo-Brewster angle increase. Plot the results derived from [3]. Although $|\Gamma_{TE}|$ increases with the loss, the shape remains similar with the incident angle θ_i. At pseudo-Brewster angle, the phase angle $\theta_{\Gamma TM}$ is close to 90°.
3) As can be seen the minimum values of $|\Gamma_{TM}|$ increases as the loss increases for the lossy medium 2. From the numerical analysis, it is concluded that $|\Gamma_{TM}|$ remains nearly constant values ranging from $0.4 < |\Gamma_{TM}| < 0.5$ for the condition $\epsilon_r'/\epsilon_r'' > 10$, this means highly lossy second medium.
4) With ϵ_r', the psudo-Brewster angle shifts to higher values.

7.5.4 Oblique Incidence: Reflection from Good Conductor

The special case of oblique incidence and reflection from good conductor has special significance in modern wireless communications, microwave biomedical engineering, and remote sensing. The theory is simple considering the attributes of the good conductor as the second lossy medium. For a good conductor $\sigma_2 \gg \omega\epsilon_2$. This results in the angle of refraction $\theta_t \to 0°$, hence, the intrinsic impedance of lossy conducting medium 2 is defined as: $\eta_2 = \sqrt{j\omega\mu_2/\sigma_2} \to 0\,\Omega$. Propagating the wave incident from a low loss medium 1 with $\eta_1 = \sqrt{\mu_1/\epsilon_1} \gg \eta_2$ at any reasonable and practical frequency range. Hence the reflection coefficients $|\Gamma_{TE}|$ and $|\Gamma_{TM}|$ are close to 1. From the special condition of the second good conducting lossy medium, we can conclude that since $cos\theta_t \to 1$, $|\Gamma_{TE}| = |\Gamma_{TM}| \cong 1$ regardless the value of the incident angle θ_i. This characteristic is evident from the analysis that the wavefront in the conducting medium is perpendicular to the plane of interface. The following example exemplifies the theory of reflection from good conductor.

Example 7.9: Dual Polarized Oblique Incidence on Good Conductor

Consider an arbitrarily polarized incident wave, which is a combination of both TE- and TM-polarizations, is launched from a chipless RFID reader antenna with an incident angle $\theta_i = 45°$. The amplitudes of the TE- and TM-polarized waves are 100 and 10 (V/m), respectively. The ultra-wideband wave has a center frequency of 24 GHz. The chipless tag is made of silver ink with conductivity $\sigma_2 = 6.2 \times 10^6$ (S/m). Calculate the followings: (i) the resultant incident wave in medium 1 (air), (ii) the complex propagation constant γ_2 in sliver ink, (iii) the true angle of refraction θ_{true},

(iv) the effective relative permittivity of medium 2, and (v) the reflection coefficients Γ_{TE} and Γ_{TM}, and (vi) the expression of the composite reflected electric field.

Answer:

Given parameters: $\theta_i = 45°$; $E_0^{TE} = 100$ (V/m); $E_0^{TM} = 10$ (V/m); $f = 24$ GHz; $\epsilon_{r1} = 1$; $\mu_{r1} = \mu_{r2} = \mu_0$; $\sigma_2 = 6.2 \times 10^6$ (S/m)

The phase constant in medium 1: $\beta_1 = (2\pi f)/c_0 = (2\pi \times 24 \times 10^9)/(3 \times 10^8) = 160\pi$ (rad/m)

i) The expression for the incident angle in medium 1:

$$\boldsymbol{E_i} = \left\{E_0^{TE}\boldsymbol{a_y} + E_0^{TM}(\boldsymbol{a_x}\cos\theta_i - \boldsymbol{a_z}\sin\theta_i)\right\}e^{-j\beta_1(x\sin\theta_i + z\cos\theta_i)}$$

$$= \left\{100\boldsymbol{a_y} + 10\left(\frac{\boldsymbol{a_x}}{\sqrt{2}} - \frac{\boldsymbol{a_z}}{\sqrt{2}}\right)\right\}e^{-j160\pi\left(\frac{x}{\sqrt{2}} + \frac{z}{\sqrt{2}}\right)} \text{ (V/m)}$$

ii) The propagation constant in medium 2:

$$\gamma_2 = \alpha_2 + j\beta_2 = (1+j)\sqrt{\frac{\omega\mu_2\sigma_2}{2}} = (1+j)\sqrt{\frac{2\pi \times 24 \times 10^9 \times 4\pi \times 10^{-7} \times 6.2 \times 10^6}{2}}$$

$$= 766.44 \times 10^3\,(1+j)\,(\text{m}^{-1})$$

iii) The true angle of refraction:

$$\theta_{true} = tan^{-1}\left(\sqrt{\frac{\omega\epsilon_0}{\sigma_2}}\right) = tan^{-1}\left(\sqrt{\frac{2\pi \times 24 \times 10^9 \times 8.854 \times 10^{-12}}{6.2 \times 10^6}}\right)$$

$$= tan^{-1}(390 \times 10^{-6}) = 0.022°$$

The ratio of intrinsic impedances of the two media: $\eta_1/\eta_2 = \sqrt{\sigma_2/(j\omega\epsilon_0)} = \sqrt{(6.2 \times 10^6)/(j2\pi \times 24 \times 10^9 \times 8.854 \times 10^{-12})} = 2.154e^{-j\pi/4} \gg 1$ and $\cos\theta_i = 0.707$

iv) Therefore, the reflection coefficients are:

$$\Gamma_{TE} = \frac{\cos\theta_i - (\eta_1/\eta_2)}{\cos\theta_i + (\eta_1/\eta_2)} \cong -1$$

$$\Gamma_{TM} = \frac{-\cos\theta_i + (\eta_1/\eta_2)}{\cos\theta_i + (\eta_1/\eta_2)} \cong 1$$

The above numerical calculation reveals that for reflections for a good conductor of an obliquely incident arbitrary polarized wave, the angle of refraction approaches zero, the magnitude of the reflection coefficient for both TE and TM polarizations approach unity. They are independent of the incident angle θ_i.

v) Finally, the expression of the reflected electric field is calculated with the help of the incident electric field and the reflection coefficients as follows:

$$\boldsymbol{E_r}(x,z) = \left\{\Gamma_{TE}E_0^{TE}\boldsymbol{a_y} + \Gamma_{TM}E_0^{TM}(\boldsymbol{a_x}\cos\theta_r + \boldsymbol{a_z}\sin\theta_r)\right\}e^{-j\beta_1(x\sin\theta_r - z\cos\theta_r)} \text{ (V/m)}$$

$$= \left\{-100\boldsymbol{a_y} + 10\left(\frac{\boldsymbol{a_x}}{\sqrt{2}} + \frac{\boldsymbol{a_z}}{\sqrt{2}}\right)\right\}e^{-j160\pi\left(\frac{x}{\sqrt{2}} - \frac{z}{\sqrt{2}}\right)} \text{ (V/m)}$$

Here, Snell's law of reflection $\theta_i = \theta_r$ is used.

7.5.5 Good Conductor to Good Conductor Interface

It is instructive to examine the theory of reflection and transmission of electromagnetic waves when the plane of incidence is formed with two different good conductive media. There are plenty of applications of the theory in remote sensing for geophysical explorations, smart packaging for communications devices, food freshness and safety, microwave and mm-wave biomedical imaging for diagnostics and hyperthermia treatments, and many more applications. For geological and mining remote sensing, the applications include measurements and imaging of underground coal seams, monitoring liquefied gases in coal seams, measuring and imaging of the cross section of boreholes deep underground mines.

The general theory of reflection and refraction at the interface of two lossy media follow the same principle of Snell's laws of reflection and reflection. As examined in the previous section for the case of a lossless medium 1 and lossy medium 2, we have experienced that the angle of refraction θ_t is complex and we need to derive the true angle of refraction θ_{true} in which direction the actual wave is propagating in medium 2. Now we extend the theory for both lossy media. For the lossy and conducting media, the angle of refraction θ_t is complex for both real incident angle θ_i for the case of uniform plane wave and complex θ_i for nonuniform plane wave.[8] In this section, we shall first examine the theory of refraction of two lossy media for: (i) the incident wave is a uniform plane wave (θ_i is real), and (ii) the incident wave is a nonuniform plane wave (θ_i is complex). Finally, we shall examine the theory of reflection for both lossy media.

The general theory of Snell's law of reflection and refraction is produced here:

$$sin\,\theta_i = sin\,\theta_r;\ \gamma_1\,sin\,\theta_i = \gamma_2\,sin\,\theta_t \tag{7.52}$$

For both lossy media, both γ_1 and γ_2 are complex and expressed as:

$$\gamma_1 = \alpha_1 + j\beta_1 = \sqrt{j\omega\mu_1(\sigma_1 + j\omega\epsilon_1)}\text{ and }\gamma_2 = \alpha_2 + j\beta_2 = \sqrt{j\omega\mu_2(\sigma_2 + j\omega\epsilon_2)} \tag{7.53}$$

To examine the properties of the lossy second medium with the interface of lossy first medium, we need to calculate the complex angle of refraction θ_t as we did in Section 7.5.1. We shall modify the propagation constant γ_2 and the angle of refraction θ_t so that complex angle is avoided as we did in the preceding section. The goal is to find the true angle of refraction and the constant amplitude and phase planes. Since the wave is nonuniform in the lossy medium 2, we can use a modified propagation in medium 2: $\gamma_2 = \alpha_2 + j\beta_2$ which is different than the intrinsic propagation constant: $\gamma_{02} = \alpha_{02} + j\beta_{02}$. Let us examine the theory of the refraction for two lossy media when the incident wave is uniform meaning the angle of incidence θ_i is considered to be real.

7.5.6 Oblique Incidence at the Interface of Two Lossy Medium with Real θ_i

Figure 7.15 illustrates the oblique incidence of a uniform plane wave incident at the interface of two lossy media. The media properties are shown in the figure with the ray trace model of the uniform plane wave in medium 1 is incident at an angle of incidence θ_i and refracted in medium 2 with an angle of refraction θ_t. Assume the polarization is TE, but the theory is equally applicable to TM polarization. The expressions of the incident, reflected and transmitted field can be expressed as:

$$\boldsymbol{E}_{i(z)} = \boldsymbol{a}_x E_0^i e^{\{-\alpha_{01}(xsin\theta_i + zcos\theta_i) - j\beta_{01}(xsin\theta_i + zcos\theta_i)\}} \tag{7.54}$$

8 The complex θ_i can be experienced if the original signal launched from a lossless dielectric medium to the first lossy medium as we have examined the theory in the preceding section.

Figure 7.15 Oblique incidence of uniform plane wave incident at the interface of two lossy media.

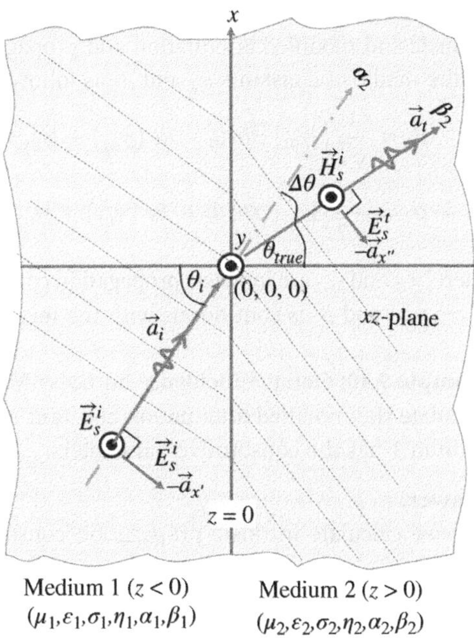

$$E_{r(z)} = \boldsymbol{a_x} E_0^r e^{\{-\alpha_{01}(x\sin\theta_i - z\cos\theta_i) - j\beta_{01}(x\sin\theta_i - z\cos\theta_i)\}} \tag{7.55}$$

$$E_{t(z)} = \boldsymbol{a_x} E_0^t e^{[-\alpha_2\{x\sin(\theta_{true} + \Delta\theta) + z\cos(\theta_{true} + \Delta\theta)\} - j\beta_2(x\sin\theta_{true} + z\cos\theta_{true})]} \tag{7.56}$$

where, α_{02} and β_{02} are the intrinsic attenuation and propagation constants, respectively; $\Delta\theta$ is the angular separations of the modified α_2 and β_2 that we need to determine. To satisfy the continuity of the tangential electric fields across the interface of the two lossy media at $z = 0$, the phases of the three field expressions in (7.54)–(7.56) must be equal. This leads to the following relations between the modified and intrinsic propagation constants:

$$\beta_{01} \sin\theta_i = \beta_2 \sin\theta_{true} \tag{7.57}$$

$$\alpha_{01} \sin\theta_i = \alpha_2 \sin(\theta_{true} + \Delta\theta) \tag{7.58}$$

According to [8] we can also write the complex vector modified propagation in medium 2 as follows:

$$\boldsymbol{\gamma_2} \cdot \boldsymbol{\gamma_2} = (\alpha_{02} + j\beta_{02})^2 \tag{7.59}$$

$$\boldsymbol{\gamma_2} = \boldsymbol{a_x}\{\alpha_2 \sin(\theta_{true} + \Delta\theta) + j\beta_2 \sin\theta_{true}\}^2 + \boldsymbol{a_z}\{\alpha_2 \cos(\theta_{true} + \Delta\theta) + j\beta_2 \cos\theta_{true}\}^2 \tag{7.60}$$

where, $\alpha_2 = |\alpha_2|$ and $\beta_2 = |\beta_2|$; the amplitudes of the modified attenuation and phase constants respectively. Substituting (7.60) into (7.59) and separating the real and imaginary parts yield the following pertinent relation between the intrinsic and the modified attenuation and propagation constants as follows:

$$\alpha_2^2 - \beta_2^2 = \alpha_{02}^2 - \beta_{02}^2 \tag{7.61}$$

$$\alpha_2\beta_2\cos\Delta\theta = \alpha_{02}^2\beta_{02}^2 \tag{7.62}$$

Using the field continuity conditions in (7.57) and (7.58) and the above relations between the intrinsic and modified attenuation and propagation constants we obtain closed form expressions for the modified constants α_2 and β_2 as follows [11]:

$$\beta_2 = \frac{1}{\sqrt{2}}\left[\,|\gamma_{01}|^2 \sin\theta_i - Re\left(\gamma_{02}^2\right) + |\gamma_{01}^2 \sin^2\theta_i - \gamma_{02}^2|\right]^{1/2} \tag{7.63}$$

$$\alpha_2 = \frac{1}{\sqrt{2}}\left[\,|\gamma_{01}|^2 \sin\theta_i + Re\left(\gamma_{02}^2\right) + |\gamma_{01}^2 \sin^2\theta_i - \gamma_{02}^2|\right]^{1/2} \tag{7.64}$$

where, γ_{01} and γ_{02} are intrinsic propagation constants of media 1 and 2 respectively. As usual, α_2 and β_2 are real and β_2 is continuous with the incident angle θ_i [11].

Example 7.10: Oblique Incidence on Lossy Media
Calculate the modified attenuation constant at $\theta_i = 45°$. The operating frequency is 6 GHz. The medium 1 has the constitutive parameters: $\epsilon_{r1} = 4$, $\sigma_1 = 0.2$ (S/m), $\epsilon_{r2} = 12$; $\sigma_2 = 0.1$ (S/m).

Answer:
First we calculate intrinsic propagation constants for medium 1 and medium 2 respectively as follows:

$$\gamma_{01} = \alpha_{01} + j\beta_{01} = \sqrt{j\omega\mu_1(\sigma_1 + j\omega\epsilon_1)}$$
$$= \sqrt{j2\pi \times 6 \times 10^9 \times 4\pi \times 10^{-7}(0.2 + j2\pi \times 10^9 \times 8.854 \times 10^{-12} \times 4)}$$
$$= \sqrt{14\,174\angle 138°} = 119\angle 69° = 42.6 + j111.1\ \left(\mathrm{m}^{-1}\right)$$

And,

$$\gamma_{02} = \alpha_{02} + j\beta_{02} = \sqrt{j\omega\mu_2(\sigma_2 + j\omega\epsilon_2)}$$
$$= \sqrt{j2\pi \times 6 \times 10^9 \times 4\pi \times 10^{-7}(0.1 + j2\pi \times 10^9 \times 8.854 \times 10^{-12} \times 12)}$$
$$= \sqrt{31\,988\angle 171°} = 179\angle 86° = 12.5 + j179\ \left(\mathrm{m}^{-1}\right)$$
$$\gamma_{02}^2 = -3.19 \times 10^4 + j4475;\ \therefore Re\left(\gamma_{02}^2\right) = -3.19 \times 10^4$$

The modified propagation constant:

$$\beta_2 = \frac{1}{\sqrt{2}}\left[\,|\gamma_{01}|^2 \sin\theta_i - Re\left(\gamma_{02}^2\right) + |\gamma_{01}^2 \sin^2\theta_i - \gamma_{02}^2|\right]^{1/2}$$

$$\text{Or, } \beta_2 = \frac{1}{\sqrt{2}}\left[|119|^2 \sin\,(60°) + 3.2 \times 10^4 + |(42.6 + j111.1)\sin^2(60°) - (12.5 + j179)^2|\right]^{\frac{1}{2}}$$

$$= \frac{1}{\sqrt{2}}\left[12\,264 + 3.19 \times 10^4 + 24\,131\right]^{\frac{1}{2}} = 184.8\ (\mathrm{rad/m})$$

[11] compares the modified and intrinsic propagation constants at the incident angle $\theta_i = 45°$ at the interface of the two lossy dielectric media of constitutive parameters $\epsilon_1 = 4$, $\sigma_1 = 0.1$ and 0.01 (S/m), $\epsilon_2 = 10$ with the function of the conductivity of medium 2 σ_2. The results are produced in Figure 7.16. As can be seen, the significant difference between the two constants as the conductivity of medium 2 increase. As σ_2 approaches σ_1 there is no significant difference in the modified and intrinsic constants. In conclusion, we can state that for a large contrast between the two lossy media the modified dielectric constant must be considered. The theory has many practical applications in mining and geological remote sensing to detect underground coal seams and the gasified coal as the two media have significant difference in conductivity.

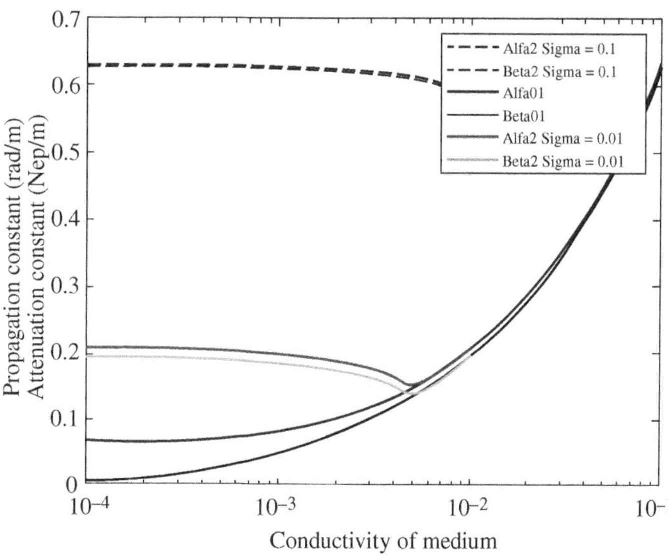

Figure 7.16 Comparison of modified and intrinsic propagation constants of a 1 MHz signal incident at $\theta_i = 45°$ at the interface of two lossy dielectric media of constitutive parameters $\epsilon_1 = 4$, $\sigma_1 = 0.1$ and 0.01 (S/m), $\epsilon_2 = 10$ [11].

The second interesting observation is the case when the second dielectric medium is a perfect dielectric meaning $\sigma_2 = 0$. In this case, [11] has done some investigation for the variation of the modified propagation constant $\beta_2(\alpha_2)$ with the incident angle θ_i as shown in Figure 7.16. As can be seen that there is a significant difference between the intrinsic and modified propagation constants with the incidence angle θ_i. According to (7.62), the angle between the phase constant, β_2 and the attenuation constant, α_2 is $\Delta\theta = 90°$ for all incident angle θ_i. In perfect medium 2, there is no net power loss and the wave propagate along β_2 direction. The real power travels along β_2 and the reactive power travels along α_2 direction [11].

As can be seen in Figure 7.17b, there is no critical angle meaning there is no incident angle for which there are transmitted wave travelling parallel to the interface $\theta_{true} = 90°$. To obtain a transmitted wave which travels parallel to the interface of the two media, we need to consider the nonuniform incident plane wave in medium 1 which is the subject of the study in the next section.

7.5.7 Refraction for Two Conductive Media

When an electromagnetic signal is refracted from two conducting media, both incident and refracted waves are nonuniform. There are phase differences between the propagation and attenuation constants of both the incident and refracted waves.

Figure 7.18 illustrates the oblique incidence of a nonuniform plane wave incident at the interface of two lossy media. The media properties with the ray trace model of the nonuniform plane wave are also shown in the figure. In medium 1, the EM wave is incident at an angle of incidence θ_i and is refracted in medium 2 with an angle of refraction θ_{true}. Assume the polarization is TE, the incident, reflected, and transmitted fields can be expressed as:

$$\boldsymbol{E}_{i(z)} = \boldsymbol{a_x} E_0^i e^{[-\alpha_{01}\{xsin(\theta_i + \Delta\theta_i) + zcos(\theta_i + \Delta\theta_i)\} - j\beta_{01}(xsin\theta_i + zcos\theta_i)]} \tag{7.65}$$

$$\boldsymbol{E}_{r(z)} = \boldsymbol{a_x} E_0^r e^{[-\alpha_{01}\{xsin(\theta_i + \Delta\theta_i) - zcos(\theta_i + \Delta\theta_i)\} - j\beta_{01}(xsin\theta_i - zcos\theta_i)]} \tag{7.66}$$

$$\boldsymbol{E}_{t(z)} = \boldsymbol{a_x} E_0^t e^{[-\alpha_2\{xsin(\theta_{true} + \Delta\theta_t) + zcos(\theta_{true} + \Delta\theta_t)\} - j\beta_2(xsin\theta_{true} + zcos\theta_{true})]} \tag{7.67}$$

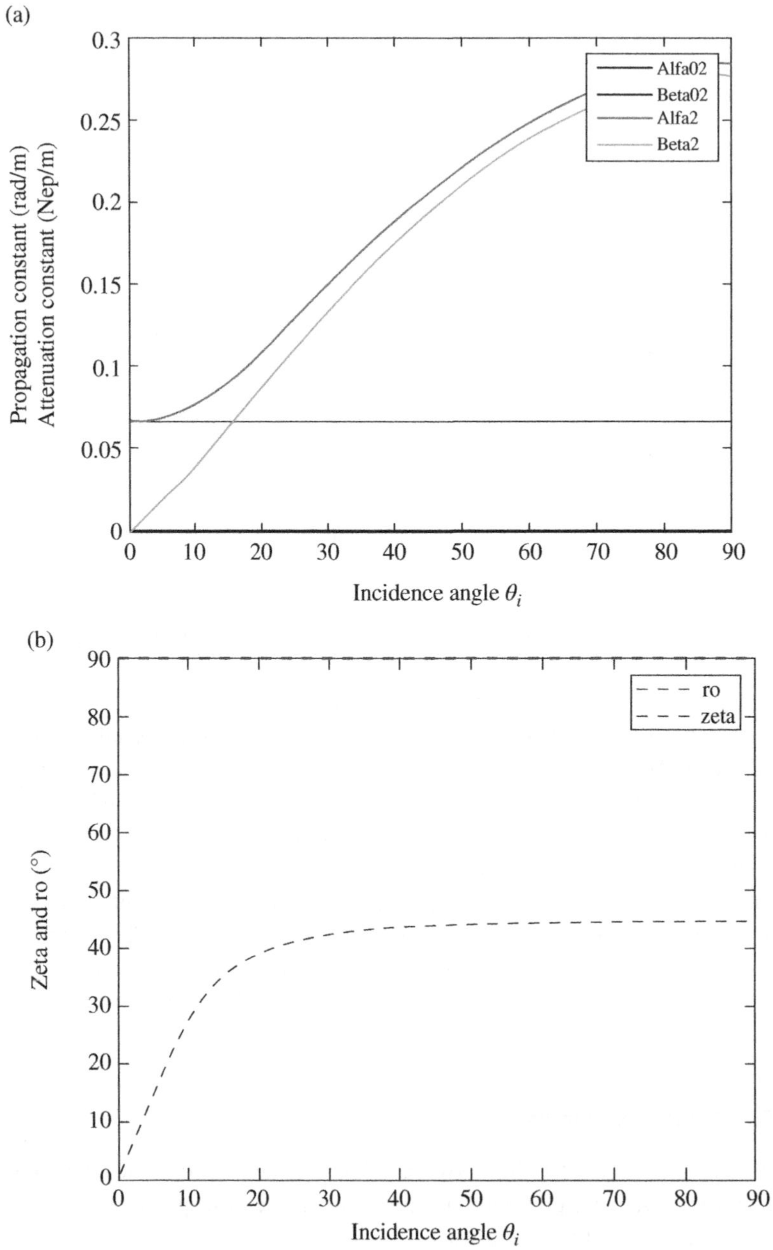

Figure 7.17 (a) Comparison of intrinsic and modified propagation constants as a function of incident angle for $\sigma_2 = 0$, and (b) variations of the phase angles transmission angles θ_{true} and $\Delta\theta$ with the incident angle θ_i [11].

where $\Delta\theta_i$ is the angular separation of the modified α_1 and β_1, $\Delta\theta_t$ is the angular separation of the modified α_2 and β_2 that we need to determine. To satisfy the continuity of the tangential electric fields across the interface of the two lossy media at $z = 0$, the phases of the three field expressions in (7.65)–(7.67) must be equal. This leads to the following relations between the modified and intrinsic propagation constants:

Figure 7.18 Oblique incidence of uniform plane wave incident at the interface of two conducting media.

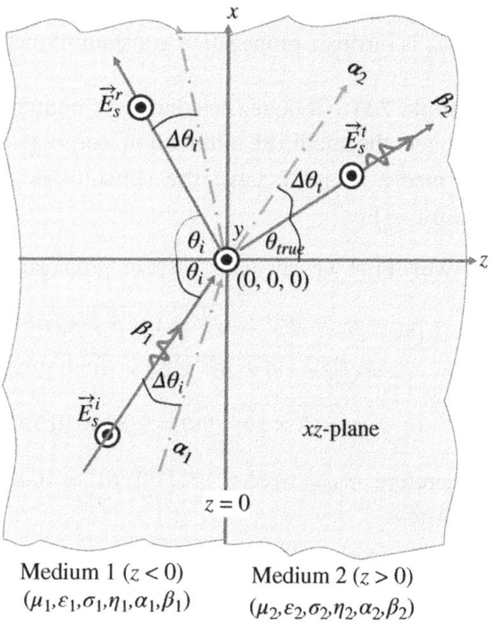

$$\beta_{01} \sin\theta_i = \beta_2 \sin\theta_{true} \tag{7.68}$$

$$\alpha_{01} \sin(\theta_i + \Delta\theta_i) = \alpha_2 \sin(\theta_{true} + \Delta\theta_t) \tag{7.69}$$

can also write the complex vector modified propagation in medium 2 [8] as follows:

$$\boldsymbol{\gamma_2} \cdot \boldsymbol{\gamma_2} = (\alpha_{02} + j\beta_{02})^2 \tag{7.70}$$

$$\boldsymbol{\gamma_2} = \boldsymbol{a_x}\{\alpha_2 \sin(\theta_{true} + \Delta\theta_t) + j\beta_2 \sin\theta_{true}\}^2 + \boldsymbol{a_z}\{\alpha_2 \cos(\theta_{true} + \Delta\theta_t) + j\beta_2 \cos\theta_{true}\}^2 \tag{7.71}$$

where, $\alpha_2 = |\alpha_2|$ and $\beta_2 = |\beta_2|$; the amplitudes of the modified attenuation and phase constants respectively. Substitution for (7.71) into (7.70) and separating the real and imaginary parts yield the following pertinent relations between the intrinsic and the modified attenuation and propagation constants as follows:

$$\alpha_2^2 - \beta_2^2 = \alpha_{02}^2 - \beta_{02}^2 \tag{7.72}$$

$$\alpha_2\beta_2 \cos\Delta\theta_t = \alpha_{02}^2\beta_{02}^2 \tag{7.73}$$

Using the field continuity conditions in (7.68) and (7.69) and the above relations between the intrinsic and modified attenuation and propagation constants, we obtain closed form expressions for the modified constants α_2 and β_2 as follows [11]:

$$\beta_2 = \frac{1}{\sqrt{2}}\left[|\gamma_1|^2 - Re\left(\gamma_{02}^2\right) + |\gamma_1^2 - \gamma_{02}^2|\right]^{1/2} \tag{7.74}$$

$$\alpha_2 = \frac{1}{\sqrt{2}}\left[|\gamma_1|^2 + Re\left(\gamma_{02}^2\right) + |\gamma_{01}^2 - \gamma_{02}^2|\right]^{1/2} \tag{7.75}$$

where, $\gamma_1 = \alpha_1 + j\beta_1 = \alpha_{01}\, sin\,(\theta_i + \Delta\theta_t) + \beta_{01}\, sin\,\theta_i$, the modified propagation constant in medium 1, and γ_{02} is intrinsic propagation constant of medium 2. As usual, α_1, β_1. α_2 and β_2 are real.

Example 7.11: Oblique Incidence on Conductor-Conductor Interface
Calculate the modified attenuation constant at $\theta_i = 45°$. The operating frequency is 6 GHz. The media 1 and 2 have the constitutive parameters: $\epsilon_{r1} = 4$, $\sigma_1 = 0.2\left(\frac{S}{m}\right), \epsilon_{r2} = 12$; and $\sigma_2 = 0.1\left(\frac{S}{m}\right)$.

Answer: First we calculate intrinsic propagation constants for medium 1 and medium 2.

$$\gamma_{01} = \alpha_{01} + j\beta_{01} = \sqrt{j\omega\mu_1(\sigma_1 + j\omega\epsilon_1)}$$
$$= \sqrt{j2\pi \times 6 \times 10^9 \times 4\pi \times 10^{-7}(12 \times 10^5 + j2\pi \times 6 \times 10^9 \times 8.854 \times 10^{-12} \times 4)}$$
$$= \sqrt{5.68 \times 10^{10}\angle 90°} = 2.38 \times 10^5\angle 45° = 16.85 \times 10^4 + j16.85 \times 10^4\,(m^{-1})$$

Therefore, $\alpha_{01} = 16.85 \times 10^4\left(\frac{Np}{m}\right)$; $\beta_{01} = 16.85 \times 10^4\left(\frac{rad}{m}\right)$
And,

$$\gamma_{02} = \alpha_{02} + j\beta_{02} = \sqrt{j\omega\mu_2(\sigma_2 + j\omega\epsilon_2)}$$
$$= \sqrt{j2\pi \times 6 \times 10^9 \times 4\pi \times 10^{-7}(14 \times 10^4 + j2\pi \times 6 \times 10^9 \times 8.854 \times 10^{-12} \times 12)}$$
$$= \sqrt{66.3 \times 10^8\angle 90°} = 81.4 \times 10^3\angle 45° = 57.6 \times 10^3 + j57.6 \times 10^3\,(m^{-1})$$
$$\therefore Re\left(\gamma_{02}^2\right) = 0$$

Therefore, $\alpha_{02} = 57.6 \times 10^3\left(\frac{Np}{m}\right)$; $\beta_{02} = 57.6 \times 10^3\left(\frac{rad}{m}\right)$
Next we calculate the modified propagation constant in medium 1:

$$\gamma_1 = \alpha_1 + j\beta_1 = \alpha_{01}\, sin\,(\theta_i + \Delta\theta_t) + \beta_{01}\, sin\,\theta_i$$
$$= 16.85 \times 10^4\, sin\,(45° + 0°) + 16.85 \times 10^4\, sin\,(45°)$$
$$= 238.3 \times 10^3 + j238.3 \times 10^3(m^{-1});$$
$$\therefore |\gamma_1|^2 = 2 \times \left(238.3 \times 10^3\right)^2 = 1.136 \times 10^{11}$$

The modified propagation constant: $\beta_2 = \frac{1}{\sqrt{2}}\left[|\gamma_1|^2 - Re\left(\gamma_{02}^2\right) + |\gamma_1^2 - \gamma_{02}^2|\right]^{\frac{1}{2}}$

$$= \frac{1}{\sqrt{2}}\left[1.136 \times 10^{11} + \left|\left(238.3 \times 10^3 + j238.3 \times 10^3\right)^2 - \left(57.6 \times 10^3 + j57.6 \times\right)^2\right|\right]^{\frac{1}{2}}$$
$$= \frac{1}{\sqrt{2}}\left[1.136 \times 10^{11} + 1.143 \times 10^{11}\right]^{\frac{1}{2}} = 3.38 \times 10^5\left(\frac{rad}{m}\right)$$

The modified attenuation constant:

$$\alpha_2 = \frac{1}{\sqrt{2}}\left[\,|\gamma_1|^2 + Re\left(\gamma_{02}^2\right) + |\gamma_{01}^2 - \gamma_{02}^2|\right]^{1/2} = 3.38 \times 10^5\left(\frac{Np}{m}\right)$$

Finally, we can draw a few concluding remarks about the significant theory of oblique incidence at the interface of two lossy media.

We have examined the characteristics of the uniform and nonuniform plane waves obliquely incident on the interface of two lossy media. We observe that when both media are lossy, we consider the nonuniform plane wave in both media and their associated phase angles between the propagation constant and attenuation constant.

Review Questions 7.1: Propagation through Conductor-conductor Interface

Q1. Draw a conclusion of a wave propagates from one lossy medium to a second lossy medium.
Q2. Draw an inference for the wave when both media are highly conductive.
Q3. State applications of the theory of wave propagation in both highly conducting media.

In the following section, we shall describe an interesting application of the wave propagation in lossy medium which is in this case the ground soil. Ground has very interesting characteristics, it can be assumed a highly conducting medium of which the dielectric constant changes with the soil moisture contents.

We shall discuss the electromagnetic wave transmission and reflection in soil for the applications of the emerging precision agriculture. Moisture, salinity, and nutrient contents of soil for a vast region can be measured with Soil Moisture Active Passive (SMAP) satellite by NASA. Since the evaporation of soil moisture needs huge energy, monitoring of soil energy flux profile on the earth surface can provide predictions of weather forecast such as floods, droughts, landslides and bushfires besides soil moisture and nutrient contents, and ocean salinity.[9] Besides SMAP satellite we can also use the ground sensors that sense soil health in a small area of a few centimeter squares. We shall describe the wireless ground sensors first followed by the SMAP satellite soil moisture measurement system.

7.6 Emerging Applications: Precision Agriculture

World population is expected to reach 81.1 billion by 2025 and 9.6 billion by 2050. This exorbitant population size will put tremendous pressure on demand for food. The United Nations (UN) Food and Agriculture Organization (FAO) estimates worldwide food production need to be raised by 70% over the next several decades to meet the expected demand by 2050. A report on global food waste shows that each year about one-third yield is lost during production in developing and emerging countries and about one-tenth is lost in developed countries while about a billion people remain hungry worldwide [1]. To meet this increasing demand of food, farmers in Australia draw exorbitant amount of water from Murray Darling river. This tremendously impacts the environment and national economy. Last year's fish deaths in the river is a contrasting example. To combat the drought condition in Australia [2] with a motto to use water resources efficiently for irrigation, farmers are adopting precision agriculture to minimize water waste during the production process. Precision agriculture ensures efficient use of water in the crop fields to reduce water waste using the innovative devices such as soil moisture measuring microwave radiometers, and ground soil moisture measuring wireless sensors. These sensors collect data for moisture, nutrient contents, and pH levels of soil in real time and feed the information to a decision support system (DSS) so that the computerized commands can control the automated water nozzles to dispense only the required amount of water in the field. Current practices are to put the expensive peg sensors which are hardwired in the edge of the paddock. Therefore, there is a need for low cost wireless ground sensors so that they can be deployed uniformly throughout the field and can be removed during the harvesting

9 https://directory.eoportal.org/web/eoportal/satellite-missions/s/smap (accessed 29 February 2020).

seasons. Most importantly, these devices drastically reduce cost of precision agriculture technology, making it affordable to farmers who find current technologies too expensive to adopt. A decision support system (DSS) is a computerized in-built algorithm that controls spray nozzles in real time to discharge only the required amount of water. It also sends accurate data for soil health and crop yields via cloud computing to farmers and government departments. These devices can track crop yields, exact only the required amount of water and fertilizers usages in different sections of the field. Preserving water has a positive influence on climate variability, river catchments and overall development of regional and national economy. In this section an innovative idea to precision agriculture using the wireless sensors and radiometer is presented. As shown in Figure 7.19, the integrated precision agriculture technology based on the advanced electromagnetics has two components that work seamlessly to provide real-time accurate data for the farmers and control effectively the water usages in the field. Figure 7.19b shows the conventional linear irrigator with a wired ground sensor. Based on the sensor data interpreted with the DSS, the irrigator dispenses required amount of water. As stated above the commercial of the shelf (COTS) sensors are hardwired and highly expensive that deter farmers to implement the sensor in the field in large

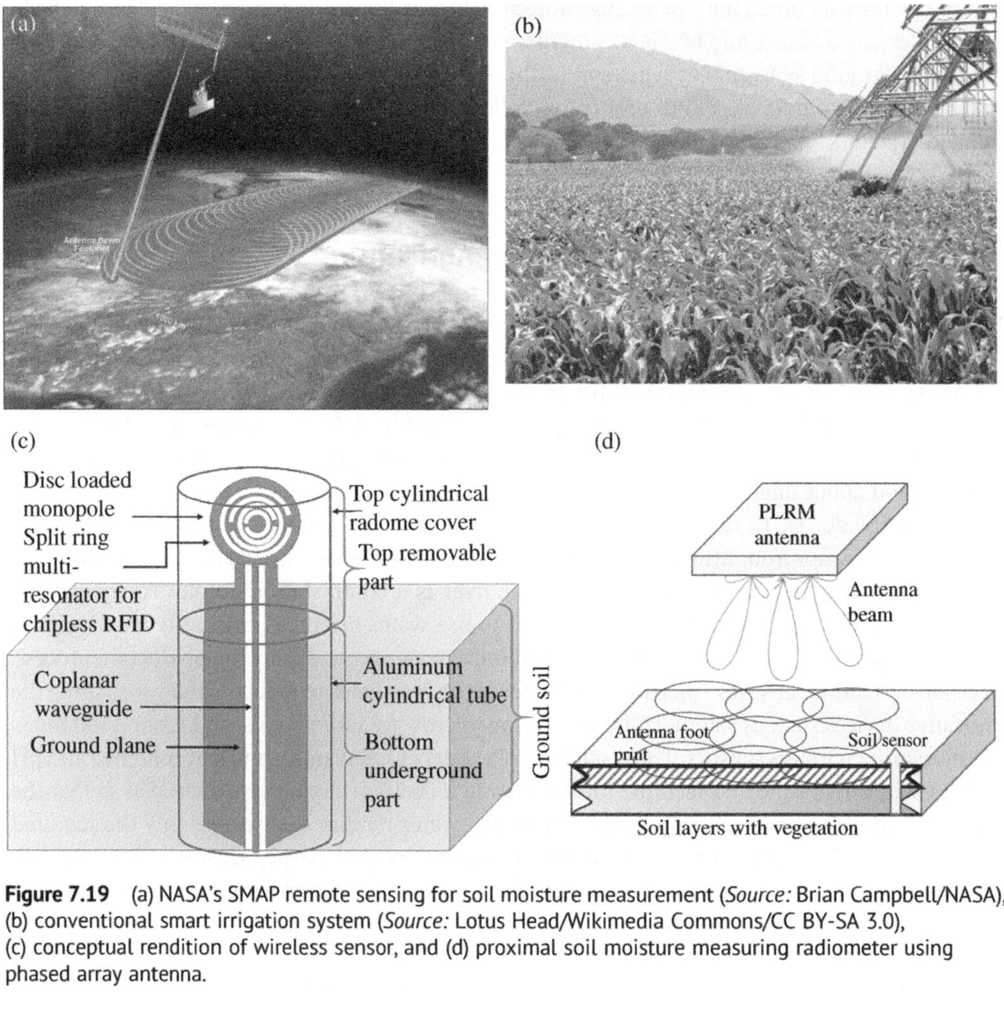

Figure 7.19 (a) NASA's SMAP remote sensing for soil moisture measurement (*Source:* Brian Campbell/NASA), (b) conventional smart irrigation system (*Source:* Lotus Head/Wikimedia Commons/CC BY-SA 3.0), (c) conceptual rendition of wireless sensor, and (d) proximal soil moisture measuring radiometer using phased array antenna.

quantities to measure the soil moisture data uniformly. The soil types and moisture contents vary with soil types, seasons, crop types, and within the paddocks. With a few expensive peg sensors in a vast field, precious measurements of soil moisture contents are not possible in the current arrangement. Only feasible option stated above is that a large number of very low-cost wireless sensors is to be deployed over the field so that data can be collected seamlessly without any wires. During the harvesting season, the top part of the sensor can be removed. In addition to instead of depending NASA' SMAP data (Figure 7.19a) which is collected with a significant latency in weeks and months intervals, the gantry-based radiometer with a multi-beam antenna (Figure 7.19d) can measure the soil moisture contents by knowing the contrast of the dielectric properties of the soil types in different part of the field in real-time.[10] Hence farmers do not need to wait for the data and then make the decision. Farmers can take instant decision to dispense correct amount of water in the crop field and can monitor the crop growth, and yields. Combining both wireless technologies of the ground-based sensor (Figure 7.19c) and gantry-based irrigator, a state-of-the-art wireless communications system can be built for the precision agriculture. In this section, the electromagnetic means of the sensors are presented. First, the ground sensor is described in the light of the electromagnetic wave propagation in lossy media followed by the radiometer theory of moisture content measurement.

7.6.1 Wireless Sensor

An innovative chipless RFID sensor for soil moisture, salinity, and nutrient measurement that utilizes the theory of advanced electromagnetics, and reflection and transmission from a lossy medium is presented in this section. As can be seen in Figure 7.19c, the sensor is made of a disc-loaded monopole antenna engraved with a concentric multi-resonant passive UWB microwave circuit. The individual concentric resonator is a split ring resonator with a gap capacitance that rings a distinctive tone (frequency) when the tag is illuminated with a UWB incident signal. The frequency signature of a resonator provides 1 : 1 correspondence of the binary identification data. Thus, multiple resonators yield distinctive multi-bit identification data stream 1010 .... This chipless RFID feature of the soil moisture probe provides distinctive identification data so that the signal received from a particular senor node can be identified. The second aspect of the disc loaded monopole antenna is the wireless communications for sensing application. The UWB antenna is the top part of the sensor. This antenna part can be detached from the bottom part during the harvesting season.

Due to the presence of the ground plane, the monopole antenna yields good match and directive radiation pattern toward the reader above the ground. Thus, less energy is wasted along the undesired direction. The bottom part of the sensor is a coplanar waveguide (CPW) based planar transmission line. The CPW is a special type of a planar transmission line with a current carrying conductor strip in the middle with gaps on both sides of the ground plane. This part is housed in an aluminum tube and is pegged in the soil so that it can come in contact with soil for moisture and nutrient information. The electromagnetic system of the bottom part is shown in Figure 7.19c. As can be seen, the current carrying CPW comes in contact with the soil after a thin protecting dielectric layer with dielectric constant ε_r and thickness h on both sides of the current carrying conductor. This thin layer protects the CPW from being oxidized when it comes in contact of wet soil. Thus, it ensures consistent electromagnetic performance. The loading effect of the soil with moisture and nutrient contents reflects back the signal to the UWB antenna via the CPW.

10 Different methods are used to measure the complex dielectric constant of soil. Examples include Topp's model, [7] Dopson's model, [9] Peplinsky's Model and Mironov's model. Out of all these model here, Mironov model will be discussed for wireless measurement of soil moisture with ground sensors.

The backscattered signal via the radiation pattern of the UWB antenna is received by the receiving antenna of the reader electronics and processed with the signal processor.

7.6.2 Soil Models

Soil is a porous complex dielectric medium with constituents of bulk soil with minerals and water contents. The porosity of soil is filled with air, bound water, and free water. Since water has the largest dielectric constant than solid soil constituents, the measurement of dielectric constant of wet soil provides an accurate measurement of the moisture content of soil.[11]

As stated above, the soil moisture contents vary between the soil types and seasons of the year. The dielectric property of soil changes based on the soil type and amount of water that it holds. If the difference between the dielectric properties (effective ε_r) of water and other constituents of a particular type of soil can be measured preciously in microwave frequency domain then it provides the moisture content data in wireless means.

Scientists try to understand the various soil compositions based on soil types and moisture contents and develop different electromagnetic models of soil. There are two different models for soil: (i) a three-phase model with water, air, and soil solid, and (ii) four phase model with bound water, free water, air, and soil solid. Usually time domain reflectometry (TDR) technique is used to measure soil moisture contents.[12] These soil dielectric models yield a relationship between the dielectric properties of different soil compositions. Different models provide different calibration equations for volumetric water content m_v which is a function of the effective dielectric constant of the soil. The volumetric water content, m_v is defined as the volume of water content as a percentage of the total volume of the soil. The volumetric water content is also measured in terms of the densities of the dry versus wet soil and is defined as:

$$m_v = \frac{\rho_d}{\rho_s} \tag{7.76}$$

where, ρ_d is the bulk density of dry soil and ρ_s is the specific density of wet soil solid. Different model provides accuracy for certain boundary conditions.

7.6.3 TDR Technique in Soil Moisture Measurements

Irrespective of wired and wireless measurements of the soil moisture contents, TDR technique of soil moisture measurement is a golden rule. A TDR technique measures the round trip of travel time of an electromagnetic signal through a medium of propagation after it reflects back from a certain discontinuity. Any impedance change in the transmission line is defined as the discontinuity. The discontinuity is usually either a short or an open circuit. The travel time depends on the velocity of the wave propagation in the medium of propagation which is in this case soil. Since the velocity is a function of the effective dielectric constant ε_r of soil, the precious measurement of the travel time t for a fixed probe length l provides the indication of the constitutive parameter ε_r hence the moisture content of soil. It is not an easy process to measure the constitutive parameter for a complex medium like soil. That is why various pseudo-empirical curve fitting approaches are used to develop various soil models. Figure 7.20 illustrates the TDR method to measure the soil moisture contents. The operating principle of a wireless soil moisture method using passive soil sensor and a proximal radiometer is shown in Figure 7.21.

---

11 This section is adopted from S. Dey [8] under supervision of the author.

12 G. Capparelli, G. Spolverino, and R. Greco, "Experimental Determination of TDR Calibration Relationship for Pyroclastic Ashes of Campania (Italy)," *Sensors* 2018, 18, 3727; doi:https://doi.org/10.3390/s18113727, pp. 1–14.

Topp et al. [7] model is a three-phase model and is universally accepted as a soil model for measurement of the volumetric moisture content of the soil. It is a third order polynomial approximation obtained experimentally for different soil textures. The volumetric moisture content as a function of the effective dielectric constant of the soil is defined as:

$$m_v = -0.053 + 29.2 \times 10^{-2}\varepsilon_r - 5.5 \times 10^{-4}\varepsilon_r^{\,2} - 4.3 \times 10^{-6}\varepsilon_r^{\,3} \tag{7.77}$$

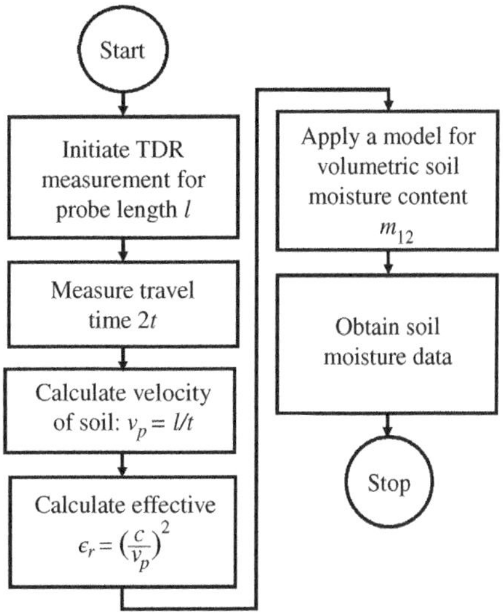

Figure 7.20 TDR measurement method for calculation of soil moisture contents.

Since Topp et al. [7] model is perceived as universal experimental approximation for all textures of soil types, it is used for TDR soil measurement with a calibration for a specific soil type. It is found by further investigations that this model is not valid for all soil types and also only works between the frequency range of 1 MHz to 1 GHz [8]. At higher microwave frequencies, it appears to be invalid. Hence for microwave UWB wireless measurement different methods should be envisaged. For accurate quantitively measurement of volumetric soil moisture contents, $m_v(\varepsilon_r)$ needs to be determined more preciously for specific soil types. Dobson et al. [9] propose a four-phase models, with complex approximations but they work for L-band frequency of interests. For UWB wireless sensors we need to consider more accurate model for a wider frequency range for 3.3–10.6 GHz. In this instance, Mironov's model is a good fit for the UWB wireless sensor applications.

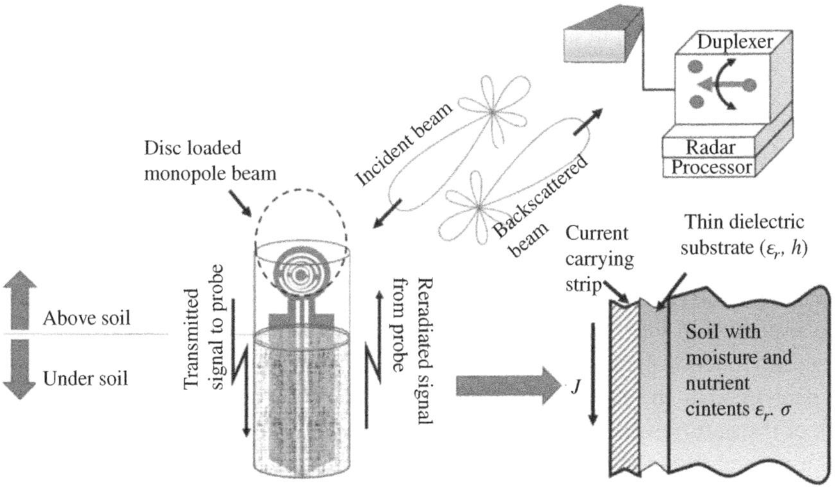

Figure 7.21 Operating principle of soil moisture sensor.

Mironov's model is based on three-phase soil moisture composition and defines the relation between the soil moisture volumetric concentrations to dielectric constant in the frequency band from 1 to 10 GHz.

$$\sqrt{\varepsilon_m} = \sqrt{\varepsilon_d} + \left(\sqrt{\varepsilon_{fw}} - 1\right) m_v \tag{7.78}$$

where ε_m is the effective dielectric constants of soil water mixture, ε_d is the effective dielectric constant of dry soil, ε_{fw} is for the wet soil and m_v is the volumetric soil moisture content. ε_d can be calculated by using the dielectric constant of soil solids ε_s and soil bulk density. Dey [8] investigated different models and after a careful consideration he selected Mironov's model for UWB sensor design.

7.6.4 Sensor Design

Dey [8] designed a UWB chipless RFID sensor for soil moisture measurement and investigated the performance in full-wave electromagnetic solver CST Microwave Studio as shown in Figure 7.22. A uniform plane wave is launched toward the sensor embedded in soil. A receiver is placed in front of the sensor to analyze the signal for soil moisture contents. Figure 7.22b shows the laboratory set up with a vector network analyzer (VNA) for complex signal analysis to determine the dielectric constant of the soil. A UWB rib horn antenna is connected with the VNA in front of the soil moisture sensor which is embedded in soil in a PVC pipe. Stevens hydra probe is inserted in the soil for measuring the soil moisture content. A thermal temperature sensor is used to monitor the temperature variation. A Getac handheld personal computer (PC) is interfaced with the moisture sensor to record the data of the soil moisture contents. The VNA is calibrated for TDR measurement of the reflected signal from the wireless sensor. The horn antenna connected with the VNA is placed 30 cm apart from the sensor. Figure 7.22c shows the measured results of the backscattered signal from the soil sensor. As can be seen in the measured result, there is a regular pattern in the reflected signal from the sensor with the moisture content. With the moisture content the backscattered signal strength reduces. A proper analysis of the signal and linking the signal with the effective dielectric constant and loss tangent *tanδ* provide accurate wireless measurement of the soil moisture content. Table 7.1 shows the comprehensive results of the soil moisture contents in terms of m_v and the corresponding dielectric constant and loss tangent of the soil. The results indicate that the innovative and very low cost wireless sensor has huge potential in emerging precision agriculture.

In the next section, we discuss the remote sensing soil moisture measurement method using a radiometer.

7.6.5 Soil Moisture Remote Sensing Radiometer

In this section, we explain the use of the theory of lossy dielectric medium for measuring the soil moisture content using a remote sensing radiometer. A radiometer is a passive highly sensitive receiver of electromagnetic waves irradiated by the molecular interaction of the soil moisture with the atmosphere. As explained in the theory of the soil moisture measuring sensor above, the dielectric constant of soil is a function of moisture contents. In the radiometer theory, the interaction of the soil moisture content with the atmospheric molecules and the resulting radiation can be explained in terms of the natural emission of soil. In a radiometer, the measuring parameter is *the emissivity* of the natural radiation due to the molecular activities of the lossy ground soil. The passive remote sensing radiometer does not use an illumination source but acts like a highly sensitive passive listener of the natural microwave emission from the soil at specific frequencies.

(a)

(b)

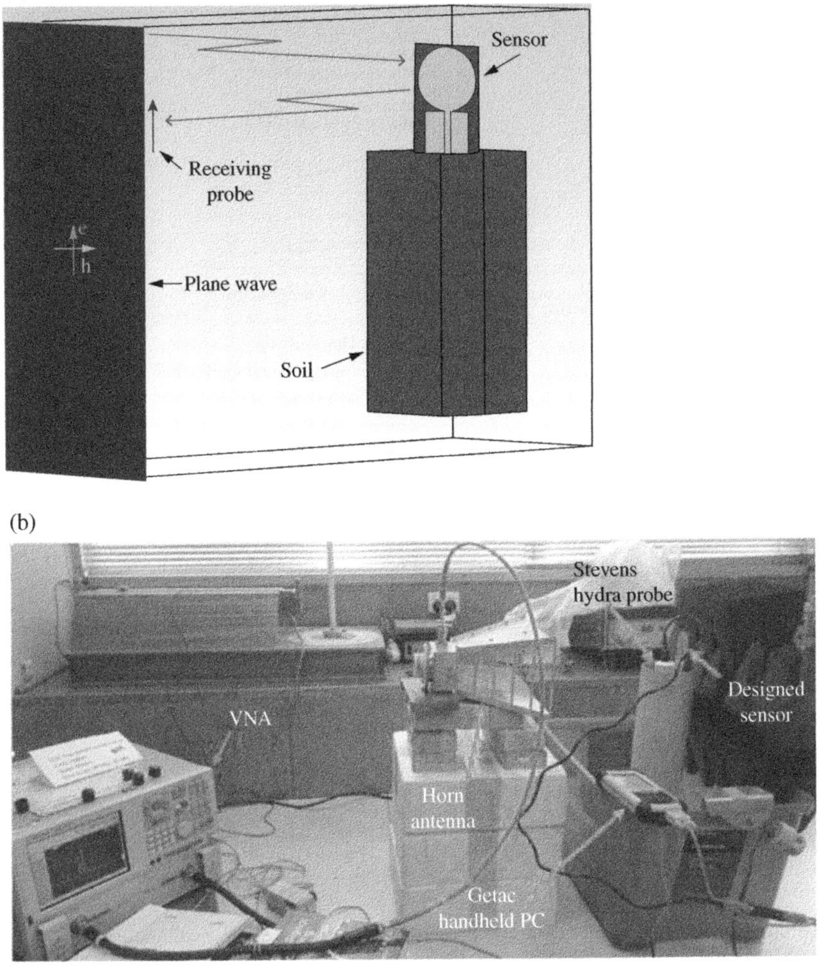

Figure 7.22 Chipless RFID soil moisture sensor. (a) CST simulations setup, (b) laboratory set up with a vector network analyser (VNA), and (c) measured results of the backscattered signal from the soil sensor. *Source:* Dey [8].

The most popular frequencies at which natural emissions from soil are measured are P-, L-, Ku and Ka-bands. The natural emission can be explained with Planck's black body concept. Planck introduced the concept in his quantum theory in 1901. In any physical object above absolute zero temperature of $-273\,^\circ$C or $0\,^\circ$K, atoms undergo in thermal motion and emits electromagnetic radiation. This radiation is called natural or black body radiation. A black body is a hypothetical ideal object at a thermal equilibrium that absorbs or emits radiation at all frequencies from all directions. Although the radiation is very weak, highly sensitive microwave and millimeter wave receivers measure the radiation. A microwave radiometer is a highly sensitive passive listener (receiver) that can detect this natural microwave emission from wet soil.

The time average power intensity of the natural electromagnetic emission from soil can be defined with Poynting theorem: $P_{ave} = |E|/2\eta$ (W/m^2); E is the received radiation electric field

(c)

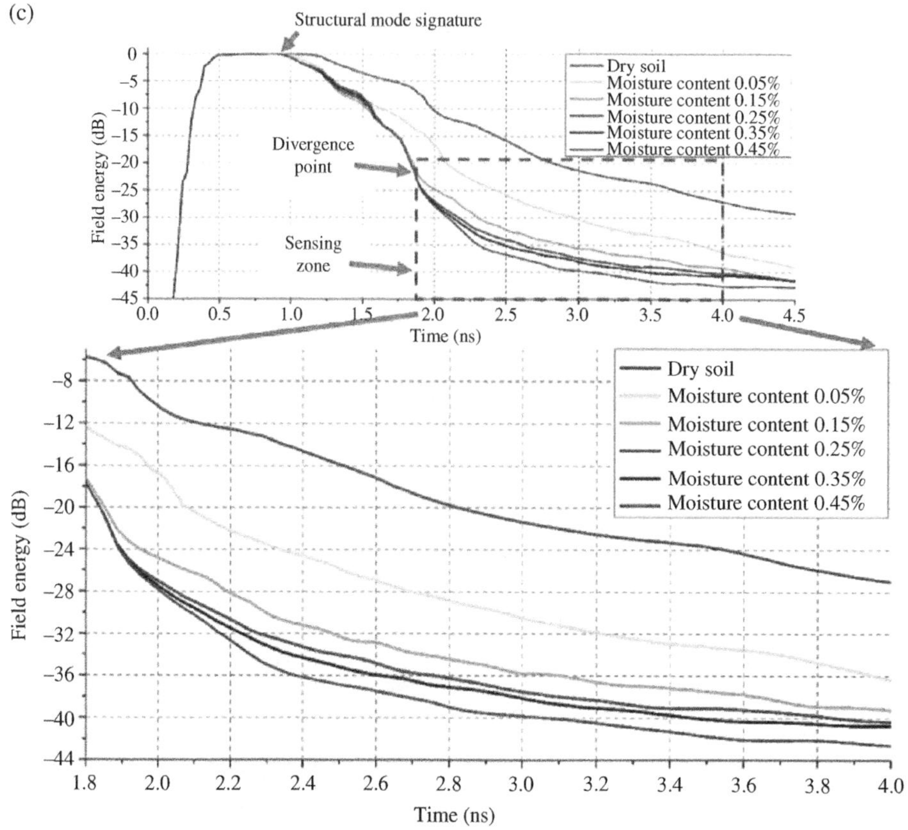

Figure 7.22 (Continued)

Table 7.1 Measured volumetric moisture content with their corresponding theoretically calculated dielectric constant for different water levels.

	Reading at one hour		**Reading at two hours**	
Soil Condition	**Volumetric moisture content (m_v)**	**Dielectric Constant (ε_r)**	**Volumetric moisture content (m_v)**	**Dielectric Constant (ε_r)**
Dry	0.01 (initial value)	2.59	0.01 (initial value)	2.59
Soil with 50 ml water	0.176	5.04	0.173	5.0
Soil with 100 ml water	0.197	5.41	0.204	5.54
Soil with 150 ml water	0.274	6.88	0.268	6.76
Soil with 250 ml water	0.284	7.08	0.280	7.0
Soil with 350 ml water	0.302	7.45	0.296	7.33
Soil with 450 ml water	0.309	7.60	0.322	7.88

intensity and η is the intrinsic impedance of the propagation medium relative to the soil moisture contents. The emission intensity or emissivity e_p of soil is signal polarization depended and is measured in term of its brightness temperature T_{bp}. The brightness temperature is defined as the physical temperature of a black-body emitting the same amount of energy. The emissivity of a black body is one. Similar to the black-body, a hypothetical white-body is a perfect reflector that reflects energy at all directions. The emissivity of a wide body is zero. Most of the practical objects are grey body which has emissivity in between those of the black-body and white-body. The brightness temperature (T_{Bp}) represents the ability of natural emission of materials and is defined in terms of the emissivity (e_p) and the physical temperature (T_{ph}) as

$$T_{Bp} = e_p.T_{ph} \tag{7.79}$$

where T_{ph} is the physical temperature of the material in °K, and the subscript "p" represents the polarization of the electromagnetic signal, which can be either horizontal or vertical. (7.79) is achieved from the Planck's black-body radiation law using the Rayleigh–Jeans approximation for microwave frequencies. For soil, the emissivity varies from around 0.95 for dry soil (with moisture content of 0.05 m^3/m^3) to around 0.6 for saturated wet soil (with moisture content of 0.4 m^3/m^3), depending on electromagnetic wavelength, surface roughness, incidence angle, and soil properties. If the soil temperature can be considered as 300 ° K, then a change from dry soil to wet soil is seen as 90 ° K difference because of different brightness temperature, whereas actual difference in temperature is very small (approximately 1 ° K). Such a large variation of brightness temperature in a sensitive receiver can be easily detected.

Example 7.12: Soil Moisture Remote Sensing
A dry soil sample with moisture contents 0.05 (m^3/m^3) has emissivity of 0.95 and 0.945 in vertical and horizontal polarization, respectively. The physical temperature is 300 ° K. Calculate the brightness temperature of the soil in both polarizations.

Answer:

The brightness temperature in vertical polarization:

$$T_{BV} = e_V.T_{ph} = 0.95 \times 300°\text{K} = 285°\text{K} = 12°\text{C}$$

The brightness temperature in horizontal polarization:

$$T_{BH} = e_H.T_{ph} = 0.945 \times 300°\text{K} = 283.5°\text{K} = 10.5°\text{C}$$

The difference in temperature in the two polarizations is 1.5 ° C.

Satellite remote sensing radiometer is the only practical way to provide high resolution soil moisture data. However, the resolution is in the order of tens of kilometers depending on the frequency of operation. Polarimetric microwave radiometer (PMR) systems are used for remote sensing of environmental parameters such as ocean salinity and soil moisture. A PMR exploits the natural emission from soil which are polarization dependent as shown in Figure 7.23. A PMR is a passive listener to natural emission at microwave frequencies and receives power intensity P_{ave}. A PMR consists of a scanhead or a multi-beam antenna as shown Figure 7.23a, with a beamforming network, the receiver electronics, and an embedded processor Figure 7.23b. Two orthogonally polarized antennas receive natural emissions of vegetation and soil in terms of *brightness temperature* ($T_{Bp} = eT_{ph}$) at both vertical (T_{Bp}^V) and horizontal (T_{Bp}^H) polarizations. As shown in Figure 7.23b, the beamforming network steers the beam in prescribed directions as controlled by switching

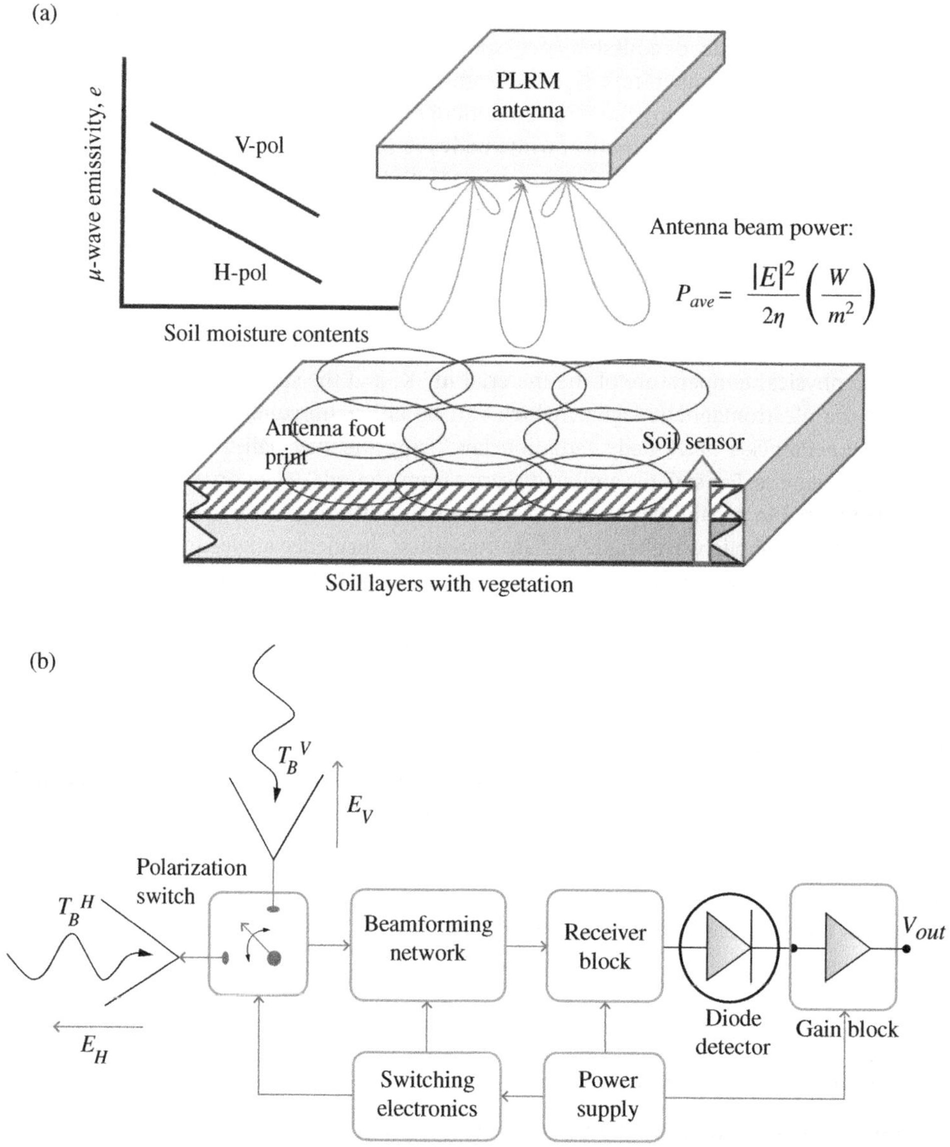

Figure 7.23 (a) Concept of multi-beam soil moisture radiometer, and (b) radiometer.

electronics. The receiver block amplifies, denoises and down-converts, and delivers the received signal to a detector that converts the AC signal to DC. The gain block amplifies the signal and generates output voltage V_{out}. The computer (not shown here) samples the analog signal V_{out}, collects digital data corresponding to *brightness temperature*, T_{Bp}, of the scanned area, and interprets the environmental parameters. The microwave bands that have been used for PMR are P-, L-, C-, X-, Ku-, Ka-, and W-bands. L-band is currently the optimal choice for soil moisture remote sensing because of its being less impacted by soil roughness, vegetation, and atmosphere.

In the next section, the experimental set up and test results of soil samples in laboratory environment is presented.

7.6.6 Test Set Up

Figure 7.24 illustrates a laboratory test set up of soil moisture measurement radiometer. A soil sample is scanned with a horn antenna. The antenna is only in the receive-only mode that direct the radiation pattern toward the soil sample with moisture contents. A high gain and low noise figure low noise amplifier (LNA) receives the signal from the horn antenna and amplifies the natural emission of the soil. As a highly sensitive receiver the spectrum analyzer processes the signal and displays the signal strength for different moisture contents.

Figure 7.25 illustrates the experimental procedure. The percentage of moisture contents is defined in terms of the normalized difference between the wet and dry soil as follows:

$$\text{moisture content (w/w)\%} = \frac{W_w - W_d}{W_d} \times 100\% \tag{7.80}$$

where W_w is the weight in kg of wet soil and W_d is the weight in kg of dry soil. As shown in Figure 7.25, phase 1 starts with soil preparation. A soil sample is taken and completely dried out in an electric oven. The sample is measured and recorded as W_d in kg. Then a known volume of water is added with the soil sample and the mixer is left for two hours to settle. Then the wet soil is measured and recorded as W_w in kg. The (w/w)% is recorded. Then Phase 2 started with the experimental set up with a known antenna configuration in a controlled environment as shown in Figure 7.24. The same temperature, air pressure, and ambient light is ensured in the controlled environment besides the anechoic environment with the electromagnetic foam absorber so that no electromagnetic signal leakage occurs for ambient interferences. The measurement is taken in the far field zone of the antenna. As stated above, the four set of antennas are used in the experiment. Figure 7.26 illustrates the signal strength received by three sets of measurements at L-, Ku-, and Ka-bands in a highly controlled environment with electromagnetic signal absorbing conical foam, a high gain antenna, a low-noise amplifier, and a spectrum analyzer as the receiver. Three set of dual polarized microstrip patch antenna arrays are used in the three frequency bands: 2 × 2-element L-band, an 8 × 8-element Ku-band and an 8 × 8-element Ka-band planar microstrip patch antenna arrays. A shared aperture 8 × 8-element Ku- and Ka-band antenna array is also used in the experiment for simultaneous measurements of signal strength in both bands and is shown in Figure 7.26d. The figures illustrate the received power versus the soil moisture contents in % for the dual polarized received signals from the antennas. The spectrum analyzer provides very sensitive and accurate received power in dBm.

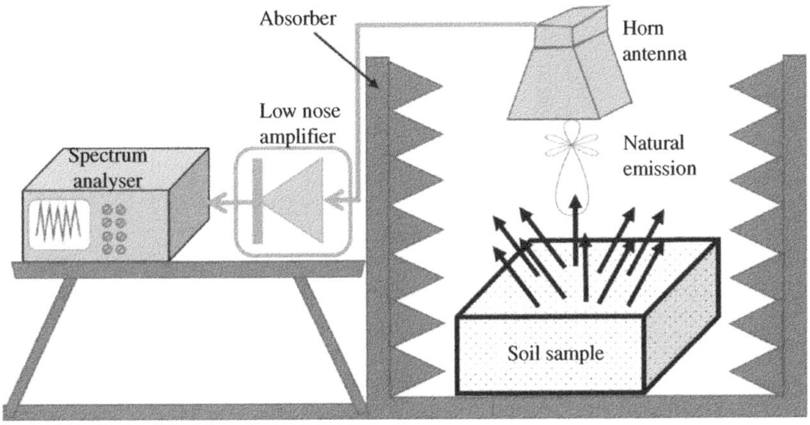

Figure 7.24 Experimental set to measure soil moisture in a laboratory setting.

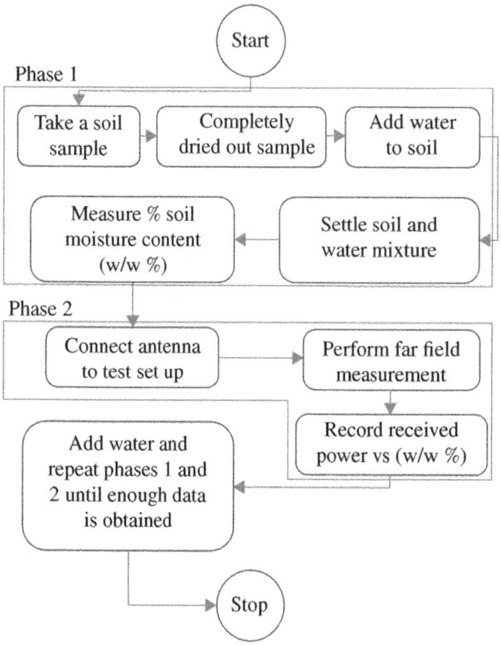

Figure 7.25 Experimental procedure of soil moisture radiometer.

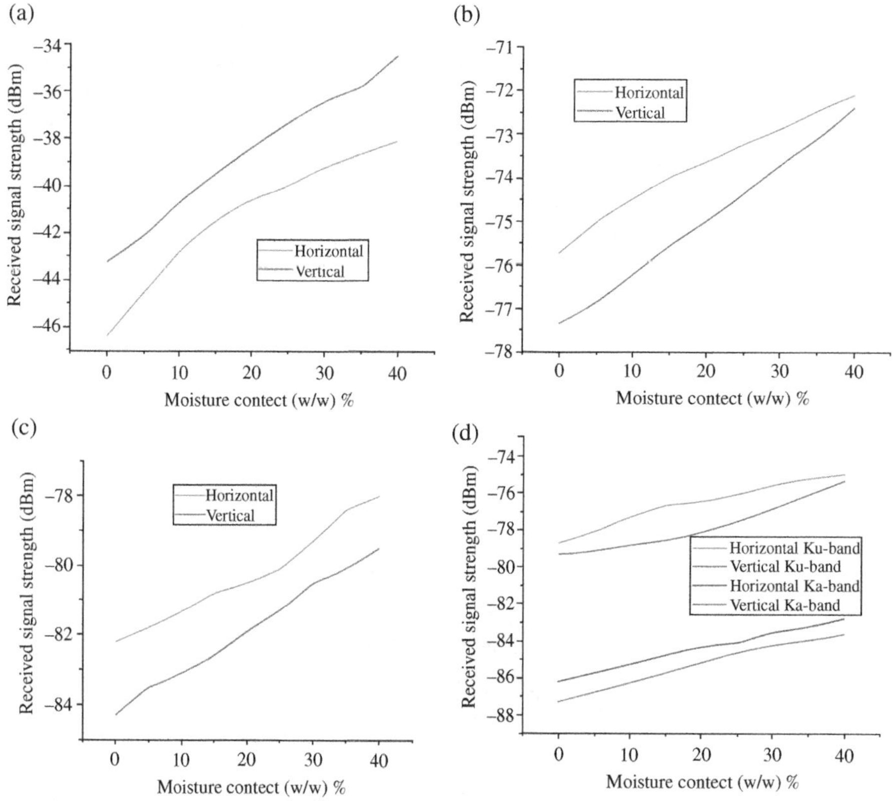

Figure 7.26 Received signal strengths with moisture content at (a) L-band, (b) Ku-band, (c) Ka-band, and (d) Ku- and Ka-band together with a shared aperture antenna array. All antennas are dual polarized antennas. *Source:* Printed with permission Hassan PhD Thesis, Monash University 2018.

From the results, it can be seen that the proof of concept experiments at different frequency bands have brought a consistent relationships with the soil moisture contents in both vertical and horizontal polarizations. Although the signal strength in dBm is a representative, illustration of the brightness temperature versus the soil moisture contents is in %, the signal levels indicates that L-band has the highest signal power and hence can provide the most penetration depth inside ground soil. In literature it states that L-band provides the highest penetration depth in the order of 5 cm whereas the millimeter wave measurements provide depth in the order of millimeters. Specific denoising signal processing algorithms are used to covert the received power in the brightness temperature profile of the soil.

7.7 Chapter Summary

Electromagnetic wave propagation through lossy media has a significant implication in modern industrial applications. This chapter has examined many important practical aspects of wave propagation is lossy media and interfaces of complex boundary value problems. The wide range applications of the theories have mesmerizing impact in modern day technologies. In this chapter only a few examples are presented. Readers are encouraged to investigate more emerging applications of the theory of plane wave propagation in lossy media and the boundary value problems at the interfaces of lossy media to enrich the knowledge in the field.

7.8 Problems

P7.1 A K-band UWB chipless RFID is printed with silver ink of finite conductivity of $\sigma_2 = 3.7 \times 10^6$ (S/m). The operating frequency is centered at 24 GHz with frequency span to 22–26 GHz. Calculate the followings: (i) the skin depth at 24 GHz, (ii) the surface resistance, (iii) the surface impedance, (iv) the reflection coefficient, and (v) percent of power reflected back in air.

P7.2 Repeat P7.1 for the lower bound 22 GHz of the K-band UWB frequency band.

P7.3 Repeat P7.1 for the lower and upper bound 26 GHz of the K-band UWB frequency band.

P7.4 For a 950 MHz RFID tag made with a 3 μm thick conducting aluminum foil (bulk metal) embedded in 70 μm paper dielectric laminate of $\varepsilon_r = 2.4$ and loss tangent $tan\delta = 0.02$. Calculate: (i) the Γ_{eff} at the air–paper interface and 100% power reflected back, and (ii) if the RFID reader transmit 3 (W/m^2), calculate the reflected power back from the air–paper interface. Conductivity for bulk Al is 36.9×10^6 (S/m).

P7.5 Refer to P7.4 now the tag is printed with a screen printer with silver ink with conductivity 5×10^5 (S/m). Assume the ink is a few sink depths thick and can be assumed semi-infinitely extended and printed on a plastic laminate. The impinging uniform plane wave induces a surface current of 130 (mA/m^2). Calculate: (i) the skin depth and the surface resistance, (ii) the equivalent surface impedance, (iii) the real power loss on the conductor, (iv) the power reflected from the conducting surface, and finally, (v) draw a conclusion on the performance of the microwave signal which are backscattered from the RFID tag to the reader's receiving antenna. (Assume $\mu_2 = 4\pi \times 10^{-7}$ (H/m).)

P7.6 A UWB microwave system uses an antenna which is printed with a nanoparticle silver ink of finite conductivity having sheet resistance 50 Ω. The operating frequency spans from 3 to

11 GHz. Calculate the followings: (i) the skin depth at the center frequency, (ii) the surface resistance, (iii) the surface impedance, (iv) the reflection coefficient, and (v) percent of power reflected back in air.

P7.7 A uniform plane wave which is defined with a magnetic field with amplitude $H_0^i = 12$ (mA/m) is incident on an air–lossy dielectric interface at 2.45 GHz. The loss tangent is 0.01 and dielectric constant is 4.2. Calculate the power flow intensity at 1 m from the interface inside the loss dielectric.

P7.8 Human skin has the layer of epidermis, Dermis, hypodermis and muscle. If the effective dielectric constant and conductivity of skin are 56 and 0.53 (S/m) at 300 MHz. Calculate the reflection coefficient and % of power reflected back from a microwave medical imaging transmitter. Assume skin is a dielectric medium so that $\mu_r = 1$.

P7.9 In a hyperthermia treatment a needle shaped antenna is inserted into a tumor. The effective permittivity and loss tangent of the tumor tissues are 55 and 0.1, respectively, at 300 MHz. Calculate equivalent impedance at 300 MHz and the power absorbed by the tumor with radius 1 cm if the antenna induces 10 mA current in the tumor.

P7.10 In P7.9, if the shrinkage rate of the tumor is 1 (μW/cm^2), calculate the new radius of the tumor.

P7.11 For microwave biomedical imaging, a simple phantom model for a human breast is developed as shown in Figure P7.11. Here the breast tissue is emulated with canola oil and a wooden log as the tumor inside the breast tissue. The allowable power transmission is 5 (mW/cm^2). The dielectric constant of canola well is 3.1 and loss tangle is 0.014 and those for the wooden straw is 36 and 0.29, respectively. The radius of the straw is 1.5 cm and overall radius is 8.5 cm. Do the followings: (i) draw an equivalent transmission line model of the electromagnetic system of breast, and calculate impedance in each layer, (ii) effective reflection coefficient, (iii) percentage power reflected, (iv) effective transmission coefficient, (v) incident electric field at the air–skin interface 1, and (vi) received power density by the receiving antenna at the air–skin interface. The operating frequency is at 20 MHz. Hints: For simplicity of calculation consider the wooden straw is of infinite extend so that we consider the impedance of wooden straw equal to its intrinsic impedance.

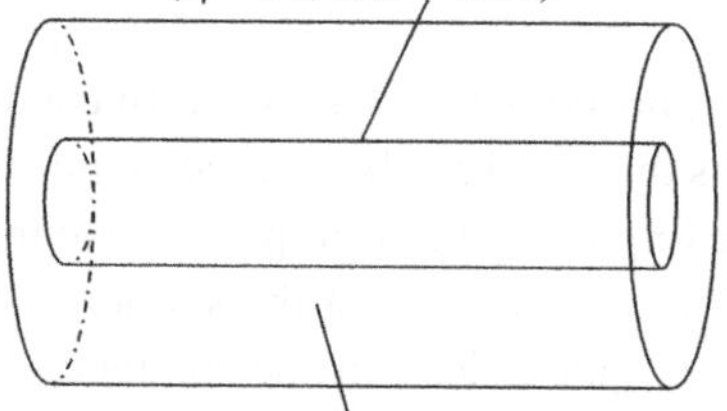

Figure P7.11 Simplified electromagnetic model of breast with canola oil as normal tissue and wooden straw as the tumor.

P7.12 Consider a plastic casing is made of Perspex with dielectric constant $\varepsilon_r = 2.7$, and loss tangent $tan\delta = 0.02$ at 1 GHz as shown in Figure P7.12. The plastic casing is 10 mm thick. The both sides of the casing are covered with nano-particle silver ink of thickness of 20 μm. The conductivity of silver ink is 15×10^5 (S/m). Assume the silver ink has $\varepsilon_{r2} = 1$ and $\mu_{r2} = 1$. (i) Calculate the skin depth and equivalent impedance of the silver ink, (ii) calculate the reflection coefficient and transmission coefficient at the Perspex and the silver ink, and (iii) comment on the coefficients and justifications if you need any extra shielding inside the Perspex casing.

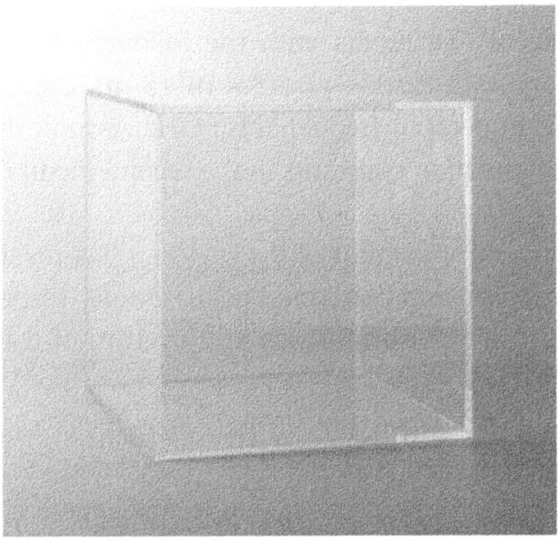

Figure P7.12 A Perspex display case.

P7.13 For P7.12 (i) calculate the reflection coefficient and transmission coefficient at the interface of air and the silver ink, and (ii) comment on the coefficients and justifications if you need any extra shielding inside the Perspex casing.

P7.14 For P7.12, determine the transmitted field in Perspex in terms of the incident field for its intensity of 25 (V/m) and the constitutive parameters of the media. (ii) Calculate the shielding effectiveness of the casing.

P7.15 For P7.14, calculate the true angle of refraction in a shield at 1 GHz. Assume $\mu_1 = \mu_2 = \mu_0$. The angle of incidence is 80°. Infer the result.

P7.16 For P7.14 calculate the true angle of refraction for the incident angle of 5° and make inference on the results.

P7.17 Calculate the true angle of refraction for seawater at 1 GHz for the incident angle of 5°. Consider sea water's conductivity of 4 (S/m) and relative permittivity of 81 as usual.

P7.18 For P7.12 calculate: (i) the refractive index $n_2(\theta_i)$ of sliver ink; (ii) the attenuation and propagation constants, and (iii) the wavelength inside silver ink. Make inferences on the incidence angle and the refractive index.

P7.19 For P7.12 consider an arbitrary polarized incident wave, which is a combination of both TE- and TM-polarizations, is launched from an antenna with an incident angle $\theta_i = 60°$. The amplitudes of the TE- and TM-polarized waves are 250 and 25 (V/m), respectively.

Calculate the followings: (i) the time domain resultant incident electric field in medium 1 (air), (ii) the complex propagation constant γ_2 in sliver ink, (iii) the true angle of refraction θ_{true}, (iv) the effective relative permittivity of medium 2, (v) reflection coefficients Γ_{TE} and Γ_{TM}, and (vi) The time domain expression of the composite reflected electric field.

P7.20 Calculate the modified attenuation constant at $\theta_i = 45°$. The operating frequency is 1 GHz. Media 1 and 2 have the following constitutive parameters: $\epsilon_{r1} = 1$, $\sigma_1 = 0.2$ (S/m), $\epsilon_{r2} = 12$; and $\sigma_2 = 0.1$ (S/m).

P7.21 A smart RFID reflective surface is made of two media with the following constitutive parameters: $\epsilon_{r1} = 1$, $\sigma_1 = 15 \times 10^5$ (S/m), $\epsilon_{r2} = 2.4$; and $\sigma_2 = 15 \times 10^4$ (S/m). Calculate the modified attenuation constant at $\theta_i = 65°$. The operating frequency is 1 GHz. Assume the angular spread between the propagation and attenuation constants in the second medium is 0°.

P7.22 A wet soil sample with the volumetric moisture content of 0.4 (m^3/m^3) has polarization sensitive emissivity of 0.595 and 0.594 in vertical and horizontal polarizations, respectively. The physical temperature is 270 ° K. Calculate the brightness temperature of the soil in both polarizations.

P7.23 A proximal radiometer probes are always calibrated pointing toward sky and then the calibration date is used to measure the brightness temperature of soil. In a clear cold morning, the physical temperature of sky was −27 ° C and the polarization sensitive emissivity in vertical and horizontal polarizations was 0.852 and 0.850, respectively. Calculate the brightness temperature of sky in both polarizations.

Acknowledgments

1) Chipless RFID research was supported with Australian Research Council (ARC) Discovery and Linkage Project Grants:
   i) DP0665523: Chipless RFID for Barcode Replacement.
   ii) LP0989652: Printable, Multi-Bit RFID for Banknotes, industry partners: Securency International Pty Ltd and SatNet Pty Ltd.
   iii) LP0991435: Back-scatter based RFID system capable of reading multiple chipless tags for regional and suburban libraries supported by Express Promotion Pty Ltd.
   iv) LP130101044: Discreet Reading of Printable Multi-bit Chipless RFID Tags on Polymer Banknotes, industry partners: Securency International Pty Ltd and SatNet Pty Ltd.
2) The Radiometer research was supported by ARC Discovery Project: DP160104233 - Airborne passive radiometer for high resolution soil moisture monitoring.

References

1 McKinsey & Company Report (2016). How big data will revolutionize the global food chain. https://www.mckinsey.com/business-functions/digital-mckinsey/our-insights/how-big-data-will-revolutionize-the-global-food-chain (accessed 25 February 2019).
2 Cai, W., Cowan, T., Briggs, P. et al. (2009). Rising temperature depletes soil moisture and exacerbates severe drought conditions across Southeast Australia. *Geophys. Res. Lett.* 36 (21): L21709.
3 Inan, U. and Inan, A. (2000). *Electromagnetic waves*. Upper Saddle River, NJ: Prentice Hall.

4 Zakavi, P. and Karmakar, N.C. (2010). Wireless orthopedic pin for bone healing and growth: antenna development. *IEEE Trans. Antennas Propag.* 58 (12): 4069–4074.
5 Islam, A. and Karmakar, N.C. (2012). A novel compact printable dual polarized chipless RFID system. *IEEE Trans. Microw. Theory Tech.* 60 (7): 2142–2151.
6 Zomorrodi, M. and Karmakar, N.C. (2016). Optimized MIMO-SAR technique for fast EM-imaging of chipless RFID system. *IEEE Trans. Microw. Theory Tech.* 64 (11): 4008–4017.
7 Topp, G.C., Davis, J.L., and Annan, A.P. (1980). Electromagnetic determination of soil water content: measurement in coaxial transmission lines. *Water Resour. Res.* 16: 574–582.
8 Dey, S. (2018). Electromagnetic transduction based chipless RFID sensors for internet of things (IoT). PhD thesis. Monash University
9 Dobson, M.C., Ulaby, F.T., Hallikainen, M.T., and El-Rayes, M.A. (1985). Microwave dielectric behavior of wet soil: part II. Dielectric mixing models. *IEEE Trans. Geosci. Remote Sens.* GE-23: 35–46.
10 Roth, K., Schulin, R., Fluhler, H., and Attinger, W. (1990). Calibration of time domain reflectometry for water content measurement using a composite dielectric approach. *Water Resour. Res.* 26: 2267–2273.
11 Radcliff. R and Balanis. C. (1981). Electromagnetic geophysical imaging incorporating refraction and reflection. *IEEE Transactions on Antennas and Propagation.* 29(2): 288–292. https://ieeexplore.ieee.org/document/1142554.

Appendix A

Useful Electromagnetic Data

It is a standard practice in professional engineering to use the electromagnetic quantities with their associated units and use them with confidence. Except ratios in normalized forms with the division of similar quantities or in decibels (dB) scales, all electromagnetic quantities are appropriately defined with the units. Without units the electromagnetic entities have no significant meaning. Therefore, students must learn these electromagnetic quantities with appropriate applications of units and apply them to solve any numerical and practical design problems with brevity. All scientific definitions of electromagnetic quantities are associated with four fundamental internationally adopted units – the length in meters (m), the mass in kilogram (kg), time in seconds (s), and charge in Coulombs (C). The system of units is known as the meter–kilogram–second–coulomb (MKSC) system of units. The MKSC system of units is also known as Giorgi system after the name of Giorgi who first introduced the system of unit in 1901 [1].

Table A.1 Electromagnetic quantities and their units.

Quantity	Symbol	Unit with symbol
Capacitance	C	Farad (F)
Electric charge	Q	Coulomb (C)
Electric charge density (volumetric)	ρ_v	$\frac{\text{Coulomb}}{\text{m}^3}\left(\frac{\text{C}}{\text{m}^3}\right)$
Electric flux density	D	$\frac{\text{Coulomb}}{\text{m}^2}\left(\frac{\text{C}}{\text{m}^2}\right)$
Magnetic charge density	B	Weber (Wb)
Conductance	G	mho (℧)
Conductivity	σ	$\frac{\text{mho}}{\text{m}}\left(\frac{℧}{\text{m}}\right)$ or (S/m)
Current	I	Ampere (A)
Current density (surface)	J	$\frac{\text{Ampere}}{\text{m}}\left(\frac{\text{A}}{\text{m}}\right)$
Current density (volume)	J	$\frac{\text{Ampere}}{\text{m}^2}\left(\frac{\text{A}}{\text{m}^2}\right)$

(*Continued*)

Fields and Waves in Electromagnetic Communications, First Edition. Nemai Chandra Karmakar.

Companion website: www.wiley.com/go/fieldsandwavesinelectromagneticcommunications

Table A.1 (Continued)

Quantity	Symbol	Unit with symbol
Electric diploe moment	p	Coulomb-m (C − m)
Electric polarization	P	$\frac{\text{Coulomb}}{\text{m}^2}\left(\frac{\text{C}}{\text{m}^2}\right)$
Electric potential or voltage	V	Volt (V)
Magnetic diploe moment	m	Ampere-m^2 (A−m^2)
Displacement vector	D	$\frac{\text{Coulomb}}{\text{m}^2}\left(\frac{\text{C}}{\text{m}^2}\right)$
Energy, work	E, W	Joule (J)
Electric field intensity (electric field)	E	$\frac{\text{Volts}}{\text{m}}\left(\frac{\text{V}}{\text{m}}\right)$
Magnetic field intensity (magnetic field)	H	$\frac{\text{Ampere}}{\text{m}}\left(\frac{\text{A}}{\text{m}}\right)$
Magnetic flux	ψ	Weber (Wb)
Magnetic flux density	B	$\frac{\text{Weber}}{\text{m}^2}\left(\frac{\text{Wb}}{\text{m}^2}\right)$
Force	F	Newton (N)
Inductance	L	Henry (H)
Intrinsic impedance	η	Ohm (Ω)
Permeability	μ	$\frac{\text{Henry}}{\text{m}}\left(\frac{\text{H}}{\text{m}}\right)$
Permittivity	ε	$\frac{\text{Farad}}{\text{m}}\left(\frac{\text{F}}{\text{m}}\right)$
Polarization vector	P	$\frac{\text{Coulomb}}{\text{m}^2}\left(\frac{\text{C}}{\text{m}^2}\right)$
Power	P	Watt (W)
Resistance	R	Ohm (Ω)
Resistivity	ρ	Ohm-m (Ω − m)
Magnetic vector potential	A	Volts-second/meter $\left(\frac{\text{V}\cdot\text{s}}{\text{m}}\right)$
Magnetic scalar potential	ψ_m	Ampere (A)
Magnetization vector	M	Ampere/meter $\left(\frac{\text{A}}{\text{m}}\right)$
Mutual inductance	M	Henry (H)
Phase constant	β	meter^{-1} (m^{-1})
Attenuation constant	α	$\frac{\text{Neper}}{\text{meter}}\left(\frac{\text{Np}}{\text{m}}\right)$

Table A.2 Definitions of electromagnetic quantities.

Quantity	Definition
Electric flux density or displacement vector	$\boldsymbol{D} = \varepsilon_0\varepsilon_r\boldsymbol{E} + \boldsymbol{P}$
Magnetic field	$\boldsymbol{H} = \frac{1}{\mu}\boldsymbol{B} - \boldsymbol{M}$
Gauss law in point from (electric)	$\nabla \cdot \boldsymbol{D} = \rho_v$
Gauss law in point from (magnetic)	$\nabla \cdot \boldsymbol{B} = 0$
Faraday's law	$\nabla \times \boldsymbol{E} = -\frac{\partial \boldsymbol{B}}{\partial t}$
Ampere's circuital law	$\nabla \times \boldsymbol{H} = \sigma\boldsymbol{E} + \frac{\partial \boldsymbol{D}}{\partial t}$
Lorentz force	$\boldsymbol{F} = \boldsymbol{F_e} + \boldsymbol{F_m} = q(\boldsymbol{E} + \nu \times \boldsymbol{B})$
Poynting vector	$\boldsymbol{P} = \boldsymbol{E} \times \boldsymbol{H}$
Poisson's equation	$\nabla^2\phi = -\frac{\rho}{\varepsilon}$
Wave equation for magnetic vector potential	$\nabla^2\boldsymbol{A} = -\mu\boldsymbol{J}$
Magnetization vector	$\boldsymbol{M} = \frac{1}{2}\int(\boldsymbol{R} \times \boldsymbol{J})\,dV$
Electric field intensity	$\boldsymbol{E} = -\nabla\phi - \frac{\partial \boldsymbol{A}}{\partial t}$

Table A.3 Relative permeability and conductivity of common conducting media.

Materials	Relative permeability	Conductivity $\left(\times\, 10^7\, \frac{\mho}{\text{m}}\right)$
Aluminum	1.00000065	3.45–3.75
Cobalt	250	
Copper	0.999994	5.8 (annealed); 5.65 (hard drawn)
Silver	0.99999981	6.139–6.17
Gold	0.999998	4.10
Zinc (with trace of iron)		1.74
Brass		1.2–1.57
Nickel	50–600	1.28
Iron	5000	1.0
Steel	100–40000	0.5–1.0
Tin	1	0.869
Wood (dry)	0.99999942	

Table A.4 Relative permittivity, loss tangent of common dielectric and non-conducting media.

Materials	Resistivity, ρ M$\Omega \cdot$m (300° K)	Relative permeability (1 GHz)	Loss tangent ($tan\delta$) (1 GHz)
Air (dry, sea level)	4×10^7	1.0005	0.000
Alumina 99.5%	$> 10^6$	9.8	0.0001–0.0002
96%	$> 10^6$	9.0	0.0006
85%	$> 10^6$	8.5	0.001 5
Aluminum nitride	106	8.9	0.001
Bakelite	1–100	4.74	0.022
Beryllium oxide (toxic)	$> 10^8$	6.7	0.004
Diamond	10^5–10^{10}	5.68	< 0.0001
Ferrite (MnZn)	0.1–10 $\Omega \cdot$m	13–16	0.0004
Ferrite (NiZn)	0.1–12.4	13–16	0.0004
FR-4 circuit board	8×10^5	4.3–4.5	0.01
GaAs	1.0	12.85	0.0006
InP	Up to 0.001	12.4	0.001
Glass	$> 1 \times 10^6\ \Omega \cdot$m	4–7	0.0003–0.002
Borosilicate	$> 2 \times 10^7\ \Omega \cdot$m	6.1	0.00090
BaO/Al_2O_3 (Schott AF45) Borosilicate	$> 1 \times 10^6\ \Omega \cdot$m	4.6	0.00037
BaO (Schott Borofloat 33) Mica	2×10^5	5.4	0.0006
Mylar	10^{10}	3.2	0.005
Paper, white	3.5×10^6	3	0.008
Polycrystalline	$> 10^6$	10.13	0.00004–0.00007
Polypropylene	$> 10^7$	2.25	0.0003
Quartz (fused)	$> 10^{11}$	3.8	0.00075
Polyethylene	$> 10^7$	2.26	0.0002
Polyimide	10^{10}	3.2	0.005
Silicon dioxide (SiO_2)	$5.8 \times \underline{10^7}$	3.7–4.1	0.001
Nitride (Si_3N_4)	10^7	7.5	0.001
Teflon	10^{10}	2.1	0.003
Vacuum	∞	1	0
Distilled water	182	80	0.1
Ice Water (273 K)	1	4.2	0.05
Wood (dry oak)	3×10^{11}	1.5–4	0.01
Zirconia (variable)	104	28	0.0009

Acknowledgment

The miscellaneous data for various materials are collected and compiled from [2–6].

References

1 Islam, M.A. (1969). *Electromagnetic Theory*. Dhaka, Bangladesh: E.P. University of Engineering and Technology.
2 Standard Reference Data Database (2022). U.S. National Institute of Standards and Technology and References, http://www.nist.gov/srd. Cited in http://onlinelibrary.wiley.com/doi/10.1002/9781118936160.app2/pdf, (accessed 03 March 2022).
3 International Council for Science (2022). Committee on Data for Science and Technology. http://www.codata.org/resources/databases/index.html. Cited in http://onlinelibrary.wiley.com/doi/10.1002/9781118936160.app2/pdf (accessed 03 March 2022).
4 Cox, J., Wagman, D., and Medvedev, V. (1989). *CODATA Key Values for Thermodynamics*. Hemisphere Publishing Corp. http://onlinelibrary.wiley.com/doi/10.1002/9781118936160.app2/pdf (accessed 03 March 2022).
5 Elert, G. (2007). The physics hypertextbook. http://hypertextbook.com http://onlinelibrary.wiley.com/doi/10.1002/9781118936160.app2/pdf (accessed 03 March 2022).
6 Uma, S., McConnell, A., Asheghi, M. et al. Temperature-dependent thermal conductivity of undoped polycrystalline silicon layers. *Int. J. Thermal Physics* 22 (2): 605–616. http://onlinelibrary.wiley.com/doi/10.1002/9781118936160.app2/pdf, (accessed 03 March 2022).

Index

Fields and Waves in Electromagnetic Communications, First Edition. Nemai Chandra Karmakar.

Companion website: www.wiley.com/go/fieldsandwavesinelectromagneticcommunications

Printed and bound by CPI Group (UK) Ltd, Croydon, CR0 4YY
16/04/2025
14658589-0005